Astro Navigation Demystified

ISBN 978-09541331-2-2
Written by Jack Case, M.A.
Published by Bookcase Learning Resources
Revised Edition. July 2012

Section 1 – The theory of Astro Navigation (Celestial Navigation).

Section 2 – The Practice of Astro Navigation (Celestial Navigation) involving the use of the Rapid Sight Reduction Tables for Navigation (published as NP303 in the UK) and the Sight Reduction Tables for Air Navigation (published as Pub. No.249 in the USA).

Copyright © Jack Case 2010
All rights reserved
Registered U.S. Copyright
Registered U.K. Copyright
A catalogue record for this book is available in the British Library.

This book contains extracts from the Nautical Almanac and the Rapid Sight Reduction Tables for Navigation. These are Crown Copyright and are reproduced by permission of the United Kingdom Hydrographic Office.

The Astro Navigation Demystified website provides a free resource for all those interested in the subject: www.astronavigationdemystified.com

Preface

Learn to navigate by the sun, moon, stars and planets with complete confidence. From novice to master in one book. Unlike some skimpy books which don't fully cover the subject, this is a comprehensive course in astro navigation ranging from simple explanations of the basic principles to clear instructions in the use of rapid sight reduction techniques for position fixing. Although it is a large book, it is thoroughly cross-referenced and it is written in a style which enables you to move from one section to another without having to read it from beginning to end.

Reviews

ReR wrote (on Amazon.com)

This is the best book on celestial navigation I've found so far. I really like the format of explanation, example, self test. There are a lot of new concepts and terms (to me) in this study and I often find, while taking the self test at the end of each chapter, that I didn't understand something as well as I thought I did so I go back and review and retest. Another feature I like is the appendix with explanations of trig formulas that are used in the chapters. You don't need to know how the formulas were derived to understand the material in the chapters but if you're into math and want to know how they were derived- It's there.

David Clark (Maritime Master and Admiralty Pilot) wrote:

There is a wealth of information in this well written and really interesting book. It is packed with useful and practical tips. There are a lot of historical facts on the mathematical principles founded by the likes of Archimedes and Eratosthenes that underpin the subject. This is an ideal book for learning the art of astro navigation and really improving your knowledge of mathematics. I have certainly learnt a great deal from the book and will use it as one of my main sources of reference on the subject.

©2009 Jack Case

Introduction

How could seafarers navigate the oceans if the global positioning system (GPS) failed? The answer is, they could revert to the tried and tested art of **astro navigation**. The problem is that we have become so reliant on automated navigation systems that traditional methods are being forgotten and yet, there is a very real danger that the GPS could be destroyed.

During periods of increased solar activity, massive amounts of material erupt from the Sun. These eruptions are known as coronal mass ejections and when they impact with the Earth they cause disturbances to its magnetic field known as magnetic storms. Major magnetic storms have been known to destroy electricity grids; shut down the Internet, blank out communications networks and wipe out satellite systems (including the global positioning system). Couple this danger with that posed by cyber terrorists who could block GPS signals at any time, then it can easily be seen that navigators who rely solely on electronic navigation systems could be faced with serious problems.

Unfortunately, many sea-goers are deterred from learning astro navigation because they perceive it to be a very difficult subject to learn. In fact, it is very interesting and easy to learn but sadly, some writers and teachers of the subject attempt to disguise its simplicity by creating an aura of mystery about it.

This book on the other hand, is written in plain language and aims to make the art of astro navigation easy and enjoyable to learn. Readers are led gently through the theoretical aspects of the subject before being helped to develop practical skills. Concepts and techniques are clearly explained with the aid of simple diagrams and each topic is re-enforced by practical examples and exercises. Suggested solutions to questions, together with relevant resource data is placed immediately following each exercise instead of being hidden away at the back of the book.

The style of the book allows those who already have some understanding of the subject to proceed straight to the practical section with the ability to dip into the theory as required. It should be emphasised however that, as with most subjects, a solid grounding in the theory makes mastery of the practice much easier.

Some of the explanations involve mathematical calculations but these are kept to a minimum so that the flow of arguments is not interrupted. For those who would like to study fuller mathematical explanations, these can be found in the appendix.

About The Author

Jack Case is an experienced navigator who became a teacher when he left the sea. He taught navigation not only to sailors but also to students of mathematics and geography and was able to demonstrate the important links between these subjects. Over many years, he developed the art of teaching navigation in an interesting way so that it could be understood by students of all ages.

Many of the author's adult students, including small boat owners and yachtsmen, expressed the need for an easily read and uncomplicated book to help them to understand astro navigation. Hence the reason for writing this book.

Acknowledgements

I would like to thank the following for their contributions to the creation of this book:

David Clark, Maritime Master and Admiralty Pilot, for his unstinting and painstaking help with proof reading and editing.

Russell for his IT expertise.

Gwenda for her patience and support.

The United Kingdom Hydrographic Office for kindly allowing me to include extracts from the Nautical Almanac and Sight-Reduction Tables in this book.

Why Trigonometry?

Much reliance has been placed on trigonometric calculations in the theoretical section of this book. You might wonder if this is really necessary in an age when calculators, computers and a host of electronic navigation aids are available to make such calculations for us. The following statement in the International Maritime Organisation regulations provides an answer to this question: "The provision of trigonometric tables onboard and regular practise with them by all officers and navigation related staff is compulsory". In the event of GPS and other electronic navigation equipment failure, it would be irresponsible of a ship's master if his ship were to go dangerously off course simply because trigonometric calculations could not be made.

For those who would like to brush up on trigonometry, a brief exposition of the topic is given in the appendix to this book on page 297.

Contents

Preface and Reviews		ii
Introduction		iii
Section 1	The Theory of Astro Navigation (Celestial Navigation)	1
Chapter 1	The Celestial Sphere	2
Chapter 2	Time	16
Chapter 3	Latitude and Longitude	29
Chapter 4	Calculating Local Hour Angle and Declination	52
Chapter 5	Azimuth and Altitude	76
Chapter 6	Meridian Passage	102
Chapter 7	The Astronomical Position Line	135
Section 2	The Practice of Astro Navigation (Celestial Navigation)	155
Chapter 8	Introduction to the Rapid Sight Reduction Tables	156
Chapter 9	Single Observation and Position Line	178
Chapter 10	Three Body Fix	189
Chapter 11	Fix with Movement of the Observer (MOO)	204
Chapter 12	Fix with Movement of the Body (MOB)	219
Chapter 13	Rapid Sight Reduction Tables Self Test	236
Appendix	Mathematical Topics Relevant to Astro Navigation	297
Index		313

Section 1
The Theory of Astro Navigation

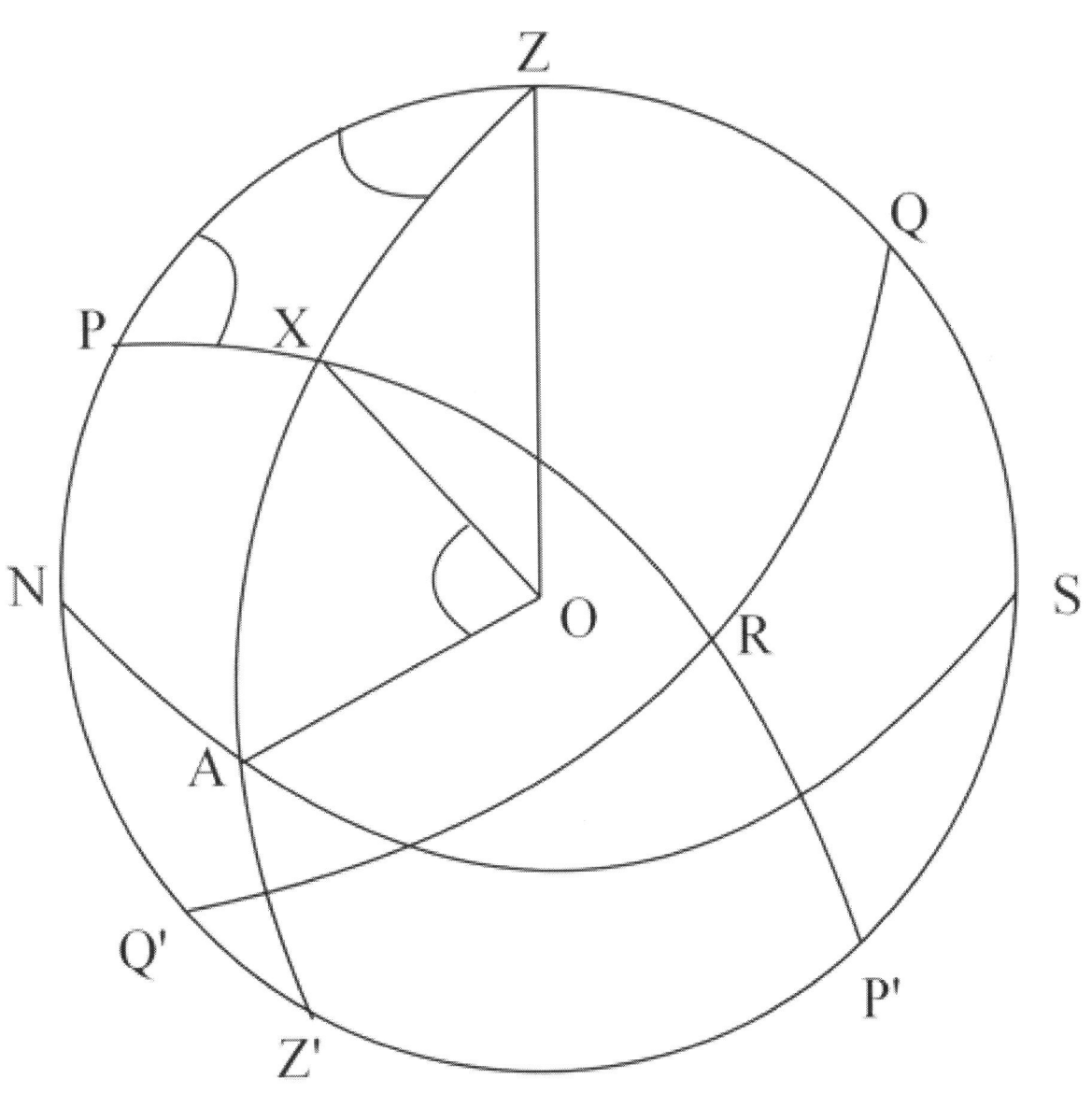

Chapter 1
The Celestial Sphere

Ptolemy was a Greek astronomer who lived in Alexandria in the first century AD. He believed that the Earth was at the centre of the universe as shown in the following diagram of his geocentric model.

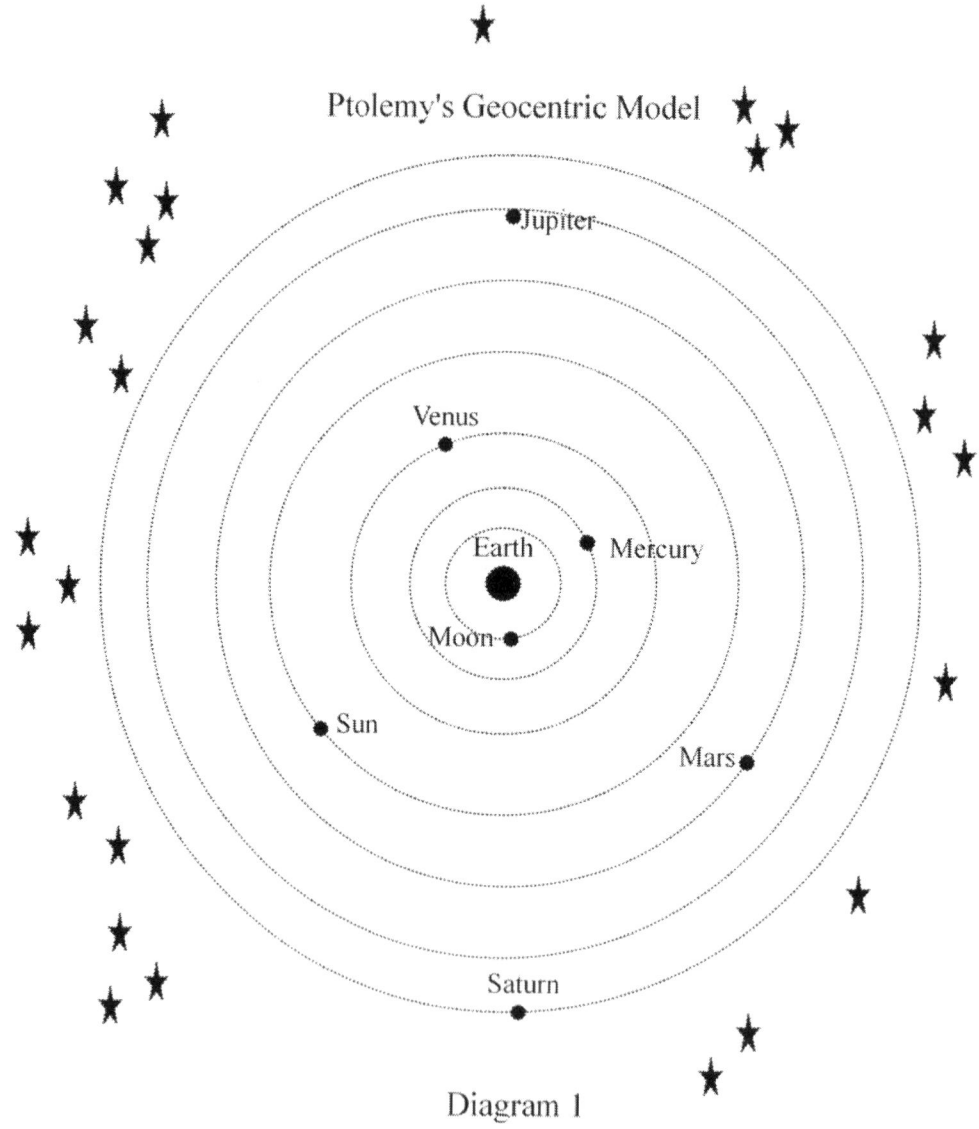

Diagram 1

For the purposes of astro navigation, we adopt a similar Ptolemaic worldview. We assume that the Earth is at the centre of a vast sphere which we call the 'Celestial Sphere'. The 'celestial bodies' such as the Sun, Moon, stars and planets are placed on the inner surface of the celestial sphere much as we would see them in the roof of a planetarium. The fact that the 'celestial bodies' are at greatly varying distances from the Earth and not actually on the inner surface of a sphere is not important since we are only concerned with the angular distances between them. Whereas, in coastal navigation, we can obtain 'position lines' by measuring the angular distances between lighthouses and other geographical features, in astro navigation we obtain position lines by measuring the angular distances between 'celestial bodies'.

In diagram 2, X and Y represent two celestial bodies and X' and Y' represent their corresponding positions on the imaginary celestial sphere. It can be seen that the angular distance between X and Y is exactly the same as that between X' and Y'. So, even though the celestial bodies are really at X and Y, no error is introduced by assuming that they are at X' and Y'.

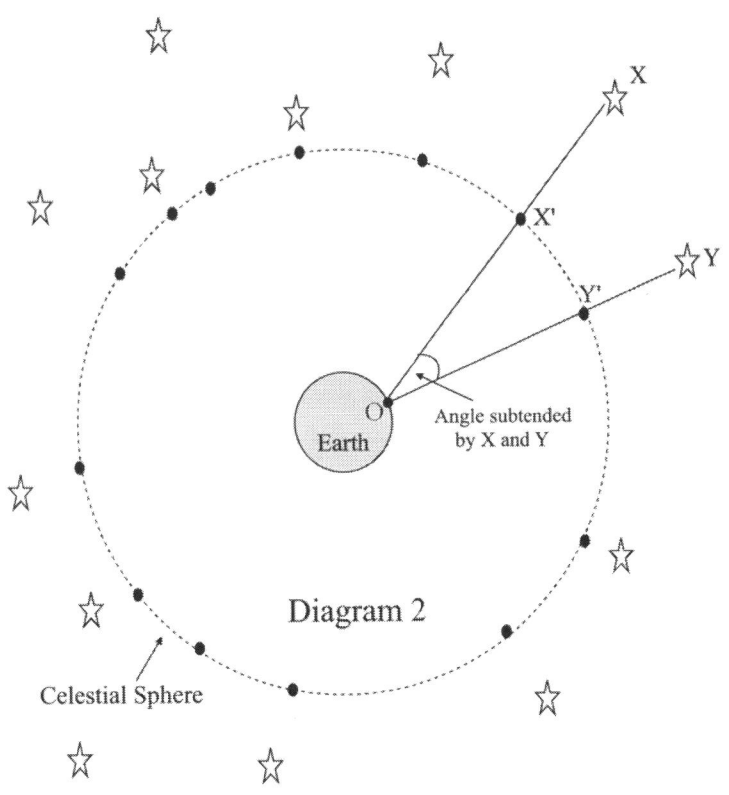

Diagram 2

As discussed above, we can obtain position lines by measuring the angular distances between X' and Y'. However, because the observer, at point O, is on the surface of a sphere and X' and Y' are on the surface of an imaginary sphere, we cannot solve the angle X'OY' by the use of 'straight line trigonometry'; instead we must resort to the use of 'spherical trigonometry'. For those who would like to improve their knowledge of spherical trigonometry, an exposition of the subject is given in the appendix.

The stars are at such an immense distance from the Earth that the movement of their relative positions in the sky, which is so slow and so small, can be discounted without any great loss of accuracy and we can assume therefore, that they are in fixed positions in the celestial sphere. There will however, be very small variations in their positions that result from the Earth's orbital motion around the Sun. We must also remember that, as the Earth rotates about its axis from West to East, the stars will appear to move in the opposite direction i.e. from East to West.

Because of the orbital motion of the Earth, the Sun appears to move around the celestial sphere taking one year to complete a revolution. This apparent movement of the Sun is called the **Ecliptic**. So, unlike the stars, the Sun's position in the celestial sphere is not fixed. Also, unlike the stars, the Sun is relatively close to the Earth so its apparent movement in the sky is comparatively great and fast and therefore must be taken into account in navigational calculations. Like the stars, the Sun appears to move westwards across the sky during the course of a day as the Earth rotates eastwards about its axis.

The planets in the solar system orbit the Sun in an anti-clockwise direction when viewed from the north pole of the celestial sphere. The apparent motions of the planets, when viewed from the Earth, are complicated by the facts that they are at varying distances from the Sun, have different orbital patterns and travel at different speeds. Like the stars and the Sun, the planets will appear to move westwards across the sky as the Earth rotates and this further complicates navigational calculations. Only those planets that are sufficiently prominent to be observed with an ordinary sextant are considered to be 'navigational planets'. These are: Venus, Mars, Jupiter and Saturn.

The following explanation will require a little imagination:
Imagine that we put a light at the centre of the Earth and that we use this light to project the surface of the Earth onto the Celestial sphere. The result would be that every point on the Earth's surface would have a corresponding point on the Celestial sphere. In exactly the same way, imaginary lines on the Earth's surface such as the Equator would have corresponding imaginary lines on the celestial sphere.

In diagram 3 on the following page, B represents the position of an observer on the Earth's surface and Z represents the projection of B onto the celestial sphere. So Z is the point on the celestial sphere directly above the observer and is called the **'Zenith**

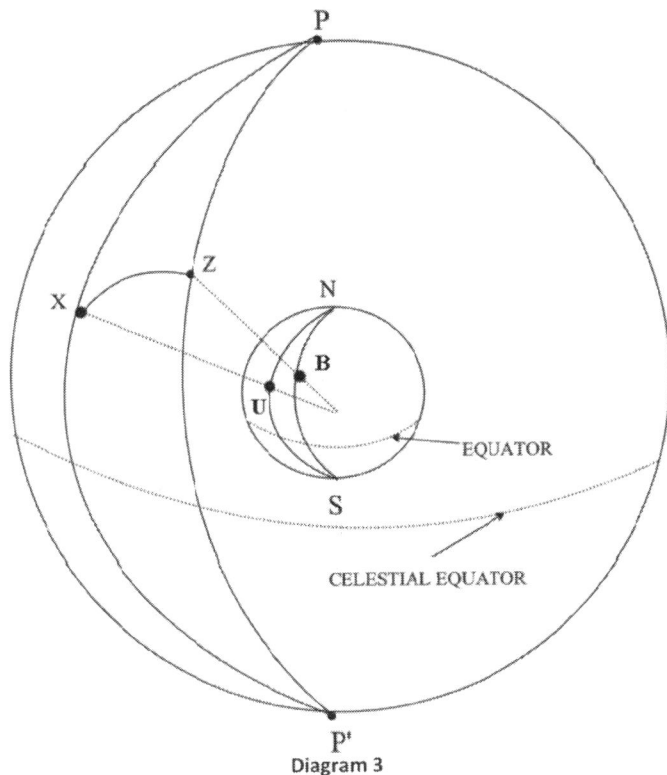

Diagram 3

X represents the position of a **celestial body** and U represents its projection onto the surface of the Earth. So U is the point on the Earth's surface immediately below the celestial body and is called the '**Geographical Position**' (GP).
Similarly,
P is the projection of N,
P' is the projection of S,
The arc ZX is the projection of the arc BU,
The Celestial Equator is the projection of the Earth's Equator,
The Celestial Meridian PZP' is the projection of the observer's meridian NBS,
The Celestial Meridian PXP' is the projection of the meridian NUS on which lies the geographical position.

Suppose yourself to be on the Earth's surface at point B. You would not be able to see the North Pole (point N) nor would you be able to see point U (the geographical position). However, you would be able to see the celestial bodies in the sky. So, although the spherical triangle NBU is inaccessible, you would be able to solve it, in effect, by solving the **spherical triangle PZX** (see diagram 3).
Our aim in, astro navigation, is to determine the geographical position of a celestial body by solving the triangle PZX and then to calculate our own position in terms of our direction and distance from the GP.

In astronomy, we need a celestial coordinate systems for fixing the positions of all celestial bodies in the celestial sphere. To this end, we express a celestial body's position in the celestial sphere in relation to its angular distances from the Celestial Equator and the celestial meridian that passes through the **'First Point of Aries'**. This is similar to the way in which we use latitude and longitude to identify a position on the Earth's surface in relation to its angular distances from the Equator and the Greenwich Meridian.

The First Point of Aries is usually represented by the 'ram's horn' symbol shown below:

Just as the Greenwich meridian has been arbitrarily chosen as the zero point for measuring longitude on the surface of the Earth, the first point of Aries has been chosen as the zero point in the celestial sphere. It is the point at which the Sun crosses the celestial equator moving from south to north (at the vernal Equinox in other words). The confusing thing is that, although this point lay in the constellation of Aries when it was chosen by the ancient astronomers, due to precession, it now lies in Pisces.

Declination. The Declination of a celestial body is its angular distance North or South of the Celestial Equator. The declinations of the stars change very slowly and can be considered to be almost constant for up to a month at a time. The declination of the Sun changes relatively fast from 23.5° North to 23.5° South and back again during the course of a year.

The Moon's declination is more difficult to predict because the rate of change is even more rapid than that of the Sun and the pattern of the changes is less uniform.

Like the Sun and the Moon, the declinations of the planets also change rapidly in comparison with the stars.

Declination can be summarised as the celestial equivalent of Latitude since it is the angular distance of a celestial body North or South of the Celestial Equator.

The Equinoxes. The Sun crosses the celestial equator on two occasions during the course of a year and these occasions are known as the equinoxes. At the equinoxes, at all places on Earth, the nights and days are of equal duration (i.e. 12 hours) hence the term equinoxes (equal nights). Because the Sun is on the celestial equator at the equinoxes, its declination is of course 0°.

The Autumnal Equinox occurs on about the 22nd September when the Sun crosses the celestial equator as it moves southwards from 23.5°N, the northernmost limit of its declination.

The Vernal Equinox occurs on about the 20th March when the Sun crosses the celestial equator as it moves northwards from 23.5°S, the southernmost limit of its declination.

The Solstices. The times when the Sun reaches the limits of its path of declination are known as the solstices. The word solstice is taken from 'solstitium', the Latin for 'sun stands still'. This is because the apparent movement of the Sun seems to stop before it changes direction

The Summer Solstice (mid-summer in the northern hemisphere) occurs on about 21st June when the Sun's declination reaches 23.5° North (the tropic of Cancer).

The Winter Solstice (mid-winter in the northern hemisphere) occurs on about 21st December when the Sun's declination is 23.5°South (the tropic of Capricorn).

Note. The latitude of the tropic of Cancer is currently drifting south at approximately 0.5" per year while the latitude of the tropic of Capricorn is drifting north at the same rate.

The dates of the equinoxes and the solstices will vary slightly during the four-year cycle between leap years for the following reason: Each year is approximately 365.25 days in length. However, for the sake of convenience, the Gregorian calendar divides three years of the cycle into 365 days and the fourth (the leap year) into 366. So, the Vernal Equinox sometimes falls on 20th March and sometimes on 21st. The Autumnal Equinox sometimes falls on 22nd September and sometimes on 23rd. Similarly, the Summer Solstice usually falls on 21st June but sometimes falls on 20th. The Winter Solstice usually falls on 21st December but sometimes falls on 22nd.

As discussed above, declination is a method of expressing a celestial body's position north or south of the celestial equator. There are several ways of expressing a celestial body's East/West position.

Right Ascension (RA). This is used by astronomers to define the position of a celestial body and is defined as the angle between the meridian of the First Point of Aries and the meridian of the celestial body measured in an Easterly direction from Aries. RA is not used in astro navigation; Sidereal Hour Angle is used instead:

Sidereal Hour Angle (SHA). This is similar to RA in as much that it is defined as the angle between the meridian of the First Point of Aries and the meridian of the celestial body. However, the difference is that SHA is measured westwards from Aries while RA is measured eastwards.

Consider Diagram 4:

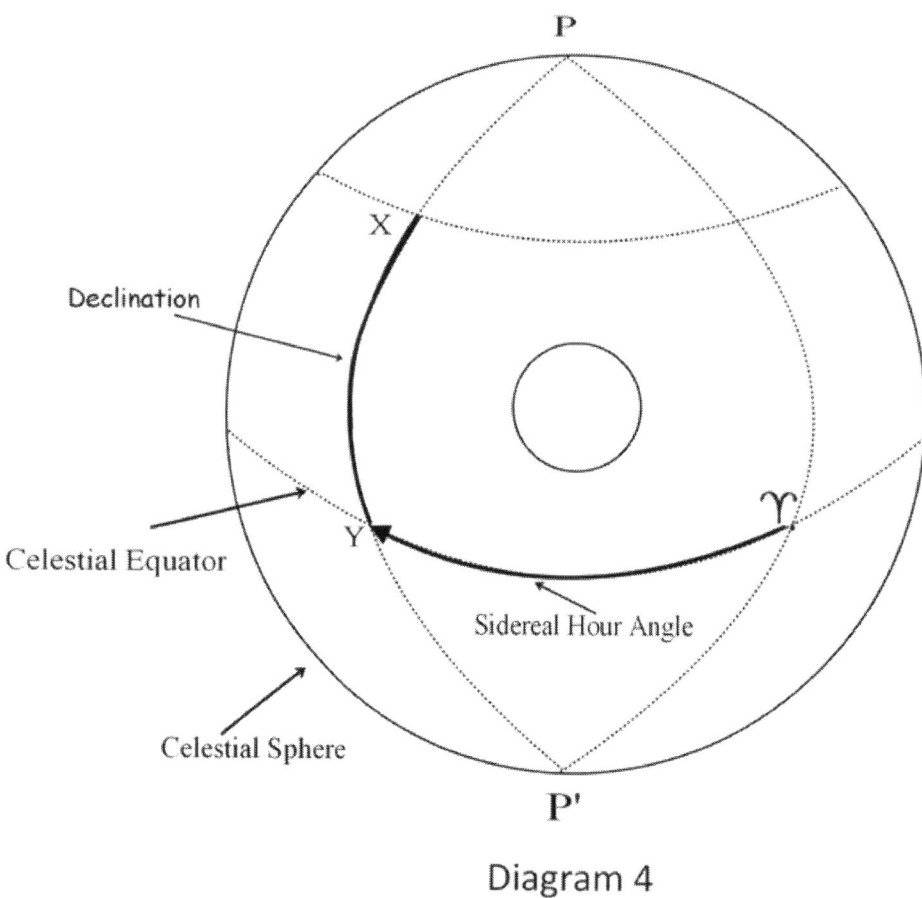

Diagram 4

X is the position of a celestial body in the celestial sphere.
PXP' is the meridian of the celestial body.
Y is the point at which the body's meridian crosses the celestial equator.
♈ is the First Point of Aries.

The **Sidereal Hour Angle** is the angle ♈PY. That is the angle between the meridian running through the First Point of Aries and the meridian running through the celestial body measured at the pole P.
 It can also be defined as the angular distance ♈Y. That is the angular distance measured **westwards** along the Celestial Equator from the meridian of the First Point of Aries to the meridian of the celestial body.

Right Ascension can also be defined as the angle between the meridian of the First Point of Aries and the meridian of the celestial body but the difference is that it is measured in an **easterly** direction from Aries.

From this, we can conclude that

RA = 360° - SHA and
SHA = 360° - RA.

Local Hour Angle (LHA). In astro navigation, we need to know the position of a celestial body relative to our own position.

Returning to Diagram 3 which is repeated below:

LHA is the angle BNU on the Earth's surface which corresponds to the angle ZPX in the Celestial sphere. In other words, it is the angle between the meridian of the observer and the meridian of the geographical position of the celestial body (GP).

Due to the Earth's rotation, the Sun moves through 15° of longitude in 1 hour and it moves through 15 minutes of arc in 1 minute of time. So the angle ZPX can be measured in terms of time and for this reason, it is know as the Local Hour Angle.

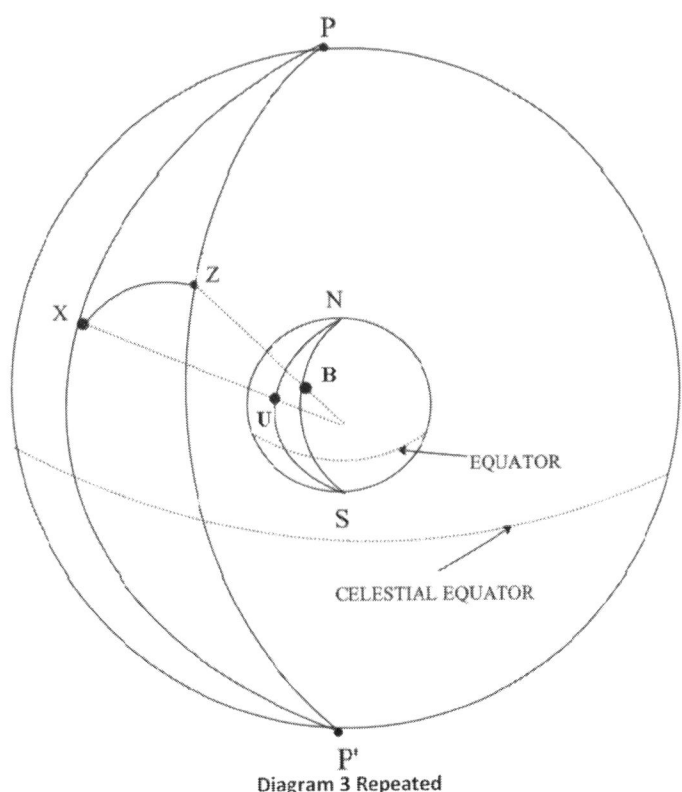

Diagram 3 Repeated

LHA is measured westwards from the observer's meridian and can be expressed in terms of either angular distance or time. For example, at noon (GMT) the Sun's GP will be on the Greenwich Meridian (0°). If the time at an observer's position is 2

hours and 3 minutes after noon, then the angular distance between the observer's meridian of longitude and the Greenwich Meridian must be
(2 ×15°) + (3× 15') = 30° 45'. Because it is after noon at the observer's position, the longitude of that position must be to the East of the Greenwich Meridian since the Earth rotates from West to East. Therefore the observer's longitude must be 30° 45' East and since LHA is measured westwards from the observer's meridian, the LHA must also be 30° 45'. However, it should be noted that as the Earth continues to rotate eastwards, the GP of the Sun will continue to move westwards so the LHA at the observer's position will be continually changing.

Greenwich Hour Angle (GHA). As discussed above, the angle between two meridians of Longitude can be expressed as an hour angle. The hour angle between the Greenwich Meridian and the meridian of a celestial body is known as the Greenwich Hour Angle.
The Local Hour Angle between an observer's position and the geographical position of a celestial body can be found by combining the observer's longitude with the GHA.
In Diagram 5, O represents the longitude of an observer;
X represents the meridian of a celestial body;
G represents the Greenwich Meridian.
Because in this case, the observer's longitude is east and because LHA is measured westwards from the observer's meridian to the meridian of the celestial body, LHA is equal to the longitude plus the GHA.

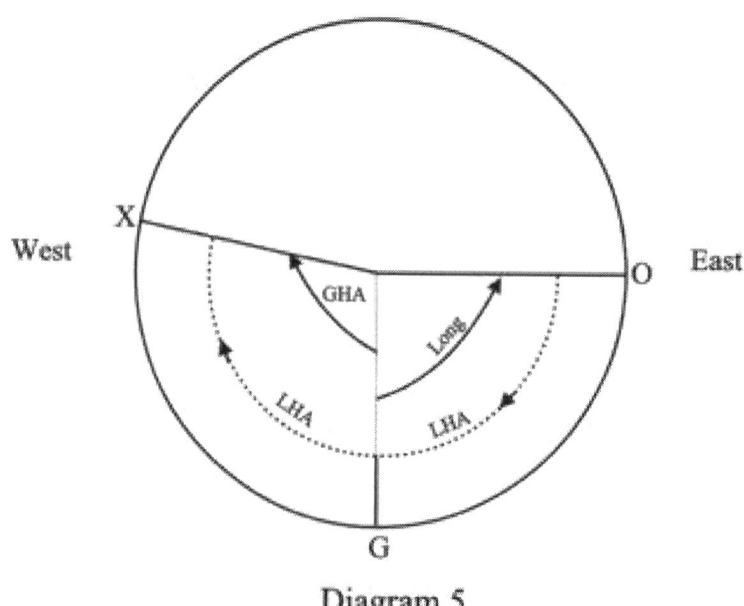

Diagram 5

So we have the rule:

Long East, LHA = GHA + LONG

However, if the observer' longitude were to be to the West of Greenwich, then LHA would be equal to the GHA minus the longitude as shown in Diagram 6:

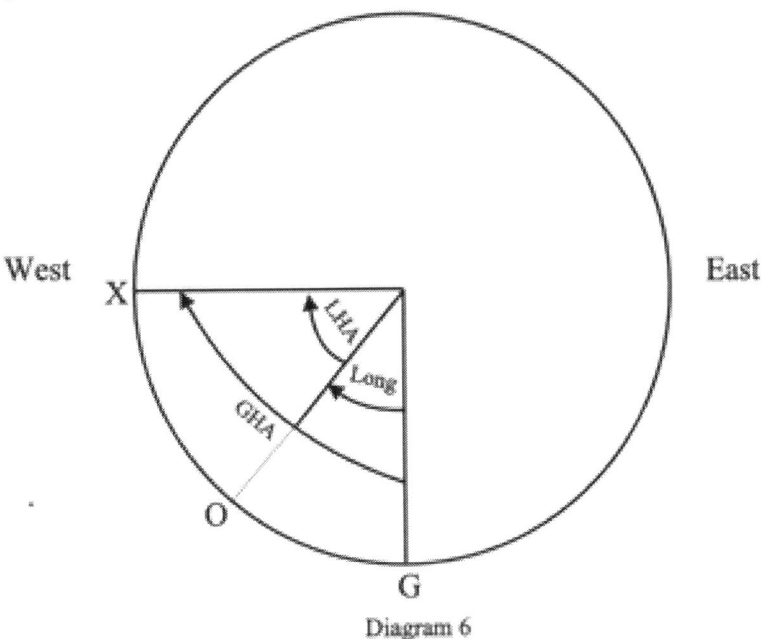

Diagram 6

In this case, the rule is:

Long West, LHA = GHA - LONG

There are two other cases to consider:

Firstly, the case shown in Diagram 7, where the longitude is east and GHA is so great that if we apply the rule **Long East, LHA = GHA + LONG**, the result will be greater than 360°.

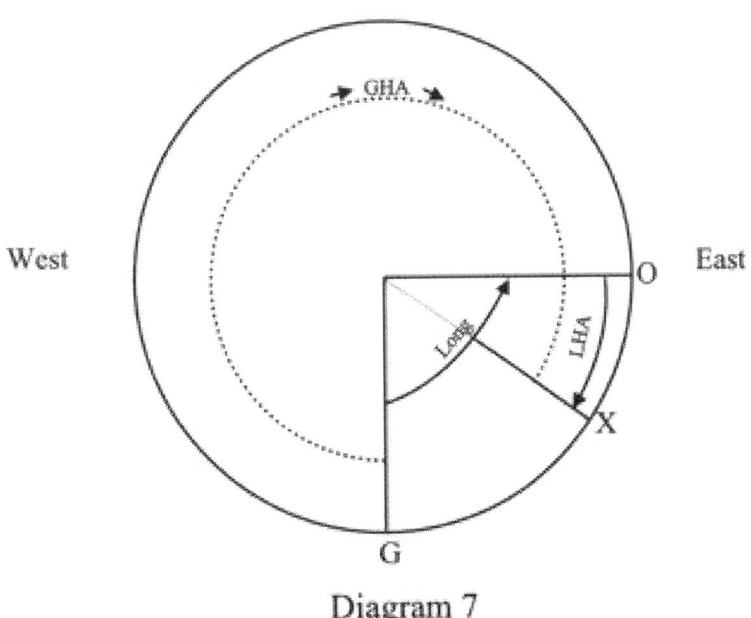

Diagram 7

©2009 Jack Case

We can see however, that the LHA is much less than 360°.
So, in this case, we modify the rule by subtracting 360° as necessary.
The rule now becomes:

Long East, LHA = GHA + LONG (-360° as necessary).

For example: If Long. is 90°E. and GHA is 300°
Then LHA = GHA + LONG -360°
= 300° + 90° = 390° - 360° = 30°

Secondly, we have the case where the longitude is west and is greater than the GHA, as shown in Diagram 8.

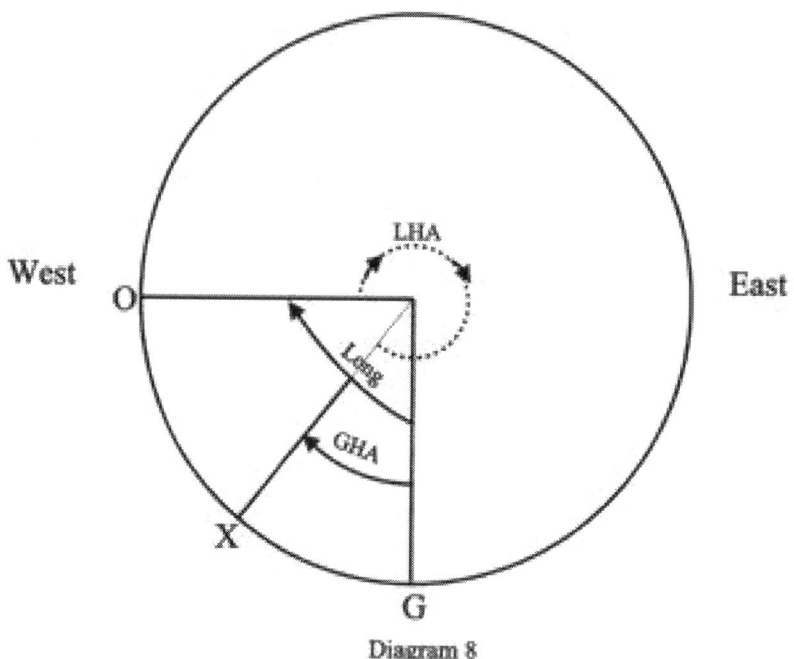

Diagram 8

In this case, we apply the rule **Long West, LHA = GHA - LONG** but because GHA is less than Longitude, we must add 360° to the result to avoid having a negative answer.
So the rule now becomes:

Long West, LHA = GHA - LONG (+360° as necessary)

For example, if Long. is 90°W. and GHA is 45,° we have:
LHA = 45° - 90° = -45° + 360° = 315°

To summarise the above rules:
Long East, LHA = GHA + LONG (- 360° as necessary)
Long West, LHA = GHA - LONG (+ 360° as necessary)

Worked Examples

Example 1.
If your longitude is 35° 46' East and the GHA of Mars is 39° 53'.8. What is the LHA?
Remember the rule:
Long East, LHA = GHA + LONG (-360°?)
GHA = 39° 53'.8
LONG = 35° 46'.0E   (+)
LHA = 75° 39'.8
(Remember 60 minutes in 1 degree)

Example 2.
Your assumed longitude = 125° 13'.0W. The GHA of the Sun is 243° 44'.7
What is the LHA?
Long West, LHA = GHA - LONG (+360°?)
GHA = 243° 44'.7
LONG = 125° 13'.0W (-)
LHA = 118° 31'.7

Example 3.
Longitude is 120°W. GHA is 70°.
What is the LHA?
LHA = GHA - LONG (+360°?)
GHA = 70° 00'.0
LONG = 120° 00'.0 W. (-)
LHA = -50° 00'.0
       360° 00'.0   (+)
LHA = 310° 00'.0

Example 4.
Longitude is 90°E. GHA is 340°
What is the LHA?
LHA = GHA + LONG (-360°?)
GHA = 340°
LONG = 90°E   (+)
LHA = 430°
       360°   (-)
LHA = 70°

Self Test

Question 1.
Position: 38° 15'N. 30° 38'.3 W,
GHA Sirius: 150° 15'.4.
What is the LHA?

Question 2.
Position: 33° 12'.8S. 70° 54'.5E,
GHA Sun: 45° 23'.4
What is the LHA?

Question 3.
Position: 52° 42'.45N. 65° 19'.67E
GHA Moon: 345° 15'.0
What is the LHA?

Question 4.
Position: 34° 28'.45S. 172° 25'.8W.
GHA Mars: 15° 30'.0
What is the LHA?

Solutions to the test questions

Q.1.
Long West, LHA = GHA - LONG
GHA = 150° 15'.4
LONG = 30° 38'.3W   (-)
LHA = 119° 37'.1

Q.2.
Long East, LHA = GHA + LONG
GHA = 45° 23'.4
LONG = 70° 54'.5E   (+)
LHA = 116° 17'.9

Q.3.
Long East, LHA = GHA + LONG (−360° ?)
GHA = 345° 15′.00
LONG = 65° 19′.67 E. (+)
LHA = 410° 34′.67
       360° 00′.00   (−)
        50° 34′.67

Q.4.
Long West, LHA = GHA − LONG (+360° ?)
GHA =   15° 30′.0
LONG = −172° 25′.8 W. (−)
LHA = −156° 55′.8
       +360° 00′.0   (+)
        203° 04′.2

Note. The calculations can be made simpler by converting minutes to fractions of a degree as follows:

To convert minutes to fractions of a degree, divide by 60.
To convert fractions of a degree to minutes, multiply by 60.
The solution to Q.4 now becomes:

GHA =   15°.5
LONG = −172°.43 W. (−)
LHA = −156°.93
       +360°.00   (+)
        203°.07
∴ LHA = 203° 04′.2

Chapter 2
Time

Definitions of Time. The apparent movement of the Sun provides the basis for our definitions of time.

The Sun as a Time-Keeper. Because the Earth's orbital motion is not uniform, there are corresponding variations in the apparent speed of the Sun along the ecliptic. For this reason, the hour angle of the True Sun does not increase at a uniform rate and therefore does not give an accurate measurement of time. To overcome this problem without losing the connection with the True Sun, the Mean Sun, as defined below, is used.

The Mean Sun is an imaginary body which is assumed to move in the celestial equator at a uniform speed round the Earth and to complete one revolution in the time taken by the True Sun to complete one revolution of the ecliptic.

The Mean Solar Day is the time taken for the Mean Sun to make one complete circuit of the Earth. In other words, it is the time taken for the Mean Sun to transit all 360 meridians of longitude. The time system based on the Mean Solar Day is known as **Mean Solar Time.**

The True Sun. Life on Earth is governed by the movement of the True Sun; that is the sun we see in the sky and not by the theoretical Mean Sun. The units of time in everyday use are defined in terms of the True Sun and are known as **Apparent Solar Time.**

Practical Units of Time. The following units of time are defined in terms of the True Sun and although they are not all used in everyday life, they are important for navigational purposes and therefore need to be defined clearly.

The Year is the time taken for the Earth to complete one orbit of the Sun.

The Day is the time taken for the Earth to complete one revolution about its own axis. In other words, it is the time taken for the True Sun to make an apparent transit of all 360 meridians of longitude.

The Hour. The day is divided into 24 hours and in 1 hour, the Sun will make an apparent transit of 15° of longitude.

The Minute. The hour is divided into 60 minutes and in 1 minute of time, the Sun will transit 15 minutes (15') of longitude or, in other words, a $\frac{1}{4}$ of 1° of longitude.

The Second. The minute is divided into 60 seconds and in 1 second of time, the Sun will transit 15 seconds (15") of longitude or, in other words, a $\frac{1}{4}$ of 1' of longitude.

Apparent Noon is when the True Sun is on an observer's meridian of longitude. It is when the Sun reaches its greatest altitude above the observer's horizon. In other words it is when the Sun is at its **zenith**.

Mean Noon occurs when the meridian of the Mean Sun coincides with the meridian of a place.

Twilight is the time between dawn and sunrise and between sunset and dusk when the Sun is just below the horizon. It is so called because scattered sunlight in the upper atmosphere illuminates the lower atmosphere so that the surface of the Earth is neither completely lit nor completely dark

Civil Twilight. Morning civil twilight begins when the geometric center of the Sun is 6° below the horizon and ends at sunrise. Evening civil twilight begins at sunset and ends when the geometric center of the Sun reaches 6° below the horizon. During civil twilight, the horizon is clearly visible and the brightest stars as well as the navigational planets Venus, Mars, Jupiter and Saturn can be seen.

Nautical twilight is the time when the center of the Sun is between 6° and 12° below the horizon. Thus, morning nautical twilight ends when morning civil twilight begins and evening nautical twilight begins when evening civil twilight ends. Nautical twilight is so named because it is when navigators are able to take reliable sights of stars and planets using a visible horizon for reference.

Star and Planet Observations. In general, the optimum conditions for taking observations of stars and planets occur during the times of civil twilight and nautical twilight when it is likely to be light enough for the horizon to be seen yet dark enough for those celestial bodies to be visible.

Local Hour Angle of the Sun is the angle between the meridian of the observer and the meridian of the geographical position of the Sun. In other words, it is the angle through which the Sun's meridian has appeared to move from its noon position.

Diagram 3 is repeated below and in this diagram, as previously explained, B is an observer's position on the Earth and NBS is his meridian of longitude.

Z is the observer's zenith and PZP' is the projection of the observer's meridian onto the celestial sphere.

X is the position of the Sun in the celestial sphere and U is its projection onto the Earth's surface (in other words, its geographical position).

PXP' is the Sun's celestial meridian and NUS is its projection onto the Earth's surface.

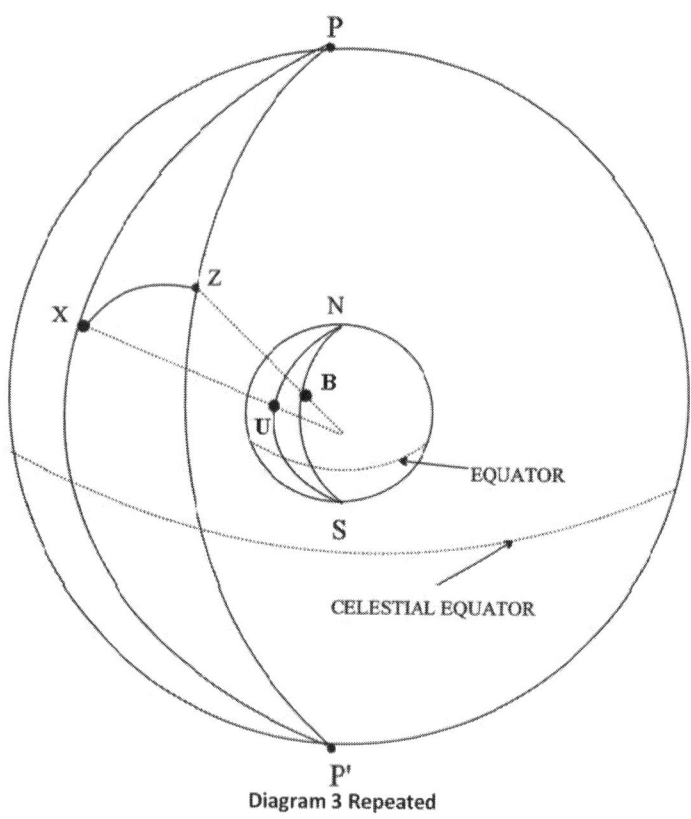

Diagram 3 Repeated

The Local Hour Angle of the Sun then is the angle BNU on the Earth's surface which corresponds to the angle ZPX in the celestial sphere. In other words, it is the angle between the meridian of the observer and the meridian of the Sun's Geographical Position. LHA is measured westwards from the observer's meridian and can be expressed as angular distance or as time.

As previously discussed, due to the Earth's rotation, the Sun appears to move through 15° of longitude in 1 hour and through 15 minutes of arc in 1 minute of time. So the angle ZPX can be measured in terms of time and for this reason, it is know as the Local Hour Angle. For example, if the angle ZPX is 30° 45'; then, in terms of time, it is 2 hrs. 3 minutes.

The Civil Day is the day that is defined for use for human activity rather than for astronomical purposes. It begins at midnight when the Local Hour Angle of the Mean Sun is 12 hours or 180° and ends the following midnight. It is divided into two periods of 12 hours each. The first period consists of the 12 hours from midnight to noon and is denoted by the abbreviation a.m. (ante meridian). The second period is the 12 hours from noon to midnight and is denoted by p.m. (post meridian).

The Astronomical Day consists of one period of 24 hours instead of two periods of 12 hours. This system is more convenient for tabulation purposes since times can be written as 4 figures and dispenses with the need for the abbreviations a.m. and p.m. For example, 11.45 a.m. is simply written as 1145 and 7.32 p.m. is written as 1932.

Local Mean Time (LMT) is the local hour angle of the Mean Sun measured westwards from the meridian of a certain place and expressed in terms of time instead of arc.

Example. In Diagram 9, imagine we are looking down on the North Pole from space.

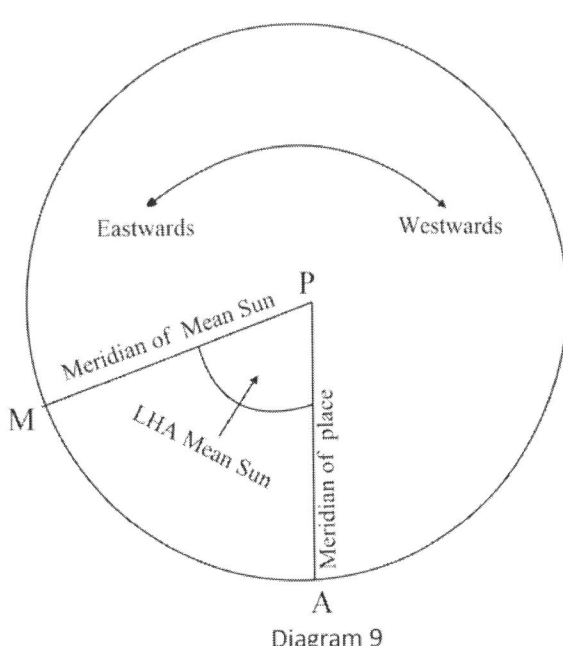

Diagram 9

Point A is a place on the Earth's surface and AP represents part of the meridian of longitude of that place.

MP is part of the meridian of longitude on which, for a brief instant, the Mean Sun lies and M is a point on the Earth's surface which lies on that meridian.

Point P is the North Pole

Angle APM is the Local Hour Angle of the Mean Sun.

Suppose the meridian of the Mean Sun is 60°W. and the longitude of point A is 15°W, then the LHA of the Mean Sun will be 45°. Since the Mean Sun moves 15° westwards in 1 hour, the time difference between point A and point M will be 3 hours. Because point A is 45° to the east of the Mean Sun's meridian, it must be 3 hours after Mean Noon and so the local mean time at point A will be 3 hours after noon or 3p.m.

Note. The difference between the LHA of the Mean Sun and the LHA of the True Sun is that the former is measured from the Mean Sun's meridian and the latter is measured from the meridian of the GP of the True Sun.

Greenwich Mean Time (GMT) is the local mean time anywhere on the meridian of Greenwich. In other words it is the Local Hour Angle of the Mean Sun on the meridian of Greenwich.

Since the Greenwich meridian is used as the base meridian from which the longitude of all places on Earth are identified, it provides the link between the LMT of a place and the LMT at Greenwich (or GMT).

Example. In Diagram 10, imagine that we are looking down on the North Pole.

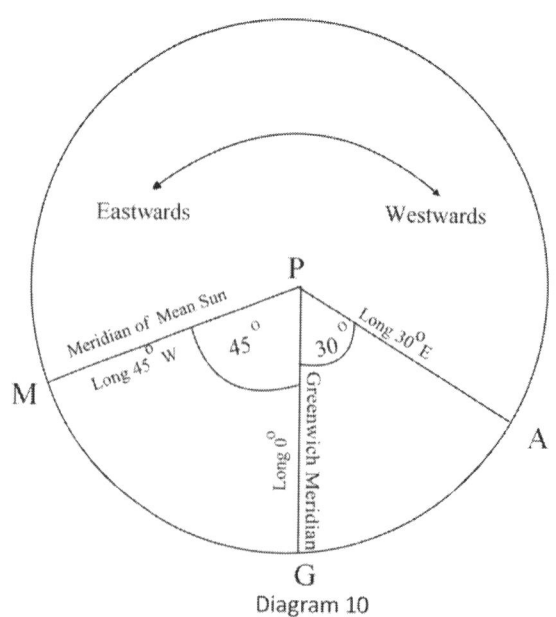

Diagram 10

P represents the North Pole

G represents the position of Greenwich on the Earth's surface.

GP represents part of the Greenwich Meridian (0°).
MP represents part of the meridian of longitude 45°W and M is a point on that meridian.
AP represents part of the meridian of longitude 30° E and A is a point on that meridian.

The meridian of the Mean Sun, for a very brief instant, coincides with the meridian 45°W. and so, at that instant, the Local Mean Time at point M is noon.
At the same instant, the Local Hour Angle of the Mean Sun at Greenwich is 45°. Therefore, the LMT at Greenwich must be 3p.m. since the time difference for 45° is 3 hours and Greenwich is to the East of M.
It follows that the Greenwich Mean Time must also be 3p.m. (since GMT is equal to the LMT at Greenwich).

The LMT at point A must be 2 hours after GMT (since the time difference for 30° is 2 hours and A is to the East of Greenwich).
Therefore, the LMT at point A must be 5p.m.

The following **Aide Memoire** is useful when trying to remember whether to add or subtract the time difference:

Long West, GMT Best.
Long East, GMT Least.

Examples.
1. When it is noon at Greenwich, the LMT at a place on longitude 15°W will be 11 a.m. (since the time difference for 15° is 1 hour and the longitude is west).
2. If Greenwich Mean Time is 4 p.m., the LMT at a place 45°E will be 7 p.m.
3. If the LMT at longitude 75°W is 0130, what is GMT? The time difference for 75° is 5 hours and since the longitude is West, GMT must be +5 hours. Therefore, GMT is 0630.
4. If the time is 2130 GMT, what is the LMT at longitude 150° 15'E.? The time difference is 10 hours, 1 minute. Since the longitude is East, LMT must be greater than GMT. Therefore, LMT is [(2130 + 1001) – 2400] = 0731 the next day.

Universal Time (UT). The term Universal Time was adopted internationally in 1928 as a more precise term than Greenwich Mean Time, because GMT can refer to either an astronomical or a civil day. However, the term Greenwich Mean Time persists in common usage to this day and is generally considered to be synonymous with the term Universal Time. It should be noted that the Nautical Almanac and other tables of astronomical data usually refer to UT instead of GMT.

Standard Time. For each place on Earth to keep its own local mean time would obviously cause a great deal of confusion and difficulty. For this reason, Standard Time is introduced so that places in the same locality can keep the same time. The time chosen is usually based on a convenient meridian running through the area and the meridian chosen usually differs from the Greenwich meridian by a whole number of hours. Some large countries, such as the U.S.A. have several different standard times.

Zone Time (ZT). It would be impossible for a ship at sea to keep to the time of its longitude because (unless it is travelling due north or south) the longitude will be constantly changing. For this reason, the sea areas of the Earth are divided into time zones which are 15° (or 1 hour) apart. The central meridian of each zone is an exact number of hours distant from the Greenwich meridian so that zone time differs from GMT by multiples of 1 hour. The time kept in each zone is the time of its central meridian and is plus or minus GMT depending on the zone's position east or west of Greenwich.

Time Zone System. The Earth is divided into 24 zones of 15° of longitude each, with the centre of the system being the Greenwich meridian. Therefore, the centre zone (zone 0) lies between 7.5° W. and 7.5° E. The zones lying to the West of Greenwich are numbered from +1 to +12 and those to the East from -1 to 12. The 12th zone is divided by the meridian 180° which is known as the International Date Line. The two halves of the 12th zone are marked + or – depending on which side of the date line they lie.

The following table shows the centre meridian of each of the 24 time zones. It should be remembered that each zone is 15° wide and extends 7.5° either side of its centre meridian. Thus, since the centre meridian of zone +4 is 60° W, it's limits are 52.5° W to 67.5°W.

0	+1	+2	+3	+4	+5	+6	+7	+8	+9	+10	+11	12
0	15w	30w	45w	60w	75w	90w	105w	120w	135w	150w	165w	180

0	-1	-2	-3	-4	-5	-6	-7	-8	-9	-10	-11	12
0	15E	30E	45E	60E	75E	90E	105E	120E	135E	150E	165E	180

To Convert Zone Time to GMT. The time kept in zone 0 is GMT. To convert any other zone time to GMT, simply apply the sign and number of the zone to the zone time. For example, if it is 1800 in zone +5, GMT will be 2300.

The Greenwich Date (G.D.) is obtained by converting the zone time and date to GMT
Examples:
ZT 0230(-8), 15 Apr = GD 1830, 14 Apr.
ZT 1915(+10), 24 Dec = GD 0515, 25 Dec.

Time-Keeping at Sea. Clocks and watches that are not used for navigational purposes are usually set to zone time to facilitate the normal routine of a ship. However, since the data in the Nautical Almanac is tabulated with reference to UT, it is necessary to keep at least one accurate time-piece constantly set to GMT/UT. Traditionally, **chronometers** are used for this purpose but **atomic radio clocks** are increasingly being used. If a chronometer is used, it should be checked daily with a universal time signal and any errors found should be noted but the chronometer should not be adjusted or disturbed. In this way, it will always be possible to have access to the exact GMT/UT for navigational purposes.

The Deck Watch. Because it is impracticable to move chronometers and atomic radio clocks, a reliable watch set to approximate GMT/UT can be used to note the time when an observation of a celestial body is made. Such a watch is known as a deck watch and its exact difference from GMT/UT should be noted daily by comparison with the chronometer or atomic radio clock.

Calculating the Greenwich Date from the deck watch time. Many deck watches and chronometers are not digital and therefore do not show time using the 24 hour clock. Although we usually automatically convert 12 hour time to 24 hour time, mistakes do sometimes happen and it is best to follow a set procedure to avoid this when making navigational calculations. The following example demonstrates the procedure:

Suppose that the zone time is 0630 in zone -9 on 15 May. The deck watch time (DWT) is 9hr. 28min. 31 sec. and the deck watch error (DWE) is 21 sec. slow.

Then:
Z.T. 0630(-9), 15 May (approx.)
G.D. 2130, 14 May (approx. GMT)

```
              h   m   s
DWT       9  28  31  (p.m.)
DWE              21  slow (+)
GMT      21  28  52   14 May
```

The Equation of Time. Although the imaginary Mean Time gives us an accurate measurement of time, it presents the navigator with a problem.

When fixing his position by an observation of the Sun, he measures the altitude of the True Sun which keeps apparent solar time. However, he notes the time of the observation from a deck watch that keeps mean solar time. To enable us to connect mean solar time with apparent solar time, we have the Equation of Time which is defined as follows:

Equation of Time = mean solar time − apparent solar time

In other words, the equation of time is the difference between apparent solar time and mean solar time taken at the same instant at one place.

Notes.
- The equation of time can be either positive or negative depending on the time of the year.
- The values range from approximately +15 to −15 mins.
- The values are positive from 15th April to 14th June and from 1st September to 24th December.
- The values are negative from 15th June to 31st August and from 25th December to 14th April.

To simplify matters for the navigator, the value of the equation of time is tabulated against GMT in the Nautical Almanac for every 12 hours.

Example.
An observation of the Sun is made at 11h 05m 22s GMT on 26 August 2009. What is the apparent solar time?
The daily page for 26 Aug. shows that the Eqn. of Time on that date is −01m 55s.
Eqn. of Time = mean solar time − apparent solar time.
∴ apparent solar time = mean solar time − Eqn. of Time.
          = 11h 05m 22s − (−)01m 55s
∴ apparent solar time = 11h 07m 17s

Converting Arc to Time and Vice Versa. As previously discussed, due to the Earth's rotation, the Sun appears to move through 360° of longitude in 1 day, 15° in 1 hour and 15' in 1 minute of time and therefore, the local hour angle of the Sun can be expressed in terms of both arc and time.

It is useful to remember the following when making conversions:

　　　　　15° ↔ 1h
　　　　　 1° ↔ 4m
　　　　　15' ↔ 1m
　　　　　 1' ↔ 4s

Conversion tables are available in publications such as the Nautical Almanac but if these are not available, it is quite easy and in fact more accurate, to make the conversions by using the following methods:

To Convert Arc Into Time.
Multiply by 4 and divide by 60
Example. Convert 65° 30' to time:

　　　　　　　　　　　　　　　h　m　s
4 × 65° ÷ 60 = 260° ÷ 60　=　4　20　00
4 × 30' ÷ 60 = 120' ÷ 60　 =　　　2　00
　　　　　　∴　65° 30' =　4　22　00

To Convert Time Into Arc.
Multiply the hours by 15 and divide the minutes and seconds by 4.
Example. Convert 12^h 32^m 15^s to arc:

　　12^h = 12 × 15　= 180° 0' 0"
　　32^m = 32 ÷ 4　 =　 8° 0' 0"
　　15^s = 15 ÷ 4　 =　 0° 3' 45"
　　∴ 12^h 32^m 15^s = 188° 3' 45"

Converting GMT to GHA.
Because GMT is measured westwards from the reciprocal of Greenwich (i.e. 180°) and GHA is measured westwards from the Greenwich meridian (i.e. 0°) we convert GMT to GHA as follows: If GMT when converted to arc is less than 180° then add 180°; if GMT is greater than 180° then subtract 180°). Examples:

Example 1. Convert 0840 GMT to GHA.
Step 1. Convert GMT to arc.
8^h = 8 × 15　　= 120° 0' 0"
40^m = 40 ÷ 4　=　10° 0' 0"
　　　　　　　　　130° 0' 0"

Step 2. Convert to GHA.
GHA =　　130° 0' 0" + 180° 0' 0" = 310° 0' 0"

Example 2. Convert 1530 GMT to GHA.
Step 1. Convert GMT to arc.
15^h = 15 × 15　= 225° 0' 0"
30^m = 30 ÷ 4　=　 7° 30' 0"
　　　　　　　　　232° 30' 0"

Step 2. Convert to GHA.
GHA = 232° 30' 0" - 180° 0' 0" = 52° 30' 0"

Note. Because GHA relates to apparent solar time and GMT relates to mean solar time, we must take the equation of time (EOT) into account when converting GMT to GHA. Therefore the next example includes a calculation for EOT.

Example 3. Convert 0415 GMT to GHA. EOT = $+1^m$

Equation of Time = mean solar time - apparent solar time
∴ apparent solar time = mean solar time - equation of time
∴ GHA = GMT - EOT.

Step 1. Convert 0415 GMT to arc.

4^h = 4 × 15   = 60° 0' 0"
15^m = 15 ÷ 4   = 3° 45' 0"
GMT =   63° 45' 0"
 - EOT   - 1' 0"   (correction for EOT)
                   63° 44' 0"

Step 2. Convert to GHA.
GHA = 63° 44' 0" + 180° 0' 0" = 243° 44' 0"

Self Test (solutions on the next page)

This test consists of general questions relating to time.

Question 1. The LMT at a place is 1830; what is the LHA of the Sun?

Question 2. If the GD is 0352, 24 Sept. what is the LMT at Longitude 162° 45' W?

Question 3. A ship's longitude is 35°, 36'.5W. If GMT is $3^h\ 35^m\ 12^s$ what is zone time?

Question 4. On 5 July, a ship's longitude is 145°, 15'.33E. If zone time is $3^h\ 20^m\ 30^s$, what is the Greenwich Date?

Question 5. A ship's longitude is 158°, 2'.38E. If the GD is 2330 3 July, what is ZT?

Question 6. Zone time is 0542 in zone -11 on 23 April. The deck watch time is 6hr. 42min. 28 sec. and the deck watch error is 34 sec. slow. What is GMT?

Question 7. If GMT is $15^h\ 15^m\ 35^s$ and a yacht's longitude is 165° 59' 30" E, what is LHA at the yacht's position? ? For this question, assume that the Equation of Time is 0.

Question 8. An observation of the Sun is made at $15^h\ 32^m\ 18^s$ GMT on 15 June 2009. N.b. The Nautical Almanac daily page for that date shows that the Eqn. of Time is $-00^m\ 30^s$.
What is the apparent solar time?

©2009 Jack Case

Test Solutions

Q.1. If the LMT is 1830, it must be 6 hours, 30 minutes after noon and since LMT is measured westwards from the place, the time difference between the place and the Sun's meridian is $6^h 30^m$.

To convert $6^h 30^m$ to arc:

6^h = 6 × 15 =	90° 0' 0"
30^m = 30 ÷ 4 =	7° 30' 0"
∴ $6^h 30^m$ =	97° 30' 0"

Therefore, the angular distance between the place and the Sun's meridian is 97° 30'.0.
∴ LHA Sun = 97° 30'.0

Q. 2.

Step 1. Convert 162° 45' to time:
4 × 162° ÷ 60 = $10^h 48^m$
4 × 45' ÷ 60 = $00^h 03^m$
∴ Time diff. = $10^h 51^m$

Step 2. Convert GMT to LMT:
The rule is: **Long West, GMT Best.**
∴ LMT = GMT - $10^h 51^m$
GD = 0352, 24 Sept.
∴ LMT = 0352 - 1051 = 1701, 23 Sept.

Q.3.

Long. 35°, 36'.5W. is in zone +2.
GMT = $3^h 35^m 12^s$
∴ ZT = $3^h 35^m 12^s - 2^h = 1^h 35^m 12^s$.

Q.4.

Date = 5 July. ZT = $3^h 20^m 30^s$
Long. 145° 15'.33E is in zone -10.
∴ GMT = $3^h 20^m 30^s - 10^h$
∴ GD = $17^h 20^m 30^s$, 4 July.

Q.5.

> Long. 158°, 2'.38E is in zone -11.
> If GD = 2330, 3 July, then ZT = 2330 + 11^h
> ∴ ZT = 1030 4 July.

Q.6.

> Z.T. 0542(-11), 23 April
> G.D. 1842, 22 April (approx. GMT)
>
> | | h | m | s | |
> |------|----|----|----|---------|
> | DWT | 6 | 42 | 28 | (p.m.) |
> | DWE | | | 34 | slow |
> | GMT | 18 | 43 | 02 | 22 April|

Q.7.

Step 1. Convert GMT to arc.
GMT = 15^h 15^m 35^s

15^h	= 15 × 15	=	225° 00' 00"
15^m	= 15 ÷ 4	=	3° 45' 00"
35^s	= 35 ÷ 4	=	0° 08' 45"
		=	228° 53' 45"

Step 2. Convert to GHA
GHA = 228° 53' 45" - 180° 00' 00" = 048° 53' 45"
(**Note.** This solution assumes that the equation of time is zero otherwise the GHA would have to be adjusted for the correct EOT).

Step 3. Calculate LHA Sun at the yacht's longitude
(Long East, LHA = GHA + LONG)
GHA = 048° 53' 45"
Long. = 165° 59' 30" E.
Therefore, LHA = 048° 53' 45" + 165° 59' 30" = 214° 53' 15"

Q.8.

> **Eqn. of Time = mean solar time − apparent solar time.**
> ∴ apparent solar time = mean solar time − Eqn. of Time.
>   = 15^h 32^m 18^s − (−)00m 30s
> ∴ Apparent solar time = 15^h 32^m 48^s

©2009 Jack Case

Chapter 3
Latitude and Longitude

A good starting point for a study of latitude and longitude is to consider the definitions of Great Circles and Small Circles:

Great Circles. A plane of a sphere which passes through the centre of the sphere is called a great circle. The paths of two great circles of a sphere will intersect at two points 180° apart on the surface of the sphere.

Small Circles. A plane of a sphere which does not pass through the centre of the sphere is called a small circle. Two small circles of a sphere need not meet.

Now that we have established these definitions, we can begin to understand the concepts of latitude and longitude.

Longitude.

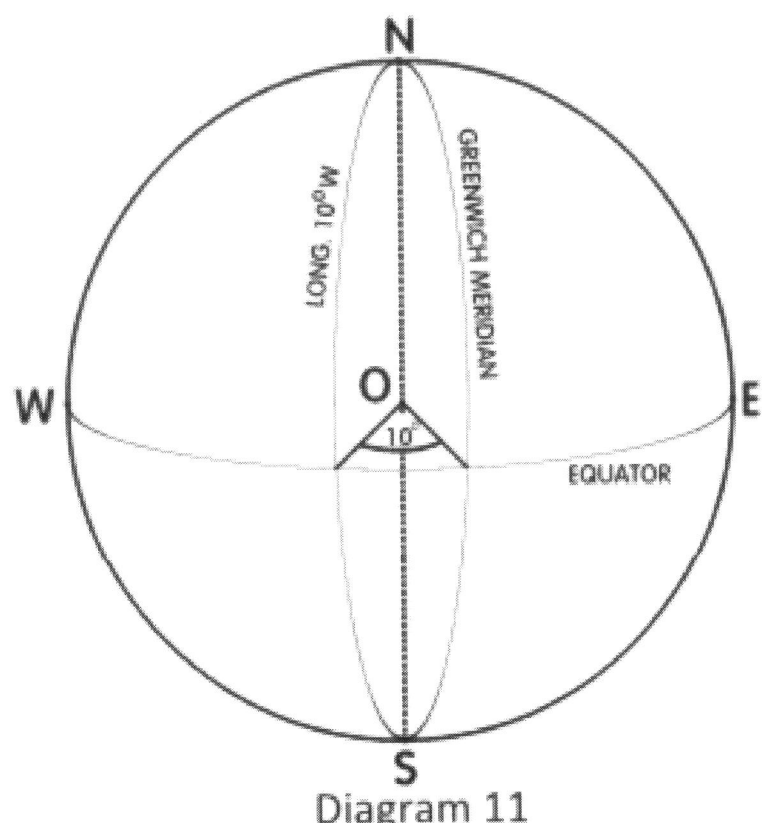

Diagram 11

In diagram 11,

©2009 Jack Case page 29

O represents the centre of the Earth.
N & S represent the North and South poles respectively.

The great circles that pass through N and S are called **meridians of longitude**. The meridian which passes through Greenwich is used as the base meridian (0°) and all other meridians are described by their angular distance east or west of 0° along the Equator from 1° to 180°. Diagram 11 shows the meridians of longitude 0° (Greenwich Meridian) and 10° West.

The Equator. The Equator is an imaginary line around the Earth forming a great circle that is equidistant from the north and south poles and as such, it forms the boundary between the northern and southern hemispheres.

Latitude. A section of the Earth's surface made by a plane parallel to the Equator is called a parallel of latitude and by the definitions established above is a small circle.

A parallel of latitude is expressed by its angular distance north or south of the Equator. As illustrated in diagram 12, all points along the parallel of Latitude 10° North have an angular distance of 10° North from the Equator.

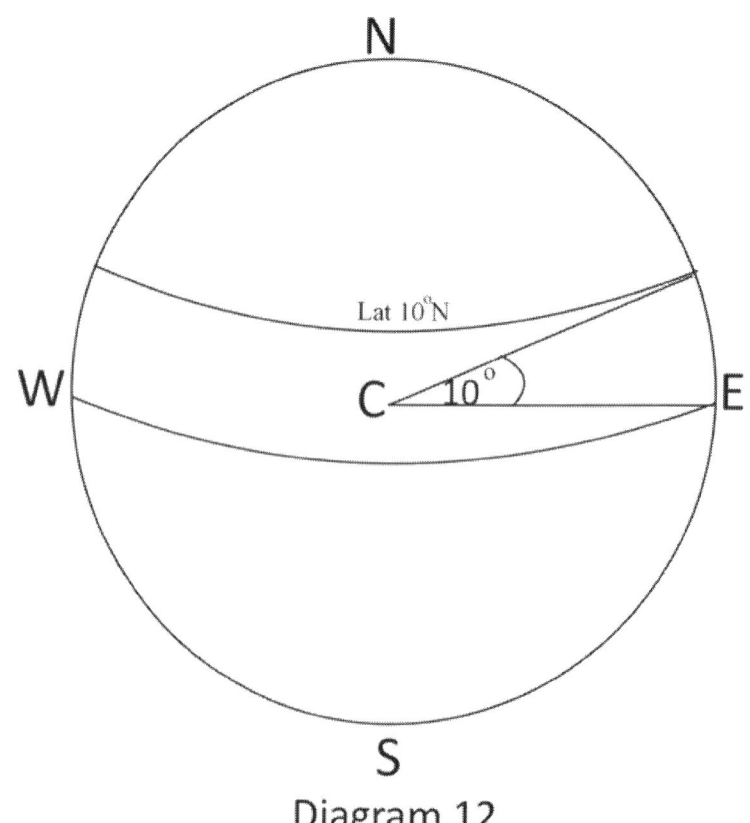

Diagram 12

Units of Length.

The standard unit of measurement used for navigation at sea is the nautical mile. However, there are other units of length associated with the word mile and this can lead to confusion for navigators. The most common of these measurements are defined below:

Statute Mile. A mile most commonly refers to the statute mile of 5,280 feet (1,760 yards, or 1,609.344 meters). The use of the statute mile as a unit of measurement is largely confined to the United States and the United Kingdom; elsewhere, it has been replaced by the kilometer as a unit of measurement on land.

Geographical Mile. The international geographical mile (g.m.) is a unit of length determined by 1 minute of arc along the Earth's equator and is defined as 1855.32 metres.

Nautical Mile. The international nautical mile (n.m.) is closely related to the geographical mile and is a unit of length corresponding approximately to one minute of arc along any meridian of longitude. It is defined as exactly 1852 metres.

Kilometre. The kilometre (km) is a unit of length in the metric system and is equal to 1000 metres. As stated above, it is used to express distances between geographical places in most countries of the world except for the United Kingdom and the United States where the statute mile is used.

Measurement Conversion Table					
	Statute Miles	Geographical Miles	Nautical Miles	Metres	Kilometres
Statute Mile	-	0.867	0.869	1609.34	1.6
Geographical Mile	1.153	-	1.002	1855.32	1.86
Nautical Mile	1.15	0.998	-	1852	1.85
Kilometre	0.62	0.539	0.54	1000	-

Earth Dimensions. The accepted value of the mean circumference of the Earth is 40041.58 kilometres. However, for navigational purposes, we need to express this measurement, sometimes in terms of nautical miles and sometimes in terms of geographical miles. We also need to understand various other dimensions of the Earth and be able to convert these from one unit of measurement to another. The following table lists the most commonly used earth dimensions expressed in terms of different units of measurement.

Earth Dimensions Table				
	Kilometres	Statute Miles	Nautical Miles	Geographical Miles
Meridional (Polar) Radius	6356.8	3941.2	3432.67	3426.3
Equatorial Radius	6378.1	3954.4	3444.17	3437.8
Mean Radius	6367.45	3947.8	3438.42	3432.06
Meridional (Polar) Circumference	40008	24805	21604.2	21564.3
Equatorial Circumference	40075.16	24846.6	21640.6	21600.5
Mean Circumference	40041.58	24825.8	21622.5	21582.4

The Relationship Between Longitude and the Nautical Mile.

As shown above, the Earth's equatorial circumference is 21640.6 n.m. Since the Equator is a great circle, 1° will subtend an arc of:

$$\frac{21640.6}{360} = 60.113 \approx 60 \text{ n.m.}$$

There are 360 meridians of Longitude so it follows that, measuring from the Earth's centre, the angular distance between adjacent meridians at the Equator is 1°. Since, as calculated above, 1° subtends an arc of 60 n.m. it follows that the distance between adjacent meridians of longitude at the Equator is 60 n.m. Returning to diagram 11, the angular difference between longitude 10° West and the Greenwich Meridian is 10°; therefore, the distance between them at the Equator is 10 x 60 = 600 n.m.

Time difference between meridians of longitude.

We know that the Earth revolves about its axis once every 24 hours. In other words, the Sun completes its apparent revolution of 360° in 24 hours. This means that the Sun crosses each of the 360 meridians of longitude once every 24 hours.

So, in 1 hour, the Sun appears to move 15°,

in 4 minutes, it appears to move 1°,

in 1 minute it appears to move 15',

in 4 seconds it appears to move 1'.

From this, it becomes obvious that there is a direct relationship between arc and time such that 1 minute of time equals 15 minutes of arc.

If we have two accurate clocks, one calibrated to GMT and the other calibrated to local time, then it is an easy matter to calculate our longitude from the difference between the two times. (In fact, we could manage with just one clock because we know that noon, local time is when the Sun is at its highest altitude).

For example, if the difference between GMT and local time is three hours, then the difference in longitude must be 3 x 15°. If local time is ahead of GMT then the local longitude must be East of the Greenwich Meridian and if local time is behind GMT the longitude must be West.

Example: If it is 18.00 GMT when it is 09.20 local time on the same day, then local time must be 8 hours and 40 minutes behind GMT.

$$\therefore \text{Long} = -[(8 \times 15°) + (40 \div 60 \times 15°)]$$
$$= -[120° + 10°]$$
$$= -130° = 130° \text{ West}$$

However, if we had no way of knowing the time in GMT, we would be in the same situation as mariners were before John Harrison invented the chronometer in the 18th. century. They had to navigate the oceans without a reliable method of calculating their longitude.

The Relationship between Latitude and the Nautical Mile.
The Earth's meridional circumference is calculated to be 40007.86 Km. which equates to 21604.2 nautical miles (n.m.). In other words, an angle of 360° at the Earth's centre subtends an arc of 21604.2 n.m. on the surface of a meridian of Longitude which is by definition a great circle.

From the above, it follows that:
1° measured along a meridian of longitude, will subtend an arc of:
$$\frac{21604.2}{360} = 60.012 \approx 60 \text{ n.m.}$$
and 1' will subtend an arc of:
$$\frac{60}{60} = 1 \text{ n.m}$$

Diagram 13 shows the parallel of latitude 1° North.

As calculated above, an angle of 1° at the Earth's centre will subtend an arc of 60 n.m. along a meridian of longitude. Therefore, any point on the parallel of latitude 1° North will have an angular distance of 60 n.m. north of the Equator.

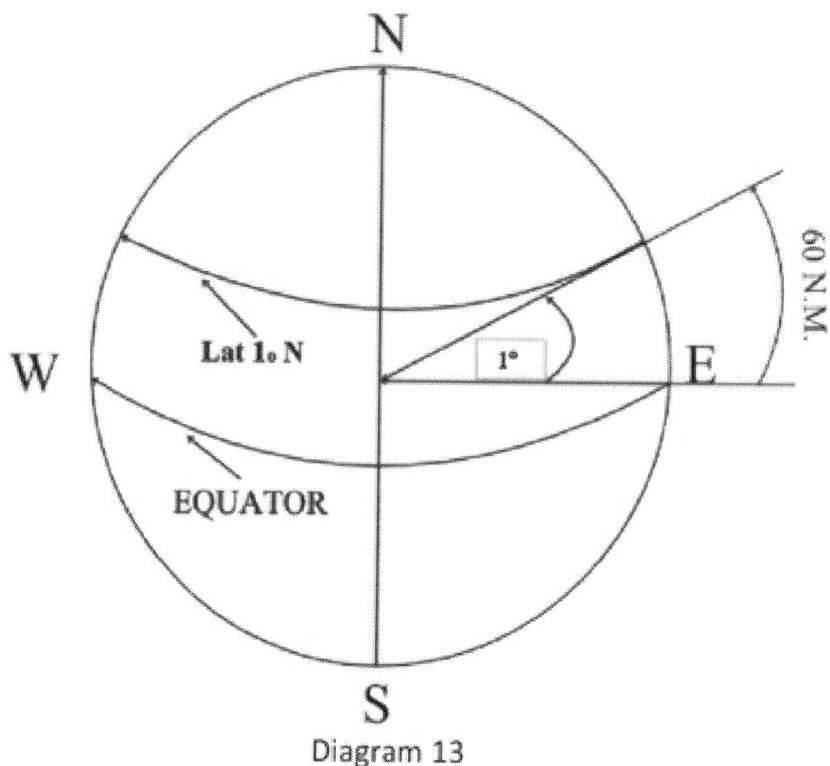

Diagram 13

Now consider diagram 14:

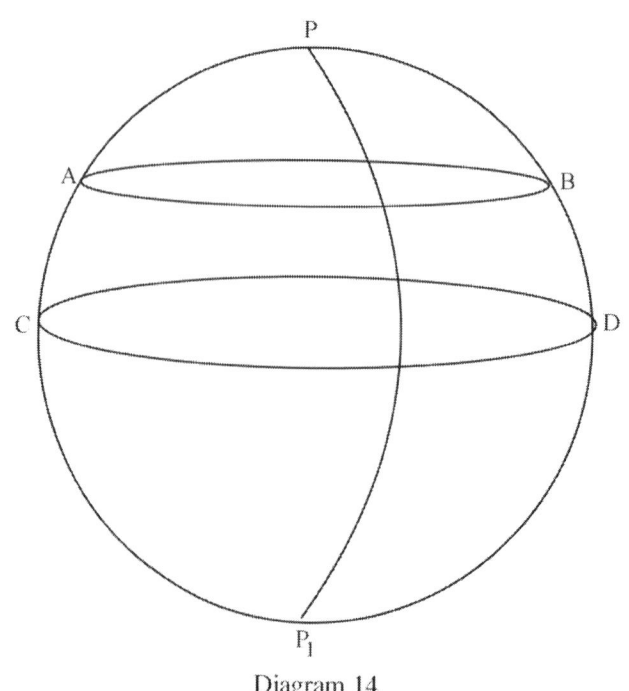

Diagram 14

If AB represents latitude 62°.8 North and CD represents latitude 48°.5 North, then the angular distance between the two lines of latitude will be 14°.3. Therefore, the distance between them, measuring due north or south along the meridian of longitude PP_1 will be: 14.3 × 60 = 858 n.m.

Measuring the Distance Between Meridians of Longitude Along a Parallel of Latitude.

We know that the distance between adjacent meridians of longitude at the Equator is 60 n.m. This is because the Equator is a great circle; parallels of latitude however, are small circles and this presents us with a problem.

In diagram 15, PBC and PAD lie on separate meridians of longitude. The arc BA is the distance between these meridians measured along a certain line of latitude. The arc CD is the distance between the same meridians measured along the Equator. Clearly, the distance CD is much greater than the distance BA.

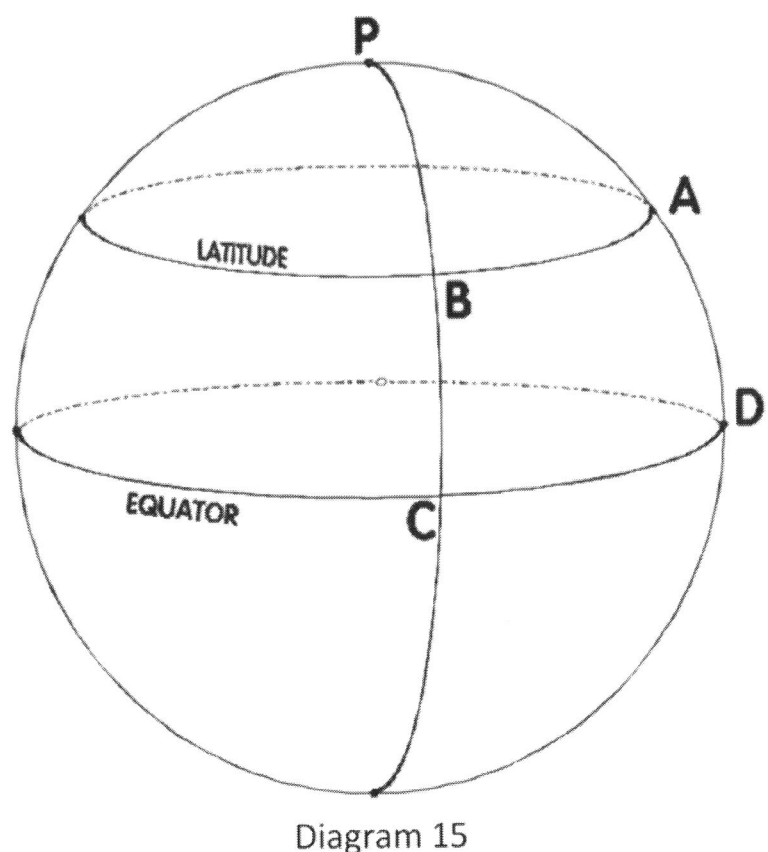

Diagram 15

To Calculate The Distance Between Two Meridians Along A Parallel Of Latitude.

The following formulae are used for calculating the difference in distance along a parallel of latitude (Ddist) corresponding to a difference in longitude (Dlong) and vice versa. (The formulae are simply stated below without explanation but a full explanation of their derivation is given in the appendix).

Ddist = Dlong × Cos Lat.   and   **Dlong = Ddist / Cos Lat.**

Since the secant is the inverse of the cosine, the formula for Dlong can be simplified to: **Dlong = Ddist × Sec Lat.**

The Rhumb Line

If a ship were to steer a steady course, that is one on which her heading remains constant, her track would cut all meridians at the same angle, as diagram 16 shows. Such a line on the Earth's surface is called a rhumb line.

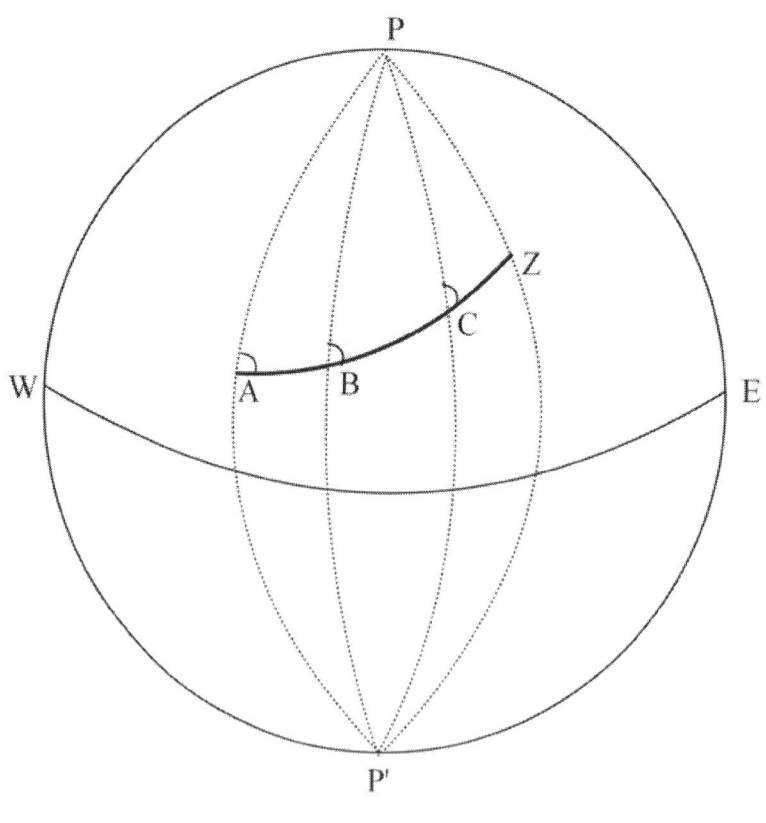

Diagram 16

When the rhumb line cuts all meridians at 90°, it will coincide with either a parallel of latitude or with the Equator. When the angle is 0°, the rhumb line will be along a meridian of longitude.

A vessel's course will always be a rhumb line; thus the course to be steered to travel from one place to another will refer to the angle between the rhumb line joining the places and any meridian.

Calculating the distance between two points along a rhumb line. In diagram 17, A, B, C, D and Z are meridians of longitude; the lines aB, bC, and cD are different parallels of latitude; and the line ABCDZ is a rhumb line. A series of right-angled triangles have been constructed along the rhumb line AZ and in each triangle, one short side lies along a meridian of longitude, one lies along a parallel of latitude and the hypotenuse lies along the rhumb line.

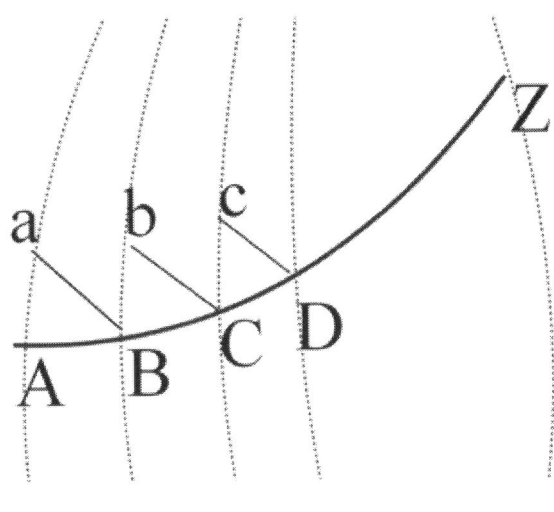

Diagram 17

It can be seen from the diagram that the east-west distance between two points along the rhumb line is the sum of the distances along the parallels of latitude corresponding to the difference in longitude in each of the right-angled triangles. This east-west distance is known as the **departure**

Middle Latitude. If we were to calculate the departure along each of the parallels of latitude aB, bC, cD, in diagram 17, we would find that they would not be equal and so the task of calculating the total departure would be complicated. In practice, the total departure is taken to be the east-west distance along the intermediate of these parallels which is known as the 'middle latitude'.
By the formula established for Ddist above, we can derive a formula to calculate departure as follows:
Departure = d.long cos(middle latitude).

Mean Latitude. In most cases, the arithmetic mean of the two latitudes can be used as the middle latitude without appreciable error, so the approximate formula **dep.= d.long cos(mean lat)** may be used.

When the difference of latitude is large (over 600 n.m.) or the latitudes are close to either of the poles, the middle latitude must be used instead of the mean latitude and in these cases, we have the more accurate formula:
Dep. = d.long cos(mid lat).

The difficulty lies in the task of calculating the middle latitude which involves finding the mean of the secants all the intermediate latitudes by integration. Such methods are obviously impracticable in situations where courses and distances have to be calculated rapidly at sea. For this reason, tables of corrections to be applied to the mean latitude are contained in various collections of nautical tables. Since, astro navigation involves short distance sailing calculations, it is not intended to copy middle latitude correction tables here; however, the following example demonstrates their use:

Suppose a ship sails from position 50°N, 32°E., to 70°N., 15°E.
The d.long is 17° and the mean latitude is 60°.
The formula for calculating departure using the mean lat. is:
dep.= d.long cos(mean lat)
Using this formula we have:
Dep. = 17° cos(60)
      = 1020' cos(60)
      = 510' or 510 n.m.

In the tables for converting mean latitude to middle latitude, the correction for a mean latitude of 60° and a difference of latitude of 20° is +1° 09'. So the middle latitude = 61°.15.
The formula for calculating departure using the middle latitude is:
Dep. = d.long cos(mid lat).
      = 1020 cos (61.15)
      = 492.17 n.m.

By comparing these results, we can see that there is a significant difference between calculations involving the mean latitude on one hand and the middle latitude on the other.

Summary of Formulae. The formulae so far derived in this chapter are summarised below:

Ddist = Dlong × Cos Lat.

Dlong = $\dfrac{\text{Ddist}}{\text{Cos Lat.}}$ = Ddist × Sec Lat.

dep.= d.long cos(mean lat) (for distances 600 n.m. or less).

Dep. = d.long cos(mid lat). (for distances over 600 n.m.).

The Rhumb Line Formulae.

With diagram 18 we expand on the work above,
- The rhumb line AZ is divided into a large number of equal parts AB, BC, CD, DZ.
- aB, bC, cD... are arcs of parallels of latitude drawn through B, C, D......
- Pa', Pb', Pc'... are meridians of longitude.
- Therefore, the angles at a, b, c...... are right angles.

If the divisions of AZ are made sufficiently small, the triangles ABa, BCb, CDc..... will be small enough to be treated as plane triangles instead of spherical triangles.
- Since the course angle is constant by the definition of a rhumb line, these small triangles are equal.

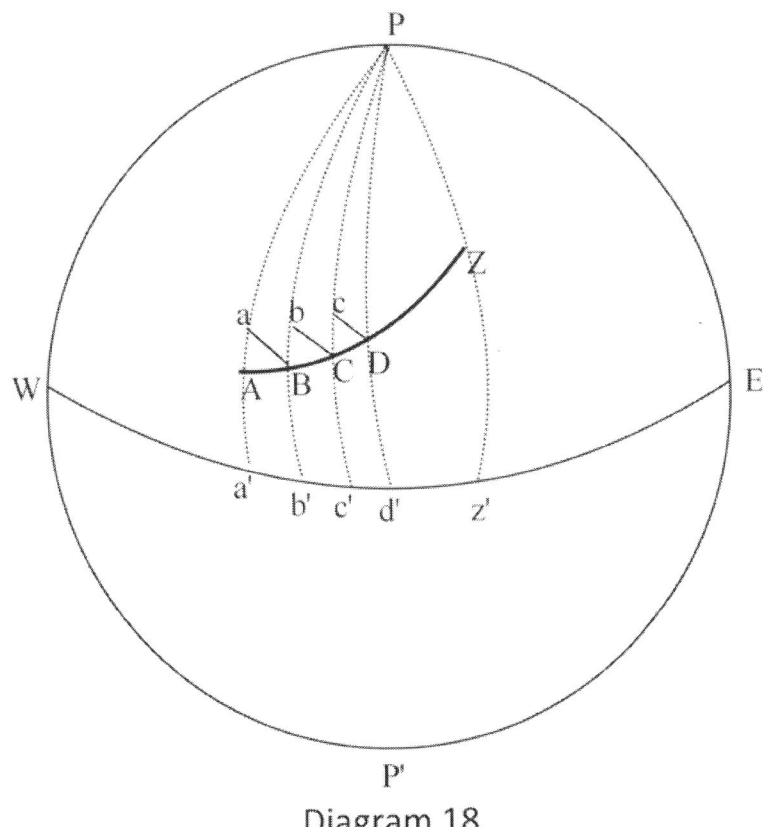

Diagram 18

Consider triangle ABa in diagram 18;
AB is the distance made good, aB is the departure along a parallel of latitude, angle aAB is the course angle.
∴ Sin(course angle) = departure ÷ dist. This formula applies to all of the small triangles since they are equal.
By transposition, the above formula becomes:
Dep = Dist x sin(course)
The departure between A and Z therefore, is the sum of the departures of all of the small triangles.
Therefore, by addition:
aB + bC + cD + ... = (AB + BC + CD + ... x sin(course)
i.e. **Dep = Dist sin(course)**

If we again consider triangle ABa, Aa = AB cos (course)
But Aa is the **difference in Latitude (D.Lat)** between A and B
So D.Lat = AB cos(course)
Again, this formula applies to all of the small triangles since they are equal.
Therefore, by addition, the total D.Lat corresponding to the total distance between A and Z becomes:
D.Lat = Dist cos(course)

We have established formulae to calculate Dep and D.Lat; we now need a formula to find the course.
If we return to triangle ABa, we can see that the course angle can be found by the formula: tan(course) = Dep ÷ D.Lat.
As before, this formula applies to all of the small equal triangles. So, by addition, the rhumb line course between A and Z can be found by the formula:
Tan(course) = Dep ÷ D.Lat

Short Distance Sailing.

Short distance sailing is a term which is applied to sailing along a rhumb-line for distances less than 600 nautical miles. From the formulae derived in this chapter, the following are used extensively in short distance sailing:
To Calculate Departure when the course is not known:
dep. = d.long cos(mean lat)

To Calculate Departure when the course is Known:
Dep = Dist x Sin(course)

To Calculate Distance when departure and course are known:

$$\text{Dist} = \frac{\text{Dep}}{\sin(\text{course})}$$

To Calculate Dlat when the distance and course are known:

$$\text{DLat} = \text{Dist} \times \cos(\text{course})$$

To Calculate Course to Steer (the rhumb line course between two points)

$$\tan(\text{course}) = \frac{\text{Dep}}{\text{D.Lat}}$$

To calculate Dlong (difference in longitude corresponding to the departure):

$$\text{DLong} = \text{Dep} \times \sec(\text{Mean.Lat}) \quad \text{or} \quad \text{Dlong} = \frac{\text{Dep}}{\cos(\text{Mean.Lat})}$$

Worked Examples

Example 1. What is the rhumb line course to steer and the distance to travel from position 40°.5N, 43°.0W to position 42°.25N 41°.8W?

Solution:

```
Dlat       = 42°.25N - 40°.5N
           = 1°.75N = 105'N
Mean Lat   = 40° 30'N + 52'.5
           = 41° 22'.5N
Dlong      = 43°.0W - 41°.8W
           = 1°.2E = 72'E
Dep        = d.long x cos(mean lat)
           = 72 cos(41.38)
           = 54'.02
Tan(course) = Dep  =  54.02
              D.Lat   105
            = 0.51
∴ course   = N27°E = 027°
Dist       =   Dep
             Sin(course)
           = 54.02
             Sin(27)
           = 120'
Course to steer = 027°
Distance to new position = 120 n.m.
```

Example 2. A vessel is in position 34° 15'.6 North, 54° 43'.5 West. What will be it's new position after steering course 065° at 12 knots for 3 hours?
Solution:

Distance travelled	= 36 n.m.
Depature	= Dist x Sin(course)
	= 36 x Sin(65) = 36 x 0.91
	= 32.6 n.m.
Dlat	= Dist x Cos(course)
	= 36 x Cos(65) = 36 x 0.42
	= 15'.21 North
New Lat	= 34° 15'.6 N. + 15'.21
	= 34° 30'.81 N.
Mean lat	= 34° 15'.6 N. + 7'.6
	= 34° 23'.2 N.
DLong.	= Dep. x Sec(Mean Lat)
	= 32.6 x Sec(34.39)
	= 32.6 x 1.21
	= 39'.4 East
New Long	= 54° 43'.5 W. - 39'.4
	= 54° 4'.1 W.
New Pos.	= 34° 30'.81 N. 54° 4'.1 W.

Example 3. Starting Pos: 55° 42'.8S. 23° 33'.8E.
Course: 225°. Speed 25 knots.
What is new position after 2.5 hours?

Solution:

Dist.	= 62.5 n.m.
Dep.	= 62.5 Sin(225) = -44.2
	= 44'.2 West(- indicates west)
Dlat	= 62.5 Cos(225) = -44'.2
	= 44'.2S. (- indicates south)
New Lat	= 55° 42'.8S. + 44'.2.
	= 56° 27'.0S.
M.lat	= 55° 42'.8S. + 22'.1
	= 56° 04'.9S.
DLong.	= 44.2 Sec(56.08)
	= 79'.2W. = 1° 19'.2W
New Long	= 23° 33'.8E - 1° 19'.2
	= 22° 14'.6E
New Pos.	= 56° 27'.0S. 22° 14'.6E

Example 4. At 0900 GMT, a life raft is reported to be in position 30° 56'.4 S, 0° 25'.6 E. A rescue ship reports that its ETA at the vicinity is 2130 GMT. The rescue ship's navigator calculates that wind and ocean currents will cause the life raft to drift in direction 345° at 3 knots. Calculate the expected position of the life raft when the rescue ship is due to arrive.

Solution:

Dist.	= 37.5 n.m.
Dep.	= 37.5 Sin(345) = 9'.7 West
Dlat	= 37.5 Cos(345) = 36'.2N
New Lat	= 30° 20'.2S.
M.lat	= 30° 38'.3S.
DLong.	= 9.7 Sec(30.64) = 11'.27W.
New Long	= 0° 14'.33E
New Pos.	= 30° 20'.2S. 0° 14'.33E

Self Test (solutions on next page)

Question 1. At 1230 GMT, a ship receives an S.O.S. signal from a vessel in position 23°.25S, 120°.5W. Her own D.R. position is 20°.4S, 118°.3W. What is the rhumb line course to steer to reach the S.O.S. position and what will be the E.T.A. if the speed is 25 knots?

Question 2. If a ship starts from position 30° 21'N, 15° 54'W and travels on a course of 030° at 12 knots for 5 hours, what will be its new position?

Question 3. At 0930, a ship's position is 48° 18'.75S, 29° 28'.30E. Course is 148° and speed is 28 knots. What will be the ship's position at 1158?

Question 4. A yacht's D.R. position at 0800 is 36° 23'.4N. 09° 15'.4W. The Skipper estimates that the course and speed made good for the next 4 hours was 075°, 6 knots. What was the estimated position at 1200?

©2009 Jack Case

Solutions To Test Questions

Solution Q.1.

Dlat	= 23°.25S - 20°.4S
	= 2°.85S = 171'S
Mean Lat	= 20°.4S + 1°.425
	= 21°.825S
Dlong	= 120°.5W - 118°.3W
	= 2°.2W = 132'W
Dep	= d.long × cos(mean lat)
	= 132 cos(21°.825)
	= 122'.5E
Tan(course)	= Dep / D.Lat = 122'.5 / 171' = 0.716
course	= S36°W = 216°
Dist	= Dep / Sin(course) = 122.5 / Sin(216)
	= 208'.41

Time to reach SOS position = 208.41 ÷ 25
= 8.3 hrs.
Course to steer = 216°
Distance to new position = 208.41 n.m.
ETA = 1230 + 0818 = 2048 GMT

Solution Q.2.

Dist.	= 60 n.m.
Dep.	= 60.Sin(30) = 30' E.
Dlat	= 60.Cos(30) = 51'.96 N.
New Lat	= 31° 12'.96 N.
M.lat	= 30° 46'.98 N. = 30°.78N.
DLong.	= 30.Sec(30.78) = 34'.92E
New Long	= 15° 54' - 34'.92 = 15° 19'.08W.
New Pos.	= 31° 12'.96N. 15° 19'.08W.

Solution Q.3.

Dist.	= 2.47 hrs @ 28 knots
	= 69.16nm
Dep.	= 69.16.Sin(148) = 36'.65 E.
Dlat	= 69.16.Cos(148) = 58'.65 S.
New Lat	= 49° 17'.4 S.
M.lat	= 48° 48'.08 S.
DLong.	= 36.65.Sec(48.8) = 55'.6E
New Long	= 30° 23'.9E
New Pos.	= 49° 17'.4 S. 30° 23'.9E

Solution Q.4.

Dist.	= 24 n.m.
Dep.	= 24.Sin(75) = 23'.18 E.
Dlat	= 24.Cos(75) = 6'.21 N.
New Lat	= 36° 29'.61 N.
M.lat	= 36° 26'.5 N.
DLong.	= 23.18.Sec(36.44) = 28'.8E
New Long	= 08° 46'.6W.
New Pos.	= 36° 29'.61 N. 08° 46'.6W.

Traverse Tables. Traverse tables have been used for about 200 years to solve the mathematical problems involved in short distance sailing. The tables simply tabulate the results of relevant trigonometric calculations and would have been extremely useful to navigators in the days before electronic calculators became available. However, it should be noted that, although traverse tables are reasonably accurate for short distance sailing, they are not accurate over long distances (over 600 nautical miles). This is because data in the traverse tables is calculated using 'straight line' trigonometry which cannot be applied to long distance, great circle sailing involving the use of 'spherical trigonometry'. (The appendix gives an exposition of trigonometry including spherical trigonometry).

Note. Recent discussions regarding the possibility that the SOS position sent by the Titanic was incorrect due to mistakes made in applying the traverse tables make interesting reading.

Traverse tables are not used so much these days because short distance sailing problems can be solved quickly and more accurately with the help of a calculator. There is also the question of expense; a book of nautical tables (containing traverse tables) can make a costly outlay for a small boat sailor whereas electronic calculators are relatively cheap to buy. For these reasons, in this book, short distance sailing problems are solved by applying the formulae for D.Lat, D.Long, departure and distance as demonstrated above. However, a brief demonstration of the use of traverse tables is given below:

Demonstration of the use of the Traverse Tables.

In this demonstration, the traverse tables are used to find the solution to test question 4 above:

i.e. Starting position = 36° 23'.4N. 09° 15'.4W.

Course and speed made good for 4 hours = 075°, 6 knots.

∴ distance covered = 24 n.m.

What was the estimated position after 4 hours?

To find new Lat:
- In the first extract from the traverse table on the following pages, the course angle 75° is at the bottom of the page.
- Opposite 24 in the Dist. column find 23.2 in the Dep. Column (use labels along the bottom row above 75°).
- Opposite 24 in the Dist. Column find 06.2 in the D.Lat column (as labelled at the bottom).

Lat	=	36° 23'.4N.
D.Lat	=	06'.2N
New Lat	=	36° 29'.6N

To find new Long:

Using the traverse tables to calculate D.Long can be confusing because the labelling is arranged for calculating D.lat. This confusion can be avoided by re-arranging the labelling in the triangle at the top of the table for 36° so that D.Long becomes the hypotenuse, Dep. becomes the adjacent and D.Lat becomes the opposite. We can now calculate the D.Long. as follows:
- In the second traverse table extract, take M.Lat 36° (nearest degree) as the course angle at the top of the page.
- Opposite 23.5, in the adjacent column (departure) find 29 in the hypotenuse column (D.Long).

Long	=	09° 15'.4 W
D.Long	=	29'.0 E
New Long	=	08° 46'.4 W

∴ New pos. = 36° 29'.6N. 08° 46'.4W. (These answers correspond reasonably closely with the answers to question 4 which were calculated by applying the formulae).

Table 3. TRAVERSE TABLE. 15° / 15°

Dist. Hyp.	D. Lat. Adj.	Dep. Opp.	Dist. Hyp.	D. Lat. Adj.	Dep. Opp.	Dist. Hyp.	D. Lat. Adj.	Dep. Opp.	Dist. Hyp.	D. Lat. Adj.	Dep. Opp.	Dist. Hyp.	D. Lat. Adj.	Dep. Opp.
1	1.0	0.3	61	58.9	15.8	121	116.9	31.3	181	174.8	46.8	241	232.8	62.4
2	1.9	0.5	62	59.9	16.0	122	117.8	31.6	182	175.8	47.1	242	233.8	62.6
3	2.9	0.8	63	60.9	16.3	123	118.8	31.8	183	176.8	47.4	243	234.7	62.9
4	3.9	1.0	64	61.8	16.6	124	119.8	32.1	184	177.7	47.6	244	235.7	63.2
5	4.8	1.3	65	62.8	16.8	125	120.7	32.4	185	178.7	47.9	245	236.7	63.4
6	5.8	1.6	66	63.8	17.1	126	121.7	32.6	186	179.7	48.1	246	237.6	63.7
7	6.8	1.8	67	64.7	17.3	127	122.7	32.9	187	180.6	48.4	247	238.6	63.9
8	7.7	2.1	68	65.7	17.6	128	123.6	33.1	188	181.6	48.7	248	239.5	64.2
9	8.7	2.3	69	66.6	17.9	129	124.6	33.4	189	182.6	48.9	249	240.5	64.4
10	9.7	2.6	70	67.6	18.1	130	125.6	33.6	190	183.5	49.2	250	241.5	64.7
11	10.6	2.8	71	68.6	18.4	131	126.5	33.9	191	184.5	49.4	251	242.4	65.0
12	11.6	3.1	72	69.5	18.6	132	127.5	34.2	192	185.5	49.7	252	243.4	65.2
13	12.6	3.4	73	70.5	18.9	133	128.5	34.4	193	186.4	50.0	253	244.4	65.5
14	13.5	3.6	74	71.5	19.2	134	129.4	34.7	194	187.4	50.2	254	245.3	65.7
15	14.5	3.9	75	72.4	19.4	135	130.4	34.9	195	188.4	50.5	255	246.3	66.0
16	15.5	4.1	76	73.4	19.7	136	131.4	35.2	196	189.3	50.7	256	247.3	66.3
17	16.4	4.4	77	74.4	19.9	137	132.3	35.5	197	190.3	51.0	257	248.2	66.5
18	17.4	4.7	78	75.3	20.2	138	133.3	35.7	198	191.3	51.2	258	249.2	66.8
19	18.4	4.9	79	76.3	20.4	139	134.3	36.0	199	192.2	51.5	259	250.2	67.0
20	19.3	5.2	80	77.3	20.7	140	135.2	36.2	200	193.2	51.8	260	251.1	67.3
21	20.3	5.4	81	78.2	21.0	141	136.2	36.5	201	194.2	52.0	261	252.1	67.6
22	21.3	5.7	82	79.2	21.2	142	137.2	36.8	202	195.1	52.3	262	253.1	67.8
23	22.2	6.0	83	80.2	21.5	143	138.1	37.0	203	196.1	52.5	263	254.0	68.1
24	23.2	6.2	84	81.1	21.7	144	139.1	37.3	204	197.0	52.8	264	255.0	68.3
25	24.1	6.5	85	82.1	22.0	145	140.1	37.5	205	198.0	53.1	265	256.0	68.6
26	25.1	6.7	86	83.1	22.3	146	141.0	37.8	206	199.0	53.3	266	256.9	68.8
27	26.1	7.0	87	84.0	22.5	147	142.0	38.0	207	199.9	53.6	267	257.9	69.1
28	27.0	7.2	88	85.0	22.8	148	143.0	38.3	208	200.9	53.8	268	258.9	69.4
29	28.0	7.5	89	86.0	23.0	149	143.9	38.6	209	201.9	54.1	269	259.8	69.6
30	29.0	7.8	90	86.9	23.3	150	144.9	38.8	210	202.8	54.4	270	260.8	69.9
31	29.9	8.0	91	87.9	23.6	151	145.9	39.1	211	203.8	54.6	271	261.8	70.1
32	30.9	8.3	92	88.9	23.8	152	146.8	39.3	212	204.8	54.9	272	262.7	70.4
33	31.9	8.5	93	89.8	24.1	153	147.8	39.6	213	205.7	55.1	273	263.7	70.7
34	32.8	8.8	94	90.8	24.3	154	148.8	39.9	214	206.7	55.4	274	264.7	70.9
35	33.8	9.1	95	91.8	24.6	155	149.7	40.1	215	207.7	55.6	275	265.6	71.2
36	34.8	9.3	96	92.7	24.8	156	150.7	40.4	216	208.6	55.9	276	266.6	71.4
37	35.7	9.6	97	93.7	25.1	157	151.7	40.6	217	209.6	56.2	277	267.6	71.7
38	36.7	9.8	98	94.7	25.4	158	152.6	40.9	218	210.6	56.4	278	268.5	72.0
39	37.7	10.1	99	95.6	25.6	159	153.6	41.2	219	211.5	56.7	279	269.5	72.2
40	38.6	10.4	100	96.6	25.9	160	154.5	41.4	220	212.5	56.9	280	270.5	72.5
41	39.6	10.6	101	97.6	26.1	161	155.5	41.7	221	213.5	57.2	281	271.4	72.7
42	40.6	10.9	102	98.5	26.4	162	156.5	41.9	222	214.4	57.5	282	272.4	73.0
43	41.5	11.1	103	99.5	26.7	163	157.4	42.2	223	215.4	57.7	283	273.4	73.2
44	42.5	11.4	104	100.5	26.9	164	158.4	42.4	224	216.4	58.0	284	274.3	73.5
45	43.5	11.6	105	101.4	27.2	165	159.4	42.7	225	217.3	58.2	285	275.3	73.8
46	44.4	11.9	106	102.4	27.4	166	160.3	43.0	226	218.3	58.5	286	276.3	74.0
47	45.4	12.2	107	103.4	27.7	167	161.3	43.2	227	219.3	58.8	287	277.2	74.3
48	46.4	12.4	108	104.3	28.0	168	162.3	43.5	228	220.2	59.0	288	278.2	74.5
49	47.3	12.7	109	105.3	28.2	169	163.2	43.7	229	221.2	59.3	289	279.2	74.8
50	48.3	12.9	110	106.3	28.5	170	164.2	44.0	230	222.2	59.5	290	280.1	75.1
51	49.3	13.2	111	107.2	28.7	171	165.2	44.3	231	223.1	59.8	291	281.1	75.3
52	50.2	13.5	112	108.2	29.0	172	166.1	44.5	232	224.1	60.0	292	282.1	75.6
53	51.2	13.7	113	109.1	29.2	173	167.1	44.8	233	225.1	60.3	293	283.0	75.8
54	52.2	14.0	114	110.1	29.5	174	168.1	45.0	234	226.0	60.6	294	284.0	76.1
55	53.1	14.2	115	111.1	29.8	175	169.0	45.3	235	227.0	60.8	295	284.9	76.4
56	54.1	14.5	116	112.0	30.0	176	170.0	45.6	236	228.0	61.1	296	285.9	76.6
57	55.1	14.8	117	113.0	30.3	177	171.0	45.8	237	228.9	61.3	297	286.9	76.9
58	56.0	15.0	118	114.0	30.5	178	171.9	46.1	238	229.9	61.6	298	287.8	77.1
59	57.0	15.3	119	114.9	30.8	179	172.9	46.3	239	230.9	61.9	299	288.8	77.4
60	58.0	15.5	120	115.9	31.1	180	173.9	46.6	240	231.8	62.1	300	289.8	77.6
Hyp.	Opp.	Adj.	Hyp.	Opp.	Adj.	Hyp.	Opp.	Adj.	Hyp.	Opp.	Adj.	Hyp.	Opp.	Adj.
Dist.	Dep.	D. Lat.	Dist.	Dep.	D. Lat.	Dist.	Dep.	D. Lat.	Dist.	Dep.	D. Lat.	Dist.	Dep.	D. Lat.

75° / 75°

Table 3. TRAVERSE TABLE.

36°  36°

Dist.	D. Lat.	Dep.	Dist.	D. Lat.	Dep.	Dist.	D. Lat.	Dep.	Dist.	D. Lat.	Dep.	Dist.	D. Lat.	Dep.
Hyp.	Adj.	Opp.	Hyp.	Adj.	Opp.	Hyp.	Adj.	Opp.	Hyp.	Adj.	Opp.	Hyp.	Adj.	Opp.
1	0.8	0.6	61	49.4	35.9	121	97.9	71.1	181	146.4	106.4	241	195.0	141.7
2	1.6	1.2	62	50.2	36.4	122	98.7	71.7	182	147.2	107.0	242	195.8	142.2
3	2.4	1.8	63	51.0	37.0	123	99.5	72.3	183	148.1	107.6	243	196.6	142.8
4	3.2	2.4	64	51.8	37.6	124	100.3	72.9	184	148.9	108.2	244	197.4	143.4
5	4.0	2.9	65	52.6	38.2	125	101.1	73.5	185	149.7	108.7	245	198.2	144.0
6	4.9	3.5	66	53.4	38.8	126	101.9	74.1	186	150.5	109.3	246	199.0	144.6
7	5.7	4.1	67	54.2	39.4	127	102.7	74.6	187	151.3	109.9	247	199.8	145.2
8	6.5	4.7	68	55.0	40.0	128	103.6	75.2	188	152.1	110.5	248	200.6	145.8
9	7.3	5.3	69	55.8	40.6	129	104.4	75.8	189	152.9	111.1	249	201.4	146.4
10	8.1	5.9	70	56.6	41.1	130	105.2	76.4	190	153.7	111.7	250	202.3	146.9
11	8.9	6.5	71	57.4	41.7	131	106.0	77.0	191	154.5	112.3	251	203.1	147.5
12	9.7	7.1	72	58.2	42.3	132	106.8	77.6	192	155.3	112.9	252	203.9	148.1
13	10.5	7.6	73	59.1	42.9	133	107.6	78.2	193	156.1	113.4	253	204.7	148.7
14	11.3	8.2	74	59.9	43.5	134	108.4	78.8	194	156.9	114.0	254	205.5	149.3
15	12.1	8.8	75	60.7	44.1	135	109.2	79.4	195	157.8	114.6	255	206.3	149.9
16	12.9	9.4	76	61.5	44.7	136	110.0	79.9	196	158.6	115.2	256	207.1	150.5
17	13.8	10.0	77	62.3	45.3	137	110.8	80.5	197	159.4	115.8	257	207.9	151.1
18	14.6	10.6	78	63.1	45.8	138	111.6	81.1	198	160.2	116.4	258	208.7	151.6
19	15.4	11.2	79	63.9	46.4	139	112.5	81.7	199	161.0	117.0	259	209.5	152.2
20	16.2	11.8	80	64.7	47.0	140	113.3	82.3	200	161.8	117.6	260	210.3	152.8
21	17.0	12.3	81	65.5	47.6	141	114.1	82.9	201	162.6	118.1	261	211.2	153.4
22	17.8	12.9	82	66.3	48.2	142	114.9	83.5	202	163.4	118.7	262	212.0	154.0
23	18.6	13.5	83	67.1	48.8	143	115.7	84.1	203	164.2	119.3	263	212.8	154.6
24	19.4	14.1	84	68.0	49.4	144	116.5	84.6	204	165.0	119.9	264	213.6	155.2
25	20.2	14.7	85	68.8	50.0	145	117.3	85.2	205	165.8	120.5	265	214.4	155.8
26	21.0	15.3	86	69.6	50.5	146	118.1	85.8	206	166.7	121.1	266	215.2	156.4
27	21.8	15.9	87	70.4	51.1	147	118.9	86.4	207	167.5	121.7	267	216.0	156.9
28	22.7	16.5	88	71.2	51.7	148	119.7	87.0	208	168.3	122.3	268	216.8	157.5
29	23.5	17.0	89	72.0	52.3	149	120.5	87.6	209	169.1	122.8	269	217.6	158.1
30	24.3	17.6	90	72.8	52.9	150	121.4	88.2	210	169.9	123.4	270	218.4	158.7
31	25.1	18.2	91	73.6	53.5	151	122.2	88.8	211	170.7	124.0	271	219.2	159.3
32	25.9	18.8	92	74.4	54.1	152	123.0	89.3	212	171.5	124.6	272	220.1	159.9
33	26.7	19.4	93	75.2	54.7	153	123.8	89.9	213	172.3	125.2	273	220.9	160.5
34	27.5	20.0	94	76.0	55.3	154	124.6	90.5	214	173.1	125.8	274	221.7	161.1
35	28.3	20.6	95	76.9	55.9	155	125.4	91.1	215	173.9	126.4	275	222.5	161.6
36	29.1	21.2	96	77.7	56.4	156	126.2	91.7	216	174.7	127.0	276	223.3	162.2
37	29.9	21.7	97	78.5	57.0	157	127.0	92.3	217	175.6	127.5	277	224.1	162.8
38	30.7	22.3	98	79.3	57.6	158	127.8	92.9	218	176.4	128.1	278	224.9	163.4
39	31.6	22.9	99	80.1	58.2	159	128.6	93.5	219	177.2	128.7	279	225.7	164.0
40	32.4	23.5	100	80.9	58.8	160	129.4	94.0	220	178.0	129.3	280	226.5	164.6
41	33.2	24.1	101	81.7	59.4	161	130.3	94.6	221	178.8	129.9	281	227.3	165.2
42	34.0	24.7	102	82.5	60.0	162	131.1	95.2	222	179.6	130.5	282	228.1	165.8
43	34.8	25.3	103	83.3	60.5	163	131.9	95.8	223	180.4	131.1	283	229.0	166.3
44	35.6	25.9	104	84.1	61.1	164	132.7	96.4	224	181.2	131.7	284	229.8	166.9
45	36.4	26.5	105	84.9	61.7	165	133.5	97.0	225	182.0	132.3	285	230.6	167.5
46	37.2	27.0	106	85.8	62.3	166	134.3	97.6	226	182.8	132.8	286	231.4	168.1
47	38.0	27.6	107	86.6	62.9	167	135.1	98.2	227	183.6	133.4	287	232.2	168.7
48	38.8	28.2	108	87.4	63.5	168	135.9	98.7	228	184.5	134.0	288	233.0	169.3
49	39.6	28.8	109	88.2	64.1	169	136.7	99.3	229	185.3	134.6	289	233.8	169.9
50	40.5	29.4	110	89.0	64.7	170	137.5	99.9	230	186.1	135.2	290	234.6	170.5
51	41.3	30.0	111	89.8	65.2	171	138.3	100.5	231	186.9	135.8	291	235.4	171.0
52	42.1	30.6	112	90.6	65.8	172	139.2	101.1	232	187.7	136.4	292	236.2	171.6
53	42.9	31.2	113	91.4	66.4	173	140.0	101.7	233	188.5	137.0	293	237.0	172.2
54	43.7	31.7	114	92.2	67.0	174	140.8	102.3	234	189.3	137.5	294	237.9	172.8
55	44.5	32.3	115	93.0	67.6	175	141.6	102.9	235	190.1	138.1	295	238.7	173.4
56	45.3	32.9	116	93.8	68.2	176	142.4	103.5	236	190.9	138.7	296	239.5	174.0
57	46.1	33.5	117	94.7	68.8	177	143.2	104.0	237	191.7	139.3	297	240.3	174.6
58	46.9	34.1	118	95.5	69.4	178	144.0	104.6	238	192.5	139.9	298	241.1	175.2
59	47.7	34.7	119	96.3	69.9	179	144.8	105.2	239	193.4	140.5	299	241.9	175.7
60	48.5	35.3	120	97.1	70.5	180	145.6	105.8	240	194.2	141.1	300	242.7	176.3
Hyp.	Opp.	Adj.	Hyp.	Opp.	Adj.	Hyp.	Opp.	Adj.	Hyp.	Opp.	Adj.	Hyp.	Opp.	Adj.
Dist.	Dep.	D. Lat.	Dist.	Dep.	D. Lat.	Dist.	Dep.	D. Lat.	Dist.	Dep.	D. Lat.	Dist.	Dep.	D. Lat.

54° 54°

Great Circle Sailing

If two points lie on the surface of the Earth, they will lie on the circumference of a great circle and the shorter arc of the great circle will be the shortest route between them.

In diagram 19, A and B are two such points and the arc AB is the shortest distance between them.

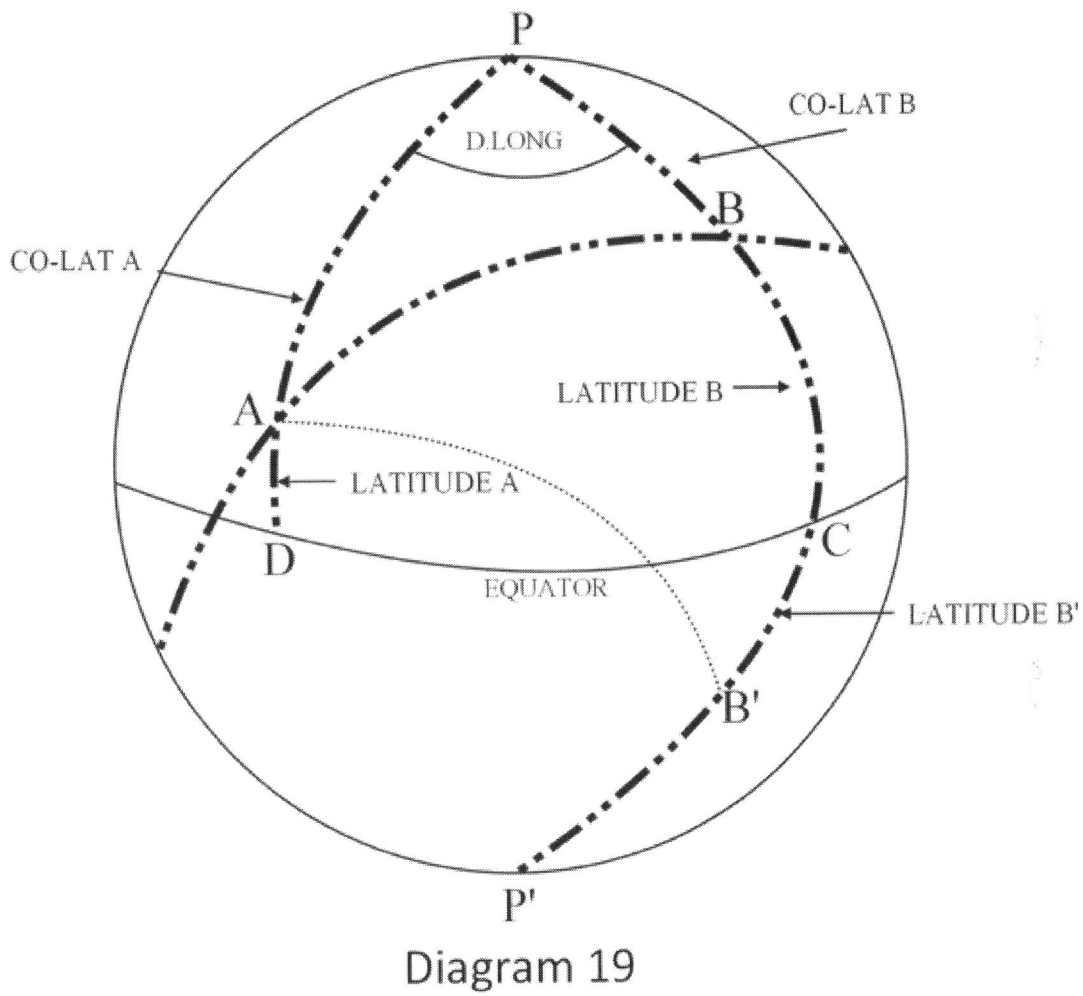

Diagram 19

PD and PC are arcs running along the meridians passing through A and B.
The arcs AD and BC are equal to the angular distances of the latitudes of points A and B respectively.

Since PA, PB and AB form the sides of the triangle PAB and since these sides are arcs of great circles, PAB must be a spherical triangle. The length of AB and the direction of one point to another can be found by solving this triangle.

©2009 Jack Case

The distance between A and B is the angular distance AB and the angle APB is the d.long. The true bearing of point B from point A is the angle between the meridian through A and the great circle joining A and B, measured clockwise from the meridian, that is, the angle PAB.

The complement of the latitude of a place (90° - Lat) is known as the 'co-latitude' (co-lat for short). PA and PB therefore, are the co-lats of points A and B respectively.

The above rule applies when the two places are in the same hemisphere. When they are in opposite hemispheres, the rule changes. B' is in the southern hemisphere, so the length of side PB' is clearly 90° + the latitude of B.

To summarise the above rules:
When latitudes have the same name:
    PA = 90° - lat A = co-lat A
    PB = 90° - lat B = co-lat B
    Angle APB = d.long.

When latitudes have opposite names:
    PA = 90° - lat A = co-lat A
    PB' = 90° + lat B'
    Angle APB = d.long.

For distances greater than 600 n.m. the rules for short distance sailing do not apply and so a ship's track has to be plotted along the path of a great circle. In practice, the busy navigator plots a great-circle track on a Mercator chart with the aid of a gnomonic chart or diagram and so avoids having to solve the spherical triangle PAB.

Gnomonic charts are constructed so that any straight line drawn on them will represent a great circle. They are formed by projecting the Earth's surface from the Earth's centre on to the tangent plane at any convenient point. Since a great circle is formed by the intersection of a plane through the Earth's centre with the Earth's surface, and as one plane will always cut another in a straight line, all great circles including the equator and meridians of longitude will appear on the chart as straight lines. Since parallels of latitude are not great circles, they form a series of curves on a gnomonic graticule. Because rhumb lines do not follow an arc of a great circle, they also will appear as curved lines.

Diagram 20 shows the great circle track AB marked on the graticule of a gnomonic chart.

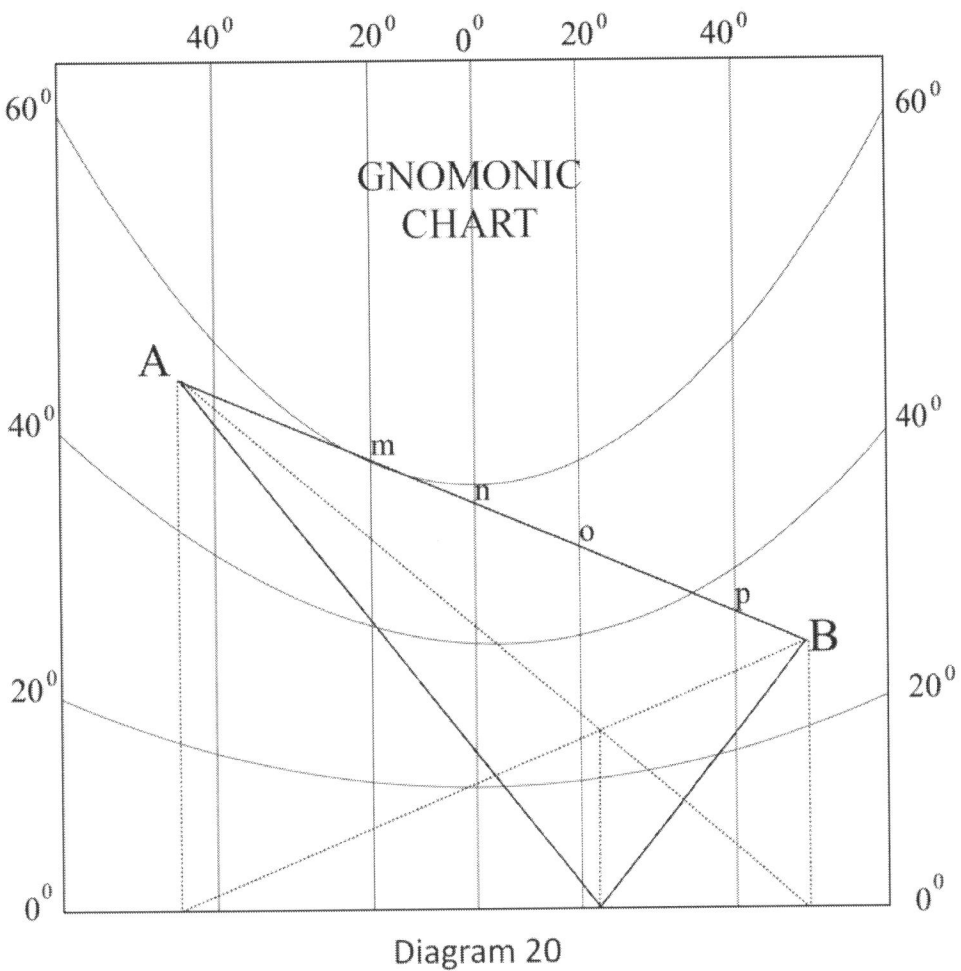

Diagram 20

The graticule is symmetrical about the meridian running through the tangent point which, in this case, is on the equator. The longitude scale can be set as convenient since the meridians on the chart do not coincide with meridians of longitude. Angles are distorted so it is impossible to take courses and distances from gnomonic charts and for this reason, they have to be used in conjunction with a Mercator chart. A great-circle track, such as AB can be transferred from a gnomonic chart by noting the latitude and longitude of convenient points such as m,n,o,p on the line AB, then marking these points on the Mercator chart and joining them with a smooth curve.

Obviously a ship cannot be steered along the continuous curve of a great circle and so the track has to be plotted as a series of rhumb lines between points along the great circle route. This is known as approximate great circle sailing or simply, great circle sailing.

Chapter 4
Calculating the Local Hour Angle and Declination of a Celestial Body.

In astro navigation, the navigator will usually begin the process of calculating a vessel's actual position by working from an approximate position. This is known as the **'assumed position'**.

In order to establish a position line from an observation of a celestial body, we first need to know its Declination and Local Hour Angle (LHA). As explained in chapter 1, we calculate the LHA by combining the vessel's longitude with the Greenwich Hour Angle (GHA) of the body.

The **Nautical Almanac** contains tables of raw data concerning the Declination and Greenwich Hour Angle for the Sun, the Moon, the navigational planets and selected stars.

The extract shown overleaf, is of one of the Nautical Almanac daily pages displaying hourly values of the Greenwich Hour Angle and the Declination of the Sun and the Moon for the 24$^{th.}$ 25$^{th.}$ and 26$^{th.}$ December. The page also shows data relating the times of rising and setting of the Sun and the Moon and the 'Equation of Time' which were explained in chapter 2. Other useful data shown concerns the meridian passage (Mer. Pas.) of the Sun and Moon (see chapter 6 for an explanation of this).

The following examples demonstrate how the daily pages are used.

Example 1. To find the LHA and Declination of the Sun.
At 04 hours, 32 minutes, 04 seconds GMT on 24 December 2009, the assumed position of your yacht is 40° 35.5' South 32° 13.8' East. Find the LHA and Declination of the Sun.

Calculating the Local Hour Angle (LHA).
Before we can calculate the LHA, we need to find the Greenwich Hour Angle of the Sun.

Step 1. Find the GHA for 0400 GMT.
From the following daily page extract, we find that the GHA of the Sun for 24 December at 0400 is 240° 07'.5. This is written as:

GHA (04^h) : 240° 07'.5

Extract from Nautical Almanac showing daily page for 24, 25 26 December.

2009 DECEMBER 24, 25, 26 (THURS., FRI., SAT.)

UT	SUN		MOON				Lat.	Twilight		Sunrise	Moonrise			
	GHA	Dec	GHA	v	Dec	d	HP		Naut.	Civil		24	25	26	27
d h	° '	° '	° '	'	° '	'	'	°	h m	h m	h m	h m	h m	h m	h m
24 00	180 08.8	S23 25.1	99 57.7	16.2	N 1 57.5	13.1	54.9	N 72	08 27	10 57	■	10 46	10 17	09 38	☐
01	195 08.4	25.1	114 32.9	16.1	2 10.6	13.2	54.9	N 70	08 07	09 55	■	10 51	10 31	10 06	09 19
02	210 08.1	25.0	129 08.0	16.0	2 23.8	13.2	54.9	68	07 51	09 20	■	10 56	10 43	10 27	10 03
03	225 07.8	.. 25.0	143 43.0	16.1	2 37.0	13.1	55.0	66	07 38	08 55	10 35	11 00	10 52	10 43	10 32
04	240 07.5	24.9	158 18.1	16.0	2 50.1	13.2	55.0	64	07 27	08 35	09 53	11 03	11 00	10 57	10 54
05	255 07.2	24.9	172 53.1	16.0	3 03.3	13.1	55.0	62	07 17	08 19	09 25	11 06	11 07	11 08	11 12
								60	07 09	08 06	09 03	11 09	11 13	11 18	11 27
06	270 06.9	S23 24.8	187 28.1	15.9	N 3 16.4	13.2	55.0	N 58	07 01	07 54	08 46	11 11	11 18	11 27	11 40
07	285 06.6	24.8	202 03.0	15.9	3 29.6	13.1	55.1	56	06 55	07 44	08 31	11 13	11 23	11 35	11 51
T 08	300 06.3	24.7	216 37.9	15.9	3 42.7	13.1	55.1	54	06 49	07 35	08 19	11 15	11 27	11 42	12 01
H 09	315 06.0	.. 24.7	231 12.8	15.8	3 55.9	13.1	55.1	52	06 43	07 27	08 08	11 17	11 31	11 48	12 09
U 10	330 05.7	24.6	245 47.6	15.8	4 09.0	13.2	55.1	50	06 38	07 19	07 58	11 18	11 35	11 54	12 17
R 11	345 05.3	24.6	260 22.4	15.8	4 22.2	13.1	55.2	45	06 26	07 03	07 37	11 22	11 42	12 06	12 34
S 12	0 05.0	S23 24.5	274 57.2	15.7	N 4 35.3	13.2	55.2	N 40	06 15	06 50	07 20	11 25	11 49	12 16	12 48
D 13	15 04.7	24.5	289 31.9	15.6	4 48.5	13.1	55.2	35	06 06	06 38	07 06	11 27	11 54	12 25	13 00
A 14	30 04.4	24.4	304 06.5	15.7	5 01.6	13.1	55.2	30	05 57	06 27	06 54	11 29	11 59	12 32	13 10
Y 15	45 04.1	.. 24.4	318 41.2	15.5	5 14.7	13.1	55.3	20	05 41	06 08	06 32	11 33	12 08	12 46	13 28
16	60 03.8	24.3	333 15.7	15.6	5 27.8	13.1	55.3	N 10	05 25	05 51	06 14	11 37	12 16	12 57	13 44
17	75 03.5	24.2	347 50.3	15.4	5 40.9	13.1	55.3	0	05 08	05 34	05 56	11 40	12 23	13 08	13 58
18	90 03.2	S23 24.2	2 24.7	15.5	N 5 54.0	13.1	55.3	S 10	04 49	05 16	05 39	11 43	12 30	13 20	14 13
19	105 02.9	24.1	16 59.2	15.3	6 07.1	13.1	55.4	20	04 26	04 55	05 20	11 47	12 38	13 32	14 29
20	120 02.6	24.1	31 33.5	15.4	6 20.2	13.0	55.4	30	03 57	04 30	04 58	11 51	12 47	13 45	14 48
21	135 02.2	.. 24.0	46 07.9	15.2	6 33.2	13.1	55.4	35	03 38	04 15	04 45	11 53	12 52	13 54	14 59
22	150 01.9	23.9	60 42.1	15.3	6 46.3	13.0	55.5	40	03 15	03 57	04 30	11 56	12 58	14 03	15 11
23	165 01.6	23.9	75 16.4	15.1	6 59.3	13.0	55.5	45	02 45	03 34	04 12	11 59	13 05	14 14	15 26
25 00	180 01.3	S23 23.8	89 50.5	15.1	N 7 12.3	13.0	55.5	S 50	01 59	03 05	03 49	12 03	13 13	14 27	15 44
01	195 01.0	23.8	104 24.6	15.1	7 25.3	13.0	55.5	52	01 31	02 49	03 38	12 05	13 17	14 33	15 53
02	210 00.7	23.7	118 58.7	15.0	7 38.3	13.0	55.6	54	00 45	02 32	03 26	12 07	13 21	14 40	16 02
03	225 00.4	.. 23.6	133 32.7	14.9	7 51.3	12.9	55.6	56	////	02 09	03 12	12 09	13 26	14 48	16 13
04	240 00.1	23.6	148 06.6	14.9	8 04.2	12.9	55.6	58	////	01 39	02 55	12 11	13 32	14 56	16 26
05	254 59.8	23.5	162 40.5	14.8	8 17.1	12.9	55.7	S 60	////	00 49	02 35	12 14	13 38	15 06	16 40
06	269 59.5	S23 23.4	177 14.3	14.7	N 8 30.0	12.9	55.7	Lat.	Sunset	Twilight		Moonset			
07	284 59.1	23.4	191 48.0	14.7	8 42.9	12.9	55.7			Civil	Naut.	24	25	26	27
08	299 58.8	23.3	206 21.7	14.6	8 55.8	12.8	55.8	°	h m	h m	h m	h m	h m	h m	h m
F 09	314 58.5	.. 23.2	220 55.3	14.5	9 08.6	12.9	55.8	N 72	■	13 03	15 33	25 38	01 38	03 51	☐
R 10	329 58.2	23.1	235 28.8	14.5	9 21.5	12.8	55.8	N 70	■	14 05	15 53	25 27	01 27	03 25	05 53
I 11	344 57.9	23.1	250 02.3	14.4	9 34.3	12.7	55.8	68	■	14 40	16 09	25 18	01 18	03 06	05 11
D 12	359 57.6	S23 23.0	264 35.7	14.3	N 9 47.0	12.8	55.9	66	13 25	15 05	16 22	25 10	01 10	02 51	04 42
A 13	14 57.3	22.9	279 09.0	14.3	9 59.8	12.7	55.9	64	14 07	15 25	16 33	25 04	01 04	02 38	04 21
Y 14	29 57.0	22.8	293 42.3	14.2	10 12.5	12.7	55.9	62	14 35	15 41	16 43	24 58	00 58	02 28	04 04
15	44 56.7	.. 22.8	308 15.5	14.1	10 25.2	12.6	56.0	60	14 57	15 55	16 51	24 53	00 53	02 19	03 50
16	59 56.4	22.7	322 48.6	14.0	10 37.8	12.7	56.0	N 58	15 14	16 06	16 59	24 49	00 49	02 11	03 38
17	74 56.1	22.6	337 21.6	14.0	10 50.5	12.6	56.0	56	15 29	16 17	17 06	24 45	00 45	02 04	03 28
18	89 55.7	S23 22.5	351 54.6	13.9	N11 03.1	12.5	56.1	54	15 42	16 26	17 12	24 42	00 42	01 58	03 19
19	104 55.4	22.5	6 27.5	13.8	11 15.6	12.6	56.1	52	15 53	16 34	17 18	24 39	00 39	01 53	03 10
20	119 55.1	22.4	21 00.3	13.7	11 28.2	12.5	56.1	50	16 03	16 41	17 23	24 36	00 36	01 48	03 03
21	134 54.8	.. 22.3	35 33.0	13.6	11 40.7	12.4	56.2	45	16 23	16 57	17 35	24 30	00 30	01 37	02 48
22	149 54.5	22.2	50 05.6	13.6	11 53.1	12.5	56.2	N 40	16 40	17 11	17 45	24 25	00 25	01 29	02 35
23	164 54.2	22.1	64 38.2	13.5	12 05.6	12.3	56.2	35	16 54	17 23	17 54	24 21	00 21	01 21	02 24
26 00	179 53.9	S23 22.0	79 10.7	13.3	N12 17.9	12.4	56.3	30	17 07	17 33	18 03	24 18	00 18	01 15	02 15
01	194 53.6	22.0	93 43.0	13.3	12 30.3	12.3	56.3	20	17 28	17 52	18 19	24 11	00 11	01 03	01 59
02	209 53.3	21.9	108 15.3	13.3	12 42.6	12.3	56.3	N 10	17 46	18 09	18 36	24 05	00 05	00 53	01 45
03	224 53.0	.. 21.8	122 47.6	13.1	12 54.9	12.2	56.4	0	18 04	18 26	18 53	24 00	00 00	00 44	01 32
04	239 52.7	21.7	137 19.7	13.0	13 07.1	12.2	56.4	S 10	18 21	18 45	19 12	23 55	24 35	00 35	01 19
05	254 52.3	21.6	151 51.7	13.0	13 19.3	12.1	56.4	20	18 40	19 05	19 34	23 49	24 25	00 25	01 05
06	269 52.0	S23 21.5	166 23.7	12.8	N13 31.4	12.1	56.5	30	19 02	19 30	20 03	23 43	24 14	00 14	00 49
07	284 51.7	21.4	180 55.5	12.8	13 43.5	12.0	56.5	35	19 15	19 45	20 22	23 39	24 08	00 08	00 40
S 08	299 51.4	21.4	195 27.3	12.6	13 55.5	12.0	56.6	40	19 30	20 03	20 45	23 35	24 01	00 01	00 30
A 09	314 51.1	.. 21.3	209 58.9	12.6	14 07.5	12.0	56.6	45	19 48	20 26	21 16	23 31	23 52	24 17	00 17
T 10	329 50.8	21.2	224 30.5	12.5	14 19.5	11.9	56.6	S 50	20 11	20 56	22 01	23 25	23 42	24 03	00 03
U 11	344 50.5	21.1	239 02.0	12.3	14 31.4	11.8	56.7	52	20 22	21 11	22 29	23 22	23 37	23 56	24 21
R 12	359 50.2	S23 21.0	253 33.3	12.3	N14 43.2	11.8	56.7	54	20 34	21 29	23 15	23 19	23 32	23 48	24 10
D 13	14 49.9	20.9	268 04.6	12.2	14 55.0	11.7	56.7	56	20 48	21 51	////	23 16	23 27	23 40	23 59
A 14	29 49.6	20.8	282 35.8	12.1	15 06.7	11.7	56.8	58	21 05	22 20	////	23 13	23 20	23 31	23 46
Y 15	44 49.3	.. 20.7	297 06.9	12.0	15 18.4	11.6	56.8	S 60	21 25	23 10	////	23 09	23 13	23 20	23 30
16	59 49.0	20.6	311 37.9	11.8	15 30.0	11.6	56.9								
17	74 48.6	20.5	326 08.7	11.8	15 41.6	11.5	56.9		SUN			MOON			
18	89 48.3	S23 20.4	340 39.5	11.7	N15 53.1	11.5	56.9	Day	Eqn. of Time		Mer.	Mer. Pass.		Age	Phase
19	104 48.0	20.3	355 10.2	11.5	16 04.6	11.3	57.0		00^h	12^h	Pass.	Upper	Lower		
20	119 47.7	20.2	9 40.7	11.5	16 15.9	11.3	57.0	d	m s	m s	h m	h m	h m	d	%
21	134 47.4	.. 20.1	24 11.2	11.3	16 27.2	11.3	57.0	24	00 36	00 21	12 00	17 50	05 29	08	48
22	149 47.1	20.0	38 41.5	11.3	16 38.5	11.2	57.1	25	00 06	00 09	12 00	18 33	06 11	09	58
23	164 46.8	19.9	53 11.8	11.1	N16 49.7	11.1	57.1	26	00 24	00 39	12 01	19 20	06 56	10	68
	SD 16.3	d 0.1	SD 15.0		15.2		15.4								

Note. Calculations for the Sun's GHA are made on the basis of an hourly difference of 15°. Although there are variations in the hourly difference, they are very small and are accounted for in the tabulated values of GHA so that there is no need for corrections to be made by the navigator.

Step 2. Calculate the increment for 32 minutes, 04 seconds:
In this step, we find the increase in GHA for minutes and seconds of GMT.
The Nautical Almanac contains tables of corrections for increments of time from 0 minutes to 59 minutes.
The extract on the next page shows the increments and corrections tables for 32 and 33 minutes.

From the extract we see that, in the table for 32^m, the increment for 04 seconds is 8° 01'.0
This is written as:
Inc. (32^m 04^s): 8° 01'.0
Note. Since GHA is always increasing, the increment correction is always added.

Step 3. Calculate GHA at 04 hr 32 min 04 sec GMT
GHA (04^h)	240° 07'.5
Inc. (32^m 04^s)	+ 8° 01'.0
GHA Sun	248° 08'.5

Step 4. Find LHA
In this step, we combine the GHA with the longitude to calculate the LHA:
(Remember the rule: Long East, LHA = GHA + LONG)
GHA	248° 08'.5
Long:	+ 32° 13'.8 East
LHA Sun	280° 22'.3

The full procedure for calculating the LHA of the Sun can be summarised in the following format:
GHA Sun (04^h)	240° 07'.5
Inc. (32^m 04^s)	8° 01'.0
GHA Sun	248° 08'.5
Long	32° 13'.8 E. (+)
LHA Sun	280° 22'.3

©2009 Jack Case

Extract from Nautical Almanac showing Increments and Corrections table for 32 and 33 minutes.

32ᵐ INCREMENTS AND CORRECTIONS

32ᵐ	SUN PLANETS	ARIES	MOON	v or d	Corrⁿ	v or d	Corrⁿ	v or d	Corrⁿ
s	° '	° '	° '	'	'	'	'	'	'
00	8 00.0	8 01.3	7 38.1	0.0	0.0	6.0	3.3	12.0	6.5
01	8 00.3	8 01.6	7 38.4	0.1	0.1	6.1	3.3	12.1	6.6
02	8 00.5	8 01.8	7 38.6	0.2	0.1	6.2	3.4	12.2	6.6
03	8 00.8	8 02.1	7 38.8	0.3	0.2	6.3	3.4	12.3	6.7
04	8 01.0	8 02.3	7 39.1	0.4	0.2	6.4	3.5	12.4	6.7
05	8 01.3	8 02.6	7 39.3	0.5	0.3	6.5	3.5	12.5	6.8
06	8 01.5	8 02.8	7 39.6	0.6	0.3	6.6	3.6	12.6	6.8
07	8 01.8	8 03.1	7 39.8	0.7	0.4	6.7	3.6	12.7	6.9
08	8 02.0	8 03.3	7 40.0	0.8	0.4	6.8	3.7	12.8	6.9
09	8 02.3	8 03.6	7 40.3	0.9	0.5	6.9	3.7	12.9	7.0
10	8 02.5	8 03.8	7 40.5	1.0	0.5	7.0	3.8	13.0	7.0
11	8 02.8	8 04.1	7 40.8	1.1	0.6	7.1	3.8	13.1	7.1
12	8 03.0	8 04.3	7 41.0	1.2	0.7	7.2	3.9	13.2	7.2
13	8 03.3	8 04.6	7 41.2	1.3	0.7	7.3	4.0	13.3	7.2
14	8 03.5	8 04.8	7 41.5	1.4	0.8	7.4	4.0	13.4	7.3
15	8 03.8	8 05.1	7 41.7	1.5	0.8	7.5	4.1	13.5	7.3
16	8 04.0	8 05.3	7 42.0	1.6	0.9	7.6	4.1	13.6	7.4
17	8 04.3	8 05.6	7 42.2	1.7	0.9	7.7	4.2	13.7	7.4
18	8 04.5	8 05.8	7 42.4	1.8	1.0	7.8	4.2	13.8	7.5
19	8 04.8	8 06.1	7 42.7	1.9	1.0	7.9	4.3	13.9	7.5
20	8 05.0	8 06.3	7 42.9	2.0	1.1	8.0	4.3	14.0	7.6
21	8 05.3	8 06.6	7 43.1	2.1	1.1	8.1	4.4	14.1	7.6
22	8 05.5	8 06.8	7 43.4	2.2	1.2	8.2	4.4	14.2	7.7
23	8 05.8	8 07.1	7 43.6	2.3	1.2	8.3	4.5	14.3	7.7
24	8 06.0	8 07.3	7 43.9	2.4	1.3	8.4	4.6	14.4	7.8
25	8 06.3	8 07.6	7 44.1	2.5	1.4	8.5	4.6	14.5	7.9
26	8 06.5	8 07.8	7 44.3	2.6	1.4	8.6	4.7	14.6	7.9
27	8 06.8	8 08.1	7 44.6	2.7	1.5	8.7	4.7	14.7	8.0
28	8 07.0	8 08.3	7 44.8	2.8	1.5	8.8	4.8	14.8	8.0
29	8 07.3	8 08.6	7 45.1	2.9	1.6	8.9	4.8	14.9	8.1
30	8 07.5	8 08.8	7 45.3	3.0	1.6	9.0	4.9	15.0	8.1
31	8 07.8	8 09.1	7 45.5	3.1	1.7	9.1	4.9	15.1	8.2
32	8 08.0	8 09.3	7 45.8	3.2	1.7	9.2	5.0	15.2	8.2
33	8 08.3	8 09.6	7 46.0	3.3	1.8	9.3	5.0	15.3	8.3
34	8 08.5	8 09.8	7 46.2	3.4	1.8	9.4	5.1	15.4	8.3
35	8 08.8	8 10.1	7 46.5	3.5	1.9	9.5	5.1	15.5	8.4
36	8 09.0	8 10.3	7 46.7	3.6	2.0	9.6	5.2	15.6	8.5
37	8 09.3	8 10.6	7 47.0	3.7	2.0	9.7	5.3	15.7	8.5
38	8 09.5	8 10.8	7 47.2	3.8	2.1	9.8	5.3	15.8	8.6
39	8 09.8	8 11.1	7 47.4	3.9	2.1	9.9	5.4	15.9	8.6
40	8 10.0	8 11.3	7 47.7	4.0	2.2	10.0	5.4	16.0	8.7
41	8 10.3	8 11.6	7 47.9	4.1	2.2	10.1	5.5	16.1	8.7
42	8 10.5	8 11.8	7 48.2	4.2	2.3	10.2	5.5	16.2	8.8
43	8 10.8	8 12.1	7 48.4	4.3	2.3	10.3	5.6	16.3	8.8
44	8 11.0	8 12.3	7 48.6	4.4	2.4	10.4	5.6	16.4	8.9
45	8 11.3	8 12.6	7 48.9	4.5	2.4	10.5	5.7	16.5	8.9
46	8 11.5	8 12.8	7 49.1	4.6	2.5	10.6	5.7	16.6	9.0
47	8 11.8	8 13.1	7 49.3	4.7	2.5	10.7	5.8	16.7	9.0
48	8 12.0	8 13.3	7 49.6	4.8	2.6	10.8	5.9	16.8	9.1
49	8 12.3	8 13.6	7 49.8	4.9	2.7	10.9	5.9	16.9	9.2
50	8 12.5	8 13.8	7 50.1	5.0	2.7	11.0	6.0	17.0	9.2
51	8 12.8	8 14.1	7 50.3	5.1	2.8	11.1	6.0	17.1	9.3
52	8 13.0	8 14.3	7 50.5	5.2	2.8	11.2	6.1	17.2	9.3
53	8 13.3	8 14.6	7 50.8	5.3	2.9	11.3	6.1	17.3	9.4
54	8 13.5	8 14.9	7 51.0	5.4	2.9	11.4	6.2	17.4	9.4
55	8 13.8	8 15.1	7 51.3	5.5	3.0	11.5	6.2	17.5	9.5
56	8 14.0	8 15.4	7 51.5	5.6	3.0	11.6	6.3	17.6	9.5
57	8 14.3	8 15.6	7 51.7	5.7	3.1	11.7	6.3	17.7	9.6
58	8 14.5	8 15.9	7 52.0	5.8	3.1	11.8	6.4	17.8	9.6
59	8 14.8	8 16.1	7 52.2	5.9	3.2	11.9	6.4	17.9	9.7
60	8 15.0	8 16.4	7 52.5	6.0	3.3	12.0	6.5	18.0	9.8

33ᵐ

33ᵐ	SUN PLANETS	ARIES	MOON	v or d	Corrⁿ	v or d	Corrⁿ	v or d	Corrⁿ
s	° '	° '	° '	'	'	'	'	'	'
00	8 15.0	8 16.4	7 52.5	0.0	0.0	6.0	3.4	12.0	6.7
01	8 15.3	8 16.6	7 52.7	0.1	0.1	6.1	3.4	12.1	6.8
02	8 15.5	8 16.9	7 52.9	0.2	0.1	6.2	3.5	12.2	6.8
03	8 15.8	8 17.1	7 53.2	0.3	0.2	6.3	3.5	12.3	6.9
04	8 16.0	8 17.4	7 53.4	0.4	0.2	6.4	3.6	12.4	6.9
05	8 16.3	8 17.6	7 53.6	0.5	0.3	6.5	3.6	12.5	7.0
06	8 16.5	8 17.9	7 53.9	0.6	0.3	6.6	3.7	12.6	7.0
07	8 16.8	8 18.1	7 54.1	0.7	0.4	6.7	3.7	12.7	7.1
08	8 17.0	8 18.4	7 54.4	0.8	0.4	6.8	3.8	12.8	7.1
09	8 17.3	8 18.6	7 54.6	0.9	0.5	6.9	3.9	12.9	7.2
10	8 17.5	8 18.9	7 54.8	1.0	0.6	7.0	3.9	13.0	7.3
11	8 17.8	8 19.1	7 55.1	1.1	0.6	7.1	4.0	13.1	7.3
12	8 18.0	8 19.4	7 55.3	1.2	0.7	7.2	4.0	13.2	7.4
13	8 18.3	8 19.6	7 55.6	1.3	0.7	7.3	4.1	13.3	7.4
14	8 18.5	8 19.9	7 55.8	1.4	0.8	7.4	4.1	13.4	7.5
15	8 18.8	8 20.1	7 56.0	1.5	0.8	7.5	4.2	13.5	7.5
16	8 19.0	8 20.4	7 56.3	1.6	0.9	7.6	4.2	13.6	7.6
17	8 19.3	8 20.6	7 56.5	1.7	0.9	7.7	4.3	13.7	7.6
18	8 19.5	8 20.9	7 56.7	1.8	1.0	7.8	4.4	13.8	7.7
19	8 19.8	8 21.1	7 57.0	1.9	1.1	7.9	4.4	13.9	7.8
20	8 20.0	8 21.4	7 57.2	2.0	1.1	8.0	4.5	14.0	7.8
21	8 20.3	8 21.6	7 57.5	2.1	1.2	8.1	4.5	14.1	7.9
22	8 20.5	8 21.9	7 57.7	2.2	1.2	8.2	4.6	14.2	7.9
23	8 20.8	8 22.1	7 57.9	2.3	1.3	8.3	4.6	14.3	8.0
24	8 21.0	8 22.4	7 58.2	2.4	1.3	8.4	4.7	14.4	8.0
25	8 21.3	8 22.6	7 58.4	2.5	1.4	8.5	4.7	14.5	8.1
26	8 21.5	8 22.9	7 58.7	2.6	1.5	8.6	4.8	14.6	8.2
27	8 21.8	8 23.1	7 58.9	2.7	1.5	8.7	4.9	14.7	8.2
28	8 22.0	8 23.4	7 59.1	2.8	1.6	8.8	4.9	14.8	8.3
29	8 22.3	8 23.6	7 59.4	2.9	1.6	8.9	5.0	14.9	8.3
30	8 22.5	8 23.9	7 59.6	3.0	1.7	9.0	5.0	15.0	8.4
31	8 22.8	8 24.1	7 59.8	3.1	1.7	9.1	5.1	15.1	8.4
32	8 23.0	8 24.4	8 00.1	3.2	1.8	9.2	5.1	15.2	8.5
33	8 23.3	8 24.6	8 00.3	3.3	1.8	9.3	5.2	15.3	8.5
34	8 23.5	8 24.9	8 00.6	3.4	1.9	9.4	5.2	15.4	8.6
35	8 23.8	8 25.1	8 00.8	3.5	2.0	9.5	5.3	15.5	8.7
36	8 24.0	8 25.4	8 01.0	3.6	2.0	9.6	5.4	15.6	8.7
37	8 24.3	8 25.6	8 01.3	3.7	2.1	9.7	5.4	15.7	8.8
38	8 24.5	8 25.9	8 01.5	3.8	2.1	9.8	5.5	15.8	8.8
39	8 24.8	8 26.1	8 01.8	3.9	2.2	9.9	5.5	15.9	8.9
40	8 25.0	8 26.4	8 02.0	4.0	2.2	10.0	5.6	16.0	8.9
41	8 25.3	8 26.6	8 02.2	4.1	2.3	10.1	5.6	16.1	9.0
42	8 25.5	8 26.9	8 02.5	4.2	2.3	10.2	5.7	16.2	9.0
43	8 25.8	8 27.1	8 02.7	4.3	2.4	10.3	5.8	16.3	9.1
44	8 26.0	8 27.4	8 02.9	4.4	2.5	10.4	5.8	16.4	9.2
45	8 26.3	8 27.6	8 03.2	4.5	2.5	10.5	5.9	16.5	9.2
46	8 26.5	8 27.9	8 03.4	4.6	2.6	10.6	5.9	16.6	9.3
47	8 26.8	8 28.1	8 03.7	4.7	2.6	10.7	6.0	16.7	9.3
48	8 27.0	8 28.4	8 03.9	4.8	2.7	10.8	6.0	16.8	9.4
49	8 27.3	8 28.6	8 04.1	4.9	2.7	10.9	6.1	16.9	9.4
50	8 27.5	8 28.9	8 04.4	5.0	2.8	11.0	6.1	17.0	9.5
51	8 27.8	8 29.1	8 04.6	5.1	2.8	11.1	6.2	17.1	9.5
52	8 28.0	8 29.4	8 04.9	5.2	2.9	11.2	6.3	17.2	9.6
53	8 28.3	8 29.6	8 05.1	5.3	3.0	11.3	6.3	17.3	9.7
54	8 28.5	8 29.9	8 05.3	5.4	3.0	11.4	6.4	17.4	9.7
55	8 28.8	8 30.1	8 05.6	5.5	3.1	11.5	6.4	17.5	9.8
56	8 29.0	8 30.4	8 05.8	5.6	3.1	11.6	6.5	17.6	9.8
57	8 29.3	8 30.6	8 06.1	5.7	3.2	11.7	6.5	17.7	9.9
58	8 29.5	8 30.9	8 06.3	5.8	3.2	11.8	6.6	17.8	9.9
59	8 29.8	8 31.1	8 06.5	5.9	3.3	11.9	6.6	17.9	10.0
60	8 30.0	8 31.4	8 06.8	6.0	3.4	12.0	6.7	18.0	10.1

Calculating Declination. The Declination is relatively simple to calculate because the hourly difference is very small in comparison with that of the GHA.

Step 1. Find the Declination for 0400 GMT. From the daily page extract for 24 December, we find that the declination of the Sun at 0400 = 23° 24'.9 South. This is written as:

Dec Sun (04^h): S23° 24'.9

At this point make a note of whether the declination is decreasing or increasing by the hour. In this case, it is decreasing.

Step 2. Calculate the correction for 32 mins.
At the bottom of the daily page beneath the column for the Sun's Dec. there is a value 'd' which is the mean hourly difference. In this case, the value of d is 0'.1. In the extract of the increments and corrections table for 32 mins, you will find three columns labelled 'v or d' and 'Corr^n'. For the correction corresponding to a value for d of 0'.1, we see that the value of the correction is also 0'.1.

Because (as we noted earlier) the declination is decreasing, the correction will have a negative value; so the correction is -0.1'.
This is written as:
d = 0'.1
Corr^n (32^m): -0'.1

Nb. *Because the rate of change of the Sun's declination is so slow, seconds can be discounted.*

The full procedure for calculating the Declination of the Sun can be summarised in the following format:

Dec Sun (04^h)	S23° 23'.6 (d = 0'.1)
Corr^n (32^m)	- 0'.1
Dec Sun (04^h 32^m)	S23° 23'.5

Example 2. Find LHA and Declination of the Moon.
On 3 December at the exact time of 16 hrs 25 mins and 30 secs. GMT, your assumed position is: 50° 12'.05N, 07° 22'.00W. Calculate the LHA and declination of the Moon.

Solution:
To find the LHA of the Moon:

Step 1. Find the GHA of the Moon at 16 hrs. From the following extract of part of the daily page of the Nautical Almanac for 3 Dec, the value of the Moon's GHA at 1600 is 222° 37'.3.
This is written as:
GHA (16^h)   222° 37'.3

UT	SUN		MOON				
	GHA	Dec	GHA	v	Dec	d	HP
d h	° '	° '	° '	'	° '	'	'
3 00	182 34.9	S22 05.1	352 47.9	2.9	N25 46.3	0.1	60.0
01	197 34.7	05.4	7 09.8	2.8	25 46.2	0.4	60.0
02	212 34.4	05.8	21 31.6	2.8	25 45.8	0.5	60.1
03	227 34.2	. . 06.2	35 53.4	2.9	25 45.3	0.8	60.1
04	242 33.9	06.5	50 15.3	2.8	25 44.5	0.9	60.1
05	257 33.7	06.9	64 37.1	2.8	25 43.6	1.1	60.1
06	272 33.5	S22 07.2	78 58.9	2.8	N25 42.5	1.3	60.1
07	287 33.2	07.6	93 20.7	2.8	25 41.2	1.5	60.1
T 08	302 33.0	07.9	107 42.5	2.8	25 39.7	1.7	60.1
H 09	317 32.7	. . 08.3	122 04.3	2.8	25 38.0	1.8	60.1
U 10	332 32.5	08.6	136 26.1	2.8	25 36.2	2.1	60.2
R 11	347 32.2	09.0	150 47.9	2.9	25 34.1	2.2	60.2
S 12	2 32.0	S22 09.3	165 09.8	2.8	N25 31.9	2.5	60.2
D 13	17 31.7	09.6	179 31.6	2.9	25 29.4	2.6	60.2
A 14	32 31.5	10.0	193 53.5	2.9	25 26.8	2.8	60.2
Y 15	47 31.2	. . 10.3	208 15.4	2.9	25 24.0	3.0	60.2
16	62 31.0	10.7	222 37.3	2.9	25 21.0	3.1	60.2
17	77 30.7	11.0	236 59.2	2.9	25 17.9	3.4	60.2

At this point, note the value of 'v' at 1600 in the column to the right of GHA. In this case the value of v is 2'.9.

Step 2. Calculate the increment for 25 mins, 30 secs.
From the following extract of part of the increments and corrections table for 25 mins, in the row for 30 secs and under the column for Moon, we find that the value of the increment is 6° 05'.1.
This is written as: **Inc. (25^m 30^s): 6° 05'.1**

©2009 Jack Case

25 m	SUN PLANETS	ARIES	MOON	v or d	Corrⁿ	v or d	Corrⁿ	v or d	Corrⁿ
s	° ′	° ′	° ′	′	′	′	′	′	′
00	6 15.0	6 16.0	5 57.9	0.0	0.0	6.0	2.6	12.0	5.1
01	6 15.3	6 16.3	5 58.2	0.1	0.0	6.1	2.6	12.1	5.1
02	6 15.5	6 16.5	5 58.4	0.2	0.1	6.2	2.6	12.2	5.2
03	6 15.8	6 16.8	5 58.6	0.3	0.1	6.3	2.7	12.3	5.2
04	6 16.0	6 17.0	5 58.9	0.4	0.2	6.4	2.7	12.4	5.3
05	6 16.3	6 17.3	5 59.1	0.5	0.2	6.5	2.8	12.5	5.3
06	6 16.5	6 17.5	5 59.3	0.6	0.3	6.6	2.8	12.6	5.4
07	6 16.8	6 17.8	5 59.6	0.7	0.3	6.7	2.8	12.7	5.4
08	6 17.0	6 18.0	5 59.8	0.8	0.3	6.8	2.9	12.8	5.4
09	6 17.3	6 18.3	6 00.1	0.9	0.4	6.9	2.9	12.9	5.5
10	6 17.5	6 18.5	6 00.3	1.0	0.4	7.0	3.0	13.0	5.5
11	6 17.8	6 18.8	6 00.5	1.1	0.5	7.1	3.0	13.1	5.6
12	6 18.0	6 19.0	6 00.8	1.2	0.5	7.2	3.1	13.2	5.6
13	6 18.3	6 19.3	6 01.0	1.3	0.6	7.3	3.1	13.3	5.7
14	6 18.5	6 19.5	6 01.3	1.4	0.6	7.4	3.1	13.4	5.7
15	6 18.8	6 19.8	6 01.5	1.5	0.6	7.5	3.2	13.5	5.7
16	6 19.0	6 20.0	6 01.7	1.6	0.7	7.6	3.2	13.6	5.8
17	6 19.3	6 20.3	6 02.0	1.7	0.7	7.7	3.3	13.7	5.8
18	6 19.5	6 20.5	6 02.2	1.8	0.8	7.8	3.3	13.8	5.9
19	6 19.8	6 20.8	6 02.5	1.9	0.8	7.9	3.4	13.9	5.9
20	6 20.0	6 21.0	6 02.7	2.0	0.9	8.0	3.4	14.0	6.0
21	6 20.3	6 21.3	6 02.9	2.1	0.9	8.1	3.4	14.1	6.0
22	6 20.5	6 21.5	6 03.2	2.2	0.9	8.2	3.5	14.2	6.0
23	6 20.8	6 21.8	6 03.4	2.3	1.0	8.3	3.5	14.3	6.1
24	6 21.0	6 22.0	6 03.6	2.4	1.0	8.4	3.6	14.4	6.1
25	6 21.3	6 22.3	6 03.9	2.5	1.1	8.5	3.6	14.5	6.2
26	6 21.5	6 22.5	6 04.1	2.6	1.1	8.6	3.7	14.6	6.2
27	6 21.8	6 22.8	6 04.4	2.7	1.1	8.7	3.7	14.7	6.2
28	6 22.0	6 23.0	6 04.6	2.8	1.2	8.8	3.7	14.8	6.3
29	6 22.3	6 23.3	6 04.8	2.9	1.2	8.9	3.8	14.9	6.3
30	6 22.5	6 23.5	6 05.1	3.0	1.3	9.0	3.8	15.0	6.4
31	6 22.8	6 23.8	6 05.3	3.1	1.3	9.1	3.9	15.1	6.4

Step 3. Calculate the 'v' correction.

As previously noted, the value of v for 16^h is 2′.9.

From the increments and corrections table for minute 25, we see that the correction corresponding to a value for v of 2′.9 is 1′.2.

This is written as: **V Corrⁿ (2′.9) 1′.2**

Explanation of v correction.

In the Nautical Almanac, the GHA of the Moon is calculated on the basis of an hourly difference of 14° 19′.0. However, unlike the Sun, the hourly difference varies greatly up to a maximum value of 15° 02′.5 and for this reason a correction value denoted by v is tabulated in the column to the right of the GHA in the daily pages. So, a further increment corresponding to the value of v must be extracted from the Increments and Corrections table.

Step 4. Calculate GHA at 16 hrs 25 mins 30 secs.

The full calculation of the Moon's GHA can be written without further explanation as follows:

GHA (16^h)   222° 37'.3
Inc. (25^m 30^s)   6° 05'.1
V Corr^n (2'.9)   1'.2
GHA Moon   228° 43'.6

Note. Since the GHA is always increasing, increment and v corrections are added to the GHA.

Step 5. Calculate LHA

GHA Moon   228° 43'.6
Long.   - 007° 22'.0 W (Remember: Long West, LHA = GHA - LONG).
LHA Moon   221° 21'.6

The full procedure for calculating the LHA of the Moon can be summarised in the following format:

GHA (16^h)   222° 37'.3
Inc. (25^m 30^s)   6° 05'.1
V Corr^n (2'.9)   1'.2
GHA Moon   228° 43'.6
Long.   007° 22'.0 W (-)
LHA Moon   221° 21'.6

Calculating Declination.

Step 1. Find the Declination of the Moon at 16 hrs.

From the daily page extract shown above, the declination of the Moon at 16 hrs is N25° 21'.0. This is written as:

Dec Moon (16^h)   N25° 21'.0

Step 2. Calculate the correction for 25 mins.

Note. The Moon's declination varies more quickly and more erratically than that of the Sun and the planets and for that reason, the d value (hourly difference) is tabulated for every hour).

From the daily page extract included at step 1 of the GHA calculation, the value of d at 16 hrs. is 3'.1. From the Increments and Corrections table for 25 mins, the correction corresponding to d = 3'.1 is 1'.3. (The declination is decreasing so the correction must be subtracted). The layout for the correction is as follows:

d = 3'.1
Corr^n (25^m) -1'.3

The full procedure for calculating the Declination of the Moon can be summarised without further explanation using the following format:

Dec Moon (16^h)   N25° 21.0'   d = 3'.1
Corr^n (25^m)   − 1'.3
Dec Moon ($16^h 25^m$)   N25° 19'.7

Example 3. To find the LHA and Declination of a planet.

On 1 January at the exact time of 08 hrs 46 mins and 12 secs. GMT, your assumed position is: 52° 45'.3S, 118° 24'.75W. Find the LHA and Declination of Mars. The following extract from the Nautical Almanac shows parts of the daily page for 1 January.

UT	ARIES	VENUS −4.4		MARS +1.3	
	GHA	GHA	Dec	GHA	Dec
d h	° ′	° ′	° ′	° ′	° ′
1 00	100 46.8	130 48.4	S13 49.1	186 55.4	S24 05.6
01	115 49.2	145 48.2	48.1	201 55.8	05.5
02	130 51.7	160 48.1	47.0	216 56.2	05.5
03	145 54.2	175 48.0 . .	45.9	231 56.6 . .	05.5
04	160 56.6	190 47.9	44.8	246 57.0	05.4
05	175 59.1	205 47.7	43.7	261 57.4	05.4
06	191 01.6	220 47.6	S13 42.6	276 57.8	S24 05.4
07	206 04.0	235 47.5	41.5	291 58.2	05.3
T 08	221 06.5	250 47.4	40.4	306 58.6	05.3
H 09	236 09.0	265 47.3 . .	39.3	321 59.0 . .	05.3
U 10	251 11.4	280 47.1	38.2	336 59.4	05.2
R 11	266 13.9	295 47.0	37.1	351 59.8	05.2
20	43 34.3	70 41.9	33.9	127 22.7	02.5
21	58 36.8	85 41.8 . .	32.8	142 23.1 . .	02.5
22	73 39.3	100 41.7	31.6	157 23.5	02.4
23	88 41.7	115 41.7	30.5	172 23.9	02.4
Mer. Pass.	h m 17 10.1	v −0.1	d 1.1	v 0.4	d 0.0

©2009 Jack Case

Note that values for v and d for the planets are to be found at the bottom of the daily page under the relevant columns

The following extract shows part of the increments and corrections table for 46 minutes.

46^m	SUN PLANETS	ARIES	MOON	v or d	Corr^n	v or d	Corr^n	v or d	Corr^n
s	° ′	° ′	° ′	′	′	′	′	′	′
00	11 30·0	11 31·9	10 58·6	0·0	0·0	6·0	4·7	12·0	9·3
01	11 30·3	11 32·1	10 58·8	0·1	0·1	6·1	4·7	12·1	9·4
02	11 30·5	11 32·4	10 59·0	0·2	0·2	6·2	4·8	12·2	9·5
03	11 30·8	11 32·6	10 59·3	0·3	0·2	6·3	4·9	12·3	9·5
04	11 31·0	11 32·9	10 59·5	0·4	0·3	6·4	5·0	12·4	9·6
05	11 31·3	11 33·1	10 59·8	0·5	0·4	6·5	5·0	12·5	9·7
06	11 31·5	11 33·4	11 00·0	0·6	0·5	6·6	5·1	12·6	9·8
07	11 31·8	11 33·6	11 00·2	0·7	0·5	6·7	5·2	12·7	9·8
08	11 32·0	11 33·9	11 00·5	0·8	0·6	6·8	5·3	12·8	9·9
09	11 32·3	11 34·1	11 00·7	0·9	0·7	6·9	5·3	12·9	10·0
10	11 32·5	11 34·4	11 01·0	1·0	0·8	7·0	5·4	13·0	10·1
11	11 32·8	11 34·6	11 01·2	1·1	0·9	7·1	5·5	13·1	10·2
12	11 33·0	11 34·9	11 01·4	1·2	0·9	7·2	5·6	13·2	10·2
13	11 33·3	11 35·1	11 01·7	1·3	1·0	7·3	5·7	13·3	10·3

Solution. Please note that from now on the solutions to the examples are given without so much explanation as the previous ones.

Step 1. Calculate LHA

GHA (08^h)   306° 58′.60
Inc. (46^m 12^s)   11° 33′.00
V Corr^n (0.′4)   0′.30
GHA Mars   318° 31′.90
Long.   118° 24′.75W (−)
LHA Mars   200° 07′.15

Step 2. Calculate Declination.

Dec Mars (08^h)   S24° 05′.3 (d = 0′.0)
Corr^n (46^m)   0′.0
Dec Mars (08^{h}46^m) S24° 05′.3

Example 4. To Find the LHA and Declination of a star.

At 05 hrs. 48 mins. 10 secs. on 4 January, your assumed position is 41° 26'.18S, 72° 17'.18E. Find the LHA and Declination of Acrux.

The following extract shows part of the Nautical Almanac daily page for 4 Jan.

	UT	ARIES	VENUS −4.5		MARS +1.3		JUPITER −1.9	
		GHA	GHA	Dec	GHA	Dec	GHA	Dec
	d h	° '	° '	° '	° '	° '	° '	° '
	4 00	103 44.2	130 41.6	S12 29.4	187 24.3	S24 02.3	161 51.7	S20 38.1
	01	118 46.7	145 41.5	28.3	202 24.7	02.2	176 53.6	38.0
	02	133 49.1	160 41.5	27.1	217 25.1	02.2	191 55.5	37.9
	03	148 51.6	175 41.4	26.0	232 25.5	02.1	206 57.3	37.8
	04	163 54.1	190 41.4	24.9	247 25.9	02.0	221 59.2	37.7
	05	178 56.5	205 41.3	23.8	262 26.3	02.0	237 01.0	37.6
	06	193 59.0	220 41.2	S12 22.6	277 26.7	S24 01.9	252 02.9	S20 37.4
	07	209 01.4	235 41.2	21.5	292 27.1	01.9	267 04.7	37.3
	08	224 03.9	250 41.1	20.4	307 27.5	01.8	282 06.6	37.2
S	09	239 06.4	265 41.1	19.3	322 27.9	01.7	297 08.5	37.1
U	10	254 08.8	280 41.0	18.1	337 28.3	01.7	312 10.3	37.0
N	11	269 11.3	295 41.0	17.0	352 28.7	01.6	327 12.2	36.8
D	12	284 13.8	310 40.9	S12 15.9	7 29.1	S24 01.5	342 14.0	S20 36.7
A	13	299 16.2	325 40.9	14.8	22 29.4	01.5	357 15.9	36.6
Y	14	314 18.7	340 40.8	13.6	37 29.8	01.4	12 17.8	36.5
	15	329 21.2	355 40.8	12.5	52 30.2	01.3	27 19.6	36.4
	16	344 23.6	10 40.7	11.4	67 30.6	01.3	42 21.5	36.2
	17	359 26.1	25 40.7	10.2	82 31.0	01.2	57 23.3	36.1
	18	14 28.6	40 40.6	S12 09.1	97 31.4	S24 01.1	72 25.2	S20 36.0
	19	29 31.0	55 40.6	08.0	112 31.8	01.0	87 27.0	35.9
	20	44 33.5	70 40.5	06.8	127 32.2	01.0	102 28.9	35.8
	21	59 35.9	85 40.5	05.7	142 32.6	00.9	117 30.8	35.6
	22	74 38.4	100 40.5	04.6	157 33.0	00.8	132 32.6	35.5
	23	89 40.9	115 40.4	03.5	172 33.4	00.8	147 34.5	35.4

The following extract (also from the daily page for 4 Jan.) shows part of the table of star data.

STARS

Name	SHA	Dec
	° '	° '
Acamar	315 20.5	S40 16.2
Achernar	335 28.9	S57 11.7
Acrux	173 13.2	S63 08.8
Adhara	255 14.8	S28 59.1
Aldebaran	290 52.9	N16 31.7
Alioth	166 23.3	N55 54.2
Alkaid	153 01.4	N49 15.7
Al Na'ir	27 48.0	S46 55.2
Alnilam	275 49.4	S 1 11.7

This extract shows part of the increments and corrections table for 48^m.

48^m	SUN PLANETS	ARIES	MOON
s	° ′	° ′	° ′
00	12 00·0	12 02·0	11 27·2
01	12 00·3	12 02·2	11 27·4
02	12 00·5	12 02·5	11 27·7
03	12 00·8	12 02·7	11 27·9
04	12 01·0	12 03·0	11 28·2
05	12 01·3	12 03·2	11 28·4
06	12 01·5	12 03·5	11 28·6
07	12 01·8	12 03·7	11 28·9
08	12 02·0	12 04·0	11 29·1
09	12 02·3	12 04·2	11 29·3
10	12 02·5	12 04·5	11 29·6
11	12 02·8	12 04·7	11 29·8
12	12 03·0	12 05·0	11 30·1
13	12 03·3	12 05·2	11 30·3
14	12 03·5	12 05·5	11 30·5

Solution to example 4:
Step 1. Calculate GHA Aries at 05 hrs. 48 mins. 10 secs. on 4 January.
The Nautical Almanac does not list the GHA of the stars; it lists their sidereal hour angle (SHA) instead. So, the GHA of a star has to be calculated from the GHA of Aries.
Remember that, as we learned in chapter 1, SHA is the angle between the meridian running through the First Point of Aries and the meridian running through the celestial body measured westwards from Aries. Therefore, the SHA of the star must be added to the GHA of Aries in order to obtain the GHA of the star.
The following calculations (using data taken from the extracts above) will make things clearer:

GHA Aries (05^h)   178° 56′.5
Inc. (48^{m}10^s)   12° 04′.5
GHA Aries   191° 01′.0

(Note. There is no v correction for Aries.)

Step 2. Calculate GHA Acrux.

GHA Aries	191° 01'.0
SHA Acrux	173° 13'.2
	364° 14'.2
	− 360°
GHA Acrux	004° 14'.2

Step 3. Calculate LHA Acrux.

GHA Acrux	004° 14'.20
Long.	72° 17'.18 E (+)
LHA Acrux	076° 31'.38

The full procedure for calculating the LHA of a star can be summarised using the following format:

GHA Aries (05^h)	178° 56'.50
Inc. (48^{m}10^s)	12° 04'.50 (+)
GHA Aries	191° 01'.00
SHA Acrux	173° 13'.20 (+)
	364° 14'.20
	360° 00'.00 (−)
GHA Acrux	004° 14'.20
Long.	72° 17'.18E (+)
LHA Acrux	076° 31'.38

Step 4. Find the Declination of Acrux.

The tabulated declination for Acrux on 4 January is S63° 08'.8

Nb. Since the declinations of the stars may be considered to be constant for up to 1 month, there is no need for any corrections to be calculated. So the declination can simply be written as:

Dec Acrux: S63° 08'.8

Self Test

Use the Nautical Almanac extracts on the following pages to answer these questions:

Question 1. Find the LHA and Declination of the Moon on 24 December at the exact time of 23 hrs 25 mins and 10 secs. GMT.
Your assumed position is: 23° 45'.05N,
18° 42'.00W.

Question 2. At 18 hrs. 48 mins. 35 secs. on 4 January, your assumed position is 38° 18'.35S 94° 33'.36E. Find the LHA and Declination of Acamar.

Question 3.
Your assumed position is 0° 05'.8 N 18° 45'.2E at 18 hrs 08 mins 15 secs on 25 September. Sunset was at 1754, and Saturn is visible above the horizon. Find the LHA and Declination of Saturn.

Question 4.
At 13 hours, 25 minutes, 18 seconds GMT on 6 December, the assumed position of your yacht is 45° 15'.6 North 18° 14'.7 West. Find the LHA and Declination of the Sun at that time.

Solutions to these questions can be found after the nautical almanac extracts on the following pages.

©2009 Jack Case   page 65

Q1. Extract of part of the daily page for 24 December.

UT	SUN		MOON				
	GHA	Dec	GHA	v	Dec	d	HP
d h	° ′	° ′	° ′	′	° ′	′	′
24 00	180 08.8	S23 25.1	99 57.7	16.2	N 1 57.5	13.1	54.9
01	195 08.4	25.1	114 32.9	16.1	2 10.6	13.2	54.9
02	210 08.1	25.0	129 08.0	16.0	2 23.8	13.2	54.9
03	225 07.8	. . 25.0	143 43.0	16.1	2 37.0	13.1	55.0
04	240 07.5	24.9	158 18.1	16.0	2 50.1	13.2	55.0
05	255 07.2	24.9	172 53.1	16.0	3 03.3	13.1	55.0
06	270 06.9	S23 24.8	187 28.1	15.9	N 3 16.4	13.2	55.0
07	285 06.6	24.8	202 03.0	15.9	3 29.6	13.1	55.1
T 08	300 06.3	24.7	216 37.9	15.9	3 42.7	13.2	55.1
H 09	315 06.0	. . 24.7	231 12.8	15.8	3 55.9	13.1	55.1
U 10	330 05.7	24.6	245 47.6	15.8	4 09.0	13.2	55.1
R 11	345 05.3	24.6	260 22.4	15.8	4 22.2	13.1	55.2
S 12	0 05.0	S23 24.5	274 57.2	15.7	N 4 35.3	13.2	55.2
D 13	15 04.7	24.5	289 31.9	15.6	4 48.5	13.1	55.2
A 14	30 04.4	24.4	304 06.5	15.7	5 01.6	13.1	55.2
Y 15	45 04.1	. . 24.4	318 41.2	15.5	5 14.7	13.1	55.3
16	60 03.8	24.3	333 15.7	15.6	5 27.8	13.1	55.3
17	75 03.5	24.2	347 50.3	15.4	5 40.9	13.1	55.3
18	90 03.2	S23 24.2	2 24.7	15.5	N 5 54.0	13.1	55.3
19	105 02.9	24.1	16 59.2	15.3	6 07.1	13.1	55.4
20	120 02.6	24.1	31 33.5	15.4	6 20.2	13.0	55.4
21	135 02.2	. . 24.0	46 07.9	15.2	6 33.2	13.1	55.4
22	150 01.9	23.9	60 42.1	15.3	6 46.3	13.0	55.5
23	165 01.6	23.9	75 16.4	15.1	6 59.3	13.0	55.5
25 00	180 01.3	S23 23.8	89 50.5	15.1	N 7 12.3	13.0	55.5
01	195 01.0	23.8	104 24.6	15.1	7 25.3	13.0	55.5
02	210 00.7	23.7	118 58.7	15.0	7 38.3	13.0	55.6
03	225 00.4	. . 23.6	133 32.7	14.9	7 51.3	12.9	55.6
04	240 00.1	23.6	148 06.6	14.9	8 04.2	12.9	55.6
05	254 59.8	23.5	162 40.5	14.8	8 17.1	12.9	55.7
06	269 59.5	S23 23.4	177 14.3	14.7	N 8 30.0	12.9	55.7
07	284 59.1	23.4	191 48.0	14.7	8 42.9	12.9	55.7
08	299 58.8	23.3	206 21.7	14.6	8 55.8	12.8	55.8

Q.1. Extract of part of the increments and corrections table for 25ᵐ.

25ᵐ s	SUN PLANETS ° ′	ARIES ° ′	MOON ° ′	v or d ′	Corrⁿ ′	v or d ′	Corrⁿ ′	v or d ′	Corrⁿ ′
00	6 15·0	6 16·0	5 57·9	0·0	0·0	6·0	2·6	12·0	5·1
01	6 15·3	6 16·3	5 58·2	0·1	0·0	6·1	2·6	12·1	5·1
02	6 15·5	6 16·5	5 58·4	0·2	0·1	6·2	2·6	12·2	5·2
03	6 15·8	6 16·8	5 58·6	0·3	0·1	6·3	2·7	12·3	5·2
04	6 16·0	6 17·0	5 58·9	0·4	0·2	6·4	2·7	12·4	5·3
05	6 16·3	6 17·3	5 59·1	0·5	0·2	6·5	2·8	12·5	5·3
06	6 16·5	6 17·5	5 59·3	0·6	0·3	6·6	2·8	12·6	5·4
07	6 16·8	6 17·8	5 59·6	0·7	0·3	6·7	2·8	12·7	5·4
08	6 17·0	6 18·0	5 59·8	0·8	0·3	6·8	2·9	12·8	5·4
09	6 17·3	6 18·3	6 00·1	0·9	0·4	6·9	2·9	12·9	5·5
10	6 17·5	6 18·5	6 00·3	1·0	0·4	7·0	3·0	13·0	5·5
11	6 17·8	6 18·8	6 00·5	1·1	0·5	7·1	3·0	13·1	5·6
12	6 18·0	6 19·0	6 00·8	1·2	0·5	7·2	3·1	13·2	5·6
13	6 18·3	6 19·3	6 01·0	1·3	0·6	7·3	3·1	13·3	5·7
14	6 18·5	6 19·5	6 01·3	1·4	0·6	7·4	3·1	13·4	5·7
15	6 18·8	6 19·8	6 01·5	1·5	0·6	7·5	3·2	13·5	5·7
16	6 19·0	6 20·0	6 01·7	1·6	0·7	7·6	3·2	13·6	5·8
17	6 19·3	6 20·3	6 02·0	1·7	0·7	7·7	3·3	13·7	5·8
18	6 19·5	6 20·5	6 02·2	1·8	0·8	7·8	3·3	13·8	5·9
19	6 19·8	6 20·8	6 02·5	1·9	0·8	7·9	3·4	13·9	5·9
20	6 20·0	6 21·0	6 02·7	2·0	0·9	8·0	3·4	14·0	6·0
21	6 20·3	6 21·3	6 02·9	2·1	0·9	8·1	3·4	14·1	6·0
22	6 20·5	6 21·5	6 03·2	2·2	0·9	8·2	3·5	14·2	6·0
23	6 20·8	6 21·8	6 03·4	2·3	1·0	8·3	3·5	14·3	6·1
24	6 21·0	6 22·0	6 03·6	2·4	1·0	8·4	3·6	14·4	6·1
25	6 21·3	6 22·3	6 03·9	2·5	1·1	8·5	3·6	14·5	6·2
26	6 21·5	6 22·5	6 04·1	2·6	1·1	8·6	3·7	14·6	6·2
27	6 21·8	6 22·8	6 04·4	2·7	1·1	8·7	3·7	14·7	6·2
28	6 22·0	6 23·0	6 04·6	2·8	1·2	8·8	3·7	14·8	6·3
29	6 22·3	6 23·3	6 04·8	2·9	1·2	8·9	3·8	14·9	6·3
30	6 22·5	6 23·5	6 05·1	3·0	1·3	9·0	3·8	15·0	6·4
31	6 22·8	6 23·8	6 05·3	3·1	1·3	9·1	3·9	15·1	6·4
32	6 23·0	6 24·0	6 05·6	3·2	1·4	9·2	3·9	15·2	6·5
33	6 23·3	6 24·3	6 05·8	3·3	1·4	9·3	4·0	15·3	6·5
34	6 23·5	6 24·5	6 06·0	3·4	1·4	9·4	4·0	15·4	6·5

©2009 Jack Case

Q2. Extracts from the daily page for 4 January showing GHA Aries and star data.

UT	ARIES	VENUS −4.5		MARS +1.3		JUPITER −1.9	
	GHA	GHA	Dec	GHA	Dec	GHA	Dec
d h	° ′	° ′	° ′	° ′	° ′	° ′	° ′
4 00	103 44.2	130 41.6	S12 29.4	187 24.3	S24 02.3	161 51.7	S20 38.1
01	118 46.7	145 41.5	28.3	202 24.7	02.2	176 53.6	38.0
02	133 49.1	160 41.5	27.1	217 25.1	02.2	191 55.5	37.9
03	148 51.6	175 41.4 ..	26.0	232 25.5 ..	02.1	206 57.3 ..	37.8
04	163 54.1	190 41.4	24.9	247 25.9	02.0	221 59.2	37.7
05	178 56.5	205 41.3	23.8	262 26.3	02.0	237 01.0	37.6
06	193 59.0	220 41.2	S12 22.6	277 26.7	S24 01.9	252 02.9	S20 37.4
07	209 01.4	235 41.2	21.5	292 27.1	01.9	267 04.7	37.3
08	224 03.9	250 41.1	20.4	307 27.5	01.8	282 06.6	37.2
S 09	239 06.4	265 41.1 ..	19.3	322 27.9 ..	01.7	297 08.5 ..	37.1
U 10	254 08.8	280 41.0	18.1	337 28.3	01.7	312 10.3	37.0
N 11	269 11.3	295 41.0	17.0	352 28.7	01.6	327 12.2	36.8
D 12	284 13.8	310 40.9	S12 15.9	7 29.1	S24 01.5	342 14.0	S20 36.7
A 13	299 16.2	325 40.9	14.8	22 29.4	01.5	357 15.9	36.6
Y 14	314 18.7	340 40.8	13.6	37 29.8	01.4	12 17.8	36.5
15	329 21.2	355 40.8 ..	12.5	52 30.2 ..	01.3	27 19.6 ..	36.4
16	344 23.6	10 40.7	11.4	67 30.6	01.3	42 21.5	36.2
17	359 26.1	25 40.7	10.2	82 31.0	01.2	57 23.3	36.1
18	14 28.6	40 40.6	S12 09.1	97 31.4	S24 01.1	72 25.2	S20 36.0
19	29 31.0	55 40.6	08.0	112 31.8	01.0	87 27.0	35.9
20	44 33.5	70 40.5	06.8	127 32.2	01.0	102 28.9	35.8
21	59 35.9	85 40.5 ..	05.7	142 32.6 ..	00.9	117 30.8 ..	35.6
22	74 38.4	100 40.5	04.6	157 33.0	00.8	132 32.6	35.5
23	89 40.9	115 40.4	03.5	172 33.4	00.8	147 34.5	35.4

STARS

Name	SHA	Dec
	° ′	° ′
Acamar	315 20.5	S40 16.2
Achernar	335 28.9	S57 11.7
Acrux	173 13.2	S63 08.8
Adhara	255 14.8	S28 59.1
Aldebaran	290 52.9	N16 31.7

Q2. Extract from part of the increments and corrections table for 48ᵐ.

48ᵐ	SUN PLANETS	ARIES	MOON	v or d	Corrⁿ	v or d	Corrⁿ	v or d	Corrⁿ
s	° ′	° ′	° ′	′	′	′	′	′	′
00	12 00·0	12 02·0	11 27·2	0·0	0·0	6·0	4·9	12·0	9·7
01	12 00·3	12 02·2	11 27·4	0·1	0·1	6·1	4·9	12·1	9·8
02	12 00·5	12 02·5	11 27·7	0·2	0·2	6·2	5·0	12·2	9·9
03	12 00·8	12 02·7	11 27·9	0·3	0·2	6·3	5·1	12·3	9·9
04	12 01·0	12 03·0	11 28·2	0·4	0·3	6·4	5·2	12·4	10·0
05	12 01·3	12 03·2	11 28·4	0·5	0·4	6·5	5·3	12·5	10·1
06	12 01·5	12 03·5	11 28·6	0·6	0·5	6·6	5·3	12·6	10·2
07	12 01·8	12 03·7	11 28·9	0·7	0·6	6·7	5·4	12·7	10·3
08	12 02·0	12 04·0	11 29·1	0·8	0·6	6·8	5·5	12·8	10·3
09	12 02·3	12 04·2	11 29·3	0·9	0·7	6·9	5·6	12·9	10·4
10	12 02·5	12 04·5	11 29·6	1·0	0·8	7·0	5·7	13·0	10·5
11	12 02·8	12 04·7	11 29·8	1·1	0·9	7·1	5·7	13·1	10·6
12	12 03·0	12 05·0	11 30·1	1·2	1·0	7·2	5·8	13·2	10·7
13	12 03·3	12 05·2	11 30·3	1·3	1·1	7·3	5·9	13·3	10·8
14	12 03·5	12 05·5	11 30·5	1·4	1·1	7·4	6·0	13·4	10·8
15	12 03·8	12 05·7	11 30·8	1·5	1·2	7·5	6·1	13·5	10·9
16	12 04·0	12 06·0	11 31·0	1·6	1·3	7·6	6·1	13·6	11·0
17	12 04·3	12 06·2	11 31·3	1·7	1·4	7·7	6·2	13·7	11·1
18	12 04·5	12 06·5	11 31·5	1·8	1·5	7·8	6·3	13·8	11·2
19	12 04·8	12 06·7	11 31·7	1·9	1·5	7·9	6·4	13·9	11·2
20	12 05·0	12 07·0	11 32·0	2·0	1·6	8·0	6·5	14·0	11·3
21	12 05·3	12 07·2	11 32·2	2·1	1·7	8·1	6·5	14·1	11·4
22	12 05·5	12 07·5	11 32·4	2·2	1·8	8·2	6·6	14·2	11·5
23	12 05·8	12 07·7	11 32·7	2·3	1·9	8·3	6·7	14·3	11·6
24	12 06·0	12 08·0	11 32·9	2·4	1·9	8·4	6·8	14·4	11·6
25	12 06·3	12 08·2	11 33·2	2·5	2·0	8·5	6·9	14·5	11·7
26	12 06·5	12 08·5	11 33·4	2·6	2·1	8·6	7·0	14·6	11·8
27	12 06·8	12 08·7	11 33·6	2·7	2·2	8·7	7·0	14·7	11·9
28	12 07·0	12 09·0	11 33·9	2·8	2·3	8·8	7·1	14·8	12·0
29	12 07·3	12 09·2	11 34·1	2·9	2·3	8·9	7·2	14·9	12·0
30	12 07·5	12 09·5	11 34·4	3·0	2·4	9·0	7·3	15·0	12·1
31	12 07·8	12 09·7	11 34·6	3·1	2·5	9·1	7·4	15·1	12·2
32	12 08·0	12 10·0	11 34·8	3·2	2·6	9·2	7·4	15·2	12·3
33	12 08·3	12 10·2	11 35·1	3·3	2·7	9·3	7·5	15·3	12·4
34	12 08·5	12 10·5	11 35·3	3·4	2·7	9·4	7·6	15·4	12·4
35	12 08·8	12 10·7	11 35·6	3·5	2·8	9·5	7·7	15·5	12·5
36	12 09·0	12 11·0	11 35·8	3·6	2·9	9·6	7·8	15·6	12·6
37	12 09·3	12 11·2	11 36·0	3·7	3·0	9·7	7·8	15·7	12·7

Q.3 Extract from parts of the daily page for 25 Sept.

UT	ARIES	MARS +0.8		JUPITER −2.7		SATURN +1.1	
	GHA	GHA	Dec	GHA	Dec	GHA	Dec
d h	° '	° '	° '	° '	° '	° '	° '
25 00	3 56.9	253 59.0	N22 48.9	43 26.9	S16 33.8	186 55.8	N 3 26.8
01	18 59.3	268 59.9	48.8	58 29.5	33.8	201 58.0	26.7
02	34 01.8	284 00.8	48.6	73 32.1	33.8	217 00.2	26.6
03	49 04.3	299 01.7 ..	48.5	88 34.7 ..	33.9	232 02.3 ..	26.5
04	64 06.7	314 02.7	48.4	103 37.3	33.9	247 04.5	26.3
05	79 09.2	329 03.6	48.2	118 39.9	34.0	262 06.7	26.2
06	94 11.7	344 04.5	N22 48.1	133 42.5	S16 34.0	277 08.9	N 3 26.1
07	109 14.1	359 05.4	47.9	148 45.1	34.1	292 11.0	26.0
08	124 16.6	14 06.4	47.8	163 47.8	34.1	307 13.2	25.8
F 09	139 19.1	29 07.3 ..	47.6	178 50.4 ..	34.1	322 15.4 ..	25.7
R 10	154 21.5	44 08.2	47.5	193 53.0	34.2	337 17.6	25.6
I 11	169 24.0	59 09.1	47.3	208 55.6	34.2	352 19.8	25.5
D 12	184 26.4	74 10.1	N22 47.2	223 58.2	S16 34.3	7 21.9	N 3 25.4
A 13	199 28.9	89 11.0	47.1	239 00.8	34.3	22 24.1	25.2
Y 14	214 31.4	104 11.9	46.9	254 03.4	34.3	37 26.3	25.1
15	229 33.8	119 12.9 ..	46.8	269 06.0 ..	34.4	52 28.5 ..	25.0
16	244 36.3	134 13.8	46.6	284 08.6	34.4	67 30.6	24.9
17	259 38.8	149 14.7	46.5	299 11.2	34.5	82 32.8	24.8
18	274 41.2	164 15.6	N22 46.3	314 13.8	S16 34.5	97 35.0	N 3 24.6
19	289 43.7	179 16.6	46.2	329 16.4	34.5	112 37.2	24.5
20	304 46.2	194 17.5	46.0	344 19.0	34.6	127 39.4	24.4
21	319 48.6	209 18.4 ..	45.9	359 21.6 ..	34.6	142 41.5 ..	24.3
22	334 51.1	224 19.4	45.7	14 24.2	34.7	157 43.7	24.2
23	349 53.6	239 20.3	45.6	29 26.8	34.7	172 45.9	24.0
26 00	4 56.0	254 21.2	N22 45.4	44 29.5	S16 34.7	187 48.1	N 3 23.9
01	19 58.5	269 22.2	45.3	59 32.1	34.8	202 50.2	23.8
02	35 00.9	284 23.1	45.1	74 34.7	34.8	217 52.4	23.7
03	50 03.4	299 24.0 ..	45.0	89 37.3 ..	34.9	232 54.6 ..	23.5
04	65 05.9	314 25.0	44.8	104 39.9	34.9	247 56.8	23.4
05	80 08.3	329 25.9	44.7	119 42.5	34.9	262 59.0	23.3
06	95 10.8	344 26.8	N22 44.5	134 45.1	S16 35.0	278 01.1	N 3 23.2
07	110 13.3	359 27.8	44.4	149 47.7	35.0	293 03.3	23.1
S 08	125 15.7	14 28.7	44.2	164 50.3	35.1	308 05.5	22.9
A 09	140 18.2	29 29.6 ..	44.1	179 52.9 ..	35.1	323 07.7 ..	22.8
T 10	155 20.7	44 30.6	43.9	194 55.5	35.1	338 09.8	22.7
U 11	170 23.1	59 31.5	43.8	209 58.1	35.2	353 12.0	22.6
R 12	185 25.6	74 32.4	N22 43.6	225 00.7	S16 35.2	8 14.2	N 3 22.5
D 13	200 28.1	89 33.4	43.5	240 03.3	35.3	23 16.4	22.3
A 14	215 30.5	104 34.3	43.3	255 05.9	35.3	38 18.6	22.2
Y 15	230 33.0	119 35.3 ..	43.2	270 08.5 ..	35.3	53 20.7 ..	22.1
16	245 35.4	134 36.2	43.0	285 11.1	35.4	68 22.9	22.0
17	260 37.9	149 37.1	42.9	300 13.7	35.4	83 25.1	21.9
18	275 40.4	164 38.1	N22 42.7	315 16.3	S16 35.4	98 27.3	N 3 21.7
19	290 42.8	179 39.0	42.6	330 18.9	35.5	113 29.5	21.6
20	305 45.3	194 40.0	42.4	345 21.5	35.5	128 31.6	21.5
21	320 47.8	209 40.9 ..	42.3	0 24.1 ..	35.6	143 33.8 ..	21.4
22	335 50.2	224 41.8	42.1	15 26.7	35.6	158 36.0	21.3
23	350 52.7	239 42.8	42.0	30 29.3	35.6	173 38.2	21.1
27 00	5 55.2	254 43.7	N22 41.8	45 31.8	S16 35.7	188 40.3	N 3 21.0
01	20 57.6	269 44.7	41.7	60 34.4	35.7	203 42.5	20.9
02	36 00.1	284 45.6	41.5	75 37.0	35.8	218 44.7	20.8
03	51 02.6	299 46.5 ..	41.4	90 39.6 ..	35.8	233 46.9 ..	20.7
04	66 05.0	314 47.5	41.2	105 42.2	35.8	248 49.1	20.5
05	81 07.5	329 48.4	41.1	120 44.8	35.9	263 51.2	20.4
06	96 09.9	344 49.4	N22 40.9	135 47.4	S16 35.9	278 53.4	N 3 20.3
	h m						
Mer. Pass.	23 36.4	v 0.9	d 0.2	v 2.6	d 0.0	v 2.2	d 0.1

©2009 Jack Case

Q.3 Extract from part of increments and corrections table for 8 mins.

m 8	SUN PLANETS	ARIES	MOON	v or d	Corrⁿ	v or d	Corrⁿ	v or d	Corrⁿ
s	° ′	° ′	° ′	′	′	′	′	′	′
00	2 00·0	2 00·3	1 54·5	0·0	0·0	6·0	0·9	12·0	1·7
01	2 00·3	2 00·6	1 54·8	0·1	0·0	6·1	0·9	12·1	1·7
02	2 00·5	2 00·8	1 55·0	0·2	0·0	6·2	0·9	12·2	1·7
03	2 00·8	2 01·1	1 55·2	0·3	0·0	6·3	0·9	12·3	1·7
04	2 01·0	2 01·3	1 55·5	0·4	0·1	6·4	0·9	12·4	1·8
05	2 01·3	2 01·6	1 55·7	0·5	0·1	6·5	0·9	12·5	1·8
06	2 01·5	2 01·8	1 56·0	0·6	0·1	6·6	0·9	12·6	1·8
07	2 01·8	2 02·1	1 56·2	0·7	0·1	6·7	0·9	12·7	1·8
08	2 02·0	2 02·3	1 56·4	0·8	0·1	6·8	1·0	12·8	1·8
09	2 02·3	2 02·6	1 56·7	0·9	0·1	6·9	1·0	12·9	1·8
10	2 02·5	2 02·8	1 56·9	1·0	0·1	7·0	1·0	13·0	1·8
11	2 02·8	2 03·1	1 57·2	1·1	0·2	7·1	1·0	13·1	1·9
12	2 03·0	2 03·3	1 57·4	1·2	0·2	7·2	1·0	13·2	1·9
13	2 03·3	2 03·6	1 57·6	1·3	0·2	7·3	1·0	13·3	1·9
14	2 03·5	2 03·8	1 57·9	1·4	0·2	7·4	1·0	13·4	1·9
15	2 03·8	2 04·1	1 58·1	1·5	0·2	7·5	1·1	13·5	1·9
16	2 04·0	2 04·3	1 58·4	1·6	0·2	7·6	1·1	13·6	1·9
17	2 04·3	2 04·6	1 58·6	1·7	0·2	7·7	1·1	13·7	1·9
18	2 04·5	2 04·8	1 58·8	1·8	0·3	7·8	1·1	13·8	2·0
19	2 04·8	2 05·1	1 59·1	1·9	0·3	7·9	1·1	13·9	2·0
20	2 05·0	2 05·3	1 59·3	2·0	0·3	8·0	1·1	14·0	2·0
21	2 05·3	2 05·6	1 59·5	2·1	0·3	8·1	1·1	14·1	2·0
22	2 05·5	2 05·8	1 59·8	2·2	0·3	8·2	1·2	14·2	2·0
23	2 05·8	2 06·1	2 00·0	2·3	0·3	8·3	1·2	14·3	2·0
24	2 06·0	2 06·3	2 00·3	2·4	0·3	8·4	1·2	14·4	2·0
25	2 06·3	2 06·6	2 00·5	2·5	0·4	8·5	1·2	14·5	2·1
26	2 06·5	2 06·8	2 00·7	2·6	0·4	8·6	1·2	14·6	2·1
27	2 06·8	2 07·1	2 01·0	2·7	0·4	8·7	1·2	14·7	2·1
28	2 07·0	2 07·3	2 01·2	2·8	0·4	8·8	1·2	14·8	2·1
29	2 07·3	2 07·6	2 01·5	2·9	0·4	8·9	1·3	14·9	2·1
30	2 07·5	2 07·8	2 01·7	3·0	0·4	9·0	1·3	15·0	2·1
31	2 07·8	2 08·1	2 01·9	3·1	0·4	9·1	1·3	15·1	2·1
32	2 08·0	2 08·4	2 02·2	3·2	0·5	9·2	1·3	15·2	2·2
33	2 08·3	2 08·6	2 02·4	3·3	0·5	9·3	1·3	15·3	2·2
34	2 08·5	2 08·9	2 02·6	3·4	0·5	9·4	1·3	15·4	2·2
35	2 08·8	2 09·1	2 02·9	3·5	0·5	9·5	1·3	15·5	2·2

Q.4 Extract from part of daily page for 6 December.

UT	SUN		MOON				
	GHA	Dec	GHA	v	Dec	d	HP
d h	° ′	° ′	° ′	′	° ′	′	′
6 00	182 16.7	S22 28.7	308 41.3	7.2	N18 06.4	11.8	60.1
01	197 16.4	29.0	323 07.5	7.2	17 54.6	11.8	60.1
02	212 16.2	29.3	337 33.7	7.3	17 42.8	12.0	60.1
03	227 15.9	29.6	352 00.0	7.4	17 30.8	12.0	60.1
04	242 15.7	29.9	6 26.4	7.6	17 18.8	12.2	60.1
05	257 15.4	30.2	20 53.0	7.6	17 06.6	12.2	60.0
06	272 15.1	S22 30.5	35 19.6	7.7	N16 54.4	12.4	60.0
07	287 14.9	30.8	49 46.3	7.8	16 42.0	12.4	60.0
08	302 14.6	31.1	64 13.1	7.9	16 29.6	12.6	60.0
S 09	317 14.3	31.4	78 40.0	8.0	16 17.0	12.6	60.0
U 10	332 14.1	31.7	93 07.0	8.1	16 04.4	12.7	60.0
N 11	347 13.8	32.0	107 34.1	8.2	15 51.7	12.8	60.0
D 12	2 13.5	S22 32.3	122 01.3	8.3	N15 38.9	12.9	59.9
A 13	17 13.3	32.6	136 28.6	8.4	15 26.0	13.0	59.9
Y 14	32 13.0	32.9	150 56.0	8.4	15 13.0	13.1	59.9
15	47 12.7	33.2	165 23.4	8.6	14 59.9	13.1	59.9
16	62 12.5	33.5	179 51.0	8.7	14 46.8	13.3	59.9
17	77 12.2	33.7	194 18.7	8.7	14 33.5	13.3	59.9
18	92 12.0	S22 34.0	208 46.4	8.9	N14 20.2	13.3	59.8
19	107 11.7	34.3	223 14.3	8.9	14 06.9	13.5	59.8
20	122 11.4	34.6	237 42.2	9.0	13 53.4	13.5	59.8
21	137 11.2	34.9	252 10.2	9.1	13 39.9	13.6	59.8
22	152 10.9	35.2	266 38.3	9.2	13 26.3	13.7	59.8
23	167 10.6	35.5	281 06.5	9.3	13 12.6	13.7	59.8
7 00	182 10.4	S22 35.7	295 34.8	9.4	N12 58.9	13.8	59.7
01	197 10.1	36.0	310 03.2	9.4	12 45.1	13.9	59.7
02	212 09.8	36.3	324 31.6	9.6	12 31.2	13.9	59.7
03	227 09.5	36.6	339 00.2	9.6	12 17.3	14.0	59.7
04	242 09.3	36.9	353 28.8	9.7	12 03.3	14.0	59.7
05	257 09.0	37.2	7 57.5	9.8	11 49.3	14.1	59.6
06	272 08.7	S22 37.4	22 26.3	9.9	N11 35.2	14.1	59.6
07	287 08.5	37.7	36 55.2	9.9	11 21.1	14.3	59.6
08	302 08.2	38.0	51 24.1	10.0	11 06.8	14.2	59.6
M 09	317 07.9	38.3	65 53.1	10.1	10 52.6	14.3	59.6
O 10	332 07.7	38.5	80 22.2	10.2	10 38.3	14.4	59.6
N 11	347 07.4	38.8	94 51.4	10.3	10 23.9	14.4	59.5
D 12	2 07.1	S22 39.1	109 20.7	10.3	N10 09.5	14.4	59.5
A 13	17 06.9	39.4	123 50.0	10.4	9 55.1	14.5	59.5
Y 14	32 06.6	39.6	138 19.4	10.5	9 40.6	14.5	59.5
15	47 06.3	39.9	152 48.9	10.5	9 26.1	14.6	59.4
16	62 06.0	40.2	167 18.4	10.7	9 11.5	14.6	59.4
17	77 05.8	40.4	181 48.1	10.6	8 56.9	14.6	59.4
18	92 05.5	S22 40.7	196 17.7	10.8	N 8 42.3	14.7	59.4
19	107 05.2	41.0	210 47.5	10.8	8 27.6	14.7	59.4
20	122 05.0	41.3	225 17.3	10.9	8 12.9	14.8	59.3
21	137 04.7	41.5	239 47.2	10.9	7 58.1	14.8	59.3
22	152 04.4	41.8	254 17.1	11.1	7 43.3	14.8	59.3
23	167 04.1	42.1	268 47.2	11.0	7 28.5	14.8	59.3
	SD 16.3	d 0.3	SD 16.3		16.2		16.1

Q.4 Extract showing part of increments and corrections table for 25^m.

25^m	SUN PLANETS	ARIES	MOON	v or d	Corr^n	v or d	Corr^n	v or d	Corr^n
s	° ′	° ′	° ′	′	′	′	′	′	′
00	6 15·0	6 16·0	5 57·9	0·0	0·0	6·0	2·6	12·0	5·1
01	6 15·3	6 16·3	5 58·2	0·1	0·0	6·1	2·6	12·1	5·1
02	6 15·5	6 16·5	5 58·4	0·2	0·1	6·2	2·6	12·2	5·2
03	6 15·8	6 16·8	5 58·6	0·3	0·1	6·3	2·7	12·3	5·2
04	6 16·0	6 17·0	5 58·9	0·4	0·2	6·4	2·7	12·4	5·3
05	6 16·3	6 17·3	5 59·1	0·5	0·2	6·5	2·8	12·5	5·3
06	6 16·5	6 17·5	5 59·3	0·6	0·3	6·6	2·8	12·6	5·4
07	6 16·8	6 17·8	5 59·6	0·7	0·3	6·7	2·8	12·7	5·4
08	6 17·0	6 18·0	5 59·8	0·8	0·3	6·8	2·9	12·8	5·4
09	6 17·3	6 18·3	6 00·1	0·9	0·4	6·9	2·9	12·9	5·5
10	6 17·5	6 18·5	6 00·3	1·0	0·4	7·0	3·0	13·0	5·5
11	6 17·8	6 18·8	6 00·5	1·1	0·5	7·1	3·0	13·1	5·6
12	6 18·0	6 19·0	6 00·8	1·2	0·5	7·2	3·1	13·2	5·6
13	6 18·3	6 19·3	6 01·0	1·3	0·6	7·3	3·1	13·3	5·7
14	6 18·5	6 19·5	6 01·3	1·4	0·6	7·4	3·1	13·4	5·7
15	6 18·8	6 19·8	6 01·5	1·5	0·6	7·5	3·2	13·5	5·7
16	6 19·0	6 20·0	6 01·7	1·6	0·7	7·6	3·2	13·6	5·8
17	6 19·3	6 20·3	6 02·0	1·7	0·7	7·7	3·3	13·7	5·8
18	6 19·5	6 20·5	6 02·2	1·8	0·8	7·8	3·3	13·8	5·9
19	6 19·8	6 20·8	6 02·5	1·9	0·8	7·9	3·4	13·9	5·9
20	6 20·0	6 21·0	6 02·7	2·0	0·9	8·0	3·4	14·0	6·0
21	6 20·3	6 21·3	6 02·9	2·1	0·9	8·1	3·4	14·1	6·0
22	6 20·5	6 21·5	6 03·2	2·2	0·9	8·2	3·5	14·2	6·0
23	6 20·8	6 21·8	6 03·4	2·3	1·0	8·3	3·5	14·3	6·1
24	6 21·0	6 22·0	6 03·6	2·4	1·0	8·4	3·6	14·4	6·1
25	6 21·3	6 22·3	6 03·9	2·5	1·1	8·5	3·6	14·5	6·2
26	6 21·5	6 22·5	6 04·1	2·6	1·1	8·6	3·7	14·6	6·2
27	6 21·8	6 22·8	6 04·4	2·7	1·1	8·7	3·7	14·7	6·2
28	6 22·0	6 23·0	6 04·6	2·8	1·2	8·8	3·7	14·8	6·3
29	6 22·3	6 23·3	6 04·8	2·9	1·2	8·9	3·8	14·9	6·3
30	6 22·5	6 23·5	6 05·1	3·0	1·3	9·0	3·8	15·0	6·4
31	6 22·8	6 23·8	6 05·3	3·1	1·3	9·1	3·9	15·1	6·4
32	6 23·0	6 24·0	6 05·6	3·2	1·4	9·2	3·9	15·2	6·5
33	6 23·3	6 24·3	6 05·8	3·3	1·4	9·3	4·0	15·3	6·5
34	6 23·5	6 24·5	6 06·0	3·4	1·4	9·4	4·0	15·4	6·5

Solutions to Test Questions

Q1.
GHA (23^h)	75° 16'.4
Inc. (25^m 10^s)	6° 00'.3
V Corr^n (15'.1)	6'.4
GHA Moon	81° 23'.1
Long.	− 18° 42'.0 W
LHA Moon	62° 41'.1
Dec Moon (23^h) (d = 13'.0)	N6° 59'.3
Corr^n (25^m)	+ 05'.5
Dec Moon (23^{h}25^m)	N7° 04'.8

Q2.
GHA Aries (18^h)	14° 28'.60
Inc. (48^{m}35^s)	12° 10'.70
GHA Aries	026° 39'.30
SHA Acamar	+ 315° 20'.50
GHA Acamar	341° 59'.80
Long.	+ 94° 33'.36 E
LHA Acamar	436° 33'.16
−360	− 360° 00'.00
LHA Acamar	076° 33'.16

Dec Acamar: S40° 16'.2 (no corrections for a star)

Q3.
GHA (18^h)	97° 35'.0
Inc. (08^m 15^s)	2° 03'.8
V Corr^n (2'.2)	0'.3
GHA Saturn	99° 39'.1
Long.	+ 18° 45'.2 E
LHA Saturn	118° 24'.3
Dec Saturn (18^h) (d = 0'.1)	N3° 24'.6
Corr^n (08^m)	0'.0
Dec Saturn (18^{h}08^m)	N3° 24'.6

©2009 Jack Case

Q4.

GHA Sun (13^h)	17° 13'.3
Inc. (25^m 18^s)	6° 19'.5
GHA Sun	23° 32'.8
Long	− 18° 14'.7 W
LHA Sun	05° 18'.1
Dec Sun (13^h)	S22° 32'.6
Corr^n (25^m) (d = 0'.3)	+ 0'.1
Dec Sun	S22° 32'.7

(Dec. increasing on 6 December).

Chapter 5
Azimuth and Altitude

We can define the position of a celestial body in relation to its local hour angle and declination. We can also state its position in relation to our own celestial meridian and celestial horizon; in other words, in terms of its azimuth and altitude from our position.

The Azimuth is similar to the bearing in that it is the angle between the observer's meridian and the direction of the celestial body. However, whereas bearings are measured clockwise from north from 0° to 360°, azimuth is measured from 0° to 180° from either north or south. If the observer is in the northern hemisphere, the azimuth is measured from north and if in the southern hemisphere, it is measured from south.

For example, in Diagram 21,

O is the position of an observer on the Earth's surface and NOS represents the observer's meridian.

X is the position of a celestial body and in terms of bearing, the direction of X from point O is 045°.

In terms of azimuth it is either N45°E for an observer is in the northern hemisphere or S135°E for an observer in the southern hemisphere.

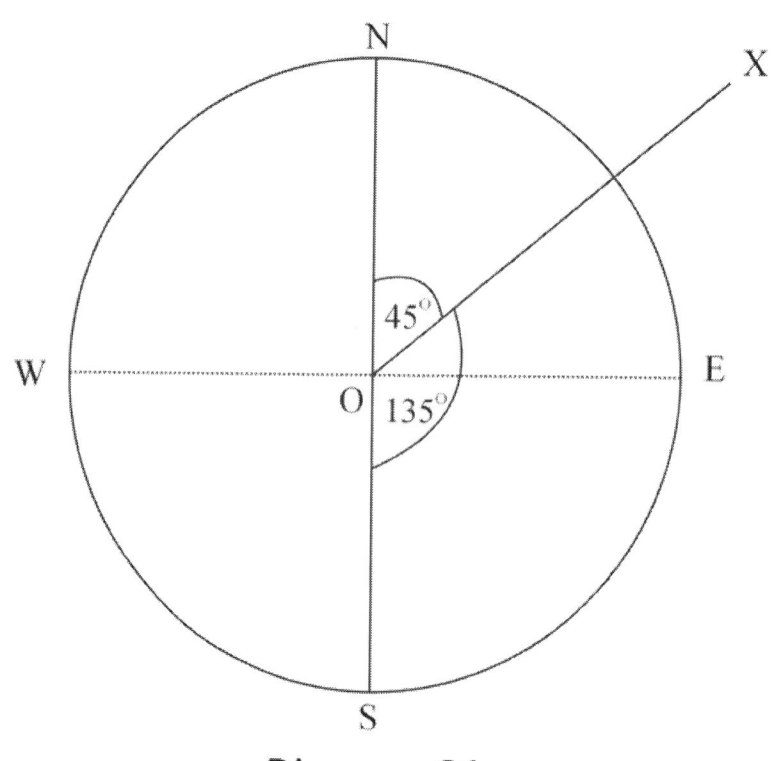

Diagram 21

The altitude is the angle between the celestial horizon and the direction of the celestial body. (The celestial horizon is the projection of the observer's horizon onto the celestial sphere).

As shown in Diagram 22, the observer measures the altitude in relation to the visible horizon from his position at O on the Earth's surface. So, the **observed altitude** is the angle HOX. However, the **true altitude** is measured from the Earth's centre in relation to the celestial horizon and is the angle RCX.

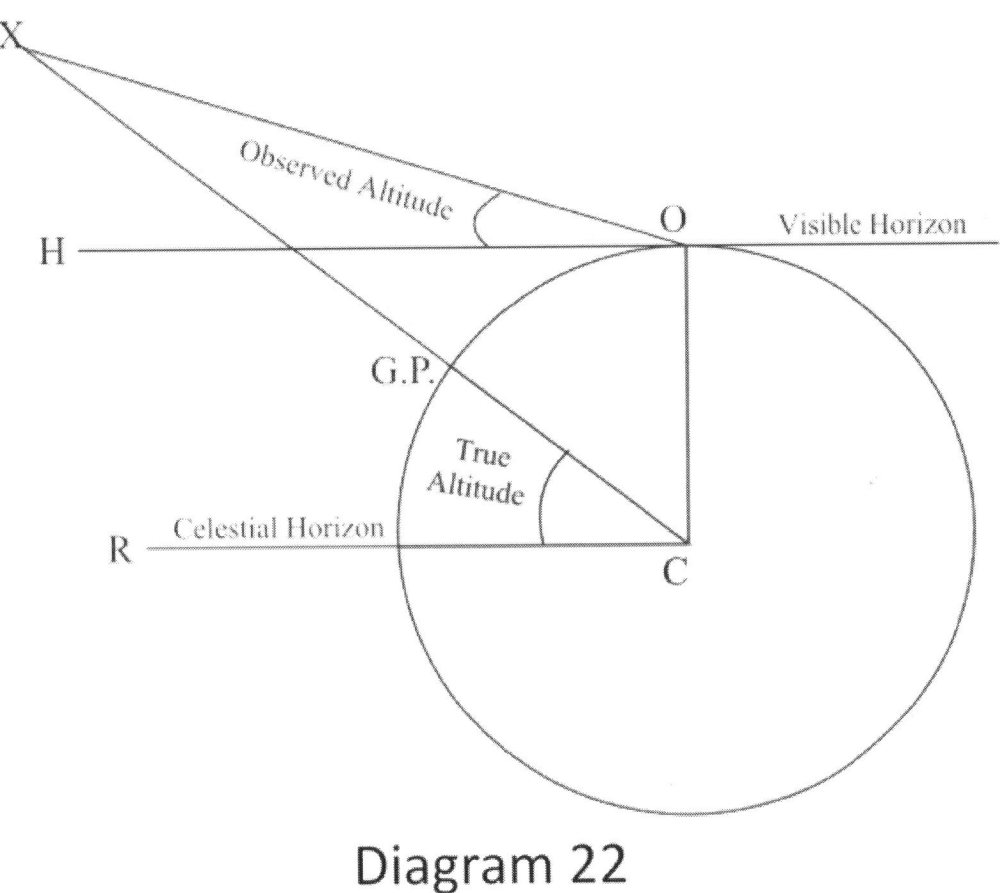

Diagram 22

Point O will be approximately 6367 Km. from the centre of the Earth and so it would seem that the visible horizon is bound to be slightly offset from the celestial horizon. Because of the vast distances of the stars and the planets from the Earth, we can assume that, in their cases, the celestial horizon and the visible horizon correspond with very little error. However, in the cases of the Sun and the Moon, which are relatively near, a correction called **Parallax** must be added. The subject of Parallax error is dealt with in greater detail later in this chapter.

So far, we have considered azimuth and altitude from a position on the surface of the Earth. To fully understand how these phenomena relate to the LHA and declination of a celestial body and hence, how they help us to establish our position, we need to consider them in relation to the celestial sphere.

Consider Diagram 23:

The celestial sphere is drawn in the plane of the observer's meridian with the observer's zenith (Z) at the top.

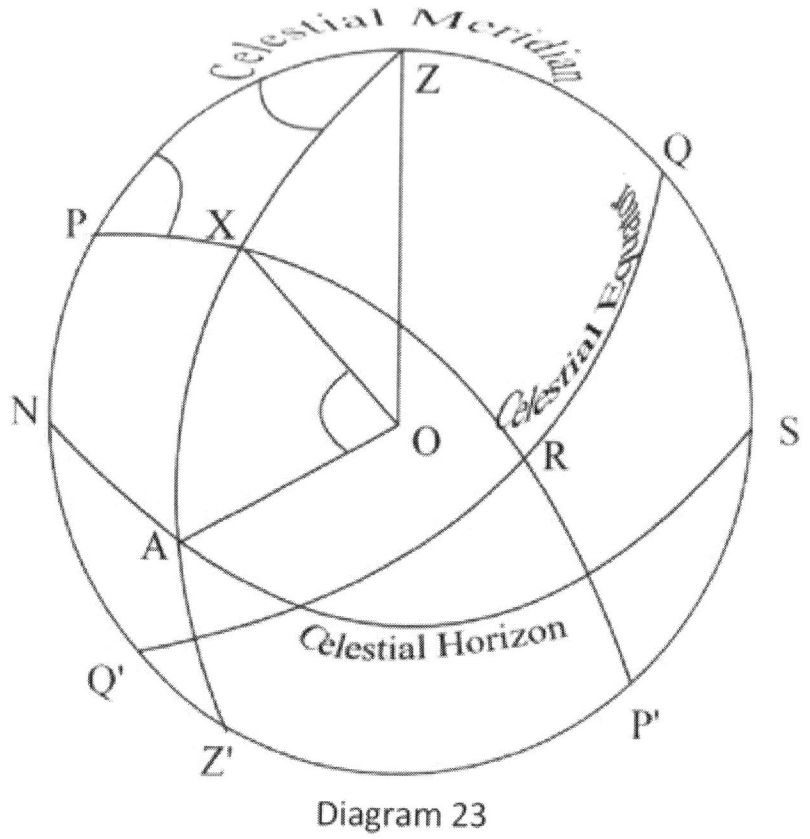

Diagram 23

Point O represents both the observer and the Earth.
The arc PZQSP' represents the observer's celestial meridian.
The arc NAS is the celestial horizon and QRQ' represents the celestial equator.
ZXAZ' is a vertical circle running through the position of the celestial body (X). (A vertical circle is a great circle that passes through the observer's zenith and is perpendicular to the celestial horizon).

The Azimuth is the angle PZX in diagram 23 (that is, the angle between the observer's celestial meridian and the vertical circle through the celestial body). It is measured from 0° to 180° east or west from the observer's meridian and depending on whether the observer is in the northern hemisphere or the southern hemisphere, it is named north or south.

The Altitude is the angle AOX in diagram 23; that is the angle from the celestial horizon to the celestial body measured along the vertical circle.

The Zenith Distance is the angular distance ZX measured along the vertical circle from the zenith to the celestial body; that is the angle XOZ.

Relationship between Altitude and Zenith Distance
Since the celestial meridian is a vertical circle and is therefore, perpendicular to the celestial horizon, it follows that angle AOZ is a right angle and angles AOX and XOZ are complementary angles. From this we can deduce that:

        Zenith Distance = 90° – Altitude

and   **Altitude = 90° – Zenith Distance**

Local Hour Angle (LHA)
LHA is the angle ZPX; that is the angle between the observer's celestial meridian and the meridian of the celestial body.

Relationship between LHA and Azimuth
Consider diagram 24:

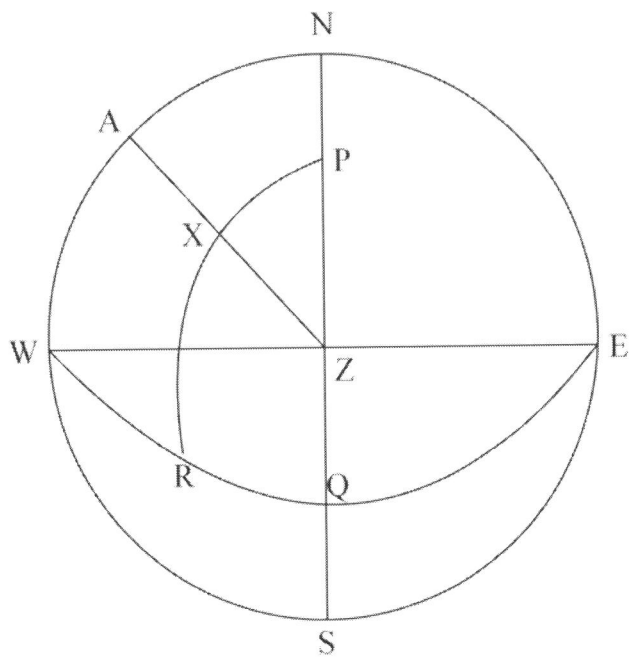

Diagram 24

Diagram 24 is drawn in the plane of the celestial horizon. Imagine that you are looking down on the celestial sphere from a position directly above the observer's zenith which is in the centre of a circle that represents the celestial horizon.

NZS represents the observer's celestial meridian.
WQE represents the celestial equator,
P is the celestial pole,
X is the position of the celestial body,
PXR represents part of the meridian of the celestial body which cuts the Equator at R.
ZPX is the LHA.
PZX is the Azimuth.

When the LHA is less than 180°, the celestial body lies to the west of the observer's meridian and when it is greater than 180° it lies to the east. (Remember LHA is measured westwards from the observer's meridian).
It follows that if the celestial body is to the west of the observer's meridian, the azimuth must be west and when to the east, the azimuth must be east.

So we have the rule:
$$\text{LHA} = 0° - 180° : \text{Azimuth West}$$
$$\text{LHA} = 180° - 360° : \text{Azimuth East}$$

The Theory of Astro navigation.
The relationships discussed above illustrate the importance of azimuth and altitude in position finding at sea. The theory of astro navigation depends on the ability to solve the spherical triangle PZX; the azimuth and altitude give us the essential data we need to do this. With this data we are able to find the LHA, declination and zenith distance of a celestial body and armed with this information, we are able to establish our position on the Earth's surface.

Measuring the Altitude. To measure the altitude of a celestial body, we use a **sextant**.
As shown in Diagram 25, the horizon is viewed directly through the sextant telescope and the celestial body is viewed via two mirrors. The upper mirror is attached to the index bar. The index bar is moved until it reflects an image of the celestial body into the lower mirror which is fixed. The position of the index bar is finely adjusted until the image of the celestial body appears to sit on the horizon.

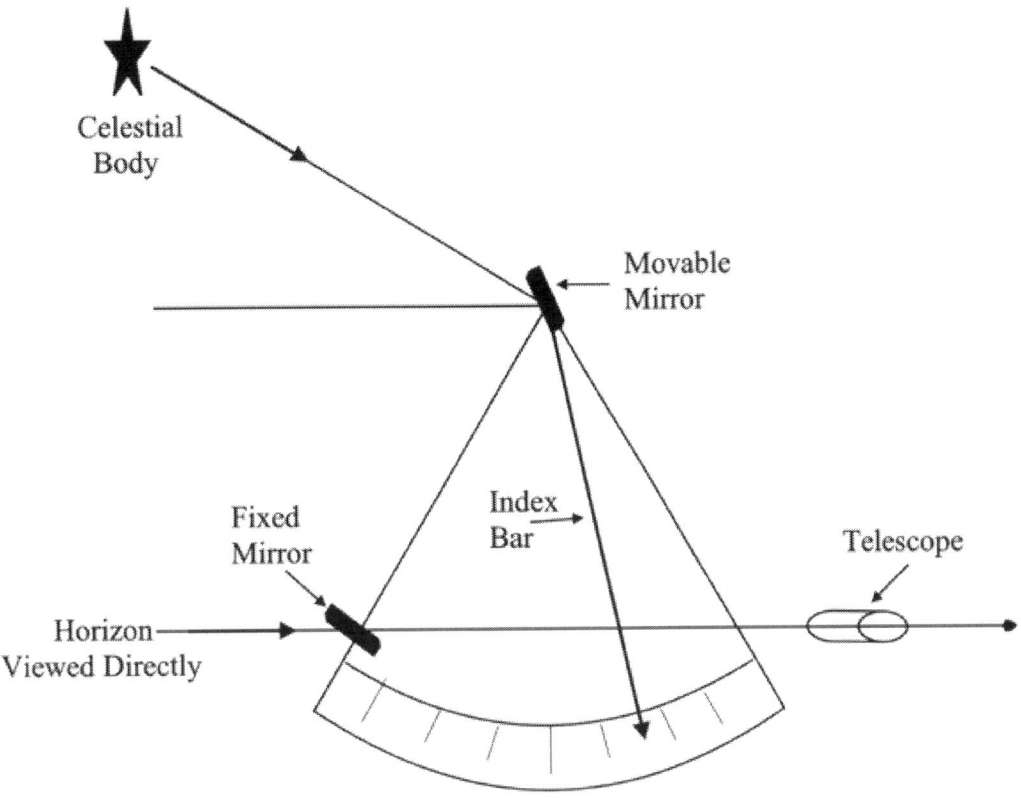

The Sextant

Diagram 25

As the index bar is adjusted, it moves a pointer over a graduated scale and when the images are made to coincide, the angle indicated by the pointer is the altitude. The altitude measured by a sextant is referred to as the **Sextant Altitude** (Sext. Alt).

Corrections. A number of corrections have to be made to the sextant altitude before we arrive at the **True Altitude**.

Index Error. No matter how carefully a sextant is manufactured, there will usually be a very small error in its reading and this is known as index error.

To calculate index error, view a single object through the sextant telescope and through the mirrors; move the index bar until the two images coincide and note the reading. If the reading is not zero, the actual reading is the index error.

For example, if the reading is 2' **too high**, it is said to be 2' **'on the arc'** and recorded as: Index Error – 2'.0. If the reading is 2' **too low**, it is said to be 2' **'off the arc'** and recorded as +2'.0.

Dip. A correction has to be made to allow for the height of the observer's eye above the horizon; this is known as Dip.
Consider Diagram 26,
O is an observer's position on the Earth's surface and E is the position of his eye. We can see that, as the observer's height of eye is raised above sea level, his visible horizon 'dips' below the true horizon and so the altitude measured at E becomes greater than that measured at O.

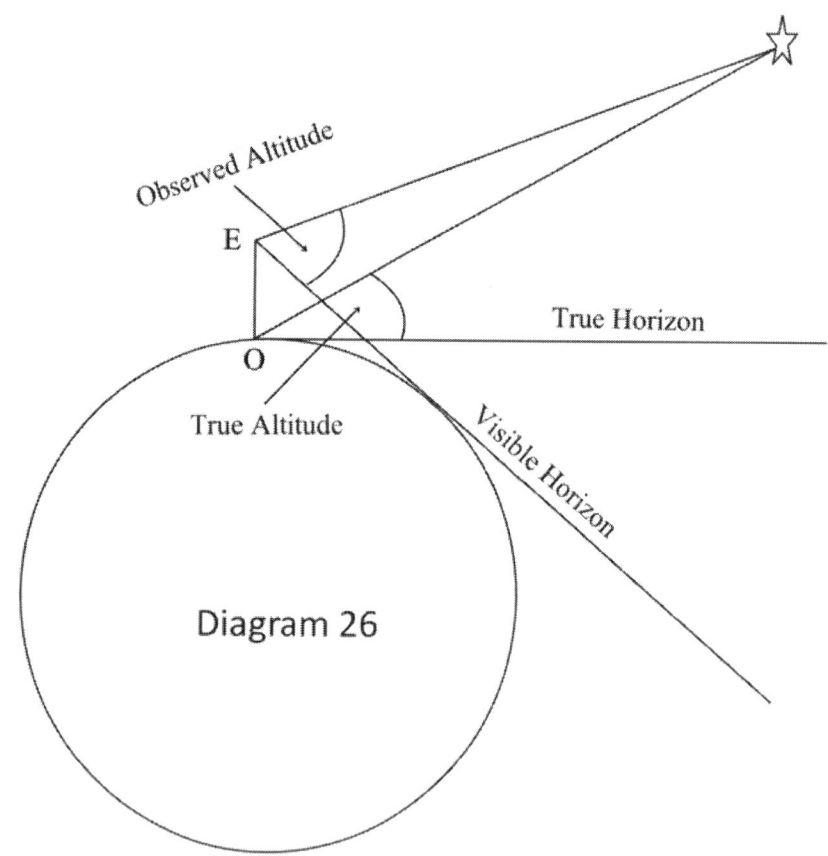

Diagram 26

Dip is the error caused by this difference and has to be subtracted from the reading.
Tables of corrections for dip are printed in the Nautical Almanac as shown in the extract opposite:

DIP				
Ht. of Eye	Corrⁿ	Ht. of Eye	Ht. of Eye	Corrⁿ
m	′	ft.	m	′
2·4	−2·8	8·0	1·0	− 1·8
2·6	−2·9	8·6	1·5	− 2·2
2·8	−3·0	9·2	2·0	− 2·5
3·0	−3·1	9·8	2·5	− 2·8
3·2	−3·2	10·5	3·0	− 3·0
3·4	−3·3	11·2	See table	
3·6	−3·4	11·9	←	
3·8	−3·5	12·6		
4·0	−3·6	13·3	m	′
4·3	−3·7	14·1	20	− 7·9
4·5	−3·8	14·9	22	− 8·3
4·7	−3·9	15·7	24	− 8·6
5·0	−4·0	16·5	26	− 9·0
5·2	−4·1	17·4	28	− 9·3
5·5	−4·2	18·3	30	− 9·6
5·8		19·1		

For example, if the height of eye is 4.6m. the correction will be 3'.8 (interpolate as necessary).

Apparent Altitude is found by applying the index error and dip to the sextant altitude.
Example:
Sextant Altitude   48° 15'.2
Index Error   +1'.3
Dip   -5'.3
Apparent Alt.   48° 11'.2

Refraction. When a ray of light from a celestial body passes through the Earth's atmosphere, it becomes bent through refraction and causes the apparent altitude to be greater than the true altitude. Since the sextant measures the apparent altitude, a correction for refraction must be applied to find the true altitude. Refraction is at its greatest when the altitude is small (i.e. when the celestial body is near the horizon) and becomes less as the altitude increases.

Corrections for refraction are included in the Altitude Correction Tables which are to be found in the Nautical Almanac and will be discussed later in this chapter.

Effect of Refraction on Dip. As well as increasing the apparent altitude of a celestial body, refraction also has an effect on the position of the visible horizon and this will in turn, have an effect on the angle of dip.
The effects of refraction are illustrated in Diagram 27.

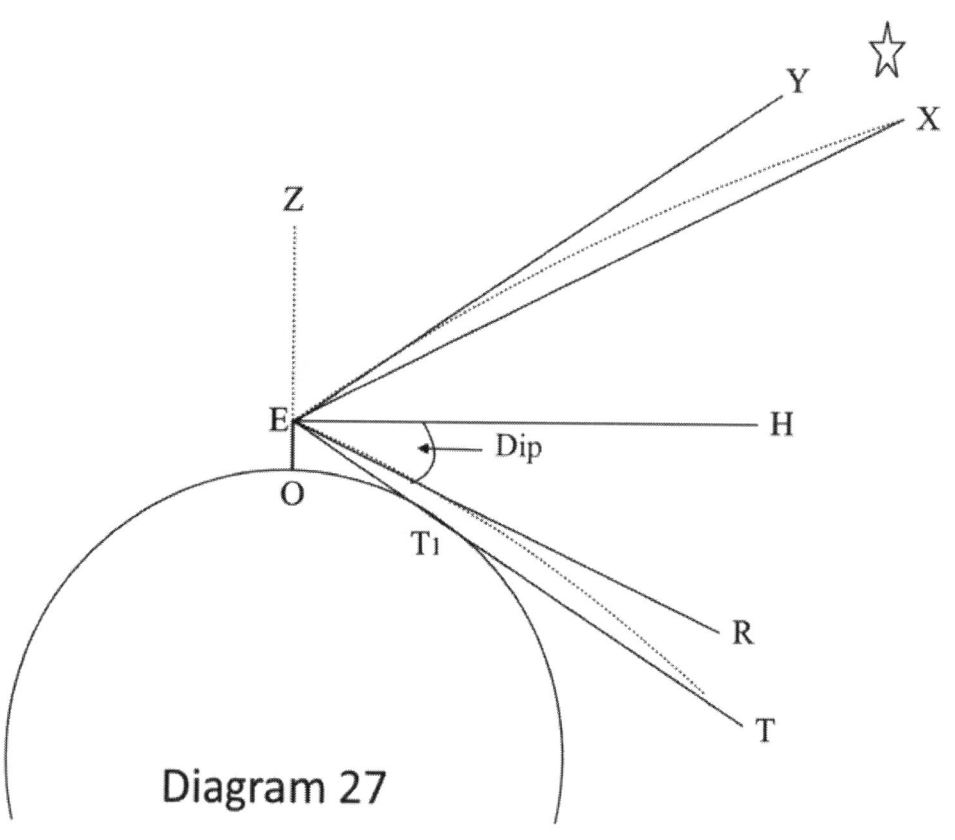

Diagram 27

XEH is the true altitude from the observer's height of eye. However, due to refraction, the celestial body appears to be at Y and so YEH becomes the apparent altitude.
ET is a tangent from the observer's eye to the Earth's surface and so T_1 should mark the position of the horizon from E.
The theoretical angle of Dip is the angle HET; however, because refraction causes the horizon to appear to be in the direction of R, angle HER becomes the angle of dip.
Terrestrial refraction, as it is known, is included in the values for dip as tabulated in the Nautical Almanac and therefore, need not be calculated by the navigator.

An **additional correction for refraction** may be needed if the temperature and atmospheric pressure are greatly different to the standard conditions which are assumed to be 10°C, 1010mb. Part of Table A4 from the Nautical Almanac is shown in the extract below. This table tabulates the additional corrections for non standard conditions:

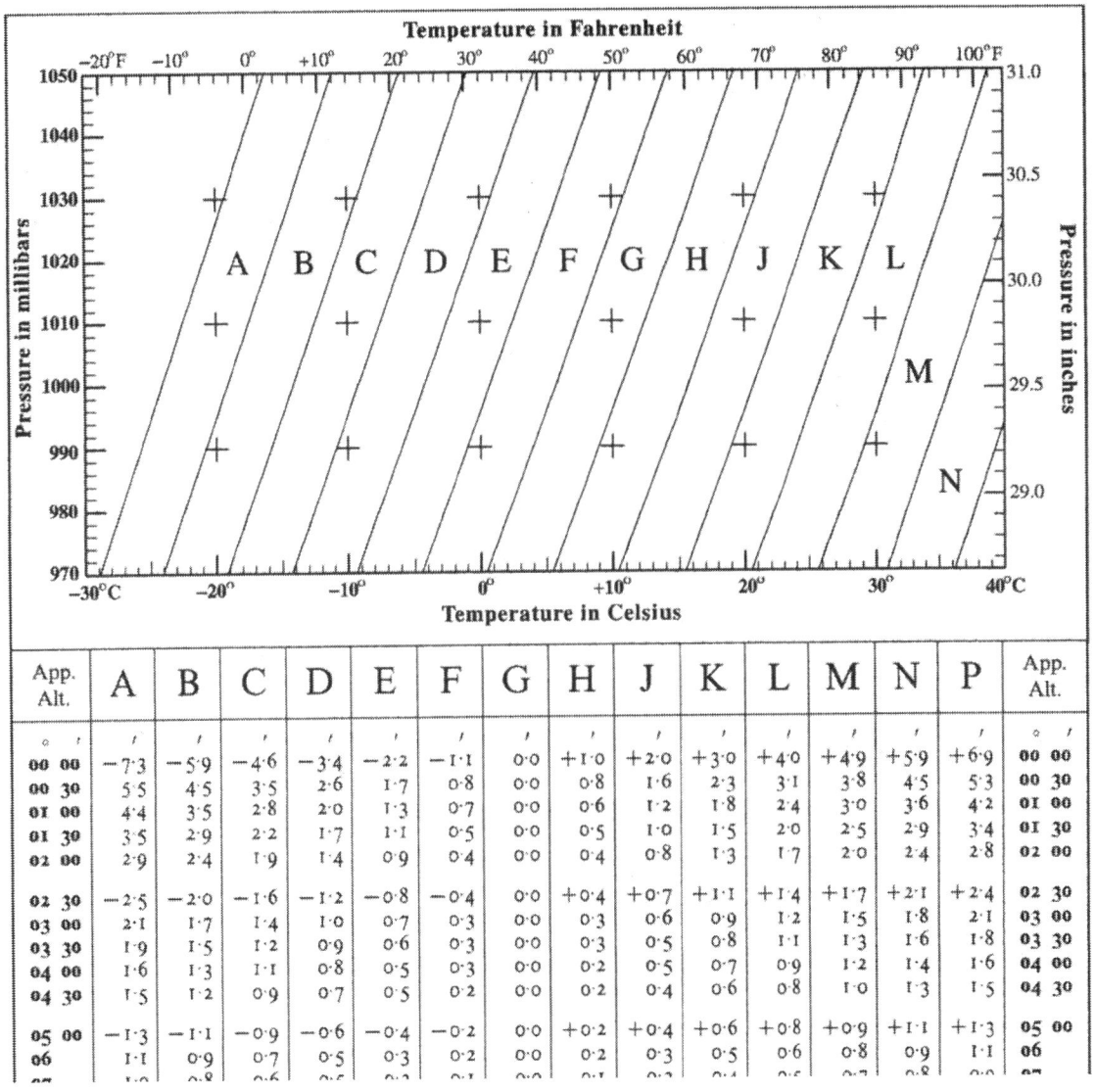

The graph above the table is entered with the temperature and pressure to find a zone letter. For example, entering with a temperature of 20°C. and a pressure of 1010mb. gives zone letter J.

The table is then entered with the apparent altitude and zone letter to find the additional correction for refraction.

Example. If apparent altitude = 4° 30', temperature = 30 °C, pressure = 1000mb, the zone letter will be L and the correction will be +0'.8.

The Sun's Semi-Diameter. The point on the Sun's circumference nearest to the horizon is called the **lower limb** (L.L.) and the point furthest from the horizon is called the **upper limb** (U.L.).

In practice, the altitude that we measure is that of the lower limb; however, what we really need is the altitude of the Sun's centre and so, as diagram 28 shows, we must add a correction for the value of its semi-diameter.

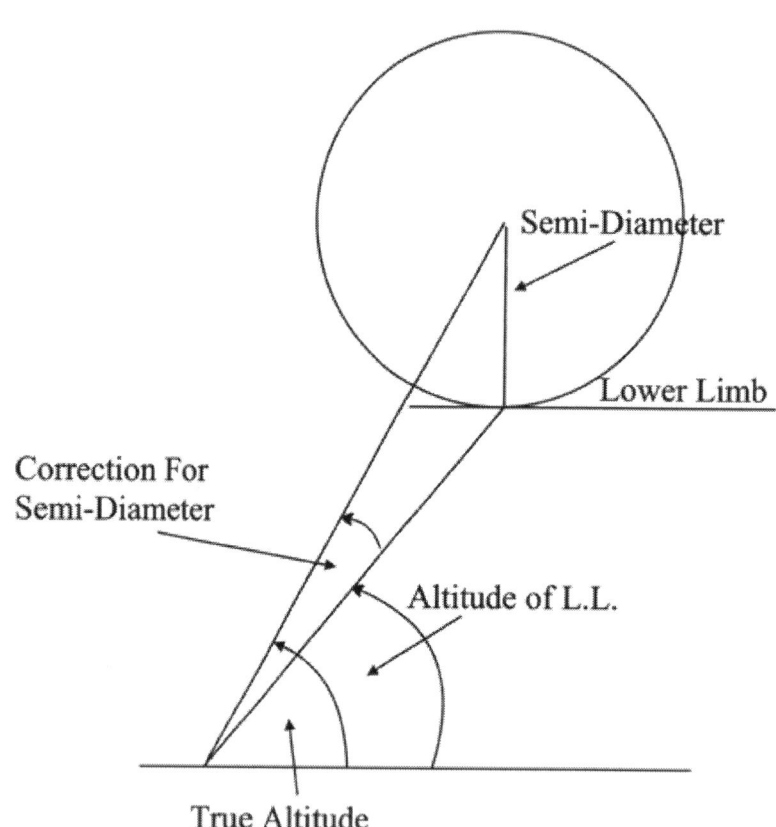

Diagram 28

The semi-diameter of the Sun is given in the daily pages of the Nautical Almanac. However, as with refraction, the correction for the Sun's semi-diameter is included in the Altitude Correction Tables and so need not be separately considered.

The Moon's Semi-Diameter. When the Moon is not full, sometimes only the upper limb will be visible and sometimes only the lower limb. In such cases, we have no choice other than to measure the altitude of the limb that we can see and to either add or 0subtract the semi-diameter accordingly.

Diagram 29 shows the moon at different phases. It should be stressed that whether the Moon's upper or lower limb is visible is dependent not only on its phase but also on the relative altitudes of the Sun and the Moon. For example, if, one morning, a crescent or gibbous moon is visible in the eastern sky and the Sun is at a higher altitude, only the upper limb will be visible but if, in the evening of the same day, the Moon is visible in the western sky and the Sun has set below the western horizon, only the lower limb will be visible.

Corrections for the Moon's semi-diameter are included in the Altitude Correction Tables and will be explained later in this chapter.

 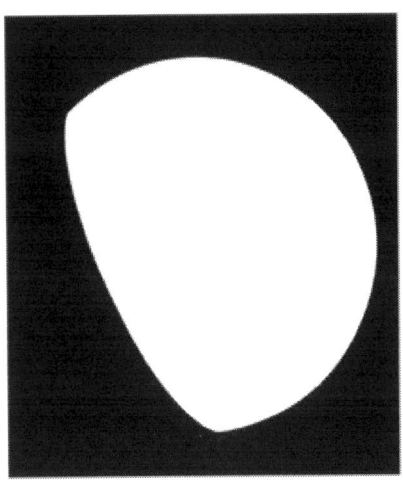

Waxing Crescent Moon   Waxing Gibbous Moon

 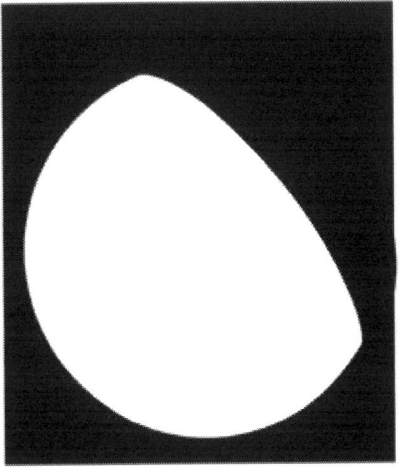

Waning Crescent Moon   Waning Gibbous Moon

Diagram 29

Semi-Diameter of Stars and Planets. To the naked eye and even through a sextant telescope, the stars and planets appear as points of light and so there is obviously no need to apply semi-diameter corrections in their cases.

Parallax. As discussed earlier in this chapter, we measure the altitude of a celestial body from our position in relation to our visible horizon; this is known as the **observed altitude**. However, when calculating the **true altitude**, measurements are made from the Earth's centre in relation to the celestial horizon. The displacement between the observed position of an object and the true position is known as **parallax**.

Parallax corrections for stars and planets. Because the stars and the planets are at such great distances from the Earth, we can assume that, in their cases, the celestial horizon and the visible horizon correspond with very little error.

However, in certain cases when extreme accuracy is needed, parallax corrections for Mars and Venus are required and these are listed in the altitude correction tables as shown in table A2 on the next page.

Parallax corrections for the Sun and the Moon. Because the Sun and the Moon are relatively close to the Earth, parallax will be significant and so a correction has to be made. These corrections are included in the altitude correction tables and therefore do not have to be applied separately.

Horizontal Parallax. Parallax error is greatest when the celestial body is close to the horizon and decreases to zero as the altitude approaches 90°. It is negligible except in the case of the Moon which is close to the Earth in comparison with the other celestial bodies. Because horizontal parallax is significant in the case of the Moon, a separate correction (abbreviated to HP) has to be applied.

Combined Correction Tables. As previously mentioned, the Altitude Correction Tables in the Nautical Almanac give combined corrections for Semi-diameter, Refraction and Parallax. There are separate tables for Dip.

Altitude Correction tables for the Sun, stars and planets. There are two tables; table A2 contains combined corrections for apparent altitudes from 9° 33' to 90° and table A3 is for apparent altitudes from 0° to 10°. Table A2 is shown in the following nautical almanac extract.

A2 ALTITUDE CORRECTION TABLES 10°–90°—SUN, STARS, PLANETS

OCT.–MAR. SUN APR.–SEPT.						STARS AND PLANETS				DIP					
App. Alt.	Lower Limb	Upper Limb	App. Alt.	Lower Limb	Upper Limb	App. Alt.	Corrn	App. Alt.	Additional Corrn	Ht. of Eye	Corrn	Ht. of Eye	Corrn	Ht. of Eye	Corrn
° ′	′	′	° ′	′	′	° ′	′			m	′	ft.		m	′
9 33	+10·8	−21·5	9 39	+10·6	−21·2	9 55	−5·3	**2009**		2·4	−2·8	8·0		1·0	−1·8
9 45	+10·9	−21·4	9 50	+10·7	−21·1	10 07	−5·2	**VENUS**		2·6	−2·9	8·6		1·5	−2·2
9 56	+11·0	−21·3	10 02	+10·8	−21·0	10 20	−5·1	Jan. 1–Jan. 28		2·8	−3·0	9·2		2·0	−2·5
10 08	+11·1	−21·2	10 14	+10·9	−20·9	10 32	−5·0	May 23–July 10		3·0	−3·1	9·8		2·5	−2·8
10 20	+11·2	−21·1	10 27	+11·0	−20·8	10 46	−4·9	°	′	3·2	−3·2	10·5		3·0	−3·0
10 33	+11·3	−21·0	10 40	+11·1	−20·7	10 59	−4·8	41	+0·2	3·4	−3·3	11·2		See table ←	
10 46	+11·4	−20·9	10 53	+11·2	−20·6	11 14	−4·7	76	+0·1	3·6	−3·4	11·9			
11 00	+11·5	−20·8	11 07	+11·3	−20·5	11 29	−4·6	Jan. 29–Feb. 21		3·8	−3·5	12·6		m	′
11 15	+11·6	−20·7	11 22	+11·4	−20·4	11 44	−4·5	May 1–May 22		4·0	−3·6	13·3		20	−7·9
11 30	+11·7	−20·6	11 37	+11·5	−20·3	12 00	−4·4	°	′	4·3	−3·7	14·1		22	−8·3
11 45	+11·8	−20·5	11 53	+11·6	−20·2	12 17	−4·3	34	+0·3	4·5	−3·8	14·9		24	−8·6
12 01	+11·9	−20·4	12 10	+11·7	−20·1	12 35	−4·2	60	+0·2	4·7	−3·9	15·7		26	−9·0
12 18	+12·0	−20·3	12 27	+11·8	−20·0	12 53	−4·1	80	+0·1	5·0	−4·0	16·5		28	−9·3
12 36	+12·1	−20·2	12 45	+11·9	−19·9	13 12	−4·0	Feb. 22–Mar. 9		5·2	−4·1	17·4			
12 54	+12·2	−20·1	13 04	+12·0	−19·8	13 32	−3·9	Apr. 15–Apr. 30		5·5	−4·2	18·3		30	−9·6
13 14	+12·3	−20·0	13 24	+12·1	−19·7	13 53	−3·8	°	′	5·8	−4·3	19·1		32	−10·0
13 34	+12·4	−19·9	13 44	+12·2	−19·6	14 16	−3·7	29	+0·4	6·1	−4·4	20·1		34	−10·3
13 55	+12·5	−19·8	14 06	+12·3	−19·5	14 39	−3·6	51	+0·3	6·3	−4·5	21·0		36	−10·6
14 17	+12·6	−19·7	14 29	+12·4	−19·4	15 03	−3·5	68	+0·2	6·6	−4·6	22·0		38	−10·8
14 41	+12·7	−19·6	14 53	+12·5	−19·3	15 29	−3·4	83	+0·1	6·9	−4·7	22·9			
15 05	+12·8	−19·5	15 18	+12·6	−19·2	15 56	−3·3	Mar. 10–Apr. 14		7·2	−4·8	23·9		40	−11·1
15 31	+12·9	−19·4	15 45	+12·7	−19·1	16 25	−3·2			7·5	−4·9	24·9		42	−11·4
15 59	+13·0	−19·3	16 13	+12·8	−19·0	16 55	−3·1	°	′	7·9	−5·0	26·0		44	−11·7
16 27	+13·1	−19·2	16 43	+12·9	−18·9	17 27	−3·0	26	+0·5	8·2	−5·1	27·1		46	−11·9
16 58	+13·2	−19·1	17 14	+13·0	−18·8	18 01	−2·9	46	+0·4	8·5	−5·2	28·1		48	−12·2
17 30	+13·3	−19·0	17 47	+13·1	−18·7	18 37	−2·8	60	+0·3	8·8	−5·3	29·2			
18 05	+13·4	−18·9	18 23	+13·2	−18·6	19 16	−2·7	73	+0·2	9·2	−5·4	30·4		ft.	′
18 41	+13·5	−18·8	19 00	+13·3	−18·5	19 56	−2·6	84	+0·1	9·5	−5·5	31·5		2	−1·4
19 20	+13·6	−18·7	19 41	+13·4	−18·4	20 40	−2·5	July 11–Dec. 31		9·9	−5·6	32·7		4	−1·9
20 02	+13·7	−18·6	20 24	+13·5	−18·3	21 27	−2·4	°	′	10·3	−5·7	33·9		6	−2·4
20 46	+13·8	−18·5	21 10	+13·6	−18·2	22 17	−2·3	60	+0·1	10·6	−5·8	35·1		8	−2·7
21 34	+13·9	−18·4	21 59	+13·7	−18·1	23 11	−2·2			11·0	−5·9	36·3		10	−3·1
22 25	+14·0	−18·3	22 52	+13·8	−18·0	24 09	−2·1	**MARS**		11·4	−6·0	37·6		See table ←	
23 20	+14·1	−18·2	23 49	+13·9	−17·9	25 12	−2·0	Jan. 1–Nov. 27		11·8	−6·1	38·9			
24 20	+14·2	−18·1	24 51	+14·0	−17·8	26 20	−1·9	°	′	12·2	−6·2	40·1		ft.	′
25 24	+14·3	−18·0	25 58	+14·1	−17·7	27 34	−1·8	60	+0·1	12·6	−6·3	41·5		70	−8·1
26 34	+14·4	−17·9	27 11	+14·2	−17·6	28 54	−1·7			13·0	−6·4	42·8		75	−8·4
27 50	+14·5	−17·8	28 31	+14·3	−17·5	30 22	−1·6	Nov. 28–Dec. 31		13·4	−6·5	44·2		80	−8·7
29 13	+14·6	−17·7	29 58	+14·4	−17·4	31 58	−1·5	°	′	13·8	−6·6	45·5		85	−8·9
30 44	+14·7	−17·6	31 33	+14·5	−17·3	33 43	−1·4	41	+0·2	14·2	−6·7	46·9		90	−9·2
32 24	+14·8	−17·5	33 18	+14·6	−17·2	35 38	−1·3	76	+0·1	14·7	−6·8	48·4		95	−9·5
34 15	+14·9	−17·4	35 15	+14·7	−17·1	37 45	−1·2			15·1	−6·9	49·8			
36 17	+15·0	−17·3	37 24	+14·8	−17·0	40 06	−1·1			15·5	−7·0	51·3		100	−9·7
38 34	+15·1	−17·2	39 48	+14·9	−16·9	42 42	−1·0			16·0	−7·1	52·8		105	−9·9
41 06	+15·2	−17·1	42 28	+15·0	−16·8	45 34	−0·9			16·5	−7·2	54·3		110	−10·2
43 56	+15·3	−17·0	45 29	+15·1	−16·7	48 45	−0·8			16·9	−7·3	55·8		115	−10·4
47 07	+15·4	−16·9	48 52	+15·2	−16·6	52 16	−0·7			17·4	−7·4	57·4		120	−10·6
50 43	+15·5	−16·8	52 41	+15·3	−16·5	56 09	−0·6			17·9	−7·5	58·9		125	−10·8
54 46	+15·6	−16·7	56 59	+15·4	−16·4	60 26	−0·5			18·4	−7·6	60·5			
59 21	+15·7	−16·6	61 50	+15·5	−16·3	65 06	−0·4			18·8	−7·7	62·1		130	−11·1
64 28	+15·8	−16·5	67 15	+15·6	−16·2	70 09	−0·3			19·3	−7·8	63·8		135	−11·3
70 10	+15·9	−16·4	73 14	+15·7	−16·1	75 32	−0·2			19·8	−7·9	65·4		140	−11·5
76 24	+16·0	−16·3	79 42	+15·8	−16·0	81 12	−0·1			20·4	−8·0	67·1		145	−11·7
83 05	+16·1	−16·2	86 31	+15·9	−15·9	87 03	0·0			20·9	−8·1	68·8		150	−11·9
90 00			90 00			90 00				21·4		70·5		155	−12·1

Altitude Correction tables for the Sun. As shown in table A2, corrections for the Sun are divided into two parts to allow for changes in the Sun's semi-diameter during the course of a year. The first part is for the period October to March and is based on a semi-diameter of 16'.15. The second part is for April to September and is based on a semi-diameter of 15'.9.

The tables are entered with the apparent altitude and the resultant correction is added in the case of the lower limb and subtracted for the upper limb.

Example: (Refer to table A2 on the previous page).

> Date = 20 May
> Apparent altitude of the Sun's lower limb = 10° 47'.0
> Correction = +11'.1 (interpolating as necessary).

Full example of altitude corrections for the Sun. The following example shows how altitude corrections for index error, dip, semi-diameter, refraction and parallax are applied to a sextant altitude of the Sun:

On 12 Feb. a reading of the Sun's lower limb was taken.
Sextant altitude = 14° 35'.5. Index error = -2'.3. Height of eye = 4.2m.
Temperature = 20°C. Pressure = 1020 mb.
Using the extracts of tables A2 above and A4 below, the calculations for the True Altitude would be as follows:

Sext. Alt.	14° 35'.5
I.E.	-2'.3
Observed Alt.	14° 33'.2
Dip	-3'.6
Apparent Alt.	14° 29'.6
Sun's correction	+12'.6 (combined correction)
Additional corr.	+ 0'.1
True Alt.	14° 42'.3

If the upper limb had been used instead of the lower limb, the result would be:

Sext. Alt.	14° 35'.5
I.E.	-2'.3
Observed Alt.	14° 33'.2
Dip	-3'.6
Apparent Alt.	14° 29'.6
Sun's correction	-19'.7
Additional corr.	+0'.1
True Alt.	14° 10'.0

A4 ALTITUDE CORRECTION TABLES—ADDITIONAL CORRECTIONS
ADDITIONAL REFRACTION CORRECTIONS FOR NON-STANDARD CONDITIONS

App. Alt.	A	B	C	D	E	F	G	H	J	K	L	M	N	P	App. Alt.
° ′	′	′	′	′	′	′	′	′	′	′	′	′	′	′	° ′
00 00	−7.3	−5.9	−4.6	−3.4	−2.2	−1.1	0.0	+1.0	+2.0	+3.0	+4.0	+4.9	+5.9	+6.9	00 00
00 30	5.5	4.5	3.5	2.6	1.7	0.8	0.0	0.8	1.6	2.3	3.1	3.8	4.5	5.3	00 30
01 00	4.4	3.5	2.8	2.0	1.3	0.7	0.0	0.6	1.2	1.8	2.4	3.0	3.6	4.2	01 00
01 30	3.5	2.9	2.2	1.7	1.1	0.5	0.0	0.5	1.0	1.5	2.0	2.5	2.9	3.4	01 30
02 00	2.9	2.4	1.9	1.4	0.9	0.4	0.0	0.4	0.8	1.3	1.7	2.0	2.4	2.8	02 00
02 30	−2.5	−2.0	−1.6	−1.2	−0.8	−0.4	0.0	+0.4	+0.7	+1.1	+1.4	+1.7	+2.1	+2.4	02 30
03 00	2.1	1.7	1.4	1.0	0.7	0.3	0.0	0.3	0.6	0.9	1.2	1.5	1.8	2.1	03 00
03 30	1.9	1.5	1.2	0.9	0.6	0.3	0.0	0.3	0.5	0.8	1.1	1.3	1.6	1.8	03 30
04 00	1.6	1.3	1.1	0.8	0.5	0.3	0.0	0.2	0.5	0.7	0.9	1.2	1.4	1.6	04 00
04 30	1.5	1.2	0.9	0.7	0.5	0.2	0.0	0.2	0.4	0.6	0.8	1.0	1.3	1.5	04 30
05 00	−1.3	−1.1	−0.9	−0.6	−0.4	−0.2	0.0	+0.2	+0.4	+0.6	+0.8	+0.9	+1.1	+1.3	05 00
06	1.1	0.9	0.7	0.5	0.3	0.2	0.0	0.2	0.3	0.5	0.6	0.8	0.9	1.1	06
07	1.0	0.8	0.6	0.5	0.3	0.1	0.0	0.1	0.3	0.4	0.5	0.7	0.8	0.9	07
08	0.8	0.7	0.5	0.4	0.3	0.1	0.0	0.1	0.2	0.4	0.5	0.6	0.7	0.8	08
09	0.7	0.6	0.5	0.4	0.2	0.1	0.0	0.1	0.2	0.3	0.4	0.5	0.6	0.7	09
10 00	−0.7	−0.5	−0.4	−0.3	−0.2	−0.1	0.0	+0.1	+0.2	+0.3	+0.4	+0.5	+0.6	+0.7	10 00
12	0.6	0.5	0.4	0.3	0.2	0.1	0.0	0.1	0.2	0.2	0.3	0.4	0.5	0.5	12
14	0.5	0.4	0.3	0.2	0.1	0.1	0.0	0.1	0.1	0.2	0.3	0.3	0.4	0.5	14
16	0.4	0.3	0.3	0.2	0.1	0.1	0.0	0.1	0.1	0.2	0.2	0.3	0.3	0.4	16
18	0.4	0.3	0.2	0.2	0.1	−0.1	0.0	+0.1	0.1	0.2	0.2	0.3	0.3	0.4	18
20 00	−0.3	−0.3	−0.2	−0.2	−0.1	0.0	0.0	0.0	+0.1	+0.1	+0.2	+0.2	+0.3	+0.3	20 00
25	0.3	0.2	0.2	0.1	0.1	0.0	0.0	0.0	0.1	0.1	0.1	0.2	0.2	0.2	25
30	0.2	0.2	0.1	0.1	0.1	0.0	0.0	0.0	+0.1	0.1	0.1	0.1	0.2	0.2	30
35	0.2	0.1	0.1	0.1	−0.1	0.0	0.0	0.0	0.0	0.1	0.1	0.1	0.1	0.2	35
40	0.1	0.1	0.1	−0.1	0.0	0.0	0.0	0.0	+0.1	0.1	0.1	0.1	0.1	0.1	40
50 00	−0.1	−0.1	−0.1	0.0	0.0	0.0	0.0	0.0	0.0	0.0	+0.1	+0.1	+0.1	+0.1	50 00

Altitude Correction Tables for Stars. As previously mentioned, because they are so far from the Earth, parallax and semi-diameter arguments for the stars are negligible and so the only corrections necessary are for dip and refraction.

Example:
Corrections for Altitude of a Star: Using the following data and the extracts from table A2 and table A4 on the preceding pages, calculate the true altitude of a star.
Data: Sextant altitude = 12° 20'.4   I.E = +1'.8   Ht. of eye = 4.8m.
Temperature = -20°C.   Pressure = 1010 mb.
(Remember, the correction for a star consists of index error, dip and refraction only since parallax and semi-diameter are negligible).

Solution:

Sext. Alt.	12° 20'.4
I.E.	+1'.8
Observed Alt.	12° 22'.2
Dip	-3'.9
Apparent Alt.	12° 18'.3
Star's correction	-4'.3
Additional corr.	-0'.6
True Alt.	12° 13'.4

Altitude Correction Tables for the Planets.
- As in the case of the stars, because they are so far from the Earth, parallax and semi-diameter arguments for the planets are negligible and so the only corrections necessary are for dip and refraction.
- For normal navigational practices, the navigational planets are treated as stars and the correction table for stars is used.
- For the very rare cases that they are needed for extreme accuracy, additional corrections for phase and parallax for Venus and parallax for Mars are given in the Nautical Almanac.

Example: Corrections for Altitude of a Planet.
Using the following data and the extracts from table A2 table A4 on the preceding pages, calculate the true altitude of a planet.
Sextant altitude = 10° 38'.6   I.E. = +2'.4   Ht. of eye = 13.0 ft.
Temperature = 10°C. Pressure = 1000 mb.

Solution:

Sext. Alt.	10° 38'.6
I.E.	+2'.4
Obs. Alt.	10° 41'.0
Dip	-3'.5
App. Alt.	10° 37'.5
Plnt's corr.	-5'.0
Additional corr.	0'.0
True Alt.	10° 32'.5

Altitude Correction Table for the Moon. There are two correction tables for the Moon in the Nautical Almanac. One is for apparent altitudes of 0° to 35° and the other is for 35° to 90°. An extract of the table for 0° to 35° is shown on the next page.

You will see from the extract, that the altitude correction table is divided into two parts. The top part of the table is entered with the apparent altitude and the correction is read from the appropriate column.

The lower part of the table is for horizontal parallax (HP) and has different columns for lower limb and upper limb. This part of the table is entered with the value of HP (which is obtained from the daily pages of the nautical almanac) and the correction is read from the same column as that for the apparent altitude correction above.

Both of the above corrections are added to the apparent altitude together with the additional correction for temperature and pressure in order to obtain the true altitude.

When the upper limb is used, 30' must be subtracted. This is because 30' is added to the upper limb corrections during compilation of the tables to keep the value small and positive.

The following example will make these explanations clearer.

Example:
At 0200 GMT on 24 December, the sextant altitude of the Moon's upper limb is measured and the following data is obtained:
Sextant Alt. = 33° 15'.0
I.E. = +1'.8
Ht. of eye = 18 ft.
H.P. = 54'.9
Temperature = -35° C. Pressure = 980 mb.
Using the following extracts, the true altitude is calculated as follows:

Sext. Alt.	33° 15'.0
I.E.	+ 1'.8
Obs. Alt.	33° 16'.8
Dip.	- 4'.1
App. Alt.	33° 12'.7
Corr. (1)	+ 57'.4 (for App. Alt.)
Corr. (2)	+ 2'.0 (for Horizontal Parallax)
Additional Corr.	+ 0'.15 (for weather)
	34° 12'.25
	- 30'.0 (for Upper Limb.)
True Alt.	33° 42'.25

Extract from part of daily page for 24 December showing the value of HP for the Moon.

UT	SUN		MOON				
	GHA	Dec	GHA	v	Dec	d	HP
d h	° '	° '	° '	'	° '	'	'
24 00	180 08.8	S23 25.1	99 57.7	16.2	N 1 57.5	13.1	54.9
01	195 08.4	25.1	114 32.9	16.1	2 10.6	13.2	54.9
02	210 08.1	25.0	129 08.0	16.0	2 23.8	13.2	54.9
03	225 07.8	.. 25.0	143 43.0	16.1	2 37.0	13.1	55.0
04	240 07.5	24.9	158 18.1	16.0	2 50.1	13.2	55.0
05	255 07.2	24.9	172 53.1	16.0	3 03.3	13.1	55.0
06	270 06.9	S23 24.8	187 28.1	15.9	N 3 16.4	13.2	55.0
07	285 06.6	24.8	202 03.0	15.9	3 29.6	13.1	55.1
T 08	300 06.3	24.7	216 37.9	15.9	3 42.7	13.2	55.1
H 09	315 06.0	.. 24.7	231 12.8	15.8	3 55.9	13.1	55.1
U 10	330 05.7	24.6	245 47.6	15.8	4 09.0	13.2	55.1
R 11	345 05.3	24.6	260 22.4	15.8	4 22.2	13.1	55.2
S 12	0 05.0	S23 24.5	274 57.2	15.7	N 4 35.3	13.2	55.2
D 13	15 04.7	24.5	289 31.9	15.6	4 48.5	13.1	55.2
A 14	30 04.4	24.4	304 06.5	15.7	5 01.6	13.1	55.2
Y 15	45 04.1	.. 24.4	318 41.2	15.5	5 14.7	13.1	55.3
16	60 03.8	24.3	333 15.7	15.6	5 27.8	13.1	55.3
17	75 03.5	24.2	347 50.3	15.4	5 40.9	13.1	55.3
18	90 03.2	S23 24.2	2 24.7	15.5	N 5 54.0	13.1	55.3
19	105 02.9	24.1	16 59.2	15.3	6 07.1	13.1	55.4
20	120 02.6	24.1	31 33.5	15.4	6 20.2	13.0	55.4
21	135 02.2	.. 24.0	46 07.9	15.2	6 33.2	13.1	55.4
22	150 01.9	23.9	60 42.1	15.3	6 46.3	13.0	55.5

ALTITUDE CORRECTION TABLES 0°–35°— MOON

App. Alt.	0°–4° Corrⁿ	5°–9° Corrⁿ	10°–14° Corrⁿ	15°–19° Corrⁿ	20°–24° Corrⁿ	25°–29° Corrⁿ	30°–34° Corrⁿ	App. Alt.
′	° ′	° ′	° ′	° ′	° ′	° ′	° ′	′
00	0 34.5	5 58.2	10 62.1	15 62.8	20 62.2	25 60.8	30 58.9	00
10	36.5	58.5	62.2	62.8	62.2	60.8	58.8	10
20	38.3	58.7	62.2	62.8	62.1	60.7	58.8	20
30	40.0	58.9	62.3	62.8	62.1	60.7	58.7	30
40	41.5	59.1	62.3	62.8	62.0	60.6	58.6	40
50	42.9	59.3	62.4	62.7	62.0	60.6	58.5	50
00	1 44.2	6 59.5	11 62.4	16 62.7	21 62.0	26 60.5	31 58.5	00
10	45.4	59.7	62.4	62.7	61.9	60.4	58.4	10
20	46.5	59.9	62.5	62.7	61.9	60.4	58.3	20
30	47.5	60.0	62.5	62.7	61.9	60.3	58.2	30
40	48.4	60.2	62.5	62.7	61.8	60.3	58.2	40
50	49.3	60.3	62.6	62.7	61.8	60.2	58.1	50
00	2 50.1	7 60.5	12 62.6	17 62.7	22 61.7	27 60.1	32 58.0	00
10	50.8	60.6	62.6	62.6	61.7	60.1	57.9	10
20	51.5	60.7	62.6	62.6	61.6	60.0	57.8	20
30	52.2	60.9	62.7	62.6	61.6	59.9	57.8	30
40	52.8	61.0	62.7	62.6	61.6	59.9	57.7	40
50	53.4	61.1	62.7	62.6	61.5	59.8	57.6	50
00	3 53.9	8 61.2	13 62.7	18 62.5	23 61.5	28 59.7	33 57.5	00
10	54.4	61.3	62.7	62.5	61.4	59.7	57.4	10
20	54.9	61.4	62.7	62.5	61.4	59.6	57.4	20
30	55.3	61.5	62.8	62.5	61.3	59.5	57.3	30
40	55.7	61.6	62.8	62.4	61.3	59.5	57.2	40
50	56.1	61.6	62.8	62.4	61.2	59.4	57.1	50
00	4 56.4	9 61.7	14 62.8	19 62.4	24 61.2	29 59.3	34 57.0	00
10	56.8	61.8	62.8	62.4	61.1	59.3	56.9	10
20	57.1	61.9	62.8	62.3	61.1	59.2	56.9	20
30	57.4	61.9	62.8	62.3	61.0	59.1	56.8	30
40	57.7	62.0	62.8	62.3	61.0	59.1	56.7	40
50	58.0	62.1	62.8	62.2	60.9	59.0	56.6	50

HP	L U	L U	L U	L U	L U	L U	L U	HP
′	′ ′	′ ′	′ ′	′ ′	′ ′	′ ′	′ ′	′
54.0	0.3 0.9	0.3 0.9	0.4 1.0	0.5 1.1	0.6 1.2	0.7 1.3	0.9 1.5	54.0
54.3	0.7 1.1	0.7 1.2	0.8 1.3	0.9 1.4	1.1 1.5	1.2 1.7	1.4 1.8	54.3
54.6	1.1 1.4	1.1 1.4	1.1 1.4	1.2 1.5	1.3 1.6	1.4 1.7	1.5 1.8	54.6
54.9	1.4 1.6	1.5 1.6	1.5 1.6	1.6 1.7	1.6 1.8	1.8 1.9	1.9 2.0	54.9
55.2	1.8 1.8	1.8 1.8	1.9 1.8	1.9 1.9	2.0 2.0	2.1 2.1	2.2 2.2	55.2
55.5	2.2 2.0	2.2 2.0	2.3 2.1	2.3 2.1	2.4 2.2	2.4 2.3	2.5 2.4	55.5
55.8	2.6 2.2	2.6 2.2	2.6 2.3	2.7 2.3	2.7 2.4	2.8 2.4	2.9 2.5	55.8
56.1	3.0 2.4	3.0 2.5	3.0 2.5	3.0 2.5	3.1 2.6	3.1 2.6	3.2 2.7	56.1
56.4	3.3 2.7	3.4 2.7	3.4 2.7	3.4 2.7	3.4 2.8	3.5 2.8	3.5 2.9	56.4
56.7	3.7 2.9	3.7 2.9	3.8 2.9	3.8 2.9	3.8 3.0	3.8 3.0	3.9 3.0	56.7
57.0	4.1 3.1	4.1 3.1	4.1 3.1	4.1 3.1	4.2 3.2	4.2 3.2	4.2 3.2	57.0
57.3	4.5 3.3	4.5 3.3	4.5 3.3	4.5 3.3	4.5 3.3	4.5 3.4	4.6 3.4	57.3
57.6	4.9 3.5	4.9 3.5	4.9 3.5	4.9 3.5	4.9 3.5	4.9 3.5	4.9 3.6	57.6
57.9	5.3 3.8	5.3 3.8	5.2 3.8	5.2 3.7	5.2 3.7	5.2 3.7	5.2 3.7	57.9
58.2	5.6 4.0	5.6 4.0	5.6 4.0	5.6 4.0	5.6 3.9	5.6 3.9	5.6 3.9	58.2
58.5	6.0 4.2	6.0 4.2	6.0 4.2	6.0 4.2	6.0 4.1	5.9 4.1	5.9 4.1	58.5
58.8	6.4 4.4	6.4 4.4	6.4 4.4	6.3 4.4	6.3 4.3	6.3 4.3	6.2 4.2	58.8
59.1	6.8 4.6	6.8 4.6	6.7 4.6	6.7 4.6	6.7 4.5	6.6 4.5	6.6 4.4	59.1
59.4	7.2 4.8	7.1 4.8	7.1 4.8	7.1 4.8	7.0 4.7	7.0 4.7	6.9 4.6	59.4
59.7	7.5 5.1	7.5 5.0	7.5 5.0	7.5 5.0	7.4 4.9	7.3 4.8	7.2 4.8	59.7
60.0	7.9 5.3	7.9 5.3	7.9 5.2	7.8 5.2	7.8 5.1	7.7 5.0	7.6 4.9	60.0
60.3	8.3 5.5	8.3 5.5	8.2 5.4	8.2 5.4	8.1 5.3	8.0 5.2	7.9 5.1	60.3
60.6	8.7 5.7	8.7 5.7	8.6 5.7	8.6 5.6	8.5 5.5	8.4 5.4	8.2 5.3	60.6
60.9	9.1 5.9	9.0 5.9	9.0 5.9	8.9 5.8	8.8 5.7	8.7 5.6	8.6 5.4	60.9
61.2	9.5 6.2	9.4 6.1	9.4 6.1	9.3 6.0	9.2 5.9	9.1 5.8	8.9 5.6	61.2
61.5	9.8 6.4	9.8 6.3	9.7 6.3	9.7 6.2	9.5 6.1	9.4 5.9	9.2 5.8	61.5

DIP

Ht. of Eye	Corrⁿ	Ht. of Eye	Ht. of Eye	Corrⁿ	Ht. of Eye
m		ft.	m		ft.
2.4	−2.8	8.0	9.5	−5.5	31.5
2.6	−2.9	8.6	9.9	−5.6	32.7
2.8	−3.0	9.2	10.3	−5.7	33.9
3.0	−3.1	9.8	10.6	−5.8	35.1
3.2	−3.2	10.5	11.0	−5.9	36.3
3.4	−3.3	11.2	11.4	−6.0	37.6
3.6	−3.4	11.9	11.8	−6.1	38.9
3.8	−3.5	12.6	12.2	−6.2	40.1
4.0	−3.6	13.3	12.6	−6.3	41.5
4.3	−3.7	14.1	13.0	−6.4	42.8
4.5	−3.8	14.9	13.4	−6.5	44.2
4.7	−3.9	15.7	13.8	−6.6	45.5
5.0	−4.0	16.5	14.2	−6.7	46.9
5.2	−4.1	17.4	14.7	−6.8	48.4
5.5	−4.2	18.3	15.1	−6.9	49.8
5.8	−4.3	19.1	15.5	−7.0	51.3
6.1	−4.4	20.1	16.0	−7.1	52.8
6.3	−4.5	21.0	16.5	−7.2	54.3
6.6	−4.6	22.0	16.9	−7.3	55.8
6.9	−4.7	22.9	17.4	−7.4	57.4
7.2	−4.8	23.9	17.9	−7.5	58.9
7.5	−4.9	24.9	18.4	−7.6	60.5
7.9	−5.0	26.0	18.8	−7.7	62.1
8.2	−5.1	27.1	19.3	−7.8	63.8
8.5	−5.2	28.1	19.8	−7.9	65.4
8.8	−5.3	29.2	20.4	−8.0	67.1
9.2	−5.4	30.4	20.9	−8.1	68.8
9.5		31.5	21.4		70.5

MOON CORRECTION TABLE

The correction is in two parts; the first correction is taken from the upper part of the table with argument apparent altitude, and the second from the lower part, with argument HP, in the same column as that from which the first correction was taken. Separate corrections are given in the lower part for lower (L) and upper(U) limbs. All corrections are to be **added** to apparent altitude, *but 30′ is to be subtracted from the altitude of the upper limb.*

For corrections for pressure and temperature see page A4.

For bubble sextant observations ignore dip, take the mean of upper and lower limb corrections and subtract 15′ from the altitude.

App. Alt. = Apparent altitude = Sextant altitude corrected for index error and dip.

A4 ALTITUDE CORRECTION TABLES—ADDITIONAL CORRECTIONS
ADDITIONAL REFRACTION CORRECTIONS FOR NON-STANDARD CONDITIONS

App. Alt.	A	B	C	D	E	F	G	H	J	K	L	M	N	P	App. Alt.
° ′	′	′	′	′	′	′	′	′	′	′	′	′	′	′	° ′
00 00	−7.3	−5.9	−4.6	−3.4	−2.2	−1.1	0.0	+1.0	+2.0	+3.0	+4.0	+4.9	+5.9	+6.9	00 00
00 30	5.5	4.5	3.5	2.6	1.7	0.8	0.0	0.8	1.6	2.3	3.1	3.8	4.5	5.3	00 30
01 00	4.4	3.5	2.8	2.0	1.3	0.7	0.0	0.6	1.2	1.8	2.4	3.0	3.6	4.2	01 00
01 30	3.5	2.9	2.2	1.7	1.1	0.5	0.0	0.5	1.0	1.5	2.0	2.5	2.9	3.4	01 30
02 00	2.9	2.4	1.9	1.4	0.9	0.4	0.0	0.4	0.8	1.3	1.7	2.0	2.4	2.8	02 00
02 30	−2.5	−2.0	−1.6	−1.2	−0.8	−0.4	0.0	+0.4	+0.7	+1.1	+1.4	+1.7	+2.1	+2.4	02 30
03 00	2.1	1.7	1.4	1.0	0.7	0.3	0.0	0.3	0.6	0.9	1.2	1.5	1.8	2.1	03 00
03 30	1.9	1.5	1.2	0.9	0.6	0.3	0.0	0.3	0.5	0.8	1.1	1.3	1.6	1.8	03 30
04 00	1.6	1.3	1.1	0.8	0.5	0.3	0.0	0.2	0.5	0.7	0.9	1.2	1.4	1.6	04 00
04 30	1.5	1.2	0.9	0.7	0.5	0.2	0.0	0.2	0.4	0.6	0.8	1.0	1.3	1.5	04 30
05 00	−1.3	−1.1	−0.9	−0.6	−0.4	−0.2	0.0	+0.2	+0.4	+0.6	+0.8	+0.9	+1.1	+1.3	05 00
06	1.1	0.9	0.7	0.5	0.3	0.2	0.0	0.2	0.3	0.5	0.6	0.8	0.9	1.1	06
07	1.0	0.8	0.6	0.5	0.3	0.1	0.0	0.1	0.3	0.4	0.5	0.7	0.8	0.9	07
08	0.8	0.7	0.5	0.4	0.3	0.1	0.0	0.1	0.2	0.4	0.5	0.6	0.7	0.8	08
09	0.7	0.6	0.5	0.4	0.2	0.1	0.0	0.1	0.2	0.3	0.4	0.5	0.6	0.7	09
10 00	−0.7	−0.5	−0.4	−0.3	−0.2	−0.1	0.0	+0.1	+0.2	+0.3	+0.4	+0.5	+0.6	+0.7	10 00
12	0.6	0.5	0.4	0.3	0.2	0.1	0.0	0.1	0.2	0.2	0.3	0.4	0.5	0.5	12
14	0.5	0.4	0.3	0.2	0.1	0.1	0.0	0.1	0.1	0.2	0.3	0.3	0.4	0.5	14
16	0.4	0.3	0.3	0.2	0.1	0.1	0.0	0.1	0.1	0.2	0.2	0.3	0.3	0.4	16
18	0.4	0.3	0.2	0.2	0.1	−0.1	0.0	+0.1	0.1	0.2	0.2	0.3	0.3	0.4	18
20 00	−0.3	−0.3	−0.2	−0.2	−0.1	0.0	0.0	0.0	+0.1	+0.1	+0.2	+0.2	+0.3	+0.3	20 00
25	0.3	0.2	0.2	0.1	0.1	0.0	0.0	0.0	0.1	0.1	0.1	0.2	0.2	0.2	25
30	0.2	0.2	0.1	0.1	0.1	0.0	0.0	0.0	+0.1	0.1	0.1	0.1	0.2	0.2	30
35	0.2	0.1	0.1	0.1	−0.1	0.0	0.0	0.0	0.0	0.1	0.1	0.1	0.1	0.2	35
40	0.1	0.1	0.1	−0.1	0.0	0.0	0.0	0.0	+0.1	0.1	0.1	0.1	0.1	0.1	40
50 00	−0.1	−0.1	−0.1	0.0	0.0	0.0	0.0	0.0	0.0	0.0	+0.1	+0.1	+0.1	+0.1	50 00

Self Test

Using the Nautical Almanac extracts on the following pages, answer these questions:

Question 1.
On 12 June, a reading of the Sun's lower limb was taken.
Sextant altitude: 50° 45'.2. Index error: +1'.8. Height of eye: 24 ft.
Temperature: 25°C. Pressure: 1000mb.
What was the true altitude?

Question 2.
Sextant altitude of Sirius: 18° 08'.5. Index error: +2'.1; Ht. of eye: 12m.
Temperature: -10°C. Pressure: 980mb.
What is the true altitude?

Question 3.
Sextant altitude of Venus: 17° 43'.3. Index error: -1'.4. Ht. of eye: 15 ft.
Temperature: 8°C. Pressure: 1025mb.
What is the true altitude?

Question 4.
Greenwich date: 1835, 6 June. Sextant altitude of Moon's lower limb: 31° 32'.8.
Index error: +2'.4. Ht. of eye: 15m. HP: 54'.8. Temperature: -18°C. Pressure: 990mb.
What is the true altitude?

A2 ALTITUDE CORRECTION TABLES 10°–90° — SUN, STARS, PLANETS

OCT.–MAR. SUN APR.–SEPT.				STARS AND PLANETS		DIP			
App. Alt.	Lower Limb	Upper Limb	App. Alt. Lower Upper Limb Limb	App. Alt.	Corr^n	App. Alt.	Additional Corr^n	Ht. of Eye	Corr^n

SUN — Oct.–Mar.

App. Alt.	Lower Limb	Upper Limb
9 33	+10.8	−21.5
9 45	+10.9	−21.4
9 56	+11.0	−21.3
10 08	+11.1	−21.2
10 20	+11.2	−21.1
10 33	+11.3	−21.0
10 46	+11.4	−20.9
11 00	+11.5	−20.8
11 15	+11.6	−20.7
11 30	+11.7	−20.6
11 45	+11.8	−20.5
12 01	+11.9	−20.4
12 18	+12.0	−20.3
12 36	+12.1	−20.2
12 54	+12.2	−20.1
13 14	+12.3	−20.0
13 34	+12.4	−19.9
13 55	+12.5	−19.8
14 17	+12.6	−19.7
14 41	+12.7	−19.6
15 05	+12.8	−19.5
15 31	+12.9	−19.4
15 59	+13.0	−19.3
16 27	+13.1	−19.2
16 58	+13.2	−19.1
17 30	+13.3	−19.0
18 05	+13.4	−18.9
18 41	+13.5	−18.8
19 20	+13.6	−18.7
20 02	+13.7	−18.6
20 46	+13.8	−18.5
21 34	+13.9	−18.4
22 25	+14.0	−18.3
23 20	+14.1	−18.2
24 20	+14.2	−18.1
25 24	+14.3	−18.0
26 34	+14.4	−17.9
27 50	+14.5	−17.8
29 13	+14.6	−17.7
30 44	+14.7	−17.6
32 24	+14.8	−17.5
34 15	+14.9	−17.4
36 17	+15.0	−17.3
38 34	+15.1	−17.2
41 06	+15.2	−17.1
43 56	+15.3	−17.0
47 07	+15.4	−16.9
50 43	+15.5	−16.8
54 46	+15.6	−16.7
59 21	+15.7	−16.6
64 28	+15.8	−16.5
70 10	+15.9	−16.4
76 24	+16.0	−16.3
83 05	+16.1	−16.2
90 00		

SUN — Apr.–Sept.

App. Alt.	Lower Limb	Upper Limb
9 39	+10.6	−21.2
9 50	+10.7	−21.1
10 02	+10.8	−21.0
10 14	+10.9	−20.9
10 27	+11.0	−20.8
10 40	+11.1	−20.7
10 53	+11.2	−20.6
11 07	+11.3	−20.5
11 22	+11.4	−20.4
11 37	+11.5	−20.3
11 53	+11.6	−20.2
12 10	+11.7	−20.1
12 27	+11.8	−20.0
12 45	+11.9	−19.9
13 04	+12.0	−19.8
13 24	+12.1	−19.7
13 44	+12.2	−19.6
14 06	+12.3	−19.5
14 29	+12.4	−19.4
14 53	+12.5	−19.3
15 18	+12.6	−19.2
15 45	+12.7	−19.1
16 13	+12.8	−19.0
16 43	+12.9	−18.9
17 14	+13.0	−18.8
17 47	+13.1	−18.7
18 23	+13.2	−18.6
19 00	+13.3	−18.5
19 41	+13.4	−18.4
20 24	+13.5	−18.3
21 10	+13.6	−18.2
21 59	+13.7	−18.1
22 52	+13.8	−18.0
23 49	+13.9	−17.9
24 51	+14.0	−17.8
25 58	+14.1	−17.7
27 11	+14.2	−17.6
28 31	+14.3	−17.5
29 58	+14.4	−17.4
31 33	+14.5	−17.3
33 18	+14.6	−17.2
35 15	+14.7	−17.1
37 24	+14.8	−17.0
39 48	+14.9	−16.9
42 28	+15.0	−16.8
45 29	+15.1	−16.7
48 52	+15.2	−16.6
52 41	+15.3	−16.5
56 59	+15.4	−16.4
61 50	+15.5	−16.3
67 15	+15.6	−16.2
73 14	+15.7	−16.1
79 42	+15.8	−16.0
86 31	+15.9	−15.9
90 00		

STARS AND PLANETS

App. Alt.	Corr^n
9 55	−5.3
10 07	−5.2
10 20	−5.1
10 32	−5.0
10 46	−4.9
10 59	−4.8
11 14	−4.7
11 29	−4.6
11 44	−4.5
12 00	−4.4
12 17	−4.3
12 35	−4.2
12 53	−4.1
13 12	−4.0
13 32	−3.9
13 53	−3.8
14 16	−3.7
14 39	−3.6
15 03	−3.5
15 29	−3.4
15 56	−3.3
16 25	−3.2
16 55	−3.1
17 27	−3.0
18 01	−2.9
18 37	−2.8
19 16	−2.7
19 56	−2.6
20 40	−2.5
21 27	−2.4
22 17	−2.3
23 11	−2.2
24 09	−2.1
25 12	−2.0
26 20	−1.9
27 34	−1.8
28 54	−1.7
30 22	−1.6
31 58	−1.5
33 43	−1.4
35 38	−1.3
37 45	−1.2
40 06	−1.1
42 42	−1.0
45 34	−0.9
48 45	−0.8
52 16	−0.7
56 09	−0.6
60 26	−0.5
65 06	−0.4
70 09	−0.3
75 32	−0.2
81 12	−0.1
87 03	0.0
90 00	

Additional Corr^n — 2009

VENUS

Jan. 1–Jan. 28
May 23–July 10

°	′
0	+0.2
41	+0.1
76	

Jan. 29–Feb. 21
May 1–May 22

°	′
0	+0.3
34	+0.2
60	+0.1
80	

Feb. 22–Mar. 9
Apr. 15–Apr. 30

°	′
0	+0.4
29	+0.3
51	+0.2
68	+0.1
83	

Mar. 10–Apr. 14

°	′
0	+0.5
26	+0.4
46	+0.3
60	+0.2
73	+0.1
84	

July 11–Dec. 31

°	′
0	+0.1
60	

MARS

Jan. 1–Nov. 27

°	′
0	+0.1
60	

Nov. 28–Dec. 31

°	′
0	+0.2
41	+0.1
76	

DIP

Ht. of Eye (m)	Corr^n
2.4	−2.8
2.6	−2.9
2.8	−3.0
3.0	−3.1
3.2	−3.2
3.4	−3.3
3.6	−3.4
3.8	−3.5
4.0	−3.6
4.3	−3.7
4.5	−3.8
4.7	−3.9
5.0	−4.0
5.2	−4.1
5.5	−4.2
5.8	−4.3
6.1	−4.4
6.3	−4.5
6.6	−4.6
6.9	−4.7
7.2	−4.8
7.5	−4.9
7.9	−5.0
8.2	−5.1
8.5	−5.2
8.8	−5.3
9.2	−5.4
9.5	−5.5
9.9	−5.6
10.3	−5.7
10.6	−5.8
11.0	−5.9
11.4	−6.0
11.8	−6.1
12.2	−6.2
12.6	−6.3
13.0	−6.4
13.4	−6.5
13.8	−6.6
14.2	−6.7
14.7	−6.8
15.1	−6.9
15.5	−7.0
16.0	−7.1
16.5	−7.2
16.9	−7.3
17.4	−7.4
17.9	−7.5
18.4	−7.6
18.8	−7.7
19.3	−7.8
19.8	−7.9
20.4	−8.0
20.9	−8.1
21.4	

Ht. of Eye (ft)	Corr^n
8.0	
8.6	
9.2	
9.8	
10.5	
11.2	
11.9	
12.6	
13.3	
14.1	
14.9	
15.7	
16.5	
17.4	
18.3	
19.1	
20.1	
21.0	
22.0	
22.9	
23.9	
24.9	
26.0	
27.1	
28.1	
29.2	
30.4	
31.5	
32.7	
33.9	
35.1	
36.3	
37.6	
38.9	
40.1	
41.5	
42.8	
44.2	
45.5	
46.9	
48.4	
49.8	
51.3	
52.8	
54.3	
55.8	
57.4	
58.9	
60.5	
62.1	
63.8	
65.4	
67.1	
68.8	
70.5	

Ht. of Eye	Corr^n
m	
1.0	−1.8
1.5	−2.2
2.0	−2.5
2.5	−2.8
3.0	−3.0
See table ←	
m	
20	−7.9
22	−8.3
24	−8.6
26	−9.0
28	−9.3
30	−9.6
32	−10.0
34	−10.3
36	−10.6
38	−10.8
40	−11.1
42	−11.4
44	−11.7
46	−11.9
48	−12.2
ft.	
2	−1.4
4	−1.9
6	−2.4
8	−2.7
10	−3.1
See table ←	
ft.	
70	−8.1
75	−8.4
80	−8.7
85	−8.9
90	−9.2
95	−9.5
100	−9.7
105	−9.9
110	−10.2
115	−10.4
120	−10.6
125	−10.8
130	−11.1
135	−11.3
140	−11.5
145	−11.7
150	−11.9
155	−12.1

ALTITUDE CORRECTION TABLES 0°–35° — MOON

App. Alt.	0°–4° Corrⁿ	5°–9° Corrⁿ	10°–14° Corrⁿ	15°–19° Corrⁿ	20°–24° Corrⁿ	25°–29° Corrⁿ	30°–34° Corrⁿ	App. Alt.
00	0° 34.5	5° 58.2	10° 62.1	15° 62.8	20° 62.2	25° 60.8	30° 58.9	00
10	36.5	58.5	62.2	62.8	62.2	60.8	58.8	10
20	38.3	58.7	62.2	62.8	62.1	60.7	58.8	20
30	40.0	58.9	62.3	62.8	62.1	60.7	58.7	30
40	41.5	59.1	62.3	62.8	62.0	60.6	58.6	40
50	42.9	59.3	62.4	62.7	62.0	60.6	58.5	50
00	1° 44.2	6° 59.5	11° 62.4	16° 62.7	21° 62.0	26° 60.5	31° 58.5	00
10	45.4	59.7	62.4	62.7	61.9	60.4	58.4	10
20	46.5	59.9	62.5	62.7	61.9	60.4	58.3	20
30	47.5	60.0	62.5	62.7	61.9	60.3	58.2	30
40	48.4	60.2	62.5	62.7	61.8	60.3	58.2	40
50	49.3	60.3	62.6	62.7	61.8	60.2	58.1	50
00	2° 50.1	7° 60.5	12° 62.6	17° 62.7	22° 61.7	27° 60.1	32° 58.0	00
10	50.8	60.6	62.6	62.6	61.7	60.1	57.9	10
20	51.5	60.7	62.6	62.6	61.6	60.0	57.8	20
30	52.2	60.9	62.7	62.6	61.6	59.9	57.8	30
40	52.8	61.0	62.7	62.6	61.6	59.9	57.7	40
50	53.4	61.1	62.7	62.6	61.5	59.8	57.6	50
00	3° 53.9	8° 61.2	13° 62.7	18° 62.5	23° 61.5	28° 59.7	33° 57.5	00
10	54.4	61.3	62.7	62.5	61.4	59.7	57.4	10
20	54.9	61.4	62.7	62.5	61.4	59.6	57.4	20
30	55.3	61.5	62.8	62.5	61.3	59.5	57.3	30
40	55.7	61.6	62.8	62.4	61.3	59.5	57.2	40
50	56.1	61.6	62.8	62.4	61.2	59.4	57.1	50
00	4° 56.4	9° 61.7	14° 62.8	19° 62.4	24° 61.2	29° 59.3	34° 57.0	00
10	56.8	61.8	62.8	62.4	61.1	59.3	56.9	10
20	57.1	61.9	62.8	62.3	61.1	59.2	56.9	20
30	57.4	61.9	62.8	62.3	61.0	59.1	56.8	30
40	57.7	62.0	62.8	62.3	61.0	59.1	56.7	40
50	58.0	62.1	62.8	62.2	60.9	59.0	56.6	50

HP	L U	L U	L U	L U	L U	L U	L U	HP
54.0	0.3 0.9	0.3 0.9	0.4 1.0	0.5 1.1	0.6 1.2	0.7 1.3	0.9 1.5	54.0
54.3	0.7 1.1	0.7 1.2	0.8 1.2	0.8 1.3	0.9 1.4	1.1 1.5	1.2 1.7	54.3
54.6	1.1 1.4	1.1 1.4	1.1 1.4	1.2 1.5	1.3 1.6	1.4 1.7	1.5 1.8	54.6
54.9	1.4 1.6	1.5 1.6	1.5 1.6	1.6 1.7	1.6 1.8	1.8 1.9	1.9 2.0	54.9
55.2	1.8 1.8	1.8 1.8	1.9 1.8	1.9 1.9	2.0 2.0	2.1 2.1	2.2 2.2	55.2
55.5	2.2 2.0	2.2 2.0	2.3 2.1	2.3 2.1	2.4 2.2	2.4 2.3	2.5 2.4	55.5
55.8	2.6 2.2	2.6 2.2	2.6 2.3	2.7 2.3	2.7 2.4	2.8 2.4	2.9 2.5	55.8
56.1	3.0 2.4	3.0 2.5	3.0 2.5	3.0 2.5	3.1 2.6	3.1 2.6	3.2 2.7	56.1
56.4	3.3 2.7	3.4 2.7	3.4 2.7	3.4 2.7	3.4 2.8	3.5 2.8	3.5 2.9	56.4
56.7	3.7 2.9	3.7 2.9	3.8 2.9	3.8 2.9	3.8 3.0	3.8 3.0	3.9 3.0	56.7
57.0	4.1 3.1	4.1 3.1	4.1 3.1	4.1 3.1	4.2 3.2	4.2 3.2	4.2 3.2	57.0
57.3	4.5 3.3	4.5 3.3	4.5 3.3	4.5 3.3	4.5 3.3	4.5 3.4	4.6 3.4	57.3
57.6	4.9 3.5	4.9 3.5	4.9 3.5	4.9 3.5	4.9 3.5	4.9 3.5	4.9 3.6	57.6
57.9	5.3 3.8	5.3 3.8	5.2 3.8	5.2 3.7	5.2 3.7	5.2 3.7	5.2 3.7	57.9
58.2	5.6 4.0	5.6 4.0	5.6 4.0	5.6 4.0	5.6 3.9	5.6 3.9	5.6 3.9	58.2
58.5	6.0 4.2	6.0 4.2	6.0 4.2	6.0 4.2	6.0 4.1	5.9 4.1	5.9 4.1	58.5
58.8	6.4 4.4	6.4 4.4	6.4 4.4	6.3 4.4	6.3 4.3	6.3 4.3	6.2 4.2	58.8
59.1	6.8 4.6	6.8 4.6	6.7 4.6	6.7 4.6	6.7 4.5	6.6 4.5	6.6 4.4	59.1
59.4	7.2 4.8	7.1 4.8	7.1 4.8	7.1 4.8	7.0 4.7	7.0 4.7	6.9 4.6	59.4
59.7	7.5 5.1	7.5 5.0	7.5 5.0	7.5 5.0	7.4 4.9	7.3 4.8	7.2 4.8	59.7
60.0	7.9 5.3	7.9 5.3	7.9 5.2	7.8 5.2	7.8 5.1	7.7 5.0	7.6 4.9	60.0
60.3	8.3 5.5	8.3 5.5	8.2 5.4	8.2 5.4	8.1 5.3	8.0 5.2	7.9 5.1	60.3
60.6	8.7 5.7	8.7 5.7	8.6 5.7	8.6 5.6	8.5 5.5	8.4 5.4	8.2 5.3	60.6
60.9	9.1 5.9	9.0 5.9	9.0 5.9	8.9 5.8	8.8 5.7	8.7 5.6	8.6 5.4	60.9
61.2	9.5 6.2	9.4 6.1	9.4 6.1	9.3 6.0	9.2 5.9	9.1 5.8	8.9 5.6	61.2
61.5	9.8 6.4	9.8 6.3	9.7 6.3	9.7 6.2	9.5 6.1	9.4 5.9	9.2 5.8	61.5

DIP

Ht. of Eye	Corrⁿ	Ht. of Eye	Ht. of Eye	Corrⁿ	Ht. of Eye
m		ft.	m		ft.
2.4	−2.8	8.0	9.5	−5.5	31.5
2.6	−2.9	8.6	9.9	−5.6	32.7
2.8	−3.0	9.2	10.3	−5.7	33.9
3.0	−3.1	9.8	10.6	−5.8	35.1
3.2	−3.2	10.5	11.0	−5.9	36.3
3.4	−3.3	11.2	11.4	−6.0	37.6
3.6	−3.4	11.9	11.8	−6.1	38.9
3.8	−3.5	12.6	12.2	−6.2	40.1
4.0	−3.6	13.3	12.6	−6.3	41.5
4.3	−3.7	14.1	13.0	−6.4	42.8
4.5	−3.8	14.9	13.4	−6.5	44.2
4.7	−3.9	15.7	13.8	−6.6	45.5
5.0	−4.0	16.5	14.2	−6.7	46.9
5.2	−4.1	17.4	14.7	−6.8	48.4
5.5	−4.2	18.3	15.1	−6.9	49.8
5.8	−4.3	19.1	15.5	−7.0	51.3
6.1	−4.4	20.1	16.0	−7.1	52.8
6.3	−4.5	21.0	16.5	−7.2	54.3
6.6	−4.6	22.0	16.9	−7.3	55.8
6.9	−4.7	22.9	17.4	−7.4	57.4
7.2	−4.8	23.9	17.9	−7.5	58.9
7.5	−4.9	24.9	18.4	−7.6	60.5
7.9	−5.0	26.0	18.8	−7.7	62.1
8.2	−5.1	27.1	19.3	−7.8	63.8
8.5	−5.2	28.1	19.8	−7.9	65.4
8.8	−5.3	29.2	20.4	−8.0	67.1
9.2	−5.4	30.4	20.9	−8.1	68.8
9.5		31.5	21.4		70.5

MOON CORRECTION TABLE

The correction is in two parts; the first correction is taken from the upper part of the table with argument apparent altitude, and the second from the lower part, with argument HP, in the same column as that from which the first correction was taken. Separate corrections are given in the lower part for lower (L) and upper (U) limbs. All corrections are to be **added** to apparent altitude, *but 30′ is to be subtracted from the altitude of the upper limb.*

For corrections for pressure and temperature see page A4.

For bubble sextant observations ignore dip, take the mean of upper and lower limb corrections and subtract 15′ from the altitude.

App. Alt. = Apparent altitude = Sextant altitude corrected for index error and dip.

A4 ALTITUDE CORRECTION TABLES—ADDITIONAL CORRECTIONS
ADDITIONAL REFRACTION CORRECTIONS FOR NON-STANDARD CONDITIONS

App. Alt.	A	B	C	D	E	F	G	H	J	K	L	M	N	P	App. Alt.
° ′	′	′	′	′	′	′	′	′	′	′	′	′	′	′	° ′
00 00	−7·3	−5·9	−4·6	−3·4	−2·2	−1·1	0·0	+1·0	+2·0	+3·0	+4·0	+4·9	+5·9	+6·9	00 00
00 30	5·5	4·5	3·5	2·6	1·7	0·8	0·0	0·8	1·6	2·3	3·1	3·8	4·5	5·3	00 30
01 00	4·4	3·5	2·8	2·0	1·3	0·7	0·0	0·6	1·2	1·8	2·4	3·0	3·6	4·2	01 00
01 30	3·5	2·9	2·2	1·7	1·1	0·5	0·0	0·5	1·0	1·5	2·0	2·5	2·9	3·4	01 30
02 00	2·9	2·4	1·9	1·4	0·9	0·4	0·0	0·4	0·8	1·3	1·7	2·0	2·4	2·8	02 00
02 30	−2·5	−2·0	−1·6	−1·2	−0·8	−0·4	0·0	+0·4	+0·7	+1·1	+1·4	+1·7	+2·1	+2·4	02 30
03 00	2·1	1·7	1·4	1·0	0·7	0·3	0·0	0·3	0·6	0·9	1·2	1·5	1·8	2·1	03 00
03 30	1·9	1·5	1·2	0·9	0·6	0·3	0·0	0·3	0·5	0·8	1·1	1·3	1·6	1·8	03 30
04 00	1·6	1·3	1·1	0·8	0·5	0·3	0·0	0·2	0·5	0·7	0·9	1·2	1·4	1·6	04 00
04 30	1·5	1·2	0·9	0·7	0·5	0·2	0·0	0·2	0·4	0·6	0·8	1·0	1·3	1·5	04 30
05 00	−1·3	−1·1	−0·9	−0·6	−0·4	−0·2	0·0	+0·2	+0·4	+0·6	+0·8	+0·9	+1·1	+1·3	05 00
06	1·1	0·9	0·7	0·5	0·3	0·2	0·0	0·2	0·3	0·5	0·6	0·8	0·9	1·1	06
07	1·0	0·8	0·6	0·5	0·3	0·1	0·0	0·1	0·3	0·4	0·5	0·7	0·8	0·9	07
08	0·8	0·7	0·5	0·4	0·3	0·1	0·0	0·1	0·2	0·4	0·5	0·6	0·7	0·8	08
09	0·7	0·6	0·5	0·4	0·2	0·1	0·0	0·1	0·2	0·3	0·4	0·5	0·6	0·7	09
10 00	−0·7	−0·5	−0·4	−0·3	−0·2	−0·1	0·0	+0·1	+0·2	+0·3	+0·4	+0·5	+0·6	+0·7	10 00
12	0·6	0·5	0·4	0·3	0·2	0·1	0·0	0·1	0·2	0·2	0·3	0·4	0·5	0·5	12
14	0·5	0·4	0·3	0·2	0·1	0·1	0·0	0·1	0·1	0·2	0·3	0·3	0·4	0·5	14
16	0·4	0·3	0·3	0·2	0·1	0·1	0·0	0·1	0·1	0·2	0·2	0·3	0·3	0·4	16
18	0·4	0·3	0·2	0·2	0·1	−0·1	0·0	+0·1	0·1	0·2	0·2	0·3	0·3	0·4	18
20 00	−0·3	−0·3	−0·2	−0·2	−0·1	0·0	0·0	0·0	+0·1	+0·1	+0·2	+0·2	+0·3	+0·3	20 00
25	0·3	0·2	0·2	0·1	0·1	0·0	0·0	0·0	0·1	0·1	0·1	0·2	0·2	0·2	25
30	0·2	0·2	0·1	0·1	0·1	0·0	0·0	0·0	+0·1	0·1	0·1	0·1	0·2	0·2	30
35	0·2	0·1	0·1	0·1	−0·1	0·0	0·0	0·0	0·0	0·1	0·1	0·1	0·1	0·2	35
40	0·1	0·1	0·1	−0·1	0·0	0·0	0·0	0·0	0·0	+0·1	0·1	0·1	0·1	0·1	40
50 00	−0·1	−0·1	−0·1	0·0	0·0	0·0	0·0	0·0	0·0	0·0	+0·1	+0·1	+0·1	+0·1	50 00

©2009 Jack Case

Solutions

Q1.
Sext. Alt.	50° 45'.2
I.E.	+1'.8
Observed Alt.	50° 47'.0
Dip	-4'.8
Apparent Alt.	50° 42'.2
Sun's Corr.	+15'.2
Add'nl. Corr.	0'.0
True Alt.	50° 57'.4

Q2.
Sext. Alt.	18° 08'.5
I.E.	+2'.1
Obs. Alt.	18° 10'.6
Dip.	-6'.1
App. Alt.	18° 04'.5
Star's Corr.	-2'.9
Add'nl. Corr.	-0'.2
True Alt.	18° 01'.4

Q.3.
Sext. Alt.	17° 43'.3
I.E.	-1'.4
Obs. Alt.	17° 41'.9
Dip	-3'.8
App. Alt.	17° 38'.1
Plnt's Corr.	-3'.0
Add'nl. Corr.	-0'.1
True Alt.	17° 35'.0

Q.4.
Sext. Alt.	31° 32'.8
I.E.	+ 2'.4
Obs. Alt.	31° 35'.2
Dip.	- 6'.8
App. Alt.	31° 28'.4
Corr. (1)	+ 58'.2 (for App. Alt.)
Corr. (2)	+ 1'.8 (for H.P. 54'.8)
Add'nl. Corr.	-0'.2
True Alt.	32° 28'.2

Chapter 6
Calculating Latitude From the Altitude of the Sun at Meridian Passage.

We learned, in previous chapters, that there is a relationship between our latitude and the Sun's altitude and declination. From this relationship, it follows that, if we were able to find the values of the Sun's altitude and declination, we would have the means to calculate our latitude. We can use the nautical almanac to find the Sun's declination for any minute of the day and we can measure its altitude by using a sextant so all we need to know is how we can use this information to calculate the latitude.

Meridian Passage (Mer. Pas.) occurs when a celestial body crosses the observer's meridian of longitude and at that instant, it will reach its greatest altitude above the observer's horizon. The method shown here involves calculating the vessel's latitude from the altitude of the Sun at meridian passage (meridian altitude) and then using the latitude so found as a position line.

The Sun's meridian Passage occurs at apparent noon when it is at its zenith. However, it should be remembered that the time of Mer. Pas. as tabulated in the nautical almanac, is in terms of Greenwich Mean Time. In other words, the event tabulated is 'mean noon' which does not necessarily coincide with 'apparent noon' which relates to the True Sun. (If you are confused by this, please refer to chapter 2).

Time of the Sun's Meridian Passage. It would be impractical for an observer to measure the meridian altitude by watching the Sun and noting when it is at its zenith. The usual method is to calculate when the Sun will be on your meridian and to measure the altitude at that moment. This calculation is made by estimating the D.R. position at the time of meridian passage. Because the altitude changes slowly when the Sun approaches its zenith, a small error in the calculated time of meridian passage will make only a negligible difference to the altitude. For this reason, it is acceptable to calculate the time of meridian passage to the nearest minute. For this reason, the Equation of Time is not usually taken into account unless extreme accuracy is required.

If the vessel is under way, its position at the time of meridian passage will not be known beforehand. However, it is necessary to calculate the time of meridian passage before it occurs. This is achieved by calculating the time of mer. pas. at the D.R. position a few hours before noon and then estimating what the vessel's position will be at that time. The time of mer. pas. can then be recalculated for the new position.

There are several methods by which the new position could be calculated such as laying off the course and speed on the chart, using the traverse tables, or relying on the GPS system. However, since we are not dealing with basic chart-work or electronic navigation systems in this book, the method demonstrated here involves the short-distance sailing technique which was explained in chapter 3. Furthermore, for the reasons also explained in chapter 3, the traverse tables are not used.

The following diagrams will help to explain how the latitude can be calculated from the altitude and declination of the Sun.

In diagram 30,
NOS   represents the horizon and O represents the position of the observer,
X   represents the position of the Sun,
Z   represents an imaginary position exactly above the observer so that
          OZ is perpendicular to NOS,
OX   represents a line from the observer to the Sun,
∠XOS   represents the altitude of the Sun,
∠XOZ   equals 90° - Altitude.

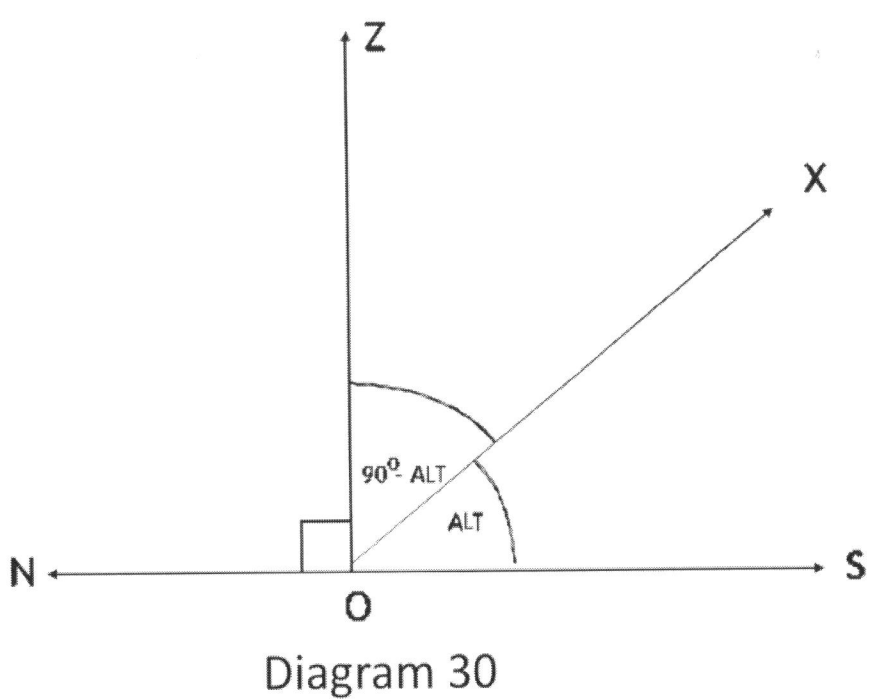

Diagram 30

Now consider Diagram 31.
Imagine the Earth to be at the centre of the imaginary celestial sphere with the positions Z and X projected onto the surface of that sphere.

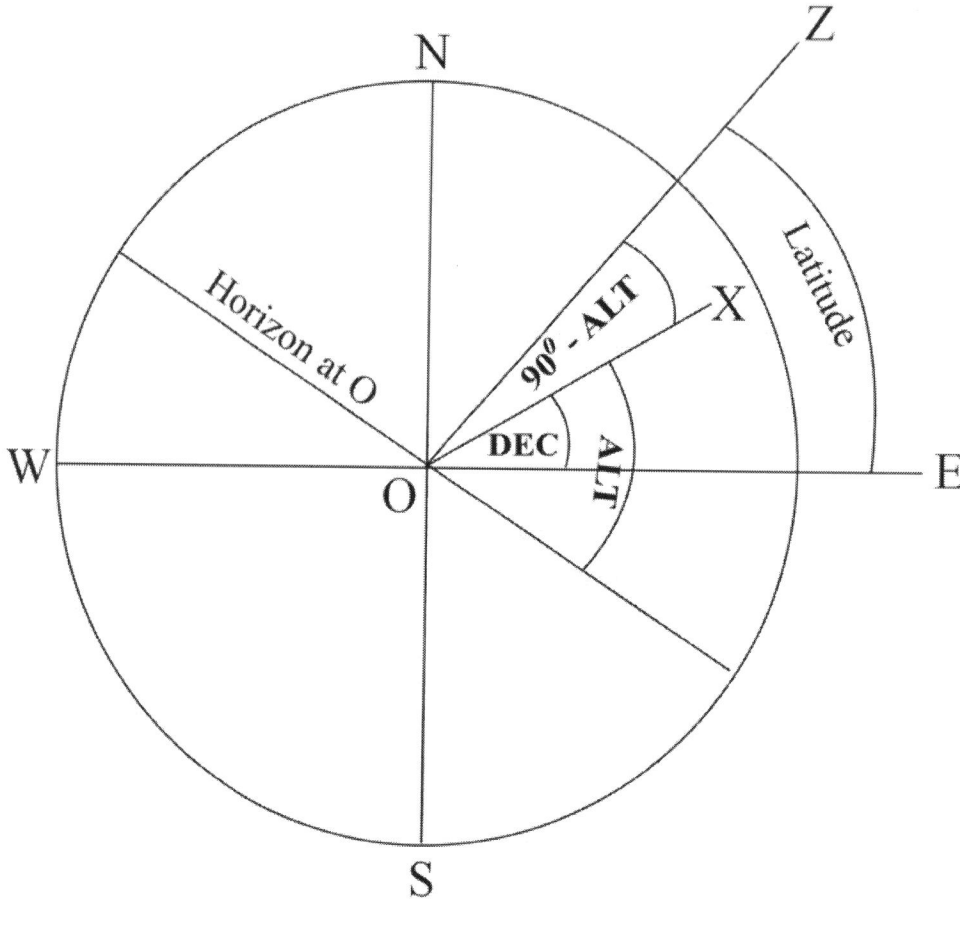

Diagram 31

WOE represents the projection of the Equator onto the surface of the sphere,

NOS represents the projection of a line joining the North and South poles (i.e. a meridian of longitude) onto the surface of the sphere,

∠EOZ equals the latitude of the observer,

∠EOX equals the declination of the Sun,

∠XOZ equals 90° - Altitude (as established in diagram 30)

∴ **Lat. = Declination + (90°-Altitude).**

In the case above, latitude and declination are in the same hemisphere but the latitude is greater than the declination.

There are two other cases to consider:
In Diagram 32, latitude and declination are in the same hemisphere but declination is greater than latitude.

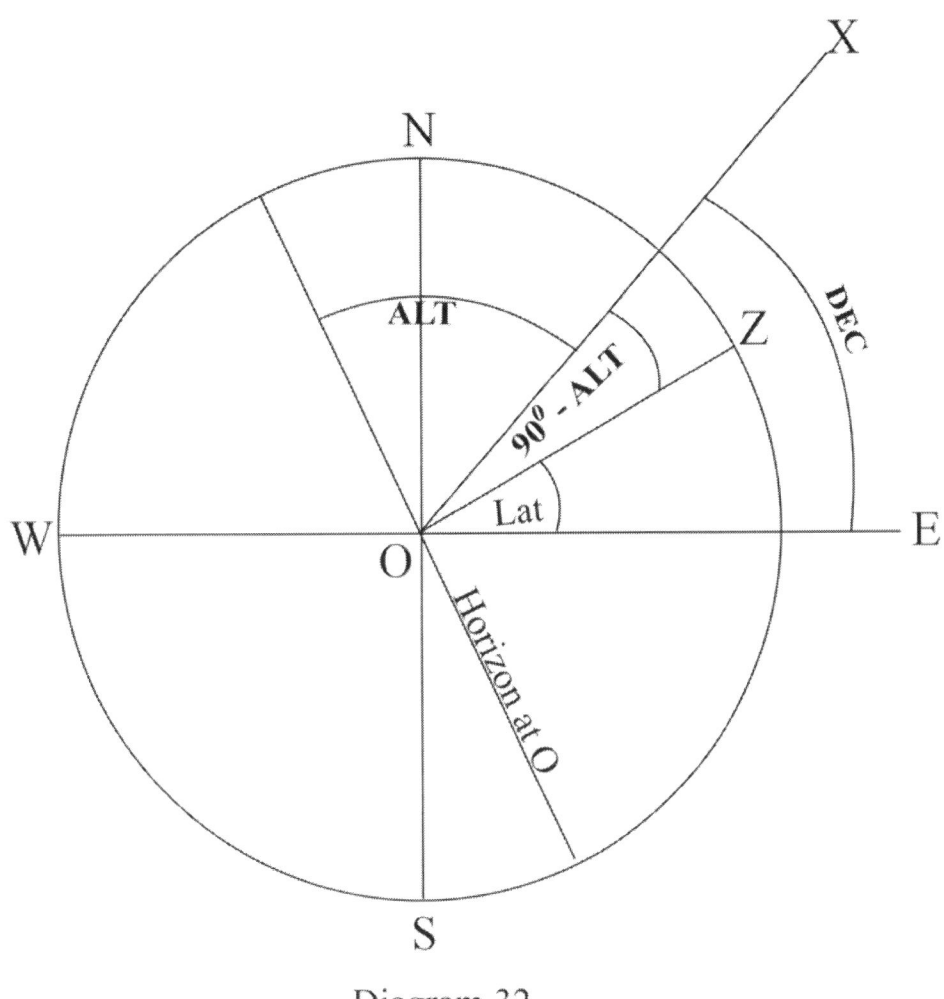

Diagram 32

In this case:
Declination = (90° - Altitude) + Latitude
∴ **Lat. = Dec. - (90° - Alt.)**

In the next case, latitude and declination are in opposite hemispheres as shown in Diagram 33:

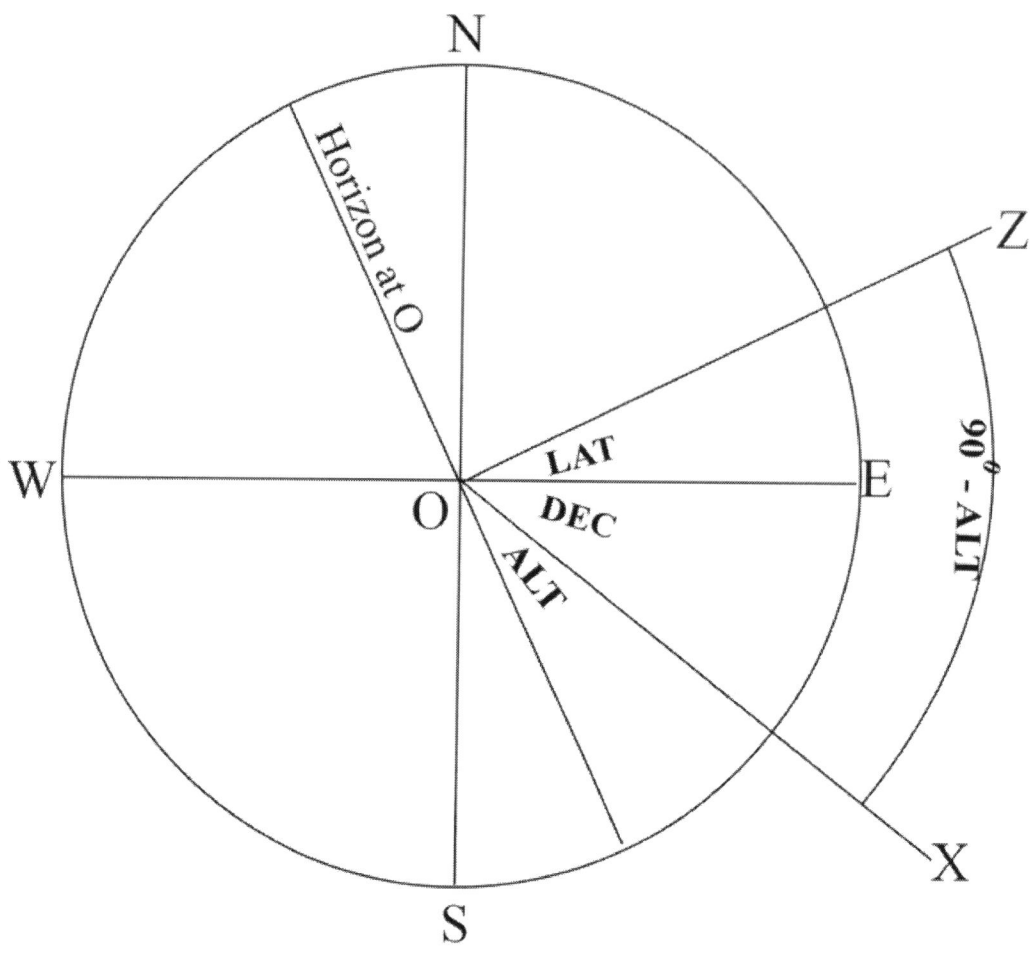

Diagram 33

In this case:
Latitude + Declination = (90° - Altitude)
∴   Lat. = (90° - Alt.) - Dec.

The rules for the three cases are summarised as follows:
(i)   Latitude and declination **same names** and **latitude greater** than declination:
      LAT = DEC + (90° - ALT)

(ii) Latitude and declination **same names** and **declination greater** than latitude:
      LAT = DEC - (90° - ALT)

(iii) Latitude and declination **contrary names**:
      LAT = (90° - ALT) - DEC

Examples

The following examples demonstrate how latitude can be calculated by applying the above rules when the altitude and declination of the Sun at meridian passage are known.

Notes.
1. Before working through these examples, it is recommended that you revise 'Short Distance Sailing' in chapter 3.
2. Since refraction is negligible when the Sun is at its zenith, additional altitude correction for non standard conditions is not necessary when calculating the true altitude at meridian passage.
3. Nautical Almanac extracts relevant to these examples are to be found following the solution to example 3.

Example 1.

At 1000 (zone -9) on 17 December 2009, a ship's D.R. Position is 41° 15'.0S. 134° 52'.0E.

The course is 030° and the speed is 15 knots.

At Mer Pas. the altitude of the Sun's lower limb is measured and found to be 72° 18'.2.   Height of eye = 12m.   Index error = +2'.1.

Tasks:

a. Calculate D.R. position at Mer. Pas.

b. Calculate the latitude from the Mer. Alt.

Solution:

Step 1. Determine Time of Mer. Pas. at Greenwich.
From the Nautical Almanac,
Mer. Pas. at Greenwich = 1156 GMT.

Step 2. Estimate Time of Mer.Pas.at first D.R. Pos.
- Convert Long. to time. (recap ch. 2)

Long. 134° 52'.0E

	h	m	s
4 × 134° ÷ 60 = 8.93^h	= 8	55	48.0
4 × 52'.0 ÷ 60 = 3.46^m	= 0	03	27.6
	= 8	59	15.6

- Estimate zone time of Mer.Pas.

	h	m	s
Mer. Pas. Greenwich	11	56	
Long. (long east GMT least)	-08	59	15.6
Local Mer. Pas (GMT)	02	56	44.4
	≈ 02	57	
Zone (-9)	+09		
Zone time Mer. Pas.	11	57	

©2009 Jack Case

Step 3. Calculate D.R. Pos. at time of Estimated Mer. Pas.

Course = 030° Speed = 15 knots.
Estimated time of Mer. Pas = 1157
Distance run in $1^h\ 57^m$ = 15 x 1.95 = 29.25 n.m.

Dep.	= Dist x Sin(course)
	= 29.25 x Sin(30) = 14'.62E
D.Lat.	= Dist x Cos(course)
	= 29.25 x Cos(30) = 25'.4N
New Lat.	= 41° 15'.0S - 25'.4N = 40° 49'.6S
Mid. Lat	= 41° 15'.0S. - 12'.7N
	= 41° 02'.3S. = 41°.04S
D.Long.	= Dep. x Sec(M.Lat)
	= 14'.62 x Sec(41°.04) = 19'.4E
New Long.	= 134° 52'.0E + 19'.4E
	= 135° 11'.4E
New Pos.	= 40° 49'.6S. 135° 11'.4E.

Step 4. Calculate time of Mer.Pas.at Second D.R. Pos.

	h	m	
Mer. Pas. Greenwich:	11	56	(GMT)
Long (135° 30'0E):	-09	02	
Local Mer. Pas (GMT)	02	54	
Zone (-9)	+ 09		
Zone time Mer. Pas.	11	54	

Step 5. Determine Declination at Time of Local Mer. Pas. (recap ch. 5)

Local Mer. Pas (GMT) $02^h\ 54^m$	
Dec Sun (02^h) (from Almanac)	S23° 21'.2
Corr^n (54^m) (d = 0'.1)	+ 0'.1
Dec Sun ($02^h\ 54^m$)	S23° 21'.3

> **Step 6. Calculate Meridian Altitude.** (recap ch. 6)
> Sext. Alt.   72° 18'.2
> I.E.   + 2'.1
> Observed Alt.   72° 20'.3
> Dip   - 6'.1
> Apparent Alt.   72° 14'.2
> Sun's Corr.   + 15'.9
> True Alt.   72° 30'.1

> **Step 7.** Determine Latitude.
> LAT = DEC + (90° - ALT) (rule i)
>   = 23° 21'.3 + (90° - 72° 30'.1)
>   = 23° 21'.3 + 17° 29'.9
>   = 40° 51'.2 S.

Example 2.
Date: 23 June 2009.
Zone time: 0930 (zone -7)
D.R. Position: 2° 18'.3 N. 104° 15'.8E.
Course: 265°. Speed: 22 knots.
Altitude of the Sun's lower limb at Mer. Pas.: 69.5°.
Height of eye: 2.5m. Index error: -1'.5.

Tasks:
a. Calculate D.R. position at Mer. Pas.
b. Calculate the latitude from the Mer. Alt.

Solution:

> **Step 1. Determine Time of Mer. Pas. at Greenwich for 23 June.**
> From the Nautical Almanac,
> Mer. Pas. at Greenwich = 1202 GMT.

Step 2. Estimate Time of Mer.Pas. at first D.R. Pos.

- Convert Long. to time. (recap ch. 2)

Long. 104° 15'.8E.

	h	m	s
4 × 104° ÷ 60 = 6.93^h =	6	55	48
4 × 15'.8 ÷ 60 = 1.05^m =		1	3
	6	56	51 ≈ 6^h 57^m

- Estimate zone time of Mer.Pas.

	h	m	s
Mer. Pas. Greenwich	12	02	
Long. (long east GMT least)	-06	56	51
Local Mer. Pas (GMT)	05	06	09
≈	05	06	
Zone (-7)	+07		
Zone time Mer. Pas.	12	06	

Step 3. Calculate D.R. Pos. at time of Estimated Mer. Pas.

Course = 265° Speed = 22 knots.
Time at first D.R. Position = 0930(-7)
Estimated time of Mer. Pas = 1206(-7)
Distance run in 2^h 36^m = 22 × 2.6
                              = 57.2 n.m.

Dep.   = Dist × Sin(course)
         = 57.2 × Sin(265)
         = -56'.99 = 56'.99W

D.Lat.   = Dist × Cos(course)
         = 57.2 × Cos(265)
         = -4'.99 = 4'.99S

New Lat. = 2° 18'.3 N. - 4'.99
         = 2° 13'.3 N.

Mid. Lat = 2° 18'.3 N. - 2'.5
         = 2° 15'.8 N. = 2°.26 N

D.Long. = Dep. × Sec(M.Lat)
         = 56.99 × Sec(2°.28)
         = 57'.04W

New Long. = 104° 15'.8E - 57'.04
          = 103° 18'.76

Step 4. Calculate time of Mer.Pas. at Second D.R. Pos.

	h	m
Mer. Pas. Greenwich:	12	02 (GMT)
Long (103° 18'.76E):	-06	53
Local Mer. Pas (GMT)	05	09
Zone (-7)	+ 07	
Zone time Mer. Pas.	12	09

Step 5. Determine Declination at Time of Local Mer. Pas.

Local Mer. Pas (GMT) $05^h\ 09^m$
Dec Sun (05^h) (from Almanac) N23° 25'.6
Corr^n (09^m) (d = 0'.0)   0'.0
Dec Sun ($05^h\ 09^m$)   N23° 25'.6

Step 6. Calculate Meridian Altitude.

Sext. Alt.	69° 30'.0
I.E.	- 1'.5
Observed Alt.	69° 28'.5
Dip	- 2'.8
Apparent Alt.	69° 25'.7
Sun's Corr.	+ 15'.6
True Alt.	69° 41'.3

Step 7. Determine Latitude.

LAT = DEC - (90° - ALT) (rule ii)
    = 23° 25'.6 – (90° – 69° 41'.3)
    = 23° 25'.6 – 20° 18'.7
    = **3° 06'.9 N.**

Example 3. A yacht is making a trans-Atlantic crossing.
At 0800 on 30 Sept. 2009 the D.R. Position was 6° 35'.13 N. 22.5°W
Estimated course and speed made good for next 5 hours = 135° at 6.5 knots.
At Mer. Pas. the altitude of the Sun's lower limb was measured and found to be 80° 38'.41. Height of eye = 2.9m.   Index error = 3'.2.

©2009 Jack Case

Tasks:
a. Calculate D.R. position at Mer. Pas.
b. Calculate the latitude from the Mer. Alt.

Solution.

Step 1. Determine Time of Mer. Pas. at Greenwich for 30 Sept.
From the Nautical Almanac,
Mer. Pas. at Greenwich = 1150 GMT.

Step 2. Estimate Time of Mer.Pas. at first D.R. Pos.
- Convert Long. to time.

Long.   22° 30'.0 W.

	h	m	s
$4 \times 22° \div 60 = 1.47^h$	= 1	28	12
$4 \times 30'.0 \div 60$	=	2	0
	= 1	30	12

- Estimate zone time of Mer.Pas.

	h	m	s
Mer. Pas. Greenwich	11	50	00
Long. (long west GMT best)	+01	30	12
Local Mer. Pas (GMT)	13	20	12
	≈ 13	20	
Zone (+1)	-01		
Zone time Mer. Pas.	12	20	

Step 3. Calculate D.R. Pos.at time of Estimated Mer. Pas.
Course = 135° Speed = 6.5 knots.
Estimated time of Mer. Pas = 1220
Distance run in $4^h 20^m$ = 6.5 x 4.33 = 28.15 n.m.
Dep.   = Dist x Sin(course) = 28.15 x Sin(135) = 19'.9 E.
D.Lat.   = Dist x Cos(course) = 28.15 x Cos(135) = -19'.9   = 19'.9S
New Lat.   = 6° 35'.13 N - 19'.9 = 6° 15'.23 N.
Mid. Lat   = 6° 35'.13 N - 9.'95 = 6° 25'.18 N = 6°.42 N
D.Long.   = Dep. x Sec(M.Lat) = 19.9 x Sec(6°.42) = 20'.02E
New Long. = 22° 30'.0W - 20'.2 = 22° 09'.8W
∴ New Pos. = 6° 15'.23 N. 22° 09'.8W.

Step 4. Calculate time of Mer.Pas. at Second D.R. Pos.

	h	m	
Mer. Pas. Greenwich:	11	50	(GMT)
Long (22° 09'.8W):	+01	29	
Local Mer. Pas (GMT)	13	19	
Zone (+1)	- 01		
Zone time Mer. Pas.	12	19	

Step 5. Determine Declination at Time of Local Mer. Pas.
Local Mer. Pas (GMT) $13^h\ 19^m$
Dec Sun (13^h) (from Almanac) S2° 58'.7
Corr^n (19^m) (d = 1'.0)   + 0'.3
Dec Sun ($05^h\ 07^m$)   S2° 59'.0

Step 6. Calculate Meridian Altitude.

Sext. Alt.	80° 38'.41
I.E.	- 3'.20
Observed Alt.	80° 35'.21
Dip	- 3'.00
Apparent Alt.	80° 32'.21
Sun's Corr.	+ 15'.80
True Alt.	80° 48'.01

Step 7. Determine Latitude.
LAT = (90° - ALT) - DEC (rule iii)
    = (90° - 80° 48'.01) - 2° 59'.0
    = 9° 11'.99 - 2° 59'.0 = 6° 12'.9N.

Nautical Almanac extracts relevant to the above examples begin on the next page.

2009 DECEMBER 15, 16, 17 (TUES., WED., THURS.)

UT	SUN		MOON					Lat.	Twilight		Sunrise	Moonrise			
	GHA	Dec	GHA	v	Dec	d	HP		Naut.	Civil		15	16	17	18
d h	° ′	° ′	° ′	′	° ′	′	′	°	h m	h m	h m	h m	h m	h m	h m
15 00	181 15.4	S23 15.7	200 18.2	9.1	S24 58.3	3.6	55.5	N 72	08 21	10 50	■	■	■	■	■
01	196 15.1	15.8	214 46.3	9.1	25 01.9	3.3	55.5	N 70	08 02	09 49	■	■	■	■	■
02	211 14.8	15.9	229 14.4	9.1	25 05.2	3.3	55.5	68	07 46	09 15	■	■	■	■	■
03	226 14.5	.. 16.1	243 42.5	9.1	25 08.5	3.1	55.5	66	07 33	08 50	10 29	■	■	■	12 04
04	241 14.2	16.2	258 10.6	9.1	25 11.6	3.0	55.5	64	07 22	08 30	09 48	10 16	11 22	11 26	11 24
05	256 13.9	16.3	272 38.7	9.1	25 14.6	2.8	55.4	62	07 12	08 14	09 20	09 14	10 13	10 43	10 56
								60	07 04	08 01	08 58	08 40	09 38	10 14	10 34
06	271 13.6	S23 16.4	287 06.8	9.1	S25 17.4	2.8	55.4	N 58	06 57	07 49	08 41	08 15	09 12	09 51	10 17
07	286 13.3	16.6	301 34.9	9.0	25 20.2	2.5	55.4	56	06 50	07 39	08 26	07 56	08 52	09 33	10 02
T 08	301 13.0	16.7	316 02.9	9.1	25 22.7	2.5	55.4	54	06 44	07 30	08 14	07 39	08 35	09 18	09 49
U 09	316 12.7	.. 16.8	330 31.0	9.1	25 25.2	2.3	55.4	52	06 38	07 22	08 03	07 25	08 21	09 05	09 38
E 10	331 12.4	16.9	344 59.1	9.0	25 27.5	2.2	55.4	50	06 33	07 14	07 53	07 12	08 08	08 53	09 27
S 11	346 12.1	17.1	359 27.1	9.1	25 29.7	2.0	55.3	45	06 21	06 58	07 32	06 47	07 42	08 28	09 06
D 12	1 11.8	S23 17.2	13 55.2	9.1	S25 31.7	1.9	55.3	N 40	06 11	06 45	07 15	06 27	07 22	08 09	08 49
A 13	16 11.5	17.3	28 23.3	9.0	25 33.6	1.8	55.3	35	06 01	06 33	07 01	06 10	07 04	07 53	08 35
Y 14	31 11.2	17.4	42 51.3	9.1	25 35.4	1.6	55.3	30	05 53	06 23	06 49	05 55	06 50	07 39	08 22
15	46 10.9	.. 17.5	57 19.4	9.1	25 37.0	1.5	55.3	20	05 36	06 04	06 28	05 31	06 24	07 15	08 00
16	61 10.6	17.7	71 47.5	9.1	25 38.5	1.4	55.2	N 10	05 20	05 46	06 09	05 09	06 03	06 54	07 42
17	76 10.3	17.8	86 15.6	9.1	25 39.9	1.2	55.2	0	05 03	05 29	05 52	04 50	05 42	06 34	07 24
18	91 10.0	S23 17.9	100 43.7	9.1	S25 41.1	1.1	55.2	S 10	04 44	05 11	05 34	04 30	05 22	06 15	07 06
19	106 09.7	18.0	115 11.8	9.1	25 42.2	1.0	55.2	20	04 22	04 51	05 16	04 09	05 00	05 54	06 48
20	121 09.4	18.1	129 39.9	9.1	25 43.2	0.8	55.2	30	03 53	04 26	04 54	03 44	04 35	05 30	06 26
21	136 09.1	.. 18.2	144 08.0	9.1	25 44.0	0.7	55.2	35	03 34	04 11	04 41	03 30	04 21	05 15	06 13
22	151 08.8	18.4	158 36.1	9.2	25 44.7	0.5	55.1	40	03 11	03 53	04 26	03 13	04 04	04 59	05 58
23	166 08.5	18.5	173 04.3	9.1	25 45.2	0.4	55.1	45	02 41	03 30	04 08	02 54	03 43	04 39	05 40
16 00	181 08.1	S23 18.6	187 32.4	9.2	S25 45.6	0.3	55.1	S 50	01 56	03 01	03 45	02 29	03 17	04 14	05 18
01	196 07.8	18.7	202 00.6	9.2	25 45.9	0.2	55.1	52	01 28	02 46	03 34	02 17	03 05	04 02	05 08
02	211 07.5	18.8	216 28.8	9.2	25 46.1	0.0	55.1	54	00 43	02 28	03 22	02 04	02 50	03 49	04 56
03	226 07.2	.. 18.9	230 57.0	9.3	25 46.1	0.1	55.1	56	////	02 06	03 08	01 48	02 34	03 33	04 42
04	241 06.9	19.0	245 25.3	9.2	25 46.0	0.3	55.1	58	////	01 36	02 51	01 30	02 14	03 14	04 26
05	256 06.6	19.1	259 53.5	9.3	25 45.7	0.4	55.0	S 60	////	00 48	02 31	01 07	01 48	02 50	04 07

UT	SUN		MOON					Lat.	Sunset	Twilight		Moonset			
	GHA	Dec	GHA	v	Dec	d	HP			Civil	Naut.	15	16	17	18
06	271 06.3	S23 19.2	274 21.8	9.3	S25 45.3	0.5	55.0	°	h m	h m	h m	h m	h m	h m	h m
W 07	286 06.0	19.4	288 50.1	9.3	25 44.8	0.6	55.0	N 72	■	13 01	15 30	■	■	■	■
E 08	301 05.7	19.5	303 18.4	9.4	25 44.2	0.8	55.0	N 70	■	14 02	15 50	■	■	■	■
D 09	316 05.4	.. 19.6	317 46.8	9.4	25 43.4	0.9	55.0	68	■	14 37	16 05	■	■	■	■
N 10	331 05.1	19.7	332 15.2	9.4	25 42.5	1.1	55.0	66	13 22	15 02	16 18	■	■	■	15 17
E 11	346 04.8	19.8	346 43.6	9.4	25 41.4	1.1	54.9	64	14 04	15 21	16 29	11 46	12 30	14 13	15 57
S 12	1 04.5	S23 19.9	1 12.0	9.5	S25 40.3	1.4	54.9	62	14 32	15 37	16 39	12 48	13 39	14 56	16 24
D 13	16 04.2	20.0	15 40.5	9.5	25 38.9	1.4	54.9	60	14 53	15 51	16 47	13 22	14 14	15 25	16 46
A 14	31 03.9	20.1	30 09.0	9.5	25 37.5	1.6	54.9	N 58	15 10	16 02	16 55	13 47	14 39	15 47	17 03
Y 15	46 03.6	.. 20.2	44 37.5	9.6	25 35.9	1.7	54.9	56	15 25	16 12	17 01	14 07	14 59	16 04	17 17
16	61 03.3	20.3	59 06.1	9.6	25 34.2	1.8	54.9	54	15 38	16 21	17 08	14 23	15 16	16 20	17 30
17	76 03.0	20.4	73 34.7	9.6	25 32.4	1.9	54.8	52	15 49	16 30	17 13	14 38	15 30	16 33	17 42
18	91 02.7	S23 20.5	88 03.3	9.7	S25 30.5	2.1	54.8	50	15 59	16 37	17 19	14 50	15 43	16 44	17 50
19	106 02.4	20.6	102 32.0	9.7	25 28.4	2.2	54.8	45	16 19	16 53	17 30	15 16	16 09	17 08	18 11
20	121 02.1	20.7	117 00.7	9.8	25 26.2	2.4	54.8	N 40	16 36	17 06	17 41	15 36	16 29	17 27	18 27
21	136 01.7	.. 20.8	131 29.5	9.8	25 23.8	2.4	54.8	35	16 50	17 18	17 50	15 53	16 46	17 43	18 41
22	151 01.4	20.9	145 58.3	9.8	25 21.4	2.6	54.8	30	17 02	17 29	17 59	16 08	17 01	17 57	18 53
23	166 01.1	21.0	160 27.1	9.9	25 18.8	2.8	54.8	20	17 24	17 48	18 15	16 33	17 26	18 20	19 13
17 00	181 00.8	S23 21.0	174 56.0	10.0	S25 16.0	2.8	54.7	N 10	17 42	18 05	18 31	16 55	17 48	18 40	19 31
01	196 00.5	21.1	189 25.0	9.9	25 13.2	3.0	54.7	0	17 59	18 22	18 48	17 15	18 08	18 59	19 47
02	211 00.2	21.2	203 53.9	10.0	25 10.2	3.1	54.7	S 10	18 17	18 40	19 07	17 35	18 28	19 17	20 03
03	225 59.9	.. 21.3	218 22.9	10.1	25 07.1	3.2	54.7	20	18 36	19 00	19 30	17 57	18 49	19 37	20 21
04	240 59.6	21.4	232 52.0	10.1	25 03.9	3.3	54.7	30	18 58	19 25	19 59	18 22	19 14	20 00	20 41
05	255 59.3	21.5	247 21.1	10.2	25 00.6	3.5	54.7	35	19 11	19 41	20 17	18 37	19 28	20 13	20 52
06	270 59.0	S23 21.6	261 50.3	10.2	S24 57.1	3.6	54.7	40	19 26	19 59	20 40	18 54	19 45	20 29	21 06
07	285 58.7	21.7	276 19.5	10.3	24 53.5	3.7	54.6	45	19 44	20 21	21 11	19 14	20 05	20 47	21 21
T 08	300 58.4	21.8	290 48.8	10.3	24 49.8	3.8	54.6	S 50	20 06	20 51	21 56	19 40	20 30	21 10	21 40
H 09	315 58.1	.. 21.8	305 18.1	10.3	24 46.0	3.9	54.6	52	20 17	21 06	22 24	19 52	20 42	21 21	21 49
U 10	330 57.8	21.9	319 47.4	10.4	24 42.1	4.1	54.6	54	20 29	21 24	23 09	20 07	20 56	21 33	21 59
R 11	345 57.5	22.0	334 16.8	10.5	24 38.0	4.2	54.6	56	20 44	21 46	////	20 23	21 12	21 47	22 11
S 12	0 57.1	S23 22.1	348 46.3	10.5	S24 33.8	4.3	54.6	58	21 00	22 16	////	20 43	21 31	22 03	22 24
D 13	15 56.8	22.2	3 15.8	10.6	24 29.5	4.4	54.6	S 60	21 21	23 05	////	21 09	21 55	22 23	22 39
A 14	30 56.5	22.3	17 45.4	10.6	24 25.1	4.5	54.6								
Y 15	45 56.2	.. 22.3	32 15.0	10.7	24 20.6	4.7	54.5	Day	SUN			MOON			
16	60 55.9	22.4	46 44.7	10.8	24 15.9	4.7	54.5		Eqn. of Time		Mer.	Mer. Pass.		Age	Phase
17	75 55.6	22.5	61 14.5	10.8	24 11.2	4.9	54.5		00ʰ	12ʰ	Pass.	Upper	Lower		
18	90 55.3	S23 22.6	75 44.3	10.8	S24 06.3	5.0	54.5	d	m s	m s	h m	h m	h m	d	%
19	105 55.0	22.7	90 14.1	10.9	24 01.3	5.1	54.5	15	05 02	04 48	11 55	11 02	23 29	29	1
20	120 54.7	22.7	104 44.0	11.0	23 56.2	5.2	54.5	16	04 33	04 19	11 56	11 55	24 21	30	0	●
21	135 54.4	.. 22.8	119 14.0	11.1	23 51.0	5.3	54.5	17	04 04	03 49	11 56	12 46	00 21	01	1
22	150 54.1	22.9	133 44.1	11.1	23 45.7	5.4	54.5								
23	165 53.8	23.0	148 14.2	11.1	S23 40.3	5.6	54.4								
	SD 16.3	d 0.1	SD 15.1		15.0		14.9								

INCREMENTS AND CORRECTIONS

54m

54 s	SUN PLANETS	ARIES	MOON	v or d	Corrn	v or d	Corrn	v or d	Corrn
	° '	° '	° '	'	'	'	'	'	'
00	13 30·0	13 32·2	12 53·1	0·0	0·0	6·0	5·5	12·0	10·9
01	13 30·3	13 32·5	12 53·3	0·1	0·1	6·1	5·5	12·1	11·0
02	13 30·5	13 32·7	12 53·6	0·2	0·2	6·2	5·6	12·2	11·1
03	13 30·8	13 33·0	12 53·8	0·3	0·3	6·3	5·7	12·3	11·2
04	13 31·0	13 33·2	12 54·1	0·4	0·4	6·4	5·8	12·4	11·3
05	13 31·3	13 33·5	12 54·3	0·5	0·5	6·5	5·9	12·5	11·4
06	13 31·5	13 33·7	12 54·5	0·6	0·5	6·6	6·0	12·6	11·4
07	13 31·8	13 34·0	12 54·8	0·7	0·6	6·7	6·1	12·7	11·5
08	13 32·0	13 34·2	12 55·0	0·8	0·7	6·8	6·2	12·8	11·6
09	13 32·3	13 34·5	12 55·2	0·9	0·8	6·9	6·3	12·9	11·7
10	13 32·5	13 34·7	12 55·5	1·0	0·9	7·0	6·4	13·0	11·8
11	13 32·8	13 35·0	12 55·7	1·1	1·0	7·1	6·4	13·1	11·9
12	13 33·0	13 35·2	12 56·0	1·2	1·1	7·2	6·5	13·2	12·0
13	13 33·3	13 35·5	12 56·2	1·3	1·2	7·3	6·6	13·3	12·1
14	13 33·5	13 35·7	12 56·4	1·4	1·3	7·4	6·7	13·4	12·2
15	13 33·8	13 36·0	12 56·7	1·5	1·4	7·5	6·8	13·5	12·3
16	13 34·0	13 36·2	12 56·9	1·6	1·5	7·6	6·9	13·6	12·4
17	13 34·3	13 36·5	12 57·2	1·7	1·5	7·7	7·0	13·7	12·4
18	13 34·5	13 36·7	12 57·4	1·8	1·6	7·8	7·1	13·8	12·5
19	13 34·8	13 37·0	12 57·6	1·9	1·7	7·9	7·2	13·9	12·6
20	13 35·0	13 37·2	12 57·9	2·0	1·8	8·0	7·3	14·0	12·7
21	13 35·3	13 37·5	12 58·1	2·1	1·9	8·1	7·4	14·1	12·8
22	13 35·5	13 37·7	12 58·3	2·2	2·0	8·2	7·4	14·2	12·9
23	13 35·8	13 38·0	12 58·6	2·3	2·1	8·3	7·5	14·3	13·0
24	13 36·0	13 38·2	12 58·8	2·4	2·2	8·4	7·6	14·4	13·1
25	13 36·3	13 38·5	12 59·1	2·5	2·3	8·5	7·7	14·5	13·2
26	13 36·5	13 38·7	12 59·3	2·6	2·4	8·6	7·8	14·6	13·3
27	13 36·8	13 39·0	12 59·5	2·7	2·5	8·7	7·9	14·7	13·4
28	13 37·0	13 39·2	12 59·8	2·8	2·5	8·8	8·0	14·8	13·4
29	13 37·3	13 39·5	13 00·0	2·9	2·6	8·9	8·1	14·9	13·5
30	13 37·5	13 39·7	13 00·3	3·0	2·7	9·0	8·2	15·0	13·6
31	13 37·8	13 40·0	13 00·5	3·1	2·8	9·1	8·3	15·1	13·7
32	13 38·0	13 40·2	13 00·7	3·2	2·9	9·2	8·4	15·2	13·8
33	13 38·3	13 40·5	13 01·0	3·3	3·0	9·3	8·4	15·3	13·9
34	13 38·5	13 40·7	13 01·2	3·4	3·1	9·4	8·5	15·4	14·0
35	13 38·8	13 41·0	13 01·5	3·5	3·2	9·5	8·6	15·5	14·1
36	13 39·0	13 41·2	13 01·7	3·6	3·3	9·6	8·7	15·6	14·2
37	13 39·3	13 41·5	13 01·9	3·7	3·4	9·7	8·8	15·7	14·3
38	13 39·5	13 41·7	13 02·2	3·8	3·5	9·8	8·9	15·8	14·4
39	13 39·8	13 42·0	13 02·4	3·9	3·5	9·9	9·0	15·9	14·4
40	13 40·0	13 42·2	13 02·6	4·0	3·6	10·0	9·1	16·0	14·5
41	13 40·3	13 42·5	13 02·9	4·1	3·7	10·1	9·2	16·1	14·6
42	13 40·5	13 42·7	13 03·1	4·2	3·8	10·2	9·3	16·2	14·7
43	13 40·8	13 43·0	13 03·4	4·3	3·9	10·3	9·4	16·3	14·8
44	13 41·0	13 43·2	13 03·6	4·4	4·0	10·4	9·4	16·4	14·9
45	13 41·3	13 43·5	13 03·8	4·5	4·1	10·5	9·5	16·5	15·0
46	13 41·5	13 43·7	13 04·1	4·6	4·2	10·6	9·6	16·6	15·1
47	13 41·8	13 44·0	13 04·3	4·7	4·3	10·7	9·7	16·7	15·2
48	13 42·0	13 44·3	13 04·6	4·8	4·4	10·8	9·8	16·8	15·3
49	13 42·3	13 44·5	13 04·8	4·9	4·5	10·9	9·9	16·9	15·4
50	13 42·5	13 44·8	13 05·0	5·0	4·5	11·0	10·0	17·0	15·4
51	13 42·8	13 45·0	13 05·3	5·1	4·6	11·1	10·1	17·1	15·5
52	13 43·0	13 45·3	13 05·5	5·2	4·7	11·2	10·2	17·2	15·6
53	13 43·3	13 45·5	13 05·7	5·3	4·8	11·3	10·3	17·3	15·7
54	13 43·5	13 45·8	13 06·0	5·4	4·9	11·4	10·4	17·4	15·8
55	13 43·8	13 46·0	13 06·2	5·5	5·0	11·5	10·4	17·5	15·9
56	13 44·0	13 46·3	13 06·5	5·6	5·1	11·6	10·5	17·6	16·0
57	13 44·3	13 46·5	13 06·7	5·7	5·2	11·7	10·6	17·7	16·1
58	13 44·5	13 46·8	13 06·9	5·8	5·3	11·8	10·7	17·8	16·2
59	13 44·8	13 47·0	13 07·2	5·9	5·4	11·9	10·8	17·9	16·3
60	13 45·0	13 47·3	13 07·4	6·0	5·5	12·0	10·9	18·0	16·4

55m

55 s	SUN PLANETS	ARIES	MOON	v or d	Corrn	v or d	Corrn	v or d	Corrn
	° '	° '	° '	'	'	'	'	'	'
00	13 45·0	13 47·3	13 07·4	0·0	0·0	6·0	5·6	12·0	11·1
01	13 45·3	13 47·5	13 07·7	0·1	0·1	6·1	5·6	12·1	11·2
02	13 45·5	13 47·8	13 07·9	0·2	0·2	6·2	5·7	12·2	11·3
03	13 45·8	13 48·0	13 08·1	0·3	0·3	6·3	5·8	12·3	11·4
04	13 46·0	13 48·3	13 08·4	0·4	0·4	6·4	5·9	12·4	11·5
05	13 46·3	13 48·5	13 08·6	0·5	0·5	6·5	6·0	12·5	11·6
06	13 46·5	13 48·8	13 08·8	0·6	0·6	6·6	6·1	12·6	11·7
07	13 46·8	13 49·0	13 09·1	0·7	0·6	6·7	6·2	12·7	11·7
08	13 47·0	13 49·3	13 09·3	0·8	0·7	6·8	6·3	12·8	11·8
09	13 47·3	13 49·5	13 09·6	0·9	0·8	6·9	6·4	12·9	11·9
10	13 47·5	13 49·8	13 09·8	1·0	0·9	7·0	6·5	13·0	12·0
11	13 47·8	13 50·0	13 10·0	1·1	1·0	7·1	6·6	13·1	12·1
12	13 48·0	13 50·3	13 10·3	1·2	1·1	7·2	6·7	13·2	12·2
13	13 48·3	13 50·5	13 10·5	1·3	1·2	7·3	6·8	13·3	12·3
14	13 48·5	13 50·8	13 10·8	1·4	1·3	7·4	6·8	13·4	12·4
15	13 48·8	13 51·0	13 11·0	1·5	1·4	7·5	6·9	13·5	12·5
16	13 49·0	13 51·3	13 11·2	1·6	1·5	7·6	7·0	13·6	12·6
17	13 49·3	13 51·5	13 11·5	1·7	1·6	7·7	7·1	13·7	12·7
18	13 49·5	13 51·8	13 11·7	1·8	1·7	7·8	7·2	13·8	12·8
19	13 49·8	13 52·0	13 12·0	1·9	1·8	7·9	7·3	13·9	12·9
20	13 50·0	13 52·3	13 12·2	2·0	1·9	8·0	7·4	14·0	13·0
21	13 50·3	13 52·5	13 12·4	2·1	1·9	8·1	7·5	14·1	13·0
22	13 50·5	13 52·8	13 12·7	2·2	2·0	8·2	7·6	14·2	13·1
23	13 50·8	13 53·0	13 12·9	2·3	2·1	8·3	7·7	14·3	13·2
24	13 51·0	13 53·3	13 13·1	2·4	2·2	8·4	7·8	14·4	13·3
25	13 51·3	13 53·5	13 13·4	2·5	2·3	8·5	7·9	14·5	13·4
26	13 51·5	13 53·8	13 13·6	2·6	2·4	8·6	8·0	14·6	13·5
27	13 51·8	13 54·0	13 13·9	2·7	2·5	8·7	8·0	14·7	13·6
28	13 52·0	13 54·3	13 14·1	2·8	2·6	8·8	8·1	14·8	13·7
29	13 52·3	13 54·5	13 14·3	2·9	2·7	8·9	8·2	14·9	13·8
30	13 52·5	13 54·8	13 14·6	3·0	2·8	9·0	8·3	15·0	13·9
31	13 52·8	13 55·0	13 14·8	3·1	2·9	9·1	8·4	15·1	14·0
32	13 53·0	13 55·3	13 15·1	3·2	3·0	9·2	8·5	15·2	14·1
33	13 53·3	13 55·5	13 15·3	3·3	3·1	9·3	8·6	15·3	14·2
34	13 53·5	13 55·8	13 15·5	3·4	3·1	9·4	8·7	15·4	14·2
35	13 53·8	13 56·0	13 15·8	3·5	3·2	9·5	8·8	15·5	14·3
36	13 54·0	13 56·3	13 16·0	3·6	3·3	9·6	8·9	15·6	14·4
37	13 54·3	13 56·5	13 16·2	3·7	3·4	9·7	9·0	15·7	14·5
38	13 54·5	13 56·8	13 16·5	3·8	3·5	9·8	9·1	15·8	14·6
39	13 54·8	13 57·0	13 16·7	3·9	3·6	9·9	9·2	15·9	14·7
40	13 55·0	13 57·3	13 17·0	4·0	3·7	10·0	9·3	16·0	14·8
41	13 55·3	13 57·5	13 17·2	4·1	3·8	10·1	9·3	16·1	14·9
42	13 55·5	13 57·8	13 17·4	4·2	3·9	10·2	9·4	16·2	15·0
43	13 55·8	13 58·0	13 17·7	4·3	4·0	10·3	9·5	16·3	15·1
44	13 56·0	13 58·3	13 17·9	4·4	4·1	10·4	9·6	16·4	15·2
45	13 56·3	13 58·5	13 18·2	4·5	4·2	10·5	9·7	16·5	15·3
46	13 56·5	13 58·8	13 18·4	4·6	4·3	10·6	9·8	16·6	15·4
47	13 56·8	13 59·0	13 18·6	4·7	4·3	10·7	9·9	16·7	15·4
48	13 57·0	13 59·3	13 18·9	4·8	4·4	10·8	10·0	16·8	15·5
49	13 57·3	13 59·5	13 19·1	4·9	4·5	10·9	10·1	16·9	15·6
50	13 57·5	13 59·8	13 19·3	5·0	4·6	11·0	10·2	17·0	15·7
51	13 57·8	14 00·0	13 19·6	5·1	4·7	11·1	10·3	17·1	15·8
52	13 58·0	14 00·3	13 19·8	5·2	4·8	11·2	10·4	17·2	15·9
53	13 58·3	14 00·5	13 20·1	5·3	4·9	11·3	10·5	17·3	16·0
54	13 58·5	14 00·8	13 20·3	5·4	5·0	11·4	10·5	17·4	16·1
55	13 58·8	14 01·0	13 20·5	5·5	5·1	11·5	10·6	17·5	16·2
56	13 59·0	14 01·3	13 20·8	5·6	5·2	11·6	10·7	17·6	16·3
57	13 59·3	14 01·5	13 21·0	5·7	5·3	11·7	10·8	17·7	16·4
58	13 59·5	14 01·8	13 21·3	5·8	5·4	11·8	10·9	17·8	16·5
59	13 59·8	14 02·0	13 21·5	5·9	5·5	11·9	11·0	17·9	16·6
60	14 00·0	14 02·3	13 21·7	6·0	5·6	12·0	11·1	18·0	16·7

2009 JUNE 21, 22, 23 (SUN., MON., TUES.)

UT	SUN GHA	SUN Dec	MOON GHA	v	MOON Dec	d	HP
d h	° '	° '	° '	'	° '	'	'
21 00	179 34.5	N23 26.3	207 54.2	3.1	N25 15.7	5.1	60.3
01	194 34.3	26.4	222 16.3	3.0	25 20.8	5.0	60.4
02	209 34.2	26.4	236 38.3	2.9	25 25.8	4.7	60.4
03	224 34.0	26.4	251 00.2	2.8	25 30.5	4.6	60.4
04	239 33.9	26.4	265 22.0	2.7	25 35.1	4.3	60.4
05	254 33.8	26.4	279 43.7	2.7	25 39.4	4.2	60.5
06	269 33.6	N23 26.4	294 05.4	2.5	N25 43.6	4.0	60.5
07	284 33.5	26.4	308 26.9	2.5	25 47.6	3.9	60.5
08	299 33.4	26.4	322 48.4	2.4	25 51.5	3.6	60.6
S 09	314 33.2	26.4	337 09.8	2.3	25 55.1	3.4	60.6
U 10	329 33.1	26.4	351 31.1	2.2	25 58.5	3.3	60.6
N 11	344 32.9	26.3	5 52.3	2.2	26 01.8	3.0	60.6
D 12	359 32.8	N23 26.3	20 13.5	2.0	N26 04.8	2.9	60.7
A 13	14 32.7	26.3	34 34.5	2.1	26 07.7	2.7	60.7
Y 14	29 32.5	26.3	48 55.6	1.9	26 10.4	2.4	60.7
15	44 32.4	26.3	63 16.5	1.9	26 12.8	2.3	60.7
16	59 32.3	26.3	77 37.4	1.9	26 15.1	2.1	60.7
17	74 32.1	26.3	91 58.3	1.7	26 17.2	1.9	60.8
18	89 32.0	N23 26.3	106 19.0	1.8	N26 19.1	1.6	60.8
19	104 31.8	26.3	120 39.8	1.6	26 20.7	1.5	60.8
20	119 31.7	26.3	135 00.4	1.7	26 22.2	1.3	60.8
21	134 31.6	26.3	149 21.1	1.6	26 23.5	1.0	60.9
22	149 31.4	26.3	163 41.7	1.5	26 24.5	0.9	60.9
23	164 31.3	26.3	178 02.2	1.5	26 25.4	0.7	60.9
22 00	179 31.2	N23 26.2	192 22.7	1.5	N26 26.1	0.4	60.9
01	194 31.0	26.2	206 43.2	1.4	26 26.5	0.3	60.9
02	209 30.9	26.2	221 03.6	1.5	26 26.8	0.0	61.0
03	224 30.8	26.2	235 24.1	1.4	26 26.8	0.1	61.0
04	239 30.6	26.2	249 44.5	1.3	26 26.7	0.4	61.0
05	254 30.5	26.2	264 04.8	1.4	26 26.3	0.5	61.0
06	269 30.3	N23 26.1	278 25.2	1.3	N26 25.8	0.8	61.0
07	284 30.2	26.1	292 45.5	1.4	26 25.0	1.0	61.0
08	299 30.1	26.1	307 05.9	1.3	26 24.0	1.2	61.0
M 09	314 29.9	26.1	321 26.2	1.3	26 22.8	1.4	61.1
O 10	329 29.8	26.1	335 46.5	1.3	26 21.4	1.6	61.1
N 11	344 29.7	26.1	350 06.8	1.3	26 19.8	1.8	61.1
D 12	359 29.5	N23 26.0	4 27.1	1.4	N26 18.0	2.0	61.1
A 13	14 29.4	26.0	18 47.5	1.3	26 16.0	2.2	61.1
Y 14	29 29.2	26.0	33 07.8	1.3	26 13.8	2.4	61.1
15	44 29.1	26.0	47 28.1	1.4	26 11.4	2.7	61.1
16	59 29.0	25.9	61 48.5	1.4	26 08.7	2.8	61.1
17	74 28.8	25.9	76 08.9	1.4	26 05.9	3.0	61.2
18	89 28.7	N23 25.9	90 29.3	1.4	N26 02.9	3.3	61.2
19	104 28.6	25.9	104 49.7	1.4	25 59.6	3.4	61.2
20	119 28.4	25.8	119 10.1	1.5	25 56.2	3.7	61.2
21	134 28.3	25.8	133 30.6	1.5	25 52.5	3.8	61.2
22	149 28.2	25.8	147 51.1	1.6	25 48.7	4.1	61.2
23	164 28.0	25.7	162 11.7	1.6	25 44.6	4.2	61.2
23 00	179 27.9	N23 25.7	176 32.3	1.6	N25 40.4	4.5	61.2
01	194 27.7	25.7	190 52.9	1.7	25 35.9	4.6	61.2
02	209 27.6	25.7	205 13.6	1.7	25 31.3	4.9	61.2
03	224 27.5	25.6	219 34.3	1.8	25 26.4	5.0	61.2
04	239 27.3	25.6	233 55.1	1.8	25 21.4	5.3	61.2
05	254 27.2	25.6	248 15.9	1.9	25 16.1	5.4	61.2
06	269 27.1	N23 25.5	262 36.8	2.0	N25 10.7	5.7	61.2
07	284 26.9	25.5	276 57.8	2.0	25 05.0	5.8	61.2
08	299 26.8	25.4	291 18.8	2.1	24 59.2	6.0	61.2
T 09	314 26.7	25.4	305 39.9	2.1	24 53.2	6.2	61.2
U 10	329 26.5	25.4	320 01.0	2.3	24 47.0	6.4	61.2
E 11	344 26.4	25.3	334 22.3	2.3	24 40.6	6.6	61.2
S 12	359 26.3	N23 25.3	348 43.6	2.3	N24 34.0	6.8	61.2
D 13	14 26.1	25.3	3 04.9	2.5	24 27.2	6.9	61.2
A 14	29 26.0	25.2	17 26.4	2.5	24 20.3	7.1	61.2
Y 15	44 25.8	25.2	31 47.9	2.6	24 13.2	7.4	61.2
16	59 25.7	25.1	46 09.5	2.7	24 05.8	7.5	61.2
17	74 25.6	25.1	60 31.2	2.8	23 58.3	7.6	61.2
18	89 25.4	N23 25.1	74 53.0	2.8	N23 50.7	7.9	61.2
19	104 25.3	25.0	89 14.8	3.0	23 42.8	8.0	61.2
20	119 25.2	25.0	103 36.8	3.0	23 34.8	8.2	61.2
21	134 25.0	24.9	117 58.8	3.2	23 26.6	8.4	61.2
22	149 24.9	24.9	132 21.0	3.2	23 18.2	8.5	61.2
23	164 24.8	24.8	146 43.2	3.3	N23 09.7	8.8	61.2
	SD 15.8	d 0.0	SD 16.5		16.6		16.7

Lat.	Twilight Naut.	Twilight Civil	Sunrise	Moonrise 21	Moonrise 22	Moonrise 23	Moonrise 24
°	h m	h m	h m	h m	h m	h m	h m
N 72	□	□	□	□	□	□	□
N 70	□	□	□	□	□	□	□
68	□	□	□	□	□	□	□
66	□	□	□	□	□	□	02 52
64	////	////	01 31	□	□	01 09	03 36
62	////	////	02 09	00 01	00 40	02 08	04 05
60	////	00 49	02 36	00 32	01 19	02 41	04 27
N 58	////	01 41	02 57	00 56	01 47	03 06	04 45
56	////	02 11	03 13	01 15	02 08	03 26	05 01
54	00 45	02 33	03 28	01 31	02 26	03 42	05 13
52	01 32	02 51	03 40	01 45	02 41	03 56	05 25
50	02 00	03 06	03 51	01 57	02 54	04 09	05 35
45	02 46	03 36	04 13	02 22	03 21	04 34	05 56
N 40	03 17	03 59	04 32	02 42	03 43	04 54	06 13
35	03 40	04 17	04 47	02 59	04 01	05 11	06 27
30	03 59	04 32	05 00	03 13	04 16	05 26	06 39
20	04 28	04 57	05 22	03 38	04 42	05 51	07 00
N 10	04 51	05 18	05 41	04 00	05 05	06 12	07 18
0	05 10	05 36	05 58	04 20	05 26	06 32	07 35
S 10	05 27	05 53	06 16	04 40	05 47	06 52	07 52
20	05 43	06 10	06 34	05 02	06 09	07 13	08 10
30	05 59	06 29	06 56	05 28	06 36	07 37	08 31
35	06 08	06 40	07 08	05 43	06 51	07 52	08 43
40	06 18	06 52	07 22	06 00	07 09	08 08	08 56
45	06 28	07 05	07 39	06 21	07 31	08 28	09 13
S 50	06 40	07 21	08 00	06 48	07 58	08 53	09 33
52	06 45	07 29	08 10	07 01	08 12	09 05	09 42
54	06 51	07 37	08 21	07 16	08 27	09 18	09 52
56	06 57	07 46	08 34	07 33	08 45	09 34	10 04
58	07 04	07 56	08 48	07 54	09 07	09 52	10 18
S 60	07 11	08 08	09 06	08 21	09 36	10 15	10 34

Lat.	Sunset	Twilight Civil	Twilight Naut.	Moonset 21	Moonset 22	Moonset 23	Moonset 24
°	h m	h m	h m	h m	h m	h m	h m
N 72	□	□	□	□	□	□	□
N 70	□	□	□	□	□	□	23 59
68	□	□	□	□	□	□	(00 00) (23 31)
66	□	□	□	□	23 26	23 15	23 08
64	22 33	////	////	□	23 26	23 15	23 08
62	21 54	////	////	21 36	22 27	22 45	22 51
60	21 28	23 14	////	20 56	21 53	22 22	22 36
N 58	21 07	22 23	////	20 29	21 28	22 03	22 23
56	20 51	21 53	////	20 08	21 08	21 47	22 12
54	20 36	21 31	23 19	19 50	20 51	21 34	22 02
52	20 24	21 13	22 31	19 35	20 37	21 22	21 53
50	20 13	20 58	22 03	19 22	20 24	21 11	21 46
45	19 51	20 28	21 18	18 55	19 58	20 49	21 29
N 40	19 33	20 05	20 47	18 33	19 38	20 31	21 15
35	19 17	19 47	20 24	18 15	19 20	20 16	21 03
30	19 04	19 32	20 05	18 00	19 05	20 03	20 53
20	18 42	19 07	19 36	17 34	18 40	19 41	20 35
N 10	18 23	18 46	19 13	17 12	18 18	19 21	20 19
0	18 06	18 28	18 54	16 51	17 57	19 02	20 04
S 10	17 48	18 11	18 37	16 30	17 36	18 44	19 49
20	17 30	17 54	18 21	16 08	17 14	18 24	19 33
30	17 08	17 35	18 05	15 42	16 48	18 00	19 14
35	16 56	17 24	17 56	15 26	16 33	17 47	19 03
40	16 42	17 12	17 47	15 09	16 15	17 31	18 50
45	16 25	16 59	17 36	14 47	15 54	17 12	18 35
S 50	16 04	16 43	17 24	14 20	15 26	16 48	18 17
52	15 54	16 35	17 19	14 07	15 13	16 36	18 08
54	15 43	16 27	17 13	13 52	14 58	16 23	17 58
56	15 30	16 18	17 07	13 34	14 40	16 08	17 47
58	15 16	16 08	17 00	13 13	14 18	15 50	17 34
S 60	14 58	15 56	16 53	12 46	13 50	15 28	17 19

Day	SUN Eqn. of Time 00h	SUN Eqn. of Time 12h	SUN Mer. Pass.	MOON Mer. Pass. Upper	MOON Mer. Pass. Lower	Age	Phase
d	m s	m s	h m	h m	h m	d	%
21	01 42	01 49	12 02	10 35	23 08	28	3
22	01 55	02 02	12 02	11 41	24 14	29	0
23	02 08	02 15	12 02	12 47	00 14	01	1

©2009 Jack Case

INCREMENTS AND CORRECTIONS

8m

s	SUN PLANETS	ARIES	MOON	v or d	Corrⁿ	v or d	Corrⁿ	v or d	Corrⁿ
	° ′	° ′	° ′	′	′	′	′	′	′
00	2 00.0	2 00.3	1 54.5	0.0	0.0	6.0	0.9	12.0	1.7
01	2 00.3	2 00.6	1 54.8	0.1	0.0	6.1	0.9	12.1	1.7
02	2 00.5	2 00.8	1 55.0	0.2	0.0	6.2	0.9	12.2	1.7
03	2 00.8	2 01.1	1 55.2	0.3	0.0	6.3	0.9	12.3	1.7
04	2 01.0	2 01.3	1 55.5	0.4	0.1	6.4	0.9	12.4	1.8
05	2 01.3	2 01.6	1 55.7	0.5	0.1	6.5	0.9	12.5	1.8
06	2 01.5	2 01.8	1 56.0	0.6	0.1	6.6	0.9	12.6	1.8
07	2 01.8	2 02.1	1 56.2	0.7	0.1	6.7	0.9	12.7	1.8
08	2 02.0	2 02.3	1 56.4	0.8	0.1	6.8	1.0	12.8	1.8
09	2 02.3	2 02.6	1 56.7	0.9	0.1	6.9	1.0	12.9	1.8
10	2 02.5	2 02.8	1 56.9	1.0	0.1	7.0	1.0	13.0	1.8
11	2 02.8	2 03.1	1 57.2	1.1	0.2	7.1	1.0	13.1	1.9
12	2 03.0	2 03.3	1 57.4	1.2	0.2	7.2	1.0	13.2	1.9
13	2 03.3	2 03.6	1 57.6	1.3	0.2	7.3	1.0	13.3	1.9
14	2 03.5	2 03.8	1 57.9	1.4	0.2	7.4	1.0	13.4	1.9
15	2 03.8	2 04.1	1 58.1	1.5	0.2	7.5	1.1	13.5	1.9
16	2 04.0	2 04.3	1 58.4	1.6	0.2	7.6	1.1	13.6	1.9
17	2 04.3	2 04.6	1 58.6	1.7	0.2	7.7	1.1	13.7	1.9
18	2 04.5	2 04.8	1 58.8	1.8	0.3	7.8	1.1	13.8	2.0
19	2 04.8	2 05.1	1 59.1	1.9	0.3	7.9	1.1	13.9	2.0
20	2 05.0	2 05.3	1 59.3	2.0	0.3	8.0	1.1	14.0	2.0
21	2 05.3	2 05.6	1 59.5	2.1	0.3	8.1	1.1	14.1	2.0
22	2 05.5	2 05.8	1 59.8	2.2	0.3	8.2	1.2	14.2	2.0
23	2 05.8	2 06.1	2 00.0	2.3	0.3	8.3	1.2	14.3	2.0
24	2 06.0	2 06.3	2 00.3	2.4	0.3	8.4	1.2	14.4	2.0
25	2 06.3	2 06.6	2 00.5	2.5	0.4	8.5	1.2	14.5	2.1
26	2 06.5	2 06.8	2 00.7	2.6	0.4	8.6	1.2	14.6	2.1
27	2 06.8	2 07.1	2 01.0	2.7	0.4	8.7	1.2	14.7	2.1
28	2 07.0	2 07.3	2 01.2	2.8	0.4	8.8	1.2	14.8	2.1
29	2 07.3	2 07.6	2 01.5	2.9	0.4	8.9	1.3	14.9	2.1
30	2 07.5	2 07.8	2 01.7	3.0	0.4	9.0	1.3	15.0	2.1
31	2 07.8	2 08.1	2 01.9	3.1	0.4	9.1	1.3	15.1	2.1
32	2 08.0	2 08.4	2 02.2	3.2	0.5	9.2	1.3	15.2	2.2
33	2 08.3	2 08.6	2 02.4	3.3	0.5	9.3	1.3	15.3	2.2
34	2 08.5	2 08.9	2 02.6	3.4	0.5	9.4	1.3	15.4	2.2
35	2 08.8	2 09.1	2 02.9	3.5	0.5	9.5	1.3	15.5	2.2
36	2 09.0	2 09.4	2 03.1	3.6	0.5	9.6	1.4	15.6	2.2
37	2 09.3	2 09.6	2 03.4	3.7	0.5	9.7	1.4	15.7	2.2
38	2 09.5	2 09.9	2 03.6	3.8	0.5	9.8	1.4	15.8	2.2
39	2 09.8	2 10.1	2 03.8	3.9	0.6	9.9	1.4	15.9	2.3
40	2 10.0	2 10.4	2 04.1	4.0	0.6	10.0	1.4	16.0	2.3
41	2 10.3	2 10.6	2 04.3	4.1	0.6	10.1	1.4	16.1	2.3
42	2 10.5	2 10.9	2 04.6	4.2	0.6	10.2	1.4	16.2	2.3
43	2 10.8	2 11.1	2 04.8	4.3	0.6	10.3	1.5	16.3	2.3
44	2 11.0	2 11.4	2 05.0	4.4	0.6	10.4	1.5	16.4	2.3
45	2 11.3	2 11.6	2 05.3	4.5	0.6	10.5	1.5	16.5	2.3
46	2 11.5	2 11.9	2 05.5	4.6	0.7	10.6	1.5	16.6	2.4
47	2 11.8	2 12.1	2 05.7	4.7	0.7	10.7	1.5	16.7	2.4
48	2 12.0	2 12.4	2 06.0	4.8	0.7	10.8	1.5	16.8	2.4
49	2 12.3	2 12.6	2 06.2	4.9	0.7	10.9	1.5	16.9	2.4
50	2 12.5	2 12.9	2 06.5	5.0	0.7	11.0	1.6	17.0	2.4
51	2 12.8	2 13.1	2 06.7	5.1	0.7	11.1	1.6	17.1	2.4
52	2 13.0	2 13.4	2 06.9	5.2	0.7	11.2	1.6	17.2	2.4
53	2 13.3	2 13.6	2 07.2	5.3	0.8	11.3	1.6	17.3	2.5
54	2 13.5	2 13.9	2 07.4	5.4	0.8	11.4	1.6	17.4	2.5
55	2 13.8	2 14.1	2 07.7	5.5	0.8	11.5	1.6	17.5	2.5
56	2 14.0	2 14.4	2 07.9	5.6	0.8	11.6	1.6	17.6	2.5
57	2 14.3	2 14.6	2 08.1	5.7	0.8	11.7	1.7	17.7	2.5
58	2 14.5	2 14.9	2 08.4	5.8	0.8	11.8	1.7	17.8	2.5
59	2 14.8	2 15.1	2 08.6	5.9	0.8	11.9	1.7	17.9	2.5
60	2 15.0	2 15.4	2 08.9	6.0	0.9	12.0	1.7	18.0	2.6

9m

s	SUN PLANETS	ARIES	MOON	v or d	Corrⁿ	v or d	Corrⁿ	v or d	Corrⁿ
	° ′	° ′	° ′	′	′	′	′	′	′
00	2 15.0	2 15.4	2 08.9	0.0	0.0	6.0	1.0	12.0	1.9
01	2 15.3	2 15.6	2 09.1	0.1	0.0	6.1	1.0	12.1	1.9
02	2 15.5	2 15.9	2 09.3	0.2	0.0	6.2	1.0	12.2	1.9
03	2 15.8	2 16.1	2 09.6	0.3	0.0	6.3	1.0	12.3	1.9
04	2 16.0	2 16.4	2 09.8	0.4	0.1	6.4	1.0	12.4	2.0
05	2 16.3	2 16.6	2 10.0	0.5	0.1	6.5	1.0	12.5	2.0
06	2 16.5	2 16.9	2 10.3	0.6	0.1	6.6	1.0	12.6	2.0
07	2 16.8	2 17.1	2 10.5	0.7	0.1	6.7	1.1	12.7	2.0
08	2 17.0	2 17.4	2 10.8	0.8	0.1	6.8	1.1	12.8	2.0
09	2 17.3	2 17.6	2 11.0	0.9	0.1	6.9	1.1	12.9	2.0
10	2 17.5	2 17.9	2 11.2	1.0	0.2	7.0	1.1	13.0	2.1
11	2 17.8	2 18.1	2 11.5	1.1	0.2	7.1	1.1	13.1	2.1
12	2 18.0	2 18.4	2 11.7	1.2	0.2	7.2	1.1	13.2	2.1
13	2 18.3	2 18.6	2 12.0	1.3	0.2	7.3	1.2	13.3	2.1
14	2 18.5	2 18.9	2 12.2	1.4	0.2	7.4	1.2	13.4	2.1
15	2 18.8	2 19.1	2 12.4	1.5	0.2	7.5	1.2	13.5	2.1
16	2 19.0	2 19.4	2 12.7	1.6	0.3	7.6	1.2	13.6	2.2
17	2 19.3	2 19.6	2 12.9	1.7	0.3	7.7	1.2	13.7	2.2
18	2 19.5	2 19.9	2 13.1	1.8	0.3	7.8	1.2	13.8	2.2
19	2 19.8	2 20.1	2 13.4	1.9	0.3	7.9	1.3	13.9	2.2
20	2 20.0	2 20.4	2 13.6	2.0	0.3	8.0	1.3	14.0	2.2
21	2 20.3	2 20.6	2 13.9	2.1	0.3	8.1	1.3	14.1	2.2
22	2 20.5	2 20.9	2 14.1	2.2	0.3	8.2	1.3	14.2	2.2
23	2 20.8	2 21.1	2 14.3	2.3	0.4	8.3	1.3	14.3	2.3
24	2 21.0	2 21.4	2 14.6	2.4	0.4	8.4	1.3	14.4	2.3
25	2 21.3	2 21.6	2 14.8	2.5	0.4	8.5	1.3	14.5	2.3
26	2 21.5	2 21.9	2 15.1	2.6	0.4	8.6	1.4	14.6	2.3
27	2 21.8	2 22.1	2 15.3	2.7	0.4	8.7	1.4	14.7	2.3
28	2 22.0	2 22.4	2 15.5	2.8	0.4	8.8	1.4	14.8	2.3
29	2 22.3	2 22.6	2 15.8	2.9	0.5	8.9	1.4	14.9	2.4
30	2 22.5	2 22.9	2 16.0	3.0	0.5	9.0	1.4	15.0	2.4
31	2 22.8	2 23.1	2 16.2	3.1	0.5	9.1	1.4	15.1	2.4
32	2 23.0	2 23.4	2 16.5	3.2	0.5	9.2	1.5	15.2	2.4
33	2 23.3	2 23.6	2 16.7	3.3	0.5	9.3	1.5	15.3	2.4
34	2 23.5	2 23.9	2 17.0	3.4	0.5	9.4	1.5	15.4	2.4
35	2 23.8	2 24.1	2 17.2	3.5	0.6	9.5	1.5	15.5	2.5
36	2 24.0	2 24.4	2 17.4	3.6	0.6	9.6	1.5	15.6	2.5
37	2 24.3	2 24.6	2 17.7	3.7	0.6	9.7	1.5	15.7	2.5
38	2 24.5	2 24.9	2 17.9	3.8	0.6	9.8	1.6	15.8	2.5
39	2 24.8	2 25.1	2 18.2	3.9	0.6	9.9	1.6	15.9	2.5
40	2 25.0	2 25.4	2 18.4	4.0	0.6	10.0	1.6	16.0	2.5
41	2 25.3	2 25.6	2 18.6	4.1	0.6	10.1	1.6	16.1	2.5
42	2 25.5	2 25.9	2 18.9	4.2	0.7	10.2	1.6	16.2	2.6
43	2 25.8	2 26.1	2 19.1	4.3	0.7	10.3	1.6	16.3	2.6
44	2 26.0	2 26.4	2 19.3	4.4	0.7	10.4	1.6	16.4	2.6
45	2 26.3	2 26.7	2 19.6	4.5	0.7	10.5	1.7	16.5	2.6
46	2 26.5	2 26.9	2 19.8	4.6	0.7	10.6	1.7	16.6	2.6
47	2 26.8	2 27.2	2 20.1	4.7	0.7	10.7	1.7	16.7	2.6
48	2 27.0	2 27.4	2 20.3	4.8	0.8	10.8	1.7	16.8	2.7
49	2 27.3	2 27.7	2 20.5	4.9	0.8	10.9	1.7	16.9	2.7
50	2 27.5	2 27.9	2 20.8	5.0	0.8	11.0	1.7	17.0	2.7
51	2 27.8	2 28.2	2 21.0	5.1	0.8	11.1	1.8	17.1	2.7
52	2 28.0	2 28.4	2 21.3	5.2	0.8	11.2	1.8	17.2	2.7
53	2 28.3	2 28.7	2 21.5	5.3	0.8	11.3	1.8	17.3	2.7
54	2 28.5	2 28.9	2 21.7	5.4	0.9	11.4	1.8	17.4	2.8
55	2 28.8	2 29.2	2 22.0	5.5	0.9	11.5	1.8	17.5	2.8
56	2 29.0	2 29.4	2 22.2	5.6	0.9	11.6	1.8	17.6	2.8
57	2 29.3	2 29.7	2 22.5	5.7	0.9	11.7	1.9	17.7	2.8
58	2 29.5	2 29.9	2 22.7	5.8	0.9	11.8	1.9	17.8	2.8
59	2 29.8	2 30.2	2 22.9	5.9	0.9	11.9	1.9	17.9	2.8
60	2 30.0	2 30.4	2 23.2	6.0	1.0	12.0	1.9	18.0	2.9

©2009 Jack Case

2009 SEPTEMBER 28, 29, 30 (MON., TUES., WED.)

UT	SUN		MOON				Lat.	Twilight		Sunrise	Moonrise			
	GHA	Dec	GHA	v	Dec	d	HP		Naut.	Civil		28	29	30	1
d h	° '	° '	° '	'	° '	'	'	°	h m	h m	h m	h m	h m	h m	h m
28 00	182 18.5	S 1 59.4	70 23.5 12.8		S21 30.6	7.9	54.2	N 72	03 39	05 03	06 10	■■■	18 14	17 31	17 00
01	197 18.7	2 00.4	84 55.3 12.9		21 22.7	7.9	54.2	N 70	03 53	05 07	06 08	18 34	17 43	17 15	16 53
02	212 18.9	01.4	99 27.2 12.9		21 14.8	8.1	54.2	68	04 03	05 11	06 06	17 46	17 20	17 02	16 48
03	227 19.1	02.4	113 59.1 13.0		21 06.7	8.1	54.2	66	04 12	05 13	06 04	17 15	17 02	16 52	16 43
04	242 19.3	03.3	128 31.1 13.0		20 58.6	8.2	54.2	64	04 19	05 16	06 03	16 52	16 47	16 43	16 38
05	257 19.5	04.3	143 03.1 13.1		20 50.4	8.3	54.2	62	04 25	05 18	06 02	16 33	16 35	16 35	16 35
06	272 19.7	S 2 05.3	157 35.2 13.1		S20 42.1	8.4	54.2	60	04 30	05 19	06 01	16 18	16 24	16 28	16 31
07	287 20.0	06.3	172 07.3 13.2		20 33.7	8.5	54.2	N 58	04 35	05 21	06 00	16 05	16 15	16 22	16 29
08	302 20.2	07.2	186 39.5 13.2		20 25.2	8.6	54.2	56	04 38	05 22	05 59	15 53	16 07	16 17	16 26
M 09	317 20.4	08.2	201 11.7 13.3		20 16.6	8.7	54.2	54	04 42	05 23	05 58	15 43	15 59	16 12	16 24
O 10	332 20.6	09.2	215 44.0 13.3		20 07.9	8.8	54.2	52	04 45	05 24	05 58	15 35	15 53	16 08	16 22
N 11	347 20.8	10.1	230 16.3 13.4		19 59.1	8.8	54.2	50	04 47	05 25	05 57	15 27	15 47	16 04	16 20
D 12	2 21.0	S 2 11.1	244 48.7 13.5		S19 50.3	8.9	54.2	45	04 52	05 26	05 56	15 09	15 34	15 56	16 16
A 13	17 21.2	12.1	259 21.2 13.5		19 41.4	9.0	54.2	N 40	04 56	05 27	05 54	14 55	15 24	15 49	16 12
Y 14	32 21.4	13.1	273 53.7 13.5		19 32.4	9.1	54.2	35	04 59	05 28	05 53	14 43	15 14	15 42	16 09
15	47 21.6	14.0	288 26.2 13.6		19 23.3	9.2	54.2	30	05 01	05 28	05 52	14 33	15 06	15 37	16 06
16	62 21.8	15.0	302 58.8 13.6		19 14.1	9.3	54.2	20	05 03	05 28	05 50	14 15	14 52	15 28	16 02
17	77 22.0	16.0	317 31.4 13.7		19 04.8	9.3	54.2	N 10	05 03	05 28	05 49	13 59	14 40	15 19	15 57
18	92 22.3	S 2 16.9	332 04.1 13.8		S18 55.5	9.4	54.2	0	05 02	05 26	05 47	13 44	14 29	15 12	15 53
19	107 22.5	17.9	346 36.9 13.8		18 46.1	9.5	54.3	S 10	05 00	05 24	05 45	13 29	14 17	15 04	15 50
20	122 22.7	18.9	1 09.7 13.8		18 36.6	9.6	54.3	20	04 56	05 21	05 43	13 14	14 05	14 55	15 45
21	137 22.9	19.9	15 42.5 13.9		18 27.0	9.7	54.3	30	04 49	05 17	05 41	12 55	13 51	14 46	15 41
22	152 23.1	20.8	30 15.4 13.9		18 17.3	9.7	54.3	35	04 44	05 14	05 39	12 44	13 42	14 40	15 38
23	167 23.3	21.8	44 48.3 14.0		18 07.6	9.8	54.3	40	04 39	05 11	05 38	12 32	13 33	14 34	15 35
29 00	182 23.5	S 2 22.8	59 21.3 14.0		S17 57.8	9.9	54.3	45	04 32	05 06	05 36	12 17	13 22	14 26	15 31
01	197 23.7	23.8	73 54.3 14.1		17 47.9	10.0	54.3	S 50	04 22	05 01	05 33	12 00	13 08	14 17	15 27
02	212 23.9	24.7	88 27.4 14.1		17 37.9	10.0	54.3	52	04 18	04 58	05 32	11 51	13 02	14 13	15 25
03	227 24.1	25.7	103 00.5 14.2		17 27.9	10.1	54.3	54	04 13	04 56	05 31	11 42	12 55	14 09	15 23
04	242 24.3	26.7	117 33.7 14.2		17 17.8	10.2	54.3	56	04 07	04 52	05 30	11 31	12 47	14 04	15 20
05	257 24.5	27.6	132 06.9 14.3		17 07.6	10.3	54.3	58	04 01	04 49	05 28	11 18	12 38	13 58	15 17
06	272 24.8	S 2 28.6	146 40.2 14.3		S16 57.3	10.3	54.3	S 60	03 53	04 44	05 27	11 04	12 28	13 51	15 14

								Lat.	Sunset	Twilight		Moonset			
										Civil	Naut.	28	29	30	1
07	287 25.0	29.6	161 13.5 14.3		16 47.0	10.4	54.3								
08	302 25.2	30.6	175 46.8 14.4		16 36.6	10.5	54.3	°	h m	h m	h m	h m	h m	h m	h m
T 09	317 25.4	31.5	190 20.2 14.4		16 26.1	10.6	54.3	N 72	17 28	18 35	19 58	■■■	23 26	25 38	01 38
U 10	332 25.6	32.5	204 53.6 14.5		16 15.5	10.6	54.3	N 70	17 31	18 31	19 45	21 32	23 55	25 52	01 52
E 11	347 25.8	33.5	219 27.1 14.5		16 04.9	10.7	54.4	68	17 33	18 28	19 34	22 19	24 17	00 17	02 03
S 12	2 26.0	S 2 34.4	234 00.6 14.6		S15 54.2	10.7	54.4	66	17 34	18 25	19 26	22 49	24 33	00 33	02 12
D 13	17 26.2	35.4	248 34.1 14.6		15 43.5	10.9	54.4	64	17 36	18 23	19 19	23 12	24 47	00 47	02 19
A 14	32 26.4	36.4	263 07.7 14.6		15 32.6	10.9	54.4	62	17 37	18 21	19 13	23 29	24 58	00 58	02 26
Y 15	47 26.6	37.4	277 41.3 14.7		15 21.7	10.9	54.4	60	17 38	18 20	19 08	23 44	25 08	01 08	02 31
16	62 26.8	38.3	292 15.0 14.7		15 10.8	11.0	54.4	N 58	17 39	18 18	19 04	23 57	25 16	01 16	02 36
17	77 27.0	39.3	306 48.7 14.7		14 59.8	11.1	54.4	56	17 40	18 17	19 01	24 07	00 07	01 24	02 40
18	92 27.2	S 2 40.3	321 22.4 14.8		S14 48.7	11.2	54.4	54	17 41	18 16	18 57	24 17	00 17	01 30	02 44
19	107 27.4	41.2	335 56.2 14.8		14 37.5	11.2	54.4	52	17 42	18 15	18 55	24 25	00 25	01 36	02 48
20	122 27.6	42.2	350 30.0 14.8		14 26.3	11.3	54.4	50	17 43	18 15	18 52	24 32	00 32	01 42	02 51
21	137 27.9	43.2	5 03.8 14.9		14 15.0	11.3	54.4	45	17 44	18 13	18 47	24 48	00 48	01 53	02 57
22	152 28.1	44.1	19 37.7 14.9		14 03.7	11.4	54.5	N 40	17 46	18 13	18 44	00 01	01 01	02 02	03 03
23	167 28.3	45.1	34 11.6 14.9		13 52.3	11.5	54.5	35	17 47	18 12	18 41	00 14	01 12	02 10	03 08
30 00	182 28.5	S 2 46.1	48 45.5 15.0		S13 40.8	11.5	54.5	30	17 48	18 12	18 39	00 26	01 22	02 17	03 12
01	197 28.7	47.1	63 19.5 15.0		13 29.3	11.6	54.5	20	17 50	18 13	18 37	00 47	01 38	02 29	03 19
02	212 28.9	48.0	77 53.5 15.0		13 17.7	11.6	54.5	N 10	17 52	18 13	18 37	01 04	01 53	02 40	03 26
03	227 29.1	49.0	92 27.5 15.1		13 06.1	11.7	54.5	0	17 54	18 14	18 38	01 21	02 06	02 49	03 32
04	242 29.3	50.0	107 01.6 15.1		12 54.4	11.8	54.5	S 10	17 55	18 16	18 41	01 37	02 19	02 59	03 38
05	257 29.5	50.9	121 35.7 15.1		12 42.6	11.8	54.5	20	17 58	18 20	18 45	01 54	02 33	03 09	03 44
06	272 29.7	S 2 51.9	136 09.8 15.2		S12 30.8	11.9	54.6	30	18 00	18 24	18 52	02 14	02 49	03 21	03 51
W 07	287 29.9	52.9	150 44.0 15.2		12 18.9	11.9	54.6	35	18 02	18 27	18 57	02 25	02 58	03 28	03 55
E 08	302 30.1	53.9	165 18.1 15.2		12 07.0	12.0	54.6	40	18 04	18 31	19 03	02 39	03 09	03 35	03 59
D 09	317 30.3	54.8	179 52.3 15.3		11 55.1	12.1	54.6	45	18 06	18 35	19 10	02 54	03 21	03 44	04 05
N 10	332 30.5	55.8	194 26.6 15.2		11 43.0	12.1	54.6	S 50	18 08	18 41	19 19	03 13	03 36	03 55	04 11
E 11	347 30.7	56.8	209 00.8 15.3		11 30.9	12.1	54.6	52	18 09	18 43	19 24	03 22	03 43	03 59	04 14
S 12	2 30.9	S 2 57.7	223 35.1 15.3		S11 18.8	12.2	54.6	54	18 11	18 46	19 29	03 32	03 50	04 05	04 17
D 13	17 31.1	58.7	238 09.4 15.3		11 06.6	12.2	54.6	56	18 12	18 50	19 35	03 43	03 59	04 11	04 20
A 14	32 31.3	2 59.7	252 43.7 15.3		10 54.4	12.3	54.7	58	18 14	18 53	19 42	03 56	04 09	04 17	04 24
Y 15	47 31.5	3 00.6	267 18.0 15.4		10 42.1	12.3	54.7	S 60	18 15	18 58	19 50	04 11	04 19	04 25	04 28
16	62 31.7	01.6	281 52.4 15.4		10 29.8	12.4	54.7								
17	77 32.0	02.6	296 26.8 15.4		10 17.4	12.4	54.7			SUN			MOON		
18	92 32.2	S 3 03.6	311 01.2 15.4		S10 05.0	12.5	54.7	Day	Eqn. of Time		Mer.	Mer. Pass.		Age	Phase
19	107 32.4	04.5	325 35.6 15.4		9 52.5	12.5	54.7		00h	12h	Pass.	Upper	Lower		
20	122 32.6	05.5	340 10.0 15.5		9 40.0	12.6	54.7	d	m s	m s	h m	h m	h m	d %	
21	137 32.8	06.5	354 44.5 15.5		9 27.4	12.6	54.8	28	09 13	09 24	11 51	19 55	07 33	10 71	
22	152 33.0	07.4	9 19.0 15.4		9 14.8	12.7	54.8	29	09 34	09 44	11 50	20 39	08 17	11 79	
23	167 33.2	08.4	23 53.4 15.5		S 9 02.1	12.7	54.8	30	09 53	10 03	11 50	21 22	09 01	12 86	
	SD 16.0	d 1.0	SD 14.8		14.8		14.9								

INCREMENTS AND CORRECTIONS

18m

s	SUN PLANETS	ARIES	MOON	v or d	Corrn	v or d	Corrn	v or d	Corrn
00	4 30.0	4 30.7	4 17.7	0.0	0.0	6.0	1.9	12.0	3.7
01	4 30.3	4 31.0	4 17.9	0.1	0.0	6.1	1.9	12.1	3.7
02	4 30.5	4 31.2	4 18.2	0.2	0.1	6.2	1.9	12.2	3.8
03	4 30.8	4 31.5	4 18.4	0.3	0.1	6.3	1.9	12.3	3.8
04	4 31.0	4 31.7	4 18.7	0.4	0.1	6.4	2.0	12.4	3.8
05	4 31.3	4 32.0	4 18.9	0.5	0.2	6.5	2.0	12.5	3.9
06	4 31.5	4 32.2	4 19.1	0.6	0.2	6.6	2.0	12.6	3.9
07	4 31.8	4 32.5	4 19.4	0.7	0.2	6.7	2.1	12.7	3.9
08	4 32.0	4 32.7	4 19.6	0.8	0.2	6.8	2.1	12.8	3.9
09	4 32.3	4 33.0	4 19.8	0.9	0.3	6.9	2.1	12.9	4.0
10	4 32.5	4 33.2	4 20.1	1.0	0.3	7.0	2.2	13.0	4.0
11	4 32.8	4 33.5	4 20.3	1.1	0.3	7.1	2.2	13.1	4.0
12	4 33.0	4 33.7	4 20.6	1.2	0.4	7.2	2.2	13.2	4.1
13	4 33.3	4 34.0	4 20.8	1.3	0.4	7.3	2.3	13.3	4.1
14	4 33.5	4 34.2	4 21.0	1.4	0.4	7.4	2.3	13.4	4.1
15	4 33.8	4 34.5	4 21.3	1.5	0.5	7.5	2.3	13.5	4.2
16	4 34.0	4 34.8	4 21.5	1.6	0.5	7.6	2.3	13.6	4.2
17	4 34.3	4 35.0	4 21.8	1.7	0.5	7.7	2.4	13.7	4.2
18	4 34.5	4 35.3	4 22.0	1.8	0.6	7.8	2.4	13.8	4.3
19	4 34.8	4 35.5	4 22.2	1.9	0.6	7.9	2.4	13.9	4.3
20	4 35.0	4 35.8	4 22.5	2.0	0.6	8.0	2.5	14.0	4.3
21	4 35.3	4 36.0	4 22.7	2.1	0.6	8.1	2.5	14.1	4.3
22	4 35.5	4 36.3	4 22.9	2.2	0.7	8.2	2.5	14.2	4.4
23	4 35.8	4 36.5	4 23.2	2.3	0.7	8.3	2.6	14.3	4.4
24	4 36.0	4 36.8	4 23.4	2.4	0.7	8.4	2.6	14.4	4.4
25	4 36.3	4 37.0	4 23.7	2.5	0.8	8.5	2.6	14.5	4.5
26	4 36.5	4 37.3	4 23.9	2.6	0.8	8.6	2.7	14.6	4.5
27	4 36.8	4 37.5	4 24.1	2.7	0.8	8.7	2.7	14.7	4.5
28	4 37.0	4 37.8	4 24.4	2.8	0.9	8.8	2.7	14.8	4.6
29	4 37.3	4 38.0	4 24.6	2.9	0.9	8.9	2.7	14.9	4.6
30	4 37.5	4 38.3	4 24.9	3.0	0.9	9.0	2.8	15.0	4.6
31	4 37.8	4 38.5	4 25.1	3.1	1.0	9.1	2.8	15.1	4.7
32	4 38.0	4 38.8	4 25.3	3.2	1.0	9.2	2.8	15.2	4.7
33	4 38.3	4 39.0	4 25.6	3.3	1.0	9.3	2.9	15.3	4.7
34	4 38.5	4 39.3	4 25.8	3.4	1.0	9.4	2.9	15.4	4.7
35	4 38.8	4 39.5	4 26.1	3.5	1.1	9.5	2.9	15.5	4.8
36	4 39.0	4 39.8	4 26.3	3.6	1.1	9.6	3.0	15.6	4.8
37	4 39.3	4 40.0	4 26.5	3.7	1.1	9.7	3.0	15.7	4.8
38	4 39.5	4 40.3	4 26.8	3.8	1.2	9.8	3.0	15.8	4.9
39	4 39.8	4 40.5	4 27.0	3.9	1.2	9.9	3.1	15.9	4.9
40	4 40.0	4 40.8	4 27.2	4.0	1.2	10.0	3.1	16.0	4.9
41	4 40.3	4 41.0	4 27.5	4.1	1.3	10.1	3.1	16.1	5.0
42	4 40.5	4 41.3	4 27.7	4.2	1.3	10.2	3.1	16.2	5.0
43	4 40.8	4 41.5	4 28.0	4.3	1.3	10.3	3.2	16.3	5.0
44	4 41.0	4 41.8	4 28.2	4.4	1.4	10.4	3.2	16.4	5.1
45	4 41.3	4 42.0	4 28.4	4.5	1.4	10.5	3.2	16.5	5.1
46	4 41.5	4 42.3	4 28.7	4.6	1.4	10.6	3.3	16.6	5.1
47	4 41.8	4 42.5	4 28.9	4.7	1.4	10.7	3.3	16.7	5.1
48	4 42.0	4 42.8	4 29.2	4.8	1.5	10.8	3.3	16.8	5.2
49	4 42.3	4 43.0	4 29.4	4.9	1.5	10.9	3.4	16.9	5.2
50	4 42.5	4 43.3	4 29.6	5.0	1.5	11.0	3.4	17.0	5.2
51	4 42.8	4 43.5	4 29.9	5.1	1.6	11.1	3.4	17.1	5.3
52	4 43.0	4 43.8	4 30.1	5.2	1.6	11.2	3.5	17.2	5.3
53	4 43.3	4 44.0	4 30.3	5.3	1.6	11.3	3.5	17.3	5.3
54	4 43.5	4 44.3	4 30.6	5.4	1.7	11.4	3.5	17.4	5.4
55	4 43.8	4 44.5	4 30.8	5.5	1.7	11.5	3.5	17.5	5.4
56	4 44.0	4 44.8	4 31.1	5.6	1.7	11.6	3.6	17.6	5.4
57	4 44.3	4 45.0	4 31.3	5.7	1.8	11.7	3.6	17.7	5.5
58	4 44.5	4 45.3	4 31.5	5.8	1.8	11.8	3.6	17.8	5.5
59	4 44.8	4 45.5	4 31.8	5.9	1.8	11.9	3.7	17.9	5.5
60	4 45.0	4 45.8	4 32.0	6.0	1.9	12.0	3.7	18.0	5.6

19m

s	SUN PLANETS	ARIES	MOON	v or d	Corrn	v or d	Corrn	v or d	Corrn
00	4 45.0	4 45.8	4 32.0	0.0	0.0	6.0	2.0	12.0	3.9
01	4 45.3	4 46.0	4 32.3	0.1	0.0	6.1	2.0	12.1	3.9
02	4 45.5	4 46.3	4 32.5	0.2	0.1	6.2	2.0	12.2	4.0
03	4 45.8	4 46.5	4 32.7	0.3	0.1	6.3	2.0	12.3	4.0
04	4 46.0	4 46.8	4 33.0	0.4	0.1	6.4	2.1	12.4	4.0
05	4 46.3	4 47.0	4 33.2	0.5	0.2	6.5	2.1	12.5	4.1
06	4 46.5	4 47.3	4 33.4	0.6	0.2	6.6	2.1	12.6	4.1
07	4 46.8	4 47.5	4 33.7	0.7	0.2	6.7	2.2	12.7	4.1
08	4 47.0	4 47.8	4 33.9	0.8	0.3	6.8	2.2	12.8	4.2
09	4 47.3	4 48.0	4 34.2	0.9	0.3	6.9	2.2	12.9	4.2
10	4 47.5	4 48.3	4 34.4	1.0	0.3	7.0	2.3	13.0	4.2
11	4 47.8	4 48.5	4 34.6	1.1	0.4	7.1	2.3	13.1	4.3
12	4 48.0	4 48.8	4 34.9	1.2	0.4	7.2	2.3	13.2	4.3
13	4 48.3	4 49.0	4 35.1	1.3	0.4	7.3	2.4	13.3	4.3
14	4 48.5	4 49.3	4 35.4	1.4	0.5	7.4	2.4	13.4	4.4
15	4 48.8	4 49.5	4 35.6	1.5	0.5	7.5	2.4	13.5	4.4
16	4 49.0	4 49.8	4 35.8	1.6	0.5	7.6	2.5	13.6	4.4
17	4 49.3	4 50.0	4 36.1	1.7	0.6	7.7	2.5	13.7	4.5
18	4 49.5	4 50.3	4 36.3	1.8	0.6	7.8	2.5	13.8	4.5
19	4 49.8	4 50.5	4 36.6	1.9	0.6	7.9	2.6	13.9	4.5
20	4 50.0	4 50.8	4 36.8	2.0	0.7	8.0	2.6	14.0	4.6
21	4 50.3	4 51.0	4 37.0	2.1	0.7	8.1	2.6	14.1	4.6
22	4 50.5	4 51.3	4 37.3	2.2	0.7	8.2	2.7	14.2	4.6
23	4 50.8	4 51.5	4 37.5	2.3	0.7	8.3	2.7	14.3	4.6
24	4 51.0	4 51.8	4 37.7	2.4	0.8	8.4	2.7	14.4	4.7
25	4 51.3	4 52.0	4 38.0	2.5	0.8	8.5	2.8	14.5	4.7
26	4 51.5	4 52.3	4 38.2	2.6	0.8	8.6	2.8	14.6	4.7
27	4 51.8	4 52.5	4 38.5	2.7	0.9	8.7	2.8	14.7	4.8
28	4 52.0	4 52.8	4 38.7	2.8	0.9	8.8	2.9	14.8	4.8
29	4 52.3	4 53.1	4 38.9	2.9	0.9	8.9	2.9	14.9	4.8
30	4 52.5	4 53.3	4 39.2	3.0	1.0	9.0	2.9	15.0	4.9
31	4 52.8	4 53.6	4 39.4	3.1	1.0	9.1	3.0	15.1	4.9
32	4 53.0	4 53.8	4 39.7	3.2	1.0	9.2	3.0	15.2	4.9
33	4 53.3	4 54.1	4 39.9	3.3	1.1	9.3	3.0	15.3	5.0
34	4 53.5	4 54.3	4 40.1	3.4	1.1	9.4	3.1	15.4	5.0
35	4 53.8	4 54.6	4 40.4	3.5	1.1	9.5	3.1	15.5	5.0
36	4 54.0	4 54.8	4 40.6	3.6	1.2	9.6	3.1	15.6	5.1
37	4 54.3	4 55.1	4 40.8	3.7	1.2	9.7	3.2	15.7	5.1
38	4 54.5	4 55.3	4 41.1	3.8	1.2	9.8	3.2	15.8	5.1
39	4 54.8	4 55.6	4 41.3	3.9	1.3	9.9	3.2	15.9	5.2
40	4 55.0	4 55.8	4 41.6	4.0	1.3	10.0	3.3	16.0	5.2
41	4 55.3	4 56.1	4 41.8	4.1	1.3	10.1	3.3	16.1	5.2
42	4 55.5	4 56.3	4 42.0	4.2	1.4	10.2	3.3	16.2	5.3
43	4 55.8	4 56.6	4 42.3	4.3	1.4	10.3	3.3	16.3	5.3
44	4 56.0	4 56.8	4 42.5	4.4	1.4	10.4	3.4	16.4	5.3
45	4 56.3	4 57.1	4 42.8	4.5	1.5	10.5	3.4	16.5	5.4
46	4 56.5	4 57.3	4 43.0	4.6	1.5	10.6	3.4	16.6	5.4
47	4 56.8	4 57.6	4 43.2	4.7	1.5	10.7	3.5	16.7	5.4
48	4 57.0	4 57.8	4 43.5	4.8	1.6	10.8	3.5	16.8	5.5
49	4 57.3	4 58.1	4 43.7	4.9	1.6	10.9	3.5	16.9	5.5
50	4 57.5	4 58.3	4 43.9	5.0	1.6	11.0	3.6	17.0	5.5
51	4 57.8	4 58.6	4 44.2	5.1	1.7	11.1	3.6	17.1	5.6
52	4 58.0	4 58.8	4 44.4	5.2	1.7	11.2	3.6	17.2	5.6
53	4 58.3	4 59.1	4 44.7	5.3	1.7	11.3	3.7	17.3	5.6
54	4 58.5	4 59.3	4 44.9	5.4	1.8	11.4	3.7	17.4	5.7
55	4 58.8	4 59.6	4 45.1	5.5	1.8	11.5	3.7	17.5	5.7
56	4 59.0	4 59.8	4 45.4	5.6	1.8	11.6	3.8	17.6	5.7
57	4 59.3	5 00.1	4 45.6	5.7	1.9	11.7	3.8	17.7	5.8
58	4 59.5	5 00.3	4 45.9	5.8	1.9	11.8	3.8	17.8	5.8
59	4 59.8	5 00.6	4 46.1	5.9	1.9	11.9	3.9	17.9	5.8
60	5 00.0	5 00.8	4 46.3	6.0	2.0	12.0	3.9	18.0	5.9

A2 ALTITUDE CORRECTION TABLES 10°–90°—SUN, STARS, PLANETS

OCT.—MAR. SUN APR.—SEPT.						STARS AND PLANETS				DIP					
App. Alt.	Lower Limb	Upper Limb	App. Alt.	Lower Limb	Upper Limb	App. Alt.	Corrⁿ	App. Alt.	Additional Corrⁿ	Ht. of Eye	Corrⁿ	Ht. of Eye		Ht. of Eye	Corrⁿ
° ′	′	′	° ′	′	′	° ′	′		**2009**	m	′	ft.		m	′
9 33	+10·8	−21·5	9 39	+10·6	−21·2	9 55	−5·3		**VENUS**	2·4	−2·8	8·0		1·0	−1·8
9 45	+10·9	−21·4	9 50	+10·7	−21·1	10 07	−5·2		Jan. 1–Jan. 28	2·6	−2·9	8·6		1·5	−2·2
9 56	+11·0	−21·3	10 02	+10·8	−21·0	10 20	−5·1		May 23–July 10	2·8	−3·0	9·2		2·0	−2·5
10 08	+11·1	−21·2	10 14	+10·9	−20·9	10 32	−5·0			3·0	−3·1	9·8		2·5	−2·8
10 20	+11·2	−21·1	10 27	+11·0	−20·8	10 46	−4·9		° ′	3·2	−3·2	10·5		3·0	−3·0
10 33	+11·3	−21·0	10 40	+11·1	−20·7	10 59	−4·8		41 +0·2	3·4	−3·3	11·2		See table	
10 46	+11·4	−20·9	10 53	+11·2	−20·6	11 14	−4·7		76 +0·1	3·6	−3·4	11·9		←	
11 00	+11·5	−20·8	11 07	+11·3	−20·5	11 29	−4·6		Jan. 29–Feb. 21	3·8	−3·5	12·6			
11 15	+11·6	−20·7	11 22	+11·4	−20·4	11 44	−4·5		May 1–May 22	4·0	−3·6	13·3		m	′
11 30	+11·7	−20·6	11 37	+11·5	−20·3	12 00	−4·4			4·3	−3·7	14·1		20	−7·9
11 45	+11·8	−20·5	11 53	+11·6	−20·2	12 17	−4·3		° ′	4·5	−3·8	14·9		22	−8·3
12 01	+11·9	−20·4	12 10	+11·7	−20·1	12 35	−4·2		34 +0·3	4·7	−3·9	15·7		24	−8·6
12 18	+12·0	−20·3	12 27	+11·8	−20·0	12 53	−4·1		60 +0·2	5·0	−4·0	16·5		26	−9·0
12 36	+12·1	−20·2	12 45	+11·9	−19·9	13 12	−4·0		80 +0·1	5·2	−4·1	17·4		28	−9·3
12 54	+12·2	−20·1	13 04	+12·0	−19·8	13 32	−3·9		Feb. 22–Mar. 9	5·5	−4·2	18·3		30	−9·6
13 14	+12·3	−20·0	13 24	+12·1	−19·7	13 53	−3·8		Apr. 15–Apr. 30	5·8	−4·3	19·1		32	−10·0
13 34	+12·4	−19·9	13 44	+12·2	−19·6	14 16	−3·7		° ′	6·1	−4·4	20·1		34	−10·3
13 55	+12·5	−19·8	14 06	+12·3	−19·5	14 39	−3·6		29 +0·4	6·3	−4·5	21·0		36	−10·6
14 17	+12·6	−19·7	14 29	+12·4	−19·4	15 03	−3·5		51 +0·3	6·6	−4·6	22·0		38	−10·8
14 41	+12·7	−19·6	14 53	+12·5	−19·3	15 29	−3·4		68 +0·2	6·9	−4·7	22·9			
15 05	+12·8	−19·5	15 18	+12·6	−19·2	15 56	−3·3		83 +0·1	7·2	−4·8	23·9		40	−11·1
15 31	+12·9	−19·4	15 45	+12·7	−19·1	16 25	−3·2		Mar. 10–Apr. 14	7·5	−4·9	24·9		42	−11·4
15 59	+13·0	−19·3	16 13	+12·8	−19·0	16 55	−3·1		° ′	7·9	−5·0	26·0		44	−11·7
16 27	+13·1	−19·2	16 43	+12·9	−18·9	17 27	−3·0		26 +0·5	8·2	−5·1	27·1		46	−11·9
16 58	+13·2	−19·1	17 14	+13·0	−18·8	18 01	−2·9		46 +0·4	8·5	−5·2	28·1		48	−12·2
17 30	+13·3	−19·0	17 47	+13·1	−18·7	18 37	−2·8		60 +0·3	8·8	−5·3	29·2		ft.	
18 05	+13·4	−18·9	18 23	+13·2	−18·6	19 16	−2·7		73 +0·2	9·2	−5·4	30·4		2	−1·4
18 41	+13·5	−18·8	19 00	+13·3	−18·5	19 56	−2·6		84 +0·1	9·5	−5·5	31·5		4	−1·9
19 20	+13·6	−18·7	19 41	+13·4	−18·4	20 40	−2·5		July 11–Dec. 31	9·9	−5·6	32·7		6	−2·4
20 02	+13·7	−18·6	20 24	+13·5	−18·3	21 27	−2·4		° ′	10·3	−5·7	33·9		8	−2·7
20 46	+13·8	−18·5	21 10	+13·6	−18·2	22 17	−2·3		60 +0·1	10·6	−5·8	35·1		10	−3·1
21 34	+13·9	−18·4	21 59	+13·7	−18·1	23 11	−2·2		**MARS**	11·0	−5·9	36·3		See table	
22 25	+14·0	−18·3	22 52	+13·8	−18·0	24 09	−2·1		Jan. 1–Nov. 27	11·4	−6·0	37·6		←	
23 20	+14·1	−18·2	23 49	+13·9	−17·9	25 12	−2·0		° ′	11·8	−6·1	38·9			
24 20	+14·2	−18·1	24 51	+14·0	−17·8	26 20	−1·9		60 +0·1	12·2	−6·2	40·1		ft.	′
25 24	+14·3	−18·0	25 58	+14·1	−17·7	27 34	−1·8		Nov. 28–Dec. 31	12·6	−6·3	41·5		70	−8·1
26 34	+14·4	−17·9	27 11	+14·2	−17·6	28 54	−1·7		° ′	13·0	−6·4	42·8		75	−8·4
27 50	+14·5	−17·8	28 31	+14·3	−17·5	30 22	−1·6		41 +0·2	13·4	−6·5	44·2		80	−8·7
29 13	+14·6	−17·7	29 58	+14·4	−17·4	31 58	−1·5		76 +0·1	13·8	−6·6	45·5		85	−8·9
30 44	+14·7	−17·6	31 33	+14·5	−17·3	33 43	−1·4			14·2	−6·7	46·9		90	−9·2
32 24	+14·8	−17·5	33 18	+14·6	−17·2	35 38	−1·3			14·7	−6·8	48·4		95	−9·5
34 15	+14·9	−17·4	35 15	+14·7	−17·1	37 45	−1·2			15·1	−6·9	49·8			
36 17	+15·0	−17·3	37 24	+14·8	−17·0	40 06	−1·1			15·5	−7·0	51·3		100	−9·7
38 34	+15·1	−17·2	39 48	+14·9	−16·9	42 42	−1·0			16·0	−7·1	52·8		105	−9·9
41 06	+15·2	−17·1	42 28	+15·0	−16·8	45 34	−0·9			16·5	−7·2	54·3		110	−10·2
43 56	+15·3	−17·0	45 29	+15·1	−16·7	48 45	−0·8			16·9	−7·3	55·8		115	−10·4
47 07	+15·4	−16·9	48 52	+15·2	−16·6	52 16	−0·7			17·4	−7·4	57·4		120	−10·6
50 43	+15·5	−16·8	52 41	+15·3	−16·5	56 09	−0·6			17·9	−7·5	58·9		125	−10·8
54 46	+15·6	−16·7	56 59	+15·4	−16·4	60 26	−0·5			18·4	−7·6	60·5			
59 21	+15·7	−16·6	61 50	+15·5	−16·3	65 06	−0·4			18·8	−7·7	62·1		130	−11·1
64 28	+15·8	−16·5	67 15	+15·6	−16·2	70 09	−0·3			19·3	−7·8	63·8		135	−11·3
70 10	+15·9	−16·4	73 14	+15·7	−16·1	75 32	−0·2			19·8	−7·9	65·4		140	−11·5
76 24	+16·0	−16·3	79 42	+15·8	−16·0	81 12	−0·1			20·4	−8·0	67·1		145	−11·7
83 05	+16·1	−16·2	86 31	+15·9	−15·9	87 03	0·0			20·9	−8·1	68·8		150	−11·9
90 00			90 00			90 00				21·4		70·5		155	−12·1

©2009 Jack Case

Self Test

Notes.
1. Before working through these questions, it is recommended that you revise 'Short Distance Sailing' in chapter 3.
2. Since refraction is negligible when the Sun is at its zenith, additional altitude correction for non standard conditions is not necessary when calculating the true altitude at meridian passage.
3. Nautical Almanac extracts relevant to these questions are to be found following the solution to question 3.

Question 1.
Date: 20 Aug. 2009.
Zone time: 1015 (zone +2)
D.R. Position: 42° 45'.6 N. 27° 01'.25W
Course 015°, speed 29 knots
Alt. Sun's L.L. at Mer. Pas: 58° 30'.15.
Height of eye = 25 ft.
Index error = -1'.7
Tasks:
a. Calculate D.R. position at Mer. Pas.
b. Calculate the latitude from the Mer. Alt.

Question 2.
At zone time 0900(-11) on 10 Sept. 2009, the approximate position of a yacht taking part in a round the world race is 36°32'.8S, 160° 25'.2E. The estimated course and speed made good for the next 4 hours was 085°, 11.5 knots. At Mer. Pas. the altitude of the Sun's lower limb was measured and found to be 48° 56'.52. Height of eye: 2.7m. Index error: +2'.4.
Tasks:
a. Calculate D.R. position at Mer. Pas.
b. Calculate the latitude from the Mer. Alt.

Question 3.
On 3 January 2009, a destroyer's D.R. position at zone time 0945(+2) is 4° 45'.7S. 30° 25'.8W. Course and speed for the next 4 hours was 175°, 32 knots.
The altitude of the Sun's lower limb at Mer. Pas was found to be 72° 15'.24. Height of eye: 18.6m. Index error: -0'.58.
Tasks:
a. Calculate D.R. position at Mer. Pas.
b. Calculate the latitude from the Mer. Alt

Solutions To Test Questions

Question 1.

Step 1. Determine Time of Mer. Pas. at Greenwich for 20 Aug.
From the Nautical Almanac,
Mer. Pas. at Greenwich = 1203 GMT.

Step 2. Estimate Time of Mer.Pas. at first D.R. Pos.
- Convert Long. to time.
Long. 27° 01'.25W

	h	m	s
$4 \times 27° \div 60 = 1.8^h$ =	1	48	00
$4 \times 1'.25 \div 60 = 0.08^m$ =		0	05
	1	48	05
≈	1	48	

- Estimate zone time of Mer.Pas.

	h	m
Mer. Pas. Greenwich	12	03
Long.	+ 1	48
Local Mer. Pas (GMT)	13	51
Zone (+2)	- 2	
Zone time Mer. Pas.	11	51

Step 3. Calculate D.R. Pos. at time of Estimated Mer. Pas.
Course = 015°   Speed = 29 knots.
Estimated time of Mer. Pas = 1151(+2)
Time of firs D.R. Pos: 1015(+2)
Distance run in $1^h 36^m$ = 29 × 1.6 = 46.4 n.m.
Dep.   = Dist × Sin(course)
          = 46.4 × Sin(15) = 12'.01E
D.Lat.   = Dist × Cos(course)
          = 46.4 × Cos(15) = 44'.8N
New Lat. = 42° 45'.6N. + 44'.8 = 43° 30'.4 N.
Mid. Lat = 42° 45'.6N + 22'.4   = 43° 08'.0N.
D.Long.   = Dep. × Sec(M.Lat)
          = 12'.01 × Sec(43.133) = 16'.46E.
New Long. = 27° 01'.25W - 16'.46 = 26° 44'.79W
∴ New Pos. = 43° 30'.4 N. 26° 44'.79W

Step 4. Calculate time of Mer. Pas. at 2nd. DR Pos.

	h	m
Mer. Pas. Greenwich:	12	03 (GMT)
Long (26° 44'.79W):	1	47
Local Mer. Pas (GMT)	13	50
Zone (+2)	- 02	
Zone time Mer. Pas.	11	50

Step 5. Determine Declination at Time of Local Mer. Pas.

Local Mer. Pas (GMT) $13^h\ 50^m$
Dec Sun (13^h) (from Almanac) N12° 17'.4
Corr^n (50^m) (d = 0'.8)   − 0'.7
Dec Sun ($13^h\ 50^m$)   N12° 16'.7

Step 6. Calculate Meridian Altitude.

Sext. Alt.	58° 30'.15
I.E.	− 1'.70
Observed Alt.	58° 28'.45
Dip	− 4'.90
Apparent Alt.	58° 23'.55
Sun's Corr.	+15'.40
True Alt.	58° 38'.95

Step 7. Determine Latitude.

LAT = DEC + (90° − ALT) (rule i)
    = 12° 16'.7 + (90° − 58° 38'.95)
    = 12° 16'.7 + 31° 21'.05
    = 43° 37'.75 N.

Question 2

Step 1. Determine Time of Mer. Pas. at Greenwich for 10 Sept.
From the Nautical Almanac,
Mer. Pas. at Greenwich = 1157 GMT.

Step 2. Estimate Time of Mer.Pas. at first D.R. Pos.

- Convert Long. to time.

Long. = 160° 25'.2E

	h	m	s
4 × 160° ÷ 60 = 10.66^h =	10	39	36
4 × 25'.2 ÷ 60 = 1.68^m =		01	40.8
	10	41	16.8
≈	10	41	

- Estimate zone time of Mer.Pas.

	h	m
Mer. Pas. Greenwich	11	57
Long. (long east GMT least)	−10	41
Local Mer. Pas (GMT)	01	16
Zone (−11)	+11	
Zone time Mer. Pas.	12	16

Step 3. Calculate D.R.Pos. at time of Estimated Mer. Pas.

Course = 085° Speed = 11.5 knots.
Estimated time of Mer. Pas = 1216(−11)
Time of firs D.R. Pos: 0900(−11)
Distance run in 3^h 16^m = 11.5 × 3.27 = 37.6 n.m.

Dep.   = Dist × Sin(course)
     = 37.6 × Sin(85)   = 37'.46E.

D.Lat.   = Dist × Cos(course)
     = 37.6 × Cos(85)   = 3'.28N.

New Lat.   = 36° 32'.8S − 3'.28 = 36° 29'.52S.

Mid. Lat   = 36° 32'.8S. − 1'.64
     = 36° 31'.16S. = 36°.52S.

D.Long.   = Dep. × Sec(M.Lat)
     = 37'.46 × Sec(36.52) = 46'.61E.

New Long. = 160° 25'.2E. + 46'.61 = 161° 11'.81E.

∴ New Pos. = 36° 29'.52S. 161° 11'.81E.

Step 4. Calculate time of Mer. Pas. at Second D.R. Pos.

	h	m
Mer. Pas. Greenwich:	11	57 (GMT)
Long (161° 11'.81E):	-10	44
Local Mer. Pas (GMT)	01	13
Zone (-11)	+ 11	
Zone time Mer. Pas.	12	13

Step 5. Determine Declination at Time of Local Mer. Pas.

Local Mer. Pas (GMT) $01^h\ 13^m$
Dec Sun (01^h) (from Almanac)   N4° 57'.1
Corr^n (13^m) (d = 1'.0)   - 0'.2
Dec Sun ($01^h\ 13^m$)   N4° 56'.9

Step 6. Calculate Meridian Altitude.

Sext. Alt.	48° 56'.52
I.E.	+ 2'.40
Observed Alt.	48° 58'.92
Dip	- 2'.90
Apparent Alt.	48° 56'.02
Sun's Corr.	+ 15'.20
True Alt.	49° 11'.22

Step 7. Determine Latitude at Mer. Pas.

LAT = (90° - ALT) - DEC (rule iii)
     = (90° - 49° 11'.22) - 4° 56'.9
     = 40° 48'.78 - 4° 56'.9
     = 35° 51'.88S.

Question 3

> **Step 1.** Determine Time of Mer. Pas. at Greenwich for 3 Jan.
> From the Nautical Almanac,
> Mer. Pas. at Greenwich = 1205 GMT.

> **Step 2.** Estimate Time of Mer.Pas. at first D.R. Pos.
> - Convert Long. to time.
> Long. = 30° 25'.8W.
>
> | | h | m | s |
> |---|---|---|---|
> | 4 × 30° ÷ 60 = | 2 | 00 | 00 |
> | 4 × 25'.8 ÷ 60 = 1.72^m = | | 01 | 43.2 |
> | | 2 | 01 | 43.2 |
> | | ≈ 2 | 02 | |
>
> - Estimate zone time of Mer.Pas.
>
> | | h | m |
> |---|---|---|
> | Mer. Pas. Greenwich | 12 | 05 |
> | Long. | + 2 | 02 |
> | Local Mer. Pas (GMT) | 14 | 07 |
> | Zone (+2) | −2 | |
> | Zone time Mer. Pas. | 12 | 07 |

> **Step 3.** Calculate D.R. Pos. at time of Estimated Mer. Pas.
> Course = 175° Speed = 32 knots.
> Estimated time of Mer. Pas = 1207
> Distance run in 2^h 22^m = 32 × 2.37 = 75.84 n.m.
> Dep.   = Dist × Sin(course)
>   = 75.84 × Sin(175)   = 6'.6E.
> D.Lat.   = Dist × Cos(course)
>   = 75.84 × Cos(175)   = 75'.55S.
> New Lat. = 4° 45'.7S. + 75'.55 = 6° 01'.25S.
> Mid. Lat = 4° 45'.7S. + 37'.8 = 5° 23'.5'S.
> D.Long.   = Dep. × Sec(M.Lat)
>   = 6'.6 × Sec(5.39) = 6'.63E.
> New Long. = 30° 25'.8W. − 6'.63
>   = 30° 19'.17W.
> ∴ New Pos. = 6° 01'.25S. 30° 19'.17W.

©2009 Jack Case

Step 4. Calculate time of Mer. Pas. at Second D.R. Pos.

	h	m
Mer. Pas. Greenwich:	12	05 (GMT)
Long (30° 19'.17W):	+2	01
Local Mer. Pas (GMT)	14	06
Zone (+2)	-02	
Zone time Mer. Pas.	12	06

Step 5. Determine Declination at Time of Local Mer. Pas.

Local Mer. Pas (GMT) $14^h\ 06^m$
Dec Sun (14^h) (from Almanac)   S22° 46'.3
Corr^n (06^m) (d = 0'.2)   0'.0
Dec Sun ($14^h\ 06^m$)   S22° 46'.3

Step 6. Calculate Meridian Altitude.

Sext. Alt.	72° 15'.24
I.E.	-0'.58
Observed Alt.	72° 14'.66
Dip	-7'.60
Apparent Alt.	72° 07'.06
Sun's Corr.	+15'.90
True Alt.	72° 22'.96

Step 7. Determine Latitude.

LAT = DEC - (90° - ALT) (rule ii)
   = S22° 46'.3 - (90° - 72° 22'.96)
   = S22° 46'.3 - 17° 37'.04
   = 5° 09'.26 S.

Note. Nautical Almanac extracts relevant to the above questions begin on the next page.

2009 AUGUST 20, 21, 22 (THURS., FRI., SAT.)

UT	SUN GHA	SUN Dec	MOON GHA	MOON v	MOON Dec	MOON d	MOON HP	Lat.	Twilight Naut.	Twilight Civil	Sunrise	Moonrise 20	Moonrise 21	Moonrise 22	Moonrise 23
d h	° '	° '	° '	'	° '	'	'	°	h m	h m	h m	h m	h m	h m	h m
20 00	179 08.1	N12 28.2	185 25.6	8.2	N12 36.6	14.9	60.9	N 72	////	////	03 03	03 08	05 43	08 05	10 26
01	194 08.2	27.4	199 52.8	8.2	12 21.7	15.1	60.9	N 70	////	01 45	03 26	03 29	05 50	08 01	10 11
02	209 08.4	26.5	214 20.0	8.4	12 06.6	15.1	60.9	68	////	02 24	03 44	03 46	05 56	07 58	09 59
03	224 08.5	.. 25.7	228 47.4	8.4	11 51.5	15.1	60.8	66	////	02 51	03 59	03 59	06 01	07 56	09 49
04	239 08.7	24.9	243 14.8	8.5	11 36.4	15.3	60.8	64	01 32	03 11	04 10	04 10	06 04	07 54	09 41
05	254 08.8	24.1	257 42.3	8.6	11 21.1	15.3	60.8	62	02 08	03 27	04 20	04 19	06 08	07 52	09 34
06	269 08.9	N12 23.2	272 09.9	8.6	N11 05.8	15.3	60.8	60	02 32	03 40	04 29	04 27	06 11	07 51	09 28
T 07	284 09.1	22.4	286 37.5	8.7	10 50.5	15.5	60.8	N 58	02 51	03 51	04 36	04 34	06 13	07 49	09 23
H 08	299 09.2	21.6	301 05.2	8.8	10 35.0	15.4	60.8	56	03 06	04 01	04 43	04 40	06 16	07 48	09 18
U 09	314 09.4	.. 20.8	315 33.0	8.9	10 19.6	15.6	60.8	54	03 19	04 09	04 49	04 45	06 18	07 47	09 14
R 10	329 09.5	19.9	330 00.9	8.9	10 04.0	15.5	60.8	52	03 30	04 17	04 54	04 50	06 20	07 46	09 10
S 11	344 09.7	19.1	344 28.8	9.1	9 48.5	15.7	60.7	50	03 40	04 24	04 59	04 55	06 21	07 45	09 07
D 12	359 09.8	N12 18.3	358 56.9	9.0	N 9 32.8	15.6	60.7	45	03 59	04 38	05 09	05 04	06 25	07 43	09 00
A 13	14 10.0	17.4	13 24.9	9.2	9 17.2	15.8	60.7	N 40	04 14	04 49	05 17	05 12	06 28	07 42	08 53
Y 14	29 10.1	16.6	27 53.1	9.2	9 01.4	15.7	60.7	35	04 27	04 58	05 25	05 19	06 31	07 40	08 48
15	44 10.3	.. 15.8	42 21.3	9.3	8 45.7	15.8	60.7	30	04 37	05 06	05 31	05 25	06 33	07 39	08 44
16	59 10.4	15.0	56 49.6	9.4	8 29.9	15.9	60.7	20	04 52	05 19	05 42	05 35	06 37	07 37	08 36
17	74 10.6	14.1	71 18.0	9.4	8 14.0	15.9	60.6	N 10	05 04	05 30	05 51	05 43	06 41	07 36	08 29
18	89 10.7	N12 13.3	85 46.4	9.5	N 7 58.1	15.9	60.6	0	05 14	05 39	06 00	05 52	06 44	07 34	08 23
19	104 10.9	12.5	100 14.9	9.5	7 42.2	16.0	60.6	S 10	05 22	05 47	06 08	06 00	06 47	07 33	08 17
20	119 11.1	11.6	114 43.4	9.6	7 26.2	15.9	60.6	20	05 29	05 55	06 17	06 09	06 51	07 31	08 10
21	134 11.2	.. 10.8	129 12.0	9.7	7 10.3	16.1	60.6	30	05 35	06 03	06 28	06 19	06 55	07 29	08 03
22	149 11.4	10.0	143 40.7	9.8	6 54.2	16.0	60.6	35	05 38	06 08	06 33	06 24	06 57	07 28	07 58
23	164 11.5	09.1	158 09.5	9.8	6 38.2	16.1	60.5	40	05 41	06 12	06 40	06 31	07 00	07 27	07 54
21 00	179 11.7	N12 08.3	172 38.3	9.8	N 6 22.1	16.1	60.5	45	05 43	06 17	06 48	06 38	07 03	07 26	07 48
01	194 11.8	07.5	187 07.1	10.0	6 06.0	16.1	60.5	S 50	05 46	06 23	06 57	06 47	07 06	07 24	07 42
02	209 12.0	06.7	201 36.1	9.9	5 49.9	16.1	60.5	52	05 47	06 26	07 01	06 51	07 08	07 23	07 39
03	224 12.1	.. 05.8	216 05.0	10.1	5 33.8	16.2	60.4	54	05 48	06 29	07 05	06 56	07 10	07 23	07 35
04	239 12.3	05.0	230 34.1	10.1	5 17.6	16.2	60.4	56	05 49	06 32	07 10	07 01	07 12	07 22	07 32
05	254 12.4	04.2	245 03.2	10.1	5 01.4	16.2	60.4	58	05 50	06 35	07 16	07 06	07 14	07 21	07 28
06	269 12.6	N12 03.3	259 32.3	10.2	N 4 45.2	16.2	60.4	S 60	05 51	06 39	07 23	07 12	07 16	07 20	07 23
07	284 12.7	02.5	274 01.5	10.2	4 29.0	16.2	60.3	Lat.	Sunset	Twilight Civil	Twilight Naut.	Moonset 20	Moonset 21	Moonset 22	Moonset 23
08	299 12.9	01.7	288 30.7	10.3	4 12.8	16.2	60.3								
F 09	314 13.0	.. 00.8	303 00.0	10.4	3 56.6	16.2	60.3	°	h m	h m	h m	h m	h m	h m	h m
R 10	329 13.2	12 00.0	317 29.4	10.4	3 40.4	16.3	60.3	N 72	20 59	23 29	////	19 54	19 21	18 50	18 14
I 11	344 13.3	11 59.1	331 58.8	10.4	3 24.1	16.2	60.2	N 70	20 36	22 13	////	19 43	19 20	18 57	18 32
D 12	359 13.5	N11 58.3	346 28.2	10.5	N 3 07.9	16.3	60.2	68	20 19	21 37	////	19 35	19 19	19 03	18 46
A 13	14 13.7	57.5	0 57.7	10.6	2 51.6	16.2	60.2	66	20 05	21 12	23 34	19 28	19 18	19 08	18 57
Y 14	29 13.8	56.6	15 27.3	10.6	2 35.4	16.3	60.2	64	19 54	20 52	22 27	19 22	19 17	19 12	19 07
15	44 14.0	.. 55.8	29 56.9	10.6	2 19.1	16.2	60.1	62	19 44	20 37	21 54	19 17	19 16	19 16	19 16
16	59 14.1	55.0	44 26.5	10.7	2 02.9	16.2	60.1	60	19 36	20 24	21 31	19 12	19 16	19 19	19 23
17	74 14.3	54.1	58 56.2	10.7	1 46.7	16.3	60.1	N 58	19 28	20 13	21 13	19 08	19 15	19 22	19 29
18	89 14.4	N11 53.3	73 25.9	10.7	N 1 30.4	16.2	60.0	56	19 22	20 04	20 58	19 04	19 15	19 24	19 35
19	104 14.6	52.5	87 55.6	10.8	1 14.2	16.2	60.0	54	19 16	19 55	20 45	19 01	19 14	19 27	19 40
20	119 14.7	51.6	102 25.4	10.9	0 58.0	16.2	60.0	52	19 11	19 48	20 34	18 58	19 14	19 29	19 45
21	134 14.9	.. 50.8	116 55.3	10.8	0 41.8	16.2	60.0	50	19 07	19 41	20 25	18 55	19 13	19 31	19 49
22	149 15.1	49.9	131 25.1	11.0	0 25.6	16.2	59.9	45	18 57	19 28	20 06	18 49	19 13	19 35	19 58
23	164 15.2	49.1	145 55.1	10.9	N 0 09.4	16.1	59.9	N 40	18 48	19 17	19 51	18 44	19 12	19 39	20 06
22 00	179 15.4	N11 48.3	160 25.0	11.0	S 0 06.7	16.1	59.9	35	18 41	19 07	19 39	18 41	19 11	19 42	20 13
01	194 15.5	47.4	174 55.0	11.0	0 22.8	16.3	59.8	30	18 35	19 00	19 29	18 36	19 11	19 44	20 18
02	209 15.7	46.6	189 25.0	11.0	0 39.0	16.1	59.8	20	18 24	18 47	19 13	18 29	19 10	19 49	20 29
03	224 15.8	.. 45.7	203 55.0	11.1	0 55.1	16.0	59.8	N 10	18 15	18 37	19 02	18 23	19 09	19 53	20 38
04	239 16.0	44.9	218 25.1	11.1	1 11.1	16.1	59.7	0	18 06	18 28	18 52	18 17	19 08	19 57	20 46
05	254 16.2	44.1	232 55.2	11.1	1 27.2	16.0	59.7	S 10	17 58	18 19	18 44	18 11	19 07	20 01	20 55
06	269 16.3	N11 43.2	247 25.3	11.2	S 1 43.2	16.0	59.7	20	17 49	18 12	18 37	18 05	19 06	20 05	21 04
07	284 16.5	42.4	261 55.5	11.2	1 59.2	16.0	59.6	30	17 39	18 03	18 31	17 57	19 05	20 10	21 14
S 08	299 16.6	41.5	276 25.7	11.2	2 15.2	15.9	59.6	35	17 33	17 59	18 29	17 53	19 04	20 13	21 20
A 09	314 16.8	.. 40.7	290 55.9	11.2	2 31.1	15.9	59.6	40	17 27	17 55	18 26	17 48	19 03	20 16	21 27
T 10	329 17.0	39.8	305 26.1	11.3	2 47.0	15.9	59.5	45	17 19	17 49	18 24	17 43	19 03	20 20	21 35
U 11	344 17.1	39.0	319 56.4	11.3	3 02.9	15.8	59.5	S 50	17 10	17 44	18 21	17 36	19 01	20 24	21 45
R 12	359 17.3	N11 38.2	334 26.7	11.3	S 3 18.7	15.8	59.5	52	17 06	17 41	18 20	17 33	19 01	20 26	21 50
D 13	14 17.4	37.3	348 57.0	11.3	3 34.5	15.8	59.4	54	17 02	17 38	18 20	17 29	19 00	20 28	21 55
A 14	29 17.6	36.5	3 27.3	11.3	3 50.3	15.7	59.4	56	16 57	17 35	18 19	17 26	19 00	20 31	22 00
Y 15	44 17.8	.. 35.6	17 57.6	11.4	4 06.0	15.7	59.4	58	16 51	17 32	18 18	17 21	18 59	20 34	22 06
16	59 17.9	34.8	32 28.0	11.4	4 21.7	15.6	59.3	S 60	16 45	17 28	18 17	17 16	18 58	20 37	22 13
17	74 18.1	33.9	46 58.4	11.4	4 37.3	15.6	59.3		SUN			MOON			
18	89 18.2	N11 33.1	61 28.8	11.4	S 4 52.9	15.5	59.3	Day	Eqn. of Time 00h	Eqn. of Time 12h	Mer. Pass.	Mer. Pass. Upper	Mer. Pass. Lower	Age	Phase
19	104 18.4	32.2	75 59.2	11.4	5 08.4	15.5	59.2								
20	119 18.6	31.4	90 29.6	11.5	5 23.9	15.5	59.2								
21	134 18.7	.. 30.6	105 00.1	11.4	5 39.4	15.4	59.2	d	m s	m s	h m	h m	h m	d	%
22	149 18.9	29.7	119 30.5	11.5	5 54.8	15.4	59.1	20	03 28	03 21	12 03	12 04	24 31	00	0
23	164 19.0	28.9	134 01.0	11.5	S 6 10.2	15.3	59.1	21	03 14	03 06	12 03	12 56	00 31	01	2
	SD 15.8	d 0.8	SD 16.5		16.4		16.2	22	02 59	02 51	12 03	13 46	01 21	02	6

INCREMENTS AND CORRECTIONS

50m

s	SUN PLANETS	ARIES	MOON	v or d	Corrⁿ	v or d	Corrⁿ	v or d	Corrⁿ
00	12 30.0	12 32.1	11 55.8	0.0	0.0	6.0	5.1	12.0	10.1
01	12 30.3	12 32.3	11 56.1	0.1	0.1	6.1	5.1	12.1	10.2
02	12 30.5	12 32.6	11 56.3	0.2	0.2	6.2	5.2	12.2	10.3
03	12 30.8	12 32.8	11 56.5	0.3	0.3	6.3	5.3	12.3	10.4
04	12 31.0	12 33.1	11 56.8	0.4	0.3	6.4	5.4	12.4	10.4
05	12 31.3	12 33.3	11 57.0	0.5	0.4	6.5	5.5	12.5	10.5
06	12 31.5	12 33.6	11 57.3	0.6	0.5	6.6	5.6	12.6	10.6
07	12 31.8	12 33.8	11 57.5	0.7	0.6	6.7	5.6	12.7	10.7
08	12 32.0	12 34.1	11 57.7	0.8	0.7	6.8	5.7	12.8	10.8
09	12 32.3	12 34.3	11 58.0	0.9	0.8	6.9	5.8	12.9	10.9
10	12 32.5	12 34.6	11 58.2	1.0	0.8	7.0	5.9	13.0	10.9
11	12 32.8	12 34.8	11 58.5	1.1	0.9	7.1	6.0	13.1	11.0
12	12 33.0	12 35.1	11 58.7	1.2	1.0	7.2	6.1	13.2	11.1
13	12 33.3	12 35.3	11 58.9	1.3	1.1	7.3	6.1	13.3	11.2
14	12 33.5	12 35.6	11 59.2	1.4	1.2	7.4	6.2	13.4	11.3
15	12 33.8	12 35.8	11 59.4	1.5	1.3	7.5	6.3	13.5	11.4
16	12 34.0	12 36.1	11 59.7	1.6	1.3	7.6	6.4	13.6	11.4
17	12 34.3	12 36.3	11 59.9	1.7	1.4	7.7	6.5	13.7	11.5
18	12 34.5	12 36.6	12 00.1	1.8	1.5	7.8	6.6	13.8	11.6
19	12 34.8	12 36.8	12 00.4	1.9	1.6	7.9	6.6	13.9	11.7
20	12 35.0	12 37.1	12 00.6	2.0	1.7	8.0	6.7	14.0	11.8
21	12 35.3	12 37.3	12 00.8	2.1	1.8	8.1	6.8	14.1	11.9
22	12 35.5	12 37.6	12 01.1	2.2	1.9	8.2	6.9	14.2	12.0
23	12 35.8	12 37.8	12 01.3	2.3	1.9	8.3	7.0	14.3	12.0
24	12 36.0	12 38.1	12 01.6	2.4	2.0	8.4	7.1	14.4	12.1
25	12 36.3	12 38.3	12 01.8	2.5	2.1	8.5	7.2	14.5	12.2
26	12 36.5	12 38.6	12 02.0	2.6	2.2	8.6	7.2	14.6	12.3
27	12 36.8	12 38.8	12 02.3	2.7	2.3	8.7	7.3	14.7	12.4
28	12 37.0	12 39.1	12 02.5	2.8	2.4	8.8	7.4	14.8	12.5
29	12 37.3	12 39.3	12 02.8	2.9	2.4	8.9	7.5	14.9	12.5
30	12 37.5	12 39.6	12 03.0	3.0	2.5	9.0	7.6	15.0	12.6
31	12 37.8	12 39.8	12 03.2	3.1	2.6	9.1	7.7	15.1	12.7
32	12 38.0	12 40.1	12 03.5	3.2	2.7	9.2	7.7	15.2	12.8
33	12 38.3	12 40.3	12 03.7	3.3	2.8	9.3	7.8	15.3	12.9
34	12 38.5	12 40.6	12 03.9	3.4	2.9	9.4	7.9	15.4	13.0
35	12 38.8	12 40.8	12 04.2	3.5	2.9	9.5	8.0	15.5	13.0
36	12 39.0	12 41.1	12 04.4	3.6	3.0	9.6	8.1	15.6	13.1
37	12 39.3	12 41.3	12 04.7	3.7	3.1	9.7	8.2	15.7	13.2
38	12 39.5	12 41.6	12 04.9	3.8	3.2	9.8	8.2	15.8	13.3
39	12 39.8	12 41.8	12 05.1	3.9	3.3	9.9	8.3	15.9	13.4
40	12 40.0	12 42.1	12 05.4	4.0	3.4	10.0	8.4	16.0	13.5
41	12 40.3	12 42.3	12 05.6	4.1	3.5	10.1	8.5	16.1	13.6
42	12 40.5	12 42.6	12 05.9	4.2	3.5	10.2	8.6	16.2	13.6
43	12 40.8	12 42.8	12 06.1	4.3	3.6	10.3	8.7	16.3	13.7
44	12 41.0	12 43.1	12 06.3	4.4	3.7	10.4	8.8	16.4	13.8
45	12 41.3	12 43.3	12 06.6	4.5	3.8	10.5	8.8	16.5	13.9
46	12 41.5	12 43.6	12 06.8	4.6	3.9	10.6	8.9	16.6	14.0
47	12 41.8	12 43.8	12 07.0	4.7	4.0	10.7	9.0	16.7	14.1
48	12 42.0	12 44.1	12 07.3	4.8	4.0	10.8	9.1	16.8	14.1
49	12 42.3	12 44.3	12 07.5	4.9	4.1	10.9	9.2	16.9	14.2
50	12 42.5	12 44.6	12 07.8	5.0	4.2	11.0	9.3	17.0	14.3
51	12 42.8	12 44.8	12 08.0	5.1	4.3	11.1	9.4	17.1	14.4
52	12 43.0	12 45.1	12 08.2	5.2	4.4	11.2	9.4	17.2	14.5
53	12 43.3	12 45.3	12 08.5	5.3	4.5	11.3	9.5	17.3	14.6
54	12 43.5	12 45.6	12 08.7	5.4	4.5	11.4	9.6	17.4	14.6
55	12 43.8	12 45.8	12 09.0	5.5	4.6	11.5	9.7	17.5	14.7
56	12 44.0	12 46.1	12 09.2	5.6	4.7	11.6	9.8	17.6	14.8
57	12 44.3	12 46.3	12 09.4	5.7	4.8	11.7	9.8	17.7	14.9
58	12 44.5	12 46.6	12 09.7	5.8	4.9	11.8	9.9	17.8	15.0
59	12 44.8	12 46.8	12 09.9	5.9	5.0	11.9	10.0	17.9	15.1
60	12 45.0	12 47.1	12 10.2	6.0	5.1	12.0	10.1	18.0	15.2

51m

s	SUN PLANETS	ARIES	MOON	v or d	Corrⁿ	v or d	Corrⁿ	v or d	Corrⁿ
00	12 45.0	12 47.1	12 10.2	0.0	0.0	6.0	5.2	12.0	10.3
01	12 45.3	12 47.3	12 10.4	0.1	0.1	6.1	5.2	12.1	10.4
02	12 45.5	12 47.6	12 10.6	0.2	0.2	6.2	5.3	12.2	10.5
03	12 45.8	12 47.8	12 10.9	0.3	0.3	6.3	5.4	12.3	10.6
04	12 46.0	12 48.1	12 11.1	0.4	0.3	6.4	5.5	12.4	10.6
05	12 46.3	12 48.3	12 11.3	0.5	0.4	6.5	5.6	12.5	10.7
06	12 46.5	12 48.6	12 11.6	0.6	0.5	6.6	5.7	12.6	10.8
07	12 46.8	12 48.8	12 11.8	0.7	0.6	6.7	5.8	12.7	10.9
08	12 47.0	12 49.1	12 12.1	0.8	0.7	6.8	5.8	12.8	11.0
09	12 47.3	12 49.4	12 12.3	0.9	0.8	6.9	5.9	12.9	11.1
10	12 47.5	12 49.6	12 12.5	1.0	0.9	7.0	6.0	13.0	11.2
11	12 47.8	12 49.9	12 12.8	1.1	0.9	7.1	6.1	13.1	11.2
12	12 48.0	12 50.1	12 13.0	1.2	1.0	7.2	6.2	13.2	11.3
13	12 48.3	12 50.4	12 13.3	1.3	1.1	7.3	6.3	13.3	11.4
14	12 48.5	12 50.6	12 13.5	1.4	1.2	7.4	6.4	13.4	11.5
15	12 48.8	12 50.9	12 13.7	1.5	1.3	7.5	6.4	13.5	11.6
16	12 49.0	12 51.1	12 14.0	1.6	1.4	7.6	6.5	13.6	11.7
17	12 49.3	12 51.4	12 14.2	1.7	1.5	7.7	6.6	13.7	11.8
18	12 49.5	12 51.6	12 14.4	1.8	1.5	7.8	6.7	13.8	11.8
19	12 49.8	12 51.9	12 14.7	1.9	1.6	7.9	6.8	13.9	11.9
20	12 50.0	12 52.1	12 14.9	2.0	1.7	8.0	6.9	14.0	12.0
21	12 50.3	12 52.4	12 15.2	2.1	1.8	8.1	7.0	14.1	12.1
22	12 50.5	12 52.6	12 15.4	2.2	1.9	8.2	7.0	14.2	12.2
23	12 50.8	12 52.9	12 15.6	2.3	2.0	8.3	7.1	14.3	12.3
24	12 51.0	12 53.1	12 15.9	2.4	2.1	8.4	7.2	14.4	12.4
25	12 51.3	12 53.4	12 16.1	2.5	2.1	8.5	7.3	14.5	12.4
26	12 51.5	12 53.6	12 16.4	2.6	2.2	8.6	7.4	14.6	12.5
27	12 51.8	12 53.9	12 16.6	2.7	2.3	8.7	7.5	14.7	12.6
28	12 52.0	12 54.1	12 16.8	2.8	2.4	8.8	7.6	14.8	12.7
29	12 52.3	12 54.4	12 17.1	2.9	2.5	8.9	7.6	14.9	12.8
30	12 52.5	12 54.6	12 17.3	3.0	2.6	9.0	7.7	15.0	12.9
31	12 52.8	12 54.9	12 17.5	3.1	2.7	9.1	7.8	15.1	13.0
32	12 53.0	12 55.1	12 17.8	3.2	2.7	9.2	7.9	15.2	13.0
33	12 53.3	12 55.4	12 18.0	3.3	2.8	9.3	8.0	15.3	13.1
34	12 53.5	12 55.6	12 18.3	3.4	2.9	9.4	8.1	15.4	13.2
35	12 53.8	12 55.9	12 18.5	3.5	3.0	9.5	8.2	15.5	13.3
36	12 54.0	12 56.1	12 18.7	3.6	3.1	9.6	8.2	15.6	13.4
37	12 54.3	12 56.4	12 19.0	3.7	3.2	9.7	8.3	15.7	13.5
38	12 54.5	12 56.6	12 19.2	3.8	3.3	9.8	8.4	15.8	13.6
39	12 54.8	12 56.9	12 19.5	3.9	3.3	9.9	8.5	15.9	13.6
40	12 55.0	12 57.1	12 19.7	4.0	3.4	10.0	8.6	16.0	13.7
41	12 55.3	12 57.4	12 19.9	4.1	3.5	10.1	8.7	16.1	13.8
42	12 55.5	12 57.6	12 20.2	4.2	3.6	10.2	8.8	16.2	13.9
43	12 55.8	12 57.9	12 20.4	4.3	3.7	10.3	8.8	16.3	14.0
44	12 56.0	12 58.1	12 20.6	4.4	3.8	10.4	8.9	16.4	14.1
45	12 56.3	12 58.4	12 20.9	4.5	3.9	10.5	9.0	16.5	14.2
46	12 56.5	12 58.6	12 21.1	4.6	3.9	10.6	9.1	16.6	14.2
47	12 56.8	12 58.9	12 21.4	4.7	4.0	10.7	9.2	16.7	14.3
48	12 57.0	12 59.1	12 21.6	4.8	4.1	10.8	9.3	16.8	14.4
49	12 57.3	12 59.4	12 21.8	4.9	4.2	10.9	9.4	16.9	14.5
50	12 57.5	12 59.6	12 22.1	5.0	4.3	11.0	9.4	17.0	14.6
51	12 57.8	12 59.9	12 22.3	5.1	4.4	11.1	9.5	17.1	14.7
52	12 58.0	13 00.1	12 22.6	5.2	4.5	11.2	9.6	17.2	14.8
53	12 58.3	13 00.4	12 22.8	5.3	4.5	11.3	9.7	17.3	14.8
54	12 58.5	13 00.6	12 23.0	5.4	4.6	11.4	9.8	17.4	14.9
55	12 58.8	13 00.9	12 23.3	5.5	4.7	11.5	9.9	17.5	15.0
56	12 59.0	13 01.1	12 23.5	5.6	4.8	11.6	10.0	17.6	15.1
57	12 59.3	13 01.4	12 23.8	5.7	4.9	11.7	10.0	17.7	15.2
58	12 59.5	13 01.6	12 24.0	5.8	5.0	11.8	10.1	17.8	15.3
59	12 59.8	13 01.9	12 24.2	5.9	5.1	11.9	10.2	17.9	15.4
60	13 00.0	13 02.1	12 24.5	6.0	5.2	12.0	10.3	18.0	15.5

© 2009 Jack Case

2009 SEPTEMBER 10, 11, 12 (THURS., FRI., SAT.)

UT	SUN GHA	SUN Dec	MOON GHA	MOON v	MOON Dec	MOON d	MOON HP	Lat.	Twilight Naut.	Twilight Civil	Sunrise	Moonrise 10	Moonrise 11	Moonrise 12	Moonrise 13
d h	° '	° '	° '	'	° '	'	'	°	h m	h m	h m	h m	h m	h m	h m
10 00	180 43.7	N 4 58.0	302 07.4	8.2	N22 33.5	7.9	57.6	N 72	01 29	03 34	04 49	▭	▭	▭	▭
01	195 43.9	57.1	316 34.6	8.1	22 41.4	7.9	57.6	N 70	02 13	03 50	04 57	▭	▭	▭	▭
02	210 44.1	56.1	331 01.7	8.0	22 49.3	7.7	57.7	68	02 41	04 03	05 03	▭	▭	▭	▭
03	225 44.3	.. 55.2	345 28.7	8.0	22 57.0	7.6	57.7	66	03 02	04 13	05 08	▭	▭	▭	19 54
04	240 44.5	54.2	359 55.7	7.8	23 04.6	7.5	57.7	64	03 18	04 22	05 12	17 56	17 39	18 56	21 16
05	255 44.8	53.3	14 22.5	7.7	23 12.1	7.3	57.7	62	03 31	04 29	05 15	18 37	19 04	20 11	21 52
								60	03 42	04 36	05 19	19 05	19 41	20 47	22 19
06	270 45.0	N 4 52.3	28 49.2	7.7	N23 19.4	7.2	57.7	N 58	03 51	04 41	05 21	19 27	20 08	21 12	22 39
07	285 45.2	51.4	43 15.9	7.6	23 26.6	7.1	57.8	56	03 59	04 46	05 24	19 45	20 28	21 33	22 56
T 08	300 45.4	50.4	57 42.5	7.4	23 33.7	6.9	57.8	54	04 06	04 50	05 26	20 00	20 46	21 50	23 10
H 09	315 45.6	.. 49.5	72 08.9	7.4	23 40.6	6.8	57.8	52	04 12	04 54	05 28	20 13	21 01	22 05	23 23
U 10	330 45.9	48.5	86 35.3	7.3	23 47.4	6.7	57.8	50	04 18	04 57	05 30	20 25	21 13	22 17	23 34
R 11	345 46.1	47.6	101 01.6	7.2	23 54.1	6.6	57.9	45	04 29	05 04	05 34	20 49	21 40	22 43	23 57
S 12	0 46.3	N 4 46.6	115 27.8	7.1	N24 00.7	6.4	57.9	N 40	04 38	05 10	05 37	21 08	22 01	23 04	24 15
D 13	15 46.5	45.7	129 53.9	7.0	24 07.1	6.2	57.9	35	04 44	05 14	05 40	21 24	22 19	23 21	24 31
A 14	30 46.7	44.7	144 19.9	7.0	24 13.3	6.2	57.9	30	04 50	05 18	05 42	21 39	22 34	23 36	24 44
Y 15	45 47.0	.. 43.8	158 45.9	6.8	24 19.5	5.9	58.0	20	04 59	05 24	05 47	22 03	23 00	24 01	00 01
16	60 47.2	42.8	173 11.7	6.8	24 25.4	5.9	58.0	N 10	05 05	05 29	05 50	22 24	23 22	24 23	00 23
17	75 47.4	41.9	187 37.5	6.7	24 31.3	5.7	58.0	0	05 09	05 33	05 53	22 43	23 43	24 44	00 44
18	90 47.6	N 4 40.9	202 03.2	6.6	N24 37.0	5.5	58.1	S 10	05 11	05 35	05 56	23 03	24 03	00 03	01 04
19	105 47.8	40.0	216 28.8	6.5	24 42.5	5.4	58.1	20	05 12	05 38	06 00	23 24	24 26	00 26	01 26
20	120 48.0	39.0	230 54.3	6.5	24 47.9	5.3	58.1	30	05 12	05 39	06 03	23 49	24 52	00 52	01 51
21	135 48.3	.. 38.1	245 19.8	6.3	24 53.2	5.1	58.1	35	05 11	05 40	06 05	24 03	00 03	01 07	02 06
22	150 48.5	37.1	259 45.1	6.3	24 58.3	4.9	58.1	40	05 09	05 40	06 07	24 20	00 20	01 25	02 23
23	165 48.7	36.2	274 10.4	6.2	25 03.2	4.9	58.1	45	05 07	05 41	06 10	24 40	00 40	01 46	02 43
11 00	180 48.9	N 4 35.2	288 35.6	6.1	N25 08.1	4.6	58.2	S 50	05 03	05 41	06 13	25 06	01 06	02 13	03 09
01	195 49.1	34.3	303 00.7	6.1	25 12.7	4.5	58.2	52	05 02	05 41	06 14	00 01	01 18	02 27	03 21
02	210 49.4	33.3	317 25.8	5.9	25 17.2	4.4	58.2	54	05 00	05 41	06 16	00 13	01 33	02 42	03 36
03	225 49.6	.. 32.4	331 50.7	5.9	25 21.6	4.1	58.2	56	04 57	05 40	06 17	00 26	01 49	03 00	03 52
04	240 49.8	31.4	346 15.6	5.9	25 25.7	4.1	58.3	58	04 55	05 40	06 19	00 42	02 09	03 22	04 12
05	255 50.0	30.5	0 40.5	5.7	25 29.8	3.9	58.3	S 60	04 52	05 40	06 21	01 01	02 34	03 50	04 37
06	270 50.2	N 4 29.5	15 05.2	5.7	N25 33.7	3.7	58.3		Sunset	Twilight Civil	Twilight Naut.	Moonset 10	Moonset 11	Moonset 12	Moonset 13
07	285 50.5	28.6	29 29.9	5.6	25 37.4	3.5	58.3								
F 08	300 50.7	27.6	43 54.5	5.6	25 40.9	3.4	58.4								
R 09	315 50.9	.. 26.7	58 19.1	5.5	25 44.3	3.3	58.4	°	h m	h m	h m	h m	h m	h m	h m
I 10	330 51.1	25.7	72 43.6	5.4	25 47.6	3.1	58.4	N 72	19 01	20 15	22 14	▭	▭	▭	▭
D 11	345 51.3	24.8	87 08.0	5.4	25 50.7	2.9	58.4	N 70	18 54	20 00	21 34	▭	▭	▭	▭
A 12	0 51.6	N 4 23.8	101 32.4	5.3	N25 53.6	2.7	58.4	68	18 48	19 47	21 08	▭	▭	▭	▭
Y 13	15 51.8	22.9	115 56.7	5.2	25 56.3	2.6	58.5	66	18 44	19 37	20 48	▭	▭	▭	19 13
14	30 52.0	21.9	130 20.9	5.2	25 58.9	2.4	58.5	64	18 40	19 29	20 32	14 52	17 12	18 03	17 51
15	45 52.2	.. 21.0	144 45.1	5.2	26 01.3	2.3	58.5	62	18 36	19 22	20 19	14 12	15 47	16 48	17 13
16	60 52.4	20.0	159 09.3	5.0	26 03.6	2.1	58.5	60	18 33	19 16	20 09	13 45	15 10	16 12	16 47
17	75 52.7	19.1	173 33.3	5.1	26 05.7	1.9	58.6								
18	90 52.9	N 4 18.1	187 57.4	4.9	N26 07.6	1.7	58.6	N 58	18 30	19 11	20 00	13 23	14 44	15 46	16 26
19	105 53.1	17.2	202 21.3	5.0	26 09.3	1.6	58.6	56	18 28	19 06	19 52	13 06	14 23	15 25	16 08
20	120 53.3	16.2	216 45.3	4.9	26 10.9	1.4	58.6	54	18 26	19 02	19 45	12 51	14 06	15 08	15 53
21	135 53.6	.. 15.3	231 09.2	4.8	26 12.3	1.3	58.7	52	18 24	18 58	19 39	12 38	13 51	14 53	15 39
22	150 53.8	14.3	245 33.0	4.8	26 13.6	1.0	58.7	50	18 22	18 55	19 34	12 27	13 38	14 40	15 29
23	165 54.0	13.4	259 56.8	4.8	26 14.6	0.9	58.7	45	18 18	18 48	19 23	12 03	13 12	14 14	15 05
12 00	180 54.2	N 4 12.4	274 20.6	4.7	N26 15.5	0.7	58.7	N 40	18 15	18 43	19 15	11 44	12 51	13 53	14 46
01	195 54.4	11.4	288 44.3	4.7	26 16.2	0.6	58.7	35	18 13	18 38	19 08	11 29	12 34	13 35	14 30
02	210 54.7	10.5	303 08.0	4.6	26 16.8	0.4	58.8	30	18 10	18 34	19 02	11 15	12 19	13 20	14 16
03	225 54.9	.. 09.5	317 31.6	4.6	26 17.2	0.2	58.8	20	18 06	18 28	18 54	10 52	11 54	12 55	13 53
04	240 55.1	08.6	331 55.2	4.6	26 17.4	0.0	58.8	N 10	18 03	18 24	18 48	10 32	11 32	12 32	13 32
05	255 55.3	07.6	346 18.8	4.5	26 17.4	0.1	58.8	0	18 00	18 21	18 45	10 14	11 11	12 12	13 13
06	270 55.5	N 4 06.7	0 42.3	4.6	N26 17.3	0.4	58.9	S 10	17 57	18 18	18 42	09 55	10 51	11 51	12 53
07	285 55.8	05.7	15 05.9	4.5	26 16.9	0.5	58.9	20	17 54	18 16	18 41	09 35	10 29	11 28	12 32
S 08	300 56.0	04.8	29 29.4	4.4	26 16.4	0.6	58.9	30	17 50	18 14	18 42	09 12	10 04	11 03	12 08
A 09	315 56.2	.. 03.8	43 52.8	4.5	26 15.8	0.9	58.9	35	17 48	18 14	18 43	08 59	09 49	10 47	11 53
T 10	330 56.4	02.9	58 16.3	4.4	26 14.9	1.0	59.0	40	17 46	18 13	18 45	08 44	09 31	10 29	11 37
U 11	345 56.6	01.9	72 39.7	4.5	26 13.9	1.2	59.0	45	17 44	18 13	18 47	08 26	09 11	10 08	11 17
R 12	0 56.9	N 4 00.9	87 03.2	4.4	N26 12.7	1.4	59.0	S 50	17 41	18 13	18 51	08 03	08 45	09 41	10 52
D 13	15 57.1	4 00.0	101 26.6	4.3	26 11.3	1.6	59.0	52	17 40	18 13	18 53	07 52	08 32	09 28	10 39
A 14	30 57.3	3 59.0	115 49.9	4.4	26 09.7	1.7	59.0	54	17 38	18 14	18 55	07 40	08 18	09 12	10 25
Y 15	45 57.5	.. 58.1	130 13.3	4.4	26 08.0	1.9	59.1	56	17 37	18 14	18 57	07 26	08 01	08 54	10 09
16	60 57.7	57.1	144 36.7	4.4	26 06.1	2.1	59.1	58	17 35	18 14	19 00	07 10	07 41	08 33	09 50
17	75 58.0	56.2	159 00.1	4.3	26 04.0	2.3	59.1	S 60	17 33	18 15	19 03	06 50	07 15	08 05	09 25
18	90 58.2	N 3 55.2	173 23.4	4.4	N26 01.7	2.4	59.1		SUN			MOON			
19	105 58.4	54.3	187 46.8	4.3	25 59.3	2.6	59.1	Day	Eqn. of Time		Mer. Pass.	Mer. Pass.		Age	Phase
20	120 58.6	53.3	202 10.1	4.4	25 56.7	2.8	59.2		00h	12h		Upper	Lower		
21	135 58.9	.. 52.3	216 33.5	4.3	25 53.9	3.0	59.2	d	m s	m s	h m	h m	h m	d	%
22	150 59.1	51.4	230 56.8	4.4	25 50.9	3.1	59.2	10	02 54	03 05	11 57	04 00	16 28	21	67
23	165 59.3	50.4	245 20.2	4.4	N25 47.8	3.4	59.2	11	03 15	03 26	11 57	04 57	17 27	22	57
	SD 15.9	d 1.0	SD 15.8		15.9		16.1	12	03 36	03 47	11 56	05 57	18 28	23	46

INCREMENTS AND CORRECTIONS

12ᵐ

12 ᵐ/s	SUN PLANETS	ARIES	MOON	v or d	Corrⁿ	v or d	Corrⁿ	v or d	Corrⁿ
s	° ′	° ′	° ′	′	′	′	′	′	′
00	3 00·0	3 00·5	2 51·8	0·0	0·0	6·0	1·3	12·0	2·5
01	3 00·3	3 00·7	2 52·0	0·1	0·0	6·1	1·3	12·1	2·5
02	3 00·5	3 01·0	2 52·3	0·2	0·0	6·2	1·3	12·2	2·5
03	3 00·8	3 01·2	2 52·5	0·3	0·1	6·3	1·3	12·3	2·6
04	3 01·0	3 01·5	2 52·8	0·4	0·1	6·4	1·3	12·4	2·6
05	3 01·3	3 01·7	2 53·0	0·5	0·1	6·5	1·4	12·5	2·6
06	3 01·5	3 02·0	2 53·2	0·6	0·1	6·6	1·4	12·6	2·6
07	3 01·8	3 02·2	2 53·5	0·7	0·1	6·7	1·4	12·7	2·6
08	3 02·0	3 02·5	2 53·7	0·8	0·2	6·8	1·4	12·8	2·7
09	3 02·3	3 02·7	2 53·9	0·9	0·2	6·9	1·4	12·9	2·7
10	3 02·5	3 03·0	2 54·2	1·0	0·2	7·0	1·5	13·0	2·7
11	3 02·8	3 03·3	2 54·4	1·1	0·2	7·1	1·5	13·1	2·7
12	3 03·0	3 03·5	2 54·7	1·2	0·3	7·2	1·5	13·2	2·8
13	3 03·3	3 03·8	2 54·9	1·3	0·3	7·3	1·5	13·3	2·8
14	3 03·5	3 04·0	2 55·1	1·4	0·3	7·4	1·5	13·4	2·8
15	3 03·8	3 04·3	2 55·4	1·5	0·3	7·5	1·6	13·5	2·8
16	3 04·0	3 04·5	2 55·6	1·6	0·3	7·6	1·6	13·6	2·8
17	3 04·3	3 04·8	2 55·9	1·7	0·4	7·7	1·6	13·7	2·9
18	3 04·5	3 05·0	2 56·1	1·8	0·4	7·8	1·6	13·8	2·9
19	3 04·8	3 05·3	2 56·3	1·9	0·4	7·9	1·6	13·9	2·9
20	3 05·0	3 05·5	2 56·6	2·0	0·4	8·0	1·7	14·0	2·9
21	3 05·3	3 05·8	2 56·8	2·1	0·4	8·1	1·7	14·1	2·9
22	3 05·5	3 06·0	2 57·0	2·2	0·5	8·2	1·7	14·2	3·0
23	3 05·8	3 06·3	2 57·3	2·3	0·5	8·3	1·7	14·3	3·0
24	3 06·0	3 06·5	2 57·5	2·4	0·5	8·4	1·8	14·4	3·0
25	3 06·3	3 06·8	2 57·8	2·5	0·5	8·5	1·8	14·5	3·0
26	3 06·5	3 07·0	2 58·0	2·6	0·5	8·6	1·8	14·6	3·0
27	3 06·8	3 07·3	2 58·2	2·7	0·6	8·7	1·8	14·7	3·1
28	3 07·0	3 07·5	2 58·5	2·8	0·6	8·8	1·8	14·8	3·1
29	3 07·3	3 07·8	2 58·7	2·9	0·6	8·9	1·9	14·9	3·1
30	3 07·5	3 08·0	2 59·0	3·0	0·6	9·0	1·9	15·0	3·1
31	3 07·8	3 08·3	2 59·2	3·1	0·6	9·1	1·9	15·1	3·1
32	3 08·0	3 08·5	2 59·4	3·2	0·7	9·2	1·9	15·2	3·2
33	3 08·3	3 08·8	2 59·7	3·3	0·7	9·3	1·9	15·3	3·2
34	3 08·5	3 09·0	2 59·9	3·4	0·7	9·4	2·0	15·4	3·2
35	3 08·8	3 09·3	3 00·2	3·5	0·7	9·5	2·0	15·5	3·2
36	3 09·0	3 09·5	3 00·4	3·6	0·8	9·6	2·0	15·6	3·3
37	3 09·3	3 09·8	3 00·6	3·7	0·8	9·7	2·0	15·7	3·3
38	3 09·5	3 10·0	3 00·9	3·8	0·8	9·8	2·0	15·8	3·3
39	3 09·8	3 10·3	3 01·1	3·9	0·8	9·9	2·1	15·9	3·3
40	3 10·0	3 10·5	3 01·3	4·0	0·8	10·0	2·1	16·0	3·3
41	3 10·3	3 10·8	3 01·6	4·1	0·9	10·1	2·1	16·1	3·4
42	3 10·5	3 11·0	3 01·8	4·2	0·9	10·2	2·1	16·2	3·4
43	3 10·8	3 11·3	3 02·1	4·3	0·9	10·3	2·1	16·3	3·4
44	3 11·0	3 11·5	3 02·3	4·4	0·9	10·4	2·2	16·4	3·4
45	3 11·3	3 11·8	3 02·5	4·5	0·9	10·5	2·2	16·5	3·4
46	3 11·5	3 12·0	3 02·8	4·6	1·0	10·6	2·2	16·6	3·5
47	3 11·8	3 12·3	3 03·0	4·7	1·0	10·7	2·2	16·7	3·5
48	3 12·0	3 12·5	3 03·3	4·8	1·0	10·8	2·3	16·8	3·5
49	3 12·3	3 12·8	3 03·5	4·9	1·0	10·9	2·3	16·9	3·5
50	3 12·5	3 13·0	3 03·7	5·0	1·0	11·0	2·3	17·0	3·5
51	3 12·8	3 13·3	3 04·0	5·1	1·1	11·1	2·3	17·1	3·6
52	3 13·0	3 13·5	3 04·2	5·2	1·1	11·2	2·3	17·2	3·6
53	3 13·3	3 13·8	3 04·4	5·3	1·1	11·3	2·4	17·3	3·6
54	3 13·5	3 14·0	3 04·7	5·4	1·1	11·4	2·4	17·4	3·6
55	3 13·8	3 14·3	3 04·9	5·5	1·1	11·5	2·4	17·5	3·6
56	3 14·0	3 14·5	3 05·2	5·6	1·2	11·6	2·4	17·6	3·7
57	3 14·3	3 14·8	3 05·4	5·7	1·2	11·7	2·4	17·7	3·7
58	3 14·5	3 15·0	3 05·6	5·8	1·2	11·8	2·5	17·8	3·7
59	3 14·8	3 15·3	3 05·9	5·9	1·2	11·9	2·5	17·9	3·7
60	3 15·0	3 15·5	3 06·1	6·0	1·3	12·0	2·5	18·0	3·8

13ᵐ

13 ᵐ/s	SUN PLANETS	ARIES	MOON	v or d	Corrⁿ	v or d	Corrⁿ	v or d	Corrⁿ
s	° ′	° ′	° ′	′	′	′	′	′	′
00	3 15·0	3 15·5	3 06·1	0·0	0·0	6·0	1·4	12·0	2·7
01	3 15·3	3 15·8	3 06·4	0·1	0·0	6·1	1·4	12·1	2·7
02	3 15·5	3 16·0	3 06·6	0·2	0·0	6·2	1·4	12·2	2·7
03	3 15·8	3 16·3	3 06·8	0·3	0·1	6·3	1·4	12·3	2·8
04	3 16·0	3 16·5	3 07·1	0·4	0·1	6·4	1·4	12·4	2·8
05	3 16·3	3 16·8	3 07·3	0·5	0·1	6·5	1·5	12·5	2·8
06	3 16·5	3 17·0	3 07·5	0·6	0·1	6·6	1·5	12·6	2·8
07	3 16·8	3 17·3	3 07·8	0·7	0·2	6·7	1·5	12·7	2·9
08	3 17·0	3 17·5	3 08·0	0·8	0·2	6·8	1·5	12·8	2·9
09	3 17·3	3 17·8	3 08·3	0·9	0·2	6·9	1·6	12·9	2·9
10	3 17·5	3 18·0	3 08·5	1·0	0·2	7·0	1·6	13·0	2·9
11	3 17·8	3 18·3	3 08·7	1·1	0·2	7·1	1·6	13·1	2·9
12	3 18·0	3 18·5	3 09·0	1·2	0·3	7·2	1·6	13·2	3·0
13	3 18·3	3 18·8	3 09·2	1·3	0·3	7·3	1·6	13·3	3·0
14	3 18·5	3 19·0	3 09·5	1·4	0·3	7·4	1·7	13·4	3·0
15	3 18·8	3 19·3	3 09·7	1·5	0·3	7·5	1·7	13·5	3·0
16	3 19·0	3 19·5	3 09·9	1·6	0·4	7·6	1·7	13·6	3·1
17	3 19·3	3 19·8	3 10·2	1·7	0·4	7·7	1·7	13·7	3·1
18	3 19·5	3 20·0	3 10·4	1·8	0·4	7·8	1·8	13·8	3·1
19	3 19·8	3 20·3	3 10·7	1·9	0·4	7·9	1·8	13·9	3·1
20	3 20·0	3 20·5	3 10·9	2·0	0·5	8·0	1·8	14·0	3·2
21	3 20·3	3 20·8	3 11·1	2·1	0·5	8·1	1·8	14·1	3·2
22	3 20·5	3 21·0	3 11·4	2·2	0·5	8·2	1·8	14·2	3·2
23	3 20·8	3 21·3	3 11·6	2·3	0·5	8·3	1·9	14·3	3·2
24	3 21·0	3 21·6	3 11·8	2·4	0·5	8·4	1·9	14·4	3·2
25	3 21·3	3 21·8	3 12·1	2·5	0·6	8·5	1·9	14·5	3·3
26	3 21·5	3 22·1	3 12·3	2·6	0·6	8·6	1·9	14·6	3·3
27	3 21·8	3 22·3	3 12·6	2·7	0·6	8·7	2·0	14·7	3·3
28	3 22·0	3 22·6	3 12·8	2·8	0·6	8·8	2·0	14·8	3·3
29	3 22·3	3 22·8	3 13·0	2·9	0·7	8·9	2·0	14·9	3·4
30	3 22·5	3 23·1	3 13·3	3·0	0·7	9·0	2·0	15·0	3·4
31	3 22·8	3 23·3	3 13·5	3·1	0·7	9·1	2·0	15·1	3·4
32	3 23·0	3 23·6	3 13·8	3·2	0·7	9·2	2·1	15·2	3·4
33	3 23·3	3 23·8	3 14·0	3·3	0·7	9·3	2·1	15·3	3·4
34	3 23·5	3 24·1	3 14·2	3·4	0·8	9·4	2·1	15·4	3·5
35	3 23·8	3 24·3	3 14·5	3·5	0·8	9·5	2·1	15·5	3·5
36	3 24·0	3 24·6	3 14·7	3·6	0·8	9·6	2·2	15·6	3·5
37	3 24·3	3 24·8	3 14·9	3·7	0·8	9·7	2·2	15·7	3·5
38	3 24·5	3 25·1	3 15·2	3·8	0·9	9·8	2·2	15·8	3·6
39	3 24·8	3 25·3	3 15·4	3·9	0·9	9·9	2·2	15·9	3·6
40	3 25·0	3 25·6	3 15·7	4·0	0·9	10·0	2·3	16·0	3·6
41	3 25·3	3 25·8	3 15·9	4·1	0·9	10·1	2·3	16·1	3·6
42	3 25·5	3 26·1	3 16·1	4·2	0·9	10·2	2·3	16·2	3·6
43	3 25·8	3 26·3	3 16·4	4·3	1·0	10·3	2·3	16·3	3·7
44	3 26·0	3 26·6	3 16·6	4·4	1·0	10·4	2·3	16·4	3·7
45	3 26·3	3 26·8	3 16·9	4·5	1·0	10·5	2·4	16·5	3·7
46	3 26·5	3 27·1	3 17·1	4·6	1·0	10·6	2·4	16·6	3·7
47	3 26·8	3 27·3	3 17·3	4·7	1·1	10·7	2·4	16·7	3·8
48	3 27·0	3 27·6	3 17·6	4·8	1·1	10·8	2·4	16·8	3·8
49	3 27·3	3 27·8	3 17·8	4·9	1·1	10·9	2·5	16·9	3·8
50	3 27·5	3 28·1	3 18·0	5·0	1·1	11·0	2·5	17·0	3·8
51	3 27·8	3 28·3	3 18·3	5·1	1·1	11·1	2·5	17·1	3·8
52	3 28·0	3 28·6	3 18·5	5·2	1·2	11·2	2·5	17·2	3·9
53	3 28·3	3 28·8	3 18·8	5·3	1·2	11·3	2·5	17·3	3·9
54	3 28·5	3 29·1	3 19·0	5·4	1·2	11·4	2·6	17·4	3·9
55	3 28·8	3 29·3	3 19·2	5·5	1·2	11·5	2·6	17·5	3·9
56	3 29·0	3 29·6	3 19·5	5·6	1·3	11·6	2·6	17·6	4·0
57	3 29·3	3 29·8	3 19·7	5·7	1·3	11·7	2·6	17·7	4·0
58	3 29·5	3 30·1	3 20·0	5·8	1·3	11·8	2·7	17·8	4·0
59	3 29·8	3 30·3	3 20·2	5·9	1·3	11·9	2·7	17·9	4·0
60	3 30·0	3 30·6	3 20·4	6·0	1·4	12·0	2·7	18·0	4·1

© 2009 Jack Case

2009 JANUARY 1, 2, 3 (THURS., FRI., SAT.)

UT	SUN		MOON				Lat.	Twilight		Sunrise	Moonrise			
	GHA	Dec	GHA	v	Dec	d	HP		Naut.	Civil		1	2	3	4
d h	° '	° '	° '	'	° '	'	'	°	h m	h m	h m	h m	h m	h m	h m
1 00	179 08.4	S23 00.5	129 33.6 15.3	S 9 50.7	13.2	55.1	N 72	08 23	10 39	■	11 16	10 45	10 16	09 42
01	194 08.1	00.3	144 07.9 15.3	9 37.5	13.2	55.2	N 70	08 04	09 48	■	11 05	10 43	10 21	09 56
02	209 07.8	23 00.1	158 42.2 15.3	9 24.3	13.3	55.2	68	07 49	09 16	■	10 57	10 41	10 25	10 08
03	224 07.5	22 59.9	173 16.5 15.4	9 11.0	13.3	55.2	66	07 37	08 52	10 26	10 49	10 39	10 28	10 17
04	239 07.2	59.7	187 50.9 15.4	8 57.7	13.3	55.2	64	07 26	08 34	09 48	10 43	10 37	10 31	10 25
05	254 06.9	59.5	202 25.3 15.4	8 44.4	13.4	55.2	62	07 17	08 18	09 22	10 38	10 36	10 34	10 32
06	269 06.6	S22 59.3	216 59.7 15.4	S 8 31.0	13.4	55.3	60	07 09	08 05	09 02	10 33	10 35	10 36	10 38
07	284 06.3	59.1	231 34.1 15.4	8 17.6	13.5	55.3	N 58	07 02	07 54	08 45	10 29	10 34	10 38	10 43
T 08	299 06.0	58.8	246 08.5 15.4	8 04.1	13.5	55.3	56	06 55	07 44	08 31	10 25	10 33	10 40	10 48
H 09	314 05.7 ..	58.6	260 42.9 15.4	7 50.6	13.5	55.3	54	06 50	07 35	08 19	10 22	10 32	10 42	10 52
U 10	329 05.5	58.4	275 17.3 15.5	7 37.1	13.6	55.3	52	06 44	07 28	08 08	10 19	10 31	10 43	10 56
R 11	344 05.2	58.2	289 51.8 15.5	7 23.5	13.7	55.4	50	06 39	07 20	07 58	10 16	10 30	10 44	11 00
S							45	06 28	07 05	07 38	10 10	10 29	10 47	11 08
D 12	359 04.9	S22 58.0	304 26.3 15.4	S 7 09.8	13.6	55.4	N 40	06 18	06 52	07 22	10 05	10 28	10 50	11 14
A 13	14 04.6	57.8	319 00.7 15.5	6 56.2	13.7	55.4	35	06 09	06 40	07 08	10 01	10 26	10 52	11 20
Y 14	29 04.3	57.6	333 35.2 15.5	6 42.5	13.8	55.4	30	06 00	06 30	06 56	09 57	10 25	10 54	11 25
15	44 04.0 ..	57.4	348 09.7 15.4	6 28.7	13.8	55.5	20	05 44	06 12	06 36	09 50	10 24	10 58	11 33
16	59 03.7	57.1	2 44.1 15.5	6 14.9	13.8	55.5	N 10	05 28	05 55	06 17	09 44	10 22	11 01	11 41
17	74 03.4	56.9	17 18.6 15.5	6 01.1	13.8	55.5	0	05 12	05 38	06 00	09 39	10 21	11 04	11 48
18	89 03.1	S22 56.7	31 53.1 15.5	S 5 47.3	13.9	55.5	S 10	04 53	05 20	05 43	09 33	10 19	11 07	11 56
19	104 02.8	56.5	46 27.6 15.5	5 33.4	13.9	55.5	20	04 31	05 00	05 25	09 27	10 18	11 10	12 04
20	119 02.5	56.3	61 02.1 15.5	5 19.5	14.0	55.6	30	04 03	04 36	05 03	09 20	10 16	11 13	12 13
21	134 02.2 ..	56.0	75 36.6 15.5	5 05.5	13.9	55.6	35	03 45	04 21	04 51	09 16	10 15	11 15	12 18
22	149 01.9	55.8	90 11.1 15.5	4 51.6	14.1	55.6	40	03 22	04 03	04 36	09 12	10 14	11 18	12 24
23	164 01.6	55.6	104 45.6 15.4	4 37.5	14.0	55.6	45	02 52	03 41	04 18	09 06	10 13	11 21	12 31
2 00	179 01.3	S22 55.4	119 20.0 15.5	S 4 23.5	14.1	55.7	S 50	02 09	03 13	03 56	09 00	10 11	11 24	12 40
01	194 01.1	55.1	133 54.5 15.5	4 09.4	14.1	55.7	52	01 43	02 58	03 46	08 57	10 11	11 26	12 44
02	209 00.8	54.9	148 29.0 15.4	3 55.3	14.1	55.7	54	01 04	02 41	03 34	08 54	10 10	11 27	12 48
03	224 00.5 ..	54.7	163 03.4 15.5	3 41.2	14.1	55.7	56	////	02 20	03 20	08 50	10 09	11 29	12 53
04	239 00.2	54.5	177 37.9 15.4	3 27.1	14.2	55.8	58	////	01 52	03 04	08 46	10 08	11 31	12 58
05	253 59.9	54.2	192 12.3 15.5	3 12.9	14.2	55.8	S 60	////	01 10	02 45	08 42	10 07	11 34	13 05
06	268 59.6	S22 54.0	206 46.8 14.5	S 2 58.7	14.2	55.8								
07	283 59.3	53.8	221 21.2 14.5	2 44.5	14.3	55.8	Lat.	Sunset	Twilight		Moonset			
08	298 59.0	53.6	235 55.6 14.5	2 30.2	14.2	55.9			Civil	Naut.	1	2	3	4
F 09	313 58.7 ..	53.3	250 30.0 15.4	2 16.0	14.3	55.9	°	h m	h m	h m	h m	h m	h m	h m
R 10	328 58.4	53.1	265 04.4 15.3	2 01.7	14.4	55.9	N 72	■	13 30	15 46	20 52	22 54	25 00	01 00
I 11	343 58.1	52.9	279 38.7 15.4	1 47.3	14.3	56.0	N 70	■	14 21	16 05	21 00	22 52	24 49	00 49
D 12	358 57.8	S22 52.6	294 13.1 15.3	S 1 33.0	14.3	56.0	68	■	14 53	16 20	21 06	22 51	24 40	00 40
A 13	13 57.6	52.4	308 47.4 15.3	1 18.7	14.4	56.0	66	13 43	15 16	16 32	21 11	22 50	24 33	00 33
Y 14	28 57.3	52.2	323 21.7 15.3	1 04.3	14.4	56.1	64	14 20	15 35	16 42	21 15	22 49	24 26	00 26
15	43 57.0 ..	51.9	337 56.0 15.2	0 49.9	14.4	56.1	62	14 46	15 50	16 51	21 19	22 49	24 21	00 21
16	58 56.7	51.7	352 30.2 15.3	0 35.5	14.4	56.1	60	15 07	16 03	16 59	21 22	22 48	24 16	00 16
17	73 56.4	51.5	7 04.5 15.2	0 21.1	14.5	56.1								
18	88 56.1	S22 51.2	21 38.7 15.2	S 0 06.6	14.4	56.1	N 58	15 24	16 15	17 07	21 25	22 48	24 12	00 12
19	103 55.8	51.0	36 12.9 15.1	N 0 07.8	14.5	56.2	56	15 38	16 24	17 13	21 28	22 47	24 09	00 09
20	118 55.5	50.7	50 47.0 15.2	0 22.3	14.4	56.2	54	15 50	16 33	17 19	21 30	22 47	24 05	00 05
21	133 55.2 ..	50.5	65 21.2 15.1	0 36.7	14.5	56.2	52	16 00	16 41	17 24	21 32	22 46	24 02	00 02
22	148 54.9	50.3	79 55.3 15.0	0 51.2	14.5	56.3	50	16 10	16 48	17 29	21 34	22 46	24 00	00 00
23	163 54.7	50.0	94 29.3 15.1	1 05.7	14.5	56.3	45	16 30	17 04	17 41	21 38	22 45	23 54	25 06
3 00	178 54.4	S22 49.8	109 03.4 15.0	N 1 20.2	14.5	56.3	N 40	16 46	17 17	17 51	21 42	22 44	23 49	24 57
01	193 54.1	49.5	123 37.4 14.9	1 34.7	14.6	56.3	35	17 00	17 28	18 00	21 44	22 44	23 45	24 49
02	208 53.8	49.3	138 11.3 15.0	1 49.3	14.5	56.4	30	17 12	17 38	18 08	21 47	22 43	23 41	24 42
03	223 53.5 ..	49.0	152 45.3 14.9	2 03.8	14.5	56.4	20	17 33	17 57	18 24	21 51	22 43	23 35	24 31
04	238 53.2	48.8	167 19.2 14.8	2 18.3	14.6	56.5	N 10	17 51	18 14	18 40	21 55	22 42	23 30	24 20
05	253 52.9	48.6	181 53.0 14.8	2 32.9	14.5	56.5	0	18 08	18 30	18 56	21 59	22 41	23 25	24 11
06	268 52.6	S22 48.3	196 26.8 14.8	N 2 47.4	14.6	56.5	S 10	18 25	18 48	19 15	22 02	22 40	23 20	24 01
07	283 52.4	48.1	211 00.6 14.7	3 02.0	14.5	56.5	20	18 43	19 08	19 37	22 06	22 39	23 14	23 51
S 08	298 52.1	47.8	225 34.3 14.7	3 16.5	14.6	56.6	30	19 05	19 32	20 05	22 10	22 39	23 08	23 40
A 09	313 51.8 ..	47.6	240 08.0 14.6	3 31.1	14.5	56.6	35	19 18	19 47	20 23	22 13	22 38	23 04	23 33
T 10	328 51.5	47.3	254 41.6 14.6	3 45.6	14.6	56.6	40	19 32	20 05	20 46	22 15	22 37	23 00	23 26
U 11	343 51.2	47.1	269 15.2 14.6	4 00.2	14.6	56.7	45	19 50	20 27	21 15	22 18	22 37	22 56	23 17
R 12	358 50.9	S22 46.8	283 48.8 14.5	N 4 14.8	14.5	56.7	S 50	20 11	20 55	21 58	22 22	22 36	22 50	23 07
D 13	13 50.6	46.6	298 22.3 14.4	4 29.3	14.6	56.7	52	20 22	21 10	22 24	22 24	22 35	22 48	23 02
A 14	28 50.4	46.3	312 55.7 14.4	4 43.9	14.6	56.7	54	20 34	21 27	23 02	22 26	22 35	22 45	22 57
Y 15	43 50.1 ..	46.0	327 29.1 14.4	4 58.4	14.6	56.8	56	20 47	21 48	////	22 28	22 35	22 42	22 51
16	58 49.8	45.8	342 02.5 14.3	5 12.9	14.6	56.8	58	21 03	22 15	////	22 30	22 34	22 39	22 44
17	73 49.5	45.5	356 35.8 14.2	5 27.5	14.5	56.8	S 60	21 23	22 56	////	22 32	22 34	22 35	22 37
18	88 49.2	S22 45.3	11 09.0 14.2	N 5 42.0	14.5	56.9		SUN			MOON			
19	103 48.9	45.0	25 42.2 14.1	5 56.5	14.5	56.9	Day	Eqn. of Time		Mer.	Mer. Pass.		Age	Phase
20	118 48.6	44.8	40 15.3 14.0	6 11.0	14.5	56.9		00ʰ	12ʰ	Pass.	Upper	Lower		
21	133 48.4 ..	44.5	54 48.3 14.0	6 25.5	14.5	57.0	d	m s	m s	h m	h m	h m	d	%
22	148 48.1	44.2	69 21.3 13.9	6 40.0	14.5	57.0	1	03 26	03 40	12 04	15 49	03 28	05	21
23	163 47.8	44.0	83 54.2 13.9	N 6 54.5	14.4	57.0	2	03 54	04 08	12 04	16 31	04 10	06	30
	SD 16.3	d 0.2	SD 15.1		15.3		15.4	3	04 22	04 36	12 05	17 14	04 52	07	40

INCREMENTS AND CORRECTIONS

6ᵐ

s	SUN PLANETS	ARIES	MOON	v or d	Corrⁿ	v or d	Corrⁿ	v or d	Corrⁿ
00	1 30·0	1 30·2	1 25·9	0·0	0·0	6·0	0·7	12·0	1·3
01	1 30·3	1 30·5	1 26·1	0·1	0·0	6·1	0·7	12·1	1·3
02	1 30·5	1 30·7	1 26·4	0·2	0·0	6·2	0·7	12·2	1·3
03	1 30·8	1 31·0	1 26·6	0·3	0·0	6·3	0·7	12·3	1·3
04	1 31·0	1 31·2	1 26·9	0·4	0·0	6·4	0·7	12·4	1·3
05	1 31·3	1 31·5	1 27·1	0·5	0·1	6·5	0·7	12·5	1·4
06	1 31·5	1 31·8	1 27·3	0·6	0·1	6·6	0·7	12·6	1·4
07	1 31·8	1 32·0	1 27·6	0·7	0·1	6·7	0·7	12·7	1·4
08	1 32·0	1 32·3	1 27·8	0·8	0·1	6·8	0·7	12·8	1·4
09	1 32·3	1 32·5	1 28·0	0·9	0·1	6·9	0·7	12·9	1·4
10	1 32·5	1 32·8	1 28·3	1·0	0·1	7·0	0·8	13·0	1·4
11	1 32·8	1 33·0	1 28·5	1·1	0·1	7·1	0·8	13·1	1·4
12	1 33·0	1 33·3	1 28·8	1·2	0·1	7·2	0·8	13·2	1·4
13	1 33·3	1 33·5	1 29·0	1·3	0·1	7·3	0·8	13·3	1·4
14	1 33·5	1 33·8	1 29·2	1·4	0·2	7·4	0·8	13·4	1·5
15	1 33·8	1 34·0	1 29·5	1·5	0·2	7·5	0·8	13·5	1·5
16	1 34·0	1 34·3	1 29·7	1·6	0·2	7·6	0·8	13·6	1·5
17	1 34·3	1 34·5	1 30·0	1·7	0·2	7·7	0·8	13·7	1·5
18	1 34·5	1 34·8	1 30·2	1·8	0·2	7·8	0·8	13·8	1·5
19	1 34·8	1 35·0	1 30·4	1·9	0·2	7·9	0·9	13·9	1·5
20	1 35·0	1 35·3	1 30·7	2·0	0·2	8·0	0·9	14·0	1·5
21	1 35·3	1 35·5	1 30·9	2·1	0·2	8·1	0·9	14·1	1·5
22	1 35·5	1 35·8	1 31·1	2·2	0·2	8·2	0·9	14·2	1·5
23	1 35·8	1 36·0	1 31·4	2·3	0·2	8·3	0·9	14·3	1·5
24	1 36·0	1 36·3	1 31·6	2·4	0·3	8·4	0·9	14·4	1·6
25	1 36·3	1 36·5	1 31·9	2·5	0·3	8·5	0·9	14·5	1·6
26	1 36·5	1 36·8	1 32·1	2·6	0·3	8·6	0·9	14·6	1·6
27	1 36·8	1 37·0	1 32·3	2·7	0·3	8·7	0·9	14·7	1·6
28	1 37·0	1 37·3	1 32·6	2·8	0·3	8·8	1·0	14·8	1·6
29	1 37·3	1 37·5	1 32·8	2·9	0·3	8·9	1·0	14·9	1·6
30	1 37·5	1 37·8	1 33·1	3·0	0·3	9·0	1·0	15·0	1·6
31	1 37·8	1 38·0	1 33·3	3·1	0·3	9·1	1·0	15·1	1·6
32	1 38·0	1 38·3	1 33·5	3·2	0·3	9·2	1·0	15·2	1·6
33	1 38·3	1 38·5	1 33·8	3·3	0·4	9·3	1·0	15·3	1·7
34	1 38·5	1 38·8	1 34·0	3·4	0·4	9·4	1·0	15·4	1·7
35	1 38·8	1 39·0	1 34·3	3·5	0·4	9·5	1·0	15·5	1·7
36	1 39·0	1 39·3	1 34·5	3·6	0·4	9·6	1·0	15·6	1·7
37	1 39·3	1 39·5	1 34·7	3·7	0·4	9·7	1·1	15·7	1·7
38	1 39·5	1 39·8	1 35·0	3·8	0·4	9·8	1·1	15·8	1·7
39	1 39·8	1 40·0	1 35·2	3·9	0·4	9·9	1·1	15·9	1·7
40	1 40·0	1 40·3	1 35·4	4·0	0·4	10·0	1·1	16·0	1·7
41	1 40·3	1 40·5	1 35·7	4·1	0·4	10·1	1·1	16·1	1·7
42	1 40·5	1 40·8	1 35·9	4·2	0·5	10·2	1·1	16·2	1·8
43	1 40·8	1 41·0	1 36·2	4·3	0·5	10·3	1·1	16·3	1·8
44	1 41·0	1 41·3	1 36·4	4·4	0·5	10·4	1·1	16·4	1·8
45	1 41·3	1 41·5	1 36·6	4·5	0·5	10·5	1·1	16·5	1·8
46	1 41·5	1 41·8	1 36·9	4·6	0·5	10·6	1·1	16·6	1·8
47	1 41·8	1 42·0	1 37·1	4·7	0·5	10·7	1·2	16·7	1·8
48	1 42·0	1 42·3	1 37·4	4·8	0·5	10·8	1·2	16·8	1·8
49	1 42·3	1 42·5	1 37·6	4·9	0·5	10·9	1·2	16·9	1·8
50	1 42·5	1 42·8	1 37·8	5·0	0·5	11·0	1·2	17·0	1·8
51	1 42·8	1 43·0	1 38·1	5·1	0·6	11·1	1·2	17·1	1·9
52	1 43·0	1 43·3	1 38·3	5·2	0·6	11·2	1·2	17·2	1·9
53	1 43·3	1 43·5	1 38·5	5·3	0·6	11·3	1·2	17·3	1·9
54	1 43·5	1 43·8	1 38·8	5·4	0·6	11·4	1·2	17·4	1·9
55	1 43·8	1 44·0	1 39·0	5·5	0·6	11·5	1·2	17·5	1·9
56	1 44·0	1 44·3	1 39·3	5·6	0·6	11·6	1·3	17·6	1·9
57	1 44·3	1 44·5	1 39·5	5·7	0·6	11·7	1·3	17·7	1·9
58	1 44·5	1 44·8	1 39·7	5·8	0·6	11·8	1·3	17·8	1·9
59	1 44·8	1 45·0	1 40·0	5·9	0·6	11·9	1·3	17·9	1·9
60	1 45·0	1 45·3	1 40·2	6·0	0·7	12·0	1·3	18·0	2·0

7ᵐ

s	SUN PLANETS	ARIES	MOON	v or d	Corrⁿ	v or d	Corrⁿ	v or d	Corrⁿ
00	1 45·0	1 45·3	1 40·2	0·0	0·0	6·0	0·8	12·0	1·5
01	1 45·3	1 45·5	1 40·5	0·1	0·0	6·1	0·8	12·1	1·5
02	1 45·5	1 45·8	1 40·7	0·2	0·0	6·2	0·8	12·2	1·5
03	1 45·8	1 46·0	1 40·9	0·3	0·0	6·3	0·8	12·3	1·5
04	1 46·0	1 46·3	1 41·2	0·4	0·1	6·4	0·8	12·4	1·6
05	1 46·3	1 46·5	1 41·4	0·5	0·1	6·5	0·8	12·5	1·6
06	1 46·5	1 46·8	1 41·6	0·6	0·1	6·6	0·8	12·6	1·6
07	1 46·8	1 47·0	1 41·9	0·7	0·1	6·7	0·8	12·7	1·6
08	1 47·0	1 47·3	1 42·1	0·8	0·1	6·8	0·9	12·8	1·6
09	1 47·3	1 47·5	1 42·4	0·9	0·1	6·9	0·9	12·9	1·6
10	1 47·5	1 47·8	1 42·6	1·0	0·1	7·0	0·9	13·0	1·6
11	1 47·8	1 48·0	1 42·8	1·1	0·1	7·1	0·9	13·1	1·6
12	1 48·0	1 48·3	1 43·1	1·2	0·2	7·2	0·9	13·2	1·7
13	1 48·3	1 48·5	1 43·3	1·3	0·2	7·3	0·9	13·3	1·7
14	1 48·5	1 48·8	1 43·6	1·4	0·2	7·4	0·9	13·4	1·7
15	1 48·8	1 49·0	1 43·8	1·5	0·2	7·5	0·9	13·5	1·7
16	1 49·0	1 49·3	1 44·0	1·6	0·2	7·6	1·0	13·6	1·7
17	1 49·3	1 49·5	1 44·3	1·7	0·2	7·7	1·0	13·7	1·7
18	1 49·5	1 49·8	1 44·5	1·8	0·2	7·8	1·0	13·8	1·7
19	1 49·8	1 50·1	1 44·8	1·9	0·2	7·9	1·0	13·9	1·7
20	1 50·0	1 50·3	1 45·0	2·0	0·3	8·0	1·0	14·0	1·8
21	1 50·3	1 50·6	1 45·2	2·1	0·3	8·1	1·0	14·1	1·8
22	1 50·5	1 50·8	1 45·5	2·2	0·3	8·2	1·0	14·2	1·8
23	1 50·8	1 51·1	1 45·7	2·3	0·3	8·3	1·0	14·3	1·8
24	1 51·0	1 51·3	1 45·9	2·4	0·3	8·4	1·1	14·4	1·8
25	1 51·3	1 51·6	1 46·2	2·5	0·3	8·5	1·1	14·5	1·8
26	1 51·5	1 51·8	1 46·4	2·6	0·3	8·6	1·1	14·6	1·8
27	1 51·8	1 52·1	1 46·7	2·7	0·3	8·7	1·1	14·7	1·8
28	1 52·0	1 52·3	1 46·9	2·8	0·4	8·8	1·1	14·8	1·9
29	1 52·3	1 52·6	1 47·1	2·9	0·4	8·9	1·1	14·9	1·9
30	1 52·5	1 52·8	1 47·4	3·0	0·4	9·0	1·1	15·0	1·9
31	1 52·8	1 53·1	1 47·6	3·1	0·4	9·1	1·1	15·1	1·9
32	1 53·0	1 53·3	1 47·9	3·2	0·4	9·2	1·2	15·2	1·9
33	1 53·3	1 53·6	1 48·1	3·3	0·4	9·3	1·2	15·3	1·9
34	1 53·5	1 53·8	1 48·3	3·4	0·4	9·4	1·2	15·4	1·9
35	1 53·8	1 54·1	1 48·6	3·5	0·4	9·5	1·2	15·5	1·9
36	1 54·0	1 54·3	1 48·8	3·6	0·5	9·6	1·2	15·6	2·0
37	1 54·3	1 54·6	1 49·0	3·7	0·5	9·7	1·2	15·7	2·0
38	1 54·5	1 54·8	1 49·3	3·8	0·5	9·8	1·2	15·8	2·0
39	1 54·8	1 55·1	1 49·5	3·9	0·5	9·9	1·2	15·9	2·0
40	1 55·0	1 55·3	1 49·8	4·0	0·5	10·0	1·3	16·0	2·0
41	1 55·3	1 55·6	1 50·0	4·1	0·5	10·1	1·3	16·1	2·0
42	1 55·5	1 55·8	1 50·2	4·2	0·5	10·2	1·3	16·2	2·0
43	1 55·8	1 56·1	1 50·5	4·3	0·5	10·3	1·3	16·3	2·0
44	1 56·0	1 56·3	1 50·7	4·4	0·6	10·4	1·3	16·4	2·1
45	1 56·3	1 56·6	1 51·0	4·5	0·6	10·5	1·3	16·5	2·1
46	1 56·5	1 56·8	1 51·2	4·6	0·6	10·6	1·3	16·6	2·1
47	1 56·8	1 57·1	1 51·4	4·7	0·6	10·7	1·3	16·7	2·1
48	1 57·0	1 57·3	1 51·7	4·8	0·6	10·8	1·4	16·8	2·1
49	1 57·3	1 57·6	1 51·9	4·9	0·6	10·9	1·4	16·9	2·1
50	1 57·5	1 57·8	1 52·1	5·0	0·6	11·0	1·4	17·0	2·1
51	1 57·8	1 58·1	1 52·4	5·1	0·6	11·1	1·4	17·1	2·1
52	1 58·0	1 58·3	1 52·6	5·2	0·7	11·2	1·4	17·2	2·2
53	1 58·3	1 58·6	1 52·9	5·3	0·7	11·3	1·4	17·3	2·2
54	1 58·5	1 58·8	1 53·1	5·4	0·7	11·4	1·4	17·4	2·2
55	1 58·8	1 59·1	1 53·3	5·5	0·7	11·5	1·4	17·5	2·2
56	1 59·0	1 59·3	1 53·6	5·6	0·7	11·6	1·5	17·6	2·2
57	1 59·3	1 59·6	1 53·8	5·7	0·7	11·7	1·5	17·7	2·2
58	1 59·5	1 59·8	1 54·1	5·8	0·7	11·8	1·5	17·8	2·2
59	1 59·8	2 00·1	1 54·3	5·9	0·7	11·9	1·5	17·9	2·2
60	2 00·0	2 00·3	1 54·5	6·0	0·8	12·0	1·5	18·0	2·3

© 2009 Jack Case

A2 ALTITUDE CORRECTION TABLES 10°–90°—SUN, STARS, PLANETS

OCT.–MAR. SUN APR.–SEPT.				STARS AND PLANETS				DIP					
App. Alt.	Lower Limb	Upper Limb	App. Alt.	Lower Limb	Upper Limb	App. Alt.	Corrⁿ	App. Alt.	Additional Corrⁿ	Ht. of Eye	Corrⁿ	Ht. of Eye	Corrⁿ	Ht. of Eye	Corrⁿ
° ′	′	′	° ′	′	′	° ′	′		**2009**	m	′	ft.	′	m	′
9 33	+10.8	−21.5	9 39	+10.6	−21.2	9 55	−5.3		**VENUS**	2.4	−2.8	8.0		1.0	− 1.8
9 45	+10.9	−21.4	9 50	+10.7	−21.1	10 07	−5.2		Jan. 1–Jan. 28	2.6	−2.9	8.6		1.5	− 2.2
9 56	+11.0	−21.3	10 02	+10.8	−21.0	10 20	−5.1		May 23–July 10	2.8	−3.0	9.2		2.0	− 2.5
10 08	+11.1	−21.2	10 14	+10.9	−20.9	10 32	−5.0		° ′	3.0	−3.1	9.8		2.5	− 2.8
10 20	+11.2	−21.1	10 27	+11.0	−20.8	10 46	−4.9		0 +0.2	3.2	−3.2	10.5		3.0	− 3.0
10 33	+11.3	−21.0	10 40	+11.1	−20.7	10 59	−4.8		41 +0.1	3.4	−3.3	11.2		See table
10 46	+11.4	−20.9	10 53	+11.2	−20.6	11 14	−4.7		76	3.6	−3.4	11.9		←
11 00	+11.5	−20.8	11 07	+11.3	−20.5	11 29	−4.6		Jan. 29–Feb. 21	3.8	−3.5	12.6		m	′
11 15	+11.6	−20.7	11 22	+11.4	−20.4	11 44	−4.5		May 1–May 22	4.0	−3.6	13.3		20	− 7.9
11 30	+11.7	−20.6	11 37	+11.5	−20.3	12 00	−4.4		° ′	4.3	−3.7	14.1		22	− 8.3
11 45	+11.8	−20.5	11 53	+11.6	−20.2	12 17	−4.3		0 +0.3	4.5	−3.8	14.9		24	− 8.6
12 01	+11.9	−20.4	12 10	+11.7	−20.1	12 35	−4.2		34 +0.2	4.7	−3.9	15.7		26	− 9.0
12 18	+12.0	−20.3	12 27	+11.8	−20.0	12 53	−4.1		60 +0.1	5.0	−4.0	16.5		28	− 9.3
12 36	+12.1	−20.2	12 45	+11.9	−19.9	13 12	−4.0		80	5.2	−4.1	17.4			
12 54	+12.2	−20.1	13 04	+12.0	−19.8	13 32	−3.9		Feb. 22–Mar. 9	5.5	−4.2	18.3		30	− 9.6
13 14	+12.3	−20.0	13 24	+12.1	−19.7	13 53	−3.8		Apr. 15–Apr. 30	5.8	−4.3	19.1		32	−10.0
13 34	+12.4	−19.9	13 44	+12.2	−19.6	14 16	−3.7		° ′	6.1	−4.4	20.1		34	−10.3
13 55	+12.5	−19.8	14 06	+12.3	−19.5	14 39	−3.6		29 +0.4	6.3	−4.5	21.0		36	−10.6
14 17	+12.6	−19.7	14 29	+12.4	−19.4	15 03	−3.5		51 +0.3	6.6	−4.6	22.0		38	−10.8
14 41	+12.7	−19.6	14 53	+12.5	−19.3	15 29	−3.4		68 +0.2	6.9	−4.7	22.9			
15 05	+12.8	−19.5	15 18	+12.6	−19.2	15 56	−3.3		83 +0.1	7.2	−4.8	23.9		40	−11.1
15 31	+12.9	−19.4	15 45	+12.7	−19.1	16 25	−3.2		Mar. 10–Apr. 14	7.5	−4.9	24.9		42	−11.4
15 59	+13.0	−19.3	16 13	+12.8	−19.0	16 55	−3.1		° ′	7.9	−5.0	26.0		44	−11.7
16 27	+13.1	−19.2	16 43	+12.9	−18.9	17 27	−3.0		26 +0.5	8.2	−5.1	27.1		46	−11.9
16 58	+13.2	−19.1	17 14	+13.0	−18.8	18 01	−2.9		46 +0.4	8.5	−5.2	28.1		48	−12.2
17 30	+13.3	−19.0	17 47	+13.1	−18.7	18 37	−2.8		60 +0.3	8.8	−5.3	29.2		ft.	′
18 05	+13.4	−18.9	18 23	+13.2	−18.6	19 16	−2.7		73 +0.2	9.2	−5.4	30.4		2	− 1.4
18 41	+13.5	−18.8	19 00	+13.3	−18.5	19 56	−2.6		84 +0.1	9.5	−5.5	31.5		4	− 1.9
19 20	+13.6	−18.7	19 41	+13.4	−18.4	20 40	−2.5		July 11–Dec. 31	9.9	−5.6	32.7		6	− 2.4
20 02	+13.7	−18.6	20 24	+13.5	−18.3	21 27	−2.4		° ′	10.3	−5.7	33.9		8	− 2.7
20 46	+13.8	−18.5	21 10	+13.6	−18.2	22 17	−2.3		60 +0.1	10.6	−5.8	35.1		10	− 3.1
21 34	+13.9	−18.4	21 59	+13.7	−18.1	23 11	−2.2			11.0	−5.9	36.3			
22 25	+14.0	−18.3	22 52	+13.8	−18.0	24 09	−2.1		**MARS**	11.4	−6.0	37.6		See table
23 20	+14.1	−18.2	23 49	+13.9	−17.9	25 12	−2.0		Jan. 1–Nov. 27	11.8	−6.1	38.9		←
24 20	+14.2	−18.1	24 51	+14.0	−17.8	26 20	−1.9		° ′	12.2	−6.2	40.1		ft.	′
25 24	+14.3	−18.0	25 58	+14.1	−17.7	27 34	−1.8		60 +0.1	12.6	−6.3	41.5		70	− 8.1
26 34	+14.4	−17.9	27 11	+14.2	−17.6	28 54	−1.7			13.0	−6.4	42.8		75	− 8.4
27 50	+14.5	−17.8	28 31	+14.3	−17.5	30 22	−1.6		Nov. 28–Dec. 31	13.4	−6.5	44.2		80	− 8.7
29 13	+14.6	−17.7	29 58	+14.4	−17.4	31 58	−1.5		° ′	13.8	−6.6	45.5		85	− 8.9
30 44	+14.7	−17.6	31 33	+14.5	−17.3	33 43	−1.4		41 +0.2	14.2	−6.7	46.9		90	− 9.2
32 24	+14.8	−17.5	33 18	+14.6	−17.2	35 38	−1.3		76 +0.1	14.7	−6.8	48.4		95	− 9.5
34 15	+14.9	−17.4	35 15	+14.7	−17.1	37 45	−1.2			15.1	−6.9	49.8			
36 17	+15.0	−17.3	37 24	+14.8	−17.0	40 06	−1.1			15.5	−7.0	51.3		100	− 9.7
38 34	+15.1	−17.2	39 48	+14.9	−16.9	42 42	−1.0			16.0	−7.1	52.8		105	− 9.9
41 06	+15.2	−17.1	42 28	+15.0	−16.8	45 34	−0.9			16.5	−7.2	54.3		110	−10.2
43 56	+15.3	−17.0	45 29	+15.1	−16.7	48 45	−0.8			16.9	−7.3	55.8		115	−10.4
47 07	+15.4	−16.9	48 52	+15.2	−16.6	52 16	−0.7			17.4	−7.4	57.4		120	−10.6
50 43	+15.5	−16.8	52 41	+15.3	−16.5	56 09	−0.6			17.9	−7.5	58.9		125	−10.8
54 46	+15.6	−16.7	56 59	+15.4	−16.4	60 26	−0.5			18.4	−7.6	60.5			
59 21	+15.7	−16.6	61 50	+15.5	−16.3	65 06	−0.4			18.8	−7.7	62.1		130	−11.1
64 28	+15.8	−16.5	67 15	+15.6	−16.2	70 09	−0.3			19.3	−7.8	63.8		135	−11.3
70 10	+15.9	−16.4	73 14	+15.7	−16.1	75 32	−0.2			19.8	−7.9	65.4		140	−11.5
76 24	+16.0	−16.3	79 42	+15.8	−16.0	81 12	−0.1			20.4	−8.0	67.1		145	−11.7
83 05	+16.1	−16.2	86 31	+15.9	−15.9	87 03	0.0			20.9	−8.1	68.8		150	−11.9
90 00			90 00			90 00				21.4		70.5		155	−12.1

Chapter 7
The Astronomical Position Line

The theory of position fixing by astro navigation depends on the ability to solve the triangle PZX (see diag. 3) by relating the observed altitude of a celestial body (which is theoretically accurate) to a D.R. position (which is only approximate). In this way, we are able to determine the geographical position of the celestial body and then calculate our true position in relation to it at the instant of the observations.

Before moving on, it should be pointed out that although the Sun is used as the 'celestial body' for this demonstration, we will see in succeeding chapters, that the Moon, stars and planets can also be used.

Suppose we are in a yacht and we measure the altitude of the Sun and find it to be 35°; what does this tell us? All that we know is that the yacht lies somewhere on the circumference of a circle centered at the geographical position of the Sun. Such a circle is known as a **'position circle'** since our position is known to lie somewhere on its circumference.

Diagram 34 shows that, at any point on the circumference of the circle, the Sun's altitude will be 35° and our distance from the GP will be equal to the radius. The problem is to establish at which precise point on the position circle the yacht lies.

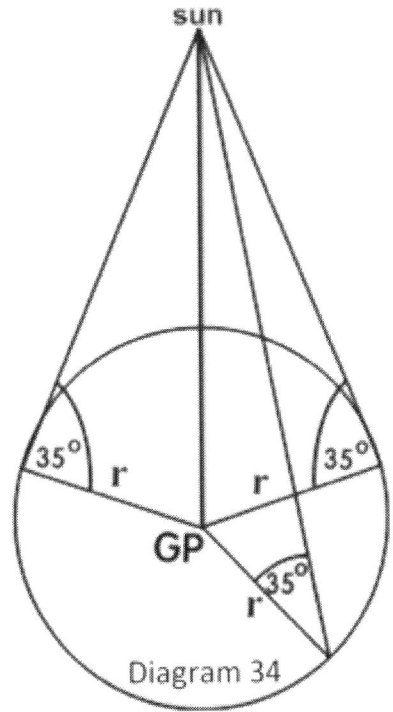

Diagram 34

Now consider diagram 35,

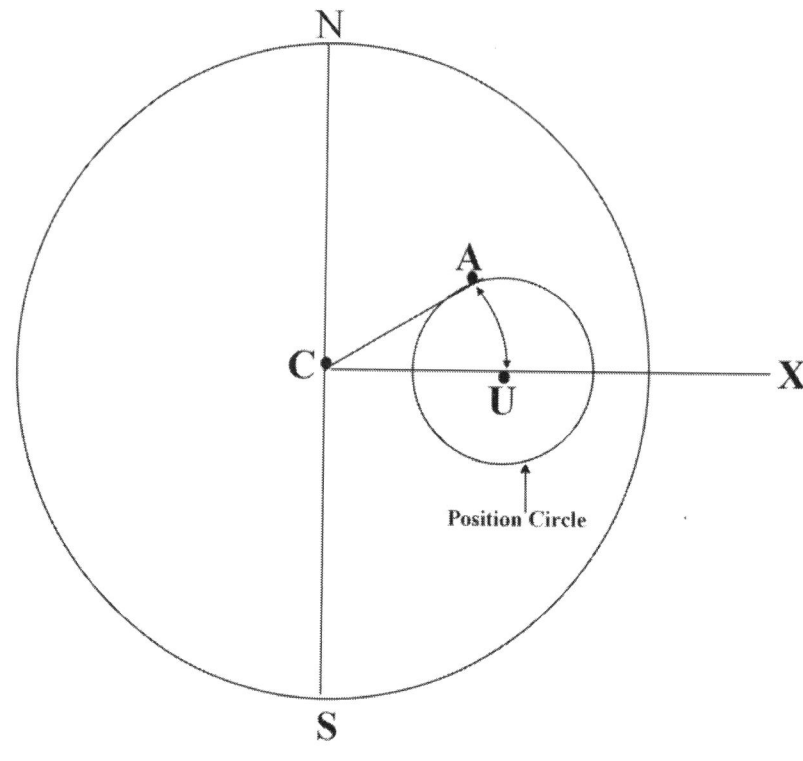

Diagram 35

A is the yacht's position,
X is the position of the Sun,
U is the GP of the Sun,
C is the Earth's centre.
Point A lies somewhere on the position circle centred at U and AU is the radius of the position circle.

At first, it might seem that all we need to do is to observe the bearing of the Sun at the same time that we measure its altitude and then draw the line of bearing on the chart along with the position circle. In this way, it would seem that our true position would correspond to the intersection of these lines on the chart. However, there is a problem with this idea which makes it impracticable. Because of the great distance of the Sun from the Earth, the radius of the position circle will be very large (approximately 3000 n.m. or so). A chart on which such a large circle could be drawn would require such a small scale that accurate position-fixing would be impossible. However, we know our D.R. position which, although approximate, should be accurate to within a degree of latitude and longitude and this may give us another way of tackling the problem.

If we could work out what the altitude would have been at the D.R. position at the time that the altitude was measured at the true position, we would then be able to compare the two altitudes and calculate the difference between them.

Since the D.R. position is only approximate, no accuracy will be lost if we establish a position in the vicinity of the D.R. position in which fractions of degrees are rounded up to make calculations easier. Such a position is known as an assumed position.

In diagram 36,
A marks the true position.
B marks the assumed position.
U marks the geographical position of the Sun (GP).

It can be seen that what we are interested in is the difference between AU and BU since this represents the distance from the assumed position to the true position which is what we need to calculate.

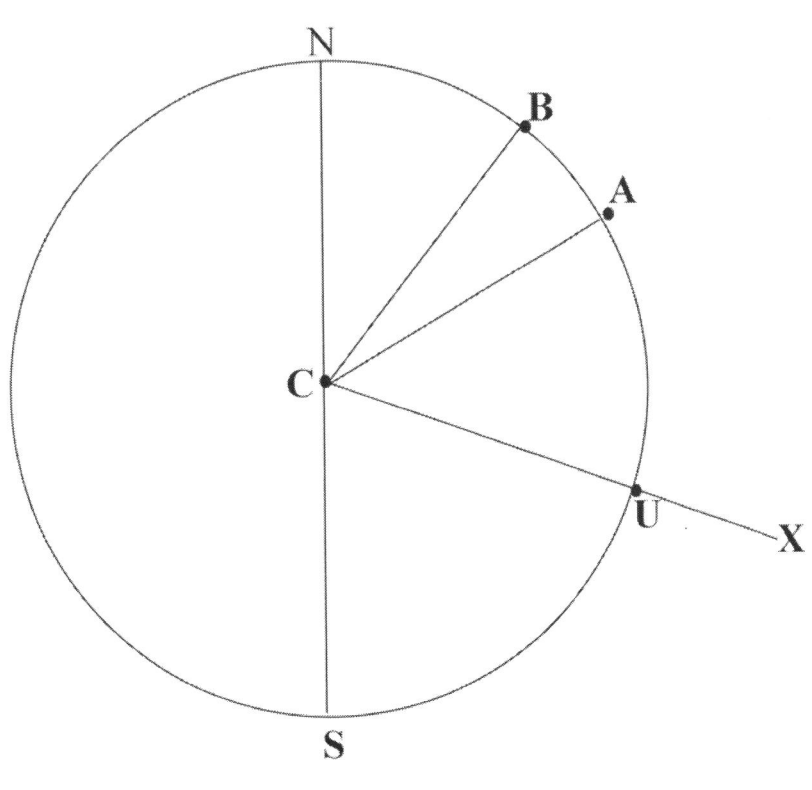

Diagram 36

We can see from the diagram that the difference between AU & BU is equal to AB. This seems obvious but it is an important point to make because what we are investigating is whether AB can be found from the difference between the altitudes of the Sun at points A and B.

We know that there is a direct relationship between the altitude of the Sun at a place and the distance of that place from the Sun's GP. The altitude can be measured from the yacht at the true position (A) and so AU can easily be calculated. The altitude cannot be measured at the assumed position but its distance from the GP can be calculated by another method. The difference between these distances will be the distance from the assumed position to the true position. That is AB = AU ~ BU.

Calculating the zenith distance at the True Position.
Diagram 36 seems to suggest that A and U lie on the same meridian of longitude but this obviously will only be the case at apparent noon. Since we are trying to discover a method of calculating position at any time, we must consider the problem with A and U lying on different meridians of longitude.
Consider diagram 37

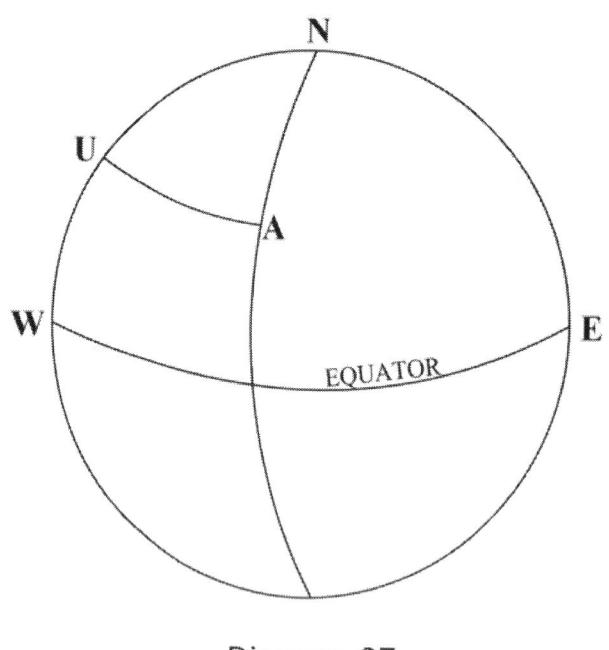

Diagram 37

In diagram 37, the GP of the Sun (U) lies on a different meridian of longitude to that of the true position (A).
N   represents the North Pole,
WE   represents the Equator,
WU   represents the declination of the Sun,
NU   represents 90° - declination,
NA   represents 90° - latitude of A,

AU represents the distance of the Sun's geographical position from the true position (A).

We can see that the task of finding the distance AU depends on solving the triangle NAU.

In Diagram 38, the true position of the yacht is represented by A.

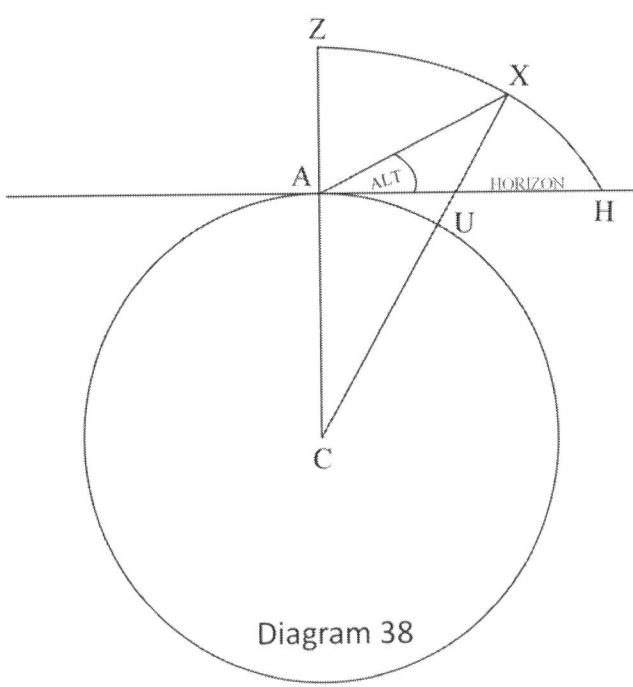

Diagram 38

For this example, we will assume that the altitude measured from the true position is 35°.
Z is the zenith of A (that is, a point immediately above A along the extension of a line joining A to the centre of the Earth).
H represents the horizon from point A (i.e. a tangent to the Earth's surface at A in the direction of AU).
X represents the position of the Sun.
U represents the GP of the Sun.
The angle ZAH is 90° and subtends the arc ZH
The angle XAH is the altitude i.e. 35°.
The arc XH is subtended by the angle XAH.
The angular distance ZX is equal to the angular distance AU since ZX and AU are both arcs subtended by the angle ZCX.

Angle ZAX = 90° – angle XAH (i.e. 90° - altitude).
Therefore ZX = 90° – altitude expressed as an angular distance.

As we established in chapter 5, we call the angular distance ZX the 'zenith distance'.

We have established that the arc AU equals the arc ZX in terms of angular distance and since 1 minute of arc along the path of a great circle on the surface of the Earth equals 1 nautical mile, the angular distance AU equals the length of the arc AU in nautical miles. So we have found a method of calculating the distance between the true position and the GP of the Sun.
i.e. AU = 90° - Altitude = 90° - 35° = 55°
and 55° = 3300' or 3300 n.m.
From this we can conclude that the distance of the true position from the GP is 3300 n.m.

Calculating the Zenith Distance of the Assumed Position.

Diagram 3 which is repeated on the next page shows that the angular distance BU on the Earth's surface is equal to the angular distance ZX in the spherical triangle PZX.

In the diagram,
- X represents the position of the Sun on the celestial sphere,
- Z represents a point on the sphere which coincides with the zenith of the assumed position (B),
- P represents the projection of the North Pole onto the celestial sphere,
- PX = NU = (90° - the declination of the Sun),
- PZ = NB = (90° - the latitude of the assumed position),
- ZX = BU = (90° - the altitude of the Sun).

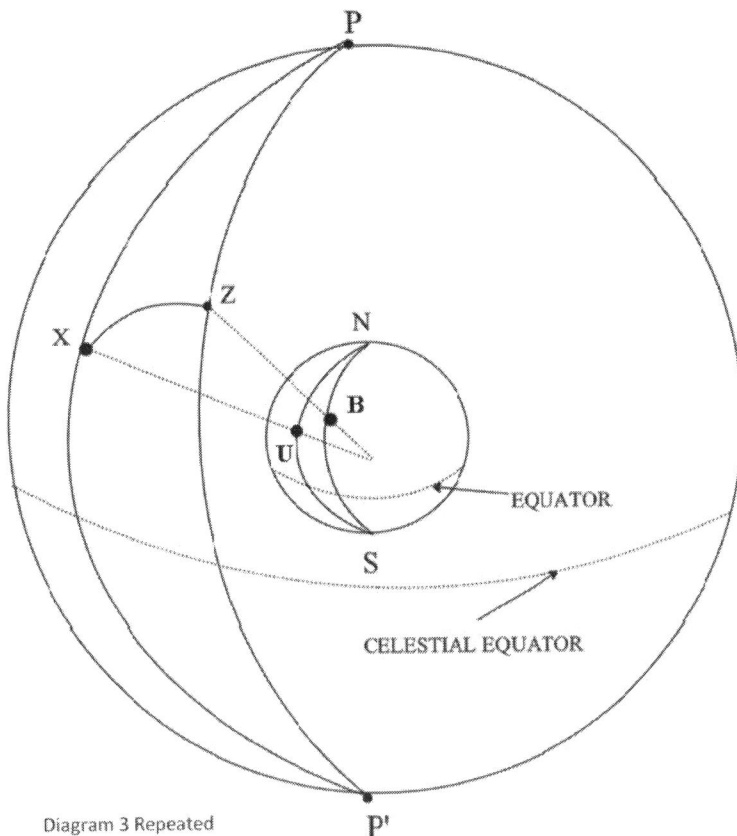

Diagram 3 Repeated

So we can see, from the diagram that the triangle NBU on the Earth's surface can be solved, in effect, by solving the triangle PZX in the celestial sphere.

We can find the Sun's declination from the Nautical Almanac at the time that the altitude was measured at the true position and so it follows that we can calculate PX (i.e. 90° - Declination).

We know the approximate latitude of the assumed position and so we can calculate PZ which, as stated above, is equal to 90° - Latitude.

We have established that we are able to calculate two sides of the spherical triangle PZX (i.e. PX = 90° – Dec. and PZ = 90° – Lat.)
We must now find a way of calculating the third side, ZX (the zenith distance) which is equal to the angular distance BU on the Earth's surface (i.e. the distance from the assumed position to the GP).

To summarise the problem, the sides PZ & PX of the spherical triangle PZX are known but the included angle ZPX is not known and that is what must be found in order to calculate the side ZX.

The angle ZPX is equal to the angle BNU on the Earth's surface (that is, the difference between the longitude of the assumed position and the longitude of the GP of the Sun). We will know the approximate longitude of the yacht from the assumed position. If we were also to know the longitude of the Sun's GP then the angle ZPX would be the difference between the two longitudes (i.e. ZPX = Long. B ~ Long. U). So, what we must now investigate, is how to calculate the longitude of the GP.

We know from chapter 1 that the angle ZPX can be measured in terms of time and for this reason, it is know as the Local Hour Angle. Suppose, from the nautical almanac, we found that the Sun's GHA at the instant of the observation at the true position was 228° 53' 45" and say the longitude of the assumed position was 165° 59' 30"E.

We could calculate the LHA as follows:
We could calculate the LHA as follows:
GHA = 228° 53' 45"
Long. = 165° 59' 30" E (+)
LHA = (394° 53' 15") - (360°)
LHA = 34° 53' 15" and therefore, angle ZPX = 34° 53' 15"

Referring back to diagram 3 above, the side ZX in the spherical triangle PZX can now be found since the other two sides and the included angle are known. Since the arc ZX in the celestial sphere equals the arc BU on the Earth's surface, it would seem that all we need to do now is to convert the angular distance 34° 53' 15" to nautical miles in order to find the distance from the assumed position to the Sun's GP.

However, in astro navigation, things are not always as simple as this because, as previously discussed, the Earth's surface is curved and therefore it follows that lines drawn on that surface must also be curved. Therefore, to find the length of ZX we must employ 'spherical trigonometry'.

To fully understand the complexities of navigating on the surface of a sphere, it would be helpful to have a knowledge of 'spherical trigonometry'. However, it would only interrupt the flow of this chapter to offer an exposition of the topic here and so we will simply resort to employing the formulas developed in the appendix instead. Those who would prefer to understand how those formulas were developed should refer to the appendix.

At this point, we need to reflect on what we have learned.
The following demonstration employs the **'Intercept method'** also known as the **'Marcq St. Hilaire'** method after the French Navigator who devised it in 1875.
We will start with a new set of data as follows:
Date: 21 June 2009.   Time: $12^h\ 41^m\ 19^s$ GMT.
Assumed Position: 45°.30' N, 10° 19'.8 W.   True altitude of Sun: 68°.06

The modified version of diagram 3 below shows that AU is equal to the angular distance ZX in the spherical triangle PZX i.e. the zenith distance at the true position.

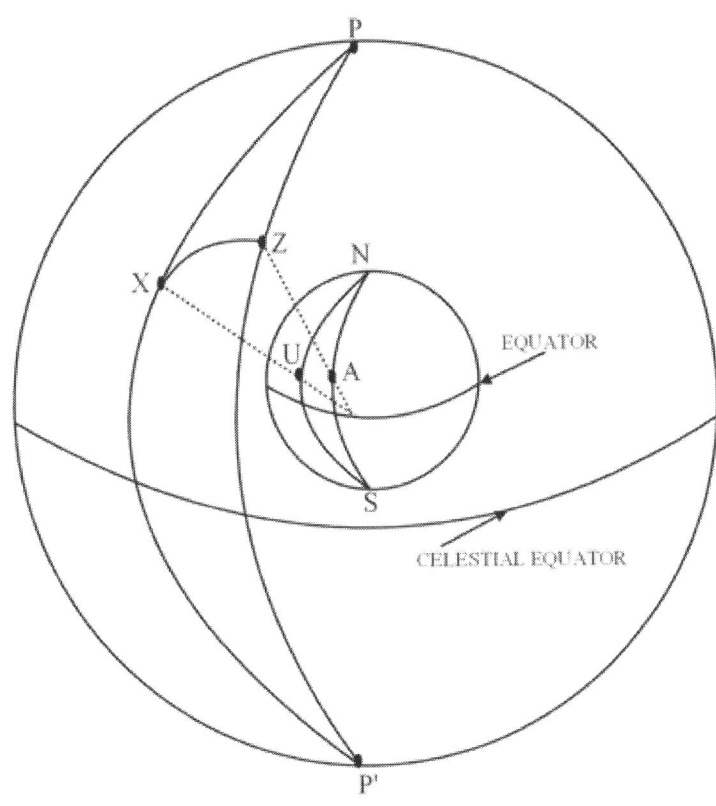

Diagram 3 Modified

To recap, our tasks are to calculate the zenith distance at the assumed position and at the true position. The difference between these distances will give us the 'intercept'.

Calculating the Zenith Distance at the True Position

As we learned previously, the 'zenith Distance' is calculated by the formula: 90° - Altitude and for this demonstration, it was stated that the Sun's altitude, as measured at the true position, was 68°.06

Using this information, the calculation for finding the zenith distance at the true position would be as shown below:

> Zenith Distance = 90° - Alt
> = 90° - 68°.06
> = 21°.914 = 1314'.84
> = 1314.84 n.m.

The next step is to calculate the zenith distance of the assumed position at the instant of measuring the altitude at the true position.

Calculating the zenith distance of the assumed position:

Note. For the sake of simplicity, the values of GHA, Declination and the increments for 41 minutes are given in the following demonstration without inclusion of the relevant Nautical Almanac extracts.

> **Step 1.** Find the Sun's Declination at the time of measuring the altitude.
> Greenwich date: June 21^d 12^h 41^m 19^s
> Dec Sun (12^h):   N23° 26'.3 (d = 0'.0)
> Corrn (41^m):   0'.0
> Dec Sun: (12^h41^m) N23° 26'.3

> **Step 2.** Calculate side PX of the triangle PZX.
> PX = 90° - Dec.
> = 90° - 23° 26'.3
> = 90° - 23°.44
> ∴ PX = 66°.56

> **Step 3. Calculate side PZ**
> PZ = 90° − lat.
>   = 90° − 45°.5
> ∴ PZ = 44°.5

> **Step 4. Calculate angle ZPX (LHA Sun)**
> (Remember the rule: Long West, LHA = GHA − LONG)
> GHA Sun (12^h)   359° 32'.8
> Inc. (41^m 19^s)   10° 19'.8
> GHA Sun   369° 52'.6
> Long   10° 19'.8 W. (−)
> LHA Sun   359° 32'.8
> ∴ ZPX = 359° 32'.8
>   = 359°.55

Note. The formula derived in the appendix for calculating the third side of a spherical triangle when the other two sides and their included angle are known is:
Cos a = [Cos b × Cos c] + [Sin b × Sin c × Cos A]. (Where a, b & c are sides of the spherical triangle and A is the included angle).
This formula is used at step 5 without explanation. A full explanation of the derivation of the formula is given in the appendix.

> **Step 5. Calculate side ZX**
> Cos ZX = [Cos PZ. Cos PX] + [Sin PZ. Sin PX. Cos ZPX]
> Substituting the values calculated above into the formula gives:
> Cos ZX = [Cos(44.5°).Cos(66.56°)] + [Sin(44.5°).Sin(66.56°).Cos(359°.55)]
>   = [0.713 × 0.398] + [0.701 × 0.917 × 0.999969]
>   = 0.284 + 0.643
> Cos ZX = 0.927
> ∴ ZX = 22°.028 = 1321'.68

Summary. From the above calculations, we have established that the zenith distance of the true position is 1314.84 n.m.
We have also established that the zenith distance of the assumed position is 1321.68 n.m.

Pause for Thought.
Before tackling step 6, we need to pause to clarify what we have so far calculated.

We know that the distance from the GP to the true position is 1314.84 n.m. and we know that the distance from the GP to the assumed position is 1321.68 n.m. n.m.

From this we can deduce that the true position is 6.84 n.m. closer to the GP than the assumed position.

Consider Diagram 39:

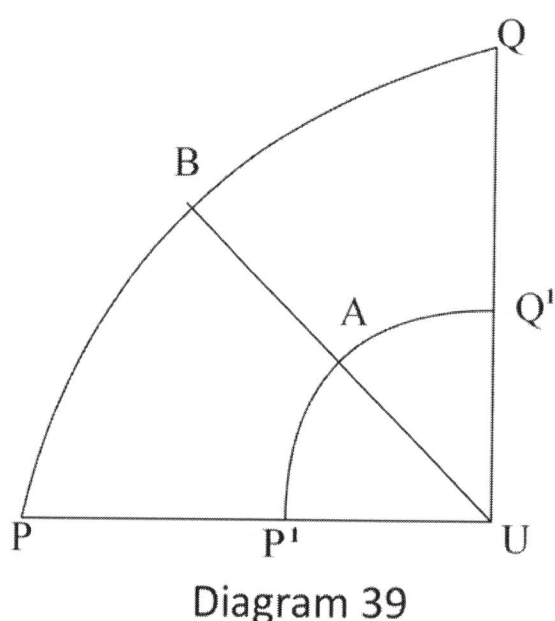

Diagram 39

B represents the Assumed. Position
A represents the true Position
U represents the Sun's GP
BU is known to be 1321.68 n.m. since this is equal to the zenith distance of B, (ZX in triangle PZX). This means that B lies on the circumference of a circle of radius 1321.68 n.m. centered at the GP. PBQ represents a small arc of this circle.
AU is known to be 1314.84 n.m. since this is equal to the zenith distance of A. Therefore, A lies somewhere on the circumference of a circle of radius 1314.84 n.m. centered at U and a small arc of this circle is represented by P^1AQ^1.

All we can conclude from this is that the true position is 6.84 nautical miles from the assumed position. We must remember that the line BA represents the distance and not the direction of A from B. We do not know the direction at this stage and to find it must be our next step.

Casting our minds back to chapter 5, we know that the angle between the observer's meridian and the direction of the celestial body is known as the Azimuth and is the angle PZX in diagram 3.

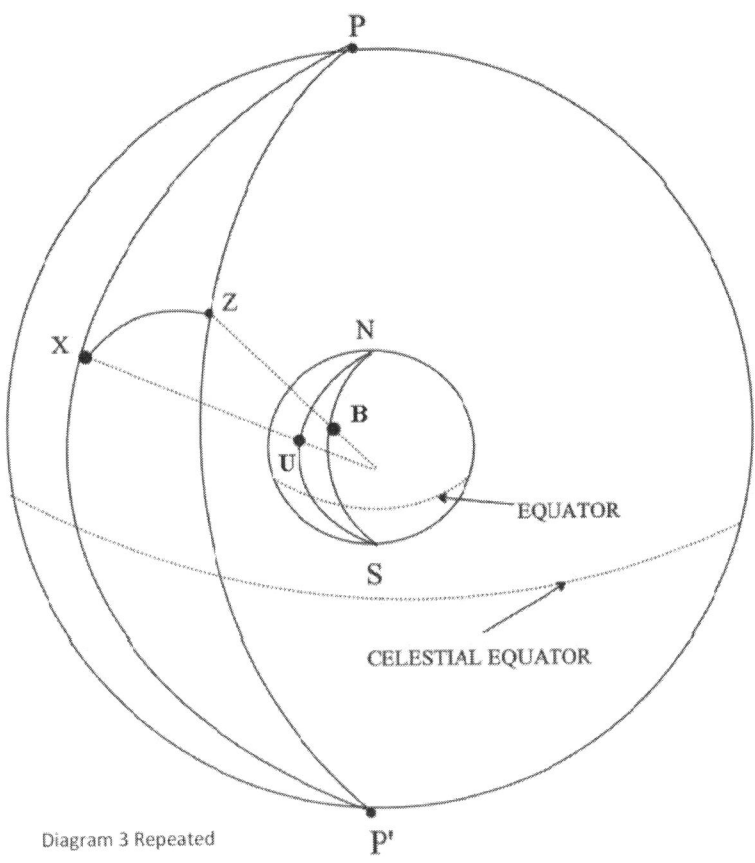

Diagram 3 Repeated

To summarize, we know that the true position is 6.84 nautical miles away from the assumed position in the direction indicated by the azimuth.
So now, in step 6, the task is to calculate angle PZX.

As discussed at step 5, the formula derived in the appendix for calculating the third side of a spherical triangle when the other two sides and their included angle are known is:
Cos a = [Cos b x Cos c] + [Sin b x Sin c x Cos A]
Transposing the formula for Cos A, we have:
Cos A = Cos a - (Cos b. Cos c) ÷ (Sin b. Sin c)
Substituting for the labels in the spherical triangle PZX, this becomes:
Cos PZX = Cos PX - (Cos ZX. Cos PZ) ÷ (Sin ZX. Sin PZ)

From our calculations in the preceding steps, we have deduced the following:

PX = 66°.56, PZ = 44°.5, ZX = 22°.028; so in order to calculate PZX, we must substitute for these values in the above formula:

Step 6. Calculate Angle PZX

Cos PZX = Cos PX − (Cos ZX . Cos PZ) ÷ (Sin ZX . Sin PZ)

= Cos(66°.55)−[Cos(22°.028)×Cos(44°.5)] ÷ [Sin(22°.028) × Sin(44°.5)]

= 0.3979 − [0.927 × 0.713] ÷ [0.375 × 0.7]

= 0.398 − 0.66095 ÷ 0.2625

= −0.2629 ÷ 0.2625

= −1

∴ PZX = 180°

∴ Azimuth = N180°E (since the latitude of the assumed position is north).

From the above, we have deduced that the azimuth of the GP from the assumed position is N180°E. In other words, the bearing of the Sun would have been due south. Since the true position is 6.84 nautical miles closer to the GP than the assumed position in the direction of the azimuth, its bearing from the assumed position must be due south.

We know that, since the assumed position is north of 23.5°N, the Sun will bear due south at noon. So, the fact that the azimuth of the Sun was due south would indicate that local time was noon at the time of the sighting. We can check that this was so by converting the GMT of the sighting to LMT:

Using the data given above:

GMT: $12^h\ 41^m\ 19^s$

Long: 10° 19'.8 W.

Convert 10° 19'.8 to time:

$4 \times 10° \div 60 = 0.(6)^h = 40^m$

$4 \times 19'.8 \div 60 = 1.32^m = 1^m\ 19^s$

$\qquad\qquad\qquad\qquad 41^m\ 19^s$

∴ Time difference = $41^m\ 19^s$ (Lat. west, GMT best)

∴ LMT = $12^h\ 41^m\ 19^s - 41^m\ 19^s$

= $12^h\ 00^m$

So we have discovered that the azimuth and the compass bearing provide exactly the same directional information albeit in different formats.

This begs the question: Why go to the trouble of calculating azimuth when it is easier to take the compass bearing? However, there are essential differences between the calculated azimuth and the bearing measured by a compass. We calculate the azimuth by finding the angle PZX from the values of the sides PZ, PX and ZX in the spherical triangle ZPX and these values are derived from data relating to the assumed position. If we take a compass bearing, obviously we can only do that from the true position. At the time of taking the altitude, we would not know where the true position is although we would know how far it is from the assumed position. There is also the point that our calculations relate to GMT and so the azimuth will relate to the mean sun whereas a compass bearing would obviously relate to the apparent sun. Therefore, our aim must be to find the direction of the true position from the assumed position and we can only do this by calculating the azimuth.

If the two positions are close together, then there will be very little difference between the calculated azimuth and the compass bearing and no accuracy will be lost by using the latter. However, at the time of taking the altitude, we do not know how close together the two positions are and so, for the sake of accuracy, it is important to calculate the azimuth from the assumed position.

So far, we have found that the true position is 6.84 n.m. from the assumed position in the direction of south. Does this tell us where the true position is? Unfortunately not; we must remember that the assumed position is approximate and therefore the position that we have found 6.84 south of it must also be an approximate position.

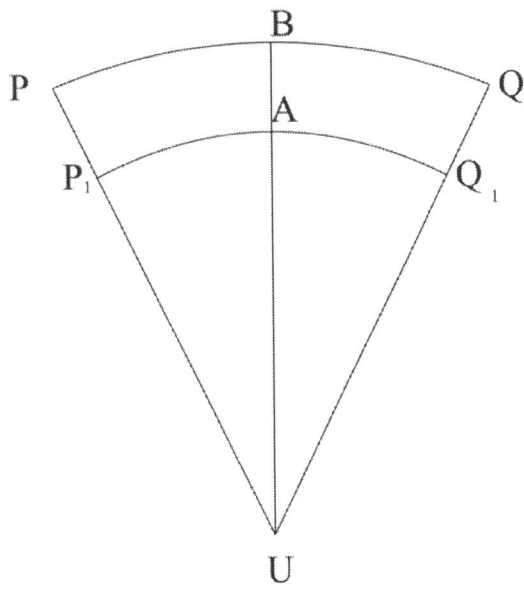

Diagram 40

In Diagram 40, we have taken Diagram 39 and rotated it so that BAU is aligned to north/south, we will see that all that we know is that the true position lies on an arc of radius 1314.84 n.m in a northerly direction from point U, somewhere in the vicinity of point A.

We also know that point B lies on an arc of radius 1321.68 n.m. from point U and that point A is 6.84 n.m. from B in a southerly direction.

In Diagram 41, we have further modified the original diagram by removing point A and replacing it with point J.

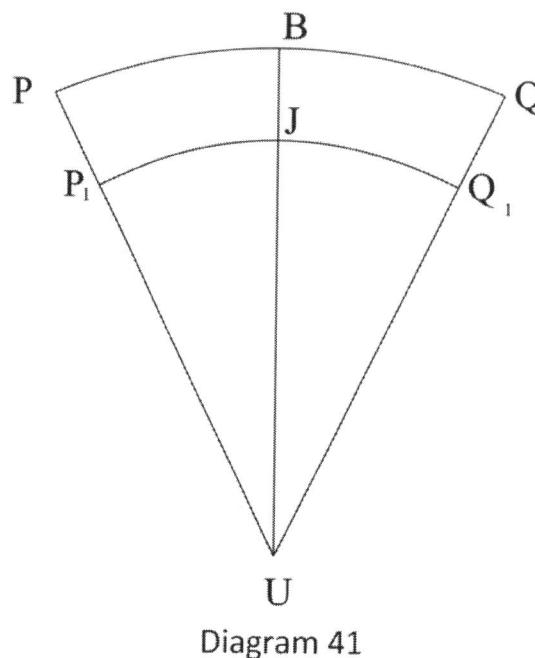

Diagram 41

The true position still lies on the arc P_1JQ_1 but at this stage, we don't know exactly where; all that is known is that it is somewhere on this arc in the vicinity of J. We can say that J is one point which is the correct distance from the GP of the Sun but this will not necessarily coincide with the true position.

At this point, it may seem that we are becoming 'bogged-down' with mathematical arguments without actually finding a solution to the problem. However, if we continue with this line of investigation for a little longer, things will begin to become clear.

To continue, we know that the assumed position lies somewhere in the vicinity of point B on the arc PBQ on a position circle of radius 1321.68 n.m. from the GP (point U) in diagram 41. We also know that the true position lies somewhere in the vicinity of point J on the arc P_1JQ_1 on a position circle of radius 1314.84 n.m. from the GP. The difference between the two position circles (the distance BJ) is known as the **'intercept'**.

Since the circumference of a circle at any point is at right-angles to the radius at that point, no accuracy will be lost by drawing the small arc P1JQ1 as a straight line through J at right-angles to the intercept. The navigator will then be able to draw a short, straight line on the chart, along which he knows his position lies; such a line is known as a position line.

Diagram 42 (not drawn to scale) shows how the results so far achieved might be represented on a navigational chart.

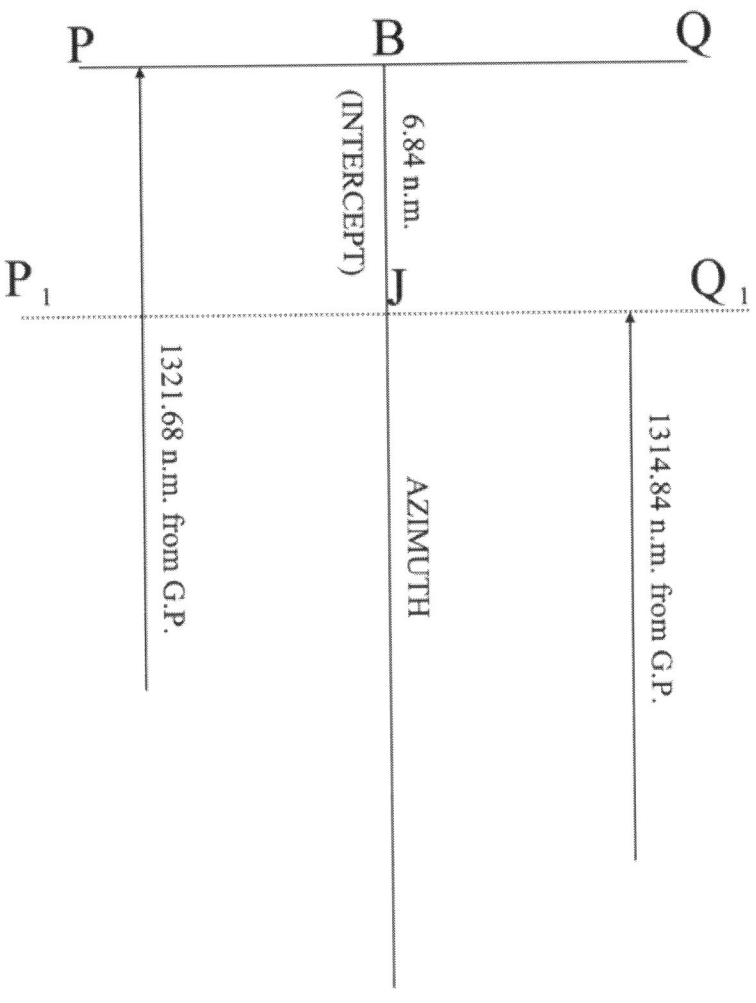

Diagram 42

We can fix a position on a chart by finding the intersection of at least two position lines. However, with only one altitude of the Sun upon which to base our calculations, we are only able to obtain one position line and therefore it would seem, at first sight, that we are unable to find an exact position from this method.

However, by measuring the altitude of the Sun at two different times of the day we could achieve two position lines and this may provide a solution to the problem.

If some way could be found to transfer one of the position lines so that it intersects with the other, a positional fix could be achieved.

Suppose that for the next hour, the vessel sails at 15 knots, on a course of 315°. This means that it will depart from some point on P_1JQ_1 and after one hour, it will arrive at some point on another position line which is parallel to P_1JQ_1.

So, if we transfer the position line P_1JQ_1 along the vessel's course of 315° for a distance of 15 n.m. we will establish a new position line P_2Q_2 as shown in Diagram 43.

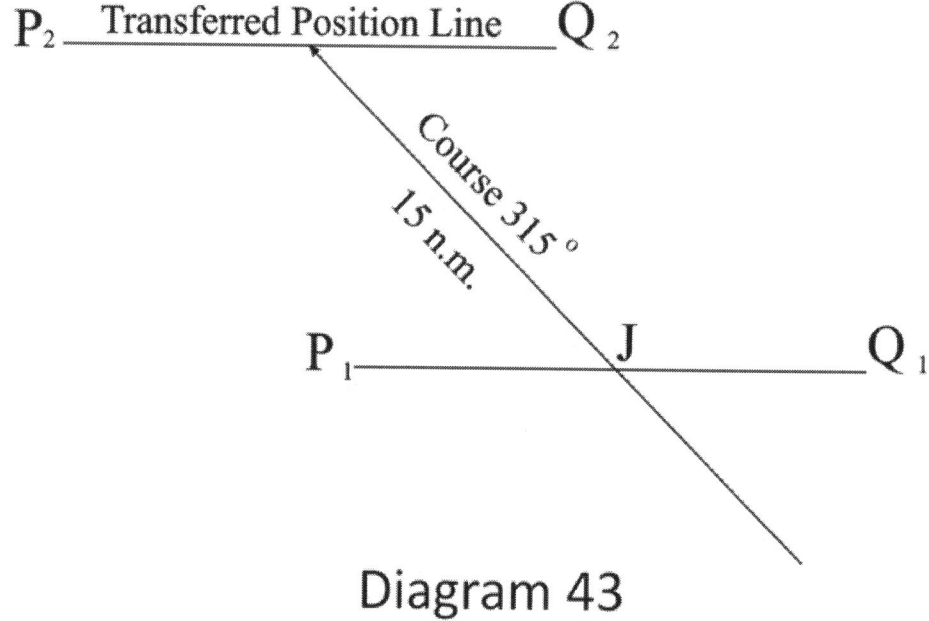

Diagram 43

We can see from the diagram that any point on the new position line will be 15 n.m. on a bearing of 315° from its corresponding point on P_1JQ_1.

At this point, we still have only one position line (i.e. the transferred position line). However, after it has sailed on course 315° for 15 miles, the vessel will arrive at a new assumed position.

If a new position line is established at the new assumed position from a fresh set of measurements and calculations, it will be known that the vessel's position lies somewhere on the new position line. At the same time, it will be known that the vessel's position also lies at some point on the transferred position line. So, it is easy to see that the exact position must be at the intersection of these two position lines.

Diagram 44 summarises our findings to date:
Point B is the original assumed position.

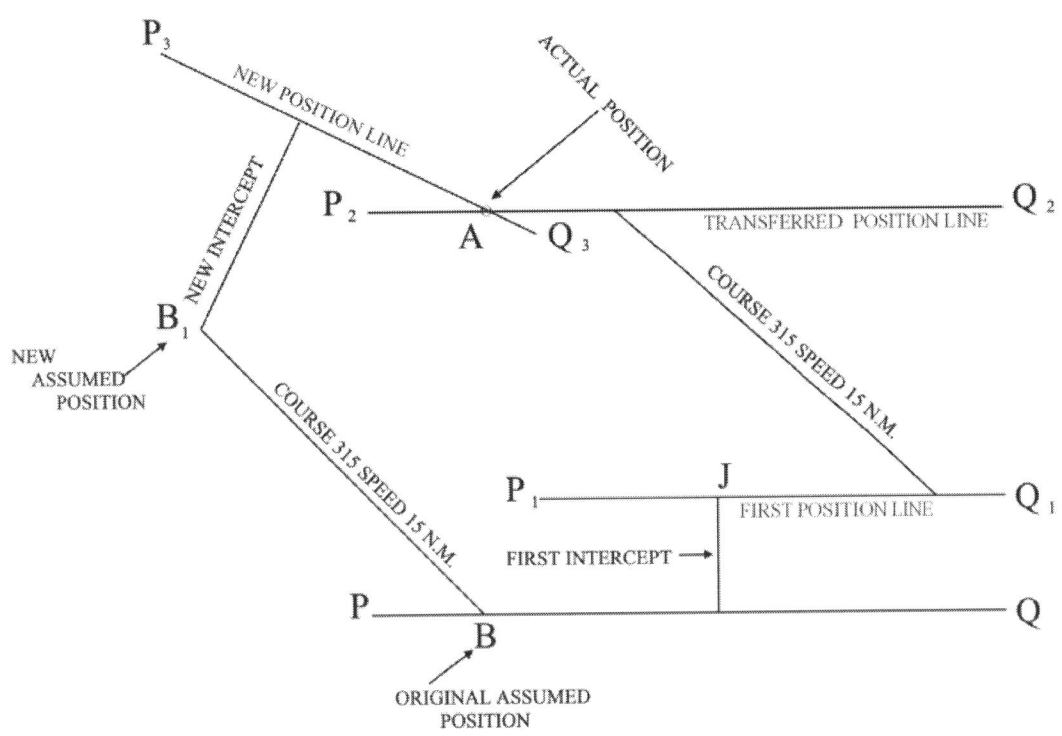

Diagram 44

P_1JQ_1 is the first position line which was established by applying the first intercept to the original assumed position.

Point B_1 is the new assumed position after travelling at 15 knots on course 315° for 1 hour.

P_2AQ_2 is the transferred position line. (i.e. the first position line transferred for 15 n.m. along bearing 315°).

P_3AQ_3 is the new position line which has been established from a new set of calculations made after arriving at the new assumed position and after applying a new intercept.

Point A is at the intersection of the new position line and the transferred position line and is therefore is the true position.

To avoid becoming swamped by facts and figures (death by numbers), we have not discussed the calculations for the second position line but suffice it to say that these would have followed the same procedures as for the first position line.

Although the intercept method involves a time-delay before an accurate position is established, this would not present a problem on an ocean passage when the position would need to be established only once or twice a day. In coastal waters where a vessel's position must be 'fixed' more frequently, position lines would normally be obtained from terrestrial objects such as points of land (assuming electronic navigational systems are not being used).

There is nothing terribly wrong with this; the limitations described above have always applied to traditional methods of astro navigation.

At this point, we should clarify the relevant terminology:

The fix: The position of a vessel established by the intersection of two or more position lines is known as a fix. As we have discussed, position lines may be obtained from a variety of sources such as visual bearings, electronic navigation aids, radar, and astronomical observations.

The Observed Position: Where position lines are derived from astronomical observations, the resultant fix is known as an observed position and is marked on the chart as 'Obs'.

As we have experienced, the mathematical solution of triangle PZX by spherical trigonometry is lengthy and gives scope for arithmetical error. Since time and accuracy are very important for the navigator, tables of computed altitude and azimuth (known as sight reduction tables) are available and these enable us to obtain solutions for all combinations of latitude, declination and hour angle. By using these tables, finding the calculated altitude and azimuth becomes a relatively simple and fast operation.

Although it has been necessary to put our minds through some torturous calculations in order to develop an understanding of the principles of astro navigation, in practice, things become much simpler as we will find in the next section which takes us from theory to practice.

Exercises. Please note that exercises have not been set for this chapter on the theory of the astronomical position line since calculations involving trigonometric ratios will rarely be used in practice. However, exercises are set in the following chapters which demonstrate practical sight reduction techniques.

Section 2 – Practical Astro Navigation

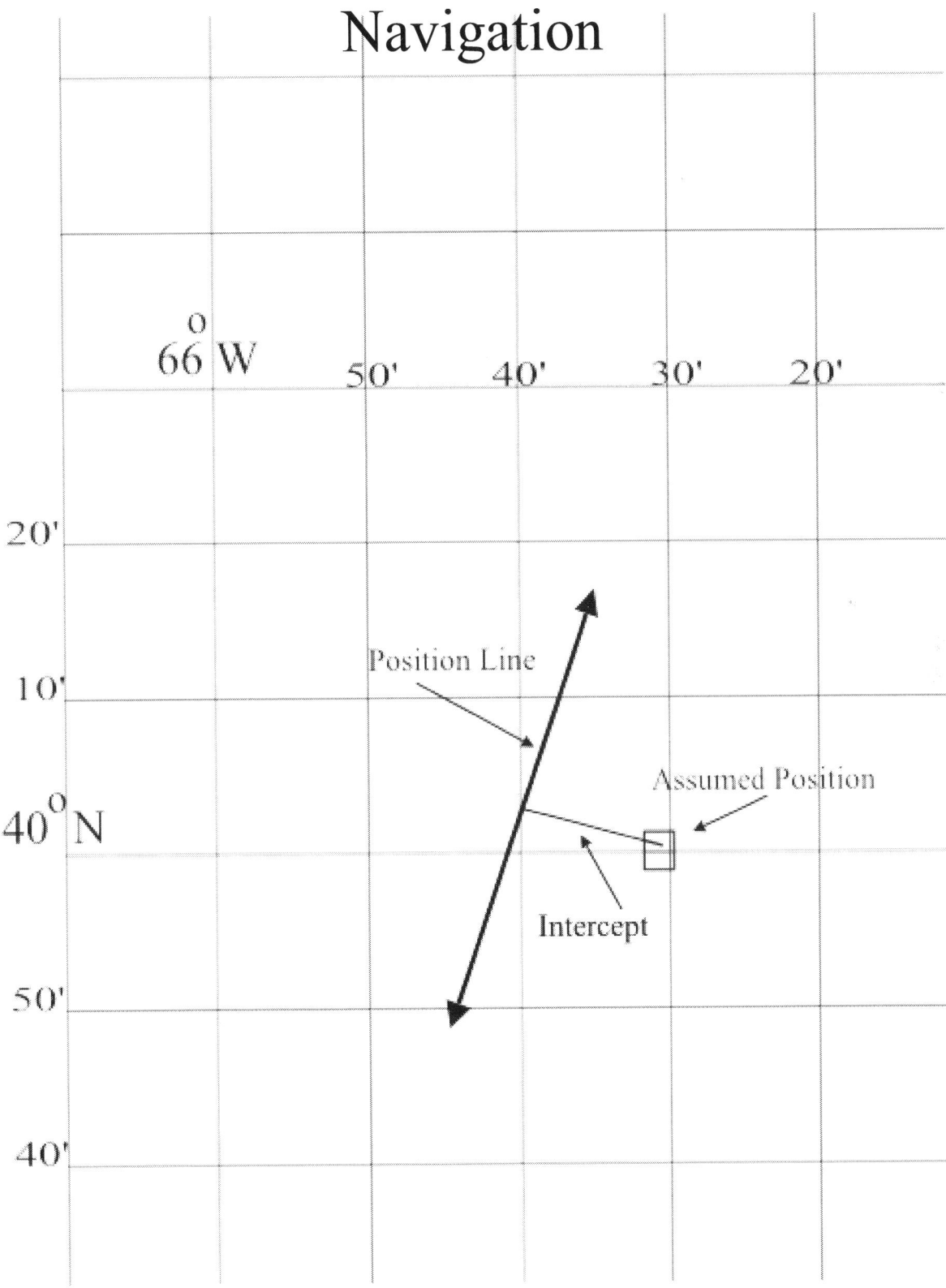

Chapter 8
Introduction to the Rapid Sight Reduction Tables for Navigation (UK) / Sight Reduction Tables for Air Navigation (USA)

The mathematical solution of the triangle PZX by the use of spherical trigonometry is time consuming and gives considerable scope for arithmetical error. Because time and accuracy are of the essence in practical navigation, tables are available that enable us to obtain the solutions of the triangle PZX for all combinations of Declination, Hour Angle and Latitude. After the mathematical torture of section 1 it will come as quite a relief to find that, in practice, we can calculate altitude and azimuth by relatively simple table operations.

In this and the following chapters, explanations and demonstrations involve the use of the Rapid Sight Reduction Tables which are contained in the following publications:
Rapid Sight Reduction Tables for Navigation (published as NP303 by the United Kingdom Hydrographic Office).
Sight Reduction Tables for Air Navigation (published as Pub. No. 249 by the Nautical Almanac Office of the US Naval Observatory).

Advantage of using the Rapid Sight Reduction Tables
The advantage of these tables is that they can be used to calculate an observed position quickly and easily to an accuracy of $\pm$ 0.5 nautical miles without reliance on spherical trigonometry. This slight level of inaccuracy is well within tolerance since the accuracy of astro navigation position finding at sea is generally considered to be $\pm$ 1 n.m. which, in the vast expanses of the oceans, is acceptable.

Before we look at the Rapid Sight Reduction Tables, we need to revise the following:
The Observed Position: As explained in chapter 7, where position lines are derived from astronomical observations, the resultant fix is known as an observed position and is marked on the chart as 'Obs'.
The Assumed Position. In section one, we based our calculations on the 'approximate position' which is the D.R. position at the time of the observation. When working with these tables, although we start from the approximate position, we use an 'assumed position' to help us to keep the number of interpolations to a minimum. We can do this because, when we calculate the intercept, we assume that the vessel is in the vicinity of the D.R. position which of course is not an accurate position. We can therefore, within reasonable limits, choose a position in the neighbourhood of the D.R. position that will not only save time and work but will also reduce the risk of making arithmetic errors.
We do this in the following way:

For the **Assumed Latitude** we use the nearest whole number of degrees to the approximate latitude.

For the **Assumed Longitude**, we take the nearest longitude to the approximate longitude that will make the Hour Angle a whole number of degrees.

GHA and Declination of the Sun. When making observations of the Sun, the Rapid Sight Reduction tables can be used without reference to the Nautical Almanac since they include tables of the Sun's GHA and declination. However, these tables have an accuracy of ±1.'6 and taking this inaccuracy into account along with the inherent inaccuracy of the main tables, it is considered preferable to take the GHA and declination from the Nautical Almanac when available.

Universal Time (UT). As in the Nautical Almanac, the time scale used in sight reduction is Universal Time (UT) and not GMT. This does lead to some confusion, especially since, traditionally, we calculate times of observations in terms of GMT. However, this confusion can be dispelled if we remember that UT is generally considered to be synonymous with GMT. (See chapter 2 for an explanation of universal time).

Star and Planet Observations. As explained in chapter 2, the optimum conditions for taking observations of stars and planets occur during the times of civil twilight and nautical twilight when it is likely to be light enough for the horizon to be seen yet dark enough for the celestial bodies to be visible. In section one, we ignored this convention because the focus was on demonstrating procedures and calculation techniques. However, to make the examples and exercises in this chapter more realistic, the convention is re-instated. Times of sunrise, sunset and nautical and civil twilights are tabulated in the daily pages of the Nautical Almanac.

Outline Procedures for Rapid Sight Reduction.

You may find these outline procedures confusing at first reading; however, it is essential to provide them in order to create a reference source for use when working through the demonstrations and exercises in the following chapters. If you do find them hard to follow, you may find it easier to move straight to the demonstrations and exercises begininning at chapter 9 and then refer back to the outline procedures as necessary. In this way, the overall rapid sight reduction procedure will become clear. The procedures for using these tables includes a 'Planning Phase' and a 'Fix Phase'. In the planning phase, the approximate azimuth and altitude of chosen celestial bodies are calculated before the process of taking the sights is undertaken in the fix phase.

Planning Phase

Step 1. Calculate the Assumed Position as follows:
  a. Begin by estimating the approximate position which will normally be the DR position at the time that the observation is made.

b. Calculate the Assumed Latitude which will be the nearest whole number of degrees to the approximate latitude.

c. Calculate the Assumed Longitude which will be the nearest longitude to the approximate longitude that will make the Hour Angle a whole number of degrees.

Step 2. Calculate the Greenwich date and UT at the time of the observation. (For this step, revise chapter 2 if necessary).

When an observed position is established from the transferred position lines of more than one observation or when special reduction techniques are used (as explained on page 170) the following should also be noted: The difference between the time of the fix, the time of each observation and whether the time of the fix is earlier or later than each of the observations.

Step 3. Calculate the GHA and declination. (For this step, you may need to revise chapter 4).

Calculate the GHA of the body being observed, adding or subtracting multiples of 360° as necessary so that the result is between 0° and 360°. Also calculate the declination of the body being observed.

Step 4. Calculate LHA. (You may need to revise chapter 4 for this step). Apply the assumed longitude to the GHA to calculate the LHA. (If assumed long. is east, add to GHA, if west, subtract). Add or subtract multiples of 360° so that LHA is between 0° and 360° if necessary.

Example:

GHA	358° 45'.0
As. long. (E)	+58° 15'.0
LHA	417°
±360°	360°
	57°

Step 5. Main table entry. From the main tables, we extract the following:
Tabulated altitude **(Hc)**,
The difference **(d)**
Azimuth angle **(Z)**.

Explanation of the above values.
Hc is the tabulated altitude (Tab. Alt.)
d is the correction for the minutes of declination (Dec').
Z is the azimuth angle.

Brief Explanation of the Use of Main Tables.
Referring to the extract from one of the 'main tables' below, the method of entering the table is as follows:

Find the page for the assumed latitude. (Select 'Same Name' or 'Contrary Name' depending on whether latitude and declination are the same or contrary).

Where the column for the value of the degrees of declination (Dec°) intersects with the row for the value of the LHA, read off the values for Hc, d, Z.

Example. Part of the main table for 40° Contrary is shown on the following page. Suppose the assumed latitude is 40°N, the declination is 10° 12'S. and the LHA is 300°. In the table, we find Dec° (the degrees part of the declination (i.e. 10°)) along the top row and the LHA (i.e. 300°) down the right hand column.

At the intersection of these values, we extract the following:

Hc = 15° 24',   d = -44',   Z = 118°.

10°			11°			12°			13°			14°			LHA
Hc	d	Z	Hc	d	Z	Hc	d	Z	Hc	d	Z	Hc	d	Z	
° '	'	°	° '	'	°	° '	'	°	° '	'	°	° '	'	°	°
09 08	41	111	08 27	42	112	07 45	41	113	07 04	42	114	06 22	42	114	291
09 51	42	112	09 09	42	113	08 27	41	114	07 46	42	114	07 04	42	115	292
10 33	42	113	09 51	42	113	09 09	42	114	08 27	42	115	07 45	42	116	293
11 15	42	113	10 33	42	114	09 51	42	115	09 09	42	116	08 27	43	116	294
11 58	-43	114	11 15	-42	115	10 33	-43	116	09 50	-42	116	09 08	-43	117	295
12 39	42	115	11 57	43	116	11 14	43	116	10 31	43	117	09 48	43	118	296
13 21	43	116	12 38	43	116	11 55	43	117	11 12	43	118	10 29	43	118	297
14 02	43	116	13 19	43	117	12 36	43	118	11 53	44	118	11 09	43	119	298
14 43	43	117	14 00	43	118	13 17	44	118	12 33	44	119	11 49	44	120	299
15 24	-44	118	14 40	-43	119	13 57	-44	119	13 13	-44	120	12 29	-44	121	300
16 05	44	119	15 21	44	119	14 37	44	120	13 53	45	121	13 08	44	121	301
16 45	44	119	16 01	45	120	15 16	44	121	14 32	45	121	13 47	44	122	302
17 25	45	120	16 40	44	121	15 56	45	121	15 11	45	122	14 26	45	123	303
18 04	44	121	17 20	45	122	16 35	45	122	15 50	45	123	15 05	46	124	304
18 44	-45	122	17 59	-45	122	17 14	-46	123	16 28	-45	124	15 43	-46	124	305
19 23	46	122	18 37	45	123	17 52	46	124	17 06	45	124	16 21	46	125	306
20 01	45	123	19 16	46	124	18 30	46	125	17 44	46	125	16 58	46	126	307
20 40	46	124	19 54	46	125	19 08	47	125	18 21	46	126	17 35	46	127	308
21 18	47	125	20 31	46	125	19 45	47	126	18 58	46	127	18 12	47	127	309
21 55	-46	126	21 09	-47	126	20 22	-47	127	19 35	-47	128	18 48	-47	128	310
22 32	47	126	21 45	47	127	20 58	47	128	20 11	47	128	19 24	48	129	311
23 09	47	127	22 22	47	128	21 35	48	129	20 47	48	129	19 59	47	130	312
23 46	48	128	22 58	48	129	22 10	48	129	21 22	48	130	20 34	48	131	313
24 21	47	129	23 34	48	130	22 46	49	130	21 57	48	131	21 09	48	132	314
24 57	-48	130	24 09	-49	130	23 20	-48	131	22 32	-49	132	21 43	-49	132	315

(right side of table labeled: LAT 40°)

Next, we use table 5 to find the correction for Dec' (the minutes part of the declination) as demonstrated below:

In the previous example, the value of Dec' was 12' and the value of d was found to be -44'.

In the extract from Table 5 below, values of d are tabulated across the top row and values of Dec' are shown down the right hand column.

Ignoring the – sign, we find 44' (the value of d) in the top row and 12' (the value of dec') in the right column.

We extract the value of the correction which is 9'. However, the value of d was -44' so we must remember to apply the same sign to the correction. Therefore, the correction now becomes -9'.

©2009 Jack Case

This correction is applied to the value of Hc which, in the example, was found to be 15° 24'.
Therefore, the corrected value of the tabulated altitude (Hc) is:
15° 24' - 9' = 15° 15'.

To convert azimuth angle (Z) to true bearing (Zn). The true bearing is usually calculated at step 7; however, it is helpful to make this calculation at an earlier stage, especially in the case of stars and planets. We calculate the approximate altitude at step 5 (i.e. the Tab. Alt.) and if we can also calculate the true bearing at this stage, we can use this information to verify the identities of the bodies chosen for our observations before proceeding too far.

The method of calculating the true bearing is dependent on the LHA which is calculated at step 4 and the azimuth angle which is calculated at step 5 so we need to have reached these stages before we can proceed further. It would seem logical therefore, to enter the calculations for true bearing at the end of step 5. However, for the sake of convenience, the examples in this book place the calculations at the end of step 4 which is shorter and less complicated than step 5.

40	41	42	43	44	45	46	47	48	49	50	51	52	53	54	55	56	57	58	59	60	d
'	'	'	'	'	'	'	'	'	'	'	'	'	'	'	'	'	'	'	'	'	'
0	0	0	0	0	0	0	0	0	0	0	0	0	0	0	0	0	0	0	0	0	0
1	1	1	1	1	1	1	1	1	1	1	1	1	1	1	1	1	1	1	1	1	1
1	1	1	1	1	2	2	2	2	2	2	2	2	2	2	2	2	2	2	2	2	2
2	2	2	2	2	2	2	2	2	2	2	3	3	3	3	3	3	3	3	3	3	3
3	3	3	3	3	3	3	3	3	3	3	3	3	4	4	4	4	4	4	4	4	4
3	3	4	4	4	4	4	4	4	4	4	4	4	4	4	5	5	5	5	5	5	5
4	4	4	4	4	4	5	5	5	5	5	5	5	5	5	6	6	6	6	6	6	6
5	5	5	5	5	5	5	5	6	6	6	6	6	6	6	6	7	7	7	7	7	7
5	5	6	6	6	6	6	6	6	7	7	7	7	7	7	7	7	8	8	8	8	8
6	6	6	6	7	7	7	7	7	7	8	8	8	8	8	8	8	9	9	9	9	9
7	7	7	7	7	8	8	8	8	8	8	8	9	9	9	9	9	10	10	10	10	10
7	8	8	8	8	8	8	9	9	9	9	9	10	10	10	10	10	10	11	11	11	11
8	8	8	9	9	9	9	9	10	10	10	10	10	11	11	11	11	11	12	12	12	12
9	9	9	9	10	10	10	10	10	11	11	11	11	11	12	12	12	12	13	13	13	13
9	10	10	10	10	10	11	11	11	11	12	12	12	12	13	13	13	13	14	14	14	14
10	10	10	11	11	11	12	12	12	12	12	13	13	13	14	14	14	14	14	15	15	15
11	11	11	11	12	12	12	13	13	13	13	14	14	14	14	15	15	15	15	16	16	16
11	12	12	12	12	13	13	13	14	14	14	14	15	15	15	16	16	16	16	17	17	17
12	12	13	13	13	14	14	14	14	15	15	15	16	16	16	17	17	17	17	18	18	18
13	13	13	14	14	14	15	15	15	16	16	16	16	17	17	17	18	18	18	19	19	19
13	14	14	14	15	15	15	16	16	16	17	17	17	18	18	18	19	19	19	20	20	20
14	14	15	15	15	16	16	16	17	17	18	18	18	19	19	19	20	20	20	21	21	21
15	15	15	16	16	16	17	17	18	18	18	19	19	19	20	20	21	21	21	22	22	22
15	16	16	16	17	17	18	18	18	19	19	20	20	20	21	21	21	22	22	23	23	23
16	16	17	17	18	18	18	19	19	20	20	20	21	21	22	22	22	23	23	24	24	24

The method of calculating the true bearing is explained below:

At step 5, we extract a value for 'Z' from the main table; this is the azimuth angle. To convert the azimuth angle to a true bearing (Zn), we make use of the following table of rules:

	Lat. North	Lat. South
LHA>180°	Zn = Z	Zn = 180° - Z
LHA<180°	Zn = 360°-Z	Zn = 180° + Z

Example. Using the values derived in the previous example, we have:
Lat 40°N, LHA 300°, Tab. alt. (Hc): 15° 15', Azimuth angle(Z): 118°.
From the table above, the true bearing is calculated as follows:
LHA>180° and latitude is north ∴ Zn = Z
∴ Zn (true bearing) = 118°.
We calculated that the tabulated altitude (Hc) is 15° 15', and we have now calculated that the true bearing (Zn) is 118°, so we are able to verify the identity of the body observed by checking that the observed altitude and bearing correspond fairly closely with these figures. **(Because we work from an assumed position, we do not expect them to correspond exactly).**

Fix Phase

Step 6. Correct the sextant altitude.

Before tackling this step, you may wish to revise chapter 5.

The following types of corrections are made depending on the body being observed:
- **Index error**
- **Dip**
- **Altitude corrections** (combined corrections for refraction, parallax and semi-diameter)
- **Refraction for stars and planets**
- **Semi-diameter and horizontal parallax for the Moon**
- **Additional refraction.** An additional correction for refraction may be needed if the temperature and atmospheric pressure are greatly different to the standard conditions which are assumed to be 10°C, 1010mb

Altitude Correction Tables

The Rapid Sight Reduction Tables publication includes supplementary tables to assist with the above corrections as follows:
- Table 6a: Dip
- Table 6b: Refraction for stars and planets
- Table 6c: Additional refraction
- Table 6d: Altitude correction tables for the Sun which give combined corrections for refraction, parallax and semi-diameter as explained in chapter 5.

Notes
1. Although these tables contain the same data as the altitude correction tables contained in the nautical almanac, their layout is different.
2. Altitude correction tables for the Moon are not included and so these must be taken from the Nautical Almanac.

Examples: Using the methods explained in chapter 5, sextant altitude (Hs) is corrected to give observed altitude (Ho) in the following examples:

Example 1. Planet observation.

Sext. Alt. (Hs)	07° 5'.32	
Index error (IE)	-1'.41	
Dip (D) (5.0m)	-3'.90	(table 6a)
Ap. Alt. (Ha)	07° 07'.01	
Alt. Cor. (R) =	-7'.40	(table 6b)
Add'l refrac.	-0'.50	(table 6c) (-5°C. 1010mb.)
Obs. Alt. (Ho)=	06° 59'.11	

Example 2. Sun lower limb observation (month of August).

Sext. Alt. (Hs)	35° 12'.45	
Index error (IE)	+0'.74	
Dip (D) (14.5 ft.)	-3'.70	(table 6a)
Ap. Alt. (Ha)	35° 09'.49	
Alt. Cor. (R) =	+14'.60	(table 6d)
Add'l refrac.	-0'.10	(table 6c) (20°C. 980mb.)
Obs. Alt. (Ho)=	35° 23'.99	

Extracts. Data used in the above examples is taken from the extracts of the relevant altitude correction tables which are shown on the following pages.

TABLE 6 — ALTITUDE CORRECTION TABLES

a. Dip of the Horizon

Ht. of Eye (m)	Corrⁿ D	Ht. of Eye (m)	Ht. of Eye (m)	Corrⁿ D	Ht. of Eye (ft.)
0.00	−0.3	0.0	13.0	−6.4	42.8
0.03	−0.4	0.1	13.4	−6.5	44.2
0.06	−0.5	0.2	13.8	−6.6	45.5
0.09	−0.6	0.3	14.2	−6.7	47.0
0.13	−0.7	0.4	14.7	−6.8	48.4
0.18	−0.8	0.5	15.1	−6.9	49.8
0.23	−0.9	0.7	15.5	−7.0	51.3
0.29	−1.0	0.9	16.0	−7.1	52.8
0.35	−1.1	1.1	16.5	−7.2	54.3
0.42	−1.2	1.4	16.9	−7.3	55.8
0.45	−1.3	1.6	17.4	−7.4	57.4
0.5	−1.4	1.9	17.9	−7.5	58.9
0.6	−1.5	2.2	18.4	−7.6	60.5
0.7	−1.6	2.5	18.8	−7.7	62.1
0.8	−1.7	2.8	19.3	−7.8	63.8
0.9	−1.8	3.2	19.8	−7.9	65.4
1.1	−1.9	3.6	20.4	−8.0	67.1
1.2	−2.0	4.0	20.9	−8.1	68.8
1.3	−2.1	4.4	21.4	−8.2	70.5
1.4	−2.2	4.9	21.9	−8.3	72.3
1.6	−2.3	5.3	22.5	−8.4	74.1
1.7	−2.4	5.8	23.0	−8.5	75.8
1.9	−2.5	6.3	23.5	−8.6	77.6
2.0	−2.6	6.9	24.1	−8.7	79.5
2.2	−2.7	7.4	24.7	−8.8	81.3
2.4	−2.8	8.0	25.2	−8.9	83.2
2.6	−2.9	8.6	25.8	−9.0	85.1
2.8	−3.0	9.2	26.4	−9.1	87.0
3.0	−3.1	9.8	27.0	−9.2	88.9
3.2	−3.2	10.5	27.6	−9.3	90.9
3.4	−3.3	11.2	28.2	−9.4	92.9
3.6	−3.4	11.9	28.8	−9.5	94.9
3.8	−3.5	12.6	29.4	−9.6	96.9
4.0	−3.6	13.3	30.0	−9.7	98.9
4.3	−3.7	14.1	30.6	−9.8	101.0
4.5	−3.8	14.9	31.3	−9.9	103.1
4.7	−3.9	15.7	31.9	−10.0	105.2
5.0	−4.0	16.5	32.6	−10.1	107.3
5.2	−4.1	17.4	33.2	−10.2	109.4
5.5	−4.2	18.3	33.9	−10.3	111.6
5.8	−4.3	19.1	34.5	−10.4	113.8
6.1	−4.4	20.1	35.2	−10.5	116.0
6.3	−4.5	21.0	35.9	−10.6	118.2
6.6	−4.6	22.0	36.6	−10.7	120.5
6.9	−4.7	22.9	37.3	−10.8	122.8
7.2	−4.8	23.9	38.0	−10.9	125.1
7.5	−4.9	25.0	38.7	−11.0	127.4
7.9	−5.0	26.0	39.4	−11.1	129.7
8.2	−5.1	27.1	40.1	−11.2	132.1
8.5	−5.2	28.1	40.8	−11.3	134.5
8.8	−5.3	29.2	41.5	−11.4	136.9
9.2	−5.4	30.4	42.3	−11.5	139.3
9.5	−5.5	31.5	43.0	−11.6	141.7
9.9	−5.6	32.7	43.8	−11.7	144.2
10.3	−5.7	33.9	44.5	−11.8	146.7
10.6	−5.8	35.1	45.3	−11.9	149.2
11.0	−5.9	36.3	46.1	−12.0	151.7
11.4	−6.0	37.6	46.8	−12.1	154.3
11.8	−6.1	38.9	47.6	−12.2	156.8
12.2	−6.2	40.1	48.4	−12.3	159.4
12.6	−6.3	41.5	49.2	−12.4	162.1
13.0		42.8	50.0		164.7

b. Refraction for Stars & Planets

App. Alt. H_a	Corrⁿ R	App. Alt. H_a	Corrⁿ R	App. Alt. H_a	Corrⁿ R
0° 00′	−33.8′	3 30	−12.9	9 55	
0 03	33.2	3 35	12.7	10 07	−5.3
0 06	32.6	3 40	12.5	10 20	−5.2
0 09	32.0	3 45	12.3	10 32	−5.1
0 12	31.5	3 50	12.1	10 46	−5.0
0 15	30.9	3 55	11.9	10 59	−4.9
0 18	−30.4	4 00	−11.7	11 14	−4.8
0 21	29.8	4 05	11.5	11 29	−4.7
0 24	29.3	4 10	11.4	11 44	−4.6
0 27	28.8	4 15	11.2	12 00	−4.5
0 30	28.3	4 20	11.0	12 17	−4.4
0 33	27.9	4 25	10.9	12 35	−4.3
0 36	−27.4	4 30	−10.7	12 53	−4.2
0 39	26.9	4 35	10.6	13 12	−4.1
0 42	26.5	4 40	10.4	13 32	−4.0
0 45	26.1	4 45	10.3	13 53	−3.9
0 48	25.7	4 50	10.1	14 16	−3.8
0 51	25.3	4 55	10.0	14 39	−3.7
0 54	−24.9	5 00	9.8	15 03	−3.6
0 57	24.5	5 05	9.7	15 29	−3.5
1 00	24.1	5 10	9.6	15 56	−3.4
1 03	23.7	5 15	9.5	16 25	−3.3
1 06	23.4	5 20	9.3	16 55	−3.2
1 09	23.0	5 25	9.2	17 27	−3.1
1 12	−22.7	5 30	−9.1	18 01	−3.0
1 15	22.3	5 35	9.0	18 37	−2.9
1 18	22.0	5 40	8.9	19 16	−2.8
1 21	21.7	5 45	8.8	19 56	−2.7
1 24	21.4	5 50	8.7	20 40	−2.6
1 27	21.1	5 55	8.6	21 27	−2.5
1 30	−20.8	6 00	−8.5	22 17	−2.4
1 35	20.3	6 10	8.3	23 11	−2.3
1 40	19.9	6 20	8.1	24 09	−2.2
1 45	19.4	6 30	7.9	25 12	−2.1
1 50	19.0	6 40	7.7	26 20	−2.0
1 55	18.6	6 50	7.6	27 34	−1.9
2 00	−18.2	7 00	−7.4	28 54	−1.8
2 05	17.8	7 10	7.2	30 22	−1.7
2 10	17.4	7 20	7.1	31 58	−1.6
2 15	17.1	7 30	6.9	33 43	−1.5
2 20	16.7	7 40	6.8	35 38	−1.4
2 25	16.4	7 50	6.7	37 45	−1.3
2 30	−16.1	8 00	−6.6	40 06	−1.2
2 35	15.8	8 10	6.4	42 42	−1.1
2 40	15.4	8 20	6.3	45 34	−1.0
2 45	15.2	8 30	6.2	48 45	−0.9
2 50	14.9	8 40	6.1	52 16	−0.8
2 55	14.6	8 50	6.0	56 09	−0.7
3 00	−14.3	9 00	−5.9	60 26	−0.6
3 05	14.1	9 10	5.8	65 06	−0.5
3 10	13.8	9 20	5.7	70 09	−0.4
3 15	13.6	9 30	5.6	75 32	−0.3
3 20	13.4	9 40	5.5	81 12	−0.2
3 25	13.1	9 50	5.4	87 03	−0.1
3 30	−12.9	10 00	−5.3	90 00	0.0

H_a = App. Alt. = Apparent Altitude
= Sextant altitude corrected for index error (IE) & dip (L
= H_s + IE + D

In critical cases ascend.

TABLE 6 — ALTITUDE CORRECTION TABLES

c. Additional Refraction for Non-standard Conditions

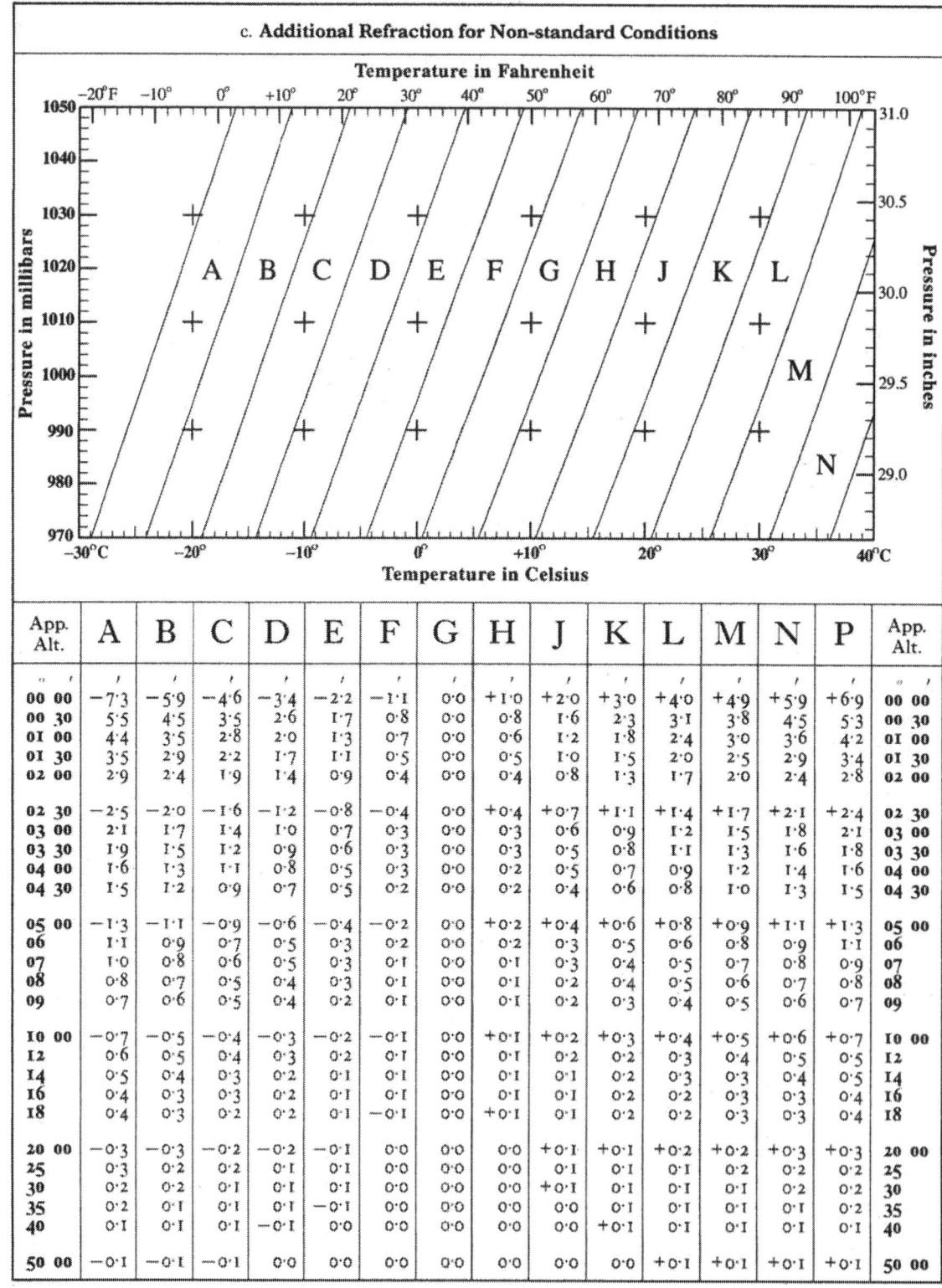

App. Alt.	A	B	C	D	E	F	G	H	J	K	L	M	N	P	App. Alt.
° ′	′	′	′	′	′	′	′	′	′	′	′	′	′	′	° ′
00 00	−7·3	−5·9	−4·6	−3·4	−2·2	−1·1	0·0	+1·0	+2·0	+3·0	+4·0	+4·9	+5·9	+6·9	00 00
00 30	5·5	4·5	3·5	2·6	1·7	0·8	0·0	0·8	1·6	2·3	3·1	3·8	4·5	5·3	00 30
01 00	4·4	3·5	2·8	2·0	1·3	0·7	0·0	0·6	1·2	1·8	2·4	3·0	3·6	4·2	01 00
01 30	3·5	2·9	2·2	1·7	1·1	0·5	0·0	0·5	1·0	1·5	2·0	2·5	2·9	3·4	01 30
02 00	2·9	2·4	1·9	1·4	0·9	0·4	0·0	0·4	0·8	1·3	1·7	2·0	2·4	2·8	02 00
02 30	−2·5	−2·0	−1·6	−1·2	−0·8	−0·4	0·0	+0·4	+0·7	+1·1	+1·4	+1·7	+2·1	+2·4	02 30
03 00	2·1	1·7	1·4	1·0	0·7	0·3	0·0	0·3	0·6	0·9	1·2	1·5	1·8	2·1	03 00
03 30	1·9	1·5	1·2	0·9	0·6	0·3	0·0	0·3	0·5	0·8	1·1	1·3	1·6	1·8	03 30
04 00	1·6	1·3	1·1	0·8	0·5	0·3	0·0	0·2	0·5	0·7	0·9	1·2	1·4	1·6	04 00
04 30	1·5	1·2	0·9	0·7	0·5	0·2	0·0	0·2	0·4	0·6	0·8	1·0	1·3	1·5	04 30
05 00	−1·3	−1·1	−0·9	−0·6	−0·4	−0·2	0·0	+0·2	+0·4	+0·6	+0·8	+0·9	+1·1	+1·3	05 00
06	1·1	0·9	0·7	0·5	0·3	0·2	0·0	0·2	0·3	0·5	0·6	0·8	0·9	1·1	06
07	1·0	0·8	0·6	0·5	0·3	0·1	0·0	0·1	0·3	0·4	0·5	0·7	0·8	0·9	07
08	0·8	0·7	0·5	0·4	0·3	0·1	0·0	0·1	0·2	0·4	0·5	0·6	0·7	0·8	08
09	0·7	0·6	0·5	0·4	0·2	0·1	0·0	0·1	0·2	0·3	0·4	0·5	0·6	0·7	09
10 00	−0·7	−0·5	−0·4	−0·3	−0·2	−0·1	0·0	+0·1	+0·2	+0·3	+0·4	+0·5	+0·6	+0·7	10 00
12	0·6	0·5	0·4	0·3	0·2	0·1	0·0	0·1	0·2	0·2	0·3	0·4	0·5	0·5	12
14	0·5	0·4	0·3	0·2	0·1	0·1	0·0	0·1	0·1	0·2	0·3	0·3	0·4	0·5	14
16	0·4	0·3	0·3	0·2	0·1	0·1	0·0	0·1	0·1	0·2	0·2	0·3	0·3	0·4	16
18	0·4	0·3	0·2	0·2	0·1	−0·1	0·0	+0·1	0·1	0·2	0·2	0·3	0·3	0·4	18
20 00	−0·3	−0·3	−0·2	−0·2	−0·1	0·0	0·0	0·0	+0·1	+0·1	+0·2	+0·2	+0·3	+0·3	20 00
25	0·3	0·2	0·2	0·1	0·1	0·0	0·0	0·0	0·1	0·1	0·1	0·2	0·2	0·2	25
30	0·2	0·2	0·1	0·1	0·1	0·0	0·0	0·0	+0·1	0·1	0·1	0·1	0·2	0·2	30
35	0·2	0·1	0·1	0·1	−0·1	0·0	0·0	0·0	0·0	0·1	0·1	0·1	0·1	0·2	35
40	0·1	0·1	0·1	−0·1	0·0	0·0	0·0	0·0	0·0	+0·1	0·1	0·1	0·1	0·1	40
50 00	−0·1	−0·1	−0·1	0·0	0·0	0·0	0·0	0·0	0·0	0·0	+0·1	+0·1	+0·1	+0·1	50 00

The graph is entered with arguments temperature and pressure to find a zone letter; using as arguments this zone letter and apparent altitude (sextant altitude corrected for index error and dip), a correction is taken from the table. This correction is to be applied to the sextant altitude in addition to the corrections for standard conditions (Table **6**b).

TABLE 6d — ALTITUDE CORRECTION TABLES FOR THE SUN

App. Alt. H_a	OCT.–MAR. Lower Limb R	OCT.–MAR. Upper Limb R	APR.–SEPT. Lower Limb R	APR.–SEPT. Upper Limb R	App. Alt. H_a	OCT.–MAR. Lower Limb R	OCT.–MAR. Upper Limb R	APR.–SEPT. Lower Limb R	APR.–SEPT. Upper Limb R	App. Alt. H_a	OCT.–MAR. Lower Limb R	OCT.–MAR. Upper Limb R	App. Alt. H_a	APR.–SEPT. Lower Limb R	APR.–SEPT. Upper Limb R
0 00	−17.5	−49.8	−17.8	−49.6	3 30	+ 3.4	−28.9	+ 3.1	−28.7	9 33	+10.8	−21.5	9 39	+10.6	−21.2
0 03	16.9	49.2	17.2	49.0	3 35	3.6	28.7	3.3	28.5	9 45	+10.9	−21.4	9 50	+10.7	−21.1
0 06	16.3	48.6	16.6	48.4	3 40	3.8	28.5	3.6	28.2	9 56	+11.0	−21.3	10 02	+10.8	−21.0
0 09	15.7	48.0	16.0	47.8	3 45	4.0	28.3	3.8	28.0	10 08	+11.1	−21.2	10 14	+10.9	−20.9
0 12	15.2	47.5	15.4	47.2	3 50	4.2	28.1	4.0	27.8	10 20	+11.2	−21.1	10 27	+11.0	−20.8
0 15	14.6	46.9	14.8	46.6	3 55	4.4	27.9	4.1	27.7	10 33	+11.3	−21.0	10 40	+11.1	−20.7
0 18	−14.1	−46.4	−14.3	−46.1	4 00	+ 4.6	−27.7	+ 4.3	−27.5	10 46	+11.4	−20.9	10 53	+11.2	−20.6
0 21	13.5	45.8	13.8	45.6	4 05	4.8	27.5	4.5	27.3	11 00	+11.5	−20.8	11 07	+11.3	−20.5
0 24	13.0	45.3	13.3	45.1	4 10	4.9	27.4	4.7	27.1	11 15	+11.6	−20.7	11 22	+11.4	−20.4
0 27	12.5	44.8	12.8	44.6	4 15	5.1	27.2	4.9	26.9	11 30	+11.7	−20.6	11 37	+11.5	−20.3
0 30	12.0	44.3	12.3	44.1	4 20	5.3	27.0	5.0	26.8	11 45	+11.8	−20.5	11 53	+11.6	−20.2
0 33	11.6	43.9	11.8	43.6	4 25	5.4	26.9	5.2	26.6	12 01	+11.9	−20.4	12 10	+11.7	−20.1
0 36	−11.1	−43.4	−11.3	−43.1	4 30	+ 5.6	−26.7	+ 5.3	−26.5	12 18	+12.0	−20.3	12 27	+11.8	−20.0
0 39	10.6	42.9	10.9	42.7	4 35	5.7	26.6	5.5	26.3	12 36	+12.1	−20.2	12 45	+11.9	−19.9
0 42	10.2	42.5	10.5	42.3	4 40	5.9	26.4	5.6	26.2	12 54	+12.2	−20.1	13 04	+12.0	−19.8
0 45	9.8	42.1	10.0	41.8	4 45	6.0	26.3	5.8	26.0	13 14	+12.3	−20.0	13 24	+12.1	−19.7
0 48	9.4	41.7	9.6	41.4	4 50	6.2	26.1	5.9	25.9	13 34	+12.4	−19.9	13 44	+12.2	−19.6
0 51	9.0	41.3	9.2	41.0	4 55	6.3	26.0	6.1	25.7	13 55	+12.5	−19.8	14 06	+12.3	−19.5
0 54	− 8.6	−40.9	− 8.8	−40.6	5 00	+ 6.4	−25.9	+ 6.2	−25.6	14 17	+12.6	−19.7	14 29	+12.4	−19.4
0 57	8.2	40.5	8.4	40.2	5 05	6.6	25.7	6.3	25.5	14 41	+12.7	−19.6	14 53	+12.5	−19.3
1 00	7.8	40.1	8.0	39.8	5 10	6.7	25.6	6.5	25.3	15 05	+12.8	−19.5	15 18	+12.6	−19.2
1 03	7.4	39.7	7.7	39.5	5 15	6.8	25.5	6.6	25.2	15 31	+12.9	−19.4	15 45	+12.7	−19.1
1 06	7.1	39.4	7.3	39.1	5 20	7.0	25.3	6.7	25.1	15 59	+13.0	−19.3	16 13	+12.8	−19.0
1 09	6.7	39.0	7.0	38.8	5 25	7.1	25.2	6.8	25.0	16 27	+13.1	−19.2	16 43	+12.9	−18.9
1 12	− 6.4	−38.7	− 6.6	−38.4	5 30	+ 7.2	−25.1	+ 6.9	−24.9	16 58	+13.2	−19.1	17 14	+13.0	−18.8
1 15	6.0	38.3	6.3	38.1	5 35	7.3	25.0	7.1	24.7	17 30	+13.3	−19.0	17 47	+13.1	−18.7
1 18	5.7	38.0	6.0	37.8	5 40	7.4	24.9	7.2	24.6	18 05	+13.4	−18.9	18 23	+13.2	−18.6
1 21	5.4	37.7	5.7	37.5	5 45	7.5	24.8	7.3	24.5	18 41	+13.5	−18.8	19 00	+13.3	−18.5
1 24	5.1	37.4	5.3	37.1	5 50	7.6	24.7	7.4	24.4	19 20	+13.6	−18.7	19 41	+13.4	−18.4
1 27	4.8	37.1	5.0	36.8	5 55	7.7	24.6	7.5	24.3	20 02	+13.7	−18.6	20 24	+13.5	−18.3
1 30	− 4.5	−36.8	− 4.7	−36.5	6 00	+ 7.8	−24.5	+ 7.6	−24.2	20 46	+13.8	−18.5	21 10	+13.6	−18.2
1 35	4.0	36.3	4.3	36.1	6 10	8.0	24.3	7.8	24.0	21 34	+13.9	−18.4	21 59	+13.7	−18.1
1 40	3.6	35.9	3.8	35.6	6 20	8.2	24.1	8.0	23.8	22 25	+14.0	−18.3	22 52	+13.8	−18.0
1 45	3.1	35.4	3.4	35.2	6 30	8.4	23.9	8.2	23.6	23 20	+14.1	−18.2	23 49	+13.9	−17.9
1 50	2.7	35.0	2.9	34.7	6 40	8.6	23.7	8.3	23.5	24 20	+14.2	−18.1	24 51	+14.0	−17.8
1 55	2.3	34.6	2.5	34.3	6 50	8.7	23.6	8.5	23.3	25 24	+14.3	−18.0	25 58	+14.1	−17.7
2 00	− 1.9	−34.2	− 2.1	−33.9	7 00	+ 8.9	−23.4	+ 8.7	−23.1	26 34	+14.4	−17.9	27 11	+14.2	−17.6
2 05	1.5	33.8	1.7	33.5	7 10	9.1	23.2	8.8	23.0	27 50	+14.5	−17.8	28 31	+14.3	−17.5
2 10	1.1	33.4	1.4	33.2	7 20	9.2	23.1	9.0	22.8	29 13	+14.6	−17.7	29 58	+14.4	−17.4
2 15	0.8	33.1	1.0	32.8	7 30	9.3	23.0	9.1	22.7	30 44	+14.7	−17.6	31 33	+14.5	−17.3
2 20	0.4	32.7	0.7	32.5	7 40	9.5	22.8	9.2	22.6	32 24	+14.8	−17.5	33 18	+14.6	−17.2
2 25	− 0.1	32.4	− 0.3	32.1	7 50	9.6	22.7	9.4	22.4	34 15	+14.9	−17.4	35 15	+14.7	−17.1
2 30	+ 0.2	−32.1	0.0	−31.8	8 00	+ 9.7	−22.6	+ 9.5	−22.3	36 17	+15.0	−17.3	37 24	+14.8	−17.0
2 35	0.5	31.8	+ 0.3	31.5	8 10	9.9	22.4	9.6	22.2	38 34	+15.1	−17.2	39 48	+14.9	−16.9
2 40	0.8	31.5	0.6	31.2	8 20	10.0	22.3	9.7	22.1	41 06	+15.2	−17.1	42 28	+15.0	−16.8
2 45	1.1	31.2	0.9	30.9	8 30	10.1	22.2	9.9	21.9	43 56	+15.3	−17.0	45 29	+15.1	−16.7
2 50	1.4	30.9	1.2	30.6	8 40	10.2	22.1	10.0	21.8	47 07	+15.4	−16.9	48 52	+15.2	−16.6
2 55	1.7	30.6	1.4	30.4	8 50	10.3	22.0	10.1	21.7	50 43	+15.5	−16.8	52 41	+15.3	−16.5
3 00	+ 2.0	−30.3	+ 1.7	−30.1	9 00	+10.4	−21.9	+10.2	−21.6	54 46	+15.6	−16.7	56 59	+15.4	−16.4
3 05	2.2	30.1	2.0	29.8	9 10	10.5	21.8	10.3	21.5	59 21	+15.7	−16.6	61 50	+15.5	−16.3
3 10	2.5	29.8	2.2	29.6	9 20	10.6	21.7	10.4	21.4	64 28	+15.8	−16.5	67 15	+15.6	−16.2
3 15	2.7	29.6	2.5	29.3	9 30	10.7	21.6	10.5	21.3	70 10	+15.9	−16.4	73 14	+15.7	−16.1
3 20	2.9	29.4	2.7	29.1	9 40	10.8	21.5	10.6	21.2	76 24	+16.0	−16.3	79 42	+15.8	−16.0
3 25	3.2	29.1	2.9	28.9	9 50	10.9	21.4	10.6	21.2	83 05	+16.1	−16.2	86 31	+15.8	−16.0
3 30	+ 3.4	−28.9	+ 3.1	−28.7	10 00	+11.0	−21.3	+10.7	−21.1	90 00			90 00	+15.9	−15.9

App. Alt. = Apparent altitude = Sextant altitude corrected for index error and dip = $H_a = H_s + IE + D$

In critical cases ascend. Additional corrections for temperature and pressure are given on the previous page.

ALTITUDE CORRECTION TABLES 0°–35° — MOON

App. Alt.	0°–4° Corrⁿ	5°–9° Corrⁿ	10°–14° Corrⁿ	15°–19° Corrⁿ	20°–24° Corrⁿ	25°–29° Corrⁿ	30°–34° Corrⁿ	App. Alt.
00	0° 34.5	5° 58.2	10° 62.1	15° 62.8	20° 62.2	25° 60.8	30° 58.9	00
10	36.5	58.5	62.2	62.8	62.2	60.8	58.8	10
20	38.3	58.7	62.2	62.8	62.1	60.7	58.8	20
30	40.0	58.9	62.3	62.8	62.1	60.7	58.7	30
40	41.5	59.1	62.3	62.8	62.0	60.6	58.6	40
50	42.9	59.3	62.4	62.7	62.0	60.6	58.5	50
00	1° 44.2	6° 59.5	11° 62.4	16° 62.7	21° 62.0	26° 60.5	31° 58.5	00
10	45.4	59.7	62.4	62.7	61.9	60.4	58.4	10
20	46.5	59.9	62.5	62.7	61.9	60.4	58.3	20
30	47.5	60.0	62.5	62.7	61.9	60.3	58.2	30
40	48.4	60.2	62.5	62.7	61.8	60.3	58.2	40
50	49.3	60.3	62.6	62.7	61.8	60.2	58.1	50
00	2° 50.1	7° 60.5	12° 62.6	17° 62.7	22° 61.7	27° 60.1	32° 58.0	00
10	50.8	60.6	62.6	62.6	61.7	60.1	57.9	10
20	51.5	60.7	62.6	62.6	61.6	60.0	57.8	20
30	52.2	60.9	62.7	62.6	61.6	59.9	57.8	30
40	52.8	61.0	62.7	62.6	61.6	59.9	57.7	40
50	53.4	61.1	62.7	62.6	61.5	59.8	57.6	50
00	3° 53.9	8° 61.2	13° 62.7	18° 62.5	23° 61.5	28° 59.7	33° 57.5	00
10	54.4	61.3	62.7	62.5	61.4	59.7	57.4	10
20	54.9	61.4	62.7	62.5	61.4	59.6	57.4	20
30	55.3	61.5	62.8	62.5	61.3	59.5	57.3	30
40	55.7	61.6	62.8	62.4	61.3	59.5	57.2	40
50	56.1	61.6	62.8	62.4	61.2	59.4	57.1	50
00	4° 56.4	9° 61.7	14° 62.8	19° 62.4	24° 61.2	29° 59.3	34° 57.0	00
10	56.8	61.8	62.8	62.4	61.1	59.3	56.9	10
20	57.1	61.9	62.8	62.3	61.1	59.2	56.9	20
30	57.4	61.9	62.8	62.3	61.0	59.1	56.8	30
40	57.7	62.0	62.8	62.3	61.0	59.1	56.7	40
50	58.0	62.1	62.8	62.2	60.9	59.0	56.6	50

HP	L U	L U	L U	L U	L U	L U	L U	HP
54.0	0.3 0.9	0.3 0.9	0.4 1.0	0.5 1.1	0.6 1.2	0.7 1.3	0.9 1.5	54.0
54.3	0.7 1.1	0.7 1.2	0.8 1.2	0.8 1.3	0.9 1.4	1.1 1.5	1.2 1.7	54.3
54.6	1.1 1.4	1.1 1.4	1.1 1.4	1.2 1.5	1.3 1.6	1.4 1.7	1.5 1.8	54.6
54.9	1.4 1.6	1.5 1.6	1.5 1.6	1.6 1.7	1.6 1.8	1.8 1.9	1.9 2.0	54.9
55.2	1.8 1.8	1.8 1.8	1.9 1.8	1.9 1.9	2.0 2.0	2.1 2.1	2.2 2.2	55.2
55.5	2.2 2.0	2.2 2.0	2.3 2.1	2.3 2.1	2.4 2.2	2.4 2.3	2.5 2.4	55.5
55.8	2.6 2.2	2.6 2.2	2.6 2.3	2.7 2.3	2.7 2.4	2.8 2.4	2.9 2.5	55.8
56.1	3.0 2.4	3.0 2.5	3.0 2.5	3.0 2.5	3.1 2.6	3.1 2.6	3.2 2.7	56.1
56.4	3.3 2.7	3.4 2.7	3.4 2.7	3.4 2.7	3.4 2.8	3.5 2.8	3.5 2.9	56.4
56.7	3.7 2.9	3.7 2.9	3.8 2.9	3.8 2.9	3.8 3.0	3.8 3.0	3.9 3.0	56.7
57.0	4.1 3.1	4.1 3.1	4.1 3.1	4.1 3.1	4.2 3.2	4.2 3.2	4.2 3.2	57.0
57.3	4.5 3.3	4.5 3.3	4.5 3.3	4.5 3.3	4.5 3.4	4.5 3.4	4.6 3.4	57.3
57.6	4.9 3.5	4.9 3.5	4.9 3.5	4.9 3.5	4.9 3.5	4.9 3.5	4.9 3.6	57.6
57.9	5.3 3.8	5.3 3.8	5.2 3.8	5.2 3.7	5.2 3.7	5.2 3.7	5.2 3.7	57.9
58.2	5.6 4.0	5.6 4.0	5.6 4.0	5.6 4.0	5.6 3.9	5.6 3.9	5.6 3.9	58.2
58.5	6.0 4.2	6.0 4.2	6.0 4.2	6.0 4.2	6.0 4.1	5.9 4.1	5.9 4.1	58.5
58.8	6.4 4.4	6.4 4.4	6.4 4.4	6.3 4.4	6.3 4.3	6.3 4.3	6.2 4.2	58.8
59.1	6.8 4.6	6.8 4.6	6.7 4.6	6.7 4.6	6.7 4.5	6.6 4.5	6.6 4.4	59.1
59.4	7.2 4.8	7.1 4.8	7.1 4.8	7.1 4.8	7.0 4.7	7.0 4.7	6.9 4.6	59.4
59.7	7.5 5.1	7.5 5.0	7.5 5.0	7.5 5.0	7.4 4.9	7.3 4.8	7.2 4.8	59.7
60.0	7.9 5.3	7.9 5.3	7.9 5.2	7.8 5.2	7.8 5.1	7.7 5.0	7.6 4.9	60.0
60.3	8.3 5.5	8.3 5.5	8.2 5.4	8.2 5.4	8.1 5.3	8.0 5.2	7.9 5.1	60.3
60.6	8.7 5.7	8.7 5.7	8.6 5.7	8.6 5.6	8.5 5.5	8.4 5.4	8.2 5.3	60.6
60.9	9.1 5.9	9.0 5.9	9.0 5.9	8.9 5.8	8.8 5.7	8.7 5.6	8.6 5.4	60.9
61.2	9.5 6.2	9.4 6.1	9.4 6.1	9.3 6.0	9.2 5.9	9.1 5.8	8.9 5.6	61.2
61.5	9.8 6.4	9.8 6.3	9.7 6.3	9.7 6.2	9.5 6.1	9.4 5.9	9.2 5.8	61.5

DIP

Ht. of Eye (m)	Corrⁿ	Ht. of Eye (ft)	Ht. of Eye (m)	Corrⁿ	Ht. of Eye (ft)
2.4	−2.8	8.0	9.5	−5.5	31.5
2.6	−2.9	8.6	9.9	−5.6	32.7
2.8	−3.0	9.2	10.3	−5.7	33.9
3.0	−3.1	9.8	10.6	−5.8	35.1
3.2	−3.2	10.5	11.0	−5.9	36.3
3.4	−3.3	11.2	11.4	−6.0	37.6
3.6	−3.4	11.9	11.8	−6.1	38.9
3.8	−3.5	12.6	12.2	−6.2	40.1
4.0	−3.6	13.3	12.6	−6.3	41.5
4.3	−3.7	14.1	13.0	−6.4	42.8
4.5	−3.8	14.9	13.4	−6.5	44.2
4.7	−3.9	15.7	13.8	−6.6	45.5
5.0	−4.0	16.5	14.2	−6.7	46.9
5.2	−4.1	17.4	14.7	−6.8	48.4
5.5	−4.2	18.3	15.1	−6.9	49.8
5.8	−4.3	19.1	15.5	−7.0	51.3
6.1	−4.4	20.1	16.0	−7.1	52.8
6.3	−4.5	21.0	16.5	−7.2	54.3
6.6	−4.6	22.0	16.9	−7.3	55.8
6.9	−4.7	22.9	17.4	−7.4	57.4
7.2	−4.8	23.9	17.9	−7.5	58.9
7.5	−4.9	24.9	18.4	−7.6	60.5
7.9	−5.0	26.0	18.8	−7.7	62.1
8.2	−5.1	27.1	19.3	−7.8	63.8
8.5	−5.2	28.1	19.8	−7.9	65.4
8.8	−5.3	29.2	20.4	−8.0	67.1
9.2	−5.4	30.4	20.9	−8.1	68.8
9.5		31.5	21.4		70.5

MOON CORRECTION TABLE

The correction is in two parts; the first correction is taken from the upper part of the table with argument apparent altitude, and the second from the lower part, with argument HP, in the same column as that from which the first correction was taken. Separate corrections are given in the lower part for lower (L) and upper (U) limbs. All corrections are to be **added** to apparent altitude, *but 30′ is to be subtracted from the altitude of the upper limb.*

For corrections for pressure and temperature see page A4.

For bubble sextant observations ignore dip, take the mean of upper and lower limb corrections and subtract 15′ from the altitude.

App. Alt. = Apparent altitude = Sextant altitude corrected for index error and dip.

ALTITUDE CORRECTION TABLES 35°–90°— MOON

App. Alt.	35°–39° Corrⁿ	40°–44° Corrⁿ	45°–49° Corrⁿ	50°–54° Corrⁿ	55°–59° Corrⁿ	60°–64° Corrⁿ	65°–69° Corrⁿ	70°–74° Corrⁿ	75°–79° Corrⁿ	80°–84° Corrⁿ	85°–89° Corrⁿ	App. Alt.
00	35° 56.5	40° 53.7	45° 50.5	50° 46.9	55° 43.1	60° 38.9	65° 34.6	70° 30.0	75° 25.3	80° 20.5	85° 15.6	00
10	56.4	53.6	50.4	46.8	42.9	38.8	34.4	29.9	25.2	20.4	15.5	10
20	56.3	53.5	50.2	46.7	42.8	38.7	34.3	29.7	25.0	20.2	15.3	20
30	56.2	53.4	50.1	46.5	42.7	38.5	34.1	29.6	24.9	20.0	15.1	30
40	56.2	53.3	50.0	46.4	42.5	38.4	34.0	29.4	24.7	19.9	15.0	40
50	56.1	53.2	49.9	46.3	42.4	38.2	33.8	29.3	24.5	19.7	14.8	50
00	36° 56.0	41° 53.1	46° 49.8	51° 46.2	56° 42.3	61° 38.1	66° 33.7	71° 29.1	76° 24.4	81° 19.6	86° 14.6	00
10	55.9	53.0	49.7	46.0	42.1	37.9	33.5	29.0	24.2	19.4	14.5	10
20	55.8	52.9	49.5	45.9	42.0	37.8	33.4	28.8	24.1	19.2	14.3	20
30	55.7	52.8	49.4	45.8	41.9	37.7	33.2	28.7	23.9	19.1	14.2	30
40	55.6	52.6	49.3	45.7	41.7	37.5	33.1	28.5	23.8	18.9	14.0	40
50	55.5	52.5	49.2	45.5	41.6	37.4	32.9	28.3	23.6	18.7	13.8	50
00	37° 55.4	42° 52.4	47° 49.1	52° 45.4	57° 41.4	62° 37.2	67° 32.8	72° 28.2	77° 23.4	82° 18.6	87° 13.7	00
10	55.3	52.3	49.0	45.3	41.3	37.1	32.6	28.0	23.3	18.4	13.5	10
20	55.2	52.2	48.8	45.2	41.2	36.9	32.5	27.9	23.1	18.2	13.3	20
30	55.1	52.1	48.7	45.0	41.0	36.8	32.3	27.7	22.9	18.1	13.2	30
40	55.0	52.0	48.6	44.9	40.9	36.6	32.2	27.6	22.8	17.9	13.0	40
50	55.0	51.9	48.5	44.8	40.8	36.5	32.0	27.4	22.6	17.8	12.8	50
00	38° 54.9	43° 51.8	48° 48.4	53° 44.6	58° 40.6	63° 36.4	68° 31.9	73° 27.2	78° 22.5	83° 17.6	88° 12.7	00
10	54.8	51.7	48.3	44.5	40.5	36.2	31.7	27.1	22.3	17.4	12.5	10
20	54.7	51.6	48.1	44.4	40.3	36.1	31.6	26.9	22.1	17.3	12.3	20
30	54.6	51.5	48.0	44.2	40.2	35.9	31.4	26.8	22.0	17.1	12.2	30
40	54.5	51.4	47.9	44.1	40.1	35.8	31.3	26.6	21.8	16.9	12.0	40
50	54.4	51.2	47.8	44.0	39.9	35.6	31.1	26.5	21.7	16.8	11.8	50
00	39° 54.3	44° 51.1	49° 47.7	54° 43.9	59° 39.8	64° 35.5	69° 31.0	74° 26.3	79° 21.5	84° 16.6	89° 11.7	00
10	54.2	51.0	47.5	43.7	39.6	35.3	30.8	26.1	21.3	16.4	11.5	10
20	54.1	50.9	47.4	43.6	39.5	35.2	30.7	26.0	21.2	16.3	11.4	20
30	54.0	50.8	47.3	43.5	39.4	35.0	30.5	25.8	21.0	16.1	11.2	30
40	53.9	50.7	47.2	43.3	39.2	34.9	30.4	25.7	20.9	16.0	11.0	40
50	53.8	50.6	47.0	43.2	39.1	34.7	30.2	25.5	20.7	15.8	10.9	50
HP	L U	L U	L U	L U	L U	L U	L U	L U	L U	L U	L U	HP
54.0	1.1 1.7	1.3 1.9	1.5 2.1	1.7 2.4	2.0 2.6	2.3 2.9	2.6 3.2	2.9 3.5	3.2 3.8	3.5 4.1	3.8 4.5	54.0
54.3	1.4 1.8	1.6 2.0	1.8 2.2	2.0 2.5	2.2 2.7	2.5 3.0	2.8 3.2	3.1 3.5	3.3 3.8	3.6 4.1	3.9 4.4	54.3
54.6	1.7 2.0	1.9 2.2	2.1 2.4	2.3 2.6	2.5 2.8	2.7 3.0	3.0 3.3	3.2 3.5	3.5 3.8	3.8 4.0	4.0 4.3	54.6
54.9	2.0 2.2	2.2 2.3	2.3 2.5	2.5 2.7	2.7 2.9	2.9 3.1	3.2 3.3	3.4 3.5	3.6 3.8	3.9 4.0	4.1 4.3	54.9
55.2	2.3 2.3	2.5 2.4	2.6 2.6	2.8 2.8	3.0 2.9	3.2 3.1	3.4 3.3	3.6 3.5	3.8 3.7	4.0 4.0	4.2 4.2	55.2
55.5	2.7 2.5	2.8 2.6	2.9 2.7	3.1 2.9	3.2 3.0	3.4 3.2	3.6 3.4	3.7 3.5	3.9 3.7	4.1 3.9	4.3 4.1	55.5
55.8	3.0 2.6	3.1 2.7	3.2 2.8	3.3 3.0	3.5 3.1	3.6 3.3	3.8 3.4	3.9 3.6	4.1 3.7	4.2 3.9	4.4 4.0	55.8
56.1	3.3 2.8	3.4 2.9	3.5 3.0	3.6 3.1	3.7 3.2	3.8 3.3	4.0 3.4	4.1 3.6	4.2 3.7	4.4 3.8	4.5 4.0	56.1
56.4	3.6 2.9	3.7 3.0	3.8 3.1	3.9 3.2	3.9 3.3	4.0 3.4	4.1 3.5	4.3 3.6	4.4 3.7	4.5 3.8	4.6 3.9	56.4
56.7	3.9 3.1	4.0 3.1	4.1 3.2	4.1 3.3	4.2 3.3	4.3 3.4	4.3 3.5	4.4 3.6	4.5 3.7	4.6 3.8	4.7 3.8	56.7
57.0	4.3 3.2	4.3 3.3	4.3 3.3	4.4 3.4	4.4 3.4	4.5 3.5	4.5 3.5	4.6 3.6	4.7 3.6	4.7 3.7	4.8 3.8	57.0
57.3	4.6 3.4	4.6 3.4	4.6 3.4	4.6 3.5	4.7 3.5	4.7 3.5	4.7 3.6	4.8 3.6	4.8 3.6	4.8 3.7	4.9 3.7	57.3
57.6	4.9 3.6	4.9 3.6	4.9 3.6	4.9 3.6	4.9 3.6	4.9 3.6	4.9 3.6	4.9 3.6	5.0 3.6	5.0 3.6	5.0 3.6	57.6
57.9	5.2 3.7	5.2 3.7	5.2 3.7	5.2 3.7	5.2 3.7	5.1 3.6	5.1 3.6	5.1 3.6	5.1 3.6	5.1 3.6	5.1 3.6	57.9
58.2	5.5 3.9	5.5 3.8	5.5 3.8	5.4 3.8	5.4 3.7	5.4 3.7	5.3 3.7	5.3 3.6	5.2 3.6	5.2 3.5	5.2 3.5	58.2
58.5	5.9 4.0	5.8 4.0	5.8 3.9	5.7 3.9	5.6 3.8	5.6 3.8	5.5 3.7	5.5 3.6	5.4 3.6	5.3 3.5	5.3 3.4	58.5
58.8	6.2 4.2	6.1 4.1	6.0 4.1	6.0 4.0	5.9 3.9	5.8 3.8	5.7 3.7	5.6 3.6	5.5 3.5	5.4 3.5	5.3 3.4	58.8
59.1	6.5 4.3	6.4 4.3	6.3 4.2	6.2 4.1	6.1 4.0	6.0 3.9	5.9 3.8	5.8 3.6	5.7 3.5	5.6 3.4	5.4 3.3	59.1
59.4	6.8 4.5	6.7 4.4	6.6 4.3	6.5 4.2	6.4 4.1	6.2 3.9	6.1 3.8	6.0 3.7	5.8 3.5	5.7 3.4	5.5 3.2	59.4
59.7	7.1 4.7	7.0 4.5	6.9 4.4	6.8 4.3	6.6 4.1	6.5 4.0	6.3 3.8	6.1 3.7	6.0 3.5	5.8 3.3	5.6 3.2	59.7
60.0	7.5 4.8	7.3 4.7	7.2 4.5	7.0 4.4	6.9 4.2	6.7 4.0	6.5 3.9	6.3 3.7	6.1 3.5	5.9 3.3	5.7 3.1	60.0
60.3	7.8 5.0	7.6 4.8	7.5 4.7	7.3 4.5	7.1 4.3	6.9 4.1	6.7 3.9	6.5 3.7	6.3 3.5	6.0 3.2	5.8 3.0	60.3
60.6	8.1 5.1	7.9 5.0	7.7 4.8	7.6 4.6	7.3 4.4	7.1 4.2	6.9 3.9	6.7 3.7	6.4 3.4	6.2 3.2	5.9 2.9	60.6
60.9	8.4 5.3	8.2 5.1	8.0 4.9	7.8 4.7	7.6 4.5	7.3 4.2	7.1 4.0	6.8 3.7	6.6 3.4	6.3 3.2	6.0 2.9	60.9
61.2	8.7 5.4	8.5 5.2	8.3 5.0	8.1 4.8	7.8 4.5	7.6 4.3	7.3 4.0	7.0 3.7	6.7 3.4	6.4 3.1	6.1 2.8	61.2
61.5	9.1 5.6	8.8 5.4	8.6 5.1	8.3 4.9	8.1 4.6	7.8 4.3	7.5 4.0	7.2 3.7	6.9 3.4	6.5 3.1	6.2 2.7	61.5

Step 7. Calculate the intercept(p).
We learned in chapter 7 that the intercept is the difference between the zenith distance of the observed altitude and the zenith distance of the calculated altitude. The calculation for this can be reduced to the difference between the observed altitude and the calculated (tabulated) altitude.

Therefore, in simple terms, to calculate the intercept (p), we find the difference between the observed altitude (Ho) and the tabulated altitude (Hc).

This can be abbreviated to p = (Ho-Hc)

Note. The intercept will be towards the direction of the observed body if the observed altitude is greater than the tabulated altitude and away from the body if less. We can abbreviate this as follows:

Intercept will be + (towards) if Ho > Hc. Intercept will be – (away) if Ho < Hc

Example:

Ho = 40° 46'.23
Hc = 40° 24'.00
p(Ho-Hc) = +22'.23 (∴ intercept is 22.23 n.m. towards the true bearing).

Step 8. Apply movement corrections to the intercept as follows:

a. Movement of observer (MOO) (Table 1)

This correction need only be made when there has been a significant change in the position of the observer between observations. For example, when a fix is made by transferring position lines from observations made at different times while a ship is under way.

Brief Explanation of the use of Table 1.

In the extract from Table 1 shown on the next page, distance (to the nearest mile) is tabulated across the top row. Rel. Zn (Zn – course) is shown in the left or right hand columns. The correction is read from the relevant intersecting cell and is used to correct the intercept as shown in the following example. (Note. To calculate the sign of the correction, use the table of rules which follows the extract.

Example. In this example, the distance made good between the observation and a later fix is 20.0 n.m. on a course of 315°. The true bearing (Zn) is 015° and the intercept (p) is -5'.58 away:

True Bearing (Zn) = 015°
±360° (to make Zn > C) = 360°
　　　　　　　　　　　　375°
Course (C) = 315°
Rel Zn (Zn – C) = 060°
Dist. = 20'
Intercept p = -5'.58 (away)
Table 1 corn. = +10'.00 (fix is later)   (see the table of rules
Corrected p = +4'.42 (∴ corrected intercept = 4.42 n.m. to 015°)

MOO TABLE 1 — ALTITUDE CORRECTION FOR CHANGE IN POSITION OF OBSERVER

Distance Made Good — nautical miles

Rel. Zn	1	2	3	4	5	6	7	8	10	15	20	25	30	35	40	45	50	75	100	150	Rel. Zn
000	+1.0	+2.0	+3.0	+4.0	+5.0	+6.0	+7.0	+8.0	+10.0	+15.0	+20.0	+25.0	+30.0	+35.0	+40.0	+45.0	+50.0	+75.0	+100.0	+150.0	000
002	1.0	2.0	3.0	4.0	5.0	6.0	7.0	8.0	10.0	15.0	20.0	25.0	30.0	35.0	40.0	45.0	50.0	75.0	99.9	149.9	358
004	1.0	2.0	3.0	4.0	5.0	6.0	7.0	8.0	10.0	15.0	20.0	24.9	29.9	34.9	39.9	44.9	49.9	74.8	99.8	149.6	356
006	1.0	2.0	3.0	4.0	5.0	6.0	7.0	8.0	9.9	14.9	19.9	24.9	29.8	34.8	39.8	44.8	49.7	74.6	99.5	149.2	354
008	1.0	2.0	3.0	4.0	5.0	5.9	6.9	7.9	9.9	14.9	19.8	24.8	29.7	34.7	39.6	44.6	49.5	74.3	99.0	148.5	352
010	+1.0	+2.0	+3.0	+3.9	+4.9	+5.9	+6.9	+7.9	+9.8	+14.8	+19.7	+24.6	+29.5	+34.5	+39.4	+44.3	+49.2	+73.9	+98.5	+147.7	350
012	1.0	2.0	2.9	3.9	4.9	5.9	6.8	7.8	9.8	14.7	19.6	24.5	29.3	34.2	39.1	44.0	48.9	73.4	97.8	146.7	348
014	1.0	1.9	2.9	3.9	4.9	5.8	6.8	7.8	9.7	14.6	19.4	24.3	29.1	34.0	38.8	43.7	48.5	72.8	97.0	145.5	346
016	1.0	1.9	2.9	3.8	4.8	5.8	6.7	7.7	9.6	14.4	19.2	24.0	28.8	33.6	38.5	43.3	48.1	72.1	96.1	144.2	344
018	1.0	1.9	2.9	3.8	4.8	5.7	6.7	7.6	9.5	14.3	19.0	23.8	28.5	33.3	38.0	42.8	47.6	71.3	95.1	142.7	342
020	+0.9	+1.9	+2.8	+3.8	+4.7	+5.6	+6.6	+7.5	+9.4	+14.1	+18.8	+23.5	+28.2	+32.9	+37.6	+42.3	+47.0	+70.5	+94.0	+141.0	340
022	0.9	1.9	2.8	3.7	4.6	5.6	6.5	7.4	9.3	13.9	18.5	23.2	27.8	32.5	37.1	41.7	46.4	69.5	92.7	139.1	338
024	0.9	1.8	2.7	3.7	4.6	5.5	6.4	7.3	9.1	13.7	18.3	22.8	27.4	32.0	36.5	41.1	45.7	68.5	91.4	137.0	336
026	0.9	1.8	2.7	3.6	4.5	5.4	6.3	7.2	9.0	13.5	18.0	22.5	27.0	31.5	36.0	40.4	44.9	67.4	89.9	134.8	334
028	0.9	1.8	2.6	3.5	4.4	5.3	6.2	7.1	8.8	13.2	17.7	22.1	26.5	30.9	35.3	39.7	44.1	66.2	88.3	132.4	332
030	+0.9	+1.7	+2.6	+3.5	+4.3	+5.2	+6.1	+6.9	+8.7	+13.0	+17.3	+21.7	+26.0	+30.3	+34.6	+39.0	+43.3	+65.0	+86.6	+129.9	330
032	0.8	1.7	2.5	3.4	4.2	5.1	5.9	6.8	8.5	12.7	17.0	21.2	25.4	29.7	33.9	38.2	42.4	63.6	84.8	127.2	328
034	0.8	1.7	2.5	3.3	4.1	5.0	5.8	6.6	8.3	12.4	16.6	20.7	24.9	29.0	33.2	37.3	41.5	62.2	82.9	124.4	326
036	0.8	1.6	2.4	3.2	4.0	4.9	5.7	6.5	8.1	12.1	16.2	20.2	24.3	28.3	32.4	36.4	40.5	60.7	80.9	121.4	324
038	0.8	1.6	2.4	3.2	3.9	4.7	5.5	6.3	7.9	11.8	15.8	19.7	23.6	27.6	31.5	35.5	39.4	59.1	78.8	118.2	322
040	+0.8	+1.5	+2.3	+3.1	+3.8	+4.6	+5.4	+6.1	+7.7	+11.5	+15.3	+19.2	+23.0	+26.8	+30.6	+34.5	+38.3	+57.5	+76.6	+114.9	320
042	0.7	1.5	2.2	3.0	3.7	4.5	5.2	5.9	7.4	11.1	14.9	18.6	22.3	26.0	29.7	33.4	37.2	55.7	74.3	111.5	318
044	0.7	1.4	2.2	2.9	3.6	4.3	5.0	5.8	7.2	10.8	14.4	18.0	21.6	25.2	28.8	32.4	36.0	54.0	71.9	107.9	316
046	0.7	1.4	2.1	2.8	3.5	4.2	4.9	5.6	6.9	10.4	13.9	17.4	20.8	24.3	27.8	31.3	34.7	52.1	69.5	104.2	314
048	0.7	1.3	2.0	2.7	3.3	4.0	4.7	5.4	6.7	10.0	13.4	16.7	20.1	23.4	26.8	30.1	33.5	50.2	66.9	100.4	312
050	+0.6	+1.3	+1.9	+2.6	+3.2	+3.9	+4.5	+5.1	+6.4	+9.6	+12.9	+16.1	+19.3	+22.5	+25.7	+28.9	+32.1	+48.2	+64.3	+96.4	310
052	0.6	1.2	1.8	2.5	3.1	3.7	4.3	4.9	6.2	9.2	12.3	15.4	18.5	21.5	24.6	27.7	30.8	46.2	61.6	92.3	308
054	0.6	1.2	1.8	2.4	2.9	3.5	4.1	4.7	5.9	8.8	11.8	14.7	17.6	20.6	23.5	26.5	29.4	44.1	58.8	88.2	306
056	0.6	1.1	1.7	2.2	2.8	3.4	3.9	4.5	5.6	8.4	11.2	14.0	16.8	19.6	22.4	25.2	28.0	41.9	55.9	83.9	304
058	0.5	1.1	1.6	2.1	2.6	3.2	3.7	4.2	5.3	7.9	10.6	13.2	15.9	18.5	21.2	23.8	26.5	39.7	53.0	79.5	302
060	+0.5	+1.0	+1.5	+2.0	+2.5	+3.0	+3.5	+4.0	+5.0	+7.5	+10.0	+12.5	+15.0	+17.5	+20.0	+22.5	+25.0	+37.5	+50.0	+75.0	300
062	0.5	0.9	1.4	1.9	2.3	2.8	3.3	3.8	4.7	7.0	9.4	11.7	14.1	16.4	18.8	21.1	23.5	35.2	46.9	70.4	298
064	0.4	0.9	1.3	1.8	2.2	2.6	3.1	3.5	4.4	6.6	8.8	11.0	13.2	15.3	17.5	19.7	21.9	32.9	43.8	65.8	296
066	0.4	0.8	1.2	1.6	2.0	2.4	2.8	3.3	4.1	6.1	8.1	10.2	12.2	14.2	16.3	18.3	20.3	30.5	40.7	61.0	294
068	0.4	0.7	1.1	1.5	1.9	2.2	2.6	3.0	3.7	5.6	7.5	9.4	11.2	13.1	15.0	16.9	18.7	28.1	37.5	56.2	292
070	+0.3	+0.7	+1.0	+1.4	+1.7	+2.1	+2.4	+2.7	+3.4	+5.1	+6.8	+8.6	+10.3	+12.0	+13.7	+15.4	+17.1	+25.7	+34.2	+51.3	290
072	0.3	0.6	0.9	1.2	1.5	1.9	2.2	2.5	3.1	4.6	6.2	7.7	9.3	10.8	12.4	13.9	15.5	23.2	30.9	46.4	288
074	0.3	0.6	0.8	1.1	1.4	1.7	1.9	2.2	2.8	4.1	5.5	6.9	8.3	9.6	11.0	12.4	13.8	20.7	27.6	41.3	286
076	0.2	0.5	0.7	1.0	1.2	1.5	1.7	1.9	2.4	3.6	4.8	6.0	7.3	8.5	9.7	10.9	12.1	18.1	24.2	36.3	284

Note. The sign of the correction is determined by application of the following rules:

Rules for application of sign from tables 1 & 2		
Time of fix	Sign from table	To intercept
Later than observation	+	Add
	-	Subtract
Earlier than observation	+	Subtract
	-	Add

b. Movement of Body (MOB) (Table 2)

This special technique may be used when observations are planned ahead so that the same assumed position, calculated altitude, azimuth and LHA are re-used in the reduction of the observations. The advantage of this technique is that the amount of calculation is considerably reduced. However, a correction for the movement of the body between the time of the fix and the time of the observation has to be made. This correction is referred to as the MOB.

Brief Explanation of the use of Table 2.

In the extract from Table 2 on the next page, Latitude (to the nearest degree) is tabulated across the top row and True Zn is shown in the left and right hand columns. The correction for 1 minute of time is read from the relevant intersecting cell and is used to correct the intercept as demonstrated in the following example. The sign of the correction is taken from the table of rules which follows the extract.

Example. In this example, the observation is made 2 minutes after the time of the fix and the intercept (p), calculated at the time of the observation, is +14'.95 (towards). The latitude was 40°N and the true bearing (Zn) was 105°. The procedure for finding the corrected intercept from Table 2 is as follows:

Lat = N40°
Zn = 105°
Fix is 2^m earlier
Intercept: +14'.95 (towards)
corr. for 1^m: +11'.1 (from table 2)
corr. for 2^m: +22'.2
Correction: -22'.2 (after applying rules)
Corrected 'cpt = +14'.95 -22'.2 = -7'.25
Therefore, corrected intercept is 7.25 n.m. away from 105°.

TABLE 2 — ALTITUDE CORRECTION FOR CHANGE IN POSITION OF BODY

Correction for 1 Minute of Time

True Zn	0°	5°	10°	15°	20°	25°	30°	35°	40°	45°	50°	55°	60°	65°	70°	75°	80°	85°	True Zn
°	′	′	′	′	′	′	′	′	′	′	′	′	′	′	′	′	′	′	°
090	+15·0	+15·0	+14·8	+14·5	+14·1	+13·6	+13·0	+12·3	+11·5	+10·6	+9·7	+8·6	+7·5	+6·4	+5·1	+3·9	+2·6	+1·3	090
093	15·0	15·0	14·8	14·5	14·1	13·6	13·0	12·3	11·5	10·6	9·7	8·6	7·5	6·3	5·1	3·9	2·6	1·3	087
096	15·0	14·9	14·7	14·4	14·1	13·6	13·0	12·3	11·5	10·6	9·6	8·6	7·5	6·3	5·1	3·9	2·6	1·3	084
099	14·9	14·8	14·6	14·3	14·0	13·5	12·9	12·2	11·4	10·5	9·5	8·5	7·4	6·3	5·1	3·8	2·6	1·3	081
102	14·7	14·7	14·5	14·2	13·8	13·3	12·7	12·1	11·3	10·4	9·5	8·4	7·4	6·2	5·0	3·8	2·6	1·3	078
105	+14·5	+14·5	+14·3	+14·0	+13·7	+13·2	+12·6	+11·9	+11·1	+10·3	+9·3	+8·3	+7·3	+6·1	+5·0	+3·8	+2·5	+1·3	075
108	14·3	14·3	14·1	13·8	13·4	13·0	12·4	11·7	11·0	10·1	9·2	8·2	7·2	6·0	4·9	3·7	2·5	1·2	072
111	14·0	14·0	13·8	13·6	13·2	12·7	12·2	11·5	10·8	9·9	9·0	8·1	7·0	5·9	4·8	3·6	2·4	1·2	069
114	13·7	13·7	13·5	13·3	12·9	12·5	11·9	11·3	10·5	9·7	8·8	7·9	6·9	5·8	4·7	3·6	2·4	1·2	066
117	13·4	13·4	13·2	12·9	12·6	12·1	11·6	11·0	10·3	9·5	8·6	7·7	6·7	5·7	4·6	3·5	2·3	1·2	063
120	+13·0	+13·0	+12·8	+12·6	+12·2	+11·8	+11·3	+10·7	+10·0	+9·2	+8·4	+7·5	+6·5	+5·5	+4·5	+3·4	+2·3	+1·1	060
123	12·6	12·6	12·4	12·2	11·9	11·4	10·9	10·3	9·7	8·9	8·1	7·2	6·3	5·3	4·3	3·3	2·2	1·1	057
126	12·2	12·1	12·0	11·8	11·4	11·0	10·5	10·0	9·3	8·6	7·8	7·0	6·1	5·1	4·2	3·1	2·1	1·1	054
129	11·7	11·6	11·5	11·3	11·0	10·6	10·1	9·6	9·0	8·3	7·5	6·7	5·8	4·9	4·0	3·0	2·0	1·0	051
132	11·2	11·1	11·0	10·8	10·5	10·1	9·7	9·2	8·6	7·9	7·2	6·4	5·6	4·7	3·8	2·9	1·9	1·0	048
135	+10·6	+10·6	+10·5	+10·3	+10·0	+9·6	+9·2	+8·7	+8·1	+7·5	+6·8	+6·1	+5·3	+4·5	+3·6	+2·8	+1·8	+0·9	045
138	10·1	10·0	9·9	9·7	9·5	9·1	8·7	8·2	7·7	7·1	6·5	5·8	5·0	4·3	3·4	2·6	1·7	0·9	042
141	9·5	9·4	9·3	9·1	8·9	8·6	8·2	7·8	7·3	6·7	6·1	5·4	4·7	4·0	3·2	2·4	1·6	0·8	039
144	8·8	8·8	8·7	8·5	8·3	8·0	7·7	7·2	6·8	6·3	5·7	5·1	4·4	3·7	3·0	2·3	1·5	0·8	036
147	8·2	8·2	8·1	7·9	7·7	7·4	7·1	6·7	6·3	5·8	5·3	4·7	4·1	3·5	2·8	2·1	1·4	0·7	033
150	+7·5	+7·5	+7·4	+7·3	+7·1	+6·8	+6·5	+6·2	+5·8	+5·3	+4·8	+4·3	+3·8	+3·2	+2·6	+1·9	+1·3	+0·7	030
153	6·8	6·8	6·7	6·6	6·4	6·2	5·9	5·6	5·2	4·8	4·4	3·9	3·4	2·9	2·3	1·8	1·2	0·6	027
156	6·1	6·1	6·0	5·9	5·7	5·5	5·3	5·0	4·7	4·3	3·9	3·5	3·1	2·6	2·1	1·6	1·1	0·5	024
159	5·4	5·4	5·3	5·2	5·1	4·9	4·7	4·4	4·1	3·8	3·5	3·1	2·7	2·3	1·8	1·4	0·9	0·5	021
162	4·6	4·6	4·6	4·5	4·4	4·2	4·0	3·8	3·6	3·3	3·0	2·7	2·3	2·0	1·6	1·2	0·8	0·4	018
165	+3·9	+3·9	+3·8	+3·8	+3·7	+3·5	+3·4	+3·2	+3·0	+2·8	+2·5	+2·2	+1·9	+1·6	+1·3	+1·0	+0·7	+0·3	015
168	3·1	3·1	3·1	3·0	2·9	2·8	2·7	2·6	2·4	2·2	2·0	1·8	1·6	1·3	1·1	0·8	0·5	0·3	012
171	2·4	2·3	2·3	2·3	2·2	2·1	2·0	1·9	1·8	1·7	1·5	1·3	1·2	1·0	0·8	0·6	0·4	0·2	009
174	1·6	1·6	1·5	1·5	1·5	1·4	1·4	1·3	1·2	1·1	1·0	0·9	0·8	0·7	0·5	0·4	0·3	0·1	006

Latitude

MOB

Note. The sign of the correction is determined by application of the following rules:

Rules for application of sign from tables 1 & 2		
Time of fix	Sign from table	To intercept
Later than observation	+	Add
	-	Subtract
Earlier than observation	+	Subtract
	-	Add

9. **Plot the position line.** Firstly, plot the assumed position using the assumed long. calculated at step 4 and the assumed lat. calculated at step 5. Secondly, using the corrected intercept data calculated at step 8, plot the position line at right angles to the bearing 105° at 7.25 n.m. from the assumed position.

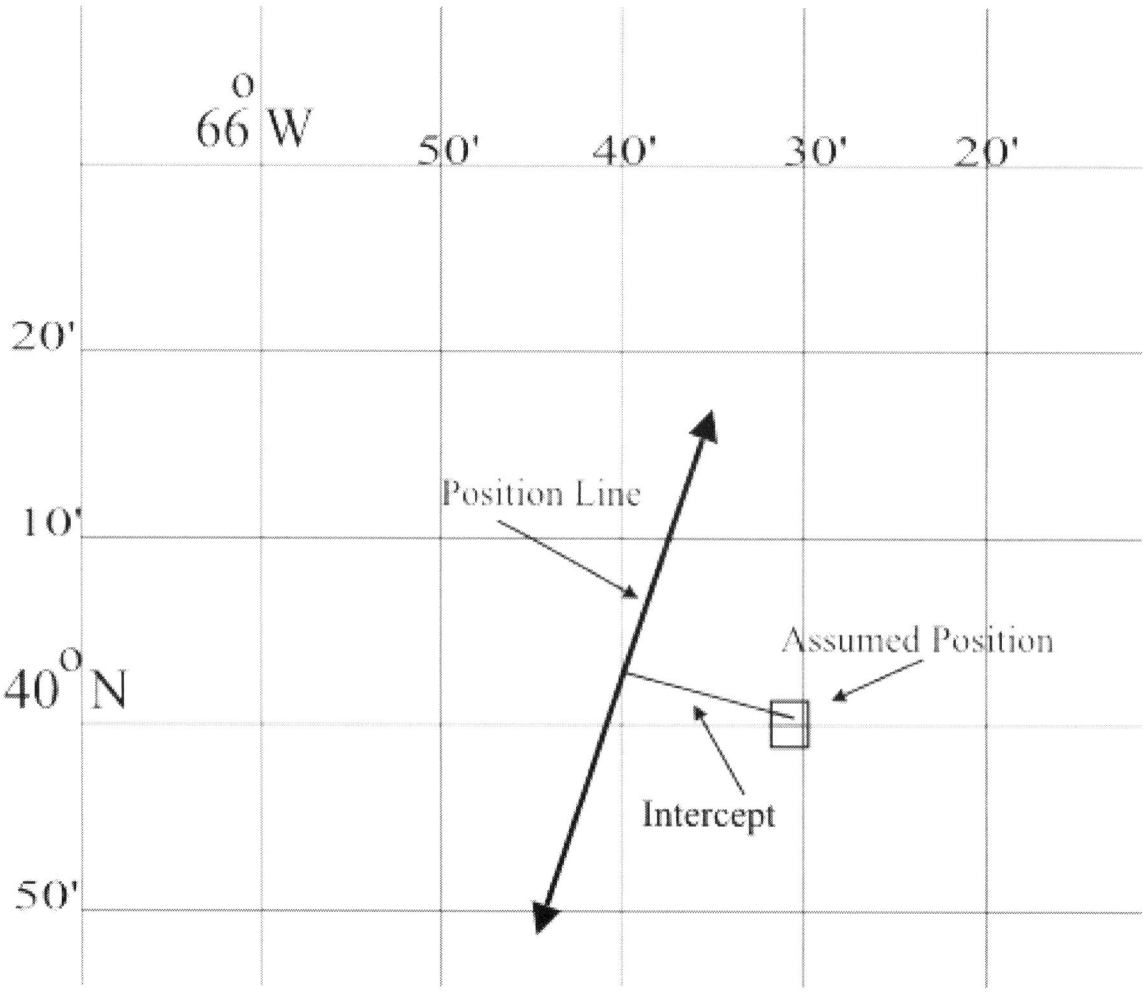

Planning Star and Planet observations.

Unlike the Sun and the Moon which are easily identified, the approximate positions of stars and planets need to be established before observations are made. The times of rising and setting of the Sun and Moon can be found in the daily pages of the nautical almanac so we know when they will be visible above the horizon. Risings and settings for stars and planets are not listed and so these need to be calculated. In fact, knowledge of the precise times of rising and setting are not always essential, all that we need to know in most cases, is which bodies are above our horizon at the times of planned observations. We can then calculate the approximate azimuth and altitude of the chosen bodies in the planning phase of the rapid sight reduction procedures. There are various techniques and devices that can be used to predict the approximate positions of celestial bodies such as 'star globes', 'star charts', computer software and the ABC tables contained in Norie's nautical tables.

However, the following 'rule of thumb' method is quite useful especially in the cramped confines of a chart-table, in a rough sea. It is emphasised that this is not intended to be a method of accurately pin-pointing the position of a star or planet but a simple and quick method of establishing whether or not the body is likely to be visible at the time an observation is required. It should also be emphasised that this method becomes less reliable as you approach the polar regions where LHA becomes less relevant and the rule for declination becomes less accurate when the body is in the same direction as the pole.

Stars. To find if a star will be above the horizon at our position at the time of the planned observations, we need to take two things into account, its local hour angle and its declination.

LHA. For a star to be visible above the horizon at our position, its meridian must be within 90° east or west of our estimated longitude at the time of the planned observations. Another way of saying this is that either the LHA must be less than 90° or, if the LHA is greater than 180°, then LHA ± 360° must be less than 90°.

The following method can be used to calculate the approximate LHA of a star before commencing the sight reduction procedure:

1. From the nautical almanac daily pages, find GHA Aries (to the nearest degree) at the planned time of the observation.
2. From the 'Index to Selected Stars' in the Nautical Almanac, find the SHA of the star (to the nearest degree). An extract of the 'Index to Selected Stars' is shown on page 177.
3. Calculate the estimated longitude (to the nearest degree) at the planned time of the observation.
4. Combine the SHA, GHA Aries and the estimated longitude to find the approximate LHA.

5. For the star to be visible, its LHA must be less than 90° or if the LHA is greater than 180°, the LHA ± 360° must be less than 90°.

Example 1:
GHA Aries   34°
SHA star   + 29°
Est. Long.   + 24°E   (east add, west subtract)
Approx LHA   87°

Example 2:
GHA Aries   233°
SHA star   +126°
Est. Long.   - 47°W   (east add, west subtract)
Approx LHA   312°
360° - LHA   48°

Example 3.
GHA Aries   197°
SHA star   +218°
Est. Long.   +16°E   (east add, west subtract)
Approx LH   431°
LHA - 360°   71°

In all three cases, since the result is less than 90°, we know that the star should be above the eastern and western horizons at the planned time of the observation. However, it should be remembered that this method does not give us an accurate LHA; this will be calculated at step 4 of the procedure for rapid sight reduction. Nor does it give us the expected altitude and bearing; these will be calculated at steps 4 & 5.

Declination. As well as the LHA, we have to take into account the declination. For a star to be visible above the horizon, it's declination must be within 90° of the latitude of our position. If our latitude is north, then the declination must be within the range 90° north to (90° - latitude) south. If the latitude is south, then the declination must be within the range 90° south to (90° - latitude) north.

We can formulate the above statements as follows:

Estimated latitude North: visible range = 90°N to (90° - Lat)S.
Estimated latitude South: visible range = 90°S to (90° - Lat)N.

Example. Estimated Lat: 40°S
Range of visible declinations = 90°S to (90° - Lat)N.
                              = 90°S to 50°N.

Notes.
1. The SHAs and declinations of the navigational stars are listed in the 'Index to Selected Stars' of which an extract is shown on page 177.

2. Tables of stars are included in the nautical almanac daily pages but these do not include the complete list of navigational stars and do not show magnitude.

Guidelines for using the above methods:
1. Choose a star with a high visual magnitude first; if calculations show that this star will not be above the horizon, then try others with lower magnitudes until a match is found.
2. Reject stars whose LHAs and declinations are found to be close to the 90° limits because it may found that, after applying increment and altitude corrections, these are just below the horizon.
3. Remember that this method becomes less reliable as you approach the polar regions.

Planets. The method of calculating which planets will be above the horizon at a vessel's estimated position at a certain time is much more straightforward than that for stars. Firstly, there are only 4 'navigational' planets compared to 57 'navigational' stars. Secondly, the planets are listed in the nautical almanac daily pages by their GHA instead of their SHA.

For a planet, we also need to know whether or not its approximate LHA will be within 90° east or west of our estimated longitude and whether or not its declination will be within 90° north or south of our estimated latitude at the planned time of the observations. In this way, we will know whether or not it will be above our horizon. The method is quite simple:

LHA
1. From the nautical almanac daily pages, find the GHA of the planet (to the nearest degree) at the planned time of the observation.
2. Calculate the estimated longitude (to the nearest degree) at the planned time of the observation.
3. Combine the GHA and the estimated longitude to find the approximate LHA.

Example 1.
GHA	78°
Estimated long.	-47°W. (east add, west subtract)
Approx LHA	31°

Example 2.
GHA	325°
Estimated long.	+48°E. (east add, west subtract)
Approx LHA	373°
LHA - 360°	13°

In both of these examples, since the result is less than 90°, we know that the planet should be visible above the eastern and western horizons at the planned time of the observation.

Declination. For declination, the rule is the same as in the case for stars:
    Estimated latitude North: visible range = 90°N to (90° - Lat)S.
    Estimated latitude South: visible range = 90°S to (90° - Lat)N.

Example 1.   Estimated lat: 45°N.
Range of visible declinations = 90°N to (90° - Lat)S.
                               = 90°N to (90° - 45°)S.
                               = 90°N to 45°S.
Example 2.   Estimated lat: 75°S.
Range of visible declinations = 90°S to (90° - Lat)N.
                               = 90°S to 15°N.

The examples in the following chapters will help to give a clear understanding of the rapid sight reduction techniques explained above.

An extract of the 'Index to Selected Stars' is shown on the following page.

INDEX TO SELECTED STARS, 2009

Name	No	Mag	SHA	Dec	No	Name	Mag	SHA	Dec
			°	°				°	°
Acamar	7	3·2	315	S 40	1	Alpheratz	2·1	358	N 29
Achernar	5	0·5	335	S 57	2	Ankaa	2·4	353	S 42
Acrux	30	1·3	173	S 63	3	Schedar	2·2	350	N 57
Adhara	19	1·5	255	S 29	4	Diphda	2·0	349	S 18
Aldebaran	10	0·9	291	N 17	5	Achernar	0·5	335	S 57
Alioth	32	1·8	166	N 56	6	Hamal	2·0	328	N 24
Alkaid	34	1·9	153	N 49	7	Acamar	3·2	315	S 40
Al Na'ir	55	1·7	28	S 47	8	Menkar	2·5	314	N 4
Alnilam	15	1·7	276	S 1	9	Mirfak	1·8	309	N 50
Alphard	25	2·0	218	S 9	10	Aldebaran	0·9	291	N 17
Alphecca	41	2·2	126	N 27	11	Rigel	0·1	281	S 8
Alpheratz	1	2·1	358	N 29	12	Capella	0·1	281	N 46
Altair	51	0·8	62	N 9	13	Bellatrix	1·6	279	N 6
Ankaa	2	2·4	353	S 42	14	Elnath	1·7	278	N 29
Antares	42	1·0	112	S 26	15	Alnilam	1·7	276	S 1
Arcturus	37	0·0	146	N 19	16	Betelgeuse	Var.*	271	N 7
Atria	43	1·9	108	S 69	17	Canopus	−0·7	264	S 53
Avior	22	1·9	234	S 60	18	Sirius	−1·5	259	S 17
Bellatrix	13	1·6	279	N 6	19	Adhara	1·5	255	S 29
Betelgeuse	16	Var.*	271	N 7	20	Procyon	0·4	245	N 5
Canopus	17	−0·7	264	S 53	21	Pollux	1·1	244	N 28
Capella	12	0·1	281	N 46	22	Avior	1·9	234	S 60
Deneb	53	1·3	50	N 45	23	Suhail	2·2	223	S 43
Denebola	28	2·1	183	N 15	24	Miaplacidus	1·7	222	S 70
Diphda	4	2·0	349	S 18	25	Alphard	2·0	218	S 9
Dubhe	27	1·8	194	N 62	26	Regulus	1·4	208	N 12
Elnath	14	1·7	278	N 29	27	Dubhe	1·8	194	N 62
Eltanin	47	2·2	91	N 51	28	Denebola	2·1	183	N 15
Enif	54	2·4	34	N 10	29	Gienah	2·6	176	S 18
Fomalhaut	56	1·2	15	S 30	30	Acrux	1·3	173	S 63
Gacrux	31	1·6	172	S 57	31	Gacrux	1·6	172	S 57
Gienah	29	2·6	176	S 18	32	Alioth	1·8	166	N 56
Hadar	35	0·6	149	S 60	33	Spica	1·0	159	S 11
Hamal	6	2·0	328	N 24	34	Alkaid	1·9	153	N 49
Kaus Australis	48	1·9	84	S 34	35	Hadar	0·6	149	S 60
Kochab	40	2·1	137	N 74	36	Menkent	2·1	148	S 36
Markab	57	2·5	14	N 15	37	Arcturus	0·0	146	N 19
Menkar	8	2·5	314	N 4	38	Rigil Kentaurus	−0·3	140	S 61
Menkent	36	2·1	148	S 36	39	Zubenelgenubi	2·8	137	S 16
Miaplacidus	24	1·7	222	S 70	40	Kochab	2·1	137	N 74
Mirfak	9	1·8	309	N 50	41	Alphecca	2·2	126	N 27
Nunki	50	2·0	76	S 26	42	Antares	1·0	112	S 26
Peacock	52	1·9	53	S 57	43	Atria	1·9	108	S 69
Pollux	21	1·1	244	N 28	44	Sabik	2·4	102	S 16
Procyon	20	0·4	245	N 5	45	Shaula	1·6	96	S 37
Rasalhague	46	2·1	96	N 13	46	Rasalhague	2·1	96	N 13
Regulus	26	1·4	208	N 12	47	Eltanin	2·2	91	N 51
Rigel	11	0·1	281	S 8	48	Kaus Australis	1·9	84	S 34
Rigil Kentaurus	38	−0·3	140	S 61	49	Vega	0·0	81	N 39
Sabik	44	2·4	102	S 16	50	Nunki	2·0	76	S 26
Schedar	3	2·2	350	N 57	51	Altair	0·8	62	N 9
Shaula	45	1·6	96	S 37	52	Peacock	1·9	53	S 57
Sirius	18	−1·5	259	S 17	53	Deneb	1·3	50	N 45
Spica	33	1·0	159	S 11	54	Enif	2·4	34	N 10
Suhail	23	2·2	223	S 43	55	Al Na'ir	1·7	28	S 47
Vega	49	0·0	81	N 39	56	Fomalhaut	1·2	15	S 30
Zubenelgenubi	39	2·8	137	S 16	57	Markab	2·5	14	N 15

©2009 Jack Case

Chapter 9
Demonstration of Single Observation and Position Line

In this chapter we demonstrate how the rapid sight reduction tables can be used to obtain a position line from a single observation.

Notes.
1. To fully understand this demonstration, it will be necessary to refer to the outline procedures in chapter 8 and it may be advisable to revise chapters 2, 4 and 5 as well.
2. In order to demonstrate the method of plotting the position line at step 9, a grid has been constructed to show the area of the plot. The scale of the drawing is not accurate enough to reliably pin-point the position of the fix; to do this, the position-line would have to be plotted on a navigational chart for the latitude of the estimated position. However, the drawing is accurate enough to demonstrate the method of plotting the position line.
3. Relevant extracts of tables are placed immediately following the solution to the example. Because of their size, some tables have been cropped so that they show only the relevant portions.
4. Self-test questions on the use of the rapid sight reduction tables can be found in chapter 13.

Aim. The aim of this demonstration is to establish a position line based on the following imaginary data:

D.R. Position: 51° 54'N 21° 55'W.
Date: 18 July 2009
Zone Time: $16^h 44^m$ (+1)
DWT: $17^h 50^m 28^s$
DWE 40^s fast
Body observed: Sun L.L.
Sextant Alt: 32° 10.'4. Bearing: 261°
Index error: +0'.54
Ht. of eye: 8m.
Temperature: 28°C. Pressure: 991mb.

Solution:
Planning Phase:

Step 1. Approx. Pos: 51° 54'N, 21° 55'W	

Step 2. Calculate GMT at time of observation.	
Date: 18 July. 2009	
Zone time:	$18^d\ 16^h\ 44^m$
Zone corrn:	$+1^h$
Greenwich date:	$18^d\ 17^h\ 44^m\ 00^s$
DWT:	$17^h\ 50^m\ 28^s$
DWE:	-40^s
UT:	$17^h\ 49^m\ 48^s$

Step 3. Calculate GHA and Dec. of Sun.

	GHA	Dec
UT 17^h	73° 26'.1	N20° 54'.7 (d:0'.5)
Inc. $49^m\ 48^s$:	12° 27'.0	-0'.4
	85° 53'.1	N20° 54'.3

Step 4. Calculate LHA

GHA:	85° 53'.1	
Assumed Long:	21° 53'.1	(West -)
LHA:	64°	(-360 if reqd.)

Convert azimuth angle to true bearing:		
Azimuth (Z)	=	98°
True bearing (Zn) (360° - Z)	=	262°
(see rules for calculating Zn below)		

Please see page 161 for an explanation of the reason for calculating true bearing here.
(Take the value of the Azimuth (Z) from step 5).

Rules for calculating true bearing (Zn)		
	Lat. North	Lat. South
LHA>180°	Zn = Z	Zn = 180° - Z
LHA<180°	Zn = 360°-Z	Zn = 180° + Z

Method for step 5: Find main table page for lat. 52°, same name and declination 20°. Where the column for Declination 20° intersects with the row for LHA 64°, read off the values for Hc, d, Z. The correct values in this case are as follows: Hc: 31° 33'   d: +45'   Z: 98°

Step 5. Main table entry. Extract values for Hc, d, Z from main table using the following data from the calculations above:	
Assumed Lat:	52° N
Dec. Sun (degrees) Dec°:	20°N (same)
Dec. Sun (minutes) Dec':	54'.3
LHA Sun:	64°
Table page:	92

Next, turn to table 5 and find the correction for Dec' under the column for 45 (the value of d) and in the row for 54 (the nearest whole number to the value of Dec'. Note the sign of d in the main table and apply the same sign to the value of the correction. The correction in this case is: +40'.

Record the findings at step 5 as follows:

	Hc	d	Z
Extracted values:	31° 33'	+45'	98°
Corrn. for Dec' 54'.0	+40'	(table 5)	
Corr. Alt. (Hc) =	32° 13'		

Fix Phase

Step 6. Correct sextant altitude (Hs).	
Sext. Alt. (Hs) =	32° 10.'4
Index error (IE) =	+0'.54
Dip (ht. 8m.) (D) =	-5'.0 (table 6a)
Apparent Alt. (Ha) =	32° 05.'94
Alt. correction (R) =	+14'.5 (table 6d)
Added refrac'n (d) =	+0'.1 (table 6c)
Observed Alt. (Ho) =	32° 20.'54

Step 7. Calculate intercept (p)		
	Ho =	32° 20'.54
	Hc =	32° 13'.00
Ho - Hc	p =	+7'.54
∴ Intercept = 7.54 n.m. towards 262° (bearing calculated at step 4)		

Step 8. Corrections to Intercept.

(No corrections required since there has been no movement between observation and plot).

Step 9. Plot the position line.

Firstly, plot the assumed position using the assumed long. calculated at step 4 and the assumed lat. calculated at step 5. Secondly, using the intercept data calculated at step 7, plot the position line at right angles to the bearing 262° at 7.54 n.m. from the assumed position.

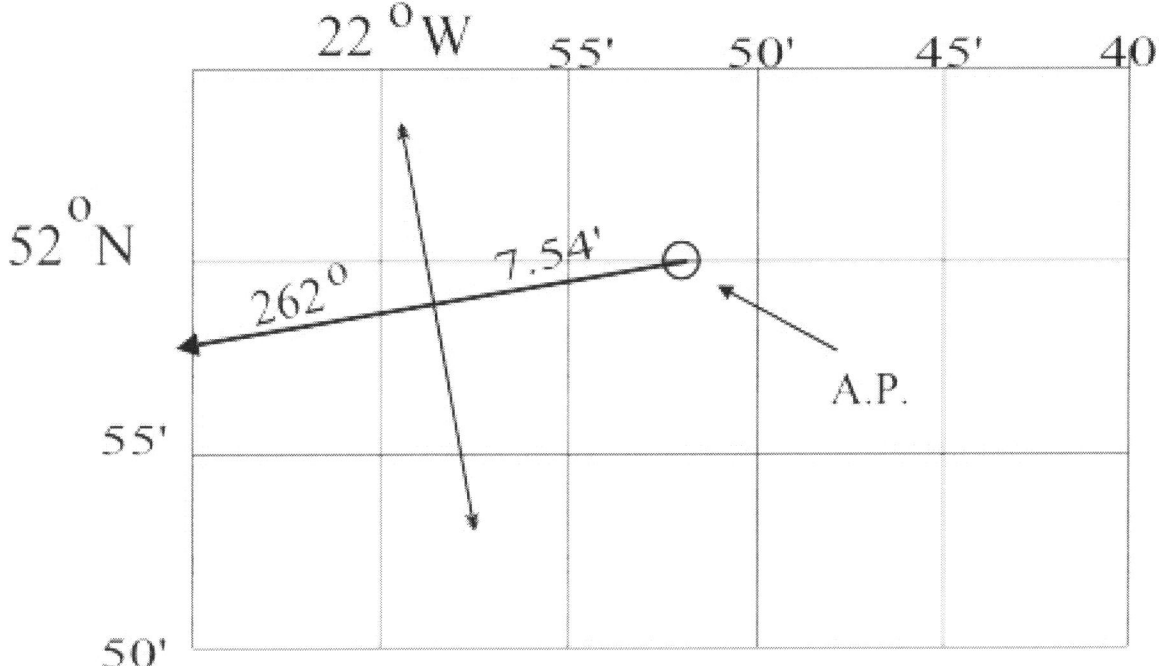

Extracts

Extracts relevant to the above example are to be found on the following pages.

2009 JULY 18, 19, 20 (SAT., SUN., MON.)

UT	SUN GHA	SUN Dec	MOON GHA	MOON v	MOON Dec	MOON d	MOON HP
d h	° '	° '	° '	'	° '	'	'
18 00	178 26.9	N21 02.2	240 54.7	5.1	N24 21.3	6.5	59.3
01	193 26.8	01.8	255 18.8	5.1	24 27.8	6.3	59.3
02	208 26.8	01.3	269 42.9	4.9	24 34.1	6.3	59.4
03	223 26.7	00.9	284 06.8	4.9	24 40.4	6.0	59.4
04	238 26.7	00.5	298 30.7	4.7	24 46.4	5.9	59.4
05	253 26.6	21 00.0	312 54.4	4.6	24 52.3	5.8	59.5
06	268 26.6	N20 59.6	327 18.0	4.5	N24 58.1	5.5	59.5
S 07	283 26.5	59.1	341 41.5	4.4	25 03.6	5.5	59.6
A 08	298 26.5	58.7	356 04.9	4.3	25 09.1	5.2	59.6
T 09	313 26.4	58.2	10 28.2	4.1	25 14.3	5.1	59.6
U 10	328 26.4	57.8	24 51.3	4.1	25 19.4	4.9	59.7
R 11	343 26.4	57.4	39 14.4	4.0	25 24.3	4.8	59.7
D 12	358 26.3	N20 56.9	53 37.4	3.9	N25 29.1	4.5	59.7
A 13	13 26.3	56.5	68 00.3	3.7	25 33.6	4.4	59.8
Y 14	28 26.2	56.0	82 23.0	3.7	25 38.0	4.3	59.8
15	43 26.2	55.6	96 45.7	3.6	25 42.3	4.0	59.8
16	58 26.1	55.1	111 08.3	3.5	25 46.3	3.9	59.9
17	73 26.1	54.7	125 30.8	3.4	25 50.2	3.7	59.9
18	88 26.0	N20 54.2	139 53.2	3.3	N25 53.9	3.5	59.9
19	103 26.0	53.8	154 15.5	3.3	25 57.4	3.4	60.0
20	118 25.9	53.3	168 37.7	3.2	26 00.8	3.1	60.0
21	133 25.9	52.9	182 59.9	3.0	26 03.9	3.0	60.0
22	148 25.9	52.4	197 21.9	3.0	26 06.9	2.8	60.1
23	163 25.8	52.0	211 43.9	2.9	26 09.7	2.6	60.1
19 00	178 25.8	N20 51.5	226 05.8	2.8	N26 12.3	2.4	60.1
01	193 25.7	51.1	240 27.6	2.8	26 14.7	2.3	60.2
02	208 25.7	50.6	254 49.4	2.7	26 17.0	2.0	60.2
03	223 25.6	50.2	269 11.1	2.6	26 19.0	1.8	60.2
04	238 25.6	49.7	283 32.7	2.5	26 20.8	1.7	60.3
05	253 25.6	49.2	297 54.2	2.5	26 22.5	1.5	60.3
06	268 25.5	N20 48.8	312 15.7	2.4	N26 24.0	1.2	60.3
07	283 25.5	48.3	326 37.1	2.4	26 25.2	1.1	60.3
S 08	298 25.4	47.9	340 58.5	2.3	26 26.3	0.9	60.4
U 09	313 25.4	47.4	355 19.8	2.3	26 27.2	0.7	60.4
N 10	328 25.4	47.0	9 41.1	2.2	26 27.9	0.5	60.4
D 11	343 25.3	46.5	24 02.3	2.2	26 28.4	0.3	60.5
A 12	358 25.3	N20 46.0	38 23.5	2.1	N26 28.7	0.1	60.5
Y 13	13 25.2	45.6	52 44.6	2.1	26 28.8	0.1	60.5
14	28 25.2	45.1	67 05.7	2.0	26 28.7	0.3	60.5
15	43 25.2	44.7	81 26.7	2.0	26 28.4	0.5	60.6
16	58 25.1	44.2	95 47.7	2.0	26 27.9	0.7	60.6
17	73 25.1	43.7	110 08.7	2.0	26 27.2	1.0	60.6
18	88 25.0	N20 43.3	124 29.7	1.9	N26 26.2	1.1	60.7
19	103 25.0	42.8	138 50.6	1.9	26 25.1	1.3	60.7
20	118 25.0	42.3	153 11.5	1.9	26 23.8	1.5	60.7
21	133 24.9	41.9	167 32.4	1.8	26 22.3	1.7	60.7
22	148 24.9	41.4	181 53.2	1.9	26 20.6	1.9	60.8
23	163 24.8	40.9	196 14.1	1.8	26 18.7	2.1	60.8
20 00	178 24.8	N20 40.5	210 34.9	1.9	N26 16.6	2.3	60.8
01	193 24.8	40.0	224 55.8	1.8	26 14.3	2.5	60.8
02	208 24.7	39.5	239 16.6	1.8	26 11.8	2.8	60.8
03	223 24.7	39.1	253 37.4	1.8	26 09.0	2.9	60.9
04	238 24.7	38.6	267 58.2	1.9	26 06.1	3.1	60.9
05	253 24.6	38.1	282 19.1	1.8	26 03.0	3.3	60.9
06	268 24.6	N20 37.7	296 39.9	1.9	N25 59.7	3.5	60.9
07	283 24.5	37.2	311 00.8	1.8	25 56.2	3.8	61.0
08	298 24.5	36.7	325 21.6	1.9	25 52.4	3.9	61.0
M 09	313 24.5	36.2	339 42.5	1.9	25 48.5	4.1	61.0
O 10	328 24.4	35.8	354 03.4	1.9	25 44.4	4.3	61.0
N 11	343 24.4	35.3	8 24.3	1.9	25 40.1	4.6	61.0
D 12	358 24.4	N20 34.8	22 45.2	2.0	N25 35.5	4.7	61.0
A 13	13 24.3	34.3	37 06.2	2.0	25 30.8	4.9	61.1
Y 14	28 24.3	33.9	51 27.2	2.0	25 25.9	5.1	61.1
15	43 24.3	33.4	65 48.2	2.1	25 20.8	5.3	61.1
16	58 24.2	32.9	80 09.3	2.1	25 15.5	5.5	61.1
17	73 24.2	32.4	94 30.4	2.2	25 10.0	5.7	61.1
18	88 24.2	N20 32.0	108 51.6	2.1	N25 04.3	5.9	61.1
19	103 24.1	31.5	123 12.7	2.3	24 58.4	6.1	61.2
20	118 24.1	31.0	137 34.0	2.3	24 52.3	6.3	61.2
21	133 24.1	30.5	151 55.3	2.3	24 46.0	6.5	61.2
22	148 24.0	30.0	166 16.6	2.4	24 39.5	6.6	61.2
23	163 24.0	29.6	180 38.0	2.4	N24 32.9	6.9	61.2
	SD 15.8	d 0.5	SD 16.3		16.5		16.6

Lat.	Twilight Naut.	Twilight Civil	Sunrise	Moonrise 18	Moonrise 19	Moonrise 20	Moonrise 21
°	h m	h m	h m	h m	h m	h m	h m
N 72	□	□	□	□	□	□	□
N 70	□	□	□	□	□	□	□
68	////	////	00 47	□	□	□	□
66	////	////	01 53	□	□	□	□
64	////	////	02 28	21 09	□	□	00 33
62	////	01 16	02 52	22 26	23 27	25 14	01 14
60	////	01 58	03 12	23 03	24 06	00 06	01 42
N 58	////	02 26	03 28	23 29	24 33	00 33	02 04
56	01 09	02 47	03 41	23 50	24 54	00 54	02 21
54	01 48	03 04	03 53	24 07	00 07	01 12	02 36
52	02 14	03 18	04 03	24 22	00 22	01 27	02 49
50	02 34	03 31	04 12	24 35	00 35	01 40	03 01
45	03 10	03 56	04 31	00 10	01 01	02 07	03 24
N 40	03 36	04 15	04 47	00 29	01 22	02 28	03 43
35	03 56	04 31	05 00	00 45	01 40	02 45	03 59
30	04 12	04 45	05 11	00 59	01 55	03 01	04 12
20	04 38	05 08	05 30	01 22	02 21	03 26	04 34
N 10	04 58	05 25	05 47	01 43	02 43	03 49	04 56
0	05 15	05 41	06 03	02 02	03 04	04 09	05 14
S 10	05 30	05 56	06 18	02 21	03 25	04 30	05 33
20	05 44	06 11	06 34	02 42	03 48	04 52	05 53
30	05 58	06 27	06 53	03 06	04 14	05 18	06 16
35	06 05	06 36	07 04	03 20	04 29	05 33	06 29
40	06 13	06 46	07 16	03 37	04 47	05 51	06 44
45	06 21	06 58	07 30	03 57	05 09	06 12	07 03
S 50	06 31	07 11	07 48	04 22	05 36	06 38	07 25
52	06 35	07 17	07 56	04 34	05 50	06 51	07 36
54	06 40	07 24	08 05	04 48	06 05	07 06	07 48
56	06 44	07 32	08 16	05 04	06 24	07 23	08 02
58	06 50	07 40	08 28	05 23	06 46	07 44	08 18
S 60	06 56	07 49	08 41	05 47	07 15	08 10	08 38

Lat.	Sunset	Twilight Civil	Twilight Naut.	Moonset 18	Moonset 19	Moonset 20	Moonset 21
°	h m	h m	h m	h m	h m	h m	h m
N 72	□	□	□	□	□	□	□
N 70	□	□	□	□	□	□	□
68	23 16	////	////	□	□	□	22 40
66	22 16	////	////	□	□	□	21 54
64	21 43	////	////	20 22	□	21 31	21 23
62	21 18	22 52	////	19 05	20 19	20 50	21 00
60	20 59	22 12	////	18 28	19 41	20 21	20 41
N 58	20 44	21 45	////	18 02	19 13	19 59	20 26
56	20 30	21 24	22 59	17 42	18 52	19 41	20 12
54	20 19	21 07	22 22	17 25	18 34	19 26	20 00
52	20 09	20 53	21 57	17 10	18 19	19 12	19 50
50	20 00	20 41	21 37	16 57	18 06	19 01	19 41
45	19 41	20 16	21 02	16 31	17 39	18 36	19 21
N 40	19 25	19 57	20 36	16 11	17 18	18 17	19 05
35	19 12	19 41	20 16	15 53	17 00	18 00	18 51
30	19 01	19 28	20 00	15 38	16 45	17 46	18 40
20	18 42	19 06	19 34	15 13	16 19	17 21	18 19
N 10	18 25	18 48	19 14	14 51	15 56	17 00	18 01
0	18 10	18 32	18 58	14 31	15 35	16 40	17 44
S 10	17 55	18 17	18 43	14 11	15 14	16 20	17 27
20	17 38	18 02	18 29	13 49	14 51	15 59	17 09
30	17 20	17 46	18 15	13 24	14 25	15 34	16 47
35	17 09	17 37	18 08	13 09	14 09	15 19	16 35
40	16 57	17 27	18 00	12 52	13 51	15 02	16 20
45	16 43	17 15	17 52	12 32	13 30	14 41	16 03
S 50	16 25	17 02	17 42	12 06	13 02	14 15	15 42
52	16 17	16 56	17 38	11 53	12 49	14 03	15 31
54	16 08	16 49	17 34	11 39	12 33	13 48	15 20
56	15 57	16 42	17 29	11 23	12 15	13 31	15 06
58	15 45	16 33	17 23	11 03	11 52	13 11	14 51
S 60	15 32	16 24	17 18	10 39	11 23	12 45	14 32

Day	Eqn. of Time 00h	Eqn. of Time 12h	Mer. Pass.	Mer. Pass. Upper	Mer. Pass. Lower	Age	Phase
d	m s	m s	h m	h m	h m	d	%
18	06 12	06 15	12 06	08 16	20 47	26	18
19	06 17	06 19	12 06	09 20	21 52	27	10
20	06 21	06 22	12 06	10 25	22 57	28	4

INCREMENTS AND CORRECTIONS

48m

48 s	SUN PLANETS ° '	ARIES ° '	MOON ° '	v or d '	Corrn '	v or d '	Corrn '	v or d '	Corrn '
00	12 00·0	12 02·0	11 27·2	0·0	0·0	6·0	4·9	12·0	9·7
01	12 00·3	12 02·2	11 27·4	0·1	0·1	6·1	4·9	12·1	9·8
02	12 00·5	12 02·5	11 27·7	0·2	0·2	6·2	5·0	12·2	9·9
03	12 00·8	12 02·7	11 27·9	0·3	0·2	6·3	5·1	12·3	9·9
04	12 01·0	12 03·0	11 28·2	0·4	0·3	6·4	5·2	12·4	10·0
05	12 01·3	12 03·2	11 28·4	0·5	0·4	6·5	5·3	12·5	10·1
06	12 01·5	12 03·5	11 28·6	0·6	0·5	6·6	5·3	12·6	10·2
07	12 01·8	12 03·7	11 28·9	0·7	0·6	6·7	5·4	12·7	10·3
08	12 02·0	12 04·0	11 29·1	0·8	0·6	6·8	5·5	12·8	10·3
09	12 02·3	12 04·2	11 29·3	0·9	0·7	6·9	5·6	12·9	10·4
10	12 02·5	12 04·5	11 29·6	1·0	0·8	7·0	5·7	13·0	10·5
11	12 02·8	12 04·7	11 29·8	1·1	0·9	7·1	5·7	13·1	10·6
12	12 03·0	12 05·0	11 30·1	1·2	1·0	7·2	5·8	13·2	10·7
13	12 03·3	12 05·2	11 30·3	1·3	1·1	7·3	5·9	13·3	10·8
14	12 03·5	12 05·5	11 30·5	1·4	1·1	7·4	6·0	13·4	10·8
15	12 03·8	12 05·7	11 30·8	1·5	1·2	7·5	6·1	13·5	10·9
16	12 04·0	12 06·0	11 31·0	1·6	1·3	7·6	6·1	13·6	11·0
17	12 04·3	12 06·2	11 31·3	1·7	1·4	7·7	6·2	13·7	11·1
18	12 04·5	12 06·5	11 31·5	1·8	1·5	7·8	6·3	13·8	11·2
19	12 04·8	12 06·7	11 31·7	1·9	1·5	7·9	6·4	13·9	11·2
20	12 05·0	12 07·0	11 32·0	2·0	1·6	8·0	6·5	14·0	11·3
21	12 05·3	12 07·2	11 32·2	2·1	1·7	8·1	6·5	14·1	11·4
22	12 05·5	12 07·5	11 32·4	2·2	1·8	8·2	6·6	14·2	11·5
23	12 05·8	12 07·7	11 32·7	2·3	1·9	8·3	6·7	14·3	11·6
24	12 06·0	12 08·0	11 32·9	2·4	1·9	8·4	6·8	14·4	11·6
25	12 06·3	12 08·2	11 33·2	2·5	2·0	8·5	6·9	14·5	11·7
26	12 06·5	12 08·5	11 33·4	2·6	2·1	8·6	7·0	14·6	11·8
27	12 06·8	12 08·7	11 33·6	2·7	2·2	8·7	7·0	14·7	11·9
28	12 07·0	12 09·0	11 33·9	2·8	2·3	8·8	7·1	14·8	12·0
29	12 07·3	12 09·2	11 34·1	2·9	2·3	8·9	7·2	14·9	12·0
30	12 07·5	12 09·5	11 34·4	3·0	2·4	9·0	7·3	15·0	12·1
31	12 07·8	12 09·7	11 34·6	3·1	2·5	9·1	7·4	15·1	12·2
32	12 08·0	12 10·0	11 34·8	3·2	2·6	9·2	7·4	15·2	12·3
33	12 08·3	12 10·2	11 35·1	3·3	2·7	9·3	7·5	15·3	12·4
34	12 08·5	12 10·5	11 35·3	3·4	2·7	9·4	7·6	15·4	12·4
35	12 08·8	12 10·7	11 35·6	3·5	2·8	9·5	7·7	15·5	12·5
36	12 09·0	12 11·0	11 35·8	3·6	2·9	9·6	7·8	15·6	12·6
37	12 09·3	12 11·2	11 36·0	3·7	3·0	9·7	7·8	15·7	12·7
38	12 09·5	12 11·5	11 36·3	3·8	3·1	9·8	7·9	15·8	12·8
39	12 09·8	12 11·7	11 36·5	3·9	3·2	9·9	8·0	15·9	12·9
40	12 10·0	12 12·0	11 36·7	4·0	3·2	10·0	8·1	16·0	12·9
41	12 10·3	12 12·2	11 37·0	4·1	3·3	10·1	8·2	16·1	13·0
42	12 10·5	12 12·5	11 37·2	4·2	3·4	10·2	8·2	16·2	13·1
43	12 10·8	12 12·8	11 37·5	4·3	3·5	10·3	8·3	16·3	13·2
44	12 11·0	12 13·0	11 37·7	4·4	3·6	10·4	8·4	16·4	13·3
45	12 11·3	12 13·3	11 37·9	4·5	3·6	10·5	8·5	16·5	13·3
46	12 11·5	12 13·5	11 38·2	4·6	3·7	10·6	8·6	16·6	13·4
47	12 11·8	12 13·8	11 38·4	4·7	3·8	10·7	8·6	16·7	13·5
48	12 12·0	12 14·0	11 38·7	4·8	3·9	10·8	8·7	16·8	13·6
49	12 12·3	12 14·3	11 38·9	4·9	4·0	10·9	8·8	16·9	13·7
50	12 12·5	12 14·5	11 39·1	5·0	4·0	11·0	8·9	17·0	13·7
51	12 12·8	12 14·8	11 39·4	5·1	4·1	11·1	9·0	17·1	13·8
52	12 13·0	12 15·0	11 39·6	5·2	4·2	11·2	9·1	17·2	13·9
53	12 13·3	12 15·3	11 39·8	5·3	4·3	11·3	9·1	17·3	14·0
54	12 13·5	12 15·5	11 40·1	5·4	4·4	11·4	9·2	17·4	14·1
55	12 13·8	12 15·8	11 40·3	5·5	4·4	11·5	9·3	17·5	14·1
56	12 14·0	12 16·0	11 40·6	5·6	4·5	11·6	9·4	17·6	14·2
57	12 14·3	12 16·3	11 40·8	5·7	4·6	11·7	9·5	17·7	14·3
58	12 14·5	12 16·5	11 41·0	5·8	4·7	11·8	9·5	17·8	14·4
59	12 14·8	12 16·8	11 41·3	5·9	4·8	11·9	9·6	17·9	14·5
60	12 15·0	12 17·0	11 41·5	6·0	4·9	12·0	9·7	18·0	14·6

49m

49 s	SUN PLANETS ° '	ARIES ° '	MOON ° '	v or d '	Corrn '	v or d '	Corrn '	v or d '	Corrn '
00	12 15·0	12 17·0	11 41·5	0·0	0·0	6·0	5·0	12·0	9·9
01	12 15·3	12 17·3	11 41·8	0·1	0·1	6·1	5·0	12·1	10·0
02	12 15·5	12 17·5	11 42·0	0·2	0·2	6·2	5·1	12·2	10·1
03	12 15·8	12 17·8	11 42·2	0·3	0·2	6·3	5·2	12·3	10·1
04	12 16·0	12 18·0	11 42·5	0·4	0·3	6·4	5·3	12·4	10·2
05	12 16·3	12 18·3	11 42·7	0·5	0·4	6·5	5·4	12·5	10·3
06	12 16·5	12 18·5	11 42·9	0·6	0·5	6·6	5·4	12·6	10·4
07	12 16·8	12 18·8	11 43·2	0·7	0·6	6·7	5·5	12·7	10·5
08	12 17·0	12 19·0	11 43·4	0·8	0·7	6·8	5·6	12·8	10·6
09	12 17·3	12 19·3	11 43·7	0·9	0·7	6·9	5·7	12·9	10·6
10	12 17·5	12 19·5	11 43·9	1·0	0·8	7·0	5·8	13·0	10·7
11	12 17·8	12 19·8	11 44·1	1·1	0·9	7·1	5·9	13·1	10·8
12	12 18·0	12 20·0	11 44·4	1·2	1·0	7·2	5·9	13·2	10·9
13	12 18·3	12 20·3	11 44·6	1·3	1·1	7·3	6·0	13·3	11·0
14	12 18·5	12 20·5	11 44·9	1·4	1·2	7·4	6·1	13·4	11·1
15	12 18·8	12 20·8	11 45·1	1·5	1·2	7·5	6·2	13·5	11·1
16	12 19·0	12 21·0	11 45·3	1·6	1·3	7·6	6·3	13·6	11·2
17	12 19·3	12 21·3	11 45·6	1·7	1·4	7·7	6·4	13·7	11·3
18	12 19·5	12 21·5	11 45·8	1·8	1·5	7·8	6·4	13·8	11·4
19	12 19·8	12 21·8	11 46·1	1·9	1·6	7·9	6·5	13·9	11·5
20	12 20·0	12 22·0	11 46·3	2·0	1·7	8·0	6·6	14·0	11·6
21	12 20·3	12 22·3	11 46·5	2·1	1·7	8·1	6·7	14·1	11·6
22	12 20·5	12 22·5	11 46·8	2·2	1·8	8·2	6·8	14·2	11·7
23	12 20·8	12 22·8	11 47·0	2·3	1·9	8·3	6·8	14·3	11·8
24	12 21·0	12 23·0	11 47·2	2·4	2·0	8·4	6·9	14·4	11·9
25	12 21·3	12 23·3	11 47·5	2·5	2·1	8·5	7·0	14·5	12·0
26	12 21·5	12 23·5	11 47·7	2·6	2·1	8·6	7·1	14·6	12·0
27	12 21·8	12 23·8	11 48·0	2·7	2·2	8·7	7·2	14·7	12·1
28	12 22·0	12 24·0	11 48·2	2·8	2·3	8·8	7·3	14·8	12·2
29	12 22·3	12 24·3	11 48·4	2·9	2·4	8·9	7·3	14·9	12·3
30	12 22·5	12 24·5	11 48·7	3·0	2·5	9·0	7·4	15·0	12·4
31	12 22·8	12 24·8	11 48·9	3·1	2·6	9·1	7·5	15·1	12·5
32	12 23·0	12 25·0	11 49·2	3·2	2·6	9·2	7·6	15·2	12·5
33	12 23·3	12 25·3	11 49·4	3·3	2·7	9·3	7·7	15·3	12·6
34	12 23·5	12 25·5	11 49·6	3·4	2·8	9·4	7·8	15·4	12·7
35	12 23·8	12 25·8	11 49·9	3·5	2·9	9·5	7·8	15·5	12·8
36	12 24·0	12 26·0	11 50·1	3·6	3·0	9·6	7·9	15·6	12·9
37	12 24·3	12 26·3	11 50·3	3·7	3·1	9·7	8·0	15·7	13·0
38	12 24·5	12 26·5	11 50·6	3·8	3·1	9·8	8·1	15·8	13·0
39	12 24·8	12 26·8	11 50·8	3·9	3·2	9·9	8·2	15·9	13·1
40	12 25·0	12 27·0	11 51·1	4·0	3·3	10·0	8·3	16·0	13·2
41	12 25·3	12 27·3	11 51·3	4·1	3·4	10·1	8·3	16·1	13·3
42	12 25·5	12 27·5	11 51·5	4·2	3·5	10·2	8·4	16·2	13·4
43	12 25·8	12 27·8	11 51·8	4·3	3·5	10·3	8·5	16·3	13·4
44	12 26·0	12 28·0	11 52·0	4·4	3·6	10·4	8·6	16·4	13·5
45	12 26·3	12 28·3	11 52·3	4·5	3·7	10·5	8·7	16·5	13·6
46	12 26·5	12 28·5	11 52·5	4·6	3·8	10·6	8·7	16·6	13·7
47	12 26·8	12 28·8	11 52·7	4·7	3·9	10·7	8·8	16·7	13·8
48	12 27·0	12 29·0	11 53·0	4·8	4·0	10·8	8·9	16·8	13·9
49	12 27·3	12 29·3	11 53·2	4·9	4·0	10·9	9·0	16·9	13·9
50	12 27·5	12 29·5	11 53·4	5·0	4·1	11·0	9·1	17·0	14·0
51	12 27·8	12 29·8	11 53·7	5·1	4·2	11·1	9·2	17·1	14·1
52	12 28·0	12 30·0	11 53·9	5·2	4·3	11·2	9·2	17·2	14·2
53	12 28·3	12 30·3	11 54·2	5·3	4·4	11·3	9·3	17·3	14·3
54	12 28·5	12 30·5	11 54·4	5·4	4·5	11·4	9·4	17·4	14·4
55	12 28·8	12 30·8	11 54·6	5·5	4·5	11·5	9·5	17·5	14·4
56	12 29·0	12 31·1	11 54·9	5·6	4·6	11·6	9·6	17·6	14·5
57	12 29·3	12 31·3	11 55·1	5·7	4·7	11·7	9·7	17·7	14·6
58	12 29·5	12 31·6	11 55·4	5·8	4·8	11·8	9·7	17·8	14·7
59	12 29·8	12 31·8	11 55·6	5·9	4·9	11·9	9·8	17·9	14·8
60	12 30·0	12 32·1	11 55·8	6·0	5·0	12·0	9·9	18·0	14·9

©2009 Jack Case

TABLE 6 — ALTITUDE CORRECTION TABLES

a. Dip of the Horizon

Ht. of Eye (m)	Corr^n D	Ht. of Eye (ft.)	Ht. of Eye (m)	Corr^n D	Ht. of Eye (ft.)
0.00	−0.3	0.0	13.0	−6.4	42.8
0.03	−0.4	0.1	13.4	−6.5	44.2
0.06	−0.5	0.2	13.8	−6.6	45.5
0.09	−0.6	0.3	14.2	−6.7	47.0
0.13	−0.7	0.4	14.7	−6.8	48.4
0.18	−0.8	0.5	15.1	−6.9	49.8
0.23	−0.9	0.7	15.5	−7.0	51.3
0.29	−1.0	0.9	16.0	−7.1	52.8
0.35	−1.1	1.1	16.5	−7.2	54.3
0.42	−1.2	1.4	16.9	−7.3	55.8
0.45	−1.3	1.6	17.4	−7.4	57.4
0.5	−1.4	1.9	17.9	−7.5	58.9
0.6	−1.5	2.2	18.4	−7.6	60.5
0.7	−1.6	2.5	18.8	−7.7	62.1
0.8	−1.7	2.8	19.3	−7.8	63.8
0.9	−1.8	3.2	19.8	−7.9	65.4
1.1	−1.9	3.6	20.4	−8.0	67.1
1.2	−2.0	4.0	20.9	−8.1	68.8
1.3	−2.1	4.4	21.4	−8.2	70.5
1.4	−2.2	4.9	21.9	−8.3	72.3
1.6	−2.3	5.3	22.5	−8.4	74.1
1.7	−2.4	5.8	23.0	−8.5	75.8
1.9	−2.5	6.3	23.5	−8.6	77.6
2.0	−2.6	6.9	24.1	−8.7	79.5
2.2	−2.7	7.4	24.7	−8.8	81.3
2.4	−2.8	8.0	25.2	−8.9	83.2
2.6	−2.9	8.6	25.8	−9.0	85.1
2.8	−3.0	9.2	26.4	−9.1	87.0
3.0	−3.1	9.8	27.0	−9.2	88.9
3.2	−3.2	10.5	27.6	−9.3	90.9
3.4	−3.3	11.2	28.2	−9.4	92.9
3.6	−3.4	11.9	28.8	−9.5	94.9
3.8	−3.5	12.6	29.4	−9.6	96.9
4.0	−3.6	13.3	30.0	−9.7	98.9
4.3	−3.7	14.1	30.6	−9.8	101.0
4.5	−3.8	14.9	31.3	−9.9	103.1
4.7	−3.9	15.7	31.9	−10.0	105.2
5.0	−4.0	16.5	32.6	−10.1	107.3
5.2	−4.1	17.4	33.2	−10.2	109.4
5.5	−4.2	18.3	33.9	−10.3	111.6
5.8	−4.3	19.1	34.5	−10.4	113.8
6.1	−4.4	20.1	35.2	−10.5	116.0
6.3	−4.5	21.0	35.9	−10.6	118.2
6.6	−4.6	22.0	36.6	−10.7	120.5
6.9	−4.7	22.9	37.3	−10.8	122.8
7.2	−4.8	23.9	38.0	−10.9	125.1
7.5	−4.9	25.0	38.7	−11.0	127.4
7.9	−5.0	26.0	39.4	−11.1	129.7
8.2	−5.1	27.1	40.1	−11.2	132.1
8.5	−5.2	28.1	40.8	−11.3	134.5
8.8	−5.3	29.2	41.5	−11.4	136.9
9.2	−5.4	30.4	42.3	−11.5	139.3
9.5	−5.5	31.5	43.0	−11.6	141.7
9.9	−5.6	32.7	43.8	−11.7	144.2
10.3	−5.7	33.9	44.5	−11.8	146.7
10.6	−5.8	35.1	45.3	−11.9	149.2
11.0	−5.9	36.3	46.1	−12.0	151.7
11.4	−6.0	37.6	46.8	−12.1	154.3
11.8	−6.1	38.9	47.6	−12.2	156.8
12.2	−6.2	40.1	48.4	−12.3	159.4
12.6	−6.3	41.5	49.2	−12.4	162.1
13.0		42.8	50.0		164.7

b. Refraction for Stars & Planets

App. Alt. H_a	Corr^n R	App. Alt. H_a	Corr^n R	App. Alt. H_a	Corr^n R
0 00	−33.8	3 30	−12.9	9 55	
0 03	33.2	3 35	12.7	10 07	−5.3
0 06	32.6	3 40	12.5	10 20	−5.2
0 09	32.0	3 45	12.3	10 32	−5.1
0 12	31.5	3 50	12.1	10 46	−5.0
0 15	30.9	3 55	11.9	10 59	−4.9
0 18	−30.4	4 00	−11.7	11 14	−4.8
0 21	29.8	4 05	11.5	11 29	−4.7
0 24	29.3	4 10	11.4	11 44	−4.6
0 27	28.8	4 15	11.2	12 00	−4.5
0 30	28.3	4 20	11.0	12 17	−4.4
0 33	27.9	4 25	10.9	12 35	−4.3
0 36	−27.4	4 30	−10.7	12 53	−4.2
0 39	26.9	4 35	10.6	13 12	−4.1
0 42	26.5	4 40	10.4	13 32	−4.0
0 45	26.1	4 45	10.3	13 53	−3.9
0 48	25.7	4 50	10.1	14 16	−3.8
0 51	25.3	4 55	10.0	14 39	−3.7
0 54	−24.9	5 00	−9.8	15 03	−3.6
0 57	24.5	5 05	9.7	15 29	−3.5
1 00	24.1	5 10	9.6	15 56	−3.4
1 03	23.7	5 15	9.5	16 25	−3.3
1 06	23.4	5 20	9.3	16 55	−3.2
1 09	23.0	5 25	9.2	17 27	−3.1
1 12	−22.7	5 30	−9.1	18 01	−3.0
1 15	22.3	5 35	9.0	18 37	−2.9
1 18	22.0	5 40	8.9	19 16	−2.8
1 21	21.7	5 45	8.8	19 56	−2.7
1 24	21.4	5 50	8.7	20 40	−2.6
1 27	21.1	5 55	8.6	21 27	−2.5
1 30	−20.8	6 00	−8.5	22 17	−2.4
1 35	20.3	6 10	8.3	23 11	−2.3
1 40	19.9	6 20	8.1	24 09	−2.2
1 45	19.4	6 30	7.9	25 12	−2.1
1 50	19.0	6 40	7.7	26 20	−2.0
1 55	18.6	6 50	7.6	27 34	−1.9
2 00	−18.2	7 00	−7.4	28 54	−1.8
2 05	17.8	7 10	7.2	30 22	−1.7
2 10	17.4	7 20	7.1	31 58	−1.6
2 15	17.1	7 30	6.9	33 43	−1.5
2 20	16.7	7 40	6.8	35 38	−1.4
2 25	16.4	7 50	6.7	37 45	−1.3
2 30	−16.1	8 00	−6.6	40 06	−1.2
2 35	15.8	8 10	6.4	42 42	−1.1
2 40	15.4	8 20	6.3	45 34	−1.0
2 45	15.2	8 30	6.2	48 45	−0.9
2 50	14.9	8 40	6.1	52 16	−0.8
2 55	14.6	8 50	6.0	56 09	−0.7
3 00	−14.3	9 00	−5.9	60 26	−0.6
3 05	14.1	9 10	5.8	65 06	−0.5
3 10	13.8	9 20	5.7	70 09	−0.4
3 15	13.6	9 30	5.6	75 32	−0.3
3 20	13.4	9 40	5.5	81 12	−0.2
3 25	13.1	9 50	5.4	87 03	−0.1
3 30	−12.9	10 00	−5.3	90 00	0.0

H_a = App. Alt. = Apparent Altitude
= Sextant altitude corrected for index error (IE) & dip (D)
= H_s + IE + D

In critical cases ascend.

TABLE 6 — ALTITUDE CORRECTION TABLES

c. Additional Refraction for Non-standard Conditions

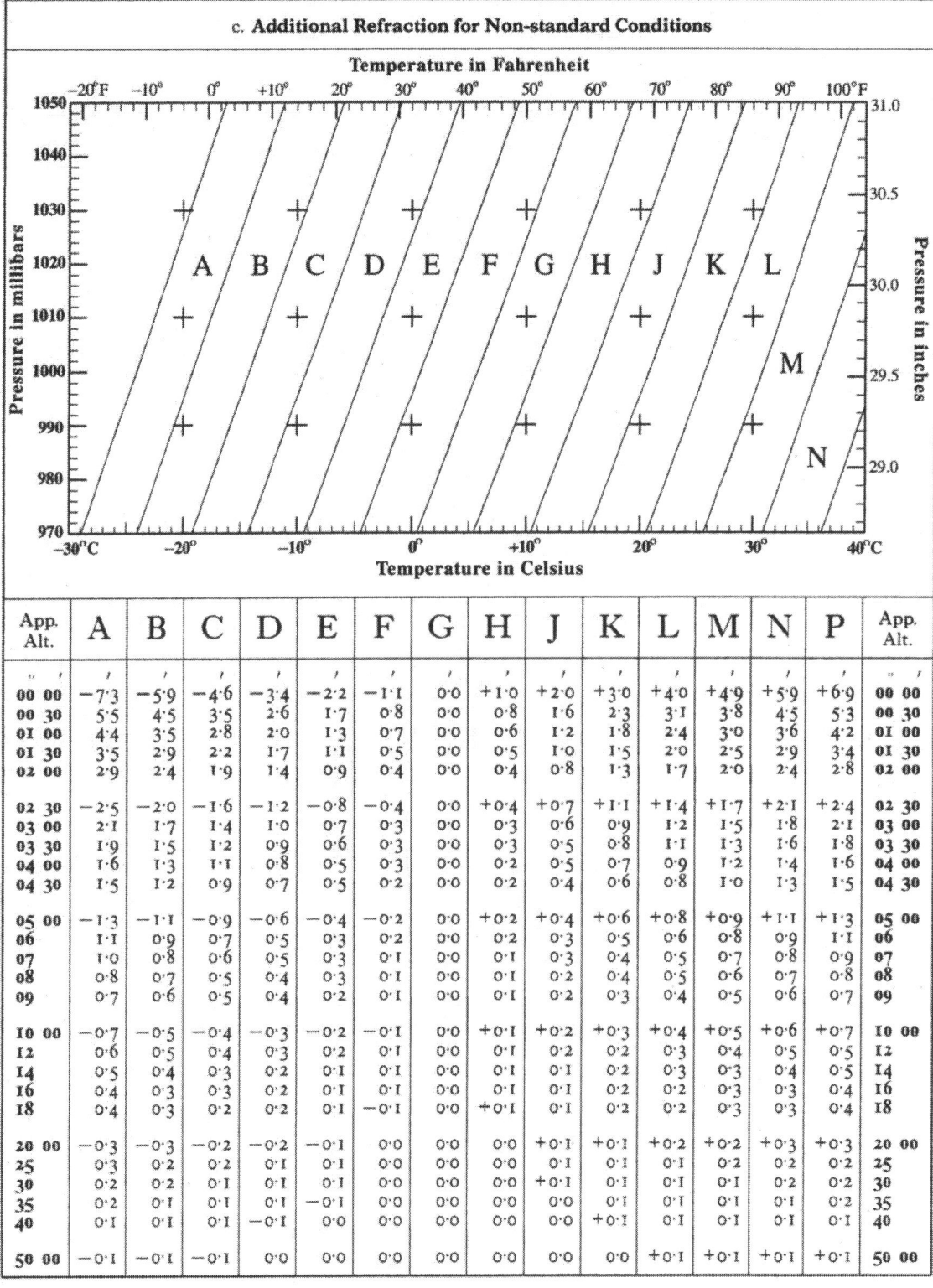

App. Alt.	A	B	C	D	E	F	G	H	J	K	L	M	N	P	App. Alt.
00 00	−7·3	−5·9	−4·6	−3·4	−2·2	−1·1	0·0	+1·0	+2·0	+3·0	+4·0	+4·9	+5·9	+6·9	00 00
00 30	5·5	4·5	3·5	2·6	1·7	0·8	0·0	0·8	1·6	2·3	3·1	3·8	4·5	5·3	00 30
01 00	4·4	3·5	2·8	2·0	1·3	0·7	0·0	0·6	1·2	1·8	2·4	3·0	3·6	4·2	01 00
01 30	3·5	2·9	2·2	1·7	1·1	0·5	0·0	0·5	1·0	1·5	2·0	2·5	2·9	3·4	01 30
02 00	2·9	2·4	1·9	1·4	0·9	0·4	0·0	0·4	0·8	1·3	1·7	2·0	2·4	2·8	02 00
02 30	−2·5	−2·0	−1·6	−1·2	−0·8	−0·4	0·0	+0·4	+0·7	+1·1	+1·4	+1·7	+2·1	+2·4	02 30
03 00	2·1	1·7	1·4	1·0	0·7	0·3	0·0	0·3	0·6	0·9	1·2	1·5	1·8	2·1	03 00
03 30	1·9	1·5	1·2	0·9	0·6	0·3	0·0	0·3	0·5	0·8	1·1	1·3	1·6	1·8	03 30
04 00	1·6	1·3	1·1	0·8	0·5	0·3	0·0	0·2	0·5	0·7	0·9	1·2	1·4	1·6	04 00
04 30	1·5	1·2	0·9	0·7	0·5	0·2	0·0	0·2	0·4	0·6	0·8	1·0	1·3	1·5	04 30
05 00	−1·3	−1·1	−0·9	−0·6	−0·4	−0·2	0·0	+0·2	+0·4	+0·6	+0·8	+0·9	+1·1	+1·3	05 00
06	1·1	0·9	0·7	0·5	0·3	0·2	0·0	0·2	0·3	0·5	0·6	0·8	0·9	1·1	06
07	1·0	0·8	0·6	0·5	0·3	0·1	0·0	0·1	0·3	0·4	0·5	0·7	0·8	0·9	07
08	0·8	0·7	0·5	0·4	0·3	0·1	0·0	0·1	0·2	0·4	0·5	0·6	0·7	0·8	08
09	0·7	0·6	0·5	0·4	0·2	0·1	0·0	0·1	0·2	0·3	0·4	0·5	0·6	0·7	09
10 00	−0·7	−0·5	−0·4	−0·3	−0·2	−0·1	0·0	+0·1	+0·2	+0·3	+0·4	+0·5	+0·6	+0·7	10 00
12	0·6	0·5	0·4	0·3	0·2	0·1	0·0	0·1	0·2	0·2	0·3	0·4	0·5	0·5	12
14	0·5	0·4	0·3	0·2	0·1	0·1	0·0	0·1	0·1	0·2	0·3	0·3	0·4	0·5	14
16	0·4	0·3	0·3	0·2	0·1	0·1	0·0	0·1	0·1	0·2	0·2	0·3	0·3	0·4	16
18	0·4	0·3	0·2	0·2	0·1	−0·1	0·0	+0·1	0·1	0·2	0·2	0·3	0·3	0·4	18
20 00	−0·3	−0·3	−0·2	−0·2	−0·1	0·0	0·0	0·0	+0·1	+0·1	+0·2	+0·2	+0·3	+0·3	20 00
25	0·3	0·2	0·2	0·1	0·1	0·0	0·0	0·0	0·1	0·1	0·1	0·2	0·2	0·2	25
30	0·2	0·2	0·1	0·1	0·1	0·0	0·0	0·0	+0·1	0·1	0·1	0·1	0·2	0·2	30
35	0·2	0·1	0·1	0·1	−0·1	0·0	0·0	0·0	0·0	0·1	0·1	0·1	0·1	0·2	35
40	0·1	0·1	0·1	−0·1	0·0	0·0	0·0	0·0	+0·1	0·1	0·1	0·1	0·1	0·1	40
50 00	−0·1	−0·1	−0·1	0·0	0·0	0·0	0·0	0·0	0·0	+0·1	+0·1	+0·1	+0·1	+0·1	50 00

The graph is entered with arguments temperature and pressure to find a zone letter; using as arguments this zone letter and apparent altitude (sextant altitude corrected for index error and dip), a correction is taken from the table. This correction is to be applied to the sextant altitude in addition to the corrections for standard conditions (Table **6**b).

TABLE 6d — ALTITUDE CORRECTION TABLES FOR THE SUN

App. Alt. H_a	OCT.–MAR. Lower Limb R	OCT.–MAR. Upper Limb R	APR.–SEPT. Lower Limb R	APR.–SEPT. Upper Limb R	App. Alt. H_a	OCT.–MAR. Lower Limb R	OCT.–MAR. Upper Limb R	APR.–SEPT. Lower Limb R	APR.–SEPT. Upper Limb R	App. Alt. H_a	OCT.–MAR. Lower Limb R	OCT.–MAR. Upper Limb R	App. Alt. H_a	APR.–SEPT. Lower Limb R	APR.–SEPT. Upper Limb R
° ′	′	′	′	′	° ′	′	′	′	′	° ′	′	′	° ′	′	′
0 00	−17·5	−49·8	−17·8	−49·6	3 30	+3·4	−28·9	+3·1	−28·7	9 33	+10·8	−21·5	9 39	+10·6	−21·2
0 03	16·9	49·2	17·2	49·0	3 35	3·6	28·7	3·3	28·5	9 45	+10·9	−21·4	9 50	+10·7	−21·1
0 06	16·3	48·6	16·6	48·4	3 40	3·8	28·5	3·6	28·2	9 56	+11·0	−21·3	10 02	+10·8	−21·0
0 09	15·7	48·0	16·0	47·8	3 45	4·0	28·3	3·8	28·0	10 08	+11·1	−21·2	10 14	+10·9	−20·9
0 12	15·2	47·5	15·4	47·2	3 50	4·2	28·1	4·0	27·8	10 20	+11·2	−21·1	10 27	+11·0	−20·8
0 15	14·6	46·9	14·8	46·6	3 55	4·4	27·9	4·1	27·7	10 33	+11·3	−21·0	10 40	+11·1	−20·7
0 18	−14·1	−46·4	−14·3	−46·1	4 00	+4·6	−27·7	+4·3	−27·5	10 46	+11·4	−20·9	10 53	+11·2	−20·6
0 21	13·5	45·8	13·8	45·6	4 05	4·8	27·5	4·5	27·3	11 00	+11·5	−20·8	11 07	+11·3	−20·5
0 24	13·0	45·3	13·3	45·1	4 10	4·9	27·4	4·7	27·1	11 15	+11·6	−20·7	11 22	+11·4	−20·4
0 27	12·5	44·8	12·8	44·6	4 15	5·1	27·2	4·9	26·9	11 30	+11·7	−20·6	11 37	+11·5	−20·3
0 30	12·0	44·3	12·3	44·1	4 20	5·3	27·0	5·0	26·8	11 45	+11·8	−20·5	11 53	+11·6	−20·2
0 33	11·6	43·9	11·8	43·6	4 25	5·4	26·9	5·2	26·6	12 01	+11·9	−20·4	12 10	+11·7	−20·1
0 36	−11·1	−43·4	−11·3	−43·1	4 30	+5·6	−26·7	+5·3	−26·5	12 18	+12·0	−20·3	12 27	+11·8	−20·0
0 39	10·6	42·9	10·9	42·7	4 35	5·7	26·6	5·5	26·3	12 36	+12·1	−20·2	12 45	+11·9	−19·9
0 42	10·2	42·5	10·5	42·3	4 40	5·9	26·4	5·6	26·2	12 54	+12·2	−20·1	13 04	+12·0	−19·8
0 45	9·8	42·1	10·0	41·8	4 45	6·0	26·3	5·8	26·0	13 14	+12·3	−20·0	13 24	+12·1	−19·7
0 48	9·4	41·7	9·6	41·4	4 50	6·2	26·1	5·9	25·9	13 34	+12·4	−19·9	13 44	+12·2	−19·6
0 51	9·0	41·3	9·2	41·0	4 55	6·3	26·0	6·1	25·7	13 55	+12·5	−19·8	14 06	+12·3	−19·5
0 54	−8·6	−40·9	−8·8	−40·6	5 00	+6·4	−25·9	+6·2	−25·6	14 17	+12·6	−19·7	14 29	+12·4	−19·4
0 57	8·2	40·5	8·4	40·2	5 05	6·6	25·7	6·3	25·5	14 41	+12·7	−19·6	14 53	+12·5	−19·3
1 00	7·8	40·1	8·0	39·8	5 10	6·7	25·6	6·5	25·3	15 05	+12·8	−19·5	15 18	+12·6	−19·2
1 03	7·4	39·7	7·7	39·5	5 15	6·8	25·5	6·6	25·2	15 31	+12·9	−19·4	15 45	+12·7	−19·1
1 06	7·1	39·4	7·3	39·1	5 20	7·0	25·3	6·7	25·1	15 59	+13·0	−19·3	16 13	+12·8	−19·0
1 09	6·7	39·0	7·0	38·8	5 25	7·1	25·2	6·8	25·0	16 27	+13·1	−19·2	16 43	+12·9	−18·9
1 12	−6·4	−38·7	−6·6	−38·4	5 30	+7·2	−25·1	+6·9	−24·9	16 58	+13·2	−19·1	17 14	+13·0	−18·8
1 15	6·0	38·3	6·3	38·1	5 35	7·3	25·0	7·1	24·7	17 30	+13·3	−19·0	17 47	+13·1	−18·7
1 18	5·7	38·0	6·0	37·8	5 40	7·4	24·9	7·2	24·6	18 05	+13·4	−18·9	18 23	+13·2	−18·6
1 21	5·4	37·7	5·7	37·5	5 45	7·5	24·8	7·3	24·5	18 41	+13·5	−18·8	19 00	+13·3	−18·5
1 24	5·1	37·4	5·3	37·1	5 50	7·6	24·7	7·4	24·4	19 20	+13·6	−18·7	19 41	+13·4	−18·4
1 27	4·8	37·1	5·0	36·8	5 55	7·7	24·6	7·5	24·3	20 02	+13·7	−18·6	20 24	+13·5	−18·3
1 30	−4·5	−36·8	−4·7	−36·5	6 00	+7·8	−24·5	+7·6	−24·2	20 46	+13·8	−18·5	21 10	+13·6	−18·2
1 35	4·0	36·3	4·3	36·1	6 10	8·0	24·3	7·8	24·0	21 34	+13·9	−18·4	21 59	+13·7	−18·1
1 40	3·6	35·9	3·8	35·6	6 20	8·2	24·1	8·0	23·8	22 25	+14·0	−18·3	22 52	+13·8	−18·0
1 45	3·1	35·4	3·4	35·2	6 30	8·4	23·9	8·2	23·6	23 20	+14·1	−18·2	23 49	+13·9	−17·9
1 50	2·7	35·0	2·9	34·7	6 40	8·6	23·7	8·3	23·5	24 20			24 51	+14·0	−17·8
1 55	2·3	34·6	2·5	34·3	6 50	8·7	23·6	8·5	23·3	25 24	+14·2	−18·1	25 58		
2 00	−1·9	−34·2	−2·1	−33·9	7 00	+8·9	−23·4	+8·7	−23·1	26 34	+14·3	−18·0	27 11	+14·1	−17·7
2 05	1·5	33·8	1·7	33·5	7 10	9·1	23·2	8·8	23·0	27 50	+14·4	−17·9	28 31	+14·2	−17·6
2 10	1·1	33·4	1·4	33·2	7 20	9·2	23·1	9·0	22·8	29 13	+14·5	−17·8	29 58	+14·3	−17·5
2 15	0·8	33·1	1·0	32·8	7 30	9·3	23·0	9·1	22·7	30 44	+14·6	−17·7	31 33	+14·4	−17·4
2 20	0·4	32·7	0·7	32·5	7 40	9·5	22·8	9·2	22·6	32 24	+14·7	−17·6	33 18	+14·5	−17·3
2 25	−0·1	32·4	−0·3	32·1	7 50	9·6	22·7	9·4	22·4	34 15	+14·8	−17·5	35 15	+14·6	−17·2
2 30	+0·2	−32·1	0·0	−31·8	8 00	+9·7	−22·6	+9·5	−22·3	36 17	+14·9	−17·4	37 24	+14·7	−17·1
2 35	0·5	31·8	+0·3	31·5	8 10	9·9	22·4	9·6	22·2	38 34	+15·0	−17·3	39 48	+14·8	−17·0
2 40	0·8	31·5	0·6	31·2	8 20	10·0	22·3	9·7	22·1	41 06	+15·1	−17·2	42 28	+14·9	−16·9
2 45	1·1	31·2	0·9	30·9	8 30	10·1	22·2	9·9	21·9	43 56	+15·2	−17·1	45 29	+15·0	−16·8
2 50	1·4	30·9	1·2	30·6	8 40	10·2	22·1	10·0	21·8	47 07	+15·3	−17·0	48 52	+15·1	−16·7
2 55	1·7	30·6	1·4	30·4	8 50	10·3	22·0	10·1	21·7	50 43	+15·4	−16·9	52 41	+15·2	−16·6
3 00	+2·0	−30·3	+1·7	−30·1	9 00	+10·4	−21·9	+10·2	−21·6	54 46	+15·5	−16·8	56 59	+15·3	−16·5
3 05	2·2	30·1	2·0	29·8	9 10	10·5	21·8	10·3	21·5	59 21	+15·6	−16·7	61 50	+15·4	−16·4
3 10	2·5	29·8	2·2	29·6	9 20	10·6	21·7	10·4	21·4	64 28	+15·7	−16·6	67 15	+15·5	−16·3
3 15	2·7	29·6	2·5	29·3	9 30	10·7	21·6	10·5	21·3	70 10	+15·8	−16·5	73 14	+15·6	−16·2
3 20	2·9	29·4	2·7	29·1	9 40	10·8	21·5	10·6	21·2	76 24	+15·9	−16·4	79 42	+15·7	−16·1
3 25	3·2	29·1	2·9	28·9	9 50	10·9	21·4	10·6	21·2	83 05	+16·0	−16·3	86 31	+15·8	−16·0
3 30	+3·4	−28·9	+3·1	−28·7	10 00	+11·0	−21·3	+10·7	−21·1	90 00	+16·1	−16·2	90 00	+15·9	−15·9

App. Alt. = Apparent altitude = Sextant altitude corrected for index error and dip = $H_a = H_s + IE + D$

In critical cases ascend. Additional corrections for temperature and pressure are given on the previous page.

Extract from main table for Lat.52, Same Name

LHA	15° Hc	d	Z	16° Hc	d	Z	17° Hc	d	Z	18° Hc	d	Z	19° Hc	d	Z	20° Hc	d	Z	21° Hc	d	Z
0	53 00	+60	180	54 00	+60	180	55 00	+60	180	56 00	+60	180	57 00	+60	180	58 00	+60	180	59 00	+60	180
1	52 59	60	178	53 59	60	178	54 59	60	178	55 59	60	178	56 59	60	178	57 59	60	178	58 59	60	178
2	52 58	60	177	53 58	60	177	54 58	60	177	55 58	60	177	56 58	60	177	57 58	60	176	58 58	60	176
3	52 55	60	175	53 55	60	175	54 55	60	175	55 55	60	175	56 55	60	175	57 55	60	175	58 55	60	175
4	52 52	60	174	53 52	59	173	54 51	60	173	55 51	60	173	56 51	60	173	57 51	60	173	58 51	59	173
5	52 47	+60	172	53 47	+60	172	54 47	+59	172	55 46	+60	172	56 46	+60	171	57 46	+59	171	58 45	+60	171
6	52 41	60	170	53 41	60	170	54 41	59	170	55 40	60	170	56 40	60	170	57 40	59	169	58 39	60	169
7	52 35	59	169	53 34	60	169	54 34	60	168	55 33	60	168	56 33	59	168	57 32	60	168	58 32	59	167
8	52 27	60	167	53 27	59	167	54 26	59	167	55 25	60	167	56 25	59	166	57 24	59	166	58 23	59	166
9	52 19	59	166	53 18	59	165	54 17	59	165	55 16	59	165	56 15	59	165	57 14	59	164	58 13	59	164
10	52 09	+59	164	53 08	+59	164	54 07	+59	164	55 06	+59	163	56 05	+59	163	57 04	+59	163	58 03	+58	162
11	51 58	59	163	52 57	59	162	53 56	59	162	54 55	58	162	55 53	59	161	56 52	59	161	57 51	58	160
12	51 47	58	161	52 45	59	161	53 44	59	160	54 43	58	160	55 41	58	160	56 39	59	159	57 38	58	159
13	51 34	59	160	52 33	58	159	53 31	58	159	54 29	59	158	55 28	58	158	56 26	58	158	57 24	58	157
14	51 21	58	158	52 19	58	158	53 17	58	157	54 15	58	157	55 13	58	156	56 11	58	156	57 09	58	155
15	51 07	+58	157	52 05	+58	156	53 03	+57	156	54 00	+58	155	54 58	+58	155	55 56	+57	154	56 53	+57	154
16	50 52	57	155	51 49	58	155	52 47	57	154	53 44	58	154	54 42	57	153	55 39	57	153	56 36	57	152
17	50 36	57	154	51 33	57	153	52 30	58	153	53 28	57	152	54 25	57	152	55 22	57	151	56 19	56	151
18	50 19	57	152	51 16	57	152	52 13	57	151	53 10	57	151	54 07	56	150	55 03	57	150	56 00	56	149
19	50 01	57	151	50 58	57	150	51 55	56	150	52 51	57	149	53 48	56	149	54 44	57	148	55 41	55	147
20	49 43	+56	149	50 39	+57	149	51 36	+56	148	52 32	+56	148	53 28	+56	147	54 24	+56	146	55 20	+56	146
21	49 23	57	148	50 20	56	147	51 16	56	147	52 12	56	146	53 08	56	146	54 04	55	145	54 59	55	144
22	49 03	56	146	49 59	56	146	50 55	56	145	51 51	56	145	52 47	55	144	53 42	55	144	54 37	55	143
23	48 43	55	145	49 38	56	145	50 34	55	144	51 29	56	143	52 25	55	143	53 20	54	142	54 14	55	141
24	48 21	56	144	49 17	55	143	50 12	55	143	51 07	55	142	52 02	55	141	52 57	54	141	53 51	54	140
25	47 59	+55	142	48 54	+55	142	49 49	+55	141	50 44	+54	141	51 38	+55	140	52 33	+54	139	53 27	+54	139
26	47 36	55	141	48 31	55	140	49 26	54	140	50 20	54	139	51 14	54	139	52 08	54	138	53 02	54	137
27	47 12	55	140	48 07	54	139	49 01	55	139	49 56	54	138	50 50	53	137	51 43	54	136	52 37	53	136
28	46 48	55	139	47 43	54	138	48 37	53	137	49 30	54	137	50 24	53	136	51 17	54	135	52 11	52	134
29	46 24	53	137	47 17	54	137	48 11	54	136	49 05	53	135	49 58	53	135	50 51	53	134	51 44	52	133
30	45 58	+54	136	46 52	+53	135	47 45	+53	135	48 38	+53	134	49 31	+53	133	50 24	+53	133	51 17	+52	132
31	45 32	54	135	46 26	53	134	47 19	53	133	48 12	52	133	49 04	53	132	49 57	52	131	50 49	51	130
32	45 06	53	134	45 59	53	133	46 52	52	132	47 44	53	131	48 37	52	131	49 29	51	130	50 20	52	129
33	44 39	52	132	45 31	53	132	46 24	52	131	47 16	52	130	48 08	52	129	49 00	51	129	49 51	52	128
34	44 11	53	131	45 04	52	130	45 56	52	130	46 48	52	129	47 40	51	128	48 31	51	128	49 22	51	127
35	43 43	+52	130	44 35	+52	129	45 27	+52	129	46 19	+51	128	47 10	+51	127	48 01	+51	126	48 52	+51	126
36	43 14	52	129	44 06	52	128	44 58	51	127	45 49	52	127	46 41	50	126	47 31	51	125	48 22	50	124
37	42 45	52	128	43 37	51	127	44 28	52	126	45 20	50	126	46 10	51	125	47 01	50	124	47 51	50	123
38	42 16	51	127	43 07	51	126	43 58	51	125	44 49	51	124	45 40	50	124	46 30	50	123	47 20	50	122
39	41 46	51	125	42 37	51	125	43 28	51	124	44 19	50	123	45 09	50	122	45 59	50	122	46 49	49	121
40	41 16	+51	124	42 07	+50	124	42 57	+51	123	43 48	+50	122	44 38	+49	121	45 27	+50	121	46 17	+49	120
41	40 45	51	123	41 36	50	123	42 26	50	122	43 16	50	121	44 06	49	120	44 55	49	119	45 44	49	119
42	40 14	50	122	41 04	50	121	41 54	50	121	42 44	50	120	43 34	49	119	44 23	49	118	45 12	48	118
43	39 42	51	121	40 33	49	120	41 22	50	120	42 12	49	119	43 01	49	118	43 50	49	117	44 39	48	116
44	39 11	50	120	40 01	49	119	40 50	50	119	41 40	49	118	42 29	48	117	43 17	49	116	44 06	48	115
45	38 39	+49	119	39 28	+50	118	40 18	+49	118	41 07	+49	117	41 56	+48	116	42 44	+48	115	43 32	+48	114
46	38 06	50	118	38 56	49	117	39 45	49	117	40 34	48	116	41 22	49	115	42 11	47	114	42 58	48	113
47	37 33	50	117	38 23	49	116	39 12	48	116	40 00	49	115	40 49	48	114	41 37	47	113	42 24	48	112
48	37 00	49	116	37 49	49	115	38 38	49	114	39 27	48	114	40 15	48	113	41 03	47	112	41 50	47	111
49	36 27	49	115	37 16	48	114	38 04	49	114	38 53	48	113	39 41	47	112	40 28	48	111	41 16	47	110
50	35 53	+49	114	36 42	+48	113	37 30	+48	113	38 18	+48	112	39 06	+48	111	39 54	+47	110	40 41	+47	109
51	35 19	49	113	36 08	48	112	36 56	48	112	37 44	48	111	38 32	47	110	39 19	47	109	40 06	47	108
52	34 45	49	112	35 34	48	111	36 22	47	111	37 09	48	110	37 57	47	109	38 44	47	108	39 31	46	108
53	34 11	48	111	34 59	48	110	35 47	48	110	36 35	47	109	37 22	47	108	38 09	47	107	38 56	46	107
54	33 36	48	110	34 24	48	110	35 12	48	109	36 00	47	108	36 47	46	107	37 33	47	106	38 20	46	106
55	33 02	+47	109	33 49	+48	109	34 37	+47	108	35 24	+47	107	36 11	+47	106	36 58	+46	106	37 44	+46	105
56	32 27	47	108	33 14	48	108	34 02	47	107	34 49	47	106	35 36	46	105	36 22	47	105	37 09	45	104
57	31 52	47	107	32 39	47	107	33 26	47	106	34 13	47	105	35 00	47	105	35 47	46	104	36 33	45	103
58	31 16	48	107	32 04	47	106	32 51	47	105	33 38	46	104	34 24	47	104	35 11	46	103	35 57	45	102
59	30 41	47	106	31 28	47	105	32 15	47	104	33 02	46	104	33 48	46	103	34 34	46	102	35 20	46	101
60	30 05	+47	105	30 52	+47	104	31 39	+47	103	32 26	+46	103	33 12	+46	102	33 58	+46	101	34 44	+46	100
61	29 29	47	104	30 16	47	103	31 03	47	102	31 50	46	102	32 36	46	101	33 22	46	100	34 08	45	99
62	28 53	47	103	29 40	47	102	30 27	46	102	31 13	47	101	32 00	46	100	32 46	45	99	33 31	45	99
63	28 17	47	102	29 04	47	101	29 51	46	101	30 37	46	100	31 23	46	99	32 09	46	99	32 55	45	98
64	27 41	47	101	28 28	46	101	29 14	47	100	30 01	46	99	30 47	46	98	31 33	45	98	32 18	45	97
65	27 05	+47	101	27 52	+46	100	28 38	+46	99	29 24	+46	98	30 10	+46	98	30 56	+45	97	31 41	+45	96
66	26 29	46	100	27 15	47	99	28 02	46	98	28 48	46	98	29 34	45	97	30 19	46	96	31 05	45	95
67	25 52	47	99	26 39	46	98	27 25	46	97	28 11	46	97	28 57	45	96	29 42	46	95	30 28	45	94

Extract from Table 5

34	35	36	37	38	39	40	41	42	43	44	45	46	47	48	49	50	51	52	53	54	55	56	57	58	59	60	d
′	′	′	′	′	′	′	′	′	′	′	′	′	′	′	′	′	′	′	′	′	′	′	′	′	′	′	′
0	0	0	0	0	0	0	0	0	0	0	0	0	0	0	0	0	0	0	0	0	0	0	0	0	0	0	0
1	1	1	1	1	1	1	1	1	1	1	1	1	1	1	1	1	1	1	1	1	1	1	1	1	1	1	1
1	1	1	1	1	1	1	1	1	1	1	2	2	2	2	2	2	2	2	2	2	2	2	2	2	2	2	2
2	2	2	2	2	2	2	2	2	2	2	2	2	2	2	2	2	3	3	3	3	3	3	3	3	3	3	3
2	2	2	2	3	3	3	3	3	3	3	3	3	3	3	3	3	3	3	4	4	4	4	4	4	4	4	4
3	3	3	3	3	3	3	3	4	4	4	4	4	4	4	4	4	4	4	4	4	5	5	5	5	5	5	5
3	4	4	4	4	4	4	4	4	4	4	4	5	5	5	5	5	5	5	5	5	6	6	6	6	6	6	6
4	4	4	4	4	5	5	5	5	5	5	5	5	5	6	6	6	6	6	6	6	6	7	7	7	7	7	7
5	5	5	5	5	5	5	5	6	6	6	6	6	6	6	7	7	7	7	7	7	7	7	8	8	8	8	8
5	5	5	6	6	6	6	6	6	6	7	7	7	7	7	7	8	8	8	8	8	8	8	9	9	9	9	9
6	6	6	6	6	6	7	7	7	7	7	8	8	8	8	8	8	8	9	9	9	9	10	10	10	10	10	10
6	6	7	7	7	7	7	8	8	8	8	8	9	9	9	9	9	9	10	10	10	10	10	10	11	11	11	11
7	7	7	7	8	8	8	8	8	9	9	9	9	9	10	10	10	10	10	11	11	11	11	11	12	12	12	12
7	8	8	8	8	8	9	9	9	9	10	10	10	10	10	11	11	11	11	11	12	12	12	12	13	13	13	13
8	8	8	9	9	9	9	10	10	10	10	10	11	11	11	11	12	12	12	12	13	13	13	13	14	14	14	14
8	9	9	9	10	10	10	10	10	11	11	11	12	12	12	12	12	13	13	13	14	14	14	14	14	15	15	15
9	9	10	10	10	10	11	11	11	11	12	12	12	13	13	13	13	14	14	14	14	15	15	15	15	16	16	16
10	10	10	10	11	11	11	12	12	12	12	13	13	13	14	14	14	14	15	15	15	16	16	16	16	17	17	17
10	10	11	11	11	12	12	12	13	13	13	14	14	14	14	15	15	15	16	16	16	16	17	17	17	18	18	18
11	11	11	12	12	12	13	13	13	14	14	14	15	15	15	16	16	16	16	17	17	17	18	18	18	19	19	19
11	12	12	12	13	13	13	14	14	14	15	15	15	16	16	17	17	17	18	18	18	19	19	19	20	20	20	20
12	12	13	13	13	14	14	14	15	15	15	16	16	16	17	17	18	18	18	19	19	19	20	20	20	21	21	21
12	13	13	14	14	14	15	15	15	16	16	16	17	17	18	18	18	19	19	19	20	20	21	21	21	22	22	22
13	13	14	14	15	15	15	16	16	16	17	17	18	18	18	19	19	20	20	20	21	21	21	22	22	23	23	23
14	14	14	15	15	16	16	16	17	17	18	18	18	19	19	20	20	20	21	21	22	22	22	23	23	24	24	24
14	15	15	15	16	16	17	17	18	18	18	19	19	20	20	20	21	21	22	22	22	23	23	24	24	25	25	25
15	15	16	16	16	17	17	18	18	19	19	20	20	20	21	21	22	22	23	23	23	24	24	25	25	26	26	26
15	16	16	17	17	18	18	18	19	19	20	20	21	21	22	22	22	23	23	24	24	25	25	26	26	27	27	27
16	16	17	17	18	18	19	19	20	20	21	21	21	22	22	23	23	24	24	25	25	26	26	27	27	28	28	28
16	17	17	18	18	19	19	20	20	21	21	22	22	23	23	24	24	25	25	26	26	27	27	28	28	29	29	29
17	18	18	18	19	20	20	20	21	22	22	22	23	24	24	24	25	26	26	26	27	28	28	28	29	30	30	30
18	18	19	19	20	20	21	21	22	22	23	23	24	24	25	25	26	26	27	27	28	28	29	29	30	30	31	31
18	19	19	20	20	21	21	22	22	23	23	24	24	25	26	26	27	27	28	28	29	29	30	30	31	31	32	32
19	19	20	20	21	21	22	23	23	24	24	25	25	26	26	27	28	28	29	29	30	30	31	31	32	32	33	33
19	20	20	21	22	22	23	23	24	24	25	26	26	27	27	28	28	29	29	30	31	31	32	32	33	33	34	34
20	20	21	22	22	23	23	24	24	25	26	26	27	27	28	29	29	30	30	31	32	32	33	33	34	34	35	35
20	21	22	22	23	23	24	25	25	26	26	27	28	28	29	29	30	31	31	32	32	33	34	34	35	35	36	36
21	22	22	23	23	24	25	25	26	27	27	28	28	29	30	30	31	31	32	33	33	34	35	35	36	36	37	37
22	22	23	23	24	25	25	26	27	27	28	28	29	30	30	31	32	32	33	34	34	35	35	36	37	37	38	38
22	23	23	24	25	25	26	27	27	28	29	29	30	31	31	32	32	33	34	34	35	36	36	37	38	38	39	39
23	23	24	25	25	26	27	27	28	29	29	30	31	31	32	33	33	34	35	35	36	37	37	38	39	39	40	40
23	24	25	25	26	27	27	28	29	29	30	31	31	32	33	33	34	35	36	36	37	38	38	39	40	40	41	41
24	24	25	26	27	27	28	29	29	30	31	32	32	33	34	34	35	36	36	37	38	38	39	40	41	41	42	42
24	25	26	27	27	28	29	29	30	31	32	32	33	34	34	35	36	37	37	38	39	39	40	41	42	42	43	43
25	26	26	27	28	29	29	30	31	32	32	33	34	34	35	36	37	37	38	39	40	40	41	42	43	43	44	44
26	26	27	28	28	29	30	31	32	32	33	34	34	35	36	37	38	38	39	40	40	41	42	43	44	44	45	45
26	27	28	28	29	30	31	31	32	33	34	34	35	36	37	38	38	39	40	41	41	42	43	44	44	45	46	46
27	27	28	29	30	31	31	32	33	34	34	35	36	37	38	38	39	40	41	42	42	43	44	45	45	46	47	47
27	28	29	30	30	31	32	33	34	34	35	36	37	38	38	39	40	41	42	42	43	44	45	46	46	47	48	48
28	29	29	30	31	32	33	33	34	35	36	37	38	38	39	40	41	42	42	43	44	45	46	47	47	48	49	49
28	29	30	31	32	32	33	34	35	36	37	38	38	39	40	41	42	42	43	44	45	46	47	48	48	49	50	50
29	30	31	31	32	33	34	35	36	37	37	38	39	40	41	42	42	43	44	45	46	47	48	48	49	50	51	51
29	30	31	32	33	34	35	36	36	37	38	39	40	41	42	42	43	44	45	46	47	48	49	49	50	51	52	52
30	31	32	33	34	34	35	36	37	38	39	40	41	42	42	43	44	45	46	47	48	49	49	50	51	52	53	53
31	32	32	33	34	35	36	37	38	39	40	40	41	42	43	44	45	46	47	48	49	50	50	51	52	53	54	54
31	32	33	34	35	36	37	38	38	39	40	41	42	43	44	45	46	47	48	49	50	50	51	52	53	54	55	55
32	33	34	35	35	36	37	38	39	40	41	42	43	44	45	46	47	48	49	49	50	51	52	53	54	55	56	56
32	33	34	35	36	37	38	39	40	41	42	43	44	45	46	47	48	48	49	50	51	52	53	54	55	56	57	57
33	34	35	36	37	38	39	40	41	42	43	44	44	45	46	47	48	49	50	51	52	53	54	55	56	57	58	58
33	34	35	36	37	38	39	40	41	42	43	44	45	46	47	48	49	50	51	52	53	54	55	56	57	58	59	59

Chapter 10
Demonstration of a Three Body Fix

In this demonstration, observations of 3 celestial bodies are taken in rapid succession. The Rapid Sight Reduction Tables are used to calculate the intercepts in order to provide 3 intersecting position lines to establish a fix.

Notes.
1. To fully understand this demonstration, it will be necessary to refer to the outline procedures in chapter 8 and it may be advisable to revise chapters 2, 4 and 5 as well.
2. Extracts of Relevant tables are placed immediately following the solution to the example. Because of their size, some tables have been cropped so that they show only the relevant portions.
3. Self-test questions on the use of the rapid sight reduction tables can be found in chapter 13.

Aim. The aim of this demonstration is to establish a 3 point fix based on the following imaginary scenario:
On 30 May at 2030, a fishing vessel stopped to haul nets. Whilst the vessel was stationary, observations of 3 celestial bodies were made in order to fix the position of the ship.
Using the following data, calculate the position of the vessel.
D.R. Position: 47° 52'N, 47° 35'W.
Zone: +3
Sunset: $19^h 49^m$
Civil twilight: $20^h 29^m$
Naut. twilight: $21^h 23^m$
Moonset: 31 May, $00^h 24^m$
Time of fix: 30 May, $21^h 00^m (+3)$
DWE 15^s fast
Index error: + 0'.53
Ht. of eye: 5.8m.

Movement. In this scenario, any movement of the ship whilst the observations are being taken is negligible and so corrections for Moo are not necessary.
Furthermore, since special techniques are not used, corrections for MOB are not required.
Observations.

Bodies selected: Using the method described in chapter 8 for planning star and planet observations, the following celestial bodies were chosen for the observations: The Moon, Saturn, and Alphecca.

Weather conditions at the time of the observations: Temperature: 16°C. Pressure: 1002mb.

Movement of the vessel due to wind and tidal stream while the observations were being conducted was negligible.

Observation 1:
DWT: $23^h 48^m 23^s$
Body observed: Moon L.L.
Sextant Alt: 34° 29.'2. Bearing: 231°

Observation 2:
DWT: $23^h 49^m 52^s$
Body observed: Saturn.
Sextant Alt: 41° 53'.50. Bearing: 224°

Observation 3.
DWT: $23^h 51^m 05^s$
Body observed: Alphecca.
Sextant Alt: 55° 02'.43. Bearing: 115°

Solution.
Planning Phase

Step 1. Approx Pos.: 47° 52'N, 47° 35'W

Step 2. Calculate Greenwich Date:
Date: 30 May 2009
Time of fix: $30^d 21^h 00^m$
Zone corrn: $+3^h$
Greenwich Date: $31^d 00^h 00^m$

Calculate UT at time of observations.

	Moon	Saturn	Alphecca
DWT:	$23^h 48^m 23^s$	$23^h 49^m 52^s$	$23^h 51^m 05^s$
DWE:	-15^s	-15^s	-15^s
UT:	$23^h 48^m 08^s$	$23^h 49^m 37^s$	$23^h 50^m 50^s$

Step 3. Calculate GHA and Dec.

	Moon	Saturn	Alphecca
Dec (23^h)	N6° 01'.8	N7° 50'.7	N26° 40'.9
d	(-15.0)	(+0.0)	(n.a.)
d corr.	-12'.1	0'.0	0'.0
Dec + corr.	N5° 49'.7	N7° 50'.7	N26° 40'.9
SHA (stars only)			126° 13'.1
GHA ϒ (23^h) (for stars)			233° 35'.2
GHA ϒ + SHA (for stars)			359° 48'.3
GHA (23^h)	75° 40'.6	66° 29'.6	
Inc (m+s)	+11° 29'.1	+12° 24'.3	+12° 44'.6
v	(+12.4)	(+2.4)	n.a.
v corr.	+10'.0	+2.0	0'.0
Sum	87° 19'.7	78° 55'.9	372° 32'.9

Step 4. Calculate LHA

	Moon	Saturn	Alphecca
GHA	87° 19'.7	78° 55'.9	372° 32'.9
As.long(W)	-47° 19'.7	-47° 55'.9	-47° 32'.9
LHA	40°	31°	325°

Convert azimuth angle to true bearing (Zn):

	Moon	Saturn	Alphecca
LHA	40°	31°	325°
Z	128°	137°	116°
Zn	232°	223°	116°

Please see page 161 for an explanation of the reason for calculating true bearings here. (Take the values of the Azimuth (Z) from step 5).

Rules for calculating true bearing (Zn):		
	Lat. North	Lat. South
LHA > 180°	Zn = Z	Zn = 180° - Z
LHA < 180°	Zn = 360° - Z	Zn = 180° + Z

Step 5. Main table entry.

	Moon	Saturn	Alphecca
As. lat.	N48°	N48°	N48°
Dec°	N5°	N7°	N26°
Same or cont.	same	same	same
Dec'	49'.7	50'.7	40'.9
LHA	40°	31°	325°
Table page:	56	56	60
Tab Hc,	35° 08'	41° 17'	54° 56'
d	+51'	+54'	+44'
Z	128°	137°	116°
Correction for mins of Dec.			
Tab Hc,	35° 08'	41° 17'	54° 56'
Cor. for Dec'	+42	+46'	+30' (table 5)
Cal. Alt. Hc= (Hc+Cor.Dec')	35° 50'	42° 03'	55° 26'

Fix Phase

Step 6. Correct sextant altitude (Hs).

	Moon	Saturn	Alphecca
Sext Alt (Hs)	34° 29'.2	41° 53'.50	55° 02'.43
I.E.	+0'.53	+0'.53	+0'.53
Dip (5.8m) (D)	-4'.2	-4'.2	-4'.2
Ap Alt (Ha)	34° 25'.53	41° 49'.83	54° 58'.76
Alt. Cor. (R)		-1'.1	-0'.7
Moon Alt Cor = (Moon HP:58'.5)	+56'.8		
Moon LL Cor =	+5'.9		
Add'l refrac.	0'.0	0'.0	0'.0
Obs Alt (Ho)	35° 28'.23	41° 48'.73	54° 58'.06

Step 7. Calculate Intercept

	Moon	Saturn	Alphecca
Ho	35° 28'.23	41° 48'.73	54° 58'.06
Hc	35° 50'	42° 03'	55° 26'
p(Ho-Hc)	-21'.77	-14'.27	-27'.94
	(away)	(away)	(away)

Therefore, intercepts are: Moon. 21'.77 from 232°
                             Saturn. 14'.27 from 223°
                             Alphecca. 27'.94 from 116°
(Note. True bearings calculated at end of step 4).

Step 8. Corrections to intercept.
(No corrections required since there has been no movement between observation and fix).

Step 9. Plot the position lines.
Firstly, for each observation, plot the assumed position using the assumed long. calculated at step 4 and the assumed lat. calculated at step 5. Secondly, plot the three position lines using the intercept data calculated at step 7.

Note. Because of the small errors inherent in astro navigation techniques, the three position lines will very rarely cross at one precise point. Usually, a small triangle known as a **'cocked-hat'** will be produced and as long as the triangle is not too large, it can safely be assumed that the ship's position is at the centre of this triangle.

Extracts relevant to the above example are to be found on the following pages.

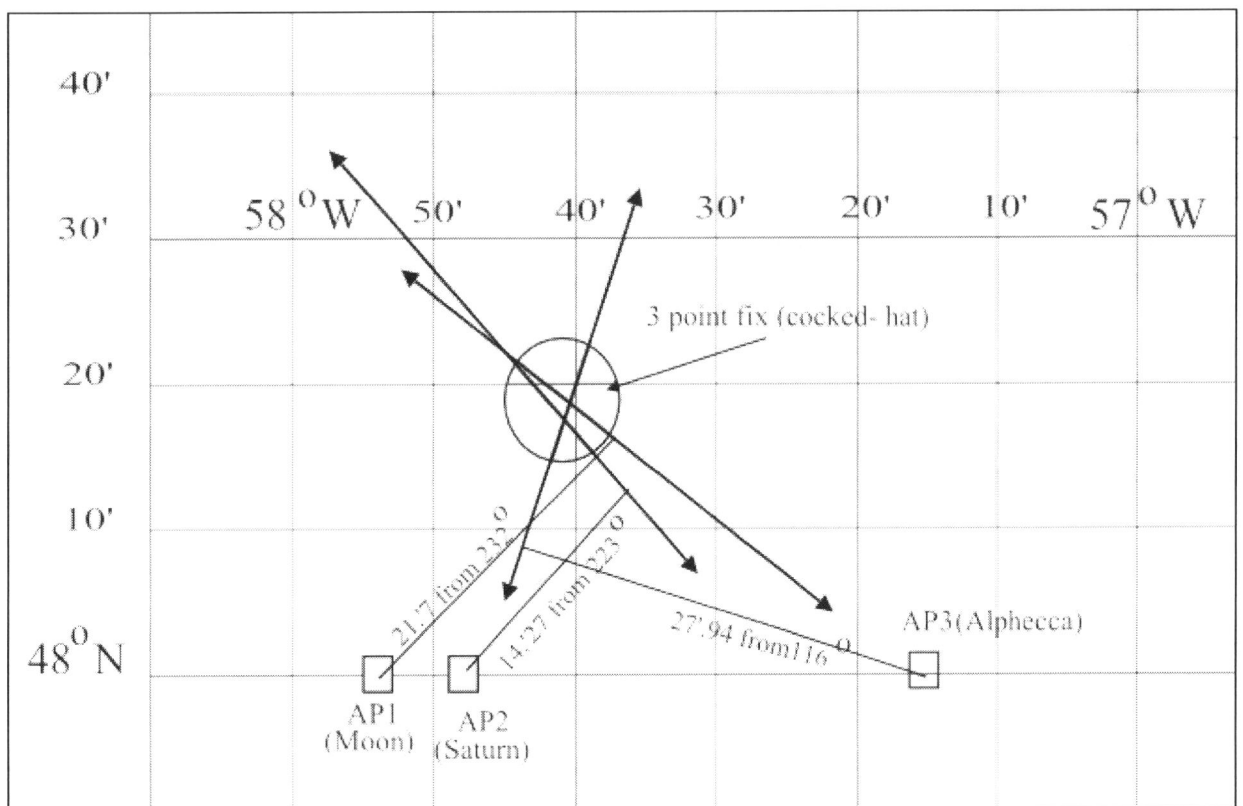

2009 MAY 28, 29, 30 (THURS., FRI., SAT.)

UT	SUN GHA	SUN Dec	MOON GHA	MOON v	MOON Dec	MOON d	MOON HP
d h	° '	° '	° '	'	° '	'	'
28 00	180 41.6	N21 26.9	128 18.4	5.5	N21 37.8	10.0	60.3
01	195 41.5	27.3	142 42.9	5.6	21 27.8	10.1	60.3
02	210 41.4	27.7	157 07.5	5.7	21 17.7	10.2	60.3
03	225 41.4	28.1	171 32.2	5.8	21 07.5	10.4	60.2
04	240 41.3	28.5	185 57.0	6.0	20 57.1	10.4	60.2
05	255 41.2	28.9	200 22.0	6.0	20 46.7	10.7	60.2
06	270 41.1	N21 29.3	214 47.0	6.2	N20 36.0	10.7	60.2
07	285 41.0	29.7	229 12.2	6.3	20 25.3	10.9	60.2
T 08	300 41.0	30.1	243 37.5	6.4	20 14.4	11.0	60.1
H 09	315 40.9	30.5	258 02.9	6.5	20 03.4	11.1	60.1
U 10	330 40.8	30.9	272 28.4	6.6	19 52.3	11.2	60.1
R 11	345 40.7	31.3	286 54.0	6.7	19 41.1	11.3	60.1
S 12	0 40.6	N21 31.7	301 19.7	6.9	N19 29.8	11.5	60.1
D 13	15 40.6	32.1	315 45.6	6.9	19 18.3	11.6	60.0
A 14	30 40.5	32.5	330 11.5	7.1	19 06.7	11.7	60.0
Y 15	45 40.4	32.8	344 37.6	7.2	18 55.0	11.7	60.0
16	60 40.3	33.2	359 03.8	7.3	18 43.3	11.9	60.0
17	75 40.2	33.6	13 30.1	7.5	18 31.4	12.0	59.9
18	90 40.2	N21 34.0	27 56.6	7.5	N18 19.4	12.1	59.9
19	105 40.1	34.4	42 23.1	7.6	18 07.3	12.2	59.9
20	120 40.0	34.8	56 49.7	7.8	17 55.1	12.3	59.9
21	135 39.9	35.2	71 16.5	7.9	17 42.8	12.4	59.8
22	150 39.8	35.6	85 43.4	8.0	17 30.4	12.5	59.8
23	165 39.8	36.0	100 10.4	8.1	17 17.9	12.6	59.8
29 00	180 39.7	N21 36.4	114 37.5	8.2	N17 05.3	12.7	59.8
01	195 39.6	36.8	129 04.7	8.3	16 52.6	12.8	59.8
02	210 39.5	37.1	143 32.0	8.4	16 39.8	12.8	59.7
03	225 39.4	37.5	157 59.4	8.6	16 27.0	13.0	59.7
04	240 39.3	37.9	172 27.0	8.6	16 14.0	13.0	59.7
05	255 39.3	38.3	186 54.6	8.8	16 01.0	13.1	59.7
06	270 39.2	N21 38.7	201 22.4	8.8	N15 47.9	13.2	59.6
07	285 39.1	39.1	215 50.2	9.0	15 34.7	13.2	59.6
08	300 39.0	39.4	230 18.2	9.1	15 21.5	13.4	59.6
F 09	315 38.9	39.8	244 46.3	9.2	15 08.1	13.4	59.5
R 10	330 38.8	40.2	259 14.5	9.2	14 54.7	13.5	59.5
I 11	345 38.8	40.6	273 42.7	9.4	14 41.2	13.5	59.5
D 12	0 38.7	N21 41.0	288 11.1	9.5	N14 27.7	13.6	59.5
A 13	15 38.6	41.4	302 39.6	9.6	14 14.1	13.7	59.4
Y 14	30 38.5	41.7	317 08.2	9.7	14 00.4	13.8	59.4
15	45 38.4	42.1	331 36.9	9.7	13 46.6	13.8	59.4
16	60 38.3	42.5	346 05.6	9.9	13 32.8	13.9	59.4
17	75 38.2	42.9	0 34.5	10.0	13 18.9	13.9	59.3
18	90 38.2	N21 43.2	15 03.5	10.1	N13 05.0	14.0	59.3
19	105 38.1	43.6	29 32.6	10.1	12 51.0	14.1	59.3
20	120 38.0	44.0	44 01.7	10.3	12 36.9	14.1	59.3
21	135 37.9	44.4	58 31.0	10.4	12 22.8	14.2	59.2
22	150 37.8	44.7	73 00.4	10.4	12 08.6	14.2	59.2
23	165 37.7	45.1	87 29.8	10.5	11 54.4	14.3	59.2
30 00	180 37.6	N21 45.5	101 59.3	10.6	N11 40.1	14.3	59.1
01	195 37.6	45.9	116 28.9	10.8	11 25.8	14.4	59.1
02	210 37.5	46.2	130 58.7	10.8	11 11.5	14.4	59.1
03	225 37.4	46.6	145 28.5	10.8	10 57.1	14.5	59.1
04	240 37.3	47.0	159 58.3	11.0	10 42.6	14.5	59.0
05	255 37.2	47.3	174 28.3	11.0	10 28.1	14.5	59.0
06	270 37.1	N21 47.7	188 58.3	11.2	N10 13.6	14.6	58.9
07	285 37.0	48.1	203 28.5	11.2	9 59.0	14.6	58.9
S 08	300 37.0	48.5	217 58.7	11.3	9 44.4	14.6	58.9
A 09	315 36.9	48.8	232 29.0	11.3	9 29.8	14.7	58.9
T 10	330 36.8	49.2	246 59.3	11.5	9 15.1	14.7	58.8
U 11	345 36.7	49.5	261 29.8	11.5	9 00.4	14.8	58.8
R 12	0 36.6	N21 49.9	276 00.3	11.6	N 8 45.6	14.7	58.8
D 13	15 36.5	50.3	290 30.9	11.7	8 30.9	14.8	58.8
A 14	30 36.4	50.6	305 01.6	11.7	8 16.1	14.9	58.7
Y 15	45 36.3	51.0	319 32.3	11.8	8 01.2	14.8	58.7
16	60 36.2	51.4	334 03.1	11.9	7 46.4	14.9	58.7
17	75 36.2	51.7	348 34.0	11.9	7 31.5	14.9	58.7
18	90 36.1	N21 52.1	3 04.9	12.0	N 7 16.6	14.9	58.6
19	105 36.0	52.4	17 35.9	12.1	7 01.7	15.0	58.6
20	120 35.9	52.8	32 07.0	12.2	6 46.7	14.9	58.6
21	135 35.8	53.2	46 38.2	12.2	6 31.8	15.0	58.5
22	150 35.7	53.5	61 09.4	12.2	6 16.8	15.0	58.5
23	165 35.6	53.9	75 40.6	12.4	N 6 01.8	15.0	58.5
	SD 15.8	d 0.4	SD 16.4		16.2		16.0

Lat.	Twilight Naut.	Twilight Civil	Sunrise	Moonrise 28	Moonrise 29	Moonrise 30	Moonrise 31
°	h m	h m	h m	h m	h m	h m	h m
N 72	▭	▭	▭	▭	06 42	09 31	11 48
N 70	▭	▭	▭	▭	07 22	09 47	11 52
68	▭	▭	▭	05 05	07 49	09 59	11 56
66	////	////	01 25	05 53	08 09	10 08	11 58
64	////	////	02 05	06 24	08 25	10 17	12 00
62	////	00 38	02 33	06 47	08 39	10 24	12 02
60	////	01 35	02 54	07 05	08 50	10 30	12 04
N 58	////	02 06	03 11	07 20	09 00	10 35	12 05
56	00 34	02 29	03 25	07 33	09 08	10 40	12 07
54	01 26	02 47	03 38	07 44	09 16	10 44	12 08
52	01 55	03 03	03 49	07 54	09 22	10 47	12 09
50	02 17	03 16	03 58	08 03	09 28	10 51	12 10
45	02 56	03 42	04 18	08 21	09 41	10 58	12 12
N 40	03 23	04 03	04 35	08 36	09 52	11 04	12 14
35	03 44	04 19	04 48	08 49	10 01	11 10	12 15
30	04 01	04 33	05 00	09 00	10 09	11 14	12 17
20	04 27	04 56	05 20	09 19	10 22	11 22	12 19
N 10	04 48	05 15	05 38	09 35	10 34	11 29	12 21
0	05 06	05 32	05 54	09 50	10 45	11 36	12 23
S 10	05 21	05 47	06 10	10 06	10 56	11 42	12 25
20	05 36	06 03	06 27	10 22	11 08	11 49	12 27
30	05 51	06 20	06 46	10 40	11 21	11 57	12 30
35	05 58	06 30	06 57	10 51	11 29	12 02	12 31
40	06 07	06 40	07 10	11 03	11 38	12 07	12 32
45	06 16	06 52	07 25	11 18	11 48	12 12	12 34
S 50	06 26	07 07	07 44	11 35	12 00	12 19	12 36
52	06 31	07 13	07 53	11 43	12 06	12 23	12 37
54	06 36	07 21	08 03	11 53	12 12	12 26	12 38
56	06 41	07 29	08 14	12 03	12 19	12 30	12 40
58	06 47	07 38	08 27	12 14	12 26	12 34	12 41
S 60	06 53	07 48	08 42	12 28	12 35	12 39	12 42

Lat.	Sunset	Twilight Civil	Twilight Naut.	Moonset 28	Moonset 29	Moonset 30	Moonset 31
°	h m	h m	h m	h m	h m	h m	h m
N 72	▭	▭	▭	▭	02 45	01 44	01 08
N 70	▭	▭	▭	▭	02 04	01 26	01 01
68	▭	▭	▭	02 22	01 35	01 12	00 55
66	22 34	////	////	01 33	01 13	01 00	00 50
64	21 52	////	////	01 01	00 56	00 51	00 46
62	21 24	23 25	////	00 38	00 41	00 42	00 42
60	21 02	22 23	////	00 19	00 29	00 35	00 39
N 58	20 45	21 51	////	00 03	00 18	00 28	00 36
56	20 30	21 27	23 28	24 09	00 09	00 22	00 33
54	20 18	21 09	22 31	24 00	00 00	00 17	00 31
52	20 07	20 53	22 01	23 53	24 12	00 12	00 28
50	19 57	20 40	21 39	23 46	24 08	00 08	00 26
45	19 37	20 13	21 00	23 32	23 59	24 22	00 22
N 40	19 21	19 53	20 33	23 19	23 51	24 18	00 18
35	19 07	19 36	20 12	23 09	23 44	24 15	00 15
30	18 55	19 22	19 54	23 00	23 38	24 12	00 12
20	18 35	18 59	19 28	22 44	23 28	24 07	00 07
N 10	18 17	18 40	19 07	22 30	23 18	24 03	00 03
0	18 01	18 23	18 49	22 17	23 10	23 59	24 45
S 10	17 45	18 08	18 34	22 04	23 01	23 54	24 45
20	17 28	17 52	18 19	21 50	22 51	23 50	24 46
30	17 09	17 34	18 04	21 33	22 40	23 44	24 46
35	16 57	17 25	17 56	21 23	22 34	23 41	24 46
40	16 44	17 14	17 48	21 12	22 27	23 38	24 46
45	16 29	17 02	17 39	20 59	22 18	23 34	24 47
S 50	16 10	16 48	17 28	20 43	22 08	23 29	24 47
52	16 02	16 41	17 24	20 36	22 03	23 27	24 47
54	15 52	16 34	17 19	20 27	21 58	23 24	24 47
56	15 41	16 26	17 14	20 18	21 52	23 21	24 47
58	15 28	16 17	17 08	20 07	21 45	23 18	24 47
S 60	15 13	16 07	17 01	19 54	21 38	23 15	24 48

Day	SUN Eqn. of Time 00h	SUN Eqn. of Time 12h	SUN Mer. Pass.	MOON Mer. Pass. Upper	MOON Mer. Pass. Lower	MOON Age	MOON Phase
d	m s	m s	h m	h m	h m	d	%
28	02 47	02 43	11 57	16 04	03 35	04	22
29	02 39	02 35	11 57	16 58	04 31	05	32
30	02 31	02 27	11 58	17 47	05 23	06	43

2009 MAY 28, 29, 30 (THURS., FRI., SAT.)

UT	ARIES GHA	VENUS −4.5 GHA	Dec	MARS +1.2 GHA	Dec	JUPITER −2.4 GHA	Dec	SATURN +0.9 GHA	Dec	STARS Name	SHA	Dec
28 00	245 40.2	225 16.9	N 6 35.5	220 10.5	N 9 30.9	276 40.7	S13 25.4	78 38.0	N 7 52.6	Acamar	315 20.9	S40 15.9
01	260 42.7	240 17.2	36.1	235 11.2	31.6	291 43.0	25.3	93 40.4	52.6	Achernar	335 29.2	S57 11.1
02	275 45.1	255 17.6	36.8	250 11.9	32.3	306 45.3	25.3	108 42.8	52.6	Acrux	173 12.6	S63 09.5
03	290 47.6	270 17.9	37.5	265 12.6	33.0	321 47.7	25.2	123 45.2	52.6	Adhara	255 15.2	S28 59.2
04	305 50.1	285 18.2	38.1	280 13.3	33.7	336 50.0	25.2	138 47.7	52.5	Aldebaran	290 53.2	N16 31.7
05	320 52.5	300 18.5	38.8	295 14.0	34.4	351 52.3	25.2	153 50.1	52.5			
06	335 55.0	315 18.8	N 6 39.5	310 14.7	N 9 35.1	6 54.6	S13 25.1	168 52.5	N 7 52.5	Alioth	166 22.8	N55 54.6
07	350 57.4	330 19.2	40.1	325 15.4	35.7	21 57.0	25.1	183 54.9	52.5	Alkaid	153 00.7	N49 16.0
T 08	5 59.9	345 19.5	40.8	340 16.1	36.4	36 59.3	25.1	198 57.4	52.4	Al Na'ir	27 47.3	S46 54.7
H 09	21 02.4	0 19.8	41.4	355 16.7	37.1	52 01.6	25.0	213 59.8	52.4	Alnilam	275 49.7	S 1 11.8
U 10	36 04.8	15 20.1	42.1	10 17.4	37.8	67 03.9	25.0	229 02.2	52.4	Alphard	217 59.2	S 8 42.1
R 11	51 07.3	30 20.4	42.8	25 18.1	38.5	82 06.3	24.9	244 04.6	52.4			
S 12	66 09.8	45 20.7	N 6 43.4	40 18.8	N 9 39.2	97 08.6	S13 24.9	259 07.0	N 7 52.3	Alphecca	126 13.1	N26 40.9
D 13	81 12.2	60 21.1	44.1	55 19.5	39.9	112 10.9	24.9	274 09.5	52.3	Alpheratz	357 46.8	N29 08.4
A 14	96 14.7	75 21.4	44.8	70 20.2	40.6	127 13.2	24.8	289 11.9	52.3	Altair	62 11.0	N 8 53.5
Y 15	111 17.2	90 21.7	45.4	85 20.9	41.3	142 15.6	24.8	304 14.3	52.3	Ankaa	353 18.7	S42 15.0
16	126 19.6	105 22.0	46.1	100 21.6	42.0	157 17.9	24.8	319 16.7	52.2	Antares	112 29.6	S26 27.3
17	141 22.1	120 22.3	46.8	115 22.3	42.7	172 20.2	24.7	334 19.1	52.2			
18	156 24.5	135 22.6	N 6 47.4	130 23.0	N 9 43.3	187 22.5	S13 24.7	349 21.6	N 7 52.2	Arcturus	145 58.1	N19 07.9
19	171 27.0	150 22.9	48.1	145 23.7	44.0	202 24.9	24.6	4 24.0	52.2	Atria	107 33.7	S69 02.7
20	186 29.5	165 23.2	48.8	160 24.4	44.7	217 27.2	24.6	19 26.4	52.1	Avior	234 19.8	S59 32.7
21	201 31.9	180 23.5	49.5	175 25.0	45.4	232 29.5	24.6	34 28.8	52.1	Bellatrix	278 35.6	N 6 21.5
22	216 34.4	195 23.8	50.1	190 25.7	46.1	247 31.9	24.5	49 31.2	52.1	Betelgeuse	271 04.9	N 7 24.5
23	231 36.9	210 24.1	50.8	205 26.4	46.8	262 34.2	24.5	64 33.7	52.0			
29 00	246 39.3	225 24.4	N 6 51.5	220 27.1	N 9 47.4	277 36.5	S13 24.5	79 36.1	N 7 52.0	Canopus	263 58.0	S52 42.1
01	261 41.8	240 24.7	52.1	235 27.8	48.1	292 38.8	24.4	94 38.5	52.0	Capella	280 39.4	N46 00.5
02	276 44.3	255 25.0	52.8	250 28.5	48.8	307 41.2	24.4	109 40.9	52.0	Deneb	49 33.4	N45 18.6
03	291 46.7	270 25.3	53.5	265 29.2	49.5	322 43.5	24.4	124 43.3	51.9	Denebola	182 36.6	N14 31.1
04	306 49.2	285 25.6	54.2	280 29.9	50.2	337 45.8	24.3	139 45.8	51.9	Diphda	348 59.0	S17 56.0
05	321 51.7	300 25.9	54.8	295 30.6	50.9	352 48.2	24.3	154 48.2	51.9			
06	336 54.1	315 26.2	N 6 55.5	310 31.3	N 9 51.6	7 50.5	S13 24.2	169 50.6	N 7 51.9	Dubhe	193 55.0	N61 42.2
07	351 56.6	330 26.5	56.2	325 32.0	52.3	22 52.8	24.2	184 53.0	51.8	Elnath	278 16.8	N28 37.0
08	6 59.1	345 26.8	56.9	340 32.7	52.9	37 55.2	24.2	199 55.4	51.8	Eltanin	90 47.0	N51 29.1
F 09	22 01.5	0 27.1	57.5	355 33.3	53.6	52 57.5	24.1	214 57.8	51.8	Enif	33 50.0	N 9 55.0
R 10	37 04.0	15 27.4	58.2	10 34.0	54.3	67 59.8	24.1	230 00.3	51.7	Fomalhaut	15 27.2	S29 34.1
I 11	52 06.4	30 27.7	58.9	25 34.7	55.0	83 02.2	24.1	245 02.7	51.7			
D 12	67 08.9	45 28.0	N 6 59.6	40 35.4	N 9 55.7	98 04.5	S13 24.0	260 05.1	N 7 51.7	Gacrux	172 04.2	S57 10.3
A 13	82 11.4	60 28.3	7 00.2	55 36.1	56.4	113 06.8	24.0	275 07.5	51.7	Gienah	175 55.3	S17 35.9
Y 14	97 13.8	75 28.6	00.9	70 36.8	57.0	128 09.2	24.0	290 09.9	51.6	Hadar	148 51.9	S60 25.4
15	112 16.3	90 28.9	01.6	85 37.5	57.7	143 11.5	23.9	305 12.3	51.6	Hamal	328 04.5	N23 30.4
16	127 18.8	105 29.2	02.3	100 38.2	58.4	158 13.8	23.9	320 14.8	51.6	Kaus Aust.	83 47.4	S34 22.8
17	142 21.2	120 29.5	02.9	115 38.9	59.1	173 16.2	23.9	335 17.2	51.6			
18	157 23.7	135 29.8	N 7 03.6	130 39.6	N 9 59.8	188 18.5	S13 23.8	350 19.6	N 7 51.5	Kochab	137 18.4	N74 07.0
19	172 26.2	150 30.1	04.3	145 40.3	10 00.5	203 20.8	23.8	5 22.0	51.5	Markab	13 41.4	N15 15.3
20	187 28.6	165 30.3	05.0	160 40.9	01.1	218 23.2	23.8	20 24.4	51.5	Menkar	314 18.5	N 4 07.6
21	202 31.1	180 30.6	05.7	175 41.6	01.8	233 25.5	23.7	35 26.8	51.4	Menkent	148 10.9	S36 25.2
22	217 33.5	195 30.9	06.3	190 42.3	02.5	248 27.8	23.7	50 29.3	51.4	Miaplacidus	221 41.0	S69 45.7
23	232 36.0	210 31.2	07.0	205 43.0	03.2	263 30.2	23.7	65 31.7	51.4			
30 00	247 38.5	225 31.5	N 7 07.7	220 43.7	N10 03.9	278 32.5	S13 23.6	80 34.1	N 7 51.4	Mirfak	308 45.3	N49 53.6
01	262 40.9	240 31.8	08.4	235 44.4	04.6	293 34.8	23.6	95 36.5	51.3	Nunki	76 01.7	S26 17.1
02	277 43.4	255 32.0	09.1	250 45.1	05.2	308 37.2	23.6	110 38.9	51.3	Peacock	53 23.5	S56 42.1
03	292 45.9	270 32.3	09.8	265 45.8	05.9	323 39.5	23.5	125 41.3	51.3	Pollux	243 31.6	N28 00.3
04	307 48.3	285 32.6	10.4	280 46.5	06.6	338 41.9	23.5	140 43.7	51.2	Procyon	245 03.1	N 5 12.0
05	322 50.8	300 32.9	11.1	295 47.2	07.3	353 44.2	23.5	155 46.2	51.2			
06	337 53.3	315 33.2	N 7 11.8	310 47.9	N10 08.0	8 46.5	S13 23.4	170 48.6	N 7 51.2	Rasalhague	96 08.9	N12 33.1
07	352 55.7	330 33.4	12.5	325 48.5	08.6	23 48.9	23.4	185 51.0	51.2	Regulus	207 46.7	N11 55.2
S 08	7 58.2	345 33.7	13.2	340 49.2	09.3	38 51.2	23.4	200 53.4	51.1	Rigel	281 15.3	S 8 11.4
A 09	23 00.7	0 34.0	13.9	355 49.9	10.0	53 53.6	23.3	215 55.8	51.1	Rigil Kent.	139 55.7	S60 52.7
T 10	38 03.1	15 34.3	14.5	10 50.6	10.7	68 55.9	23.3	230 58.2	51.1	Sabik	102 15.7	S15 44.3
U 11	53 05.6	30 34.5	15.2	25 51.3	11.4	83 58.2	23.3	246 00.6	51.0			
R 12	68 08.0	45 34.8	N 7 15.9	40 52.0	N10 12.0	99 00.6	S13 23.2	261 03.1	N 7 51.0	Schedar	349 44.5	N56 35.2
D 13	83 10.5	60 35.1	16.6	55 52.7	12.7	114 02.9	23.2	276 05.5	51.0	Shaula	96 25.6	S37 06.7
A 14	98 13.0	75 35.4	17.3	70 53.4	13.4	129 05.3	23.2	291 07.9	51.0	Sirius	258 36.7	S16 43.8
Y 15	113 15.4	90 35.6	18.0	85 54.1	14.1	144 07.6	23.1	306 10.3	50.9	Spica	158 34.2	S11 12.8
16	128 17.9	105 35.9	18.7	100 54.8	14.8	159 09.9	23.1	321 12.7	50.9	Suhail	222 54.9	S43 28.5
17	143 20.4	120 36.2	19.4	115 55.4	15.4	174 12.3	23.1	336 15.1	50.9			
18	158 22.8	135 36.4	N 7 20.0	130 56.1	N10 16.1	189 14.6	S13 23.0	351 17.5	N 7 50.8	Vega	80 40.6	N38 47.4
19	173 25.3	150 36.7	20.7	145 56.8	16.8	204 17.0	23.0	6 19.9	50.8	Zuben'ubi	137 08.5	S16 05.0
20	188 27.8	165 37.0	21.4	160 57.5	17.5	219 19.3	23.0	21 22.4	50.8		SHA	Mer. Pass.
21	203 30.2	180 37.2	22.1	175 58.2	18.1	234 21.6	22.9	36 24.8	50.7	Venus	338 45.1	8 58
22	218 32.7	195 37.5	22.8	190 58.9	18.8	249 24.0	22.9	51 27.2	50.7	Mars	333 47.8	9 18
23	233 35.2	210 37.8	23.5	205 59.6	19.5	264 26.3	22.9	66 29.6	50.7	Jupiter	30 57.2	5 29
Mer. Pass. 7h 32.1m	v 0.3	d 0.7	v 0.7	d 0.7	v 2.3	d 0.0	v 2.4	d 0.0	Saturn	192 56.7	18 39

48ᵐ INCREMENTS AND CORRECTIONS 49ᵐ

48	SUN PLANETS	ARIES	MOON	v or d	Corrⁿ	v or d	Corrⁿ	v or d	Corrⁿ	49	SUN PLANETS	ARIES	MOON	v or d	Corrⁿ	v or d	Corrⁿ	v or d	Corrⁿ
s	° ′	° ′	° ′	′	′	′	′	′	′	s	° ′	° ′	° ′	′	′	′	′	′	′
00	12 00·0	12 02·0	11 27·2	0·0	0·0	6·0	4·9	12·0	9·7	00	12 15·0	12 17·0	11 41·5	0·0	0·0	6·0	5·0	12·0	9·9
01	12 00·3	12 02·2	11 27·4	0·1	0·1	6·1	4·9	12·1	9·8	01	12 15·3	12 17·3	11 41·8	0·1	0·1	6·1	5·0	12·1	10·0
02	12 00·5	12 02·5	11 27·7	0·2	0·2	6·2	5·0	12·2	9·9	02	12 15·5	12 17·5	11 42·0	0·2	0·2	6·2	5·1	12·2	10·1
03	12 00·8	12 02·7	11 27·9	0·3	0·2	6·3	5·1	12·3	9·9	03	12 15·8	12 17·8	11 42·2	0·3	0·2	6·3	5·2	12·3	10·1
04	12 01·0	12 03·0	11 28·2	0·4	0·3	6·4	5·2	12·4	10·0	04	12 16·0	12 18·0	11 42·5	0·4	0·3	6·4	5·3	12·4	10·2
05	12 01·3	12 03·2	11 28·4	0·5	0·4	6·5	5·3	12·5	10·1	05	12 16·3	12 18·3	11 42·7	0·5	0·4	6·5	5·4	12·5	10·3
06	12 01·5	12 03·5	11 28·6	0·6	0·5	6·6	5·3	12·6	10·2	06	12 16·5	12 18·5	11 42·9	0·6	0·5	6·6	5·4	12·6	10·4
07	12 01·8	12 03·7	11 28·9	0·7	0·6	6·7	5·4	12·7	10·3	07	12 16·8	12 18·8	11 43·2	0·7	0·6	6·7	5·5	12·7	10·5
08	12 02·0	12 04·0	11 29·1	0·8	0·6	6·8	5·5	12·8	10·3	08	12 17·0	12 19·0	11 43·4	0·8	0·7	6·8	5·6	12·8	10·6
09	12 02·3	12 04·2	11 29·3	0·9	0·7	6·9	5·6	12·9	10·4	09	12 17·3	12 19·3	11 43·7	0·9	0·7	6·9	5·7	12·9	10·6
10	12 02·5	12 04·5	11 29·6	1·0	0·8	7·0	5·7	13·0	10·5	10	12 17·5	12 19·5	11 43·9	1·0	0·8	7·0	5·8	13·0	10·7
11	12 02·8	12 04·7	11 29·8	1·1	0·9	7·1	5·7	13·1	10·6	11	12 17·8	12 19·8	11 44·1	1·1	0·9	7·1	5·9	13·1	10·8
12	12 03·0	12 05·0	11 30·1	1·2	1·0	7·2	5·8	13·2	10·7	12	12 18·0	12 20·0	11 44·4	1·2	1·0	7·2	5·9	13·2	10·9
13	12 03·3	12 05·2	11 30·3	1·3	1·1	7·3	5·9	13·3	10·8	13	12 18·3	12 20·3	11 44·6	1·3	1·1	7·3	6·0	13·3	11·0
14	12 03·5	12 05·5	11 30·5	1·4	1·1	7·4	6·0	13·4	10·8	14	12 18·5	12 20·5	11 44·9	1·4	1·2	7·4	6·1	13·4	11·1
15	12 03·8	12 05·7	11 30·8	1·5	1·2	7·5	6·1	13·5	10·9	15	12 18·8	12 20·8	11 45·1	1·5	1·2	7·5	6·2	13·5	11·1
16	12 04·0	12 06·0	11 31·0	1·6	1·3	7·6	6·1	13·6	11·0	16	12 19·0	12 21·0	11 45·3	1·6	1·3	7·6	6·3	13·6	11·2
17	12 04·3	12 06·2	11 31·3	1·7	1·4	7·7	6·2	13·7	11·1	17	12 19·3	12 21·3	11 45·6	1·7	1·4	7·7	6·4	13·7	11·3
18	12 04·5	12 06·5	11 31·5	1·8	1·5	7·8	6·3	13·8	11·2	18	12 19·5	12 21·5	11 45·8	1·8	1·5	7·8	6·4	13·8	11·4
19	12 04·8	12 06·7	11 31·7	1·9	1·5	7·9	6·4	13·9	11·2	19	12 19·8	12 21·8	11 46·1	1·9	1·6	7·9	6·5	13·9	11·5
20	12 05·0	12 07·0	11 32·0	2·0	1·6	8·0	6·5	14·0	11·3	20	12 20·0	12 22·0	11 46·3	2·0	1·7	8·0	6·6	14·0	11·6
21	12 05·3	12 07·2	11 32·2	2·1	1·7	8·1	6·5	14·1	11·4	21	12 20·3	12 22·3	11 46·5	2·1	1·7	8·1	6·7	14·1	11·6
22	12 05·5	12 07·5	11 32·4	2·2	1·8	8·2	6·6	14·2	11·5	22	12 20·5	12 22·5	11 46·8	2·2	1·8	8·2	6·8	14·2	11·7
23	12 05·8	12 07·7	11 32·7	2·3	1·9	8·3	6·7	14·3	11·6	23	12 20·8	12 22·8	11 47·0	2·3	1·9	8·3	6·8	14·3	11·8
24	12 06·0	12 08·0	11 32·9	2·4	1·9	8·4	6·8	14·4	11·6	24	12 21·0	12 23·0	11 47·2	2·4	2·0	8·4	6·9	14·4	11·9
25	12 06·3	12 08·2	11 33·2	2·5	2·0	8·5	6·9	14·5	11·7	25	12 21·3	12 23·3	11 47·5	2·5	2·1	8·5	7·0	14·5	12·0
26	12 06·5	12 08·5	11 33·4	2·6	2·1	8·6	7·0	14·6	11·8	26	12 21·5	12 23·5	11 47·7	2·6	2·1	8·6	7·1	14·6	12·0
27	12 06·8	12 08·7	11 33·6	2·7	2·2	8·7	7·0	14·7	11·9	27	12 21·8	12 23·8	11 48·0	2·7	2·2	8·7	7·2	14·7	12·1
28	12 07·0	12 09·0	11 33·9	2·8	2·3	8·8	7·1	14·8	12·0	28	12 22·0	12 24·0	11 48·2	2·8	2·3	8·8	7·3	14·8	12·2
29	12 07·3	12 09·2	11 34·1	2·9	2·3	8·9	7·2	14·9	12·0	29	12 22·3	12 24·3	11 48·4	2·9	2·4	8·9	7·3	14·9	12·3
30	12 07·5	12 09·5	11 34·4	3·0	2·4	9·0	7·3	15·0	12·1	30	12 22·5	12 24·5	11 48·7	3·0	2·5	9·0	7·4	15·0	12·4
31	12 07·8	12 09·7	11 34·6	3·1	2·5	9·1	7·4	15·1	12·2	31	12 22·8	12 24·8	11 48·9	3·1	2·6	9·1	7·5	15·1	12·5
32	12 08·0	12 10·0	11 34·8	3·2	2·6	9·2	7·4	15·2	12·3	32	12 23·0	12 25·0	11 49·2	3·2	2·6	9·2	7·6	15·2	12·5
33	12 08·3	12 10·2	11 35·1	3·3	2·7	9·3	7·5	15·3	12·4	33	12 23·3	12 25·3	11 49·4	3·3	2·7	9·3	7·7	15·3	12·6
34	12 08·5	12 10·5	11 35·3	3·4	2·7	9·4	7·6	15·4	12·4	34	12 23·5	12 25·5	11 49·6	3·4	2·8	9·4	7·8	15·4	12·7
35	12 08·8	12 10·7	11 35·6	3·5	2·8	9·5	7·7	15·5	12·5	35	12 23·8	12 25·8	11 49·9	3·5	2·9	9·5	7·8	15·5	12·8
36	12 09·0	12 11·0	11 35·8	3·6	2·9	9·6	7·8	15·6	12·6	36	12 24·0	12 26·0	11 50·1	3·6	3·0	9·6	7·9	15·6	12·9
37	12 09·3	12 11·2	11 36·0	3·7	3·0	9·7	7·8	15·7	12·7	37	12 24·3	12 26·3	11 50·3	3·7	3·1	9·7	8·0	15·7	13·0
38	12 09·5	12 11·5	11 36·3	3·8	3·1	9·8	7·9	15·8	12·8	38	12 24·5	12 26·5	11 50·6	3·8	3·1	9·8	8·1	15·8	13·0
39	12 09·8	12 11·7	11 36·5	3·9	3·2	9·9	8·0	15·9	12·9	39	12 24·8	12 26·8	11 50·8	3·9	3·2	9·9	8·2	15·9	13·1
40	12 10·0	12 12·0	11 36·7	4·0	3·2	10·0	8·1	16·0	12·9	40	12 25·0	12 27·0	11 51·1	4·0	3·3	10·0	8·3	16·0	13·2
41	12 10·3	12 12·2	11 37·0	4·1	3·3	10·1	8·2	16·1	13·0	41	12 25·3	12 27·3	11 51·3	4·1	3·4	10·1	8·3	16·1	13·3
42	12 10·5	12 12·5	11 37·2	4·2	3·4	10·2	8·2	16·2	13·1	42	12 25·5	12 27·5	11 51·5	4·2	3·5	10·2	8·4	16·2	13·4
43	12 10·8	12 12·8	11 37·5	4·3	3·5	10·3	8·3	16·3	13·2	43	12 25·8	12 27·8	11 51·8	4·3	3·5	10·3	8·5	16·3	13·4
44	12 11·0	12 13·0	11 37·7	4·4	3·6	10·4	8·4	16·4	13·3	44	12 26·0	12 28·0	11 52·0	4·4	3·6	10·4	8·6	16·4	13·5
45	12 11·3	12 13·3	11 37·9	4·5	3·6	10·5	8·5	16·5	13·3	45	12 26·3	12 28·3	11 52·3	4·5	3·7	10·5	8·7	16·5	13·6
46	12 11·5	12 13·5	11 38·2	4·6	3·7	10·6	8·6	16·6	13·4	46	12 26·5	12 28·5	11 52·5	4·6	3·8	10·6	8·7	16·6	13·7
47	12 11·8	12 13·8	11 38·4	4·7	3·8	10·7	8·6	16·7	13·5	47	12 26·8	12 28·8	11 52·7	4·7	3·9	10·7	8·8	16·7	13·8
48	12 12·0	12 14·0	11 38·7	4·8	3·9	10·8	8·7	16·8	13·6	48	12 27·0	12 29·0	11 53·0	4·8	4·0	10·8	8·9	16·8	13·9
49	12 12·3	12 14·3	11 38·9	4·9	4·0	10·9	8·8	16·9	13·7	49	12 27·3	12 29·3	11 53·2	4·9	4·0	10·9	9·0	16·9	13·9
50	12 12·5	12 14·5	11 39·1	5·0	4·0	11·0	8·9	17·0	13·7	50	12 27·5	12 29·5	11 53·4	5·0	4·1	11·0	9·1	17·0	14·0
51	12 12·8	12 14·8	11 39·4	5·1	4·1	11·1	9·0	17·1	13·8	51	12 27·8	12 29·8	11 53·7	5·1	4·2	11·1	9·2	17·1	14·1
52	12 13·0	12 15·0	11 39·6	5·2	4·2	11·2	9·1	17·2	13·9	52	12 28·0	12 30·0	11 53·9	5·2	4·3	11·2	9·2	17·2	14·2
53	12 13·3	12 15·3	11 39·8	5·3	4·3	11·3	9·1	17·3	14·0	53	12 28·3	12 30·3	11 54·2	5·3	4·4	11·3	9·3	17·3	14·3
54	12 13·5	12 15·5	11 40·1	5·4	4·4	11·4	9·2	17·4	14·1	54	12 28·5	12 30·5	11 54·4	5·4	4·5	11·4	9·4	17·4	14·4
55	12 13·8	12 15·8	11 40·3	5·5	4·4	11·5	9·3	17·5	14·1	55	12 28·8	12 30·8	11 54·6	5·5	4·5	11·5	9·5	17·5	14·4
56	12 14·0	12 16·0	11 40·6	5·6	4·5	11·6	9·4	17·6	14·2	56	12 29·0	12 31·1	11 54·9	5·6	4·6	11·6	9·6	17·6	14·5
57	12 14·3	12 16·3	11 40·8	5·7	4·6	11·7	9·5	17·7	14·3	57	12 29·3	12 31·3	11 55·1	5·7	4·7	11·7	9·7	17·7	14·6
58	12 14·5	12 16·5	11 41·0	5·8	4·7	11·8	9·5	17·8	14·4	58	12 29·5	12 31·6	11 55·4	5·8	4·8	11·8	9·7	17·8	14·7
59	12 14·8	12 16·8	11 41·3	5·9	4·8	11·9	9·6	17·9	14·5	59	12 29·8	12 31·8	11 55·6	5·9	4·9	11·9	9·8	17·9	14·8
60	12 15·0	12 17·0	11 41·5	6·0	4·9	12·0	9·7	18·0	14·6	60	12 30·0	12 32·1	11 55·8	6·0	5·0	12·0	9·9	18·0	14·9

INCREMENTS AND CORRECTIONS

50ᵐ

s	SUN PLANETS	ARIES	MOON	v or d	Corrⁿ	v or d	Corrⁿ	v or d	Corrⁿ
00	12 30·0	12 32·1	11 55·8	0·0	0·0	6·0	5·1	12·0	10·1
01	12 30·3	12 32·3	11 56·1	0·1	0·1	6·1	5·1	12·1	10·2
02	12 30·5	12 32·6	11 56·3	0·2	0·2	6·2	5·2	12·2	10·3
03	12 30·8	12 32·8	11 56·5	0·3	0·3	6·3	5·3	12·3	10·4
04	12 31·0	12 33·1	11 56·8	0·4	0·3	6·4	5·4	12·4	10·4
05	12 31·3	12 33·3	11 57·0	0·5	0·4	6·5	5·5	12·5	10·5
06	12 31·5	12 33·6	11 57·3	0·6	0·5	6·6	5·6	12·6	10·6
07	12 31·8	12 33·8	11 57·5	0·7	0·6	6·7	5·6	12·7	10·7
08	12 32·0	12 34·1	11 57·7	0·8	0·7	6·8	5·7	12·8	10·8
09	12 32·3	12 34·3	11 58·0	0·9	0·8	6·9	5·8	12·9	10·9
10	12 32·5	12 34·6	11 58·2	1·0	0·8	7·0	5·9	13·0	10·9
11	12 32·8	12 34·8	11 58·5	1·1	0·9	7·1	6·0	13·1	11·0
12	12 33·0	12 35·1	11 58·7	1·2	1·0	7·2	6·1	13·2	11·1
13	12 33·3	12 35·3	11 58·9	1·3	1·1	7·3	6·1	13·3	11·2
14	12 33·5	12 35·6	11 59·2	1·4	1·2	7·4	6·2	13·4	11·3
15	12 33·8	12 35·8	11 59·4	1·5	1·3	7·5	6·3	13·5	11·4
16	12 34·0	12 36·1	11 59·7	1·6	1·3	7·6	6·4	13·6	11·4
17	12 34·3	12 36·3	11 59·9	1·7	1·4	7·7	6·5	13·7	11·5
18	12 34·5	12 36·6	12 00·1	1·8	1·5	7·8	6·6	13·8	11·6
19	12 34·8	12 36·8	12 00·4	1·9	1·6	7·9	6·6	13·9	11·7
20	12 35·0	12 37·1	12 00·6	2·0	1·7	8·0	6·7	14·0	11·8
21	12 35·3	12 37·3	12 00·8	2·1	1·8	8·1	6·8	14·1	11·9
22	12 35·5	12 37·6	12 01·1	2·2	1·9	8·2	6·9	14·2	12·0
23	12 35·8	12 37·8	12 01·3	2·3	1·9	8·3	7·0	14·3	12·0
24	12 36·0	12 38·1	12 01·6	2·4	2·0	8·4	7·1	14·4	12·1
25	12 36·3	12 38·3	12 01·8	2·5	2·1	8·5	7·2	14·5	12·2
26	12 36·5	12 38·6	12 02·0	2·6	2·2	8·6	7·2	14·6	12·3
27	12 36·8	12 38·8	12 02·3	2·7	2·3	8·7	7·3	14·7	12·4
28	12 37·0	12 39·1	12 02·5	2·8	2·4	8·8	7·4	14·8	12·5
29	12 37·3	12 39·3	12 02·8	2·9	2·4	8·9	7·5	14·9	12·5
30	12 37·5	12 39·6	12 03·0	3·0	2·5	9·0	7·6	15·0	12·6
31	12 37·8	12 39·8	12 03·2	3·1	2·6	9·1	7·7	15·1	12·7
32	12 38·0	12 40·1	12 03·5	3·2	2·7	9·2	7·7	15·2	12·8
33	12 38·3	12 40·3	12 03·7	3·3	2·8	9·3	7·8	15·3	12·9
34	12 38·5	12 40·6	12 03·9	3·4	2·9	9·4	7·9	15·4	13·0
35	12 38·8	12 40·8	12 04·2	3·5	2·9	9·5	8·0	15·5	13·0
36	12 39·0	12 41·1	12 04·4	3·6	3·0	9·6	8·1	15·6	13·1
37	12 39·3	12 41·3	12 04·7	3·7	3·1	9·7	8·2	15·7	13·2
38	12 39·5	12 41·6	12 04·9	3·8	3·2	9·8	8·2	15·8	13·3
39	12 39·8	12 41·8	12 05·1	3·9	3·3	9·9	8·3	15·9	13·4
40	12 40·0	12 42·1	12 05·4	4·0	3·4	10·0	8·4	16·0	13·5
41	12 40·3	12 42·3	12 05·6	4·1	3·5	10·1	8·5	16·1	13·6
42	12 40·5	12 42·6	12 05·9	4·2	3·5	10·2	8·6	16·2	13·6
43	12 40·8	12 42·8	12 06·1	4·3	3·6	10·3	8·7	16·3	13·7
44	12 41·0	12 43·1	12 06·3	4·4	3·7	10·4	8·8	16·4	13·8
45	12 41·3	12 43·3	12 06·6	4·5	3·8	10·5	8·8	16·5	13·9
46	12 41·5	12 43·6	12 06·8	4·6	3·9	10·6	8·9	16·6	14·0
47	12 41·8	12 43·8	12 07·0	4·7	4·0	10·7	9·0	16·7	14·1
48	12 42·0	12 44·1	12 07·3	4·8	4·0	10·8	9·1	16·8	14·1
49	12 42·3	12 44·3	12 07·5	4·9	4·1	10·9	9·2	16·9	14·2
50	12 42·5	12 44·6	12 07·8	5·0	4·2	11·0	9·3	17·0	14·3
51	12 42·8	12 44·8	12 08·0	5·1	4·3	11·1	9·3	17·1	14·4
52	12 43·0	12 45·1	12 08·2	5·2	4·4	11·2	9·4	17·2	14·5
53	12 43·3	12 45·3	12 08·5	5·3	4·5	11·3	9·5	17·3	14·6
54	12 43·5	12 45·6	12 08·7	5·4	4·5	11·4	9·6	17·4	14·6
55	12 43·8	12 45·8	12 09·0	5·5	4·6	11·5	9·7	17·5	14·7
56	12 44·0	12 46·1	12 09·2	5·6	4·7	11·6	9·8	17·6	14·8
57	12 44·3	12 46·3	12 09·4	5·7	4·8	11·7	9·8	17·7	14·9
58	12 44·5	12 46·6	12 09·7	5·8	4·9	11·8	9·9	17·8	15·0
59	12 44·8	12 46·8	12 09·9	5·9	5·0	11·9	10·0	17·9	15·1
60	12 45·0	12 47·1	12 10·2	6·0	5·1	12·0	10·1	18·0	15·2

51ᵐ

s	SUN PLANETS	ARIES	MOON	v or d	Corrⁿ	v or d	Corrⁿ	v or d	Corrⁿ
00	12 45·0	12 47·1	12 10·2	0·0	0·0	6·0	5·2	12·0	10·3
01	12 45·3	12 47·3	12 10·4	0·1	0·1	6·1	5·2	12·1	10·4
02	12 45·5	12 47·6	12 10·6	0·2	0·2	6·2	5·3	12·2	10·5
03	12 45·8	12 47·8	12 10·9	0·3	0·3	6·3	5·4	12·3	10·6
04	12 46·0	12 48·1	12 11·1	0·4	0·3	6·4	5·5	12·4	10·6
05	12 46·3	12 48·3	12 11·3	0·5	0·4	6·5	5·6	12·5	10·7
06	12 46·5	12 48·6	12 11·6	0·6	0·5	6·6	5·7	12·6	10·8
07	12 46·8	12 48·8	12 11·8	0·7	0·6	6·7	5·8	12·7	10·9
08	12 47·0	12 49·1	12 12·1	0·8	0·7	6·8	5·8	12·8	11·0
09	12 47·3	12 49·4	12 12·3	0·9	0·8	6·9	5·9	12·9	11·1
10	12 47·5	12 49·6	12 12·5	1·0	0·9	7·0	6·0	13·0	11·2
11	12 47·8	12 49·9	12 12·8	1·1	0·9	7·1	6·1	13·1	11·2
12	12 48·0	12 50·1	12 13·0	1·2	1·0	7·2	6·2	13·2	11·3
13	12 48·3	12 50·4	12 13·3	1·3	1·1	7·3	6·3	13·3	11·4
14	12 48·5	12 50·6	12 13·5	1·4	1·2	7·4	6·4	13·4	11·5
15	12 48·8	12 50·9	12 13·7	1·5	1·3	7·5	6·4	13·5	11·6
16	12 49·0	12 51·1	12 14·0	1·6	1·4	7·6	6·5	13·6	11·7
17	12 49·3	12 51·4	12 14·2	1·7	1·5	7·7	6·6	13·7	11·8
18	12 49·5	12 51·6	12 14·4	1·8	1·5	7·8	6·7	13·8	11·8
19	12 49·8	12 51·9	12 14·7	1·9	1·6	7·9	6·8	13·9	11·9
20	12 50·0	12 52·1	12 14·9	2·0	1·7	8·0	6·9	14·0	12·0
21	12 50·3	12 52·4	12 15·2	2·1	1·8	8·1	7·0	14·1	12·1
22	12 50·5	12 52·6	12 15·4	2·2	1·9	8·2	7·0	14·2	12·2
23	12 50·8	12 52·9	12 15·6	2·3	2·0	8·3	7·1	14·3	12·3
24	12 51·0	12 53·1	12 15·9	2·4	2·1	8·4	7·2	14·4	12·4
25	12 51·3	12 53·4	12 16·1	2·5	2·1	8·5	7·3	14·5	12·4
26	12 51·5	12 53·6	12 16·4	2·6	2·2	8·6	7·4	14·6	12·5
27	12 51·8	12 53·9	12 16·6	2·7	2·3	8·7	7·5	14·7	12·6
28	12 52·0	12 54·1	12 16·8	2·8	2·4	8·8	7·6	14·8	12·7
29	12 52·3	12 54·4	12 17·1	2·9	2·5	8·9	7·6	14·9	12·8
30	12 52·5	12 54·6	12 17·3	3·0	2·6	9·0	7·7	15·0	12·9
31	12 52·8	12 54·9	12 17·5	3·1	2·7	9·1	7·8	15·1	13·0
32	12 53·0	12 55·1	12 17·8	3·2	2·7	9·2	7·9	15·2	13·0
33	12 53·3	12 55·4	12 18·0	3·3	2·8	9·3	8·0	15·3	13·1
34	12 53·5	12 55·6	12 18·3	3·4	2·9	9·4	8·1	15·4	13·2
35	12 53·8	12 55·9	12 18·5	3·5	3·0	9·5	8·2	15·5	13·3
36	12 54·0	12 56·1	12 18·7	3·6	3·1	9·6	8·2	15·6	13·4
37	12 54·3	12 56·4	12 19·0	3·7	3·2	9·7	8·3	15·7	13·5
38	12 54·5	12 56·6	12 19·2	3·8	3·3	9·8	8·4	15·8	13·6
39	12 54·8	12 56·9	12 19·5	3·9	3·3	9·9	8·5	15·9	13·6
40	12 55·0	12 57·1	12 19·7	4·0	3·4	10·0	8·6	16·0	13·7
41	12 55·3	12 57·4	12 19·9	4·1	3·5	10·1	8·7	16·1	13·8
42	12 55·5	12 57·6	12 20·2	4·2	3·6	10·2	8·8	16·2	13·9
43	12 55·8	12 57·9	12 20·4	4·3	3·7	10·3	8·8	16·3	14·0
44	12 56·0	12 58·1	12 20·6	4·4	3·8	10·4	8·9	16·4	14·1
45	12 56·3	12 58·4	12 20·9	4·5	3·9	10·5	9·0	16·5	14·2
46	12 56·5	12 58·6	12 21·1	4·6	3·9	10·6	9·1	16·6	14·2
47	12 56·8	12 58·9	12 21·4	4·7	4·0	10·7	9·2	16·7	14·3
48	12 57·0	12 59·1	12 21·6	4·8	4·1	10·8	9·3	16·8	14·4
49	12 57·3	12 59·4	12 21·8	4·9	4·2	10·9	9·4	16·9	14·5
50	12 57·5	12 59·6	12 22·1	5·0	4·3	11·0	9·4	17·0	14·6
51	12 57·8	12 59·9	12 22·3	5·1	4·4	11·1	9·5	17·1	14·7
52	12 58·0	13 00·1	12 22·6	5·2	4·5	11·2	9·6	17·2	14·8
53	12 58·3	13 00·4	12 22·8	5·3	4·5	11·3	9·7	17·3	14·8
54	12 58·5	13 00·6	12 23·0	5·4	4·6	11·4	9·8	17·4	14·9
55	12 58·8	13 00·9	12 23·3	5·5	4·7	11·5	9·9	17·5	15·0
56	12 59·0	13 01·1	12 23·5	5·6	4·8	11·6	10·0	17·6	15·1
57	12 59·3	13 01·4	12 23·8	5·7	4·9	11·7	10·0	17·7	15·2
58	12 59·5	13 01·6	12 24·0	5·8	5·0	11·8	10·1	17·8	15·3
59	12 59·8	13 01·9	12 24·2	5·9	5·1	11·9	10·2	17·9	15·4
60	13 00·0	13 02·1	12 24·5	6·0	5·2	12·0	10·3	18·0	15·5

TABLE 6 — ALTITUDE CORRECTION TABLES

a. Dip of the Horizon

Ht. of Eye (m)	Corr^n D	Ht. of Eye (ft.)	Ht. of Eye (m)	Corr^n D	Ht. of Eye (ft.)
0.00	−0.3	0.0	13.0	−6.4	42.8
0.03	−0.4	0.1	13.4	−6.5	44.2
0.06	−0.5	0.2	13.8	−6.6	45.5
0.09	−0.6	0.3	14.2	−6.7	47.0
0.13	−0.7	0.4	14.7	−6.8	48.4
0.18	−0.8	0.5	15.1	−6.9	49.8
0.23	−0.9	0.7	15.5	−7.0	51.3
0.29	−1.0	0.9	16.0	−7.1	52.8
0.35	−1.1	1.1	16.5	−7.2	54.3
0.42	−1.2	1.4	16.9	−7.3	55.8
0.45	−1.3	1.6	17.4	−7.4	57.4
0.5	−1.4	1.9	17.9	−7.5	58.9
0.6	−1.5	2.2	18.4	−7.6	60.5
0.7	−1.6	2.5	18.8	−7.7	62.1
0.8	−1.7	2.8	19.3	−7.8	63.8
0.9	−1.8	3.2	19.8	−7.9	65.4
1.1	−1.9	3.6	20.4	−8.0	67.1
1.2	−2.0	4.0	20.9	−8.1	68.8
1.3	−2.1	4.4	21.4	−8.2	70.5
1.4	−2.2	4.9	21.9	−8.3	72.3
1.6	−2.3	5.3	22.5	−8.4	74.1
1.7	−2.4	5.8	23.0	−8.5	75.8
1.9	−2.5	6.3	23.5	−8.6	77.6
2.0	−2.6	6.9	24.1	−8.7	79.5
2.2	−2.7	7.4	24.7	−8.8	81.3
2.4	−2.8	8.0	25.2	−8.9	83.2
2.6	−2.9	8.6	25.8	−9.0	85.1
2.8	−3.0	9.2	26.4	−9.1	87.0
3.0	−3.1	9.8	27.0	−9.2	88.9
3.2	−3.2	10.5	27.6	−9.3	90.9
3.4	−3.3	11.2	28.2	−9.4	92.9
3.6	−3.4	11.9	28.8	−9.5	94.9
3.8	−3.5	12.6	29.4	−9.6	96.9
4.0	−3.6	13.3	30.0	−9.7	98.9
4.3	−3.7	14.1	30.6	−9.8	101.0
4.5	−3.8	14.9	31.3	−9.9	103.1
4.7	−3.9	15.7	31.9	−10.0	105.2
5.0	−4.0	16.5	32.6	−10.1	107.3
5.2	−4.1	17.4	33.2	−10.2	109.4
5.5	−4.2	18.3	33.9	−10.3	111.6
5.8	−4.3	19.1	34.5	−10.4	113.8
6.1	−4.4	20.1	35.2	−10.5	116.0
6.3	−4.5	21.0	35.9	−10.6	118.2
6.6	−4.6	22.0	36.6	−10.7	120.5
6.9	−4.7	22.9	37.3	−10.8	122.8
7.2	−4.8	23.9	38.0	−10.9	125.1
7.5	−4.9	25.0	38.7	−11.0	127.4
7.9	−5.0	26.0	39.4	−11.1	129.7
8.2	−5.1	27.1	40.1	−11.2	132.1
8.5	−5.2	28.1	40.8	−11.3	134.5
8.8	−5.3	29.2	41.5	−11.4	136.9
9.2	−5.4	30.4	42.3	−11.5	139.3
9.5	−5.5	31.5	43.0	−11.6	141.7
9.9	−5.6	32.7	43.8	−11.7	144.2
10.3	−5.7	33.9	44.5	−11.8	146.7
10.6	−5.8	35.1	45.3	−11.9	149.2
11.0	−5.9	36.3	46.1	−12.0	151.7
11.4	−6.0	37.6	46.8	−12.1	154.3
11.8	−6.1	38.9	47.6	−12.2	156.8
12.2	−6.2	40.1	48.4	−12.3	159.4
12.6	−6.3	41.5	49.2	−12.4	162.1
13.0		42.8	50.0		164.7

b. Refraction for Stars & Planets

App. Alt. H_a	Corr^n R	App. Alt. H_a	Corr^n R	App. Alt. H_a	Corr^n R
0 00	−33.8	3 30	−12.9	9 55	−5.3
0 03	33.2	3 35	12.7	10 07	−5.2
0 06	32.6	3 40	12.5	10 20	−5.1
0 09	32.0	3 45	12.3	10 32	−5.0
0 12	31.5	3 50	12.1	10 46	−4.9
0 15	30.9	3 55	11.9	10 59	−4.8
0 18	−30.4	4 00	−11.7	11 14	−4.7
0 21	29.8	4 05	11.5	11 29	−4.6
0 24	29.3	4 10	11.4	11 44	−4.5
0 27	28.8	4 15	11.2	12 00	−4.4
0 30	28.3	4 20	11.0	12 17	−4.3
0 33	27.9	4 25	10.9	12 35	−4.2
0 36	−27.4	4 30	−10.7	12 53	−4.1
0 39	26.9	4 35	10.6	13 12	−4.0
0 42	26.5	4 40	10.4	13 32	−3.9
0 45	26.1	4 45	10.3	13 53	−3.8
0 48	25.7	4 50	10.1	14 16	−3.7
0 51	25.3	4 55	10.0	14 39	−3.6
0 54	−24.9	5 00	−9.8	15 03	−3.5
0 57	24.5	5 05	9.7	15 29	−3.4
1 00	24.1	5 10	9.6	15 56	−3.3
1 03	23.7	5 15	9.5	16 25	−3.2
1 06	23.4	5 20	9.3	16 55	−3.1
1 09	23.0	5 25	9.2	17 27	−3.0
1 12	−22.7	5 30	−9.1	18 01	−2.9
1 15	22.3	5 35	9.0	18 37	−2.8
1 18	22.0	5 40	8.9	19 16	−2.7
1 21	21.7	5 45	8.8	19 56	−2.6
1 24	21.4	5 50	8.7	20 40	−2.5
1 27	21.1	5 55	8.6	21 27	−2.4
1 30	−20.8	6 00	−8.5	22 17	−2.3
1 35	20.3	6 10	8.3	23 11	−2.2
1 40	19.9	6 20	8.1	24 09	−2.1
1 45	19.4	6 30	7.9	25 12	−2.0
1 50	19.0	6 40	7.7	26 20	−1.9
1 55	18.6	6 50	7.6	27 34	−1.8
2 00	−18.2	7 00	−7.4	28 54	−1.7
2 05	17.8	7 10	7.2	30 22	−1.6
2 10	17.4	7 20	7.1	31 58	−1.5
2 15	17.1	7 30	6.9	33 43	−1.4
2 20	16.7	7 40	6.8	35 38	−1.3
2 25	16.4	7 50	6.7	37 45	−1.2
2 30	−16.1	8 00	−6.6	40 06	−1.1
2 35	15.8	8 10	6.4	42 42	−1.0
2 40	15.4	8 20	6.3	45 34	−0.9
2 45	15.2	8 30	6.2	48 45	−0.8
2 50	14.9	8 40	6.1	52 16	−0.7
2 55	14.6	8 50	6.0	56 09	−0.6
3 00	−14.3	9 00	−5.9	60 26	−0.5
3 05	14.1	9 10	5.8	65 06	−0.4
3 10	13.8	9 20	5.7	70 09	−0.3
3 15	13.6	9 30	5.6	75 32	−0.2
3 20	13.4	9 40	5.5	81 12	−0.1
3 25	13.1	9 50	5.4	87 03	0.0
3 30	−12.9	10 00	−5.3	90 00	

H_a = App. Alt. = Apparent Altitude
 = Sextant altitude corrected for index error (IE) & dip (
 = H_s + IE + D

In critical cases ascend.

TABLE 6 — ALTITUDE CORRECTION TABLES

c. Additional Refraction for Non-standard Conditions

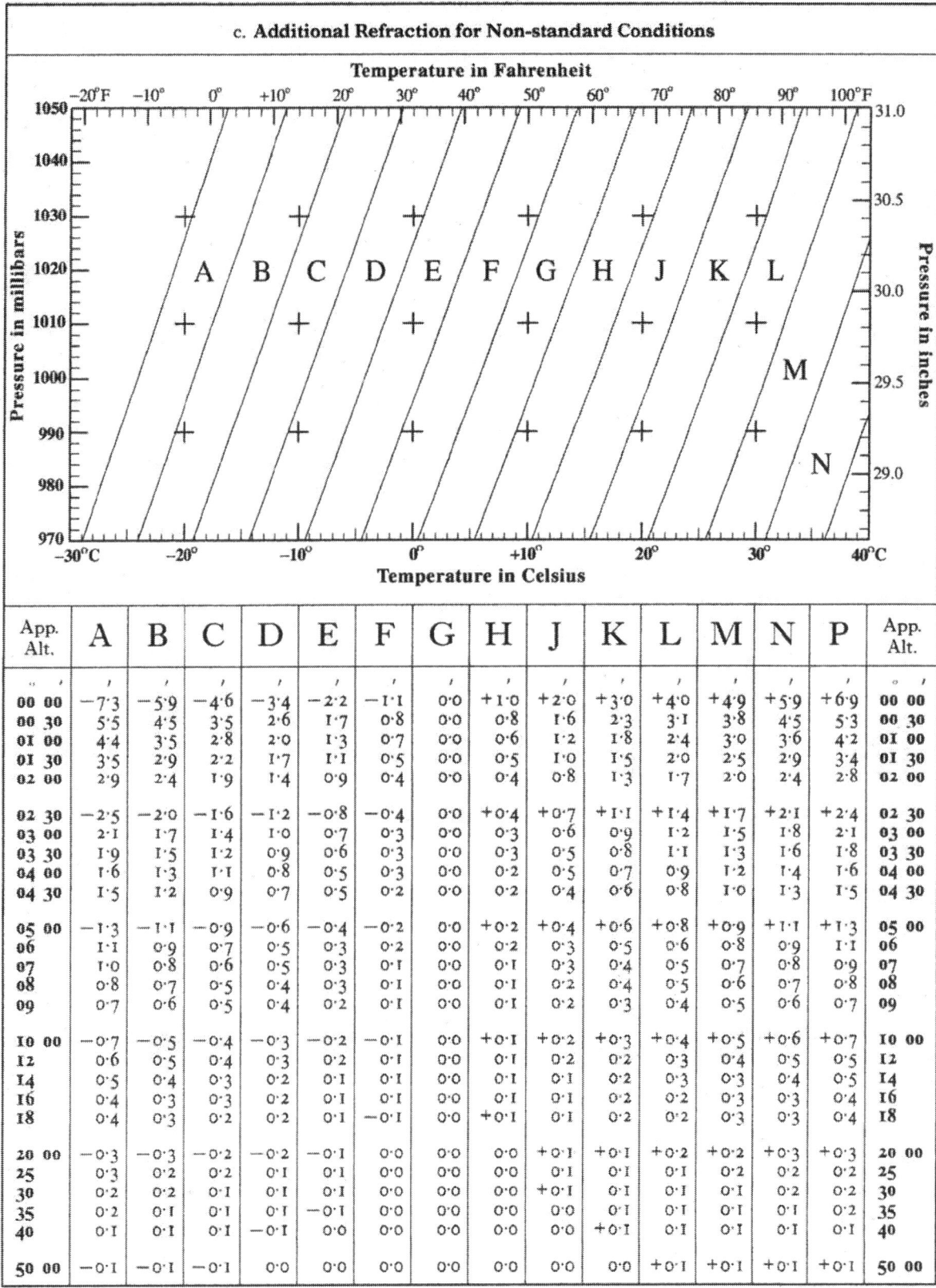

App. Alt.	A	B	C	D	E	F	G	H	J	K	L	M	N	P	App. Alt.
° ′	′	′	′	′	′	′	′	′	′	′	′	′	′	′	° ′
00 00	−7·3	−5·9	−4·6	−3·4	−2·2	−1·1	0·0	+1·0	+2·0	+3·0	+4·0	+4·9	+5·9	+6·9	00 00
00 30	5·5	4·5	3·5	2·6	1·7	0·8	0·0	0·8	1·6	2·3	3·1	3·8	4·5	5·3	00 30
01 00	4·4	3·5	2·8	2·0	1·3	0·7	0·0	0·6	1·2	1·8	2·4	3·0	3·6	4·2	01 00
01 30	3·5	2·9	2·2	1·7	1·1	0·5	0·0	0·5	1·0	1·5	2·0	2·5	2·9	3·4	01 30
02 00	2·9	2·4	1·9	1·4	0·9	0·4	0·0	0·4	0·8	1·3	1·7	2·0	2·4	2·8	02 00
02 30	−2·5	−2·0	−1·6	−1·2	−0·8	−0·4	0·0	+0·4	+0·7	+1·1	+1·4	+1·7	+2·1	+2·4	02 30
03 00	2·1	1·7	1·4	1·0	0·7	0·3	0·0	0·3	0·6	0·9	1·2	1·5	1·8	2·1	03 00
03 30	1·9	1·5	1·2	0·9	0·6	0·3	0·0	0·3	0·5	0·8	1·1	1·3	1·6	1·8	03 30
04 00	1·6	1·3	1·1	0·8	0·5	0·3	0·0	0·2	0·5	0·7	0·9	1·2	1·4	1·6	04 00
04 30	1·5	1·2	0·9	0·7	0·5	0·2	0·0	0·2	0·4	0·6	0·8	1·0	1·3	1·5	04 30
05 00	−1·3	−1·1	−0·9	−0·6	−0·4	−0·2	0·0	+0·2	+0·4	+0·6	+0·8	+0·9	+1·1	+1·3	05 00
06	1·1	0·9	0·7	0·5	0·3	0·2	0·0	0·2	0·3	0·5	0·6	0·8	0·9	1·1	06
07	1·0	0·8	0·6	0·5	0·3	0·1	0·0	0·1	0·3	0·4	0·5	0·7	0·8	0·9	07
08	0·8	0·7	0·5	0·4	0·3	0·1	0·0	0·1	0·2	0·4	0·5	0·6	0·7	0·8	08
09	0·7	0·6	0·5	0·4	0·2	0·1	0·0	0·1	0·2	0·3	0·4	0·5	0·6	0·7	09
10 00	−0·7	−0·5	−0·4	−0·3	−0·2	−0·1	0·0	+0·1	+0·2	+0·3	+0·4	+0·5	+0·6	+0·7	10 00
12	0·6	0·5	0·4	0·3	0·2	0·1	0·0	0·1	0·2	0·2	0·3	0·4	0·5	0·5	12
14	0·5	0·4	0·3	0·2	0·1	0·1	0·0	0·1	0·1	0·2	0·3	0·3	0·4	0·5	14
16	0·4	0·3	0·3	0·2	0·1	0·1	0·0	0·1	0·1	0·2	0·2	0·3	0·3	0·4	16
18	0·4	0·3	0·2	0·2	0·1	−0·1	0·0	+0·1	0·1	0·2	0·2	0·3	0·3	0·4	18
20 00	−0·3	−0·3	−0·2	−0·2	−0·1	0·0	0·0	0·0	+0·1	+0·1	+0·2	+0·2	+0·3	+0·3	20 00
25	0·3	0·2	0·2	0·1	0·1	0·0	0·0	0·0	0·1	0·1	0·1	0·2	0·2	0·2	25
30	0·2	0·2	0·1	0·1	0·1	0·0	0·0	0·0	+0·1	0·1	0·1	0·1	0·2	0·2	30
35	0·2	0·1	0·1	0·1	−0·1	0·0	0·0	0·0	0·0	0·1	0·1	0·1	0·1	0·2	35
40	0·1	0·1	0·1	−0·1	0·0	0·0	0·0	0·0	+0·1	0·1	0·1	0·1	0·1	0·1	40
50 00	−0·1	−0·1	−0·1	0·0	0·0	0·0	0·0	0·0	0·0	0·0	+0·1	+0·1	+0·1	+0·1	50 00

The graph is entered with arguments temperature and pressure to find a zone letter; using as arguments this zone letter and apparent altitude (sextant altitude corrected for index error and dip), a correction is taken from the table. This correction is to be applied to the sextant altitude in addition to the corrections for standard conditions (Table **6**b).

©2009 Jack Case

ALTITUDE CORRECTION TABLES 0°–35° — MOON

App. Alt.	0°–4° Corrⁿ	5°–9° Corrⁿ	10°–14° Corrⁿ	15°–19° Corrⁿ	20°–24° Corrⁿ	25°–29° Corrⁿ	30°–34° Corrⁿ	App. Alt.
00	0° 34.5	5° 58.2	10° 62.1	15° 62.8	20° 62.2	25° 60.8	30° 58.9	00
10	36.5	58.5	62.2	62.8	62.2	60.8	58.8	10
20	38.3	58.7	62.2	62.8	62.1	60.7	58.8	20
30	40.0	58.9	62.3	62.8	62.1	60.7	58.7	30
40	41.5	59.1	62.3	62.8	62.0	60.6	58.6	40
50	42.9	59.3	62.4	62.7	62.0	60.6	58.5	50
00	1° 44.2	6° 59.5	11° 62.4	16° 62.7	21° 62.0	26° 60.5	31° 58.5	00
10	45.4	59.7	62.4	62.7	61.9	60.4	58.4	10
20	46.5	59.9	62.5	62.7	61.9	60.4	58.3	20
30	47.5	60.0	62.5	62.7	61.9	60.3	58.2	30
40	48.4	60.2	62.5	62.7	61.8	60.3	58.2	40
50	49.3	60.3	62.6	62.7	61.8	60.2	58.1	50
00	2° 50.1	7° 60.5	12° 62.6	17° 62.7	22° 61.7	27° 60.1	32° 58.0	00
10	50.8	60.6	62.6	62.6	61.7	60.1	57.9	10
20	51.5	60.7	62.6	62.6	61.6	60.0	57.8	20
30	52.2	60.9	62.7	62.6	61.6	59.9	57.8	30
40	52.8	61.0	62.7	62.6	61.6	59.9	57.7	40
50	53.4	61.1	62.7	62.6	61.5	59.8	57.6	50
00	3° 53.9	8° 61.2	13° 62.7	18° 62.5	23° 61.5	28° 59.7	33° 57.5	00
10	54.4	61.3	62.7	62.5	61.4	59.7	57.4	10
20	54.9	61.4	62.7	62.5	61.4	59.6	57.4	20
30	55.3	61.5	62.8	62.5	61.3	59.5	57.3	30
40	55.7	61.6	62.8	62.4	61.3	59.5	57.2	40
50	56.1	61.6	62.8	62.4	61.2	59.4	57.1	50
00	4° 56.4	9° 61.7	14° 62.8	19° 62.4	24° 61.2	29° 59.3	34° 57.0	00
10	56.8	61.8	62.8	62.4	61.1	59.3	56.9	10
20	57.1	61.9	62.8	62.3	61.1	59.2	56.9	20
30	57.4	61.9	62.8	62.3	61.0	59.1	56.8	30
40	57.7	62.0	62.8	62.3	61.0	59.1	56.7	40
50	58.0	62.1	62.8	62.2	60.9	59.0	56.6	50

HP	L U	L U	L U	L U	L U	L U	L U	HP
54.0	0.3 0.9	0.3 0.9	0.4 1.0	0.5 1.1	0.6 1.2	0.7 1.3	0.9 1.5	54.0
54.3	0.7 1.1	0.7 1.2	0.8 1.2	0.8 1.3	0.9 1.4	1.1 1.5	1.2 1.7	54.3
54.6	1.1 1.4	1.1 1.4	1.1 1.4	1.2 1.5	1.3 1.6	1.4 1.7	1.5 1.8	54.6
54.9	1.4 1.6	1.5 1.6	1.5 1.6	1.6 1.7	1.6 1.8	1.8 1.9	1.9 2.0	54.9
55.2	1.8 1.8	1.8 1.8	1.9 1.8	1.9 1.9	2.0 2.0	2.1 2.1	2.2 2.2	55.2
55.5	2.2 2.0	2.2 2.0	2.3 2.1	2.3 2.1	2.4 2.2	2.4 2.3	2.5 2.4	55.5
55.8	2.6 2.2	2.6 2.2	2.6 2.3	2.7 2.3	2.7 2.4	2.8 2.4	2.9 2.5	55.8
56.1	3.0 2.4	3.0 2.5	3.0 2.5	3.0 2.5	3.1 2.6	3.1 2.6	3.2 2.7	56.1
56.4	3.3 2.7	3.4 2.7	3.4 2.7	3.4 2.7	3.4 2.8	3.5 2.8	3.5 2.9	56.4
56.7	3.7 2.9	3.7 2.9	3.8 2.9	3.8 2.9	3.8 3.0	3.8 3.0	3.9 3.0	56.7
57.0	4.1 3.1	4.1 3.1	4.1 3.1	4.1 3.1	4.2 3.2	4.2 3.2	4.2 3.2	57.0
57.3	4.5 3.3	4.5 3.3	4.5 3.3	4.5 3.3	4.5 3.3	4.5 3.4	4.6 3.4	57.3
57.6	4.9 3.5	4.9 3.5	4.9 3.5	4.9 3.5	4.9 3.5	4.9 3.5	4.9 3.6	57.6
57.9	5.3 3.8	5.3 3.8	5.2 3.8	5.2 3.7	5.2 3.7	5.2 3.7	5.2 3.7	57.9
58.2	5.6 4.0	5.6 4.0	5.6 4.0	5.6 4.0	5.6 3.9	5.6 3.9	5.6 3.9	58.2
58.5	6.0 4.2	6.0 4.2	6.0 4.2	6.0 4.2	6.0 4.1	5.9 4.1	5.9 4.1	58.5
58.8	6.4 4.4	6.4 4.4	6.4 4.4	6.3 4.4	6.3 4.3	6.3 4.3	6.2 4.2	58.8
59.1	6.8 4.6	6.8 4.6	6.7 4.6	6.7 4.6	6.7 4.5	6.6 4.5	6.6 4.4	59.1
59.4	7.2 4.8	7.1 4.8	7.1 4.8	7.1 4.8	7.0 4.7	7.0 4.7	6.9 4.6	59.4
59.7	7.5 5.1	7.5 5.0	7.5 5.0	7.5 5.0	7.4 4.9	7.3 4.8	7.2 4.8	59.7
60.0	7.9 5.3	7.9 5.3	7.9 5.2	7.8 5.2	7.8 5.1	7.7 5.0	7.6 4.9	60.0
60.3	8.3 5.5	8.3 5.5	8.2 5.4	8.2 5.4	8.1 5.3	8.0 5.2	7.9 5.1	60.3
60.6	8.7 5.7	8.7 5.7	8.6 5.7	8.6 5.6	8.5 5.5	8.4 5.4	8.2 5.3	60.6
60.9	9.1 5.9	9.0 5.9	9.0 5.9	8.9 5.8	8.8 5.7	8.7 5.6	8.6 5.4	60.9
61.2	9.5 6.2	9.4 6.1	9.4 6.1	9.3 6.0	9.2 5.9	9.1 5.8	8.9 5.6	61.2
61.5	9.8 6.4	9.8 6.3	9.7 6.3	9.7 6.2	9.5 6.1	9.4 5.9	9.2 5.8	61.5

DIP

Ht. of Eye	Corrⁿ	Ht. of Eye	Ht. of Eye	Corrⁿ	Ht. of Eye
m		ft.	m		ft.
2.4	−2.8	8.0	9.5	−5.5	31.5
2.6	−2.9	8.6	9.9	−5.6	32.7
2.8	−3.0	9.2	10.3	−5.7	33.9
3.0	−3.1	9.8	10.6	−5.8	35.1
3.2	−3.2	10.5	11.0	−5.9	36.3
3.4	−3.3	11.2	11.4	−6.0	37.6
3.6	−3.4	11.9	11.8	−6.1	38.9
3.8	−3.5	12.6	12.2	−6.2	40.1
4.0	−3.6	13.3	12.6	−6.3	41.5
4.3	−3.7	14.1	13.0	−6.4	42.8
4.5	−3.8	14.9	13.4	−6.5	44.2
4.7	−3.9	15.7	13.8	−6.6	45.5
5.0	−4.0	16.5	14.2	−6.7	46.9
5.2	−4.1	17.4	14.7	−6.8	48.4
5.5	−4.2	18.3	15.1	−6.9	49.8
5.8	−4.3	19.1	15.5	−7.0	51.3
6.1	−4.4	20.1	16.0	−7.1	52.8
6.3	−4.5	21.0	16.5	−7.2	54.3
6.6	−4.6	22.0	16.9	−7.3	55.8
6.9	−4.7	22.9	17.4	−7.4	57.4
7.2	−4.8	23.9	17.9	−7.5	58.9
7.5	−4.9	24.9	18.4	−7.6	60.5
7.9	−5.0	26.0	18.8	−7.7	62.1
8.2	−5.1	27.1	19.3	−7.8	63.8
8.5	−5.2	28.1	19.8	−7.9	65.4
8.8	−5.3	29.2	20.4	−8.0	67.1
9.2	−5.4	30.4	20.9	−8.1	68.8
9.5		31.5	21.4		70.5

MOON CORRECTION TABLE

The correction is in two parts; the first correction is taken from the upper part of the table with argument apparent altitude, and the second from the lower part, with argument HP, in the same column as that from which the first correction was taken. Separate corrections are given in the lower part for lower (L) and upper (U) limbs. All corrections are to be **added** to apparent altitude, *but 30′ is to be subtracted from the altitude of the upper limb.*

For corrections for pressure and temperature see page A4.

For bubble sextant observations ignore dip, take the mean of upper and lower limb corrections and subtract 15′ from the altitude.

App. Alt. = Apparent altitude = Sextant altitude corrected for index error and dip.

Extract of part of Main Table for Lat. 48°, Same Name (Dec 22° - 29°)

22°			23°			24°			25°			26°			27°			28°			29°			LHA
Hc	d	Z	Hc	d	Z	Hc	d	Z	Hc	d	Z	Hc	d	Z	Hc	d	Z	Hc	d	Z	Hc	d	Z	
64 00	+60	180	65 00	+60	180	66 00	+60	180	67 00	+60	180	68 00	+60	180	69 00	+60	180	70 00	+60	180	71 00	+60	180	360
63 59	60	178	64 59	60	178	65 59	60	178	66 59	60	178	67 59	60	178	68 59	60	178	69 59	60	177	70 59	60	177	359
63 57	60	176	64 57	60	176	65 57	60	176	66 57	60	175	67 57	60	175	68 57	59	175	69 56	60	175	70 56	60	175	358
63 53	60	174	64 53	60	173	65 53	60	173	66 53	59	173	67 52	60	173	68 52	60	173	69 52	60	172	70 52	59	172	357
63 48	60	172	64 48	59	171	65 47	60	171	66 47	60	171	67 47	59	170	68 46	60	170	69 46	59	170	70 45	59	169	356
63 42	+59	169	64 41	+59	169	65 40	+60	169	66 40	+59	168	67 39	+59	168	68 38	+60	168	69 38	+59	167	70 37	+59	167	355
63 34	59	167	64 33	59	167	65 32	59	167	66 31	59	166	67 30	59	166	68 29	59	165	69 28	59	165	70 27	58	164	354
63 24	59	165	64 23	59	165	65 22	59	165	66 21	58	164	67 19	59	163	68 18	58	163	69 16	59	162	70 15	58	162	353
63 13	59	163	64 12	59	163	65 11	58	162	66 09	58	162	67 07	58	161	68 05	58	161	69 03	58	160	70 01	58	159	352
63 01	58	161	63 59	58	161	64 58	58	160	65 56	58	160	66 54	57	159	67 51	58	158	68 49	57	158	69 46	57	157	351
62 48	+58	159	63 46	+57	159	64 43	+58	158	65 41	+58	158	66 39	+57	157	67 36	+57	156	68 33	+57	155	69 30	+56	154	350
62 33	57	157	63 30	58	157	64 28	57	156	65 25	57	155	66 22	57	155	67 19	56	154	68 15	56	153	69 11	56	152	349
62 17	57	156	63 14	57	155	64 11	57	154	65 08	56	153	66 04	56	153	67 00	56	152	67 56	56	151	68 52	55	150	348
62 00	56	154	62 56	57	153	63 53	56	152	64 49	56	151	65 45	56	151	66 41	55	150	67 36	55	149	68 31	54	148	347
61 41	56	152	62 37	56	151	63 33	56	150	64 29	56	149	65 25	55	148	66 20	54	148	67 14	55	146	68 09	54	145	346
61 22	+55	150	62 17	+56	149	63 13	+55	148	64 08	+55	147	65 03	+55	147	65 58	+54	146	66 52	+53	144	67 45	+53	143	345
61 01	55	148	61 56	55	147	62 51	55	146	63 46	54	146	64 40	54	145	65 34	54	144	66 28	53	142	67 21	52	141	344
60 39	55	146	61 34	55	146	62 29	54	145	63 23	54	144	64 17	53	143	65 10	53	142	66 03	52	141	66 55	52	139	343
60 17	54	145	61 11	54	144	62 05	54	143	62 59	53	142	63 52	53	141	64 45	52	140	65 37	51	139	66 28	51	137	342
59 53	54	143	60 47	53	142	61 40	53	141	62 33	53	140	63 26	52	139	64 18	52	138	65 10	51	137	66 01	50	136	341
59 28	+54	141	60 22	+53	140	61 15	+52	139	62 07	+52	138	62 59	+52	137	63 51	+51	136	64 42	+50	135	65 32	+50	134	340
59 03	53	140	59 56	52	139	60 48	52	138	61 40	52	137	62 32	51	136	63 23	50	135	64 13	50	133	65 03	49	132	339
58 36	53	138	59 29	52	137	60 21	51	136	61 12	51	135	62 03	51	134	62 54	49	133	63 43	50	132	64 33	48	130	338
58 09	52	137	59 01	52	136	59 53	51	135	60 44	50	134	61 34	50	132	62 24	49	131	63 13	49	130	64 02	47	129	337
57 41	52	135	58 33	51	134	59 24	50	133	60 14	50	132	61 04	49	131	61 53	49	130	62 42	48	128	63 30	47	127	336
57 13	+51	134	58 04	+50	133	58 54	+50	132	59 44	+49	131	60 33	+49	129	61 22	+48	128	62 10	+48	127	62 58	+46	126	335
56 43	51	132	57 34	50	131	58 24	49	130	59 13	49	129	60 02	48	128	60 50	48	127	61 38	47	125	62 25	46	124	334
56 13	50	131	57 03	50	130	57 53	49	129	58 42	48	128	59 30	48	126	60 18	47	125	61 05	46	124	61 51	46	123	333
55 42	50	129	56 32	49	128	57 21	49	127	58 10	47	126	58 57	48	125	59 45	46	124	60 31	46	123	61 17	45	121	332
55 11	49	128	56 00	49	127	56 49	48	126	57 37	47	125	58 24	47	124	59 11	46	123	59 57	46	121	60 43	44	120	331
54 39	+49	127	55 28	+48	126	56 16	+48	125	57 04	+47	124	57 51	+46	122	58 37	+46	121	59 23	+44	120	60 07	+45	119	330
54 07	48	125	54 55	48	124	55 43	47	123	56 30	47	122	57 17	45	121	58 02	46	120	58 48	44	119	59 32	44	117	329
53 34	48	124	54 22	47	123	55 09	47	122	55 56	46	121	56 42	45	120	57 27	45	119	58 12	44	117	58 56	43	116	328
53 00	48	123	53 48	47	122	54 35	46	121	55 21	46	120	56 07	45	119	56 52	44	117	57 36	44	116	58 20	43	115	327
52 26	48	122	53 14	46	121	54 00	46	120	54 46	45	119	55 31	45	117	56 16	44	116	57 00	43	115	57 43	43	114	326
51 52	+47	121	52 39	+46	120	53 25	+46	118	54 11	+45	117	54 56	+44	116	55 40	+44	115	56 24	+42	114	57 06	+42	113	325
51 17	47	119	52 04	46	118	52 50	45	117	53 35	44	116	54 19	44	115	55 03	44	114	55 47	42	113	56 29	42	111	324
50 42	46	118	51 28	46	117	52 14	45	116	52 59	44	115	53 43	44	114	54 27	42	113	55 09	43	112	55 52	41	110	323
50 07	45	117	50 52	45	116	51 37	45	115	52 22	44	114	53 06	43	113	53 49	43	112	54 32	42	110	55 14	41	109	322
49 31	45	116	50 16	45	115	51 01	44	114	51 45	44	113	52 29	43	112	53 12	42	111	53 54	42	109	54 36	41	108	321
48 54	+46	115	49 40	+44	114	50 24	+44	113	51 08	+43	112	51 51	+43	111	52 34	+42	110	53 16	+42	108	53 58	+40	107	320
48 18	45	114	49 03	44	113	49 47	44	112	50 31	43	111	51 14	42	110	51 56	42	109	52 38	41	107	53 19	40	106	319
47 41	45	113	48 26	44	112	49 10	43	111	49 53	43	110	50 36	42	109	51 18	42	108	52 00	40	106	52 40	41	105	318
47 04	44	112	47 48	44	111	48 32	43	110	49 15	43	109	49 58	42	108	50 40	41	107	51 21	41	105	52 02	39	104	317
46 26	45	111	47 11	43	110	47 54	43	109	48 37	42	108	49 19	42	107	50 01	41	106	50 42	41	104	51 23	39	103	316
45 49	+44	110	46 33	+43	109	47 16	+43	108	47 59	+42	107	48 41	+41	106	49 22	+41	105	50 03	+40	103	50 43	+40	102	315
45 11	44	109	45 55	43	108	46 38	42	107	47 20	42	106	48 02	41	105	48 43	41	104	49 24	40	103	50 04	39	101	314
44 33	43	108	45 16	43	107	45 59	42	106	46 41	42	105	47 23	41	104	48 04	41	103	48 45	40	102	49 25	39	101	313
43 55	43	107	44 38	42	106	45 20	43	105	46 03	41	104	46 44	41	103	47 25	40	102	48 05	40	101	48 45	39	100	312
43 16	43	106	43 59	43	105	44 42	42	104	45 24	41	103	46 05	41	102	46 46	40	101	47 26	40	100	48 06	39	99	311
42 37	+43	105	43 20	+43	104	44 03	+41	103	44 44	+42	102	45 26	+40	101	46 06	+40	100	46 46	+40	99	47 26	+39	98	310
41 59	42	104	42 41	42	103	43 23	42	102	44 05	41	101	44 46	41	100	45 27	40	99	46 07	39	98	46 46	39	97	309
41 20	42	103	42 02	42	102	42 44	42	101	43 26	41	100	44 07	40	99	44 47	40	98	45 27	39	97	46 06	39	96	308
40 40	43	102	41 23	42	102	42 05	41	101	42 46	41	100	43 27	40	99	44 07	40	98	44 47	39	97	45 26	39	95	307
40 01	42	102	40 43	42	101	41 25	41	100	42 06	41	99	42 47	40	98	43 27	40	97	44 07	39	96	44 46	39	95	306
39 22	+42	101	40 04	+42	100	40 46	+41	99	41 27	+40	98	42 07	+41	97	42 48	+39	96	43 27	+39	95	44 06	+39	94	305
38 42	42	100	39 24	42	99	40 06	41	98	40 47	41	97	41 28	40	96	42 08	39	95	42 47	39	94	43 26	39	93	304
38 03	42	99	38 45	41	98	39 26	41	97	40 07	41	96	40 48	40	95	41 28	39	94	42 07	39	93	42 46	38	92	303
37 23	42	98	38 05	41	97	38 46	41	96	39 27	41	96	40 08	40	95	40 48	39	94	41 27	39	93	42 06	38	92	302
36 43	42	97	37 25	41	97	38 06	41	96	38 47	41	95	39 28	39	94	40 07	40	93	40 47	39	92	41 26	38	91	301
36 03	+42	97	36 45	+41	96	37 26	+41	95	38 07	+40	94	38 47	+40	93	39 27	+40	92	40 07	+39	91	40 46	+38	90	300
35 24	41	96	36 05	41	95	36 46	41	94	37 27	40	93	38 07	40	92	38 47	40	91	39 27	38	90	40 05	39	89	299
34 44	41	95	35 25	41	94	36 06	40	93	36 47	40	92	37 27	40	91	38 07	39	91	38 46	39	90	39 25	39	89	298
34 04	41	94	34 45	41	93	35 26	41	93	36 07	40	92	36 47	40	91	37 27	39	90	38 06	39	89	38 45	39	88	297
33 23	42	94	34 05	41	93	34 46	41	92	35 27	40	91	36 07	40	90	36 47	39	89	37 26	39	88	38 05	39	87	296
32 43	+42	93	33 25	+41	92	34 06	+41	91	34 47	+40	90	35 27	+40	89	36 07	+39	88	36 46	+39	87	37 25	+39	86	295
32 03	42	92	32 45	41	91	33 26	40	90	34 06	41	89	34 47	40	88	35 27	39	88	36 06	39	87	36 45	39	86	294
31 23	42	91	32 05	41	90	32 46	40	90	33 26	41	89	34 07	39	88	34 46	39	87	35 26	39	86	36 05	39	85	293
30 43	41	90	31 24	41	90	32 05	41	89	32 46	40	88	33 26	40	87	34 06	40	86	34 46	39	85	35 25	39	84	292
30 03	41	90	30 44	41	89	31 25	41	88	32 06	40	87	32 46	40	86	33 26	40	85	34 06	39	85	34 45	39	84	291
22°			23°			24°			25°			26°			27°			28°			29°			

©2009 Jack Case

Extract of part of Main Table for Lat. 48° Same Name Dec 5° - 7°)

	3°			4°			5°			6°			7°		
LHA	Hc	d	Z	Hc	d	Z	Hc	d	Z	Hc	d	Z	Hc	d	Z
	° ′	′	°	° ′	′	°	° ′	′	°	° ′	′	°	° ′	′	°
0	45 00	+60	180	46 00	+60	180	47 00	+60	180	48 00	+60	180	49 00	+60	180
1	45 00	59	179	45 59	60	179	46 59	60	179	47 59	60	179	48 59	60	178
2	44 58	60	177	45 58	60	177	46 58	60	177	47 58	60	177	48 58	60	177
3	44 56	59	176	45 55	60	176	46 55	60	176	47 55	60	176	48 55	60	175
4	44 52	60	174	45 52	60	174	46 52	60	174	47 52	60	174	48 52	59	174
5	44 48	+59	173	45 47	+60	173	46 47	+60	173	47 47	+60	173	48 47	+60	172
6	44 42	60	172	45 42	60	171	46 42	59	171	47 41	60	171	48 41	60	171
7	44 36	59	170	45 35	60	170	46 35	60	170	47 35	59	170	48 34	60	169
8	44 29	59	169	45 28	59	169	46 27	60	168	47 27	59	168	48 26	60	168
9	44 20	60	167	45 20	59	167	46 19	59	167	47 18	59	167	48 17	60	167
10	44 11	+59	166	45 10	+59	166	46 09	+59	166	47 08	+60	165	48 08	+59	165
11	44 01	59	165	45 00	59	164	45 59	59	164	46 58	59	164	47 57	59	164
12	43 50	59	163	44 49	58	163	45 47	59	163	46 46	59	162	47 45	59	162
13	43 38	58	162	44 36	59	162	45 35	59	161	46 34	58	161	47 32	59	161
14	43 25	58	161	44 23	59	160	45 22	58	160	46 20	58	160	47 18	59	159
15	43 11	+58	159	44 09	+58	159	45 07	+59	159	46 06	+58	158	47 04	+58	158
16	42 56	58	158	43 54	58	158	44 52	58	157	45 50	58	157	46 48	58	156
17	42 41	58	157	43 39	57	156	44 36	58	156	45 34	58	155	46 32	57	155
18	42 24	58	155	43 22	58	155	44 20	57	155	45 17	57	154	46 14	57	154
19	42 07	58	154	43 05	57	154	44 02	57	153	44 59	57	153	45 56	57	152
20	41 49	+57	153	42 46	+57	152	43 43	+57	152	44 40	+57	151	45 37	+57	151
21	41 30	57	151	42 27	57	151	43 24	57	151	44 21	56	150	45 17	56	150
22	41 11	56	150	42 07	57	150	43 04	56	149	44 00	56	149	44 56	56	148
23	40 51	56	149	41 47	56	148	42 43	56	148	43 39	56	148	44 35	56	147
24	40 29	56	148	41 25	56	147	42 21	56	147	43 17	56	146	44 13	55	146
25	40 08	+55	146	41 03	+56	146	41 59	+55	146	42 54	+56	145	43 50	+55	144
26	39 45	56	145	40 41	55	145	41 36	55	144	42 31	55	144	43 26	55	143
27	39 22	55	144	40 17	55	144	41 12	55	143	42 07	55	143	43 02	54	142
28	38 58	55	143	39 53	55	142	40 48	54	142	41 42	54	141	42 36	55	141
29	38 34	54	142	39 28	54	141	40 22	55	141	41 17	54	140	42 11	54	140
30	38 08	+55	141	39 03	+54	140	39 57	+54	139	40 51	+53	139	41 44	+54	138
31	37 43	54	139	38 37	53	139	39 30	54	138	40 24	53	138	41 17	54	137
32	37 16	54	138	38 10	53	138	39 03	54	137	39 57	53	137	40 50	53	136
33	36 49	54	137	37 43	53	137	38 36	53	136	39 29	52	135	40 21	53	135
34	36 22	53	136	37 15	53	136	38 08	52	135	39 00	53	134	39 53	52	134
35	35 54	+52	135	36 46	+53	134	37 39	+52	134	38 31	+52	133	39 23	+52	133
36	35 25	52	134	36 17	53	133	37 10	52	133	38 02	52	132	38 54	51	131
37	34 56	52	133	35 48	52	132	36 40	52	132	37 32	51	131	38 23	52	130
38	34 26	52	132	35 18	52	131	36 10	51	131	37 01	51	130	37 52	51	129
39	33 56	51	131	34 47	52	130	35 39	51	130	36 30	51	129	37 21	51	128
40	33 25	+52	130	34 17	+51	129	35 08	+51	128	35 59	+50	128	36 49	+51	127
41	32 54	51	129	33 45	51	128	34 36	51	127	35 27	50	127	36 17	50	126
42	32 23	50	128	33 13	51	127	34 04	50	126	34 54	50	126	35 44	50	125
43	31 51	50	127	32 41	50	126	33 31	51	125	34 22	49	125	35 11	50	124
44	31 18	50	126	32 08	50	125	32 58	50	124	33 48	50	124	34 38	49	123

Extract of part of Table 5

31	32	33	34	35	36	37	38	39	40	41	42	43	44	45	46	47	48	49	50	51	52	53	54	55	56	57	58	59	60	d
′	′	′	′	′	′	′	′	′	′	′	′	′	′	′	′	′	′	′	′	′	′	′	′	′	′	′	′	′	′	′
0	0	0	0	0	0	0	0	0	0	0	0	0	0	0	0	0	0	0	0	0	0	0	0	0	0	0	0	0	0	0
1	1	1	1	1	1	1	1	1	1	1	1	1	1	1	1	1	1	1	1	1	1	1	1	1	1	1	1	1	1	1
1	1	1	1	1	1	1	1	1	1	1	1	1	1	2	2	2	2	2	2	2	2	2	2	2	2	2	2	2	2	2
2	2	2	2	2	2	2	2	2	2	2	2	2	2	2	2	2	2	2	2	3	3	3	3	3	3	3	3	3	3	3
2	2	2	2	2	2	2	3	3	3	3	3	3	3	3	3	3	3	3	3	3	3	4	4	4	4	4	4	4	4	4
3	3	3	3	3	3	3	3	3	3	3	4	4	4	4	4	4	4	4	4	4	4	4	4	5	5	5	5	5	5	5
3	3	3	3	4	4	4	4	4	4	4	4	4	4	4	5	5	5	5	5	5	5	5	5	6	6	6	6	6	6	6
4	4	4	4	4	4	4	4	5	5	5	5	5	5	5	5	5	6	6	6	6	6	6	6	7	7	7	7	7	7	7
4	4	4	5	5	5	5	5	5	5	5	6	6	6	6	6	6	6	7	7	7	7	7	7	7	7	8	8	8	8	8
5	5	5	5	5	5	6	6	6	6	6	6	6	7	7	7	7	7	7	8	8	8	8	8	8	8	9	9	9	9	9
5	5	6	6	6	6	6	6	6	7	7	7	7	7	8	8	8	8	8	8	8	9	9	9	9	9	10	10	10	10	10
6	6	6	6	6	7	7	7	7	7	7	8	8	8	8	8	9	9	9	9	9	10	10	10	10	10	10	11	11	11	11
6	6	7	7	7	7	7	8	8	8	8	8	9	9	9	9	9	10	10	10	10	10	11	11	11	11	11	12	12	12	12
7	7	7	7	8	8	8	8	8	9	9	9	9	10	10	10	10	11	11	11	11	11	11	12	12	12	12	13	13	13	13
7	7	8	8	8	8	9	9	9	9	10	10	10	10	10	11	11	11	11	12	12	12	12	13	13	13	13	14	14	14	14
8	8	8	8	9	9	9	10	10	10	10	10	11	11	11	12	12	12	12	12	13	13	13	14	14	14	14	14	15	15	15
8	9	9	9	9	10	10	10	11	11	11	11	11	12	12	12	13	13	13	13	14	14	14	14	15	15	15	15	16	16	16
9	9	9	10	10	11	10	11	11	11	12	12	12	13	13	13	13	14	14	14	14	15	15	15	16	16	16	16	17	17	17
9	10	10	10	10	11	11	11	12	12	12	13	13	13	14	14	14	14	15	15	15	16	16	16	16	17	17	17	18	18	18
10	10	10	11	11	11	11	11	12	12	13	13	13	13	14	14	15	15	15	15	15	16	16	16	17	17	17	18	19	19	19
10	11	11	11	12	12	12	13	13	13	14	14	14	15	15	15	16	16	17	18	18	18	19	19	19	20	20	20	20	20	20
11	11	12	12	12	13	13	13	14	14	14	15	15	15	16	16	16	17	17	18	18	18	19	19	19	20	20	20	21	21	21
11	12	12	12	13	13	14	14	14	15	15	15	16	16	16	17	17	18	18	18	19	19	19	20	20	21	21	21	22	22	22
12	12	13	13	13	14	14	15	15	15	16	16	16	17	17	18	18	18	19	19	20	20	20	21	21	21	22	22	23	23	23
12	13	13	14	14	14	15	15	16	16	16	17	17	18	18	18	19	19	20	20	20	21	21	22	22	22	23	23	24	24	24
13	13	14	14	15	15	15	16	16	17	17	18	18	18	19	19	20	20	20	21	21	22	22	22	23	23	24	24	25	25	25
13	14	14	15	15	16	16	16	17	17	18	18	19	19	20	20	20	21	21	22	22	23	23	23	24	24	25	25	26	26	26
14	14	15	15	16	16	17	17	18	18	18	19	19	20	20	21	21	22	22	22	23	23	24	24	25	25	26	26	27	27	27
14	15	15	16	16	17	17	18	18	19	19	20	20	21	21	21	22	22	23	23	24	24	25	25	26	26	27	27	28	28	28
15	15	16	16	17	17	18	18	19	19	20	20	21	21	22	22	23	23	24	24	25	25	26	26	27	27	28	28	29	29	29
16	16	16	17	18	18	18	19	20	20	20	21	22	22	22	23	24	24	24	25	26	26	26	27	28	28	28	29	30	30	30
16	17	17	18	18	19	19	20	20	21	21	22	22	23	23	24	24	25	25	26	26	27	27	28	28	29	29	30	30	31	31
17	17	18	18	19	19	20	20	21	21	22	22	23	23	24	25	25	26	26	27	27	28	28	29	29	30	30	31	31	32	32
17	18	18	19	19	20	20	21	21	22	23	23	23	24	25	25	26	26	27	28	28	29	29	30	30	31	31	32	32	33	33
18	18	19	19	20	20	21	22	22	22	23	23	24	25	26	26	27	27	28	28	29	29	30	31	31	32	32	33	33	34	34
18	19	19	20	20	21	22	22	23	23	24	24	25	26	26	27	27	28	29	29	30	30	31	32	32	33	33	34	34	35	35
19	19	20	20	21	22	22	23	23	24	25	25	26	26	27	28	28	29	29	30	31	31	32	32	33	34	34	35	35	36	36
19	20	20	21	22	22	23	23	24	25	25	26	26	27	28	28	29	30	30	31	31	32	33	33	34	35	35	36	36	37	37
20	20	21	22	22	23	23	24	25	25	26	27	27	28	28	29	30	30	31	32	32	33	34	34	35	35	36	37	37	38	38
20	21	21	22	23	23	24	25	25	26	27	27	28	29	29	30	31	31	32	32	33	34	34	35	36	36	37	38	38	39	39
21	21	22	23	23	24	25	25	26	27	27	28	29	29	30	31	31	32	33	33	34	35	35	36	37	37	38	39	39	40	40
21	22	23	23	24	25	25	26	27	27	28	29	29	30	31	31	32	33	33	34	35	36	36	37	38	38	39	40	40	41	41
22	22	23	24	24	25	26	27	27	28	29	29	30	31	31	32	33	34	34	35	36	36	37	38	38	39	40	41	41	42	42
22	23	24	24	25	26	27	27	28	29	29	30	31	32	32	33	34	34	35	36	37	37	38	39	39	40	41	42	42	43	43
23	23	24	25	26	26	27	28	29	29	30	31	32	32	33	34	34	35	36	37	37	38	39	40	40	41	42	43	43	44	44
23	24	25	26	26	27	28	28	29	30	31	32	32	33	34	34	35	36	37	38	38	39	40	40	41	42	43	44	44	45	45
24	25	25	26	27	28	28	29	30	31	31	32	33	34	34	35	36	37	38	39	39	40	41	41	42	43	44	44	45	46	46
24	25	26	27	27	28	29	30	31	31	32	33	34	34	35	36	37	38	38	39	40	41	42	42	43	44	45	45	46	47	47
25	26	26	27	28	29	30	30	31	32	33	34	34	35	36	37	38	38	39	40	41	42	42	43	44	45	46	46	47	48	48
25	26	27	28	29	29	30	31	32	33	33	34	35	36	37	38	38	39	40	41	42	42	43	44	45	46	47	47	48	49	49
26	27	28	28	29	30	31	32	32	33	34	35	36	37	38	38	39	40	41	42	42	43	44	45	46	47	48	48	49	50	50
26	27	28	29	30	31	31	32	33	34	35	36	37	37	38	39	40	41	42	42	43	44	45	46	47	48	48	49	50	51	51
27	28	29	29	30	31	32	33	34	35	36	36	37	38	39	40	41	42	42	43	44	45	46	47	48	49	49	50	51	52	52
27	28	29	30	31	32	33	34	34	35	36	37	38	39	40	41	42	42	43	44	45	46	47	48	49	49	50	51	52	53	53
28	29	30	31	32	32	33	34	35	36	37	38	39	40	40	41	42	43	44	45	46	47	48	49	50	50	51	52	53	54	54
28	29	30	31	32	33	34	35	36	37	38	38	39	40	41	42	43	44	45	46	47	48	49	50	50	51	52	53	54	55	55
29	30	31	32	33	34	35	36	37	38	39	40	40	41	42	43	44	45	46	47	48	49	50	50	51	52	53	54	55	56	56
29	30	31	32	33	34	35	36	37	38	39	40	41	42	43	44	45	46	47	48	49	50	51	52	52	53	54	55	56	57	57
30	31	32	33	34	35	36	37	38	39	40	41	42	43	44	45	46	47	48	49	50	51	52	53	54	55	56	56	57	58	58
30	31	32	33	34	35	36	37	38	39	40	41	42	43	44	45	46	47	48	49	50	51	52	53	54	55	56	57	58	59	59

Chapter 11
Demonstration of a Fix with Movement of the Observer (MOO).

In this demonstration, observations of 3 bodies are taken at different times in order to provide position lines for a fix. Because of the time lapses, considerable movement of the vessel occurs between the observations and so corrections have to be applied to the intercepts at step 8(a) as explained in the outline procedures. This correction is referred to as MOO (movement of observer) and involves the use of Table 1.

Notes.

1. To fully understand this demonstration, it will be necessary to refer to the outline procedures in chapter 8 and it may be advisable to revise chapters 2, 4 and 5 as well.
2. Relevant extracts of tables are placed immediately following the solution to the example. Because of their size, some tables have been cropped so that they show only the relevant portions.
3. Self-test questions on the use of the rapid sight reduction tables can be found in chapter 13.

Aim. The aim of this demonstration is to establish a 3 point fix based on the following imaginary scenario:

D.R. Position: 42° 53'.1S, 58° 15'E.
Date: 23 July 2009
Time of Fix: $12^h \, 00^m$ (-4)
DWE 35^s fast
Index error: -1'.41
Ht. of eye: 5m.
Temperature: -5°C. Pressure: 992mb.
Course 068° speed 12 knots maintained throughout the day.
Fix the position of the ship at $12^h \, 00^m$(-4) using position lines obtained from the following 3 observations of the Sun's lower limb.

Observation 1:
DWT: $04^h \, 35^m \, 10^s$
Body observed: Sun L.L.
Sextant Alt: 09° 12.'4. Bearing: 050°

Observation 2:
DWT: $08^h \, 01^m \, 20^s$
Body observed: Sun L.L.
Sextant Alt: 26° 55'.49. Bearing: 004°

Observation 3:
DWT: $12^h \, 02^m \, 05^s$

Body observed: Sun L.L.
Sextant Alt: 07° 20'.6. Bearing: 307°

UT of fix: 23^d 08^h 00^m
Movement of vessel:
Between observation 1 and fix: 3.48hrs. @ 12 knots = 41.76 n.m., course 068°
Between fix and observation 3: 4 hrs. @ 12 knots = 48 n.m., course 068°.

Solution.
Planning Phase

Step 1. Approx Pos:	42° 53'.1S, 58° 15'E.		
Step 2. Calculate UT of observations:			
Date: 23 July 2009			
Time of fix:	23^d 12^h 00^m		
Zone corrn:	-4^h		
Greenwich Date:	23^d 08^h 00^m		
	Obs.1	Obs.2	Obs.3
DWT:	04^h 35^m 10^s	08^h 01^m 20^s	12^h 02^m 05^s
DWE:	-35^s	-35^s	-35^s
a. UT obs:	04^h 34^m 35^s	08^h 00^m 45^s	12^h 01^m 30^s
b. UT fix:	08^h 00^m 00^s	08^h 00^m 00^s	08^h 00^m 00^s
	+ 03^h 25^m 25^s	- 00^h 00^m 45^s	- 04^h 01^m 30^s
Fix is	later	earlier	earlier
Step 3. Calculate GHA and Dec at UT of obs.			
UT obs:	04^h 34^m 35^s	08^h 00^m 45^s	12^h 01^m 30^s
Dec at UT^h	N20° 03'.2	N20° 01'.2	N19° 59'.1
d	(0'.5)	(0'.5)	(0'.5)
d cor at UT^m	-0'.3	0'.0	0'.0
Dec + corr.	N20° 02'.9	N20° 01'.2	N19° 59'.1
GHA at UT^h	238° 22'.6	298° 22'.6	358° 22'.5
Inc (m+s)	+8° 38'.8	+0° 11'.3	+0° 22'.5
Sum = GHA	247° 01'.4	298° 33'.9	358° 45'.0
Step 4. Calculate LHA			
	Obs.1	Obs.2	Obs.3
GHA	247° 01'.4	298° 33'.9	358° 45'.0

As. long. (E)	+ 57° 58'.6	+ 58° 26'.1	+ 58° 15'.0
LHA (E+)	305°	357°	417°
±360°			57°
Convert azimuth angle to true bearing (Zn):			
LHA =	305°	357°	057°
Z =	129°	177°	127°
Bearing Zn	051°	003°	307°

Rules for calculating true bearing (ZN):		
	Lat. North	Lat. South
LHA>180°	Zn = Z	Zn = 180° - Z
LHA<180°	Zn = 360°-Z	Zn = 180° + Z

Step 5. Main table entry.

	Obs.1	Obs.2	Obs.3
As. lat.	S43°	S43°	S43°
Dec°	N20°	N20°	N19°
	Contr.	Contr.	Contr.
Dec'	02'.9	01'.2	59'.1
LHA	305°	357°	57°
Table page:	31	31	31
Tab Hc,	09° 16'	26° 56'	08° 54'
d	-48'	-60'	-48'
Z	129°	177°	127°
Correction for mins of Dec. (Table 5)			
Tab Hc,	09° 16'	26° 56'	08° 54'
Cor for Dec'	-2'	-1'	-47'
Cal Alt(Hc)	09° 14'	26° 55'	08° 07'

Fix Phase

Step 6. Correct sextant altitude (Hs).

	Obs.1	Obs.2	Obs.3
Sext Alt(Hs)	09° 12'.4	26° 55'.49	07° 20'.6
I.E.	-1'.41	-1'.41	-1'.41
Dip (5.0m) (D	-3'.9	-3'.9	-3'.9
Ap Alt (Ha)	09° 07'.09	26° 50'.18	07°15'.29
Alt Cor (R)	+10'.3	+14'.1	+8'.9
Add'l refrac.	-0'.2	-0'.1	-0'.3
Obs Alt (Ho)	09° 17'.19	27° 04'.18	07° 23'.89

Step 7. Calculate Intercept			
	Obs.1	**Obs.2**	**Obs.3**
Ho =	09° 17'.9	27° 04'.18	07° 23'.89
Hc =	09° 14'.0	26° 55'.00	08° 07'.00
p(Ho-Hc) =	+3'.9	+9'.18	- 43'.11
Incpt (p) =	3'.9 towards	9'.18 towards	43'.11 from

Therefore, intercepts are:
Obs.1: 3.9 n.m. towards 051°, Obs.2: 9.18 n.m. towards 003°,
Obs.2: 43.11 from 307°.
Please see page 161 for an explanation of the reason for calculating true bearings after step 4.

Movement of vessel:
Between observation 1 and fix: 3.48hrs. @ 12 knots = 41.76 n.m., course 068°
Between fix and observation 3: 4 hrs. @ 12 knots = 48 n.m., course 068°.

Step 8. Corrections to intercept.(MOO)			
	Obs.1	**Obs.2**	**Obs.3**
Bearing (Zn)	051°	003°	307°
±360° (to make Zn > C)	360°	360°	
Zn	411°	363°	307°
Course (C)	068°	068°	068°
Rel Zn (Zn – C)	343°	295°	239°
Dist. (DMG)	41'.8	neg.	48'.0
Table 1	+40'.15		-25'.32
Fix is	later		earlier
Intercept (p) =	+ 3'.9	+9'.18	43'.11
correction =	+40'.15	0'.00	+25'.32
Corrected p =	+44'.05	+9'.18	-17'.79
	towards	towards	from
Azimuth (Zn) =	051°	003°	307°

Therefore, corrected intercepts are:
Obs1: 44'.05 towards 051°, Obs2: 9'.18 towards 003°, Obs3: 17'.79 from 307°.

Rules for application of sign from table 1		
Time of fix	Sign from table	To intercept
Later than observation	+	Add
	-	Subtract
Earlier than observation	+	Subtract
	-	Add

Reminder of the use of Table 1.
In Table 1, distance (to the nautical nearest mile) is tabulated across the top row. Rel. Zn (Zn – course) is shown in the left and right hand columns. The correction is read from the relevant intersecting cell and is used to correct the intercept as demonstrated in the above example.

Step 9. Plot the position lines (shown on next page).
Firstly, for each observation, plot the assumed position using the assumed long. calculated at step 4 and the assumed lat. calculated at step 5. Secondly, plot the three position lines using the corrected intercept data calculated at step 8.

Notes.
1. As in the last example, a 'cocked-hat' is produced and the position of the fix is assumed to be at the centre of the resultant triangle.

2. Some confusion might arise from the fact that AP2 is further to the east than AP3. This is a result of the need to choose an assumed longitude that, when combined with the GHA, will produce an LHA with a whole number of degrees. Such confusion can be avoided if it is remembered that an assumed position is just an approximate position.

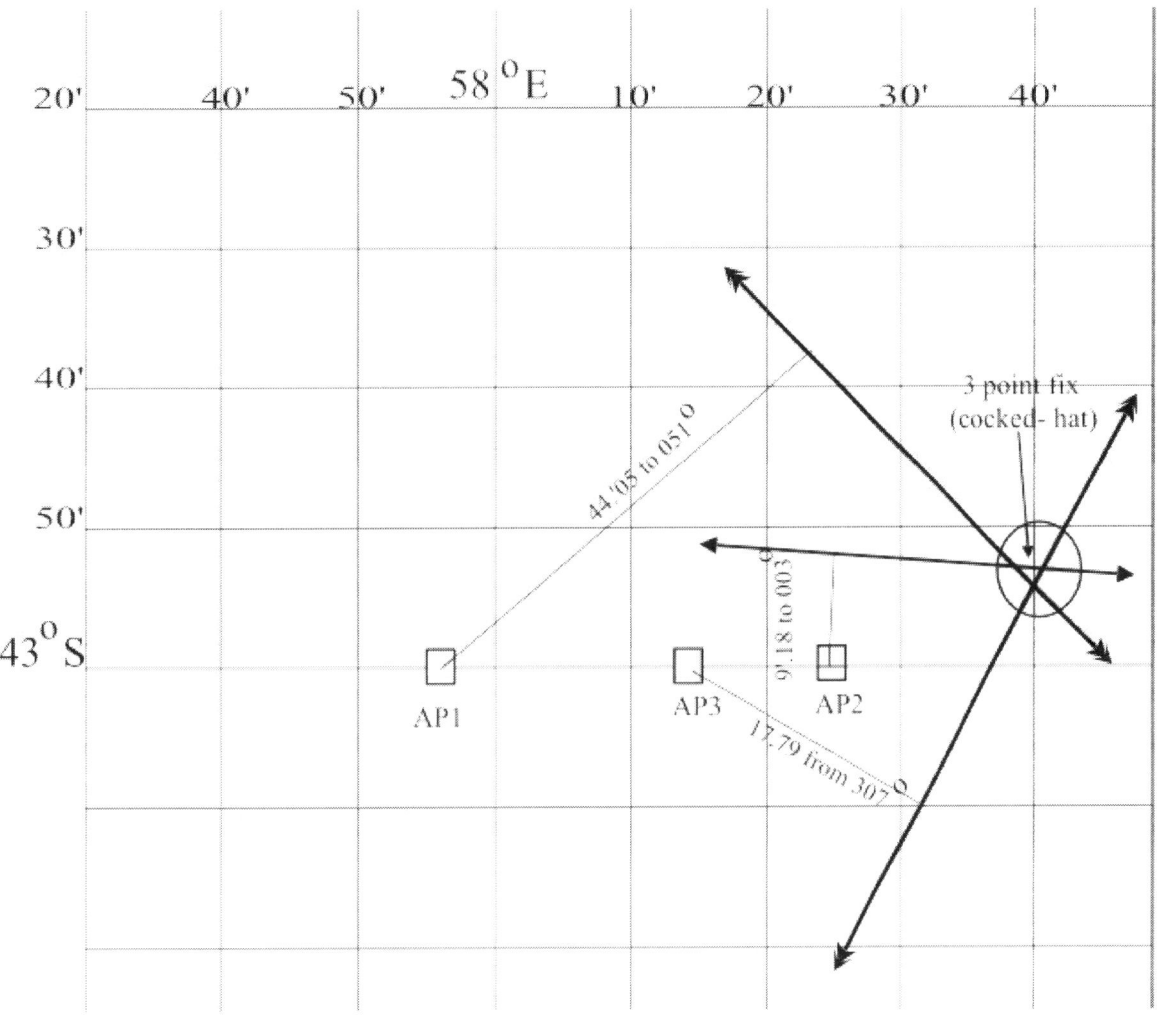

Extracts. Extracts relevant purely to this example begin on the next page. Due to their sizes, some tables have been cropped to show only the required portions.

2009 JULY 21, 22, 23 (TUES., WED., THURS.)

UT	SUN GHA	SUN Dec	MOON GHA	MOON v	MOON Dec	MOON d	MOON HP
d h	° '	° '	° '	'	° '	'	'
21 00	178 24.0	N20 29.1	194 59.4	2.5	N24 26.0	7.0	61.2
01	193 23.9	28.6	209 20.9	2.6	24 19.0	7.2	61.2
02	208 23.9	28.1	223 42.5	2.6	24 11.8	7.4	61.2
03	223 23.9	27.6	238 04.1	2.7	24 04.4	7.6	61.3
04	238 23.8	27.1	252 25.8	2.8	23 56.8	7.8	61.3
05	253 23.8	26.7	266 47.6	2.8	23 49.0	7.9	61.3
06	268 23.8	N20 26.2	281 09.4	2.9	N23 41.1	8.2	61.3
07	283 23.8	25.7	295 31.3	3.0	23 32.9	8.3	61.3
08	298 23.7	25.2	309 53.3	3.0	23 24.6	8.4	61.3
09	313 23.7	24.7	324 15.3	3.2	23 16.2	8.7	61.3
10	328 23.7	24.2	338 37.5	3.2	23 07.5	8.8	61.3
11	343 23.6	23.7	352 59.7	3.3	22 58.7	9.0	61.3
12	358 23.6	N20 23.2	7 22.0	3.3	N22 49.7	9.2	61.3
13	13 23.6	22.8	21 44.3	3.5	22 40.5	9.3	61.3
14	28 23.5	22.3	36 06.8	3.5	22 31.2	9.5	61.3
15	43 23.5	21.8	50 29.3	3.6	22 21.7	9.7	61.3
16	58 23.5	21.3	64 51.9	3.7	22 12.0	9.8	61.3
17	73 23.5	20.8	79 14.6	3.8	22 02.2	10.0	61.3
18	88 23.4	N20 20.3	93 37.4	3.9	N21 52.2	10.1	61.3
19	103 23.4	19.8	108 00.3	4.0	21 42.1	10.3	61.3
20	118 23.4	19.3	122 23.3	4.1	21 31.8	10.5	61.3
21	133 23.4	18.8	136 46.4	4.2	21 21.3	10.6	61.3
22	148 23.3	18.3	151 09.6	4.2	21 10.7	10.8	61.3
23	163 23.3	17.8	165 32.8	4.4	N20 59.9	10.9	61.3
22 00	178 23.3	N20 17.3					
01	193 23.2	16.8					
02	208 23.2	16.3	A total eclipse of the Sun occurs on this date. See page 5.				
03	223 23.2	15.8					
04	238 23.2	15.3					
05	253 23.1	14.8					
06	268 23.1	N20 14.3	266 18.5	5.0	N19 40.6	11.9	61.3
07	283 23.1	13.8	280 42.5	5.2	19 28.7	12.0	61.3
08	298 23.1	13.3	295 06.7	5.3	19 16.7	12.2	61.3
09	313 23.1	12.8	309 31.0	5.4	19 04.5	12.3	61.3
10	328 23.0	12.3	323 55.4	5.4	18 52.2	12.4	61.3
11	343 23.0	11.8	338 19.8	5.6	18 39.8	12.5	61.3
12	358 23.0	N20 11.3	352 44.4	5.7	N18 27.3	12.7	61.3
13	13 23.0	10.8	7 09.1	5.8	18 14.6	12.7	61.3
14	28 22.9	10.3	21 33.9	5.9	18 01.9	12.9	61.3
15	43 22.9	09.8	35 58.8	6.0	17 49.0	13.0	61.2
16	58 22.9	09.3	50 23.8	6.1	17 36.0	13.1	61.2
17	73 22.9	08.8	64 48.9	6.2	17 22.9	13.2	61.2
18	88 22.8	N20 08.3	79 14.1	6.3	N17 09.7	13.4	61.2
19	103 22.8	07.8	93 39.4	6.4	16 56.3	13.4	61.2
20	118 22.8	07.3	108 04.8	6.5	16 42.9	13.5	61.2
21	133 22.8	06.8	122 30.3	6.6	16 29.4	13.7	61.2
22	148 22.8	06.3	136 55.9	6.8	16 15.7	13.7	61.2
23	163 22.7	05.8	151 21.7	6.8	16 02.0	13.9	61.1
23 00	178 22.7	N20 05.3	165 47.5	6.9	N15 48.1	13.9	61.1
01	193 22.7	04.7	180 13.4	7.1	15 34.2	14.0	61.1
02	208 22.7	04.2	194 39.5	7.1	15 20.2	14.1	61.1
03	223 22.7	03.7	209 05.6	7.3	15 06.1	14.2	61.1
04	238 22.6	03.2	223 31.9	7.3	14 51.9	14.3	61.1
05	253 22.6	02.7	237 58.2	7.5	14 37.6	14.4	61.0
06	268 22.6	N20 02.2	252 24.7	7.5	N14 23.2	14.4	61.0
07	283 22.6	01.7	266 51.2	7.7	14 08.8	14.6	61.0
08	298 22.6	01.2	281 17.9	7.7	13 54.2	14.6	61.0
09	313 22.6	00.6	295 44.6	7.9	13 39.6	14.7	61.0
10	328 22.5	20 00.1	310 11.5	7.9	13 24.9	14.7	60.9
11	343 22.5	19 59.6	324 38.4	8.0	13 10.2	14.8	60.9
12	358 22.5	N19 59.1	339 05.4	8.2	N12 55.4	14.9	60.9
13	13 22.5	58.6	353 32.6	8.2	12 40.5	15.0	60.9
14	28 22.5	58.1	7 59.8	8.3	12 25.5	15.0	60.9
15	43 22.5	57.5	22 27.1	8.5	12 10.5	15.1	60.8
16	58 22.4	57.0	36 54.6	8.5	11 55.4	15.1	60.8
17	73 22.4	56.5	51 22.1	8.6	11 40.3	15.2	60.8
18	88 22.4	N19 56.0	65 49.7	8.7	N11 25.1	15.3	60.8
19	103 22.4	55.5	80 17.4	8.8	11 09.8	15.3	60.7
20	118 22.4	54.9	94 45.2	8.9	10 54.5	15.3	60.7
21	133 22.4	54.4	109 13.1	9.0	10 39.2	15.5	60.7
22	148 22.3	53.9	123 41.1	9.0	10 23.7	15.4	60.7
23	163 22.3	53.4	138 09.1	9.2	N10 08.3	15.5	60.6
	SD 15.8	d 0.5	SD 16.7		16.7		16.6

Lat.	Twilight Naut.	Twilight Civil	Sunrise	Moonrise 21	Moonrise 22	Moonrise 23	Moonrise 24
°	h m	h m	h m	h m	h m	h m	h m
N 72	▢	▢	▢			03 27	06 20
N 70	▢	▢	▢			04 02	06 32
68	////	////	01 15		01 38	04 27	06 43
66	////	////	02 06		02 23	04 46	06 51
64	////	////	02 37	00 33	02 53	05 01	06 58
62	////	01 31	03 00	01 14	03 15	05 13	07 04
60	////	02 08	03 18	01 42	03 33	05 24	07 10
N 58	////	02 33	03 33	02 04	03 48	05 33	07 14
56	01 23	02 53	03 46	02 21	04 00	05 41	07 18
54	01 57	03 09	03 57	02 36	04 11	05 48	07 22
52	02 21	03 23	04 07	02 49	04 21	05 55	07 25
50	02 40	03 35	04 16	03 01	04 30	06 00	07 28
45	03 14	03 59	04 34	03 24	04 48	06 13	07 34
N 40	03 39	04 18	04 49	03 43	05 03	06 23	07 40
35	03 59	04 34	05 02	03 59	05 15	06 31	07 44
30	04 15	04 47	05 13	04 12	05 26	06 39	07 48
20	04 40	05 08	05 32	04 36	05 45	06 52	07 55
N 10	04 59	05 25	05 48	04 56	06 01	07 03	08 02
0	05 15	05 41	06 03	05 14	06 16	07 14	08 07
S 10	05 30	05 55	06 18	05 33	06 31	07 24	08 13
20	05 43	06 10	06 33	05 53	06 47	07 35	08 19
30	05 57	06 26	06 51	06 16	07 05	07 48	08 26
35	06 04	06 35	07 02	06 29	07 16	07 55	08 30
40	06 11	06 44	07 14	06 44	07 28	08 04	08 34
45	06 19	06 55	07 28	07 03	07 42	08 13	08 39
S 50	06 28	07 08	07 45	07 25	07 59	08 25	08 45
52	06 32	07 14	07 52	07 36	08 07	08 30	08 48
54	06 36	07 21	08 01	07 48	08 16	08 36	08 51
56	06 41	07 28	08 11	08 02	08 26	08 43	08 55
58	06 46	07 36	08 23	08 18	08 38	08 50	08 58
S 60	06 51	07 45	08 36	08 38	08 51	08 58	09 03

Lat.	Sunset	Twilight Civil	Twilight Naut.	Moonset 21	Moonset 22	Moonset 23	Moonset 24
°	h m	h m	h m	h m	h m	h m	h m
N 72	▢	▢	▢		22 58	22 00	21 24
N 70	▢	▢	▢		22 21	21 44	21 19
68	22 52	////	////	22 40	21 55	21 32	21 14
66	22 04	////	////	21 54	21 34	21 21	21 10
64	21 34	////	////	21 23	21 17	21 12	21 07
62	21 11	22 37	////	21 00	21 03	21 04	21 04
60	20 53	22 02	////	20 41	20 52	20 58	21 02
N 58	20 38	21 37	////	20 26	20 41	20 52	21 00
56	20 26	21 18	22 46	20 12	20 32	20 47	20 58
54	20 15	21 02	22 13	20 00	20 24	20 42	20 56
52	20 05	20 49	21 50	19 50	20 17	20 37	20 54
50	19 56	20 37	21 32	19 41	20 11	20 33	20 53
45	19 38	20 13	20 58	19 21	19 56	20 25	20 49
N 40	19 23	19 54	20 33	19 05	19 45	20 18	20 46
35	19 10	19 39	20 14	18 51	19 35	20 11	20 44
30	19 00	19 26	19 58	18 40	19 26	20 06	20 42
20	18 41	19 05	19 33	18 19	19 10	19 56	20 38
N 10	18 25	18 47	19 14	18 01	18 57	19 48	20 35
0	18 10	18 32	18 58	17 44	18 44	19 40	20 31
S 10	17 55	18 18	18 43	17 27	18 31	19 31	20 28
20	17 40	18 03	18 30	17 09	18 17	19 23	20 24
30	17 22	17 47	18 17	16 47	18 01	19 12	20 20
35	17 11	17 39	18 10	16 35	17 52	19 07	20 18
40	17 00	17 29	18 02	16 20	17 41	19 00	20 15
45	16 46	17 18	17 54	16 03	17 28	18 52	20 12
S 50	16 29	17 05	17 45	15 42	17 13	18 42	20 08
52	16 21	16 59	17 41	15 31	17 05	18 38	20 07
54	16 12	16 53	17 37	15 20	16 57	18 33	20 05
56	16 02	16 46	17 32	15 06	16 48	18 27	20 02
58	15 51	16 38	17 28	14 51	16 37	18 21	20 00
S 60	15 38	16 29	17 22	14 32	16 25	18 14	19 57

Day	Eqn. of Time 00h	Eqn. of Time 12h	Mer. Pass.	Mer. Pass. Upper	Mer. Pass. Lower	Age	Phase
d	m s	m s	h m	h m	h m	d	%
21	06 24	06 26	12 06	11 29	24 00	29	1
22	06 27	06 28	12 06	12 30	00 00	00	0
23	06 29	06 30	12 06	13 27	00 59	01	3

©2009 Jack Case

INCREMENTS AND CORRECTIONS

0ᵐ

s	SUN PLANETS	ARIES	MOON	v or d	Corrⁿ	v or d	Corrⁿ	v or d	Corrⁿ
00	0 00·0	0 00·0	0 00·0	0·0	0·0	6·0	0·1	12·0	0·1
01	0 00·3	0 00·3	0 00·2	0·1	0·0	6·1	0·1	12·1	0·1
02	0 00·5	0 00·5	0 00·5	0·2	0·0	6·2	0·1	12·2	0·1
03	0 00·8	0 00·8	0 00·7	0·3	0·0	6·3	0·1	12·3	0·1
04	0 01·0	0 01·0	0 01·0	0·4	0·0	6·4	0·1	12·4	0·1
05	0 01·3	0 01·3	0 01·2	0·5	0·0	6·5	0·1	12·5	0·1
06	0 01·5	0 01·5	0 01·4	0·6	0·0	6·6	0·1	12·6	0·1
07	0 01·8	0 01·8	0 01·7	0·7	0·0	6·7	0·1	12·7	0·1
08	0 02·0	0 02·0	0 01·9	0·8	0·0	6·8	0·1	12·8	0·1
09	0 02·3	0 02·3	0 02·1	0·9	0·0	6·9	0·1	12·9	0·1
10	0 02·5	0 02·5	0 02·4	1·0	0·0	7·0	0·1	13·0	0·1
11	0 02·8	0 02·8	0 02·6	1·1	0·0	7·1	0·1	13·1	0·1
12	0 03·0	0 03·0	0 02·9	1·2	0·0	7·2	0·1	13·2	0·1
13	0 03·3	0 03·3	0 03·1	1·3	0·0	7·3	0·1	13·3	0·1
14	0 03·5	0 03·5	0 03·3	1·4	0·0	7·4	0·1	13·4	0·1
15	0 03·8	0 03·8	0 03·6	1·5	0·0	7·5	0·1	13·5	0·1
16	0 04·0	0 04·0	0 03·8	1·6	0·0	7·6	0·1	13·6	0·1
17	0 04·3	0 04·3	0 04·1	1·7	0·0	7·7	0·1	13·7	0·1
18	0 04·5	0 04·5	0 04·3	1·8	0·0	7·8	0·1	13·8	0·1
19	0 04·8	0 04·8	0 04·5	1·9	0·0	7·9	0·1	13·9	0·1
20	0 05·0	0 05·0	0 04·8	2·0	0·0	8·0	0·1	14·0	0·1
21	0 05·3	0 05·3	0 05·0	2·1	0·0	8·1	0·1	14·1	0·1
22	0 05·5	0 05·5	0 05·2	2·2	0·0	8·2	0·1	14·2	0·1
23	0 05·8	0 05·8	0 05·5	2·3	0·0	8·3	0·1	14·3	0·1
24	0 06·0	0 06·0	0 05·7	2·4	0·0	8·4	0·1	14·4	0·1
25	0 06·3	0 06·3	0 06·0	2·5	0·0	8·5	0·1	14·5	0·1
26	0 06·5	0 06·5	0 06·2	2·6	0·0	8·6	0·1	14·6	0·1
27	0 06·8	0 06·8	0 06·4	2·7	0·0	8·7	0·1	14·7	0·1
28	0 07·0	0 07·0	0 06·7	2·8	0·0	8·8	0·1	14·8	0·1
29	0 07·3	0 07·3	0 06·9	2·9	0·0	8·9	0·1	14·9	0·1
30	0 07·5	0 07·5	0 07·2	3·0	0·0	9·0	0·1	15·0	0·1
31	0 07·8	0 07·8	0 07·4	3·1	0·0	9·1	0·1	15·1	0·1
32	0 08·0	0 08·0	0 07·6	3·2	0·0	9·2	0·1	15·2	0·1
33	0 08·3	0 08·3	0 07·9	3·3	0·0	9·3	0·1	15·3	0·1
34	0 08·5	0 08·5	0 08·1	3·4	0·0	9·4	0·1	15·4	0·1
35	0 08·8	0 08·8	0 08·4	3·5	0·0	9·5	0·1	15·5	0·1
36	0 09·0	0 09·0	0 08·6	3·6	0·0	9·6	0·1	15·6	0·1
37	0 09·3	0 09·3	0 08·8	3·7	0·0	9·7	0·1	15·7	0·1
38	0 09·5	0 09·5	0 09·1	3·8	0·0	9·8	0·1	15·8	0·1
39	0 09·8	0 09·8	0 09·3	3·9	0·0	9·9	0·1	15·9	0·1
40	0 10·0	0 10·0	0 09·5	4·0	0·0	10·0	0·1	16·0	0·1
41	0 10·3	0 10·3	0 09·8	4·1	0·0	10·1	0·1	16·1	0·1
42	0 10·5	0 10·5	0 10·0	4·2	0·0	10·2	0·1	16·2	0·1
43	0 10·8	0 10·8	0 10·3	4·3	0·0	10·3	0·1	16·3	0·1
44	0 11·0	0 11·0	0 10·5	4·4	0·0	10·4	0·1	16·4	0·1
45	0 11·3	0 11·3	0 10·7	4·5	0·0	10·5	0·1	16·5	0·1
46	0 11·5	0 11·5	0 11·0	4·6	0·0	10·6	0·1	16·6	0·1
47	0 11·8	0 11·8	0 11·2	4·7	0·0	10·7	0·1	16·7	0·1
48	0 12·0	0 12·0	0 11·5	4·8	0·0	10·8	0·1	16·8	0·1
49	0 12·3	0 12·3	0 11·7	4·9	0·0	10·9	0·1	16·9	0·1
50	0 12·5	0 12·5	0 11·9	5·0	0·0	11·0	0·1	17·0	0·1
51	0 12·8	0 12·8	0 12·2	5·1	0·0	11·1	0·1	17·1	0·1
52	0 13·0	0 13·0	0 12·4	5·2	0·0	11·2	0·1	17·2	0·1
53	0 13·3	0 13·3	0 12·6	5·3	0·0	11·3	0·1	17·3	0·1
54	0 13·5	0 13·5	0 12·9	5·4	0·0	11·4	0·1	17·4	0·1
55	0 13·8	0 13·8	0 13·1	5·5	0·0	11·5	0·1	17·5	0·1
56	0 14·0	0 14·0	0 13·4	5·6	0·0	11·6	0·1	17·6	0·1
57	0 14·3	0 14·3	0 13·6	5·7	0·0	11·7	0·1	17·7	0·1
58	0 14·5	0 14·5	0 13·8	5·8	0·0	11·8	0·1	17·8	0·1
59	0 14·8	0 14·8	0 14·1	5·9	0·0	11·9	0·1	17·9	0·1
60	0 15·0	0 15·0	0 14·3	6·0	0·1	12·0	0·1	18·0	0·2

1ᵐ

s	SUN PLANETS	ARIES	MOON	v or d	Corrⁿ	v or d	Corrⁿ	v or d	Corrⁿ
00	0 15·0	0 15·0	0 14·3	0·0	0·0	6·0	0·2	12·0	0·3
01	0 15·3	0 15·3	0 14·6	0·1	0·0	6·1	0·2	12·1	0·3
02	0 15·5	0 15·5	0 14·8	0·2	0·0	6·2	0·2	12·2	0·3
03	0 15·8	0 15·8	0 15·0	0·3	0·0	6·3	0·2	12·3	0·3
04	0 16·0	0 16·0	0 15·3	0·4	0·0	6·4	0·2	12·4	0·3
05	0 16·3	0 16·3	0 15·5	0·5	0·0	6·5	0·2	12·5	0·3
06	0 16·5	0 16·5	0 15·7	0·6	0·0	6·6	0·2	12·6	0·3
07	0 16·8	0 16·8	0 16·0	0·7	0·0	6·7	0·2	12·7	0·3
08	0 17·0	0 17·0	0 16·2	0·8	0·0	6·8	0·2	12·8	0·3
09	0 17·3	0 17·3	0 16·5	0·9	0·0	6·9	0·2	12·9	0·3
10	0 17·5	0 17·5	0 16·7	1·0	0·0	7·0	0·2	13·0	0·3
11	0 17·8	0 17·8	0 16·9	1·1	0·0	7·1	0·2	13·1	0·3
12	0 18·0	0 18·0	0 17·2	1·2	0·0	7·2	0·2	13·2	0·3
13	0 18·3	0 18·3	0 17·4	1·3	0·0	7·3	0·2	13·3	0·3
14	0 18·5	0 18·6	0 17·7	1·4	0·0	7·4	0·2	13·4	0·3
15	0 18·8	0 18·8	0 17·9	1·5	0·0	7·5	0·2	13·5	0·3
16	0 19·0	0 19·1	0 18·1	1·6	0·0	7·6	0·2	13·6	0·3
17	0 19·3	0 19·3	0 18·4	1·7	0·0	7·7	0·2	13·7	0·3
18	0 19·5	0 19·6	0 18·6	1·8	0·0	7·8	0·2	13·8	0·3
19	0 19·8	0 19·8	0 18·9	1·9	0·0	7·9	0·2	13·9	0·3
20	0 20·0	0 20·1	0 19·1	2·0	0·1	8·0	0·2	14·0	0·4
21	0 20·3	0 20·3	0 19·3	2·1	0·1	8·1	0·2	14·1	0·4
22	0 20·5	0 20·6	0 19·6	2·2	0·1	8·2	0·2	14·2	0·4
23	0 20·8	0 20·8	0 19·8	2·3	0·1	8·3	0·2	14·3	0·4
24	0 21·0	0 21·1	0 20·0	2·4	0·1	8·4	0·2	14·4	0·4
25	0 21·3	0 21·3	0 20·3	2·5	0·1	8·5	0·2	14·5	0·4
26	0 21·5	0 21·6	0 20·5	2·6	0·1	8·6	0·2	14·6	0·4
27	0 21·8	0 21·8	0 20·8	2·7	0·1	8·7	0·2	14·7	0·4
28	0 22·0	0 22·1	0 21·0	2·8	0·1	8·8	0·2	14·8	0·4
29	0 22·3	0 22·3	0 21·2	2·9	0·1	8·9	0·2	14·9	0·4
30	0 22·5	0 22·6	0 21·5	3·0	0·1	9·0	0·2	15·0	0·4
31	0 22·8	0 22·8	0 21·7	3·1	0·1	9·1	0·2	15·1	0·4
32	0 23·0	0 23·1	0 22·0	3·2	0·1	9·2	0·2	15·2	0·4
33	0 23·3	0 23·3	0 22·2	3·3	0·1	9·3	0·2	15·3	0·4
34	0 23·5	0 23·6	0 22·4	3·4	0·1	9·4	0·2	15·4	0·4
35	0 23·8	0 23·8	0 22·7	3·5	0·1	9·5	0·2	15·5	0·4
36	0 24·0	0 24·1	0 22·9	3·6	0·1	9·6	0·2	15·6	0·4
37	0 24·3	0 24·3	0 23·1	3·7	0·1	9·7	0·2	15·7	0·4
38	0 24·5	0 24·6	0 23·4	3·8	0·1	9·8	0·2	15·8	0·4
39	0 24·8	0 24·8	0 23·6	3·9	0·1	9·9	0·2	15·9	0·4
40	0 25·0	0 25·1	0 23·9	4·0	0·1	10·0	0·3	16·0	0·4
41	0 25·3	0 25·3	0 24·1	4·1	0·1	10·1	0·3	16·1	0·4
42	0 25·5	0 25·6	0 24·3	4·2	0·1	10·2	0·3	16·2	0·4
43	0 25·8	0 25·8	0 24·6	4·3	0·1	10·3	0·3	16·3	0·4
44	0 26·0	0 26·1	0 24·8	4·4	0·1	10·4	0·3	16·4	0·4
45	0 26·3	0 26·3	0 25·1	4·5	0·1	10·5	0·3	16·5	0·4
46	0 26·5	0 26·6	0 25·3	4·6	0·1	10·6	0·3	16·6	0·4
47	0 26·8	0 26·8	0 25·5	4·7	0·1	10·7	0·3	16·7	0·4
48	0 27·0	0 27·1	0 25·8	4·8	0·1	10·8	0·3	16·8	0·4
49	0 27·3	0 27·3	0 26·0	4·9	0·1	10·9	0·3	16·9	0·4
50	0 27·5	0 27·6	0 26·2	5·0	0·1	11·0	0·3	17·0	0·4
51	0 27·8	0 27·8	0 26·5	5·1	0·1	11·1	0·3	17·1	0·4
52	0 28·0	0 28·1	0 26·7	5·2	0·1	11·2	0·3	17·2	0·4
53	0 28·3	0 28·3	0 27·0	5·3	0·1	11·3	0·3	17·3	0·4
54	0 28·5	0 28·6	0 27·2	5·4	0·1	11·4	0·3	17·4	0·4
55	0 28·8	0 28·8	0 27·4	5·5	0·1	11·5	0·3	17·5	0·4
56	0 29·0	0 29·1	0 27·7	5·6	0·1	11·6	0·3	17·6	0·4
57	0 29·3	0 29·3	0 27·9	5·7	0·1	11·7	0·3	17·7	0·4
58	0 29·5	0 29·6	0 28·2	5·8	0·1	11·8	0·3	17·8	0·4
59	0 29·8	0 29·8	0 28·4	5·9	0·1	11·9	0·3	17·9	0·4
60	0 30·0	0 30·1	0 28·6	6·0	0·2	12·0	0·3	18·0	0·5

© 2009 Jack Case

INCREMENTS AND CORRECTIONS

34ᵐ

34 s	SUN PLANETS	ARIES	MOON	v or d	Corrⁿ	v or d	Corrⁿ	v or d	Corrⁿ
	° ′	° ′	° ′	′	′	′	′	′	′
00	8 30.0	8 31.4	8 06.8	0.0	0.0	6.0	3.5	12.0	6.9
01	8 30.3	8 31.6	8 07.0	0.1	0.1	6.1	3.5	12.1	7.0
02	8 30.5	8 31.9	8 07.2	0.2	0.1	6.2	3.6	12.2	7.0
03	8 30.8	8 32.1	8 07.5	0.3	0.2	6.3	3.6	12.3	7.1
04	8 31.0	8 32.4	8 07.7	0.4	0.2	6.4	3.7	12.4	7.1
05	8 31.3	8 32.6	8 08.0	0.5	0.3	6.5	3.7	12.5	7.2
06	8 31.5	8 32.9	8 08.2	0.6	0.3	6.6	3.8	12.6	7.2
07	8 31.8	8 33.2	8 08.4	0.7	0.4	6.7	3.9	12.7	7.3
08	8 32.0	8 33.4	8 08.7	0.8	0.5	6.8	3.9	12.8	7.4
09	8 32.3	8 33.7	8 08.9	0.9	0.5	6.9	4.0	12.9	7.4
10	8 32.5	8 33.9	8 09.2	1.0	0.6	7.0	4.0	13.0	7.5
11	8 32.8	8 34.2	8 09.4	1.1	0.6	7.1	4.1	13.1	7.5
12	8 33.0	8 34.4	8 09.6	1.2	0.7	7.2	4.1	13.2	7.6
13	8 33.3	8 34.7	8 09.9	1.3	0.7	7.3	4.2	13.3	7.6
14	8 33.5	8 34.9	8 10.1	1.4	0.8	7.4	4.3	13.4	7.7
15	8 33.8	8 35.2	8 10.3	1.5	0.9	7.5	4.3	13.5	7.8
16	8 34.0	8 35.4	8 10.6	1.6	0.9	7.6	4.4	13.6	7.8
17	8 34.3	8 35.7	8 10.8	1.7	1.0	7.7	4.4	13.7	7.9
18	8 34.5	8 35.9	8 11.1	1.8	1.0	7.8	4.5	13.8	7.9
19	8 34.8	8 36.2	8 11.3	1.9	1.1	7.9	4.5	13.9	8.0
20	8 35.0	8 36.4	8 11.5	2.0	1.2	8.0	4.6	14.0	8.1
21	8 35.3	8 36.7	8 11.8	2.1	1.2	8.1	4.7	14.1	8.1
22	8 35.5	8 36.9	8 12.0	2.2	1.3	8.2	4.7	14.2	8.2
23	8 35.8	8 37.2	8 12.3	2.3	1.3	8.3	4.8	14.3	8.2
24	8 36.0	8 37.4	8 12.5	2.4	1.4	8.4	4.8	14.4	8.3
25	8 36.3	8 37.7	8 12.7	2.5	1.4	8.5	4.9	14.5	8.3
26	8 36.5	8 37.9	8 13.0	2.6	1.5	8.6	4.9	14.6	8.4
27	8 36.8	8 38.2	8 13.2	2.7	1.6	8.7	5.0	14.7	8.5
28	8 37.0	8 38.4	8 13.4	2.8	1.6	8.8	5.1	14.8	8.5
29	8 37.3	8 38.7	8 13.7	2.9	1.7	8.9	5.1	14.9	8.6
30	8 37.5	8 38.9	8 13.9	3.0	1.7	9.0	5.2	15.0	8.6
31	8 37.8	8 39.2	8 14.2	3.1	1.8	9.1	5.2	15.1	8.7
32	8 38.0	8 39.4	8 14.4	3.2	1.8	9.2	5.3	15.2	8.7
33	8 38.3	8 39.7	8 14.6	3.3	1.9	9.3	5.3	15.3	8.8
34	8 38.5	8 39.9	8 14.9	3.4	2.0	9.4	5.4	15.4	8.9
35	8 38.8	8 40.2	8 15.1	3.5	2.0	9.5	5.5	15.5	8.9
36	8 39.0	8 40.4	8 15.4	3.6	2.1	9.6	5.5	15.6	9.0
37	8 39.3	8 40.7	8 15.6	3.7	2.1	9.7	5.6	15.7	9.0
38	8 39.5	8 40.9	8 15.8	3.8	2.2	9.8	5.6	15.8	9.1
39	8 39.8	8 41.2	8 16.1	3.9	2.2	9.9	5.7	15.9	9.1
40	8 40.0	8 41.4	8 16.3	4.0	2.3	10.0	5.8	16.0	9.2
41	8 40.3	8 41.7	8 16.5	4.1	2.4	10.1	5.8	16.1	9.3
42	8 40.5	8 41.9	8 16.8	4.2	2.4	10.2	5.9	16.2	9.3
43	8 40.8	8 42.2	8 17.0	4.3	2.5	10.3	5.9	16.3	9.4
44	8 41.0	8 42.4	8 17.3	4.4	2.5	10.4	6.0	16.4	9.4
45	8 41.3	8 42.7	8 17.5	4.5	2.6	10.5	6.0	16.5	9.5
46	8 41.5	8 42.9	8 17.7	4.6	2.6	10.6	6.1	16.6	9.5
47	8 41.8	8 43.2	8 18.0	4.7	2.7	10.7	6.2	16.7	9.6
48	8 42.0	8 43.4	8 18.2	4.8	2.8	10.8	6.2	16.8	9.7
49	8 42.3	8 43.7	8 18.5	4.9	2.8	10.9	6.3	16.9	9.7
50	8 42.5	8 43.9	8 18.7	5.0	2.9	11.0	6.3	17.0	9.8
51	8 42.8	8 44.2	8 18.9	5.1	2.9	11.1	6.4	17.1	9.8
52	8 43.0	8 44.4	8 19.2	5.2	3.0	11.2	6.4	17.2	9.9
53	8 43.3	8 44.7	8 19.4	5.3	3.0	11.3	6.5	17.3	9.9
54	8 43.5	8 44.9	8 19.7	5.4	3.1	11.4	6.6	17.4	10.0
55	8 43.8	8 45.2	8 19.9	5.5	3.2	11.5	6.6	17.5	10.1
56	8 44.0	8 45.4	8 20.1	5.6	3.2	11.6	6.7	17.6	10.1
57	8 44.3	8 45.7	8 20.4	5.7	3.3	11.7	6.7	17.7	10.2
58	8 44.5	8 45.9	8 20.6	5.8	3.3	11.8	6.8	17.8	10.2
59	8 44.8	8 46.2	8 20.8	5.9	3.4	11.9	6.8	17.9	10.3
60	8 45.0	8 46.4	8 21.1	6.0	3.5	12.0	6.9	18.0	10.4

35ᵐ

35 s	SUN PLANETS	ARIES	MOON	v or d	Corrⁿ	v or d	Corrⁿ	v or d	Corrⁿ
	° ′	° ′	° ′	′	′	′	′	′	′
00	8 45.0	8 46.4	8 21.1	0.0	0.0	6.0	3.6	12.0	7.1
01	8 45.3	8 46.7	8 21.3	0.1	0.1	6.1	3.6	12.1	7.2
02	8 45.5	8 46.9	8 21.6	0.2	0.1	6.2	3.7	12.2	7.2
03	8 45.8	8 47.2	8 21.8	0.3	0.2	6.3	3.7	12.3	7.3
04	8 46.0	8 47.4	8 22.0	0.4	0.2	6.4	3.8	12.4	7.3
05	8 46.3	8 47.7	8 22.3	0.5	0.3	6.5	3.8	12.5	7.4
06	8 46.5	8 47.9	8 22.5	0.6	0.4	6.6	3.9	12.6	7.5
07	8 46.8	8 48.2	8 22.8	0.7	0.4	6.7	4.0	12.7	7.5
08	8 47.0	8 48.4	8 23.0	0.8	0.5	6.8	4.0	12.8	7.6
09	8 47.3	8 48.7	8 23.2	0.9	0.5	6.9	4.1	12.9	7.6
10	8 47.5	8 48.9	8 23.5	1.0	0.6	7.0	4.1	13.0	7.7
11	8 47.8	8 49.2	8 23.7	1.1	0.7	7.1	4.2	13.1	7.8
12	8 48.0	8 49.4	8 23.9	1.2	0.7	7.2	4.3	13.2	7.8
13	8 48.3	8 49.7	8 24.2	1.3	0.8	7.3	4.3	13.3	7.9
14	8 48.5	8 49.9	8 24.4	1.4	0.8	7.4	4.4	13.4	7.9
15	8 48.8	8 50.2	8 24.7	1.5	0.9	7.5	4.4	13.5	8.0
16	8 49.0	8 50.4	8 24.9	1.6	0.9	7.6	4.5	13.6	8.0
17	8 49.3	8 50.7	8 25.1	1.7	1.0	7.7	4.6	13.7	8.1
18	8 49.5	8 50.9	8 25.4	1.8	1.1	7.8	4.6	13.8	8.2
19	8 49.8	8 51.2	8 25.6	1.9	1.1	7.9	4.7	13.9	8.2
20	8 50.0	8 51.5	8 25.9	2.0	1.2	8.0	4.7	14.0	8.3
21	8 50.3	8 51.7	8 26.1	2.1	1.2	8.1	4.8	14.1	8.3
22	8 50.5	8 52.0	8 26.3	2.2	1.3	8.2	4.9	14.2	8.4
23	8 50.8	8 52.2	8 26.6	2.3	1.4	8.3	4.9	14.3	8.5
24	8 51.0	8 52.5	8 26.8	2.4	1.4	8.4	5.0	14.4	8.5
25	8 51.3	8 52.7	8 27.0	2.5	1.5	8.5	5.0	14.5	8.6
26	8 51.5	8 53.0	8 27.3	2.6	1.5	8.6	5.1	14.6	8.6
27	8 51.8	8 53.2	8 27.5	2.7	1.6	8.7	5.1	14.7	8.7
28	8 52.0	8 53.5	8 27.8	2.8	1.7	8.8	5.2	14.8	8.8
29	8 52.3	8 53.7	8 28.0	2.9	1.7	8.9	5.3	14.9	8.8
30	8 52.5	8 54.0	8 28.2	3.0	1.8	9.0	5.3	15.0	8.9
31	8 52.8	8 54.2	8 28.5	3.1	1.8	9.1	5.4	15.1	8.9
32	8 53.0	8 54.5	8 28.7	3.2	1.9	9.2	5.4	15.2	9.0
33	8 53.3	8 54.7	8 29.0	3.3	2.0	9.3	5.5	15.3	9.1
34	8 53.5	8 55.0	8 29.2	3.4	2.0	9.4	5.6	15.4	9.1
35	8 53.8	8 55.2	8 29.4	3.5	2.1	9.5	5.6	15.5	9.2
36	8 54.0	8 55.5	8 29.7	3.6	2.1	9.6	5.7	15.6	9.2
37	8 54.3	8 55.7	8 29.9	3.7	2.2	9.7	5.7	15.7	9.3
38	8 54.5	8 56.0	8 30.2	3.8	2.2	9.8	5.8	15.8	9.3
39	8 54.8	8 56.2	8 30.4	3.9	2.3	9.9	5.9	15.9	9.4
40	8 55.0	8 56.5	8 30.6	4.0	2.4	10.0	5.9	16.0	9.5
41	8 55.3	8 56.7	8 30.9	4.1	2.4	10.1	6.0	16.1	9.5
42	8 55.5	8 57.0	8 31.1	4.2	2.5	10.2	6.0	16.2	9.6
43	8 55.8	8 57.2	8 31.3	4.3	2.5	10.3	6.1	16.3	9.6
44	8 56.0	8 57.5	8 31.6	4.4	2.6	10.4	6.2	16.4	9.7
45	8 56.3	8 57.7	8 31.8	4.5	2.7	10.5	6.2	16.5	9.8
46	8 56.5	8 58.0	8 32.1	4.6	2.7	10.6	6.3	16.6	9.8
47	8 56.8	8 58.2	8 32.3	4.7	2.8	10.7	6.3	16.7	9.9
48	8 57.0	8 58.5	8 32.5	4.8	2.8	10.8	6.4	16.8	9.9
49	8 57.3	8 58.7	8 32.8	4.9	2.9	10.9	6.4	16.9	10.0
50	8 57.5	8 59.0	8 33.0	5.0	3.0	11.0	6.5	17.0	10.1
51	8 57.8	8 59.2	8 33.3	5.1	3.0	11.1	6.6	17.1	10.1
52	8 58.0	8 59.5	8 33.5	5.2	3.1	11.2	6.6	17.2	10.2
53	8 58.3	8 59.7	8 33.7	5.3	3.1	11.3	6.7	17.3	10.2
54	8 58.5	9 00.0	8 34.0	5.4	3.2	11.4	6.7	17.4	10.3
55	8 58.8	9 00.2	8 34.2	5.5	3.3	11.5	6.8	17.5	10.4
56	8 59.0	9 00.5	8 34.4	5.6	3.3	11.6	6.9	17.6	10.4
57	8 59.3	9 00.7	8 34.7	5.7	3.4	11.7	6.9	17.7	10.5
58	8 59.5	9 01.0	8 34.9	5.8	3.4	11.8	7.0	17.8	10.5
59	8 59.8	9 01.2	8 35.2	5.9	3.5	11.9	7.0	17.9	10.6
60	9 00.0	9 01.5	8 35.4	6.0	3.6	12.0	7.1	18.0	10.7

© 2009 Jack Case

TABLE 6 — ALTITUDE CORRECTION TABLES

a. Dip of the Horizon

Ht. of Eye (m)	Corr^n D	Ht. of Eye (ft.)	Ht. of Eye (m)	Corr^n D	Ht. of Eye (ft.)
0·00	−0·3	0·0	13·0	−6·4	42·8
0·03	−0·4	0·1	13·4	−6·5	44·2
0·06	−0·5	0·2	13·8	−6·6	45·5
0·09	−0·6	0·3	14·2	−6·7	47·0
0·13	−0·7	0·4	14·7	−6·8	48·4
0·18	−0·8	0·5	15·1	−6·9	49·8
0·23	−0·9	0·7	15·5	−7·0	51·3
0·29	−1·0	0·9	16·0	−7·1	52·8
0·35	−1·1	1·1	16·5	−7·2	54·3
0·42	−1·2	1·4	16·9	−7·3	55·8
0·45	−1·3	1·6	17·4	−7·4	57·4
0·5	−1·4	1·9	17·9	−7·5	58·9
0·6	−1·5	2·2	18·4	−7·6	60·5
0·7	−1·6	2·5	18·8	−7·7	62·1
0·8	−1·7	2·8	19·3	−7·8	63·8
0·9	−1·8	3·2	19·8	−7·9	65·4
1·1	−1·9	3·6	20·4	−8·0	67·1
1·2	−2·0	4·0	20·9	−8·1	68·8
1·3	−2·1	4·4	21·4	−8·2	70·5
1·4	−2·2	4·9	21·9	−8·3	72·3
1·6	−2·3	5·3	22·5	−8·4	74·1
1·7	−2·4	5·8	23·0	−8·5	75·8
1·9	−2·5	6·3	23·5	−8·6	77·6
2·0	−2·6	6·9	24·1	−8·7	79·5
2·2	−2·7	7·4	24·7	−8·8	81·3
2·4	−2·8	8·0	25·2	−8·9	83·2
2·6	−2·9	8·6	25·8	−9·0	85·1
2·8	−3·0	9·2	26·4	−9·1	87·0
3·0	−3·1	9·8	27·0	−9·2	88·9
3·2	−3·2	10·5	27·6	−9·3	90·9
3·4	−3·3	11·2	28·2	−9·4	92·9
3·6	−3·4	11·9	28·8	−9·5	94·9
3·8	−3·5	12·6	29·4	−9·6	96·9
4·0	−3·6	13·3	30·0	−9·7	98·9
4·3	−3·7	14·1	30·6	−9·8	101·0
4·5	−3·8	14·9	31·3	−9·9	103·1
4·7	−3·9	15·7	31·9	−10·0	105·2
5·0	−4·0	16·5	32·6	−10·1	107·3
5·2	−4·1	17·4	33·2	−10·2	109·4
5·5	−4·2	18·3	33·9	−10·3	111·6
5·8	−4·3	19·1	34·5	−10·4	113·8
6·1	−4·4	20·1	35·2	−10·5	116·0
6·3	−4·5	21·0	35·9	−10·6	118·2
6·6	−4·6	22·0	36·6	−10·7	120·5
6·9	−4·7	22·9	37·3	−10·8	122·8
7·2	−4·8	23·9	38·0	−10·9	125·1
7·5	−4·9	25·0	38·7	−11·0	127·4
7·9	−5·0	26·0	39·4	−11·1	129·7
8·2	−5·1	27·1	40·1	−11·2	132·1
8·5	−5·2	28·1	40·8	−11·3	134·5
8·8	−5·3	29·2	41·5	−11·4	136·9
9·2	−5·4	30·4	42·3	−11·5	139·3
9·5	−5·5	31·5	43·0	−11·6	141·7
9·9	−5·6	32·7	43·8	−11·7	144·2
10·3	−5·7	33·9	44·5	−11·8	146·7
10·6	−5·8	35·1	45·3	−11·9	149·2
11·0	−5·9	36·3	46·1	−12·0	151·7
11·4	−6·0	37·6	46·8	−12·1	154·3
11·8	−6·1	38·9	47·6	−12·2	156·8
12·2	−6·2	40·1	48·4	−12·3	159·4
12·6	−6·3	41·5	49·2	−12·4	162·1
13·0		42·8	50·0		164·7

b. Refraction for Stars & Planets

App. Alt. H_a	Corr^n R	App. Alt. H_a	Corr^n R	App. Alt. H_a	Corr^n R
0 00	−33·8	3 30	−12·9	9 55	
0 03	33·2	3 35	12·7	10 07	−5·3
0 06	32·6	3 40	12·5	10 20	−5·2
0 09	32·0	3 45	12·3	10 32	−5·1
0 12	31·5	3 50	12·1	10 46	−5·0
0 15	30·9	3 55	11·9	10 59	−4·9
0 18	−30·4	4 00	−11·7	11 14	−4·8
0 21	29·8	4 05	11·5	11 29	−4·7
0 24	29·3	4 10	11·4	11 44	−4·6
0 27	28·8	4 15	11·2	12 00	−4·5
0 30	28·3	4 20	11·0	12 17	−4·4
0 33	27·9	4 25	10·9	12 35	−4·3
0 36	−27·4	4 30	−10·7	12 53	−4·2
0 39	26·9	4 35	10·6	13 12	−4·1
0 42	26·5	4 40	10·4	13 32	−4·0
0 45	26·1	4 45	10·3	13 53	−3·9
0 48	25·7	4 50	10·1	14 16	−3·8
0 51	25·3	4 55	10·0	14 39	−3·7
0 54	−24·9	5 00	−9·8	15 03	−3·6
0 57	24·5	5 05	9·7	15 29	−3·5
1 00	24·1	5 10	9·6	15 56	−3·4
1 03	23·7	5 15	9·5	16 25	−3·3
1 06	23·4	5 20	9·3	16 55	−3·2
1 09	23·0	5 25	9·2	17 27	−3·1
1 12	−22·7	5 30	−9·1	18 01	−3·0
1 15	22·3	5 35	9·0	18 37	−2·9
1 18	22·0	5 40	8·9	19 16	−2·8
1 21	21·7	5 45	8·8	19 56	−2·7
1 24	21·4	5 50	8·7	20 40	−2·6
1 27	21·1	5 55	8·6	21 27	−2·5
1 30	−20·8	6 00	−8·5	22 17	−2·4
1 35	20·3	6 10	8·3	23 11	−2·3
1 40	19·9	6 20	8·1	24 09	−2·2
1 45	19·4	6 30	7·9	25 12	−2·1
1 50	19·0	6 40	7·7	26 20	−2·0
1 55	18·6	6 50	7·6	27 34	−1·9
2 00	−18·2	7 00	−7·4	28 54	−1·8
2 05	17·8	7 10	7·2	30 22	−1·7
2 10	17·4	7 20	7·1	31 58	−1·6
2 15	17·1	7 30	6·9	33 43	−1·5
2 20	16·7	7 40	6·8	35 38	−1·4
2 25	16·4	7 50	6·7	37 45	−1·3
2 30	−16·1	8 00	−6·6	40 06	−1·2
2 35	15·8	8 10	6·4	42 42	−1·1
2 40	15·4	8 20	6·3	45 34	−1·0
2 45	15·2	8 30	6·2	48 45	−0·9
2 50	14·9	8 40	6·1	52 16	−0·8
2 55	14·6	8 50	6·0	56 09	−0·7
3 00	−14·3	9 00	−5·9	60 26	−0·6
3 05	14·1	9 10	5·8	65 06	−0·5
3 10	13·8	9 20	5·7	70 09	−0·4
3 15	13·6	9 30	5·6	75 32	−0·3
3 20	13·4	9 40	5·5	81 12	−0·2
3 25	13·1	9 50	5·4	87 03	−0·1
3 30	−12·9	10 00	−5·3	90 00	0·0

H_a = App. Alt. = Apparent Altitude
= Sextant altitude corrected for index error (IE) & dip (*d*)
= H_s + IE + D

In critical cases ascend.

TABLE 6 — ALTITUDE CORRECTION TABLES

c. Additional Refraction for Non-standard Conditions

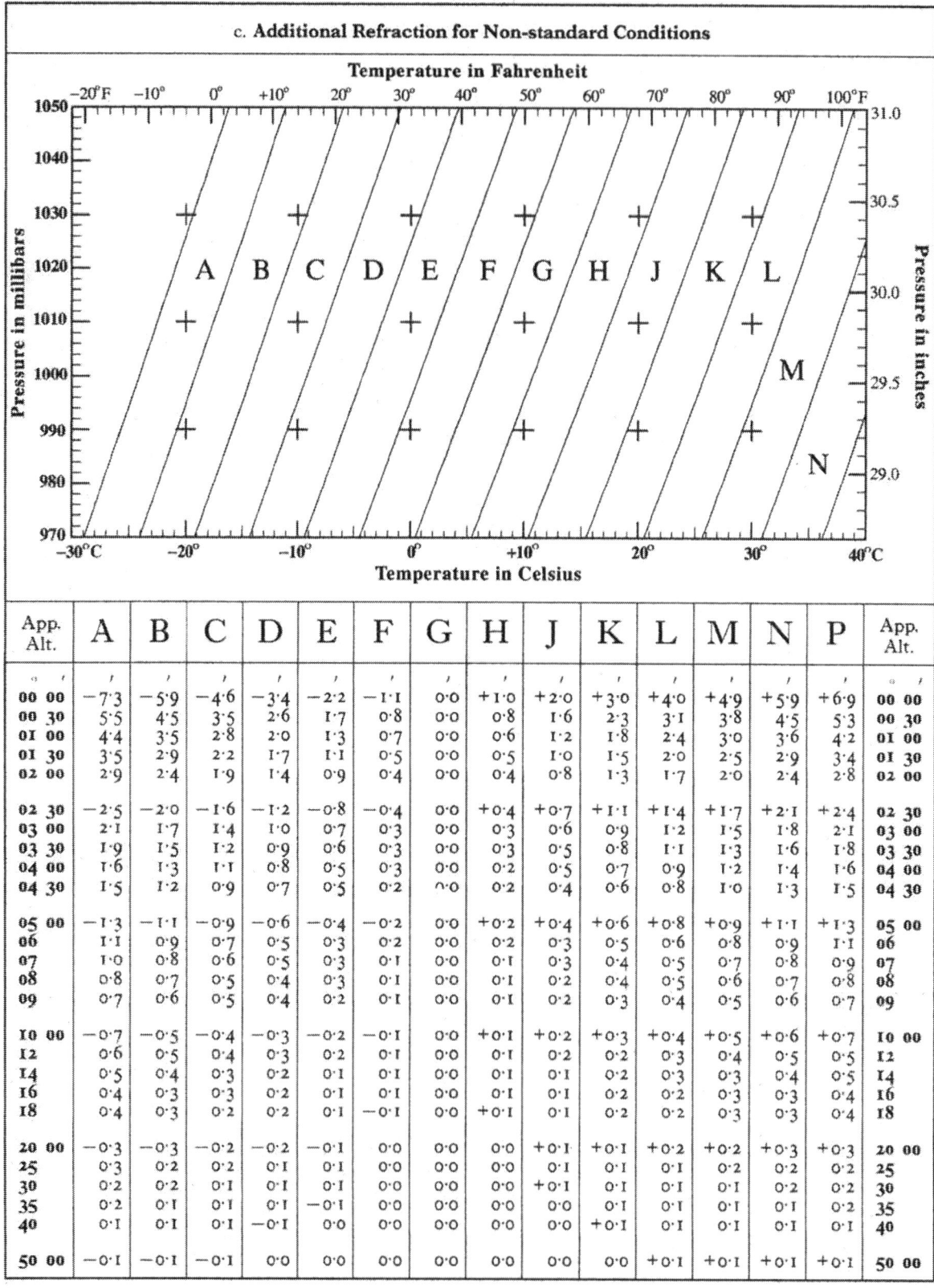

App. Alt.	A	B	C	D	E	F	G	H	J	K	L	M	N	P	App. Alt.
° ′	′	′	′	′	′	′	′	′	′	′	′	′	′	′	° ′
00 00	−7·3	−5·9	−4·6	−3·4	−2·2	−1·1	0·0	+1·0	+2·0	+3·0	+4·0	+4·9	+5·9	+6·9	00 00
00 30	5·5	4·5	3·5	2·6	1·7	0·8	0·0	0·8	1·6	2·3	3·1	3·8	4·5	5·3	00 30
01 00	4·4	3·5	2·8	2·0	1·3	0·7	0·0	0·6	1·2	1·8	2·4	3·0	3·6	4·2	01 00
01 30	3·5	2·9	2·2	1·7	1·1	0·5	0·0	0·5	1·0	1·5	2·0	2·5	2·9	3·4	01 30
02 00	2·9	2·4	1·9	1·4	0·9	0·4	0·0	0·4	0·8	1·3	1·7	2·0	2·4	2·8	02 00
02 30	−2·5	−2·0	−1·6	−1·2	−0·8	−0·4	0·0	+0·4	+0·7	+1·1	+1·4	+1·7	+2·1	+2·4	02 30
03 00	2·1	1·7	1·4	1·0	0·7	0·3	0·0	0·3	0·6	0·9	1·2	1·5	1·8	2·1	03 00
03 30	1·9	1·5	1·2	0·9	0·6	0·3	0·0	0·3	0·5	0·8	1·1	1·3	1·6	1·8	03 30
04 00	1·6	1·3	1·1	0·8	0·5	0·3	0·0	0·2	0·5	0·7	0·9	1·2	1·4	1·6	04 00
04 30	1·5	1·2	0·9	0·7	0·5	0·2	0·0	0·2	0·4	0·6	0·8	1·0	1·3	1·5	04 30
05 00	−1·3	−1·1	−0·9	−0·6	−0·4	−0·2	0·0	+0·2	+0·4	+0·6	+0·8	+0·9	+1·1	+1·3	05 00
06	1·1	0·9	0·7	0·5	0·3	0·2	0·0	0·2	0·3	0·5	0·6	0·8	0·9	1·1	06
07	1·0	0·8	0·6	0·5	0·3	0·1	0·0	0·1	0·3	0·4	0·5	0·7	0·8	0·9	07
08	0·8	0·7	0·5	0·4	0·3	0·1	0·0	0·1	0·2	0·4	0·5	0·6	0·7	0·8	08
09	0·7	0·6	0·5	0·4	0·2	0·1	0·0	0·1	0·2	0·3	0·4	0·5	0·6	0·7	09
10 00	−0·7	−0·5	−0·4	−0·3	−0·2	−0·1	0·0	+0·1	+0·2	+0·3	+0·4	+0·5	+0·6	+0·7	10 00
12	0·6	0·5	0·4	0·3	0·2	0·1	0·0	0·1	0·2	0·2	0·3	0·4	0·5	0·5	12
14	0·5	0·4	0·3	0·2	0·1	0·1	0·0	0·1	0·1	0·2	0·3	0·3	0·4	0·5	14
16	0·4	0·3	0·3	0·2	0·1	0·1	0·0	0·1	0·1	0·2	0·2	0·3	0·3	0·4	16
18	0·4	0·3	0·2	0·2	0·1	−0·1	0·0	+0·1	0·1	0·2	0·2	0·3	0·3	0·4	18
20 00	−0·3	−0·3	−0·2	−0·2	−0·1	0·0	0·0	0·0	+0·1	+0·1	+0·2	+0·2	+0·3	+0·3	20 00
25	0·3	0·2	0·2	0·1	0·1	0·0	0·0	0·0	0·1	0·1	0·1	0·2	0·2	0·2	25
30	0·2	0·2	0·1	0·1	0·1	0·0	0·0	+0·1	0·1	0·1	0·1	0·2	0·2	0·2	30
35	0·2	0·1	0·1	0·1	−0·1	0·0	0·0	0·0	0·1	0·1	0·1	0·1	0·1	0·2	35
40	0·1	0·1	0·1	−0·1	0·0	0·0	0·0	0·0	+0·1	0·1	0·1	0·1	0·1	0·1	40
50 00	−0·1	−0·1	−0·1	0·0	0·0	0·0	0·0	0·0	0·0	+0·1	+0·1	+0·1	+0·1	+0·1	50 00

The graph is entered with arguments temperature and pressure to find a zone letter; using as arguments this zone letter and apparent altitude (sextant altitude corrected for index error and dip), a correction is taken from the table. This correction is to be applied to the sextant altitude in addition to the corrections for standard conditions (Table **6**b).

TABLE 6d — ALTITUDE CORRECTION TABLES FOR THE SUN

App. Alt. H_a	OCT.–MAR. Lower Limb R	OCT.–MAR. Upper Limb R	APR.–SEPT. Lower Limb R	APR.–SEPT. Upper Limb R	App. Alt. H_a	OCT.–MAR. Lower Limb R	OCT.–MAR. Upper Limb R	APR.–SEPT. Lower Limb R	APR.–SEPT. Upper Limb R	App. Alt. H_a	OCT.–MAR. Lower Limb R	OCT.–MAR. Upper Limb R	App. Alt. H_a	APR.–SEPT. Lower Limb R	APR.–SEPT. Upper Limb R
0 00	−17.5	−49.8	−17.8	−49.6	3 30	+3.4	−28.9	+3.1	−28.7	9 33	+10.8	−21.5	9 39	+10.6	−21.2
0 03	16.9	49.2	17.2	49.0	3 35	3.6	28.7	3.3	28.5	9 45	+10.9	−21.4	9 50	+10.7	−21.1
0 06	16.3	48.6	16.6	48.4	3 40	3.8	28.5	3.6	28.2	9 56	+11.0	−21.3	10 02	+10.8	−21.0
0 09	15.7	48.0	16.0	47.8	3 45	4.0	28.3	3.8	28.0	10 08	+11.1	−21.2	10 14	+10.9	−20.9
0 12	15.2	47.5	15.4	47.2	3 50	4.2	28.1	4.0	27.8	10 20	+11.2	−21.1	10 27	+11.0	−20.8
0 15	14.6	46.9	14.8	46.6	3 55	4.4	27.9	4.1	27.7	10 33	+11.3	−21.0	10 40	+11.1	−20.7
0 18	−14.1	−46.4	−14.3	−46.1	4 00	+4.6	−27.7	+4.3	−27.5	10 46	+11.4	−20.9	10 53	+11.2	−20.6
0 21	13.5	45.8	13.8	45.6	4 05	4.8	27.5	4.5	27.3	11 00	+11.5	−20.8	11 07	+11.3	−20.5
0 24	13.0	45.3	13.3	45.1	4 10	4.9	27.4	4.7	27.1	11 15	+11.6	−20.7	11 22	+11.4	−20.4
0 27	12.5	44.8	12.8	44.6	4 15	5.1	27.2	4.9	26.9	11 30	+11.7	−20.6	11 37	+11.5	−20.3
0 30	12.0	44.3	12.3	44.1	4 20	5.3	27.0	5.0	26.8	11 45	+11.8	−20.5	11 53	+11.6	−20.2
0 33	11.6	43.9	11.8	43.6	4 25	5.4	26.9	5.2	26.6	12 01	+11.9	−20.4	12 10	+11.7	−20.1
0 36	−11.1	−43.4	−11.3	−43.1	4 30	+5.6	−26.7	+5.3	−26.5	12 18	+12.0	−20.3	12 27	+11.8	−20.0
0 39	10.6	42.9	10.9	42.7	4 35	5.7	26.6	5.5	26.3	12 36	+12.1	−20.2	12 45	+11.9	−19.9
0 42	10.2	42.5	10.5	42.3	4 40	5.9	26.4	5.6	26.2	12 54	+12.2	−20.1	13 04	+12.0	−19.8
0 45	9.8	42.1	10.0	41.8	4 45	6.0	26.3	5.8	26.0	13 14	+12.3	−20.0	13 24	+12.1	−19.7
0 48	9.4	41.7	9.6	41.4	4 50	6.2	26.1	5.9	25.9	13 34	+12.4	−19.9	13 44	+12.2	−19.6
0 51	9.0	41.3	9.2	41.0	4 55	6.3	26.0	6.1	25.7	13 55	+12.5	−19.8	14 06	+12.3	−19.5
0 54	−8.6	−40.9	−8.8	−40.6	5 00	+6.4	−25.9	+6.2	−25.6	14 17	+12.6	−19.7	14 29	+12.4	−19.4
0 57	8.2	40.5	8.4	40.2	5 05	6.6	25.7	6.3	25.5	14 41	+12.7	−19.6	14 53	+12.5	−19.3
1 00	7.8	40.1	8.0	39.8	5 10	6.7	25.6	6.5	25.3	15 05	+12.8	−19.5	15 18	+12.6	−19.2
1 03	7.4	39.7	7.7	39.5	5 15	6.8	25.5	6.6	25.2	15 31	+12.9	−19.4	15 45	+12.7	−19.1
1 06	7.1	39.4	7.3	39.1	5 20	7.0	25.3	6.7	25.1	15 59	+13.0	−19.3	16 13	+12.8	−19.0
1 09	6.7	39.0	7.0	38.8	5 25	7.1	25.2	6.8	25.0	16 27	+13.1	−19.2	16 43	+12.9	−18.9
1 12	−6.4	−38.7	−6.6	−38.4	5 30	+7.2	−25.1	+6.9	−24.9	16 58	+13.2	−19.1	17 14	+13.0	−18.8
1 15	6.0	38.3	6.3	38.1	5 35	7.3	25.0	7.1	24.7	17 30	+13.3	−19.0	17 47	+13.1	−18.7
1 18	5.7	38.0	6.0	37.8	5 40	7.4	24.9	7.2	24.6	18 05	+13.4	−18.9	18 23	+13.2	−18.6
1 21	5.4	37.7	5.7	37.5	5 45	7.5	24.8	7.3	24.5	18 41	+13.5	−18.8	19 00	+13.3	−18.5
1 24	5.1	37.4	5.3	37.1	5 50	7.6	24.7	7.4	24.4	19 20	+13.6	−18.7	19 41	+13.4	−18.4
1 27	4.8	37.1	5.0	36.8	5 55	7.7	24.6	7.5	24.3	20 02	+13.7	−18.6	20 24	+13.5	−18.3
1 30	−4.5	−36.8	−4.7	−36.5	6 00	+7.8	−24.5	+7.6	−24.2	20 46	+13.8	−18.5	21 10	+13.6	−18.2
1 35	4.0	36.3	4.3	36.1	6 10	8.0	24.3	7.8	24.0	21 34	+13.9	−18.4	21 59	+13.7	−18.1
1 40	3.6	35.9	3.8	35.6	6 20	8.2	24.1	8.0	23.8	22 25	+14.0	−18.3	22 52	+13.8	−18.0
1 45	3.1	35.4	3.4	35.2	6 30	8.4	23.9	8.2	23.6	23 20	+14.1	−18.2	23 49	+13.9	−17.9
1 50	2.7	35.0	2.9	34.7	6 40	8.6	23.7	8.3	23.5	24 20	+14.2	−18.1	24 51	+14.0	−17.8
1 55	2.3	34.6	2.5	34.3	6 50	8.7	23.6	8.5	23.3	25 24	+14.3	−18.0	25 58	+14.1	−17.7
2 00	−1.9	−34.2	−2.1	−33.9	7 00	+8.9	−23.4	+8.7	−23.1	26 34	+14.4	−17.9	27 11	+14.2	−17.6
2 05	1.5	33.8	1.7	33.5	7 10	9.1	23.2	8.8	23.0	27 50	+14.5	−17.8	28 31	+14.3	−17.5
2 10	1.1	33.4	1.4	33.2	7 20	9.2	23.1	9.0	22.8	29 13	+14.6	−17.7	29 58	+14.4	−17.4
2 15	0.8	33.1	1.0	32.8	7 30	9.3	23.0	9.1	22.7	30 44	+14.7	−17.6	31 33	+14.5	−17.3
2 20	0.4	32.7	0.7	32.5	7 40	9.5	22.8	9.2	22.6	32 24	+14.8	−17.5	33 18	+14.6	−17.2
2 25	−0.1	32.4	−0.3	32.1	7 50	9.6	22.7	9.4	22.4	34 15	+14.9	−17.4	35 15	+14.7	−17.1
2 30	+0.2	−32.1	0.0	−31.8	8 00	+9.7	−22.6	+9.5	−22.3	36 17	+15.0	−17.3	37 24	+14.8	−17.0
2 35	0.5	31.8	+0.3	31.5	8 10	9.9	22.4	9.6	22.2	38 34	+15.1	−17.2	39 48	+14.9	−16.9
2 40	0.8	31.5	0.6	31.2	8 20	10.0	22.3	9.7	22.1	41 06	+15.2	−17.1	42 28	+15.0	−16.8
2 45	1.1	31.2	0.9	30.9	8 30	10.1	22.2	9.9	21.9	43 56	+15.3	−17.0	45 29	+15.1	−16.7
2 50	1.4	30.9	1.2	30.6	8 40	10.2	22.1	10.0	21.8	47 07	+15.4	−16.9	48 52	+15.2	−16.6
2 55	1.7	30.6	1.4	30.4	8 50	10.3	22.0	10.1	21.7	50 43	+15.5	−16.8	52 41	+15.3	−16.5
3 00	+2.0	−30.3	+1.7	−30.1	9 00	+10.4	−21.9	+10.2	−21.6	54 46	+15.6	−16.7	56 59	+15.4	−16.4
3 05	2.2	30.1	2.0	29.8	9 10	10.5	21.8	10.3	21.5	59 21	+15.7	−16.6	61 50	+15.5	−16.3
3 10	2.5	29.8	2.2	29.6	9 20	10.6	21.7	10.4	21.4	64 28	+15.8	−16.5	67 15	+15.6	−16.2
3 15	2.7	29.6	2.5	29.3	9 30	10.7	21.6	10.5	21.3	70 10	+15.9	−16.4	73 14	+15.7	−16.1
3 20	2.9	29.4	2.7	29.1	9 40	10.8	21.5	10.6	21.2	76 24	+16.0	−16.3	79 42	+15.8	−16.0
3 25	3.2	29.1	2.9	28.9	9 50	10.9	21.4	10.6	21.2	83 05	+16.1	−16.2	86 31	+15.9	−15.9
3 30	+3.4	−28.9	+3.1	−28.7	10 00	+11.0	−21.3	+10.7	−21.1	90 00			90 00		

Main table lat. 43°, contrary declinations 19°, 20°; LHAs 57°, 305° and 357°.

LHA	15° Hc d Z	16° Hc d Z	17° Hc d Z	18° Hc d Z	19° Hc d Z	20° Hc d Z	LHA
69	04 24 44 115	03 40 44 116	02 56 44 117	02 12 43 117	01 29 44 118	00 45 44 119	291
68	05 03 44 116	04 19 44 117	03 35 44 117	02 51 44 118	02 07 44 119	01 23 44 119	292
67	05 43 45 117	04 58 44 117	04 14 44 118	03 30 44 119	02 46 45 119	02 01 44 120	293
66	06 22 45 117	05 37 44 118	04 53 45 119	04 08 44 119	03 24 45 120	02 39 44 121	294
65	07 01 −45 118	06 16 −45 119	05 31 −45 119	04 46 −44 120	04 02 −45 121	03 17 −45 121	295
64	07 39 45 119	06 54 45 120	06 09 45 120	05 24 45 121	04 39 45 122	03 54 45 122	296
63	08 17 45 120	07 32 45 120	06 47 45 121	06 02 45 122	05 16 45 122	04 31 45 123	297
62	08 55 45 120	08 10 45 121	07 25 46 122	06 39 46 122	05 53 45 123	05 08 46 124	298
61	09 33 45 121	08 48 46 122	08 02 46 122	07 16 46 123	06 30 46 124	05 44 46 124	299
60	10 11 −46 122	09 25 −46 122	08 39 −46 123	07 53 −47 124	07 06 −46 124	06 20 −46 125	300
59	10 48 46 123	10 02 47 123	09 15 46 124	08 29 47 124	07 42 46 125	06 56 47 126	301
58	11 25 47 123	10 38 46 124	09 52 47 125	09 05 47 125	08 18 47 126	07 31 46 127	302
57	12 01 47 124	11 14 46 125	10 28 47 125	09 41 47 126	08 54 48 127	08 06 47 127	303
56	12 37 47 125	11 50 47 125	11 03 47 126	10 16 47 127	09 29 48 127	08 41 47 128	304
55	13 13 −47 126	12 26 −48 126	11 38 −47 127	10 51 −48 128	10 03 −47 128	09 16 −48 129	305
54	13 49 48 126	13 01 48 127	12 13 47 128	11 26 48 128	10 38 48 129	09 50 48 130	306
53	14 24 48 127	13 36 48 128	12 48 48 128	12 00 48 129	11 12 49 130	10 23 49 130	307
52	14 59 49 128	14 10 48 129	13 22 48 129	12 34 49 130	11 45 49 130	10 57 49 131	308
51	15 33 49 129	14 44 48 129	13 56 49 130	13 07 49 131	12 18 48 131	11 30 49 132	309
50	16 07 −49 130	15 18 −49 130	14 29 −49 131	13 40 −49 131	12 51 −49 132	12 02 −49 133	310
49	16 41 50 130	15 51 49 131	15 02 49 132	14 13 49 132	13 24 50 133	12 34 49 133	311
48	17 14 50 131	16 24 49 132	15 35 50 132	14 45 49 133	13 56 50 134	13 06 50 134	312
47	17 46 49 132	16 57 50 133	16 07 50 133	15 17 50 134	14 27 50 134	13 37 50 135	313
46	18 19 50 133	17 29 50 134	16 39 50 134	15 49 51 135	14 58 50 135	14 08 51 136	314
45	18 51 −51 134	18 00 −50 134	17 10 −51 135	16 19 −50 136	15 29 −51 136	14 38 −51 137	315
44	19 22 50 135	18 32 51 135	17 41 51 136	16 50 51 136	15 59 51 137	15 08 51 137	316
43	19 53 51 136	19 02 51 136	18 11 51 137	17 20 51 137	16 29 51 138	15 38 52 138	317
42	20 24 52 136	19 32 51 137	18 41 51 138	17 50 52 138	16 58 51 139	16 07 52 139	318
41	20 54 52 137	20 02 52 138	19 10 51 138	18 19 52 139	17 27 52 139	16 35 52 140	319
40	21 23 −52 138	20 31 −52 139	19 39 −52 139	18 47 −52 140	17 55 −52 140	17 03 −52 141	320
39	21 52 52 139	21 00 52 140	20 08 53 140	19 15 52 141	18 23 53 141	17 30 52 142	321
38	22 21 53 140	21 28 52 141	20 36 53 141	19 43 53 142	18 50 53 142	17 57 52 143	322
37	22 49 53 141	21 56 53 141	21 03 53 142	20 10 53 142	19 17 53 143	18 24 53 143	323
36	23 16 53 142	22 23 53 142	21 30 54 143	20 36 53 143	19 43 53 144	18 50 54 144	324
35	23 43 −54 143	22 49 −53 143	21 56 −54 144	21 02 −53 144	20 09 −54 145	19 15 −54 145	325
34	24 09 54 144	23 15 53 144	22 22 54 145	21 28 54 145	20 34 54 146	19 40 54 146	326
33	24 35 54 145	23 41 54 145	22 47 54 146	21 53 55 146	20 58 54 147	20 04 54 147	327
32	25 00 54 146	24 06 55 146	23 11 54 147	22 17 55 147	21 22 54 147	20 28 55 148	328
31	25 24 54 147	24 30 54 147	23 35 55 147	22 40 54 148	21 46 55 148	20 51 55 149	329
30	25 48 −55 148	24 53 −55 148	23 58 −55 148	23 03 −55 149	22 08 −55 149	21 13 −55 150	330
29	26 11 55 149	25 16 55 149	24 21 55 149	23 26 56 150	22 30 55 150	21 35 56 151	331
28	26 34 55 150	25 39 56 150	24 43 56 150	23 47 55 151	22 52 56 151	21 56 56 152	332
27	26 56 56 151	26 00 56 151	25 04 55 151	24 09 56 152	23 13 56 152	22 17 56 153	333
26	27 17 56 152	26 21 56 152	25 25 56 152	24 29 56 153	23 33 56 153	22 37 57 153	334
25	27 38 −57 153	26 41 −56 153	25 45 −56 153	24 49 −57 154	23 52 −56 154	22 56 −57 154	335
24	27 58 57 154	27 01 57 154	26 04 56 154	25 08 57 155	24 11 57 155	23 14 56 155	336
23	28 17 57 155	27 20 57 155	26 23 57 155	25 26 57 156	24 29 57 156	23 32 57 156	337
22	28 35 57 156	27 38 57 156	26 41 57 156	25 44 57 157	24 47 57 157	23 50 57 157	338
21	28 53 57 157	27 56 58 157	26 58 57 157	26 01 57 158	25 04 58 158	24 06 57 158	339
20	29 10 −58 158	28 12 −57 158	27 15 −58 158	26 17 −57 159	25 20 −58 159	24 22 −58 159	340
19	29 26 58 159	28 28 57 159	27 31 58 159	26 33 58 160	25 35 58 160	24 37 58 160	341
18	29 42 58 160	28 44 58 160	27 46 58 160	26 48 58 161	25 50 59 161	24 51 58 161	342
17	29 56 58 161	28 58 58 161	28 00 58 162	27 02 59 162	26 03 58 162	25 05 58 162	343
16	30 10 58 162	29 12 59 162	28 13 58 163	27 15 58 163	26 17 59 163	25 18 58 163	344
15	30 23 −58 163	29 25 −59 163	28 26 −58 164	27 28 −59 164	26 29 −59 164	25 30 −58 164	345
14	30 36 59 164	29 37 59 164	28 38 59 165	27 39 58 165	26 41 59 165	25 42 59 165	346
13	30 47 59 165	29 48 59 166	28 49 59 166	27 50 59 166	26 51 59 166	25 52 59 166	347
12	30 58 59 166	29 59 59 167	29 00 59 167	28 01 60 167	27 01 59 167	26 02 59 167	348
11	31 08 60 168	30 08 59 168	29 09 59 168	28 10 59 168	27 11 60 168	26 11 59 168	349
10	31 17 −60 169	30 17 −59 169	29 18 −59 169	28 19 −60 169	27 19 −59 169	26 20 −60 170	350
9	31 25 59 170	30 25 59 170	29 26 60 170	28 26 59 170	27 27 60 170	26 27 59 171	351
8	31 32 59 171	30 33 60 171	29 33 60 171	28 33 59 171	27 34 59 171	26 34 59 172	352
7	31 39 60 172	30 39 60 172	29 39 59 172	28 40 60 172	27 40 60 173	26 40 59 173	353
6	31 44 59 173	30 45 60 173	29 45 60 173	28 45 60 173	27 45 60 174	26 45 59 174	354
5	31 49 −60 174	30 49 −60 174	29 49 −59 174	28 50 −60 175	27 50 −60 175	26 50 −60 175	355
4	31 53 60 175	30 53 60 176	29 53 60 176	28 53 60 176	27 53 60 176	26 54 60 176	356
3	31 56 60 177	30 56 60 177	29 56 60 177	28 56 60 177	27 56 60 177	26 56 60 177	357
2	31 58 60 178	30 58 60 178	29 58 60 178	28 58 60 178	27 58 60 178	26 58 60 178	358
1	32 00 60 179	31 00 60 179	30 00 60 179	29 00 60 179	28 00 60 179	27 00 60 179	359
0	32 00 −60 180	31 00 −60 180	30 00 −60 180	29 00 −60 180	28 00 −60 180	27 00 −60 180	360
	15°	16°	17°	18°	19°	20°	

Extract of part of Table 5.

31	32	33	34	35	36	37	38	39	40	41	42	43	44	45	46	47	48	49	50	51	52	53	54	55	56	57	58	59	60	d
′	′	′	′	′	′	′	′	′	′	′	′	′	′	′	′	′	′	′	′	′	′	′	′	′	′	′	′	′	′	′
0	0	0	0	0	0	0	0	0	0	0	0	0	0	0	0	0	0	0	0	0	0	0	0	0	0	0	0	0	0	0
1	1	1	1	1	1	1	1	1	1	1	1	1	1	1	1	1	1	1	1	1	1	1	1	1	1	1	1	1	1	1
1	1	1	1	1	1	1	1	1	1	1	1	1	1	2	2	2	2	2	2	2	2	2	2	2	2	2	2	2	2	2
2	2	2	2	2	2	2	2	2	2	2	2	2	2	2	2	2	2	2	2	3	3	3	3	3	3	3	3	3	3	3
2	2	2	2	2	2	2	3	3	3	3	3	3	3	3	3	3	3	3	3	3	3	4	4	4	4	4	4	4	4	4
3	3	3	3	3	3	3	3	3	3	3	4	4	4	4	4	4	4	4	4	4	4	4	4	5	5	5	5	5	5	5
3	3	3	3	4	4	4	4	4	4	4	4	5	5	5	5	5	5	5	5	5	5	5	5	6	6	6	6	6	6	6
4	4	4	4	4	4	4	4	5	5	5	5	5	5	5	5	6	6	6	6	6	6	6	6	6	7	7	7	7	7	7
4	4	4	4	4	4	5	5	5	5	5	6	6	6	6	6	6	6	7	7	7	7	7	7	7	7	8	8	8	8	8
5	5	5	5	5	5	6	6	6	6	6	6	7	7	7	7	7	7	7	8	8	8	8	8	8	8	9	9	9	9	9
5	5	6	6	6	6	6	7	7	7	7	8	7	7	8	8	8	8	9	9	9	9	9	10	10	10	10	10	10	10	10
6	6	6	6	6	7	7	7	8	8	8	8	8	8	8	8	9	9	9	9	9	10	10	10	10	10	10	11	11	11	11
6	6	7	7	7	7	7	8	8	8	8	8	9	9	9	10	10	10	10	11	11	11	11	12	12	12	12	12	12	12	12
7	7	7	7	8	8	8	8	8	9	9	9	9	10	10	10	10	10	11	11	11	11	12	12	12	12	13	13	13	13	13
7	7	8	8	8	8	9	9	9	9	10	10	10	10	10	11	11	11	11	12	12	12	13	13	13	13	13	14	14	14	14
8	8	8	8	9	9	9	10	10	10	10	10	11	11	11	12	12	12	12	12	13	13	13	14	14	14	14	14	15	15	15
8	9	9	9	9	10	10	10	11	11	11	11	12	12	13	13	13	14	13	13	14	14	14	14	15	15	15	15	16	16	16
9	9	9	10	10	10	10	11	11	11	12	12	12	12	13	13	13	14	14	14	14	15	15	15	16	16	16	16	17	17	17
9	10	10	10	10	11	11	11	12	12	12	13	13	13	14	14	14	14	15	15	15	16	16	16	16	17	17	17	18	18	18
10	10	10	11	11	11	12	12	12	13	13	13	14	14	14	15	15	15	16	16	16	16	17	17	17	18	18	18	19	19	19
10	11	11	11	12	12	12	13	13	13	14	14	14	15	15	15	16	16	17	17	17	17	18	18	18	19	19	19	20	20	20
11	11	12	12	12	13	13	13	14	14	14	15	15	15	16	16	16	17	17	18	18	18	19	19	19	20	20	20	21	21	21
11	12	12	12	13	13	13	14	14	14	15	15	15	16	16	16	17	17	18	18	19	19	19	20	20	21	21	21	22	22	22
12	12	13	13	13	14	14	15	15	15	16	16	16	17	17	18	18	18	19	19	20	20	20	21	21	21	22	22	23	23	23
12	13	13	14	14	14	15	15	16	16	16	17	17	18	18	18	19	19	20	20	20	21	21	22	22	22	23	23	24	24	24
13	13	14	14	15	15	15	16	16	17	17	18	18	18	19	19	20	20	20	21	21	22	22	22	23	23	24	24	25	25	25
13	14	14	15	15	16	16	16	17	17	18	18	19	19	20	20	20	21	21	22	22	23	23	23	24	24	25	25	26	26	26
14	14	15	15	16	16	17	17	18	18	18	19	19	20	20	21	21	22	22	22	23	23	24	24	25	25	26	26	27	27	27
14	15	15	16	16	17	17	18	18	19	19	20	20	21	21	21	22	22	23	23	24	24	25	25	26	26	27	27	28	28	28
15	15	16	16	17	17	18	18	19	19	20	20	21	21	22	22	23	23	24	24	25	25	26	26	27	27	28	28	29	29	29
16	16	16	17	18	18	18	19	20	20	20	21	22	22	22	23	24	24	24	25	26	26	26	27	28	28	28	29	30	30	30
16	17	17	18	18	19	19	20	20	21	21	22	22	23	23	24	24	25	25	26	26	27	27	28	28	29	29	30	30	31	31
17	17	18	18	19	19	20	20	21	21	22	22	23	23	24	25	25	26	26	27	27	28	28	29	29	30	30	31	31	32	32
17	18	18	19	19	20	20	21	21	22	23	23	24	24	25	25	26	26	27	28	28	29	29	30	30	31	31	32	33	33	33
18	18	19	19	20	20	21	22	22	23	23	24	24	25	26	26	27	27	28	28	29	29	30	31	31	32	32	33	33	34	34
18	19	19	20	20	21	22	22	23	23	24	24	25	26	26	27	27	28	29	29	30	30	31	32	32	33	33	34	34	35	35
19	19	20	20	21	22	22	23	23	24	25	25	26	26	27	28	28	29	29	30	31	31	32	32	33	34	34	35	35	36	36
19	20	20	21	22	22	23	23	24	25	25	26	27	27	28	28	29	30	30	31	31	32	33	33	34	35	35	36	36	37	37
20	20	21	22	22	23	23	24	25	25	26	27	27	28	28	29	30	30	31	31	32	33	34	34	35	35	36	37	37	38	38
20	21	21	22	23	23	24	25	25	26	26	27	28	29	29	30	30	31	32	32	33	34	34	35	36	36	37	38	38	39	39
21	21	22	23	23	24	25	25	26	27	27	28	29	29	30	31	31	32	33	33	34	35	35	36	37	37	38	39	39	40	40
21	22	23	23	24	25	25	26	27	27	28	29	29	30	31	31	32	33	33	34	35	36	36	37	38	38	39	40	40	41	41
22	22	23	24	24	25	26	27	27	28	29	29	30	31	32	32	33	34	34	35	36	37	37	38	38	39	40	41	41	42	42
22	23	24	24	25	26	27	27	28	29	29	30	31	32	32	33	34	34	35	36	37	37	38	39	39	40	41	42	42	43	43
23	23	24	25	26	26	27	28	29	29	30	31	32	32	33	34	34	35	36	37	37	38	39	40	40	41	42	43	43	44	44
23	24	25	26	26	27	28	28	29	30	31	32	32	33	34	35	36	36	37	38	38	39	40	40	41	42	43	44	44	45	45
24	25	25	26	27	28	28	29	30	31	31	32	33	34	34	35	36	37	38	38	39	40	41	41	42	43	44	44	45	46	46
24	25	26	27	27	28	29	30	31	31	32	33	34	34	35	36	37	38	38	39	40	41	42	42	43	44	45	45	46	47	47
25	26	26	27	28	29	30	30	31	32	33	34	34	35	36	37	38	38	39	40	41	42	42	43	44	45	46	46	47	48	48
25	26	27	28	29	29	30	31	32	33	33	34	35	36	37	38	38	39	40	41	42	42	43	44	45	46	47	47	48	49	49
26	27	28	28	29	30	31	32	32	33	34	35	36	37	38	38	39	40	41	42	42	43	44	45	46	47	48	48	49	50	50
26	27	28	29	30	31	31	32	33	34	35	36	37	37	38	39	40	41	42	42	43	44	45	46	47	48	48	49	50	51	51
27	28	29	29	30	31	32	33	34	35	36	36	37	38	39	40	41	42	42	43	44	45	46	47	48	49	49	50	51	52	52
27	28	29	30	31	32	32	33	34	35	36	37	38	39	40	41	42	43	43	44	45	46	47	48	49	50	50	51	52	53	53
28	29	30	31	32	32	33	34	35	36	37	38	39	40	40	41	42	43	44	45	46	47	48	49	50	50	51	52	53	54	54
28	29	30	31	32	33	34	35	36	37	38	38	39	40	41	42	43	44	45	46	47	48	49	50	50	51	52	53	54	55	55
29	30	31	32	33	34	34	35	36	37	38	39	40	41	42	43	44	45	46	47	48	49	50	51	51	52	53	54	55	56	56
29	30	31	32	33	34	35	36	37	38	39	40	41	42	43	44	45	46	47	48	49	50	51	52	53	54	55	56	57	57	57
30	31	32	33	34	35	36	37	38	39	40	41	42	43	44	45	46	47	48	49	50	51	52	53	54	55	56	57	58	58	58
30	31	32	33	34	35	36	37	38	39	40	41	42	43	44	45	46	47	48	49	50	51	52	53	54	55	56	57	58	59	59

Extract of part of Table 1 (MOO)

— ALTITUDE CORRECTION FOR CHANGE IN POSITION OF OBSERVER

Distance Made Good — nautical miles																Rel. Zn
4	5	6	7	8	10	15	20	25	30	35	40	45	50	75	100	150	°
+4.0	+5.0	+6.0	+7.0	+8.0	+10.0	+15.0	+20.0	+25.0	+30.0	+35.0	+40.0	+45.0	+50.0	+75.0	+100.0	+150.0	000
4.0	5.0	6.0	7.0	8.0	10.0	15.0	20.0	25.0	30.0	35.0	40.0	45.0	50.0	75.0	99.9	149.9	358
4.0	5.0	6.0	7.0	8.0	10.0	15.0	20.0	24.9	29.9	34.9	39.9	44.9	49.9	74.8	99.8	149.6	356
4.0	5.0	6.0	7.0	8.0	9.9	14.9	19.9	24.9	29.8	34.8	39.8	44.8	49.7	74.6	99.5	149.2	354
4.0	5.0	5.9	6.9	7.9	9.9	14.9	19.8	24.8	29.7	34.7	39.6	44.6	49.5	74.3	99.0	148.5	352
+3.9	+4.9	+5.9	+6.9	+7.9	+9.8	+14.8	+19.7	+24.6	+29.5	+34.5	+39.4	+44.3	+49.2	+73.9	+98.5	+147.7	350
3.9	4.9	5.9	6.8	7.8	9.8	14.7	19.6	24.5	29.3	34.2	39.1	44.0	48.9	73.4	97.8	146.7	348
3.9	4.9	5.8	6.8	7.8	9.7	14.6	19.4	24.3	29.1	34.0	38.8	43.7	48.5	72.8	97.0	145.5	346
3.8	4.8	5.8	6.7	7.7	9.6	14.4	19.2	24.0	28.8	33.6	38.5	43.3	48.1	72.1	96.1	144.2	344
3.8	4.8	5.7	6.7	7.6	9.5	14.3	19.0	23.8	28.5	33.3	38.0	42.8	47.6	71.3	95.1	142.7	342
+3.8	+4.7	+5.6	+6.6	+7.5	+9.4	+14.1	+18.8	+23.5	+28.2	+32.9	+37.6	+42.3	+47.0	+70.5	+94.0	+141.0	340
3.7	4.6	5.6	6.5	7.4	9.3	13.9	18.5	23.2	27.8	32.5	37.1	41.7	46.4	69.5	92.7	139.1	338
3.7	4.6	5.5	6.4	7.3	9.1	13.7	18.3	22.8	27.4	32.0	36.5	41.1	45.7	68.5	91.4	137.0	336
3.6	4.5	5.4	6.3	7.2	9.0	13.5	18.0	22.5	27.0	31.5	36.0	40.4	44.9	67.4	89.9	134.8	334
3.5	4.4	5.3	6.2	7.1	8.8	13.2	17.7	22.1	26.5	30.9	35.3	39.7	44.1	66.2	88.3	132.4	332
+3.5	+4.3	+5.2	+6.1	+6.9	+8.7	+13.0	+17.3	+21.7	+26.0	+30.3	+34.6	+39.0	+43.3	+65.0	+86.6	+129.9	330
3.4	4.2	5.1	5.9	6.8	8.5	12.7	17.0	21.2	25.4	29.7	33.9	38.2	42.4	63.6	84.8	127.2	328
3.3	4.1	5.0	5.8	6.6	8.3	12.4	16.6	20.7	24.9	29.0	33.2	37.3	41.5	62.2	82.9	124.4	326
3.2	4.0	4.9	5.7	6.5	8.1	12.1	16.2	20.2	24.3	28.3	32.4	36.4	40.5	60.7	80.9	121.4	324
3.2	3.9	4.7	5.5	6.3	7.9	11.8	15.8	19.7	23.6	27.6	31.5	35.5	39.4	59.1	78.8	118.2	322
+3.1	+3.8	+4.6	+5.4	+6.1	+7.7	+11.5	+15.3	+19.2	+23.0	+26.8	+30.6	+34.5	+38.3	+57.5	+76.6	+114.9	320
3.0	3.7	4.5	5.2	5.9	7.4	11.1	14.9	18.6	22.3	26.0	29.7	33.4	37.2	55.7	74.3	111.5	318
2.9	3.6	4.3	5.0	5.8	7.2	10.8	14.4	18.0	21.6	25.2	28.8	32.4	36.0	54.0	71.9	107.9	316
2.8	3.5	4.2	4.9	5.6	6.9	10.4	13.9	17.4	20.8	24.3	27.8	31.3	34.7	52.1	69.5	104.2	314
2.7	3.3	4.0	4.7	5.4	6.7	10.0	13.4	16.7	20.1	23.4	26.8	30.1	33.5	50.2	66.9	100.4	312
+2.6	+3.2	+3.9	+4.5	+5.1	+6.4	+9.6	+12.9	+16.1	+19.3	+22.5	+25.7	+28.9	+32.1	+48.2	+64.3	+96.4	310
2.5	3.1	3.7	4.3	4.9	6.2	9.2	12.3	15.4	18.5	21.5	24.6	27.7	30.8	46.2	61.6	92.3	308
2.4	2.9	3.5	4.1	4.7	5.9	8.8	11.8	14.7	17.6	20.6	23.5	26.5	29.4	44.1	58.8	88.2	306
2.2	2.8	3.4	3.9	4.5	5.6	8.4	11.2	14.0	16.8	19.6	22.4	25.2	28.0	41.9	55.9	83.9	304
2.1	2.6	3.2	3.7	4.2	5.3	7.9	10.6	13.2	15.9	18.5	21.2	23.8	26.5	39.7	53.0	79.5	302
+2.0	+2.5	+3.0	+3.5	+4.0	+5.0	+7.5	+10.0	+12.5	+15.0	+17.5	+20.0	+22.5	+25.0	+37.5	+50.0	+75.0	300
1.9	2.3	2.8	3.3	3.8	4.7	7.0	9.4	11.7	14.1	16.4	18.8	21.1	23.5	35.2	46.9	70.4	298
1.8	2.2	2.6	3.1	3.5	4.4	6.6	8.8	11.0	13.2	15.3	17.5	19.7	21.9	32.9	43.8	65.8	296
1.6	2.0	2.4	2.8	3.3	4.1	6.1	8.1	10.2	12.2	14.2	16.3	18.3	20.3	30.5	40.7	61.0	294
1.5	1.9	2.2	2.6	3.0	3.7	5.6	7.5	9.4	11.2	13.1	15.0	16.9	18.7	28.1	37.5	56.2	292
+1.4	+1.7	+2.1	+2.4	+2.7	+3.4	+5.1	+6.8	+8.6	+10.3	+12.0	+13.7	+15.4	+17.1	+25.7	+34.2	+51.3	290
1.2	1.5	1.9	2.2	2.5	3.1	4.6	6.2	7.7	9.3	10.8	12.4	13.9	15.5	23.2	30.9	46.4	288
1.1	1.4	1.7	1.9	2.2	2.8	4.1	5.5	6.9	8.3	9.6	11.0	12.4	13.8	20.7	27.6	41.3	286
1.0	1.2	1.5	1.7	1.9	2.4	3.6	4.8	6.0	7.3	8.5	9.7	10.9	12.1	18.1	24.2	36.3	284
0.8	1.0	1.2	1.5	1.7	2.1	3.1	4.2	5.2	6.2	7.3	8.3	9.4	10.4	15.6	20.8	31.2	282
+0.7	+0.9	+1.0	+1.2	+1.4	+1.7	+2.6	+3.5	+4.3	+5.2	+6.1	+6.9	+7.8	+8.7	+13.0	+17.4	+26.0	280
0.6	0.7	0.8	1.0	1.1	1.4	2.1	2.8	3.5	4.2	4.9	5.6	6.3	7.0	10.4	13.9	20.9	278
0.4	0.5	0.6	0.7	0.8	1.0	1.6	2.1	2.6	3.1	3.7	4.2	4.7	5.2	7.8	10.5	15.7	276
0.3	0.3	0.4	0.5	0.6	0.7	1.0	1.4	1.7	2.1	2.4	2.8	3.1	3.5	5.2	7.0	10.5	274
0.1	0.2	0.2	0.2	0.3	0.3	0.5	0.7	0.9	1.0	1.2	1.4	1.6	1.7	2.6	3.5	5.2	272
0.0	0.0	0.0	0.0	0.0	0.0	0.0	0.0	0.0	0.0	0.0	0.0	0.0	0.0	0.0	0.0	0.0	270
−0.1	−0.2	−0.2	−0.2	−0.3	−0.3	−0.5	−0.7	−0.9	−1.0	−1.2	−1.4	−1.6	−1.7	−2.6	−3.5	−5.2	268
0.3	0.3	0.4	0.5	0.6	0.7	1.0	1.4	1.7	2.1	2.4	2.8	3.1	3.5	5.2	7.0	10.5	266
0.4	0.5	0.6	0.7	0.8	1.0	1.6	2.1	2.6	3.1	3.7	4.2	4.7	5.2	7.8	10.5	15.7	264
0.6	0.7	0.8	1.0	1.1	1.4	2.1	2.8	3.5	4.2	4.9	5.6	6.3	7.0	10.4	13.9	20.9	262
0.7	0.9	1.0	1.2	1.4	1.7	2.6	3.5	4.3	5.2	6.1	6.9	7.8	8.7	13.0	17.4	26.0	260
−0.8	−1.0	−1.2	−1.5	−1.7	−2.1	−3.1	−4.2	−5.2	−6.2	−7.3	−8.3	−9.4	−10.4	−15.6	−20.8	−31.2	258
1.0	1.2	1.5	1.7	1.9	2.4	3.6	4.8	6.0	7.3	8.5	9.7	10.9	12.1	18.1	24.2	36.3	256
1.1	1.4	1.7	1.9	2.2	2.8	4.1	5.5	6.9	8.3	9.6	11.0	12.4	13.8	20.7	27.6	41.3	254
1.2	1.5	1.9	2.2	2.5	3.1	4.6	6.2	7.7	9.3	10.8	12.4	13.9	15.5	23.2	30.9	46.4	252
1.4	1.7	2.1	2.4	2.7	3.4	5.1	6.8	8.6	10.3	12.0	13.7	15.4	17.1	25.7	34.2	51.3	250
−1.5	−1.9	−2.2	−2.6	−3.0	−3.7	−5.6	−7.5	−9.4	−11.2	−13.1	−15.0	−16.9	−18.7	−28.1	−37.5	−56.2	248
1.6	2.0	2.4	2.8	3.3	4.1	6.1	8.1	10.2	12.2	14.2	16.3	18.3	20.3	30.5	40.7	61.0	246
1.8	2.2	2.6	3.1	3.5	4.4	6.6	8.8	11.0	13.2	15.3	17.5	19.7	21.9	32.9	43.8	65.8	244
1.9	2.3	2.8	3.3	3.8	4.7	7.0	9.4	11.7	14.1	16.4	18.8	21.1	23.5	35.2	46.9	70.4	242
2.0	2.5	3.0	3.5	4.0	5.0	7.5	10.0	12.5	15.0	17.5	20.0	22.5	25.0	37.5	50.0	75.0	240
−2.1	−2.6	−3.2	−3.7	−4.2	−5.3	−7.9	−10.6	−13.2	−15.9	−18.5	−21.2	−23.8	−26.5	−39.7	−53.0	−79.5	238
2.2	2.8	3.4	3.9	4.5	5.6	8.4	11.2	14.0	16.8	19.6	22.4	25.2	28.0	41.9	55.9	83.9	236
2.4	2.9	3.5	4.1	4.7	5.9	8.8	11.8	14.7	17.6	20.6	23.5	26.5	29.4	44.1	58.8	88.2	234
2.5	3.1	3.7	4.3	4.9	6.2	9.2	12.3	15.4	18.5	21.5	24.6	27.7	30.8	46.2	61.6	92.3	232
2.6	3.2	3.9	4.5	5.1	6.4	9.6	12.9	16.1	19.3	22.5	25.7	28.9	32.1	48.2	64.3	96.4	230
−2.7	−3.3	−4.0	−4.7	−5.4	−6.7	−10.0	−13.4	−16.7	−20.1	−23.4	−26.8	−30.1	−33.5	−50.2	−66.9	−100.4	228
2.8	3.5	4.2	4.9	5.6	6.9	10.4	13.9	17.4	20.8	24.3	27.8	31.3	34.7	52.1	69.5	104.2	226

Chapter 12
Demonstration of a Fix with Movement of the Body (MOB).

This demonstration shows the special technique that is used for planning observations to be calculated at a certain time as explained in the outline procedures at step 8(b). The advantage of this technique is that the assumed position, LHA, calculated altitude and azimuth from the planning stage are re-used when calculations for the reduction of the observations is carried out. However, a correction for the movement of the body between the time of the fix and the time of the observation has to be made. This correction is referred to as the MOB and involves the use of Table 2.

Notes.
1. To fully understand this demonstration, it will be necessary to refer to the outline procedures in chapter 8 and it may be advisable to revise chapters 2, 4 and 5 as well.
2. Relevant extracts of tables are placed immediately following the solution to the example. Because of their size, some tables have been cropped so that they show only the relevant portions.
3. Self-test questions on the use of the rapid sight reduction tables can be found in chapter 13.

Aim. The aim of this demonstration is to establish a 3 point fix based on the following imaginary scenario:

D.R. Position at $12^h\ 00^m$ (+4): 40° 12'.8N, 65° 03'W.
Date: 7 January 2009
Time of planning:
Observations are planned for the following bodies: Venus, Moon LL, Jupiter.
Time of Fix: $17^h\ 00^m$
Course 268° speed 8 knots throughout.
Sunset: $16^h\ 52^m$ (nautical almanac daily page)
Civil twilight: $17^h\ 22^m$
Nautical twilight: $17^h\ 56^m$
E.P. at time of fix: 40° 12'.7N, 65° 48'W.
DWE 15^s slow.
Index error: +0'.75
Ht. of eye: 4.1m.
The task is to calculate the position of the ship at $17^h\ 00^m$(+4).

Note. The time of the fix is slightly outside the optimum time period for star and planet observations (during civil and nautical twilight) but at 4 minutes after sunset, it is expected that, with clear skies, the celestial bodies chosen will be clearly seen. The advantage of making the fix exactly on the hour outweighs the small disadvantage of making the observations slightly before the optimum time period.

Observations.

Bodies selected: Where appropriate, the method described in chapter 8 for planning star and planet observations was used to select the following celestial bodies: The Moon, Venus, and Jupiter.

Solution:

Planning Phase

At this stage, calculations are based on the UT of the planned fix. Details of the actual observations are added later, when they have been made.

Step 1. EP at $17^h 00^m$ (+4): 40° 12'.7N, 65° 48'W.

Step 2. Calculate UT.

Z.T. of fix:	$07^d 17^h 00^m$
Zone corrn:	$+4^h$
Greenwich Date:	$07^d 21^h 00^m$
UT of fix:	$21^h 00^m$

Note. The UT of the observations are not calculated in the planning phase.

Step 3. Calculate GHA and Dec.

Note. Calculations for the planned observations are made using the UT of the planned fix ($21^h 00^m$).

	Venus	Moon(LL)	Jupiter
Dec. 21^h	S10° 43'.4	N25° 15'.8	S20° 26'.9
d	(1'.2)	(6'.4)	(0'.1)
d cor at UT^m	0'.0	+0'.1	0'.0
Dec + corr.	S10° 43'.4	N25° 15'.9	S20° 26'.9
GHA at UT^h	85° 39'.8	4° 01'.3	119° 44'.4
Inc (m+s)	0'.0	0'.0	0'.0
v	(0.0)	(+3.4)	(+1.9)
v corr.	0'.0	0'.0	0'.0
Sum = GHA	85° 39'.8	4° 01'.3	119° 44'.4

Please see page 161 for an explanation of the reason for calculating true bearings

Step 4. Calculate LHA

	Venus	Moon(LL)	Jupiter
GHA	85° 39'.8	4° 01'.3	119° 44'.4
As. long. (W)	- 65° 39'.8	- 66° 01'.3	- 65° 44'.4
LHA	20°	- 62°	54°
±360°		298°	
Convert azimuth angle to true bearing (Zn):			
LHA =	20°	298°	54°
Z =	155°	86°	129°
Bearing Zn =	205°	086°	231°

at step 4. (Take the values of the Azimuth (Z) from step 5).

	Rules for calculating true bearing	
	Lat. North	Lat. South
LHA>180°	Zn = Z	Zn = 180° - Z
LHA<180°	Zn = 360°-Z	Zn = 180° + Z

Step 5. Main table entry.

	Venus	Moon(LL)	Jupiter
As. lat.	N40°	N40°	N40°
Dec°	S10°	N25°	S20°
	contr.	same	contr.
Dec'	43'.4	15'.9	26'.9
LHA =	20°	298°	54°
Table page:	10	11	13
Tab Hc,	36° 41'	36° 42'	11° 44'
d	-57'	+32'	-47'
Z	155°	86°	129°
Correction for mins of Dec. (Table 5)			
Tab Hc,	36° 41'	36° 42'	11° 44'
Cor for Dec'	-41'	+9'	-21'
Cal. Alt. Hc=	36° 00'	36° 51'	11° 23'
(Hc+Cor.Dec'			

The fix phase

Details of the observations made at the fix phase:

Observation 1:
DWT: $20^h\ 58^m\ 45^s$
Body observed: Venus.
Sextant Alt: 36° 19.'1 Bearing: 206°

Observation 2:
DWT: $21^h\ 00^m\ 30^s$
Body observed: Moon L.L.
Sextant Alt: 35° 47'.95 Bearing: 085°

Observation 3.
DWT: $21^h\ 01^m\ 15^s$
Body observed: Jupiter.
Sextant Alt: 11° 12'.9 Bearing: 230°

Weather Conditions at the time of the observations:
Temperature: 15°C. Pressure: 1004mb.

The UT of each observation is now calculated to determine whether it is earlier or later than the UT of the fix and by how much.

Note. The interval will be positive when the fix is later than the observation and negative when it is earlier.

	Venus	**Moon**	**Jupiter**
DWT:	$20^h\ 58^m\ 45^s$	$21^h\ 00^m\ 30^s$	$21^h\ 01^m\ 15^s$
DWE:	$+15^s$	$+15^s$	$+15^s$
a. UT obs:	$20^h\ 59^m\ 00^s$	$21^h\ 00^m\ 45^s$	$21^h\ 01^m\ 30^s$
b. UT fix:	$21^h\ 00^m$	$21^h\ 00^m$	$21^h\ 00^m$
b ~ a	$+01^m$	-45^s	$-01^m\ 30^s$
∴ Fix is	1^m later	45^s earlier	$1^m\ 30^s$ earlier

Next continue from step 6:

Step 6. Correct sextant altitude (Hs).
(Sextant altitudes are now added to previous calculations).

	Venus	**Moon(LL)**	**Jupiter**
Sext. Alt (Hs)	36° 19'.1	35° 47'.95	11° 12'.9
Index error	+0'.75	+0'.75	+0'.75
Dip (4.1m) (D)	− 3'.6	− 3'.6	− 3'.6
Ap Alt (Ha)	36° 16'.25	35° 45'.10	11° 10'.05
Alt. Cor. (R)	−1'.3	+56'.15	−4'.8
(interpolate between 35° 40' & 50' for Moon)			
HPMoon = 60.3			
Corr. For HP =		+ 7'.8	
Add'l refrac. =	0'.0	0'.0	+0'.1
Obs Alt (Ho) =	36° 14'.95	36° 49'.05	11° 05'.35

Step 7. Calculate Intercept

	Venus	**Moon(LL)**	**Jupiter**
Ho =	36° 14'.95	36° 49'.05	11° 05'.35
Hc =	36° 00'	36° 51'	11° 23'
p(Ho−Hc) =	+14'.95	−1'.95	−17'.65

Therefore, intercepts are:
Venus: 14.95 n.m. to 205°, Moon: 1.95 n.m. from 086°,
Jupiter: 17.65 n.m. from 231°.
(Note. True bearings calculated at step 4).

	Venus	**Moon(LL)**	**Jupiter**
Step 8(a) MOO:	nil	nil	nil
(b) MOB (table 2)			
Lat =	N40°	N40°	N40°
Zn =	205°	086°	231°
Fix is	1ᵐ(later)	0.75ᵐ(earlier)	1.5ᵐ(earlier)
corr. for 1ᵐ =	−4'.86	+11'.5	−9'.0
Actual correction =	−4'.86	+8'.63	−13'.5
Apply rules below =	−4'.86	−8'.63	+13'.5
intercept =	+14'.95	−1'.95	−17'.65
correction =	−4'.86	−8'.63	+13'.5
Corrected 'cpt =	+10'.09	−10'.58	−4'.15
	to 205°	from 086°	from 231°

Therefore, the corrected intercepts are:
Venus: 10.09 n.m. to 205°, Moon: 10.58 n.m. from 086°,
Jupiter: 4.15 n.m. from 231°.

Rules for applying the sign from table 2		
Time of fix	Sign from table	To intercept
Later than observation	+	Add
	−	Subtract
Earlier than observation	+	Subtract
	−	Add

Step 9. Plot the position lines.

Firstly, for each observation, plot the assumed position using the assumed long. calculated at step 4 and the assumed lat. calculated at step 5. Secondly, plot the three position lines using the corrected intercept data calculated at step 8.

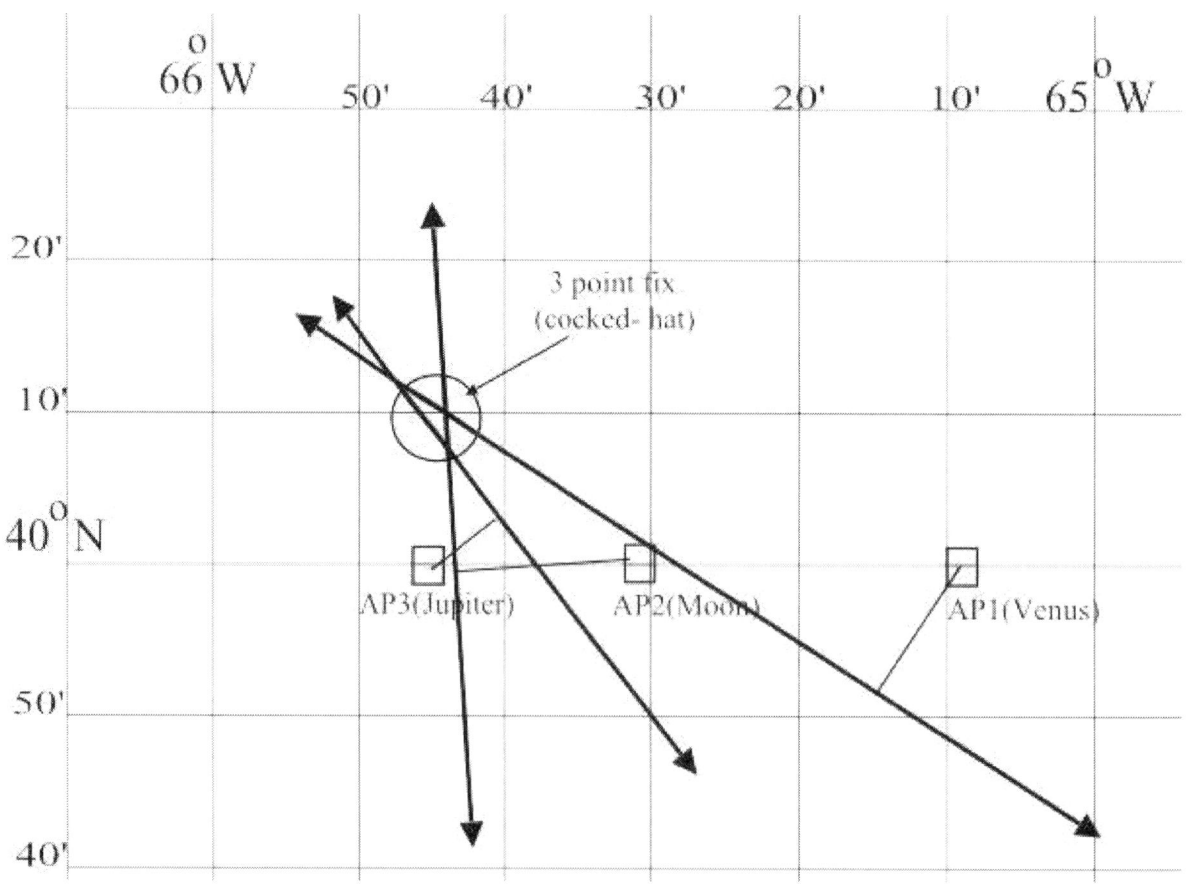

As in the previous example, a small 'cocked-hat' is produced when the position lines are plotted.

Relevant Extracts begin on the next page.

2009 JANUARY 7, 8, 9 (WED., THURS., FRI.)

UT	SUN GHA	SUN Dec	MOON GHA	MOON v	MOON Dec	MOON d	MOON HP	Lat.	Twilight Naut.	Twilight Civil	Sunrise	Moonrise 7	Moonrise 8	Moonrise 9	Moonrise 10
d h	° '	° '	° '	'	° '	'	'	°	h m	h m	h m	h m	h m	h m	h m
7 00	178 27.6	S22 22.9	61 43.0	6.0	N22 26.1	9.5	59.7	N 72	08 15	10 18	■	□	□	□	□
01	193 27.3	22.6	76 08.0	5.9	22 35.6	9.5	59.7	N 70	07 58	09 36	■	□	□	□	□
02	208 27.0	22.3	90 32.9	5.7	22 45.1	9.2	59.7	68	07 44	09 07	11 08	□	□	□	□
03	223 26.7	21.9	104 57.6	5.7	22 54.3	9.2	59.8	66	07 32	08 46	10 12	08 58	□	□	□
04	238 26.5	21.6	119 22.3	5.4	23 03.5	9.0	59.8	64	07 23	08 28	09 40	09 59	□	□	12 32
05	253 26.2	21.3	133 46.7	5.4	23 12.5	8.8	59.8	62	07 14	08 14	09 16	10 33	10 46	11 37	13 26
06	268 25.9	S22 21.0	148 11.1	5.2	N23 21.3	8.8	59.9	60	07 07	08 02	08 57	10 59	11 25	12 23	13 59
W 07	283 25.7	20.7	162 35.3	5.1	23 30.1	8.5	59.9	N 58	07 00	07 51	08 41	11 19	11 52	12 53	14 23
E 08	298 25.4	20.3	176 59.4	5.0	23 38.6	8.5	59.9	56	06 54	07 42	08 28	11 36	12 14	13 15	14 42
D 09	313 25.1	20.0	191 23.4	4.8	23 47.1	8.2	60.0	54	06 48	07 34	08 16	11 50	12 32	13 34	14 58
N 10	328 24.9	19.7	205 47.2	4.7	23 55.3	8.1	60.0	52	06 43	07 26	08 06	12 03	12 47	13 50	15 12
E 11	343 24.6	19.4	220 10.9	4.6	24 03.4	8.0	60.0	50	06 38	07 19	07 57	12 14	13 00	14 04	15 24
S 12	358 24.3	S22 19.0	234 34.5	4.5	N24 11.4	7.8	60.1	45	06 28	07 04	07 38	12 37	13 27	14 32	15 49
D 13	13 24.1	18.7	248 58.0	4.3	24 19.2	7.7	60.1	N 40	06 18	06 52	07 22	12 56	13 49	14 54	16 09
A 14	28 23.8	18.4	263 21.3	4.2	24 26.9	7.5	60.1	35	06 09	06 41	07 09	13 12	14 06	15 12	16 26
Y 15	43 23.5	18.1	277 44.5	4.1	24 34.4	7.3	60.1	30	06 01	06 31	06 57	13 25	14 22	15 28	16 40
16	58 23.3	17.7	292 07.6	4.0	24 41.7	7.2	60.2	20	05 46	06 13	06 37	13 49	14 48	15 54	17 05
17	73 23.0	17.4	306 30.6	3.8	24 48.9	7.0	60.2	N 10	05 31	05 57	06 20	14 09	15 11	16 17	17 25
								0	05 15	05 41	06 03	14 28	15 32	16 39	17 45
18	88 22.7	S22 17.1	320 53.4	3.8	N24 55.9	6.8	60.2	S 10	04 57	05 24	05 46	14 48	15 54	17 00	18 05
19	103 22.5	16.7	335 16.2	3.6	25 02.7	6.7	60.3	20	04 35	05 04	05 29	15 08	16 16	17 23	18 25
20	118 22.2	16.4	349 38.8	3.5	25 09.4	6.4	60.3	30	04 08	04 41	05 08	15 33	16 43	17 50	18 49
21	133 21.9	16.1	4 01.3	3.4	25 15.8	6.4	60.3	35	03 50	04 26	04 56	15 47	16 59	18 06	19 03
22	148 21.7	15.7	18 23.7	3.3	25 22.2	6.1	60.4	40	03 29	04 09	04 41	16 03	17 17	18 24	19 20
23	163 21.4	15.4	32 46.0	3.1	25 28.3	6.0	60.4	45	03 00	03 48	04 25	16 23	17 40	18 46	19 39
8 00	178 21.2	S22 15.1	47 08.1	3.1	N25 34.3	5.8	60.4	S 50	02 19	03 21	04 04	16 48	18 08	19 14	20 03
01	193 20.9	14.7	61 30.2	3.0	25 40.1	5.6	60.4	52	01 56	03 07	03 54	17 01	18 22	19 28	20 15
02	208 20.6	14.4	75 52.2	2.8	25 45.7	5.4	60.5	54	01 23	02 51	03 42	17 14	18 38	19 43	20 28
03	223 20.4	14.1	90 14.0	2.8	25 51.1	5.2	60.5	56	////	02 31	03 29	17 31	18 57	20 02	20 43
04	238 20.1	13.7	104 35.8	2.6	25 56.3	5.1	60.5	58	////	02 06	03 14	17 50	19 21	20 25	21 00
05	253 19.8	13.4	118 57.4	2.6	26 01.4	4.9	60.6	S 60	////	01 30	02 56	18 14	19 52	20 54	21 22
06	268 19.6	S22 13.0	133 19.0	2.4	N26 06.3	4.7	60.6	Lat.	Sunset	Twilight Civil	Twilight Naut.	Moonset 7	Moonset 8	Moonset 9	Moonset 10
07	283 19.3	12.7	147 40.4	2.4	26 11.0	4.4	60.6								
T 08	298 19.0	12.4	162 01.8	2.3	26 15.4	4.4	60.6								
H 09	313 18.8	12.0	176 23.1	2.2	26 19.8	4.1	60.7	°	h m	h m	h m	h m	h m	h m	h m
U 10	328 18.5	11.7	190 44.3	2.1	26 23.9	3.9	60.7								
R 11	343 18.3	11.3	205 05.4	2.0	26 27.8	3.7	60.7	N 72	■	13 56	16 00	□	□	□	□
S 12	358 18.0	S22 11.0	219 26.4	2.0	N26 31.5	3.5	60.7	N 70	■	14 38	16 17	□	□	□	□
D 13	13 17.7	10.6	233 47.4	1.9	26 35.0	3.3	60.8	68	13 06	15 07	16 30	□	□	□	□
A 14	28 17.5	10.3	248 08.3	1.8	26 38.3	3.2	60.8	66	14 02	15 28	16 42	07 15	□	□	□
Y 15	43 17.2	09.9	262 29.1	1.7	26 41.5	2.9	60.8	64	14 34	15 46	16 52	06 14	□	□	10 33
16	58 17.0	09.6	276 49.8	1.6	26 44.4	2.7	60.8	62	14 58	16 00	17 00	05 41	07 39	09 07	09 38
17	73 16.7	09.3	291 10.4	1.6	26 47.1	2.5	60.9	60	15 17	16 12	17 07	05 16	07 00	08 21	09 05
18	88 16.4	S22 08.9	305 31.0	1.6	N26 49.6	2.3	60.9	N 58	15 33	16 23	17 14	04 56	06 33	07 51	08 41
19	103 16.2	08.6	319 51.6	1.4	26 51.9	2.2	60.9	56	15 46	16 32	17 20	04 40	06 12	07 28	08 21
20	118 15.9	08.2	334 12.0	1.4	26 54.1	1.9	60.9	54	15 58	16 40	17 26	04 26	05 54	07 10	08 04
21	133 15.7	07.9	348 32.4	1.4	26 56.0	1.7	60.9	52	16 08	16 48	17 31	04 14	05 39	06 54	07 50
22	148 15.4	07.5	2 52.8	1.3	26 57.7	1.4	61.0	50	16 17	16 55	17 36	04 03	05 27	06 40	07 38
23	163 15.1	07.2	17 13.1	1.3	26 59.1	1.3	61.0	45	16 36	17 10	17 46	03 41	05 00	06 12	07 12
9 00	178 14.9	S22 06.8	31 33.4	1.2	N27 00.4	1.1	61.0	N 40	16 52	17 22	17 56	03 23	04 39	05 50	06 51
01	193 14.6	06.4	45 53.6	1.2	27 01.5	0.9	61.0	35	17 05	17 33	18 04	03 09	04 21	05 31	06 34
02	208 14.4	06.1	60 13.8	1.1	27 02.4	0.6	61.0	30	17 17	17 43	18 12	02 56	04 06	05 15	06 19
03	223 14.1	05.7	74 33.9	1.1	27 03.0	0.4	61.1	20	17 37	18 00	18 28	02 34	03 41	04 49	05 53
04	238 13.8	05.4	88 54.0	1.1	27 03.4	0.3	61.1	N 10	17 54	18 17	18 43	02 15	03 19	04 25	05 31
05	253 13.6	05.0	103 14.1	1.1	27 03.7	0.0	61.1	0	18 10	18 33	18 59	01 57	02 58	04 04	05 11
06	268 13.3	S22 04.7	117 34.2	1.0	N27 03.7	0.2	61.1	S 10	18 27	18 50	19 17	01 39	02 38	03 42	04 50
07	283 13.1	04.3	131 54.2	1.0	27 03.5	0.4	61.1	20	18 45	19 09	19 38	01 21	02 16	03 19	04 28
08	298 12.8	03.9	146 14.2	1.0	27 03.1	0.6	61.1	30	19 06	19 33	20 05	00 59	01 51	02 52	04 02
F 09	313 12.6	03.6	160 34.2	1.0	27 02.5	0.9	61.1	35	19 18	19 47	20 23	00 46	01 36	02 36	03 46
R 10	328 12.3	03.2	174 54.2	0.9	27 01.6	1.0	61.2	40	19 32	20 04	20 44	00 32	01 18	02 17	03 28
I 11	343 12.1	02.9	189 14.1	1.0	27 00.6	1.3	61.2	45	19 49	20 25	21 13	00 15	00 58	01 55	03 07
D 12	358 11.8	S22 02.5	203 34.1	1.0	N26 59.3	1.4	61.2	S 50	20 09	20 52	21 53	24 32	00 32	01 26	02 39
A 13	13 11.5	02.1	217 54.1	0.9	26 57.9	1.7	61.2	52	20 19	21 06	22 16	24 19	00 19	01 12	02 25
Y 14	28 11.3	01.8	232 14.0	1.0	26 56.2	1.9	61.2	54	20 31	21 22	22 48	24 05	00 05	00 56	02 10
15	43 11.0	01.4	246 34.0	1.0	26 54.3	2.1	61.2	56	20 43	21 41	////	23 49	24 37	00 37	01 52
16	58 10.8	01.0	260 54.0	0.9	26 52.2	2.4	61.2	58	20 58	22 03	////	23 29	24 13	00 13	01 29
17	73 10.5	00.7	275 13.9	1.0	26 49.8	2.5	61.2	S 60	21 16	22 41	////	23 04	23 42	25 00	01 00
18	88 10.3	S22 00.3	289 33.9	1.1	N26 47.3	2.8	61.3		SUN			MOON			
19	103 10.0	21 59.9	303 54.0	1.0	26 44.5	2.9	61.3	Day	Eqn. of Time 00ʰ	Eqn. of Time 12ʰ	Mer. Pass.	Mer. Pass. Upper	Mer. Pass. Lower	Age	Phase
20	118 09.8	59.6	318 14.0	1.1	26 41.6	3.2	61.3								
21	133 09.5	59.2	332 34.1	1.0	26 38.4	3.4	61.3	d	m s	m s	h m	h m	h m	d %	
22	148 09.3	58.8	346 54.1	1.2	26 35.0	3.6	61.3	7	06 09	06 22	12 06	20 43	08 13	11 81	
23	163 09.0	58.5	1 14.3	1.1	N26 31.4	3.8	61.3	8	06 35	06 47	12 07	21 48	09 15	12 90	☾
	SD 16.3	d 0.3	SD 16.4	16.5		16.7		9	07 00	07 12	12 07	22 55	10 21	13 96	

©2009 Jack Case   page 225

2009 JANUARY 7, 8, 9 (WED., THURS., FRI.)

UT	ARIES GHA	VENUS −4.5 GHA	VENUS Dec	MARS +1.3 GHA	MARS Dec	JUPITER −1.9 GHA	JUPITER Dec	SATURN +0.9 GHA	SATURN Dec	Star Name	SHA	Dec
d h	° ′	° ′	° ′	° ′	° ′	° ′	° ′	° ′	° ′		° ′	° ′
7 00	106 41.6	130 39.5	S11 07.6	187 52.9	S23 56.7	164 05.5	S20 29.5	293 28.0	N 5 11.2	Acamar	315 20.5	S40 16.2
01	121 44.1	145 39.5	06.4	202 53.3	56.6	179 07.3	29.4	308 30.5	11.3	Achernar	335 28.9	S57 11.7
02	136 46.5	160 39.5	05.3	217 53.7	56.5	194 09.2	29.2	323 33.0	11.3	Acrux	173 13.2	S63 08.8
03	151 49.0	175 39.5	.. 04.1	232 54.1	.. 56.4	209 11.0	.. 29.1	338 35.5	.. 11.3	Adhara	255 14.8	S28 59.1
04	166 51.5	190 39.5	03.0	247 54.5	56.3	224 12.9	29.0	353 37.9	11.3	Aldebaran	290 52.9	N16 31.7
05	181 53.9	205 39.5	01.8	262 54.9	56.2	239 14.7	28.9	8 40.4	11.3			
06	196 56.4	220 39.6	S11 00.7	277 55.3	S23 56.1	254 16.6	S20 28.8	23 42.9	N 5 11.4	Alioth	166 23.3	N55 54.2
W 07	211 58.9	235 39.6	10 59.5	292 55.7	56.0	269 18.5	28.6	38 45.4	11.4	Alkaid	153 01.4	N49 15.7
E 08	227 01.3	250 39.6	58.4	307 56.1	55.9	284 20.3	28.5	53 47.9	11.4	Al Na'ir	27 48.1	S46 55.2
D 09	242 03.8	265 39.6	.. 57.2	322 56.5	.. 55.8	299 22.2	.. 28.4	68 50.4	.. 11.4	Alnilam	275 49.4	S 1 11.8
N 10	257 06.3	280 39.6	56.0	337 56.9	55.7	314 24.0	28.3	83 52.9	11.4	Alphard	217 59.1	S 8 41.9
E 11	272 08.7	295 39.6	54.9	352 57.3	55.6	329 25.9	28.1	98 55.4	11.5			
S 12	287 11.2	310 39.6	S10 53.7	7 57.7	S23 55.5	344 27.7	S20 28.0	113 57.8	N 5 11.5	Alphecca	126 14.0	N26 40.8
D 13	302 13.7	325 39.6	52.6	22 58.1	55.4	359 29.6	27.9	129 00.3	11.5	Alpheratz	357 47.1	N29 08.6
A 14	317 16.1	340 39.7	51.4	37 58.5	55.3	14 31.4	27.8	144 02.8	11.5	Altair	62 11.8	N 8 53.5
Y 15	332 18.6	355 39.7	.. 50.3	52 58.8	.. 55.2	29 33.3	.. 27.7	159 05.3	.. 11.6	Ankaa	353 18.9	S42 15.6
16	347 21.0	10 39.7	49.1	67 59.2	55.1	44 35.2	27.5	174 07.8	11.6	Antares	112 30.6	S26 27.1
17	2 23.5	25 39.7	48.0	82 59.6	55.0	59 37.0	27.4	189 10.3	11.6			
18	17 26.0	40 39.7	S10 46.8	98 00.0	S23 54.9	74 38.9	S20 27.3	204 12.8	N 5 11.6	Arcturus	145 58.8	N19 07.9
19	32 28.4	55 39.8	45.7	113 00.4	54.8	89 40.7	27.2	219 15.3	11.7	Atria	107 36.0	S69 02.5
20	47 30.9	70 39.8	44.5	128 00.8	54.7	104 42.6	27.0	234 17.8	11.7	Avior	234 18.9	S59 32.2
21	62 33.4	85 39.8	.. 43.4	143 01.2	.. 54.6	119 44.4	.. 26.9	249 20.2	.. 11.7	Bellatrix	278 35.2	N 6 21.5
22	77 35.8	100 39.8	42.2	158 01.6	54.5	134 46.3	26.8	264 22.7	11.7	Betelgeuse	271 04.5	N 7 24.5
23	92 38.3	115 39.9	41.0	173 02.0	54.4	149 48.2	26.7	279 25.2	11.7			
8 00	107 40.8	130 39.9	S10 39.9	188 02.4	S23 54.3	164 50.0	S20 26.6	294 27.7	N 5 11.8	Canopus	263 57.1	S52 42.0
01	122 43.2	145 39.9	38.7	203 02.8	54.2	179 51.9	26.4	309 30.2	11.8	Capella	280 38.9	N46 00.6
02	137 45.7	160 39.9	37.6	218 03.2	54.1	194 53.7	26.3	324 32.7	11.8	Deneb	49 34.3	N45 18.8
03	152 48.1	175 40.0	.. 36.4	233 03.6	.. 54.0	209 55.6	.. 26.2	339 35.2	.. 11.8	Denebola	182 36.9	N14 31.1
04	167 50.6	190 40.0	35.3	248 04.0	53.8	224 57.4	26.1	354 37.7	11.9	Diphda	348 59.2	S17 56.3
05	182 53.1	205 40.0	34.1	263 04.4	53.7	239 59.3	25.9	9 40.2	11.9			
06	197 55.5	220 40.1	S10 32.9	278 04.8	S23 53.6	255 01.1	S20 25.8	24 42.7	N 5 11.9	Dubhe	193 55.1	N61 41.8
07	212 58.0	235 40.1	31.8	293 05.2	53.5	270 03.0	25.7	39 45.2	11.9	Elnath	278 16.5	N28 37.0
T 08	228 00.5	250 40.1	30.6	308 05.6	53.4	285 04.8	25.6	54 47.6	12.0	Eltanin	90 48.2	N51 29.1
H 09	243 02.9	265 40.2	.. 29.5	323 06.0	.. 53.3	300 06.7	.. 25.4	69 50.1	.. 12.0	Enif	33 50.7	N 9 55.0
U 10	258 05.4	280 40.2	28.3	338 06.4	53.2	315 08.5	25.3	84 52.6	12.0	Fomalhaut	15 27.7	S29 34.6
R 11	273 07.9	295 40.2	27.1	353 06.8	53.1	330 10.4	25.2	99 55.1	12.0			
S 12	288 10.3	310 40.3	S10 26.0	8 07.2	S23 53.0	345 12.3	S20 25.1	114 57.6	N 5 12.1	Gacrux	172 04.7	S57 09.7
D 13	303 12.8	325 40.3	24.8	23 07.6	52.9	0 14.1	25.0	130 00.1	12.1	Gienah	175 55.7	S17 35.6
A 14	318 15.3	340 40.3	23.7	38 08.0	52.8	15 16.0	24.8	145 02.6	12.1	Hadar	148 53.0	S60 24.8
Y 15	333 17.7	355 40.4	.. 22.5	53 08.4	.. 52.6	30 17.8	.. 24.7	160 05.1	.. 12.1	Hamal	328 04.5	N23 30.5
16	348 20.2	10 40.4	21.3	68 08.7	52.5	45 19.7	24.6	175 07.6	12.2	Kaus Aust.	83 48.6	S34 22.9
17	3 22.6	25 40.5	20.2	83 09.1	52.4	60 21.5	24.5	190 10.1	12.2			
18	18 25.1	40 40.5	S10 19.0	98 09.5	S23 52.3	75 23.4	S20 24.3	205 12.6	N 5 12.2	Kochab	137 20.0	N74 06.7
19	33 27.6	55 40.5	17.9	113 09.9	52.2	90 25.3	24.2	220 15.1	12.2	Markab	13 41.8	N15 15.3
20	48 30.0	70 40.6	16.7	128 10.3	52.1	105 27.1	24.1	235 17.6	12.3	Menkar	314 18.3	N 4 07.6
21	63 32.5	85 40.6	.. 15.5	143 10.7	.. 52.0	120 29.0	.. 24.0	250 20.0	.. 12.3	Menkent	148 11.7	S36 24.8
22	78 35.0	100 40.7	14.4	158 11.1	51.8	135 30.8	23.8	265 22.5	12.3	Miaplacidus	221 39.9	S69 45.1
23	93 37.4	115 40.7	13.2	173 11.5	51.7	150 32.7	23.7	280 25.0	12.3			
9 00	108 39.9	130 40.8	S10 12.0	188 11.9	S23 51.6	165 34.5	S20 23.6	295 27.5	N 5 12.4	Mirfak	308 44.9	N49 53.9
01	123 42.4	145 40.8	10.9	203 12.3	51.5	180 36.4	23.5	310 30.0	12.4	Nunki	76 02.8	S26 17.2
02	138 44.8	160 40.9	09.7	218 12.7	51.4	195 38.2	23.3	325 32.5	12.4	Peacock	53 24.9	S56 42.5
03	153 47.3	175 40.9	.. 08.5	233 13.1	.. 51.3	210 40.1	.. 23.2	340 35.0	.. 12.4	Pollux	243 31.3	N28 00.2
04	168 49.8	190 41.0	07.4	248 13.5	51.1	225 42.0	23.1	355 37.5	12.5	Procyon	245 02.8	N 5 12.1
05	183 52.2	205 41.0	06.2	263 13.9	51.0	240 43.8	23.0	10 40.0	12.5			
06	198 54.7	220 41.1	S10 05.1	278 14.3	S23 50.9	255 45.7	S20 22.9	25 42.5	N 5 12.5	Rasalhague	96 09.8	N12 33.1
07	213 57.1	235 41.1	03.9	293 14.7	50.8	270 47.5	22.7	40 45.0	12.5	Regulus	207 44.9	N11 55.2
08	228 59.6	250 41.2	02.7	308 15.1	50.7	285 49.4	22.6	55 47.5	12.6	Rigel	281 14.9	S 8 11.5
F 09	244 02.1	265 41.2	.. 01.6	323 15.5	.. 50.5	300 51.2	.. 22.5	70 50.0	.. 12.6	Rigil Kent.	139 56.7	S60 52.2
R 10	259 04.5	280 41.3	10 00.4	338 15.9	50.4	315 53.1	22.4	85 52.5	12.6	Sabik	102 16.6	S15 44.2
I 11	274 07.0	295 41.4	9 59.2	353 16.3	50.3	330 54.9	22.2	100 55.0	12.6			
D 12	289 09.5	310 41.4	S 9 58.1	8 16.7	S23 50.2	345 56.8	S20 22.1	115 57.5	N 5 12.7	Schedar	349 44.6	N56 35.6
A 13	304 11.9	325 41.5	56.9	23 17.1	50.1	0 58.6	22.0	131 00.0	12.7	Shaula	96 26.8	S37 06.6
Y 14	319 14.4	340 41.5	55.7	38 17.5	49.9	16 00.5	21.9	146 02.5	12.7	Sirius	258 36.3	S16 43.7
15	334 16.9	355 41.6	.. 54.6	53 17.8	.. 49.8	31 02.4	.. 21.7	161 05.0	.. 12.7	Spica	158 34.8	S11 12.6
16	349 19.3	10 41.7	53.4	68 18.2	49.7	46 04.2	21.6	176 07.5	12.8	Suhail	222 54.6	S43 28.1
17	4 21.8	25 41.7	52.2	83 18.6	49.6	61 06.1	21.5	191 10.0	12.8			
18	19 24.3	40 41.8	S 9 51.0	98 19.0	S23 49.4	76 07.9	S20 21.4	206 12.5	N 5 12.8	Vega	80 41.6	N38 47.4
19	34 26.7	55 41.9	49.9	113 19.4	49.3	91 09.8	21.2	221 15.0	12.8	Zuben'ubi	137 09.3	S16 04.8
20	49 29.2	70 41.9	48.7	128 19.8	49.2	106 11.6	21.1	236 17.5	12.9		SHA	Mer.Pass.
21	64 31.6	85 42.0	.. 47.5	143 20.2	.. 49.1	121 13.5	.. 21.0	251 19.9	.. 12.9		° ′	h m
22	79 34.1	100 42.1	46.4	158 20.6	48.9	136 15.3	20.9	266 22.4	12.9	Venus	22 59.1	15 17
23	94 36.6	115 42.1	45.2	173 21.0	48.8	151 17.2	20.7	281 24.9	12.9	Mars	80 21.7	11 28
	h m									Jupiter	57 09.2	12 59
Mer.Pass.	16 46.5	v 0.0	d 1.2	v 0.4	d 0.1	v 1.9	d 0.1	v 2.5	d 0.0	Saturn	186 47.0	4 21

INCREMENTS AND CORRECTIONS

0ᵐ

s	SUN PLANETS	ARIES	MOON	v or d	Corrⁿ	v or d	Corrⁿ	v or d	Corrⁿ
	° ′	° ′	° ′	′	′	′	′	′	′
00	0 00·0	0 00·0	0 00·0	0·0	0·0	6·0	0·1	12·0	0·1
01	0 00·3	0 00·3	0 00·2	0·1	0·0	6·1	0·1	12·1	0·1
02	0 00·5	0 00·5	0 00·5	0·2	0·0	6·2	0·1	12·2	0·1
03	0 00·8	0 00·8	0 00·7	0·3	0·0	6·3	0·1	12·3	0·1
04	0 01·0	0 01·0	0 01·0	0·4	0·0	6·4	0·1	12·4	0·1
05	0 01·3	0 01·3	0 01·2	0·5	0·0	6·5	0·1	12·5	0·1
06	0 01·5	0 01·5	0 01·4	0·6	0·0	6·6	0·1	12·6	0·1
07	0 01·8	0 01·8	0 01·7	0·7	0·0	6·7	0·1	12·7	0·1
08	0 02·0	0 02·0	0 01·9	0·8	0·0	6·8	0·1	12·8	0·1
09	0 02·3	0 02·3	0 02·1	0·9	0·0	6·9	0·1	12·9	0·1
10	0 02·5	0 02·5	0 02·4	1·0	0·0	7·0	0·1	13·0	0·1
11	0 02·8	0 02·8	0 02·6	1·1	0·0	7·1	0·1	13·1	0·1
12	0 03·0	0 03·0	0 02·9	1·2	0·0	7·2	0·1	13·2	0·1
13	0 03·3	0 03·3	0 03·1	1·3	0·0	7·3	0·1	13·3	0·1
14	0 03·5	0 03·5	0 03·3	1·4	0·0	7·4	0·1	13·4	0·1
15	0 03·8	0 03·8	0 03·6	1·5	0·0	7·5	0·1	13·5	0·1
16	0 04·0	0 04·0	0 03·8	1·6	0·0	7·6	0·1	13·6	0·1
17	0 04·3	0 04·3	0 04·1	1·7	0·0	7·7	0·1	13·7	0·1
18	0 04·5	0 04·5	0 04·3	1·8	0·0	7·8	0·1	13·8	0·1
19	0 04·8	0 04·8	0 04·5	1·9	0·0	7·9	0·1	13·9	0·1
20	0 05·0	0 05·0	0 04·8	2·0	0·0	8·0	0·1	14·0	0·1
21	0 05·3	0 05·3	0 05·0	2·1	0·0	8·1	0·1	14·1	0·1
22	0 05·5	0 05·5	0 05·2	2·2	0·0	8·2	0·1	14·2	0·1
23	0 05·8	0 05·8	0 05·5	2·3	0·0	8·3	0·1	14·3	0·1
24	0 06·0	0 06·0	0 05·7	2·4	0·0	8·4	0·1	14·4	0·1
25	0 06·3	0 06·3	0 06·0	2·5	0·0	8·5	0·1	14·5	0·1
26	0 06·5	0 06·5	0 06·2	2·6	0·0	8·6	0·1	14·6	0·1
27	0 06·8	0 06·8	0 06·4	2·7	0·0	8·7	0·1	14·7	0·1
28	0 07·0	0 07·0	0 06·7	2·8	0·0	8·8	0·1	14·8	0·1
29	0 07·3	0 07·3	0 06·9	2·9	0·0	8·9	0·1	14·9	0·1
30	0 07·5	0 07·5	0 07·2	3·0	0·0	9·0	0·1	15·0	0·1
31	0 07·8	0 07·8	0 07·4	3·1	0·0	9·1	0·1	15·1	0·1
32	0 08·0	0 08·0	0 07·6	3·2	0·0	9·2	0·1	15·2	0·1
33	0 08·3	0 08·3	0 07·9	3·3	0·0	9·3	0·1	15·3	0·1
34	0 08·5	0 08·5	0 08·1	3·4	0·0	9·4	0·1	15·4	0·1
35	0 08·8	0 08·8	0 08·4	3·5	0·0	9·5	0·1	15·5	0·1
36	0 09·0	0 09·0	0 08·6	3·6	0·0	9·6	0·1	15·6	0·1
37	0 09·3	0 09·3	0 08·8	3·7	0·0	9·7	0·1	15·7	0·1
38	0 09·5	0 09·5	0 09·1	3·8	0·0	9·8	0·1	15·8	0·1
39	0 09·8	0 09·8	0 09·3	3·9	0·0	9·9	0·1	15·9	0·1
40	0 10·0	0 10·0	0 09·5	4·0	0·0	10·0	0·1	16·0	0·1
41	0 10·3	0 10·3	0 09·8	4·1	0·0	10·1	0·1	16·1	0·1
42	0 10·5	0 10·5	0 10·0	4·2	0·0	10·2	0·1	16·2	0·1
43	0 10·8	0 10·8	0 10·3	4·3	0·0	10·3	0·1	16·3	0·1
44	0 11·0	0 11·0	0 10·5	4·4	0·0	10·4	0·1	16·4	0·1
45	0 11·3	0 11·3	0 10·7	4·5	0·0	10·5	0·1	16·5	0·1
46	0 11·5	0 11·5	0 11·0	4·6	0·0	10·6	0·1	16·6	0·1
47	0 11·8	0 11·8	0 11·2	4·7	0·0	10·7	0·1	16·7	0·1
48	0 12·0	0 12·0	0 11·5	4·8	0·0	10·8	0·1	16·8	0·1
49	0 12·3	0 12·3	0 11·7	4·9	0·0	10·9	0·1	16·9	0·1
50	0 12·5	0 12·5	0 11·9	5·0	0·0	11·0	0·1	17·0	0·1
51	0 12·8	0 12·8	0 12·2	5·1	0·0	11·1	0·1	17·1	0·1
52	0 13·0	0 13·0	0 12·4	5·2	0·0	11·2	0·1	17·2	0·1
53	0 13·3	0 13·3	0 12·6	5·3	0·0	11·3	0·1	17·3	0·1
54	0 13·5	0 13·5	0 12·9	5·4	0·0	11·4	0·1	17·4	0·1
55	0 13·8	0 13·8	0 13·1	5·5	0·0	11·5	0·1	17·5	0·1
56	0 14·0	0 14·0	0 13·4	5·6	0·0	11·6	0·1	17·6	0·1
57	0 14·3	0 14·3	0 13·6	5·7	0·0	11·7	0·1	17·7	0·1
58	0 14·5	0 14·5	0 13·8	5·8	0·0	11·8	0·1	17·8	0·1
59	0 14·8	0 14·8	0 14·1	5·9	0·0	11·9	0·1	17·9	0·1
60	0 15·0	0 15·0	0 14·3	6·0	0·1	12·0	0·1	18·0	0·2

1ᵐ

s	SUN PLANETS	ARIES	MOON	v or d	Corrⁿ	v or d	Corrⁿ	v or d	Corrⁿ
	° ′	° ′	° ′	′	′	′	′	′	′
00	0 15·0	0 15·0	0 14·3	0·0	0·0	6·0	0·2	12·0	0·3
01	0 15·3	0 15·3	0 14·6	0·1	0·0	6·1	0·2	12·1	0·3
02	0 15·5	0 15·5	0 14·8	0·2	0·0	6·2	0·2	12·2	0·3
03	0 15·8	0 15·8	0 15·0	0·3	0·0	6·3	0·2	12·3	0·3
04	0 16·0	0 16·0	0 15·3	0·4	0·0	6·4	0·2	12·4	0·3
05	0 16·3	0 16·3	0 15·5	0·5	0·0	6·5	0·2	12·5	0·3
06	0 16·5	0 16·5	0 15·7	0·6	0·0	6·6	0·2	12·6	0·3
07	0 16·8	0 16·8	0 16·0	0·7	0·0	6·7	0·2	12·7	0·3
08	0 17·0	0 17·0	0 16·2	0·8	0·0	6·8	0·2	12·8	0·3
09	0 17·3	0 17·3	0 16·5	0·9	0·0	6·9	0·2	12·9	0·3
10	0 17·5	0 17·5	0 16·7	1·0	0·0	7·0	0·2	13·0	0·3
11	0 17·8	0 17·8	0 16·9	1·1	0·0	7·1	0·2	13·1	0·3
12	0 18·0	0 18·0	0 17·2	1·2	0·0	7·2	0·2	13·2	0·3
13	0 18·3	0 18·3	0 17·4	1·3	0·0	7·3	0·2	13·3	0·3
14	0 18·5	0 18·6	0 17·7	1·4	0·0	7·4	0·2	13·4	0·3
15	0 18·8	0 18·8	0 17·9	1·5	0·0	7·5	0·2	13·5	0·3
16	0 19·0	0 19·1	0 18·1	1·6	0·0	7·6	0·2	13·6	0·3
17	0 19·3	0 19·3	0 18·4	1·7	0·0	7·7	0·2	13·7	0·3
18	0 19·5	0 19·6	0 18·6	1·8	0·0	7·8	0·2	13·8	0·3
19	0 19·8	0 19·8	0 18·9	1·9	0·0	7·9	0·2	13·9	0·3
20	0 20·0	0 20·1	0 19·1	2·0	0·1	8·0	0·2	14·0	0·4
21	0 20·3	0 20·3	0 19·3	2·1	0·1	8·1	0·2	14·1	0·4
22	0 20·5	0 20·6	0 19·6	2·2	0·1	8·2	0·2	14·2	0·4
23	0 20·8	0 20·8	0 19·8	2·3	0·1	8·3	0·2	14·3	0·4
24	0 21·0	0 21·1	0 20·0	2·4	0·1	8·4	0·2	14·4	0·4
25	0 21·3	0 21·3	0 20·3	2·5	0·1	8·5	0·2	14·5	0·4
26	0 21·5	0 21·6	0 20·5	2·6	0·1	8·6	0·2	14·6	0·4
27	0 21·8	0 21·8	0 20·8	2·7	0·1	8·7	0·2	14·7	0·4
28	0 22·0	0 22·1	0 21·0	2·8	0·1	8·8	0·2	14·8	0·4
29	0 22·3	0 22·3	0 21·2	2·9	0·1	8·9	0·2	14·9	0·4
30	0 22·5	0 22·6	0 21·5	3·0	0·1	9·0	0·2	15·0	0·4
31	0 22·8	0 22·8	0 21·7	3·1	0·1	9·1	0·2	15·1	0·4
32	0 23·0	0 23·1	0 22·0	3·2	0·1	9·2	0·2	15·2	0·4
33	0 23·3	0 23·3	0 22·2	3·3	0·1	9·3	0·2	15·3	0·4
34	0 23·5	0 23·6	0 22·4	3·4	0·1	9·4	0·2	15·4	0·4
35	0 23·8	0 23·8	0 22·7	3·5	0·1	9·5	0·2	15·5	0·4
36	0 24·0	0 24·1	0 22·9	3·6	0·1	9·6	0·2	15·6	0·4
37	0 24·3	0 24·3	0 23·1	3·7	0·1	9·7	0·2	15·7	0·4
38	0 24·5	0 24·6	0 23·4	3·8	0·1	9·8	0·2	15·8	0·4
39	0 24·8	0 24·8	0 23·6	3·9	0·1	9·9	0·2	15·9	0·4
40	0 25·0	0 25·1	0 23·9	4·0	0·1	10·0	0·3	16·0	0·4
41	0 25·3	0 25·3	0 24·1	4·1	0·1	10·1	0·3	16·1	0·4
42	0 25·5	0 25·6	0 24·3	4·2	0·1	10·2	0·3	16·2	0·4
43	0 25·8	0 25·8	0 24·6	4·3	0·1	10·3	0·3	16·3	0·4
44	0 26·0	0 26·1	0 24·8	4·4	0·1	10·4	0·3	16·4	0·4
45	0 26·3	0 26·3	0 25·1	4·5	0·1	10·5	0·3	16·5	0·4
46	0 26·5	0 26·6	0 25·3	4·6	0·1	10·6	0·3	16·6	0·4
47	0 26·8	0 26·8	0 25·5	4·7	0·1	10·7	0·3	16·7	0·4
48	0 27·0	0 27·1	0 25·8	4·8	0·1	10·8	0·3	16·8	0·4
49	0 27·3	0 27·3	0 26·0	4·9	0·1	10·9	0·3	16·9	0·4
50	0 27·5	0 27·6	0 26·2	5·0	0·1	11·0	0·3	17·0	0·4
51	0 27·8	0 27·8	0 26·5	5·1	0·1	11·1	0·3	17·1	0·4
52	0 28·0	0 28·1	0 26·7	5·2	0·1	11·2	0·3	17·2	0·4
53	0 28·3	0 28·3	0 27·0	5·3	0·1	11·3	0·3	17·3	0·4
54	0 28·5	0 28·6	0 27·2	5·4	0·1	11·4	0·3	17·4	0·4
55	0 28·8	0 28·8	0 27·4	5·5	0·1	11·5	0·3	17·5	0·4
56	0 29·0	0 29·1	0 27·7	5·6	0·1	11·6	0·3	17·6	0·4
57	0 29·3	0 29·3	0 27·9	5·7	0·1	11·7	0·3	17·7	0·4
58	0 29·5	0 29·6	0 28·2	5·8	0·1	11·8	0·3	17·8	0·4
59	0 29·8	0 29·8	0 28·4	5·9	0·1	11·9	0·3	17·9	0·4
60	0 30·0	0 30·1	0 28·6	6·0	0·2	12·0	0·3	18·0	0·5

© 2009 Jack Case

TABLE 6 — ALTITUDE CORRECTION TABLES

a. Dip of the Horizon

Ht. of Eye (m)	Corr^n D	Ht. of Eye (ft.)	Ht. of Eye (m)	Corr^n D	Ht. of Eye (ft.)
0·00	−0·3	0·0	13·0	−6·4	42·8
0·03	−0·4	0·1	13·4	−6·5	44·2
0·06	−0·5	0·2	13·8	−6·6	45·5
0·09	−0·6	0·3	14·2	−6·7	47·0
0·13	−0·7	0·4	14·7	−6·8	48·4
0·18	−0·8	0·5	15·1	−6·9	49·8
0·23	−0·9	0·7	15·5	−7·0	51·3
0·29	−1·0	0·9	16·0	−7·1	52·8
0·35	−1·1	1·1	16·5	−7·2	54·3
0·42	−1·2	1·4	16·9	−7·3	55·8
0·45	−1·3	1·6	17·4	−7·4	57·4
0·5	−1·4	1·9	17·9	−7·5	58·9
0·6	−1·5	2·2	18·4	−7·6	60·5
0·7	−1·6	2·5	18·8	−7·7	62·1
0·8	−1·7	2·8	19·3	−7·8	63·8
0·9	−1·8	3·2	19·8	−7·9	65·4
1·1	−1·9	3·6	20·4	−8·0	67·1
1·2	−2·0	4·0	20·9	−8·1	68·8
1·3	−2·1	4·4	21·4	−8·2	70·5
1·4	−2·2	4·9	21·9	−8·3	72·3
1·6	−2·3	5·3	22·5	−8·4	74·1
1·7	−2·4	5·8	23·0	−8·5	75·8
1·9	−2·5	6·3	23·5	−8·6	77·6
2·0	−2·6	6·9	24·1	−8·7	79·5
2·2	−2·7	7·4	24·7	−8·8	81·3
2·4	−2·8	8·0	25·2	−8·9	83·2
2·6	−2·9	8·6	25·8	−9·0	85·1
2·8	−3·0	9·2	26·4	−9·1	87·0
3·0	−3·1	9·8	27·0	−9·2	88·9
3·2	−3·2	10·5	27·6	−9·3	90·9
3·4	−3·3	11·2	28·2	−9·4	92·9
3·6	−3·4	11·9	28·8	−9·5	94·9
3·8	−3·5	12·6	29·4	−9·6	96·9
4·0	−3·6	13·3	30·0	−9·7	98·9
4·3	−3·7	14·1	30·6	−9·8	101·0
4·5	−3·8	14·9	31·3	−9·9	103·1
4·7	−3·9	15·7	31·9	−10·0	105·2
5·0	−4·0	16·5	32·6	−10·1	107·3
5·2	−4·1	17·4	33·2	−10·2	109·4
5·5	−4·2	18·3	33·9	−10·3	111·6
5·8	−4·3	19·1	34·5	−10·4	113·8
6·1	−4·4	20·1	35·2	−10·5	116·0
6·3	−4·5	21·0	35·9	−10·6	118·2
6·6	−4·6	22·0	36·6	−10·7	120·5
6·9	−4·7	22·9	37·3	−10·8	122·8
7·2	−4·8	23·9	38·0	−10·9	125·1
7·5	−4·9	25·0	38·7	−11·0	127·4
7·9	−5·0	26·0	39·4	−11·1	129·7
8·2	−5·1	27·1	40·1	−11·2	132·1
8·5	−5·2	28·1	40·8	−11·3	134·5
8·8	−5·3	29·2	41·5	−11·4	136·9
9·2	−5·4	30·4	42·3	−11·5	139·3
9·5	−5·5	31·5	43·0	−11·6	141·7
9·9	−5·6	32·7	43·8	−11·7	144·2
10·3	−5·7	33·9	44·5	−11·8	146·7
10·6	−5·8	35·1	45·3	−11·9	149·2
11·0	−5·9	36·3	46·1	−12·0	151·7
11·4	−6·0	37·6	46·8	−12·1	154·3
11·8	−6·1	38·9	47·6	−12·2	156·8
12·2	−6·2	40·1	48·4	−12·3	159·4
12·6	−6·3	41·5	49·2	−12·4	162·1
13·0		42·8	50·0		164·7

b. Refraction for Stars & Planets

App. Alt. H_a (° ′)	Corr^n R (′)	App. Alt. H_a (° ′)	Corr^n R (′)	App. Alt. H_a (° ′)	Corr^n R (′)
0 00	−33·8	3 30	−12·9	9 55	
0 03	33·2	3 35	12·7	10 07	−5·3
0 06	32·6	3 40	12·5	10 20	−5·2
0 09	32·0	3 45	12·3	10 32	−5·1
0 12	31·5	3 50	12·1	10 46	−5·0
0 15	30·9	3 55	11·9	10 59	−4·9
0 18	−30·4	4 00	−11·7	11 14	−4·8
0 21	29·8	4 05	11·5	11 29	−4·7
0 24	29·3	4 10	11·4	11 44	−4·6
0 27	28·8	4 15	11·2	12 00	−4·5
0 30	28·3	4 20	11·0	12 17	−4·4
0 33	27·9	4 25	10·9	12 35	−4·3
0 36	−27·4	4 30	−10·7	12 53	−4·2
0 39	26·9	4 35	10·6	13 12	−4·1
0 42	26·5	4 40	10·4	13 32	−4·0
0 45	26·1	4 45	10·3	13 53	−3·9
0 48	25·7	4 50	10·1	14 16	−3·8
0 51	25·3	4 55	10·0	14 39	−3·7
0 54	−24·9	5 00	−9·8	15 03	−3·6
0 57	24·5	5 05	9·7	15 29	−3·5
1 00	24·1	5 10	9·6	15 56	−3·4
1 03	23·7	5 15	9·5	16 25	−3·3
1 06	23·4	5 20	9·3	16 55	−3·2
1 09	23·0	5 25	9·2	17 27	−3·1
1 12	−22·7	5 30	−9·1	18 01	−3·0
1 15	22·3	5 35	9·0	18 37	−2·9
1 18	22·0	5 40	8·9	19 16	−2·8
1 21	21·7	5 45	8·8	19 56	−2·7
1 24	21·4	5 50	8·7	20 40	−2·6
1 27	21·1	5 55	8·6	21 27	−2·5
1 30	−20·8	6 00	−8·5	22 17	−2·4
1 35	20·3	6 10	8·3	23 11	−2·3
1 40	19·9	6 20	8·1	24 09	−2·2
1 45	19·4	6 30	7·9	25 12	−2·1
1 50	19·0	6 40	7·7	26 20	−2·0
1 55	18·6	6 50	7·6	27 34	−1·9
2 00	−18·2	7 00	−7·4	28 54	−1·8
2 05	17·8	7 10	7·2	30 22	−1·7
2 10	17·4	7 20	7·1	31 58	−1·6
2 15	17·1	7 30	6·9	33 43	−1·5
2 20	16·7	7 40	6·8	35 38	−1·4
2 25	16·4	7 50	6·7	37 45	−1·3
2 30	−16·1	8 00	−6·6	40 06	−1·2
2 35	15·8	8 10	6·4	42 42	−1·1
2 40	15·4	8 20	6·3	45 34	−1·0
2 45	15·2	8 30	6·2	48 45	−0·9
2 50	14·9	8 40	6·1	52 16	−0·8
2 55	14·6	8 50	6·0	56 09	−0·7
3 00	−14·3	9 00	−5·9	60 26	−0·6
3 05	14·1	9 10	5·8	65 06	−0·5
3 10	13·8	9 20	5·7	70 09	−0·4
3 15	13·6	9 30	5·6	75 32	−0·3
3 20	13·4	9 40	5·5	81 12	−0·2
3 25	13·1	9 50	5·4	87 03	−0·1
3 30	−12·9	10 00	−5·3	90 00	0·0

H_a = App. Alt. = Apparent Altitude
= Sextant altitude corrected for index error (IE) & dip (D)
= H_s + IE + D

In critical cases ascend.

©2009 Jack Case

TABLE 6 — ALTITUDE CORRECTION TABLES
ALTITUDE CORRECTION TABLES 35°–90° — MOON

App. Alt.	35°–39° Corrⁿ	40°–44° Corrⁿ	45°–49° Corrⁿ	50°–54° Corrⁿ	55°–59° Corrⁿ	60°–64° Corrⁿ	65°–69° Corrⁿ	70°–74° Corrⁿ	75°–79° Corrⁿ	80°–84° Corrⁿ	85°–89° Corrⁿ	App. Alt.
00	35 56.5	40 53.7	45 50.5	50 46.9	55 43.1	60 38.9	65 34.6	70 30.0	75 25.3	80 20.5	85 15.6	00
10	56.4	53.6	50.4	46.8	42.9	38.8	34.4	29.9	25.2	20.4	15.5	10
20	56.3	53.5	50.2	46.7	42.8	38.7	34.3	29.7	25.0	20.2	15.3	20
30	56.2	53.4	50.1	46.5	42.7	38.5	34.1	29.6	24.9	20.0	15.1	30
40	56.2	53.3	50.0	46.4	42.5	38.4	34.0	29.4	24.7	19.9	15.0	40
50	56.1	53.2	49.9	46.3	42.4	38.2	33.8	29.3	24.5	19.7	14.8	50
00	36 56.0	41 53.1	46 49.8	51 46.2	56 42.3	61 38.1	66 33.7	71 29.1	76 24.4	81 19.6	86 14.6	00
10	55.9	53.0	49.7	46.0	42.1	37.9	33.5	29.0	24.2	19.4	14.5	10
20	55.8	52.9	49.5	45.9	42.0	37.8	33.4	28.8	24.1	19.2	14.3	20
30	55.7	52.8	49.4	45.8	41.9	37.7	33.2	28.7	23.9	19.1	14.2	30
40	55.6	52.6	49.3	45.7	41.7	37.5	33.1	28.5	23.8	18.9	14.0	40
50	55.5	52.5	49.2	45.5	41.6	37.4	32.9	28.3	23.6	18.7	13.8	50
00	37 55.4	42 52.4	47 49.1	52 45.4	57 41.4	62 37.2	67 32.8	72 28.2	77 23.4	82 18.6	87 13.7	00
10	55.3	52.3	49.0	45.3	41.3	37.1	32.6	28.0	23.3	18.4	13.5	10
20	55.2	52.2	48.8	45.2	41.2	36.9	32.5	27.9	23.1	18.2	13.3	20
30	55.1	52.1	48.7	45.0	41.0	36.8	32.3	27.7	22.9	18.1	13.2	30
40	55.0	52.0	48.6	44.9	40.9	36.6	32.2	27.6	22.8	17.9	13.0	40
50	55.0	51.9	48.5	44.8	40.8	36.5	32.0	27.4	22.6	17.8	12.8	50
00	38 54.9	43 51.8	48 48.4	53 44.6	58 40.6	63 36.4	68 31.9	73 27.2	78 22.5	83 17.6	88 12.7	00
10	54.8	51.7	48.3	44.5	40.5	36.2	31.7	27.1	22.3	17.4	12.5	10
20	54.7	51.6	48.1	44.4	40.3	36.1	31.6	26.9	22.1	17.3	12.3	20
30	54.6	51.5	48.0	44.2	40.2	35.9	31.4	26.8	22.0	17.1	12.2	30
40	54.5	51.4	47.9	44.1	40.1	35.8	31.3	26.6	21.8	16.9	12.0	40
50	54.4	51.2	47.8	44.0	39.9	35.6	31.1	26.5	21.7	16.8	11.8	50
00	39 54.3	44 51.1	49 47.7	54 43.9	59 39.8	64 35.5	69 31.0	74 26.3	79 21.5	84 16.6	89 11.7	00
10	54.2	51.0	47.5	43.7	39.6	35.3	30.8	26.1	21.3	16.4	11.5	10
20	54.1	50.9	47.4	43.6	39.5	35.2	30.7	26.0	21.2	16.3	11.4	20
30	54.0	50.8	47.3	43.5	39.4	35.0	30.5	25.8	21.0	16.1	11.2	30
40	53.9	50.7	47.2	43.3	39.2	34.9	30.4	25.7	20.9	16.0	11.0	40
50	53.8	50.6	47.0	43.2	39.1	34.7	30.2	25.5	20.7	15.8	10.9	50

HP	L U	L U	L U	L U	L U	L U	L U	L U	L U	L U	L U	HP
54.0	1.1 1.7	1.3 1.9	1.5 2.1	1.7 2.4	2.0 2.6	2.3 2.9	2.6 3.2	2.9 3.5	3.2 3.8	3.5 4.1	3.8 4.5	54.0
54.3	1.4 1.8	1.6 2.0	1.8 2.2	2.0 2.5	2.2 2.7	2.5 3.0	2.8 3.2	3.1 3.5	3.3 3.8	3.6 4.1	3.9 4.4	54.3
54.6	1.7 2.0	1.9 2.2	2.1 2.4	2.3 2.6	2.5 2.8	2.7 3.0	3.0 3.3	3.2 3.5	3.5 3.8	3.8 4.0	4.0 4.3	54.6
54.9	2.0 2.2	2.2 2.3	2.3 2.5	2.5 2.7	2.7 2.9	2.9 3.1	3.2 3.3	3.4 3.5	3.6 3.8	3.9 4.0	4.1 4.3	54.9
55.2	2.3 2.3	2.5 2.4	2.6 2.6	2.8 2.8	3.0 2.9	3.2 3.1	3.4 3.3	3.6 3.5	3.8 3.7	4.0 4.0	4.2 4.2	55.2
55.5	2.7 2.5	2.8 2.6	2.9 2.7	3.1 2.9	3.2 3.0	3.4 3.2	3.6 3.4	3.7 3.5	3.9 3.7	4.1 3.9	4.3 4.1	55.5
55.8	3.0 2.6	3.1 2.7	3.2 2.8	3.3 3.0	3.5 3.1	3.6 3.3	3.8 3.4	3.9 3.6	4.1 3.7	4.2 3.9	4.4 4.0	55.8
56.1	3.3 2.8	3.4 2.9	3.5 3.0	3.6 3.1	3.7 3.2	3.8 3.3	4.0 3.4	4.1 3.6	4.3 3.7	4.4 3.8	4.5 4.0	56.1
56.4	3.6 2.9	3.7 3.0	3.8 3.1	3.9 3.2	3.9 3.3	4.0 3.4	4.1 3.5	4.3 3.6	4.4 3.7	4.5 3.8	4.6 3.9	56.4
56.7	3.9 3.1	4.0 3.1	4.1 3.2	4.1 3.3	4.2 3.3	4.3 3.4	4.3 3.5	4.4 3.6	4.5 3.7	4.6 3.8	4.7 3.8	56.7
57.0	4.3 3.2	4.3 3.3	4.3 3.3	4.4 3.4	4.4 3.4	4.5 3.5	4.5 3.5	4.6 3.6	4.7 3.6	4.7 3.7	4.8 3.8	57.0
57.3	4.6 3.4	4.6 3.4	4.6 3.4	4.6 3.5	4.7 3.5	4.7 3.5	4.7 3.6	4.8 3.6	4.8 3.6	4.8 3.7	4.9 3.7	57.3
57.6	4.9 3.6	4.9 3.6	4.9 3.6	4.9 3.6	4.9 3.6	4.9 3.6	4.9 3.6	4.9 3.6	5.0 3.6	5.0 3.6	5.0 3.6	57.6
57.9	5.2 3.7	5.2 3.7	5.2 3.7	5.2 3.7	5.2 3.7	5.1 3.6	5.1 3.6	5.1 3.6	5.1 3.6	5.1 3.6	5.1 3.6	57.9
58.2	5.5 3.9	5.5 3.8	5.5 3.8	5.4 3.8	5.4 3.7	5.4 3.7	5.3 3.7	5.3 3.6	5.2 3.6	5.2 3.5	5.2 3.5	58.2
58.5	5.9 4.0	5.8 4.0	5.8 3.9	5.7 3.9	5.6 3.8	5.6 3.8	5.5 3.7	5.5 3.6	5.4 3.6	5.3 3.5	5.3 3.4	58.5
58.8	6.2 4.2	6.1 4.1	6.0 4.1	6.0 4.0	5.9 3.9	5.8 3.8	5.7 3.7	5.6 3.6	5.5 3.5	5.4 3.5	5.3 3.4	58.8
59.1	6.5 4.3	6.4 4.3	6.3 4.2	6.2 4.1	6.1 4.0	6.0 3.9	5.9 3.8	5.8 3.6	5.7 3.5	5.6 3.4	5.4 3.3	59.1
59.4	6.8 4.5	6.7 4.4	6.6 4.3	6.5 4.2	6.4 4.1	6.2 3.9	6.1 3.8	6.0 3.7	5.8 3.5	5.7 3.4	5.5 3.2	59.4
59.7	7.1 4.7	7.0 4.5	6.9 4.4	6.8 4.3	6.6 4.1	6.5 4.0	6.3 3.8	6.1 3.7	6.0 3.5	5.8 3.3	5.6 3.2	59.7
60.0	7.5 4.8	7.3 4.7	7.2 4.5	7.0 4.4	6.9 4.2	6.7 4.0	6.5 3.9	6.3 3.7	6.1 3.5	5.9 3.3	5.7 3.1	60.0
60.3	7.8 5.0	7.6 4.8	7.5 4.7	7.3 4.5	7.1 4.3	6.9 4.1	6.7 3.9	6.5 3.7	6.3 3.5	6.0 3.2	5.8 3.0	60.3
60.6	8.1 5.1	7.9 5.0	7.7 4.8	7.6 4.6	7.3 4.4	7.1 4.2	6.9 3.9	6.7 3.7	6.4 3.4	6.2 3.2	5.9 2.9	60.6
60.9	8.4 5.3	8.2 5.1	8.0 4.9	7.8 4.7	7.6 4.5	7.3 4.2	7.1 4.0	6.8 3.7	6.6 3.4	6.3 3.2	6.0 2.9	60.9
61.2	8.7 5.4	8.5 5.2	8.3 5.0	8.1 4.8	7.8 4.5	7.6 4.3	7.3 4.0	7.0 3.7	6.7 3.4	6.4 3.1	6.1 2.8	61.2
61.5	9.1 5.6	8.8 5.4	8.6 5.1	8.3 4.9	8.1 4.6	7.8 4.3	7.5 4.0	7.2 3.7	6.9 3.4	6.5 3.1	6.2 2.7	61.5

TABLE 6 — ALTITUDE CORRECTION TABLES

c. Additional Refraction for Non-standard Conditions

App. Alt.	A	B	C	D	E	F	G	H	J	K	L	M	N	P	App. Alt.
° ′	′	′	′	′	′	′	′	′	′	′	′	′	′	′	° ′
00 00	−7.3	−5.9	−4.6	−3.4	−2.2	−1.1	0.0	+1.0	+2.0	+3.0	+4.0	+4.9	+5.9	+6.9	00 00
00 30	5.5	4.5	3.5	2.6	1.7	0.8	0.0	0.8	1.6	2.3	3.1	3.8	4.5	5.3	00 30
01 00	4.4	3.5	2.8	2.0	1.3	0.7	0.0	0.6	1.2	1.8	2.4	3.0	3.6	4.2	01 00
01 30	3.5	2.9	2.2	1.7	1.1	0.5	0.0	0.5	1.0	1.5	2.0	2.5	2.9	3.4	01 30
02 00	2.9	2.4	1.9	1.4	0.9	0.4	0.0	0.4	0.8	1.3	1.7	2.0	2.4	2.8	02 00
02 30	−2.5	−2.0	−1.6	−1.2	−0.8	−0.4	0.0	+0.4	+0.7	+1.1	+1.4	+1.7	+2.1	+2.4	02 30
03 00	2.1	1.7	1.4	1.0	0.7	0.3	0.0	0.3	0.6	0.9	1.2	1.5	1.8	2.1	03 00
03 30	1.9	1.5	1.2	0.9	0.6	0.3	0.0	0.3	0.5	0.8	1.1	1.3	1.6	1.8	03 30
04 00	1.6	1.3	1.1	0.8	0.5	0.3	0.0	0.2	0.5	0.7	0.9	1.2	1.4	1.6	04 00
04 30	1.5	1.2	0.9	0.7	0.5	0.2	0.0	0.2	0.4	0.6	0.8	1.0	1.3	1.5	04 30
05 00	−1.3	−1.1	−0.9	−0.6	−0.4	−0.2	0.0	+0.2	+0.4	+0.6	+0.8	+0.9	+1.1	+1.3	05 00
06	1.1	0.9	0.7	0.5	0.3	0.2	0.0	0.2	0.3	0.5	0.6	0.8	0.9	1.1	06
07	1.0	0.8	0.6	0.5	0.3	0.1	0.0	0.1	0.3	0.4	0.5	0.7	0.8	0.9	07
08	0.8	0.7	0.5	0.4	0.3	0.1	0.0	0.1	0.2	0.4	0.5	0.6	0.7	0.8	08
09	0.7	0.6	0.5	0.4	0.2	0.1	0.0	0.1	0.2	0.3	0.4	0.5	0.6	0.7	09
10 00	−0.7	−0.5	−0.4	−0.3	−0.2	−0.1	0.0	+0.1	+0.2	+0.3	+0.4	+0.5	+0.6	+0.7	10 00
12	0.6	0.5	0.4	0.3	0.2	0.1	0.0	0.1	0.2	0.2	0.3	0.4	0.5	0.5	12
14	0.5	0.4	0.3	0.2	0.1	0.1	0.0	0.1	0.1	0.2	0.3	0.3	0.4	0.5	14
16	0.4	0.3	0.3	0.2	0.1	0.1	0.0	0.1	0.1	0.2	0.2	0.3	0.3	0.4	16
18	0.4	0.3	0.2	0.2	0.1	−0.1	0.0	+0.1	0.1	0.2	0.2	0.3	0.3	0.4	18
20 00	−0.3	−0.3	−0.2	−0.2	−0.1	0.0	0.0	0.0	+0.1	+0.1	+0.2	+0.2	+0.3	+0.3	20 00
25	0.3	0.2	0.2	0.1	0.1	0.0	0.0	0.0	0.1	0.1	0.1	0.2	0.2	0.2	25
30	0.2	0.2	0.1	0.1	0.1	0.0	0.0	0.0	+0.1	0.1	0.1	0.1	0.2	0.2	30
35	0.2	0.1	0.1	0.1	−0.1	0.0	0.0	0.0	0.0	0.1	0.1	0.1	0.1	0.2	35
40	0.1	0.1	0.1	−0.1	0.0	0.0	0.0	0.0	+0.1	0.1	0.1	0.1	0.1	0.1	40
50 00	−0.1	−0.1	−0.1	0.0	0.0	0.0	0.0	0.0	0.0	0.0	+0.1	+0.1	+0.1	+0.1	50 00

The graph is entered with arguments temperature and pressure to find a zone letter; using as arguments this zone letter and apparent altitude (sextant altitude corrected for index error and dip), a correction is taken from the table. This correction is to be applied to the sextant altitude in addition to the corrections for standard conditions (Table **6**b).

Extract of part of main table for Lat 40° (contr) showing Dec 10°, LHA 20°

LHA	6° Hc	d	Z	7° Hc	d	Z	8° Hc	d	Z	9° Hc	d	Z	10° Hc	d	Z	11° Hc	d	Z
69	11 53	41	108	11 12	41	109	10 31	42	110	09 49	41	111	09 08	41	111	08 27	42	112
68	12 36	41	109	11 55	41	110	11 14	42	111	10 32	41	111	09 51	42	112	09 09	42	113
67	13 20	42	110	12 38	41	111	11 57	42	111	11 15	42	112	10 33	42	113	09 51	42	113
66	14 03	42	111	13 21	42	111	12 39	42	112	11 57	42	113	11 15	42	113	10 33	42	114
65	14 46	−42	111	14 04	−42	112	13 22	−42	113	12 40	−42	113	11 58	−43	114	11 15	−42	115
64	15 28	42	112	14 46	42	113	14 04	42	113	13 22	43	114	12 39	42	115	11 57	43	116
63	16 11	42	113	15 29	43	113	14 46	42	114	14 04	43	115	13 21	43	116	12 38	43	116
62	16 53	42	113	16 11	43	114	15 28	43	115	14 45	43	116	14 02	43	116	13 19	43	117
61	17 35	43	114	16 52	42	115	16 10	43	116	15 27	44	116	14 43	43	117	14 00	43	118
60	18 17	−43	115	17 34	−43	116	16 51	−43	116	16 08	−44	117	15 24	−44	118	14 40	−43	119
59	18 59	44	116	18 15	43	116	17 32	44	117	16 48	43	118	16 05	44	119	15 21	44	119
58	19 40	44	116	18 56	43	117	18 13	44	118	17 29	44	119	16 45	44	119	16 01	45	120
57	20 21	44	117	19 37	44	118	18 53	44	119	18 09	44	119	17 25	45	120	16 40	44	121
56	21 02	44	118	20 18	45	119	19 33	44	119	18 49	45	120	18 04	44	121	17 20	45	122
55	21 42	−44	119	20 58	−45	119	20 13	−44	120	19 29	−45	121	18 44	−45	122	17 59	−45	122
54	22 22	44	120	21 38	45	120	20 53	45	121	20 08	45	122	19 23	46	122	18 37	45	123
53	23 02	45	120	22 17	45	121	21 32	45	122	20 47	46	122	20 01	45	123	19 16	46	124
52	23 42	46	121	22 56	45	122	22 11	46	123	21 25	45	123	20 40	46	124	19 54	46	125
51	24 21	46	122	23 35	45	123	22 50	46	123	22 04	46	124	21 18	47	125	20 31	46	125
50	25 00	−46	123	24 14	−46	124	23 28	−46	124	22 42	−47	125	21 55	−46	126	21 09	−47	126
49	25 38	46	124	24 52	46	124	24 06	47	125	23 19	47	126	22 32	47	126	21 45	47	127
48	26 16	46	124	25 30	47	125	24 43	47	126	23 56	47	127	23 09	47	127	22 22	47	128
47	26 54	47	125	26 07	47	126	25 20	47	127	24 33	47	127	23 46	48	128	22 58	48	129
46	27 31	47	126	26 44	47	127	25 57	48	128	25 09	48	128	24 21	47	129	23 34	48	130
45	28 08	−47	127	27 21	−48	128	26 33	−48	128	25 45	−48	129	24 57	−48	130	24 09	−49	130
44	28 44	47	128	27 57	48	129	27 09	49	129	26 20	48	130	25 32	48	131	24 44	49	131
43	29 20	48	129	28 32	48	130	27 44	49	130	26 55	48	131	26 07	49	132	25 18	49	132
42	29 56	49	130	29 07	48	131	28 19	49	131	27 30	49	132	26 41	49	132	25 52	50	133
41	30 31	49	131	29 42	49	131	28 53	49	132	28 04	50	133	27 14	49	133	26 25	50	134
40	31 06	−50	132	30 16	−49	132	29 27	−50	133	28 37	−49	134	27 48	−50	134	26 58	−50	135
39	31 40	50	133	30 50	50	133	30 00	50	134	29 10	50	135	28 20	50	135	27 30	50	136
38	32 13	50	134	31 23	50	134	30 33	50	135	29 43	51	136	28 52	50	136	28 02	51	137
37	32 46	50	135	31 56	51	135	31 05	50	136	30 15	51	137	29 24	51	137	28 33	51	138
36	33 19	51	136	32 28	51	136	31 37	51	137	30 46	51	137	29 55	51	138	29 04	52	139
35	33 50	−51	137	32 59	−51	137	32 08	−51	138	31 17	−52	138	30 25	−51	139	29 34	−52	140
34	34 22	52	138	33 30	51	138	32 39	52	139	31 47	52	139	30 55	52	140	30 03	52	141
33	34 52	51	139	34 01	52	139	33 09	52	140	32 17	53	140	31 24	52	141	30 32	53	142
32	35 22	52	140	34 30	52	140	33 38	53	141	32 45	52	142	31 53	53	142	31 00	53	143
31	35 52	53	141	34 59	52	141	34 07	53	142	33 14	53	143	32 21	53	143	31 28	54	144
30	36 20	−52	142	35 28	−53	142	34 35	−54	143	33 41	−53	144	32 48	−53	144	31 55	−54	145
29	36 48	53	143	35 55	53	144	35 02	54	144	34 08	53	145	33 15	54	145	32 21	54	146
28	37 16	54	144	36 22	54	145	35 28	53	145	34 35	54	146	33 41	55	146	32 46	54	147
27	37 42	54	145	36 48	54	146	35 54	54	146	35 00	54	147	34 06	55	147	33 11	54	148
26	38 08	54	146	37 14	55	147	36 19	54	147	35 25	55	148	34 30	55	148	33 35	55	149
25	38 33	−54	147	37 39	−55	148	36 44	−55	149	35 49	−55	149	34 54	−55	150	33 59	−56	150
24	38 58	55	149	38 03	56	149	37 07	55	150	36 12	55	150	35 17	56	151	34 21	55	151
23	39 21	55	150	38 26	56	150	37 30	55	151	36 35	56	151	35 39	56	152	34 43	56	152
22	39 44	56	151	38 48	56	152	37 52	56	152	36 56	56	152	36 00	56	153	35 04	56	153
21	40 06	56	152	39 10	57	153	38 13	56	153	37 17	56	154	36 21	57	154	35 24	56	154
20	40 27	−57	153	39 30	−56	154	38 34	−57	154	37 37	−56	155	36 41	−57	155	35 44	−57	156
19	40 47	57	155	39 50	57	155	38 53	57	156	37 56	57	156	36 59	57	156	36 02	57	157
18	41 06	57	156	40 09	57	156	39 12	57	157	38 15	58	157	37 17	57	158	36 20	57	158
17	41 24	57	157	40 27	57	158	39 30	58	158	38 32	57	158	37 35	58	159	36 37	58	159
16	41 42	58	158	40 44	58	159	39 46	57	159	38 49	58	160	37 51	58	160	36 53	58	160
15	41 58	−58	160	41 00	−58	160	40 02	−58	160	39 04	−58	161	38 06	−58	161	37 08	−58	161

Extract of part of main table for Lat 40° (same) showing Dec 25° LHA 298°

	23°			24°			25°			26°			27°			28°			29°			
	Hc	d	Z	Hc	d	Z	Hc	d	Z	Hc	d	Z	Hc	d	Z	Hc	d	Z	Hc	d	Z	LHA
	° ′	′	°	° ′	′	°	° ′	′	°	° ′	′	°	° ′	′	°	° ′	′	°	° ′	′	°	°
0	73 00	+60	180	74 00	+60	180	75 00	+60	180	76 00	+60	180	77 00	+60	180	78 00	+60	180	79 00	+60	180	360
7	72 59	60	177	73 59	60	177	74 59	60	177	75 59	59	176	76 58	60	176	77 58	60	176	78 58	60	175	359
4	72 55	60	174	73 55	59	173	74 54	60	173	75 54	60	173	76 54	59	172	77 53	60	172	78 53	59	171	358
1	72 49	59	171	73 48	59	170	74 47	60	170	75 47	59	169	76 46	59	168	77 45	59	167	78 44	59	166	357
8	72 40	59	168	73 39	59	167	74 38	58	166	75 36	59	165	76 35	58	164	77 33	58	163	78 31	58	162	356
5	72 29	+58	165	73 27	+59	164	74 26	+58	163	75 24	+57	162	76 21	+58	161	77 19	+57	159	78 16	+56	158	355
2	72 16	57	162	73 13	58	161	74 11	57	160	75 08	57	159	76 05	56	157	77 01	56	156	77 57	55	154	354
0	72 00	57	159	72 57	57	158	73 54	56	157	74 50	56	155	75 46	55	154	76 41	54	152	77 35	54	150	353
7	71 42	56	156	72 38	56	155	73 34	56	154	74 30	54	152	75 24	54	151	76 18	53	149	77 11	52	147	352
4	71 22	56	153	72 18	55	152	73 13	54	151	74 07	54	149	75 01	52	147	75 53	52	145	76 45	50	143	351
2	71 01	+54	151	71 55	+54	149	72 49	+53	148	73 42	+53	146	74 35	+51	144	75 26	+51	142	76 17	+49	140	350
9	70 37	54	148	71 31	53	147	72 24	52	145	73 16	51	143	74 07	50	142	74 57	49	140	75 46	48	137	349
7	70 12	53	146	71 05	52	144	71 57	52	143	72 48	50	141	73 38	49	139	74 27	47	137	75 14	46	134	348
5	69 45	52	143	70 37	51	142	71 28	50	140	72 18	49	138	73 07	47	136	73 54	47	134	74 41	46	132	347
2	69 17	51	141	70 08	50	139	70 58	49	138	71 47	47	136	72 34	47	134	73 21	45	132	74 06	43	129	346
0	68 48	+49	139	69 37	+49	137	70 26	+48	136	71 14	+47	134	72 01	+45	132	72 46	+44	130	73 30	+42	127	345
8	68 17	48	137	69 05	48	135	69 53	47	133	70 40	46	132	71 26	44	130	72 10	43	127	72 53	40	125	344
6	67 45	47	135	68 32	47	133	69 19	46	131	70 05	45	130	70 50	43	128	71 33	42	125	72 15	39	123	343
4	67 11	47	133	67 58	46	131	68 44	45	129	69 29	44	128	70 13	42	126	70 55	41	123	71 36	38	121	342
2	66 37	46	131	67 23	45	129	68 08	44	128	68 52	43	126	69 35	41	124	70 16	40	122	70 56	38	119	341
1	66 02	+45	129	66 47	+45	128	67 32	+43	126	68 15	+41	124	68 56	+41	122	69 37	+38	120	70 15	+37	118	340
9	65 26	44	127	66 10	44	126	66 54	42	124	67 36	41	122	68 17	40	120	68 57	37	118	69 34	37	116	339
7	64 49	44	126	65 33	43	124	66 16	41	123	66 57	40	121	67 37	39	119	68 16	37	117	68 53	35	115	338
6	64 11	43	124	64 54	42	123	65 36	41	121	66 17	40	119	66 57	37	117	67 34	37	115	68 11	35	113	337
4	63 33	42	123	64 15	42	121	64 57	40	119	65 37	38	118	66 15	38	116	66 53	35	114	67 28	34	112	336
3	62 54	+42	121	63 36	+40	120	64 16	+40	118	64 56	+38	116	65 34	+36	114	66 10	+36	113	66 46	+33	110	335
1	62 15	41	120	62 56	40	118	63 36	38	117	64 14	38	115	64 52	36	113	65 28	34	111	66 02	33	109	334
0	61 35	40	119	62 15	39	117	62 54	38	115	63 32	37	114	64 09	36	112	64 45	34	110	65 19	32	108	333
9	60 54	40	117	61 34	39	116	62 13	37	114	62 50	36	112	63 26	35	111	64 01	34	109	64 35	32	107	332
8	60 13	39	116	60 52	38	115	61 30	37	113	62 07	36	111	62 43	35	110	63 18	33	108	63 51	31	106	331
6	59 31	+39	115	60 10	+38	113	60 48	+36	112	61 24	+36	110	62 00	+34	108	62 34	+33	107	63 07	+31	105	330
5	58 49	39	114	59 28	37	112	60 05	36	111	60 41	35	109	61 16	34	107	61 50	32	106	62 22	31	104	329
4	58 07	38	113	58 45	37	111	59 22	36	110	59 58	34	108	60 32	33	106	61 05	32	105	61 37	31	103	328
3	57 25	37	111	58 02	36	110	58 38	35	108	59 14	34	107	59 48	33	105	60 21	31	104	60 52	31	102	327
2	56 42	37	110	57 19	36	109	57 55	35	107	58 30	33	106	59 03	32	104	59 36	31	103	60 07	30	101	326
1	55 58	+37	109	56 35	+36	108	57 11	+34	106	57 45	+34	105	58 19	+32	103	58 51	+32	102	59 22	+30	100	325
0	55 15	36	108	55 51	36	107	56 27	34	105	57 01	33	104	57 34	32	102	58 06	31	101	58 37	29	99	324
9	54 31	36	107	55 07	35	106	55 42	34	105	56 16	33	103	56 49	32	102	57 21	30	100	57 51	30	98	323
8	53 47	36	106	54 23	35	105	54 58	33	104	55 31	33	102	56 04	31	101	56 35	31	99	57 06	29	98	322
7	53 03	35	105	53 38	35	104	54 13	33	103	54 46	32	101	55 19	31	100	55 50	30	98	56 20	29	97	321
6	52 19	+35	105	52 54	+34	103	53 28	+33	102	54 01	+32	100	54 33	+32	99	55 05	+30	98	55 35	+28	96	320
5	51 34	35	104	52 09	34	102	52 43	33	101	53 16	32	100	53 48	31	98	54 19	30	97	54 49	29	95	319
4	50 49	35	103	51 24	34	102	51 58	33	100	52 31	31	99	53 02	31	97	53 33	30	96	54 03	29	95	318
3	50 04	35	102	50 39	33	101	51 12	33	99	51 45	32	98	52 17	30	97	52 47	30	95	53 17	29	94	317
2	49 19	35	101	49 54	33	100	50 27	33	99	51 00	31	97	51 31	31	96	52 02	29	95	52 31	29	93	316
2	48 34	+34	100	49 08	+34	99	49 42	+32	98	50 14	+31	97	50 45	+31	95	51 16	+29	94	51 45	+29	92	315
1	47 49	34	100	48 23	33	98	48 56	32	97	49 28	32	96	50 00	30	94	50 30	29	93	50 59	29	92	314
0	47 04	33	99	47 37	33	98	48 10	32	96	48 42	32	95	49 14	30	94	49 44	30	92	50 14	28	91	313
9	46 18	34	98	46 52	33	97	47 25	32	96	47 57	31	94	48 28	30	93	48 58	30	92	49 28	28	90	312
8	45 33	33	97	46 06	33	96	46 39	32	95	47 11	31	94	47 42	30	92	48 12	30	91	48 42	28	90	311
8	44 47	+33	97	45 20	+33	95	45 53	+32	94	46 25	+31	93	46 56	+30	92	47 26	+30	90	47 56	+28	89	310
7	44 01	34	96	44 35	32	95	45 07	32	93	45 39	31	92	46 10	30	91	46 40	30	90	47 10	28	89	309
6	43 16	33	95	43 49	32	94	44 21	32	93	44 53	31	92	45 24	30	90	45 54	30	89	46 24	28	88	308
6	42 30	33	94	43 03	32	93	43 35	32	92	44 07	31	91	44 38	30	90	45 08	30	89	45 38	28	87	307
5	41 44	33	94	42 17	32	93	42 49	32	91	43 21	31	90	43 52	30	89	44 22	30	88	44 52	29	87	306
4	40 58	+33	93	41 31	+32	92	42 03	+32	91	42 35	+31	90	43 06	+30	88	43 36	+30	87	44 06	+29	86	305
3	40 12	33	92	40 45	32	91	41 17	32	90	41 49	31	89	42 20	31	88	42 51	29	87	43 20	29	86	304
3	39 26	33	92	39 59	33	91	40 32	31	90	41 03	31	88	41 34	31	87	42 05	29	86	42 34	29	85	303
2	38 40	33	91	39 13	33	90	39 46	31	89	40 17	31	88	40 48	31	87	41 19	30	86	41 49	29	84	302
1	37 54	33	90	38 27	33	89	39 00	31	88	39 31	32	87	40 03	30	86	40 33	30	85	41 03	29	84	301
1	37 08	+33	90	37 41	+33	89	38 14	+32	88	38 46	+31	87	39 17	+30	85	39 47	+30	84	40 17	+30	83	300
0	36 22	33	89	36 55	33	88	37 28	32	87	38 00	31	86	38 31	31	85	39 02	30	84	39 32	29	83	299
9	35 36	33	88	36 09	33	87	36 42	32	86	37 14	31	85	37 45	31	84	38 16	30	83	38 46	29	82	298
9	34 50	33	88	35 23	33	87	35 56	32	86	36 28	31	85	36 59	31	84	37 30	31	83	38 01	29	82	297
8	34 04	34	87	34 38	32	86	35 10	32	85	35 42	32	84	36 14	31	83	36 45	30	82	37 15	30	81	296
8	33 19	+33	87	33 52	+32	86	34 24	+33	85	34 57	+31	84	35 28	+31	83	35 59	+31	81	36 30	+30	80	295
7	32 33	33	86	33 06	33	85	33 39	32	84	34 11	32	83	34 43	31	82	35 14	31	81	35 45	30	80	294
6	31 47	33	85	32 20	33	84	32 53	33	83	33 25	32	82	33 57	32	81	34 29	30	80	34 59	31	79	293
6	31 01	33	85	31 34	33	84	32 07	33	83	32 40	32	82	33 12	31	81	33 43	31	80	34 14	31	79	292
5	30 15	34	84	30 49	33	83	31 22	32	82	31 54	32	81	32 26	32	80	32 58	31	79	33 29	31	78	291

Extract of part of main table for Lat 40° (contr) showing Dec 20° LHA 54°

LHA	15° Hc d Z	16° Hc d Z	17° Hc d Z	18° Hc d Z	19° Hc d Z	20° Hc d Z	21° Hc d Z
69	05 40 42 115	04 58 41 116	04 17 42 116	03 35 42 117	02 53 42 118	02 11 42 119	01 29 42 119
68	06 22 42 116	05 40 42 116	04 58 42 117	04 16 42 118	03 33 42 119	02 51 42 119	02 09 42 120
67	07 03 42 116	06 21 43 117	05 38 42 118	04 56 42 119	04 14 43 119	03 31 42 120	02 49 43 121
66	07 44 42 117	07 02 43 118	06 19 43 118	05 36 42 119	04 54 43 120	04 11 43 121	03 28 43 121
65	08 25 −43 118	07 42 −43 118	06 59 −43 119	06 16 −43 120	05 33 −43 121	04 50 −43 121	04 07 −43 122
64	09 05 43 118	08 22 43 119	07 39 43 120	06 56 43 121	06 13 43 121	05 30 44 122	04 46 43 123
63	09 46 44 119	09 02 43 120	08 19 43 121	07 36 44 121	06 52 44 122	06 08 43 123	05 25 44 123
62	10 26 44 120	09 42 44 121	08 58 43 121	08 15 44 122	07 31 44 123	06 47 44 123	06 03 44 124
61	11 05 43 121	10 22 44 121	09 38 44 122	08 54 45 123	08 09 44 123	07 25 44 124	06 41 44 125
60	11 45 −44 121	11 01 −45 122	10 16 −44 123	09 32 −44 123	08 48 −45 124	08 03 −44 125	07 19 −45 125
59	12 24 44 122	11 40 45 123	10 55 45 123	10 10 44 124	09 26 45 125	08 41 45 125	07 56 45 126
58	13 03 45 123	12 18 45 123	11 33 45 124	10 48 45 125	10 03 45 125	09 18 45 126	08 33 45 127
57	13 41 45 124	12 56 45 124	12 11 45 125	11 26 46 126	10 40 45 126	09 55 46 127	09 09 45 128
56	14 19 45 124	13 34 45 125	12 49 46 126	12 03 46 126	11 17 45 127	10 32 46 128	09 46 46 128
55	14 57 −45 125	14 12 −46 126	13 26 −46 126	12 40 −46 127	11 54 −46 128	11 08 −46 128	10 22 −47 129
54	15 35 46 126	14 49 46 126	14 03 47 127	13 16 46 128	12 30 46 128	11 44 47 129	10 57 46 130
53	16 12 46 127	15 26 47 127	14 39 46 128	13 53 47 129	13 06 47 129	12 19 47 130	11 32 47 130
52	16 49 47 127	16 02 47 128	15 15 47 129	14 28 47 129	13 41 47 130	12 54 47 131	12 07 47 131
51	17 25 47 128	16 38 47 129	15 51 47 129	15 04 48 130	14 16 47 131	13 29 47 131	12 42 48 132
50	18 01 −47 129	17 14 −48 130	16 26 −47 130	15 39 −48 131	14 51 −48 131	14 03 −47 132	13 16 −48 133
49	18 36 47 130	17 49 48 130	17 01 48 131	16 13 48 132	15 25 48 132	14 37 48 133	13 49 48 133
48	19 12 48 131	18 24 48 131	17 36 49 132	16 47 48 132	15 59 48 133	15 11 49 134	14 22 48 134
47	19 46 48 131	18 58 48 132	18 10 49 133	17 21 49 133	16 32 48 134	15 44 49 134	14 55 49 135
46	20 21 49 132	19 32 49 133	18 43 49 133	17 54 49 134	17 05 49 135	16 16 49 135	15 27 49 136
45	20 54 −49 133	20 05 −49 134	19 16 −49 134	18 27 −49 135	17 38 −50 135	16 48 −49 136	15 59 −50 137
44	21 28 49 134	20 39 50 134	19 49 49 135	19 00 50 136	18 10 50 136	17 20 50 137	16 30 50 137
43	22 01 50 135	21 11 50 135	20 21 50 136	19 31 50 137	18 41 50 137	17 51 50 138	17 01 50 138
42	22 33 50 135	21 43 50 136	20 53 50 137	20 03 51 137	19 12 50 138	18 22 51 139	17 31 50 139
41	23 05 50 136	22 15 51 137	21 24 50 138	20 34 51 138	19 43 51 139	18 52 51 139	18 01 51 140
40	23 36 −50 137	22 46 −51 138	21 55 −51 139	21 04 −51 139	20 13 −51 140	19 22 −51 140	18 31 −52 141
39	24 07 51 138	23 16 51 139	22 25 51 139	21 34 51 140	20 43 52 140	19 51 51 141	19 00 52 142
38	24 38 52 139	23 46 51 140	22 55 52 140	22 03 51 141	21 12 52 141	20 20 52 142	19 28 52 142
37	25 07 51 140	24 16 52 140	23 24 52 141	22 32 52 142	21 40 52 142	20 48 52 143	19 56 52 143
36	25 37 52 141	24 45 53 142	23 52 52 142	23 00 52 143	22 08 53 143	21 15 52 144	20 23 53 144
35	26 05 −52 142	25 13 −53 142	24 20 −52 143	23 28 −53 144	22 35 −53 144	21 42 −53 145	20 49 −53 145
34	26 33 52 143	25 41 53 143	24 48 53 144	23 55 53 144	23 02 53 145	22 09 54 145	21 15 53 146
33	27 01 53 144	26 08 53 144	25 15 54 145	24 21 53 145	23 28 54 146	22 34 53 146	21 41 54 147
32	27 28 54 145	26 34 53 145	25 41 54 146	24 47 54 146	23 53 53 147	22 59 54 147	22 06 54 148
31	27 54 54 146	27 00 54 146	26 06 54 147	25 12 54 147	24 18 54 148	23 24 54 148	22 30 54 149
30	28 19 −54 147	27 25 −54 147	26 31 −54 148	25 37 −55 148	24 42 −54 149	23 48 −54 149	22 54 −55 150
29	28 44 54 148	27 50 55 148	26 55 54 149	26 01 55 149	25 06 55 150	24 11 55 150	23 16 54 150
28	29 08 54 149	28 14 55 149	27 19 55 150	26 24 55 150	25 29 55 151	24 34 55 151	23 39 55 151
27	29 32 55 150	28 37 55 150	27 42 55 151	26 47 56 151	25 51 55 152	24 56 55 152	24 00 55 152
26	29 55 56 151	28 59 55 151	28 04 56 152	27 08 55 152	26 13 56 152	25 17 56 153	24 21 55 153
25	30 17 −56 152	29 21 −56 152	28 25 −55 153	27 30 −56 153	26 34 −56 153	25 38 −56 154	24 42 −56 154
24	30 38 56 153	29 42 56 153	28 46 56 154	27 50 56 154	26 54 56 154	25 58 57 155	25 01 56 155
23	30 59 56 154	30 03 57 154	29 06 56 155	28 10 57 155	27 13 56 155	26 17 57 156	25 20 56 156
22	31 19 57 155	30 22 56 155	29 26 57 156	28 29 57 156	27 32 57 156	26 35 57 157	25 38 57 157
21	31 38 57 156	30 41 57 156	29 44 57 157	28 47 57 157	27 50 57 157	26 53 57 158	25 56 57 158
20	31 56 −57 157	30 59 −57 157	30 02 −57 158	29 05 −58 158	28 07 −57 158	27 10 −57 159	26 13 −58 159
19	32 14 58 158	31 16 57 159	30 19 58 159	29 21 57 159	28 24 58 160	27 26 57 160	26 29 58 160
18	32 30 57 159	31 33 58 160	30 35 58 160	29 37 57 160	28 40 58 161	27 42 58 161	26 44 58 161
17	32 46 58 160	31 48 58 161	30 50 58 161	29 52 58 161	28 54 58 162	27 56 58 162	26 58 58 162
16	33 01 58 161	32 03 58 162	31 05 58 162	30 07 59 162	29 09 59 163	28 10 58 163	27 12 58 163
15	33 15 −58 163	32 17 −58 163	31 19 −59 163	30 20 −58 163	29 22 −59 164	28 23 −58 164	27 25 −59 164
14	33 29 59 164	32 30 58 164	31 32 59 164	30 33 59 165	29 34 58 165	28 36 59 165	27 37 59 165
13	33 41 59 165	32 42 58 165	31 44 59 165	30 45 59 166	29 46 59 166	28 47 59 166	27 48 58 166
12	33 53 59 166	32 54 59 166	31 55 59 166	30 56 59 167	29 57 59 167	28 58 59 167	27 59 59 167
11	34 03 59 167	33 04 59 167	32 05 59 168	31 06 59 168	30 07 59 168	29 08 59 168	28 09 60 168
10	34 13 −59 168	33 14 −59 168	32 15 −60 169	31 15 −59 169	30 16 −59 169	29 17 −60 169	28 17 −59 169
9	34 22 59 169	33 23 60 170	32 23 59 170	31 24 60 170	30 24 59 170	29 25 60 170	28 25 59 170
8	34 30 60 171	33 30 59 171	32 31 60 171	31 31 59 171	30 32 60 171	29 32 59 171	28 33 60 171
7	34 37 60 172	33 37 59 172	32 38 60 172	31 38 60 172	30 38 59 172	29 39 60 172	28 39 60 173
6	34 43 60 173	33 43 60 173	32 44 60 173	31 44 60 173	30 44 60 173	29 44 59 174	28 45 60 174
5	34 48 −60 174	33 48 −59 174	32 49 −60 174	31 49 −60 174	30 49 −60 174	29 49 −60 175	28 49 −60 175
4	34 52 59 175	33 53 60 175	32 53 60 175	31 53 60 176	30 53 60 176	29 53 60 176	28 53 60 176
3	34 56 60 176	33 56 60 177	32 56 60 177	31 56 60 177	30 56 60 177	29 56 60 177	28 56 60 177
2	34 58 60 178	33 58 60 178	32 58 60 178	31 58 60 178	30 58 60 178	29 58 60 178	28 58 60 178
1	35 00 60 179	34 00 60 179	33 00 60 179	32 00 60 179	31 00 60 179	30 00 60 179	29 00 60 179
0	35 00 −60 180	34 00 −60 180	33 00 −60 180	32 00 −60 180	31 00 −60 180	30 00 −60 180	29 00 −60 180

©2009 Jack Case

Extract from table 5 showing d 31 to 60.

31	32	33	34	35	36	37	38	39	40	41	42	43	44	45	46	47	48	49	50	51	52	53	54	55	56	57	58	59	60	d
′	′	′	′	′	′	′	′	′	′	′	′	′	′	′	′	′	′	′	′	′	′	′	′	′	′	′	′	′	′	′
0	0	0	0	0	0	0	0	0	0	0	0	0	0	0	0	0	0	0	0	0	0	0	0	0	0	0	0	0	0	0
1	1	1	1	1	1	1	1	1	1	1	1	1	1	1	1	1	1	1	1	1	1	1	1	1	1	1	1	1	1	1
1	1	1	1	1	1	1	1	1	1	1	1	1	1	2	2	2	2	2	2	2	2	2	2	2	2	2	2	2	2	2
2	2	2	2	2	2	2	2	2	2	2	2	2	2	2	2	2	2	2	2	3	3	3	3	3	3	3	3	3	3	3
2	2	2	2	2	2	2	3	3	3	3	3	3	3	3	3	3	3	3	3	3	3	4	4	4	4	4	4	4	4	4
3	3	3	3	3	3	3	3	3	3	4	4	4	4	4	4	4	4	4	4	4	4	4	4	5	5	5	5	5	5	5
3	3	3	3	4	4	4	4	4	4	4	4	4	4	4	5	5	5	5	5	5	5	5	5	6	6	6	6	6	6	6
4	4	4	4	4	4	4	4	5	5	5	5	5	5	5	5	5	6	6	6	6	6	6	6	7	7	7	7	7	7	7
4	4	4	5	5	5	5	5	5	6	6	6	6	6	6	6	6	6	7	7	7	7	7	7	7	7	8	8	8	8	8
5	5	5	5	5	5	6	6	6	6	6	6	6	7	7	7	7	7	7	8	8	8	8	8	8	8	9	9	9	9	9
5	5	6	6	6	6	6	6	6	7	7	7	7	7	8	8	8	8	8	8	9	9	9	9	9	9	10	10	10	10	10
6	6	6	6	6	7	7	7	8	7	8	8	8	8	8	8	9	9	9	9	9	10	10	10	10	10	10	11	11	11	11
6	6	7	7	7	7	7	8	8	8	8	8	9	9	9	9	9	10	10	10	10	10	11	11	11	11	11	12	12	12	12
7	7	7	7	8	8	8	8	8	9	9	9	9	10	10	10	10	10	11	11	11	11	11	12	12	12	12	13	13	13	13
7	7	8	8	8	8	9	9	9	9	10	10	10	10	10	11	11	11	11	12	12	12	12	13	13	13	13	14	14	14	14
8	8	8	8	9	9	9	10	10	10	10	10	11	11	11	12	12	12	12	12	13	13	13	14	14	14	14	14	15	15	15
8	9	9	9	9	10	10	10	11	11	11	12	12	13	13	13	13	14	14	14	14	15	15	15	15	16	16	15	16	16	16
9	9	9	10	10	11	10	11	11	12	12	13	12	13	14	14	14	14	15	15	15	16	16	16	16	17	17	16	17	17	17
9	10	10	10	10	11	11	11	12	12	13	13	13	14	14	15	15	15	15	16	16	16	17	17	17	18	18	17	18	18	18
10	10	10	11	11	11	12	12	12	13	13	14	14	14	14	15	16	16	16	16	16	17	17	18	18	18	18	18	19	19	19
10	11	11	11	12	12	12	13	13	14	15	15	15	16	16	15	16	17	17	18	18	18	19	19	19	20	20	19	20	20	20
11	11	12	12	12	13	13	14	14	14	15	15	15	16	17	16	17	18	18	19	19	19	20	20	20	21	21	20	21	21	21
11	12	12	12	13	13	14	14	14	15	15	16	16	16	16	17	17	18	18	18	19	19	20	20	20	21	21	21	22	22	22
12	12	13	13	13	14	14	15	15	16	16	16	17	17	17	18	18	18	19	19	20	20	20	21	21	21	22	22	23	23	23
12	13	13	14	14	14	15	15	16	16	16	17	17	18	18	18	19	19	20	20	20	21	21	22	22	22	23	23	24	24	24
13	13	14	14	15	15	15	16	16	17	17	18	18	18	19	19	20	20	20	21	21	22	22	22	23	23	24	24	25	25	25
13	14	14	15	15	16	16	16	17	17	18	18	19	19	20	20	20	21	21	22	22	23	23	23	24	24	25	25	26	26	26
14	14	15	15	16	16	17	17	18	18	19	19	19	20	20	21	21	22	22	22	23	23	24	24	25	25	26	26	27	27	27
14	15	15	16	16	17	17	18	18	19	19	20	20	21	21	22	22	22	23	23	24	24	25	25	26	26	27	27	28	28	28
15	15	16	16	17	17	18	18	19	19	20	20	21	21	22	22	23	23	24	24	25	25	26	26	27	27	28	28	29	29	29
16	16	16	17	18	18	18	19	20	20	20	21	22	22	22	23	24	24	24	25	26	26	26	27	28	28	28	29	30	30	30
16	17	17	18	18	19	19	20	20	21	21	22	22	23	23	24	24	25	25	26	26	27	27	28	28	29	29	30	30	31	31
17	17	18	18	19	19	20	20	21	21	22	22	23	23	24	25	25	26	26	27	27	28	28	29	29	30	30	31	31	32	32
17	18	18	19	19	20	20	21	21	22	23	23	24	24	25	25	26	27	27	28	28	29	29	30	30	31	31	32	32	33	33
18	18	19	19	20	20	21	22	22	23	23	24	24	25	26	26	27	27	28	28	29	29	30	31	31	32	32	33	33	34	34
18	19	19	20	20	21	22	22	23	23	24	24	25	26	26	27	27	28	29	29	30	30	31	32	32	33	33	34	34	35	35
19	19	20	20	21	22	22	23	23	24	25	25	26	26	27	28	28	29	29	30	31	31	32	32	33	34	34	35	35	36	36
19	20	20	21	22	22	23	23	24	25	25	26	27	27	28	28	29	30	30	31	31	32	33	33	34	35	35	36	36	37	37
20	20	21	22	22	23	23	24	25	25	26	27	27	28	28	29	30	30	31	32	32	33	34	34	35	35	36	37	37	38	38
20	21	21	22	23	23	24	25	25	26	27	27	28	29	29	30	31	31	32	32	33	34	34	35	36	36	37	38	38	39	39
21	21	22	23	23	24	25	25	26	27	27	28	29	29	30	31	31	32	33	33	34	35	35	36	37	37	38	39	39	40	40
21	22	23	23	24	25	25	26	27	27	28	29	29	30	31	31	32	33	33	34	35	36	36	37	38	38	39	40	40	41	41
22	22	23	24	24	25	26	27	27	28	29	29	30	31	32	32	33	34	34	35	36	36	37	38	38	39	40	41	41	42	42
22	23	24	24	25	26	27	27	28	29	29	30	31	32	32	33	34	34	35	36	37	37	38	39	39	40	41	42	42	43	43
23	23	24	25	26	26	27	28	29	29	30	31	32	32	33	34	34	35	36	37	37	38	39	40	40	41	42	43	43	44	44
23	24	25	26	26	27	28	28	29	30	31	32	32	33	34	34	35	36	37	38	38	39	40	40	41	42	43	44	44	45	45
24	25	25	26	27	28	28	29	30	31	31	32	33	34	34	35	36	37	38	38	39	40	41	41	42	43	44	44	45	46	46
24	25	26	27	27	28	29	30	31	31	32	33	34	34	35	36	37	38	38	39	40	41	42	42	43	44	45	45	46	47	47
25	26	26	27	28	29	30	30	31	32	33	34	34	35	36	37	38	38	39	40	41	42	42	43	44	45	46	46	47	48	48
25	26	27	28	29	29	30	31	32	33	33	34	35	36	37	38	38	39	40	41	42	42	43	44	45	46	47	47	48	49	49
26	27	28	28	29	30	31	32	32	33	34	35	36	37	38	38	39	40	41	42	42	43	44	45	46	47	48	48	49	50	50
26	27	28	29	30	31	31	32	33	34	35	36	37	37	38	39	40	41	42	42	43	44	45	46	47	48	48	49	50	51	51
27	28	29	29	30	31	32	33	34	35	36	36	37	38	39	40	41	42	42	43	44	45	46	47	48	49	49	50	51	52	52
27	28	29	30	31	32	33	34	34	35	36	37	38	39	40	41	42	42	43	44	45	46	47	48	49	49	50	51	52	53	53
28	29	30	31	32	32	33	34	35	36	37	38	39	40	40	41	42	43	44	45	46	47	48	49	50	50	51	52	53	54	54
28	29	30	31	32	33	34	35	36	37	38	38	39	40	41	42	43	44	45	46	47	48	49	50	50	51	52	53	54	55	55
29	30	31	32	33	34	35	35	36	37	38	39	40	41	42	43	44	45	46	47	48	49	50	50	51	52	53	54	55	56	56
29	30	31	32	33	34	35	36	37	38	39	40	41	42	43	44	45	46	47	48	49	50	51	52	52	53	54	55	56	57	57
30	31	32	33	34	35	36	37	38	39	40	41	42	43	44	45	46	47	48	49	50	51	52	53	53	54	55	56	57	58	58
30	31	32	33	34	35	36	37	38	39	40	41	42	43	44	45	46	47	48	49	50	51	52	53	54	55	56	57	58	59	59

TABLE 2 — ALTITUDE CORRECTION FOR CHANGE IN POSITION OF BODY

Correction for 1 Minute of Time   MOB

True Zn	\multicolumn{17}{c}{Latitude}	True Zn
	0°	5°	10°	15°	20°	25°	30°	35°	40°	45°	50°	55°	60°	65°	70°	75°	80°	85°	
°	′	′	′	′	′	′	′	′	′	′	′	′	′	′	′	′	′	′	°
090	+15·0	+15·0	+14·8	+14·5	+14·1	+13·6	+13·0	+12·3	+11·5	+10·6	+9·7	+8·6	+7·5	+6·4	+5·1	+3·9	+2·6	+1·3	090
093	15·0	15·0	14·8	14·5	14·1	13·6	13·0	12·3	11·5	10·6	9·7	8·6	7·5	6·3	5·1	3·9	2·6	1·3	087
096	15·0	14·9	14·7	14·4	14·1	13·6	13·0	12·3	11·5	10·6	9·6	8·6	7·5	6·3	5·1	3·9	2·6	1·3	084
099	14·9	14·8	14·6	14·3	14·0	13·5	12·9	12·2	11·4	10·5	9·5	8·5	7·4	6·3	5·1	3·8	2·6	1·3	081
102	14·7	14·7	14·5	14·2	13·8	13·3	12·7	12·1	11·3	10·4	9·5	8·4	7·4	6·2	5·0	3·8	2·6	1·3	078
105	+14·5	+14·5	+14·3	+14·0	+13·7	+13·2	+12·6	+11·9	+11·1	+10·3	+9·3	+8·3	+7·3	+6·1	+5·0	+3·8	+2·5	+1·3	075
108	14·3	14·3	14·1	13·8	13·4	13·0	12·4	11·7	11·0	10·1	9·2	8·2	7·2	6·0	4·9	3·7	2·5	1·2	072
111	14·0	14·0	13·8	13·6	13·2	12·7	12·2	11·5	10·8	9·9	9·0	8·1	7·0	5·9	4·8	3·6	2·4	1·2	069
114	13·7	13·7	13·5	13·3	12·9	12·5	11·9	11·3	10·5	9·7	8·8	7·9	6·9	5·8	4·7	3·6	2·4	1·2	066
117	13·4	13·4	13·2	12·9	12·6	12·1	11·6	11·0	10·3	9·5	8·6	7·7	6·7	5·7	4·6	3·5	2·3	1·2	063
120	+13·0	+13·0	+12·8	+12·6	+12·2	+11·8	+11·3	+10·7	+10·0	+9·2	+8·4	+7·5	+6·5	+5·5	+4·5	+3·4	+2·3	+1·1	060
123	12·6	12·6	12·4	12·2	11·9	11·4	10·9	10·3	9·7	8·9	8·1	7·2	6·3	5·3	4·3	3·3	2·2	1·1	057
126	12·2	12·1	12·0	11·8	11·4	11·0	10·5	10·0	9·3	8·6	7·8	7·0	6·1	5·1	4·2	3·1	2·1	1·1	054
129	11·7	11·6	11·5	11·3	11·0	10·6	10·1	9·6	9·0	8·3	7·5	6·7	5·8	4·9	4·0	3·0	2·0	1·0	051
132	11·2	11·1	11·0	10·8	10·5	10·1	9·7	9·2	8·6	7·9	7·2	6·4	5·6	4·7	3·8	2·9	1·9	1·0	048
135	+10·6	+10·6	+10·5	+10·3	+10·0	+9·6	+9·2	+8·7	+8·1	+7·5	+6·8	+6·1	+5·3	+4·5	+3·6	+2·8	+1·8	+0·9	045
138	10·1	10·0	9·9	9·7	9·5	9·1	8·7	8·2	7·7	7·1	6·5	5·8	5·0	4·3	3·4	2·6	1·7	0·9	042
141	9·5	9·4	9·3	9·1	8·9	8·6	8·2	7·8	7·3	6·7	6·1	5·4	4·7	4·0	3·2	2·4	1·6	0·8	039
144	8·8	8·8	8·7	8·5	8·3	8·0	7·7	7·2	6·8	6·3	5·7	5·1	4·4	3·7	3·0	2·3	1·5	0·8	036
147	8·2	8·2	8·1	7·9	7·7	7·4	7·1	6·7	6·3	5·8	5·3	4·7	4·1	3·5	2·8	2·1	1·4	0·7	033
150	+7·5	+7·5	+7·4	+7·3	+7·1	+6·8	+6·5	+6·2	+5·8	+5·3	+4·8	+4·3	+3·8	+3·2	+2·6	+1·9	+1·3	+0·7	030
153	6·8	6·8	6·7	6·6	6·4	6·2	5·9	5·6	5·2	4·8	4·4	3·9	3·4	2·9	2·3	1·8	1·2	0·6	027
156	6·1	6·1	6·0	5·9	5·7	5·5	5·3	5·0	4·7	4·3	3·9	3·5	3·1	2·6	2·1	1·6	1·1	0·5	024
159	5·4	5·4	5·3	5·2	5·1	4·9	4·7	4·4	4·1	3·8	3·5	3·1	2·7	2·3	1·8	1·4	0·9	0·5	021
162	4·6	4·6	4·6	4·5	4·4	4·2	4·0	3·8	3·6	3·3	3·0	2·7	2·3	2·0	1·6	1·2	0·8	0·4	018
165	+3·9	+3·9	+3·8	+3·8	+3·7	+3·5	+3·4	+3·2	+3·0	+2·8	+2·5	+2·2	+1·9	+1·6	+1·3	+1·0	+0·7	+0·3	015
168	3·1	3·1	3·1	3·0	2·9	2·8	2·7	2·6	2·4	2·2	2·0	1·8	1·6	1·3	1·1	0·8	0·5	0·3	012
171	2·4	2·3	2·3	2·3	2·2	2·1	2·0	1·9	1·8	1·7	1·5	1·3	1·2	1·0	0·8	0·6	0·4	0·2	009
174	1·6	1·6	1·5	1·5	1·5	1·4	1·4	1·3	1·2	1·1	1·0	0·9	0·8	0·7	0·5	0·4	0·3	0·1	006
177	0·8	0·8	0·8	0·8	0·7	0·7	0·7	0·6	0·6	0·6	0·5	0·5	0·4	0·3	0·3	0·2	0·1	0·1	003
180	0·0	0·0	0·0	0·0	0·0	0·0	0·0	0·0	0·0	0·0	0·0	0·0	0·0	0·0	0·0	0·0	0·0	0·0	000
183	−0·8	−0·8	−0·8	−0·8	−0·7	−0·7	−0·7	−0·6	−0·6	−0·6	−0·5	−0·5	−0·4	−0·3	−0·3	−0·2	−0·1	−0·1	357
186	1·6	1·6	1·5	1·5	1·5	1·4	1·4	1·3	1·2	1·1	1·0	0·9	0·8	0·7	0·5	0·4	0·3	0·1	354
189	2·4	2·3	2·3	2·3	2·2	2·1	2·0	1·9	1·8	1·7	1·5	1·3	1·2	1·0	0·8	0·6	0·4	0·2	351
192	3·1	3·1	3·1	3·0	2·9	2·8	2·7	2·6	2·4	2·2	2·0	1·8	1·6	1·3	1·1	0·8	0·5	0·3	348
195	3·9	3·9	3·8	3·8	3·7	3·5	3·4	3·2	3·0	2·8	2·5	2·2	1·9	1·6	1·3	1·0	0·7	0·3	345
198	−4·6	−4·6	−4·6	−4·5	−4·4	−4·2	−4·0	−3·8	−3·6	−3·3	−3·0	−2·7	−2·3	−2·0	−1·6	−1·2	−0·8	−0·4	342
201	5·4	5·4	5·3	5·2	5·1	4·9	4·7	4·4	4·1	3·8	3·5	3·1	2·7	2·3	1·8	1·4	0·9	0·5	339
204	6·1	6·1	6·0	5·9	5·7	5·5	5·3	5·0	4·7	4·3	3·9	3·5	3·1	2·6	2·1	1·6	1·1	0·5	336
207	6·8	6·8	6·7	6·6	6·4	6·2	5·9	5·6	5·2	4·8	4·4	3·9	3·4	2·9	2·3	1·8	1·2	0·6	333
210	7·5	7·5	7·4	7·3	7·1	6·8	6·5	6·2	5·8	5·3	4·8	4·3	3·8	3·2	2·6	1·9	1·3	0·7	330
213	−8·2	−8·2	−8·1	−7·9	−7·7	−7·4	−7·1	−6·7	−6·3	−5·8	−5·3	−4·7	−4·1	−3·5	−2·8	−2·1	−1·4	−0·7	327
216	8·8	8·8	8·7	8·5	8·3	8·0	7·7	7·2	6·8	6·3	5·7	5·1	4·4	3·7	3·0	2·3	1·5	0·8	324
219	9·5	9·4	9·3	9·1	8·9	8·6	8·2	7·8	7·3	6·7	6·1	5·4	4·7	4·0	3·2	2·4	1·6	0·8	321
222	10·1	10·0	9·9	9·7	9·5	9·1	8·7	8·2	7·7	7·1	6·5	5·8	5·0	4·3	3·4	2·6	1·7	0·9	318
225	10·6	10·6	10·5	10·3	10·0	9·6	9·2	8·7	8·1	7·5	6·8	6·1	5·3	4·5	3·6	2·8	1·8	0·9	315
228	−11·2	−11·1	−11·0	−10·8	−10·5	−10·1	−9·7	−9·2	−8·6	−7·9	−7·2	−6·4	−5·6	−4·7	−3·8	−2·9	−1·9	−1·0	312
231	11·7	11·6	11·5	11·3	11·0	10·6	10·1	9·6	9·0	8·3	7·5	6·7	5·8	4·9	4·0	3·0	2·0	1·0	309
234	12·2	12·1	12·0	11·8	11·4	11·0	10·5	10·0	9·3	8·6	7·8	7·0	6·1	5·1	4·2	3·1	2·1	1·1	306
237	12·6	12·6	12·4	12·2	11·9	11·4	10·9	10·3	9·7	8·9	8·1	7·2	6·3	5·3	4·3	3·3	2·2	1·1	303
240	13·0	13·0	12·8	12·6	12·2	11·8	11·3	10·7	10·0	9·2	8·4	7·5	6·5	5·5	4·5	3·4	2·3	1·1	300
243	−13·4	−13·4	−13·2	−12·9	−12·6	−12·1	−11·6	−11·0	−10·3	−9·5	−8·6	−7·7	−6·7	−5·7	−4·6	−3·5	−2·3	−1·2	297
246	13·7	13·7	13·5	13·3	12·9	12·5	11·9	11·3	10·5	9·7	8·8	7·9	6·9	5·8	4·7	3·6	2·4	1·2	294
249	14·0	14·0	13·8	13·6	13·2	12·7	12·2	11·5	10·8	9·9	9·0	8·1	7·0	5·9	4·8	3·6	2·4	1·2	291
252	14·3	14·3	14·1	13·8	13·4	13·0	12·4	11·7	11·0	10·1	9·2	8·2	7·2	6·0	4·9	3·7	2·5	1·2	288
255	14·5	14·5	14·3	14·0	13·7	13·2	12·6	11·9	11·1	10·3	9·3	8·3	7·3	6·1	5·0	3·8	2·5	1·3	285
258	−14·7	−14·7	−14·5	−14·2	−13·8	−13·3	−12·7	−12·1	−11·3	−10·4	−9·5	−8·4	−7·4	−6·2	−5·0	−3·8	−2·6	−1·3	282
261	14·9	14·8	14·6	14·3	14·0	13·5	12·9	12·2	11·4	10·5	9·5	8·5	7·4	6·3	5·1	3·8	2·6	1·3	279
264	15·0	14·9	14·7	14·4	14·1	13·6	13·0	12·3	11·5	10·6	9·6	8·6	7·5	6·3	5·1	3·9	2·6	1·3	276
267	15·0	15·0	14·8	14·5	14·1	13·6	13·0	12·3	11·5	10·6	9·7	8·6	7·5	6·3	5·1	3·9	2·6	1·3	273
270	15·0	15·0	14·8	14·5	14·1	13·6	13·0	12·3	11·5	10·6	9·7	8·6	7·5	6·4	5·1	3·9	2·6	1·3	270

©2009 Jack Case

Chapter 13
Rapid Sight Reduction Tables
Self Test

To be able to confidently apply the knowledge acquired from chapters 8 to 12, it is essential to have a firm grasp of the theoretical aspects of the subject of astro navigation. For this reason, it is recommended that, before tackling this test, readers revise the theory section of this book paying particular attention to chapters 2, 4 and 5.

Notes.
1. Suggested solutions follow each question.
2. Relevant extracts of tables are placed immediately after the solutions to each question. Because of their size, some tables have been cropped so that they show only the relevant portions.

Test Questions

Question 1. Referring to the outline procedures in chapter 8 and the demonstration in chapter 9 as necessary, calculate and plot a single position line using the information detailed in the following scenario.

Scenario:
D.R. Position: 39° 55'N 45° 15'W.
Date: 30 May 2009
Zone Time: $05^h 48^m$ (+3)
DWT: $08^h 50^m 28^s$
DWE 30^s fast
Body observed: Sun L.L.
Sunrise: $04^h 35^m$
Sextant Alt: 12° 14.'4   Bearing: 071°
Index error: +1'.4
Ht. of eye: 4m.
Temperature: 18°C. Pressure: 981mb.

Solution. The solution to this example is to be found on the following page.
Extracts. Relevant extracts of navigational data follow the solution.

Solution To Question 1

Step 1. Approx Pos: 39° 55'N, 45° 15'W

Step 2. Calculate GMT at time of observation.

Date:	30 May 2009
Zone time:	$30^d\ 05^h\ 48^m$
Zone corrn:	$+3^h$
GMT:	$30^d\ 08^h\ 48^m\ 00^s$
DWT:	$08^h\ 50^m\ 28^s$
DWE:	-30^s
Greenwich date:	$30^d\ 08^h\ 49^m\ 58^s$

Step 3. Calculate GHA and Dec. of Sun 30 May.

	GHA	Dec
08^h UT:	300° 37'.0	N21° 48'.5 (d:0'.4)
Inc. $49^m\ 58^s$:	12° 29'.5	+0'.3
	313° 06'.5	N21° 48'.8

Step 4. Calculate LHA

GHA:	313° 06'.5
Assumed Long:	45° 06'.5 (West -)
LHA:	268° (-360 if reqd.)

Convert azimuth angle to true bearing (Zn):

Z =	72°
LHA =	268°
True bearing (Zn) =	072°

Please see page 161 for an explanation of the reason for calculating true bearing at step 4. (Take the value of the Azimuth (Z) from step 5).

Rules for calculating true bearing (Zn)

	Lat. North	Lat. South
LHA > 180°	Zn = Z	Zn = 180° - Z
LHA < 180°	Zn = 360° - Z	Zn = 180° + Z

©2009 Jack Case

Step 5. Main table entry.

Assumed Lat:	40° N
Dec. Sun (degrees) Dec°:	21° N (same)
Dec. Sun (minutes) Dec':	48'.8
LHA Sun:	268°
Table page:	12

	Hc	d	Z	
Extracted values:	11° 51'	+37'	72°	
Corrn. for Dec'	+30'			(table 5)
Corr. Alt.	Hc = 12° 21'			

Step 6. Correct sextant altitude (Hs).

Sext. Alt.	Hs =	12° 14.'4	
Index error	IE =	+1'.4	
Dip (ht. 4m.)	D =	−3.55	(table 6a)
Apparent Alt.	Ha =	12° 12'.25	
Alt. correction	R =	+11'.7	(table 6d)
Added refrac'n	d =	+0'.2	(table 6c)
Observed Alt.	Ho =	12° 24'.15	

Step 7. Calculate intercept (p)

	Ho =	12° 24'.15
	Hc =	12° 21'.0
Ho − Hc	p =	+ 3'.15

Intercept = 3.15 n.m. towards 072°. (True bearing calculated at step 4).

Step 8. Corrections to intercept.
(No corrections required since there has been no movement between observation and fix).

Step 9. Plot the position line. (See the demonstration in chapter 9 to revise the method).

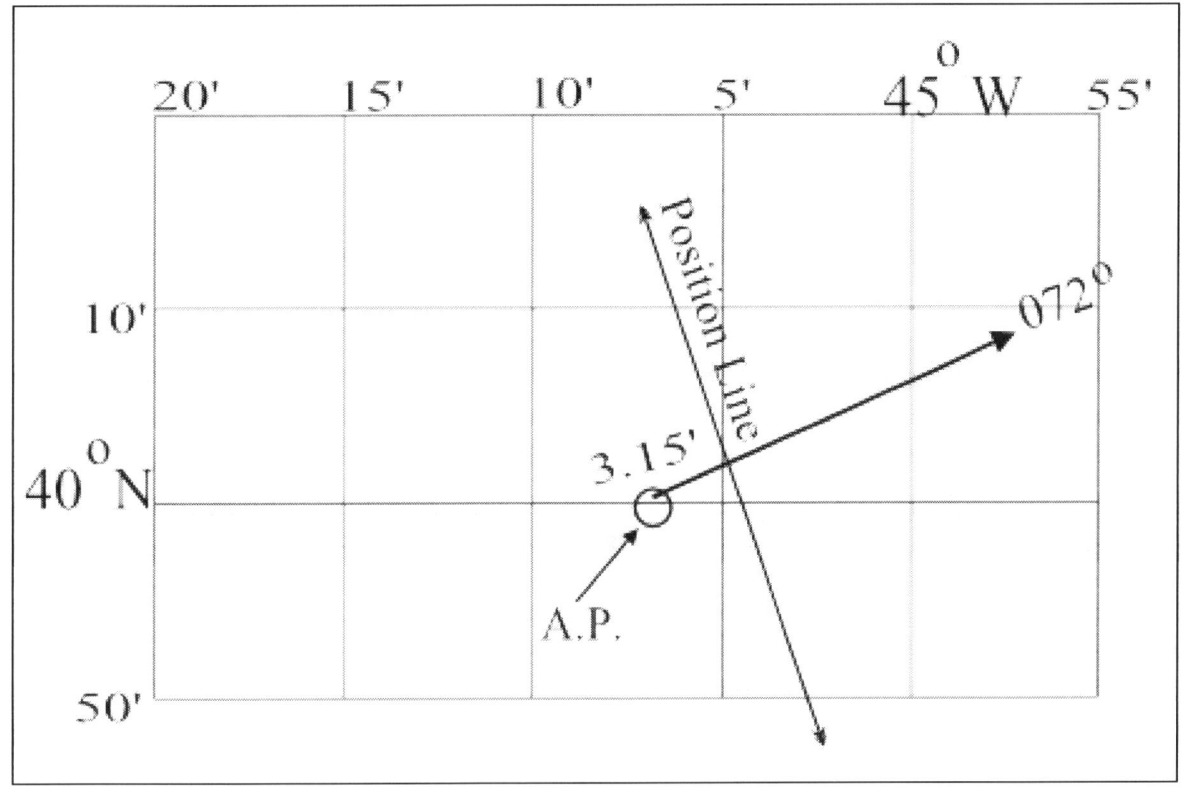

Relevant extracts follow on the next page.

2009 MAY 28, 29, 30 (THURS., FRI., SAT.)

UT	SUN		MOON					Lat.	Twilight		Sunrise	Moonrise			
	GHA	Dec	GHA	v	Dec	d	HP		Naut.	Civil		28	29	30	31
d h	° ′	° ′	° ′	′	° ′	′	′	°	h m	h m	h m	h m	h m	h m	h m
28 00	180 41.6	N21 26.9	128 18.4	5.5	N21 37.8	10.0	60.3	N 72	▭	▭	▭	▭	06 42	09 31	11 48
01	195 41.5	27.3	142 42.9	5.6	21 27.8	10.1	60.3	N 70	▭	▭	▭	▭	07 22	09 47	11 52
02	210 41.4	27.7	157 07.5	5.7	21 17.7	10.2	60.3	68	▭	▭	▭	05 05	07 49	09 59	11 56
03	225 41.4	28.1	171 32.2	5.8	21 07.5	10.4	60.2	66	////	////	01 25	05 53	08 09	10 08	11 58
04	240 41.3	28.5	185 57.0	6.0	20 57.1	10.4	60.2	64	////	////	02 05	06 24	08 25	10 17	12 00
05	255 41.2	28.9	200 22.0	6.0	20 46.7	10.7	60.2	62	////	00 38	02 33	06 47	08 39	10 24	12 02
								60	////	01 35	02 54	07 05	08 50	10 30	12 04
06	270 41.1	N21 29.3	214 47.0	6.2	N20 36.0	10.7	60.2	N 58	////	02 06	03 11	07 20	09 00	10 35	12 05
07	285 41.0	29.7	229 12.2	6.3	20 25.3	10.9	60.2	56	00 34	02 29	03 25	07 33	09 08	10 40	12 07
T 08	300 41.0	30.1	243 37.5	6.4	20 14.4	11.0	60.1	54	01 26	02 47	03 38	07 44	09 16	10 44	12 08
H 09	315 40.9	30.5	258 02.9	6.5	20 03.4	11.1	60.1	52	01 55	03 03	03 49	07 54	09 22	10 47	12 09
U 10	330 40.8	30.9	272 28.4	6.6	19 52.3	11.2	60.1	50	02 17	03 16	03 58	08 03	09 28	10 51	12 10
R 11	345 40.7	31.3	286 54.0	6.7	19 41.1	11.3	60.1	45	02 56	03 42	04 18	08 21	09 41	10 58	12 12
S 12	0 40.6	N21 31.7	301 19.7	6.9	N19 29.8	11.5	60.1	N 40	03 23	04 03	04 35	08 36	09 52	11 04	12 14
D 13	15 40.6	32.1	315 45.6	6.9	19 18.3	11.6	60.0	35	03 44	04 19	04 48	08 49	10 01	11 10	12 15
A 14	30 40.5	32.5	330 11.5	7.1	19 06.7	11.7	60.0	30	04 01	04 33	05 00	09 00	10 09	11 14	12 17
Y 15	45 40.4	32.8	344 37.6	7.2	18 55.0	11.7	60.0	20	04 27	04 56	05 20	09 19	10 22	11 22	12 19
16	60 40.3	33.2	359 03.8	7.3	18 43.3	11.9	60.0	N 10	04 48	05 15	05 38	09 35	10 34	11 29	12 21
17	75 40.2	33.6	13 30.1	7.5	18 31.4	12.0	59.9	0	05 06	05 32	05 54	09 50	10 45	11 36	12 23
18	90 40.2	N21 34.0	27 56.6	7.5	N18 19.4	12.1	59.9	S 10	05 21	05 47	06 10	10 06	10 56	11 42	12 25
19	105 40.1	34.4	42 23.1	7.6	18 07.3	12.2	59.9	20	05 36	06 03	06 27	10 22	11 08	11 49	12 27
20	120 40.0	34.8	56 49.7	7.8	17 55.1	12.3	59.9	30	05 51	06 20	06 46	10 40	11 21	11 57	12 30
21	135 39.9	35.2	71 16.5	7.9	17 42.8	12.4	59.8	35	05 58	06 30	06 57	10 51	11 29	12 02	12 31
22	150 39.8	35.6	85 43.4	8.0	17 30.4	12.5	59.8	40	06 07	06 40	07 10	11 03	11 38	12 07	12 32
23	165 39.8	36.0	100 10.4	8.1	17 17.9	12.6	59.8	45	06 16	06 52	07 25	11 18	11 48	12 12	12 34
29 00	180 39.7	N21 36.4	114 37.5	8.2	N17 05.3	12.7	59.8	S 50	06 26	07 07	07 44	11 35	12 00	12 19	12 36
01	195 39.6	36.8	129 04.7	8.3	16 52.6	12.8	59.8	52	06 31	07 13	07 53	11 43	12 06	12 23	12 37
02	210 39.5	37.1	143 32.0	8.4	16 39.8	12.8	59.7	54	06 36	07 21	08 03	11 53	12 12	12 26	12 38
03	225 39.4	37.5	157 59.4	8.6	16 27.0	13.0	59.7	56	06 41	07 29	08 14	12 03	12 19	12 30	12 40
04	240 39.3	37.9	172 27.0	8.6	16 14.0	13.0	59.7	58	06 47	07 38	08 27	12 14	12 26	12 34	12 41
05	255 39.3	38.3	186 54.6	8.8	16 01.0	13.1	59.7	S 60	06 53	07 48	08 42	12 28	12 35	12 39	12 42
06	270 39.2	N21 38.7	201 22.4	8.8	N15 47.9	13.2	59.6								
07	285 39.1	39.1	215 50.2	9.0	15 34.7	13.2	59.6	Lat.	Sunset	Twilight		Moonset			
F 08	300 39.0	39.4	230 18.2	9.1	15 21.5	13.4	59.6			Civil	Naut.	28	29	30	31
R 09	315 38.9	39.8	244 46.3	9.2	15 08.1	13.4	59.5								
I 10	330 38.8	40.2	259 14.5	9.2	14 54.7	13.5	59.5	°	h m	h m	h m	h m	h m	h m	h m
11	345 38.8	40.6	273 42.7	9.4	14 41.2	13.5	59.5	N 72	▭	▭	▭	▭	02 45	01 44	01 08
D 12	0 38.7	N21 41.0	288 11.1	9.5	N14 27.7	13.6	59.5	N 70	▭	▭	▭	▭	02 04	01 26	01 01
A 13	15 38.6	41.4	302 39.6	9.6	14 14.1	13.7	59.4	68	▭	▭	▭	02 22	01 35	01 12	00 55
Y 14	30 38.5	41.7	317 08.2	9.7	14 00.4	13.8	59.4	66	22 34	////	////	01 33	01 13	01 00	00 50
15	45 38.4	42.1	331 36.9	9.7	13 46.6	13.8	59.4	64	21 52	////	////	01 01	00 56	00 51	00 46
16	60 38.3	42.5	346 05.6	9.9	13 32.8	13.9	59.4	62	21 24	23 25	////	00 38	00 41	00 42	00 42
17	75 38.2	42.9	0 34.5	10.0	13 18.9	13.9	59.3	60	21 02	22 23	////	00 19	00 29	00 35	00 39
18	90 38.2	N21 43.2	15 03.5	10.1	N13 05.0	14.0	59.3	N 58	20 45	21 51	////	00 03	00 18	00 28	00 36
19	105 38.1	43.6	29 32.6	10.1	12 51.0	14.1	59.3	56	20 30	21 27	23 28	24 09	00 09	00 22	00 33
20	120 38.0	44.0	44 01.7	10.3	12 36.9	14.1	59.2	54	20 18	21 09	22 31	24 00	00 00	00 17	00 31
21	135 37.9	44.4	58 31.0	10.4	12 22.8	14.2	59.2	52	20 07	20 53	22 01	23 53	24 12	00 12	00 28
22	150 37.8	44.7	73 00.4	10.4	12 08.6	14.2	59.2	50	19 57	20 40	21 39	23 46	24 08	00 08	00 26
23	165 37.7	45.1	87 29.8	10.5	11 54.4	14.3	59.2	45	19 37	20 13	21 00	23 32	23 59	24 22	00 22
30 00	180 37.6	N21 45.5	101 59.3	10.6	N11 40.1	14.3	59.1	N 40	19 21	19 53	20 33	23 19	23 51	24 18	00 18
01	195 37.6	45.9	116 28.9	10.8	11 25.8	14.3	59.1	35	19 07	19 36	20 12	23 09	23 44	24 15	00 15
02	210 37.5	46.2	130 58.7	10.8	11 11.5	14.4	59.1	30	18 55	19 22	19 54	23 00	23 38	24 12	00 12
03	225 37.4	46.6	145 28.5	10.8	10 57.1	14.5	59.1	20	18 35	18 59	19 28	22 44	23 28	24 07	00 07
04	240 37.3	47.0	159 58.3	11.0	10 42.6	14.5	59.0	N 10	18 17	18 40	19 07	22 30	23 18	24 03	00 03
05	255 37.2	47.3	174 28.3	11.0	10 28.1	14.5	59.0	0	18 01	18 23	18 49	22 17	23 10	23 59	24 45
06	270 37.1	N21 47.7	188 58.3	11.2	N10 13.6	14.6	59.0	S 10	17 45	18 08	18 34	22 04	23 01	23 54	24 45
S 07	285 37.0	48.1	203 28.5	11.2	9 59.0	14.6	58.9	20	17 28	17 52	18 19	21 50	22 51	23 50	24 46
A 08	300 37.0	48.5	217 58.7	11.3	9 44.4	14.6	58.9	30	17 09	17 34	18 04	21 33	22 40	23 46	24 46
T 09	315 36.9	48.8	232 29.0	11.3	9 29.8	14.7	58.9	35	16 57	17 25	17 56	21 23	22 34	23 41	24 46
U 10	330 36.8	49.2	246 59.3	11.5	9 15.1	14.7	58.9	40	16 44	17 14	17 48	21 12	22 27	23 38	24 46
R 11	345 36.7	49.5	261 29.8	11.4	9 00.4	14.8	58.8	45	16 29	17 02	17 39	20 59	22 18	23 34	24 47
D 12	0 36.6	N21 49.9	276 00.3	11.6	N 8 45.6	14.7	58.8	S 50	16 10	16 48	17 28	20 43	22 08	23 29	24 47
A 13	15 36.5	50.3	290 30.9	11.7	8 30.9	14.8	58.8	52	16 02	16 41	17 24	20 36	22 03	23 27	24 47
Y 14	30 36.4	50.6	305 01.6	11.7	8 16.1	14.9	58.7	54	15 52	16 34	17 19	20 27	21 58	23 24	24 47
15	45 36.3	51.0	319 32.3	11.8	8 01.2	14.8	58.7	56	15 41	16 26	17 14	20 18	21 52	23 21	24 47
16	60 36.2	51.4	334 03.1	11.9	7 46.4	14.9	58.7	58	15 28	16 17	17 08	20 07	21 45	23 18	24 47
17	75 36.2	51.7	348 34.0	11.9	7 31.5	14.9	58.7	S 60	15 13	16 07	17 01	19 54	21 38	23 15	24 48
18	90 36.1	N21 52.1	3 04.9	12.0	N 7 16.6	14.9	58.6			SUN			MOON		
19	105 36.0	52.4	17 35.9	12.1	7 01.7	15.0	58.6	Day	Eqn. of Time		Mer.	Mer. Pass.		Age	Phase
20	120 35.9	52.8	32 07.0	12.2	6 46.7	14.9	58.6		00ʰ	12ʰ	Pass.	Upper	Lower		
21	135 35.8	53.2	46 38.2	12.2	6 31.8	15.0	58.5	d	m s	m s	h m	h m	h m	d %	
22	150 35.7	53.5	61 09.4	12.2	6 16.8	15.0	58.5	28	02 47	02 43	11 57	16 04	03 35	04 22	
23	165 35.6	53.9	75 40.6	12.4	N 6 01.8	15.0	58.5	29	02 39	02 35	11 57	16 58	04 31	05 32	◐
	SD 15.8	d 0.4	SD 16.4		16.2		16.0	30	02 31	02 27	11 58	17 47	05 23	06 43	

INCREMENTS AND CORRECTIONS

48m

s	SUN PLANETS	ARIES	MOON	v or d	Corrn	v or d	Corrn	v or d	Corrn
	° ′	° ′	° ′	′	′	′	′	′	′
00	12 00·0	12 02·0	11 27·2	0·0	0·0	6·0	4·9	12·0	9·7
01	12 00·3	12 02·2	11 27·4	0·1	0·1	6·1	4·9	12·1	9·8
02	12 00·5	12 02·5	11 27·7	0·2	0·2	6·2	5·0	12·2	9·9
03	12 00·8	12 02·7	11 27·9	0·3	0·2	6·3	5·1	12·3	9·9
04	12 01·0	12 03·0	11 28·2	0·4	0·3	6·4	5·2	12·4	10·0
05	12 01·3	12 03·2	11 28·4	0·5	0·4	6·5	5·3	12·5	10·1
06	12 01·5	12 03·5	11 28·6	0·6	0·5	6·6	5·3	12·6	10·2
07	12 01·8	12 03·7	11 28·9	0·7	0·6	6·7	5·4	12·7	10·3
08	12 02·0	12 04·0	11 29·1	0·8	0·6	6·8	5·5	12·8	10·3
09	12 02·3	12 04·2	11 29·3	0·9	0·7	6·9	5·6	12·9	10·4
10	12 02·5	12 04·5	11 29·6	1·0	0·8	7·0	5·7	13·0	10·5
11	12 02·8	12 04·7	11 29·8	1·1	0·9	7·1	5·7	13·1	10·6
12	12 03·0	12 05·0	11 30·1	1·2	1·0	7·2	5·8	13·2	10·7
13	12 03·3	12 05·2	11 30·3	1·3	1·1	7·3	5·9	13·3	10·8
14	12 03·5	12 05·5	11 30·5	1·4	1·1	7·4	6·0	13·4	10·8
15	12 03·8	12 05·7	11 30·8	1·5	1·2	7·5	6·1	13·5	10·9
16	12 04·0	12 06·0	11 31·0	1·6	1·3	7·6	6·1	13·6	11·0
17	12 04·3	12 06·2	11 31·3	1·7	1·4	7·7	6·2	13·7	11·1
18	12 04·5	12 06·5	11 31·5	1·8	1·5	7·8	6·3	13·8	11·2
19	12 04·8	12 06·7	11 31·7	1·9	1·5	7·9	6·4	13·9	11·2
20	12 05·0	12 07·0	11 32·0	2·0	1·6	8·0	6·5	14·0	11·3
21	12 05·3	12 07·2	11 32·2	2·1	1·7	8·1	6·5	14·1	11·4
22	12 05·5	12 07·5	11 32·4	2·2	1·8	8·2	6·6	14·2	11·5
23	12 05·8	12 07·7	11 32·7	2·3	1·9	8·3	6·7	14·3	11·6
24	12 06·0	12 08·0	11 32·9	2·4	1·9	8·4	6·8	14·4	11·6
25	12 06·3	12 08·2	11 33·2	2·5	2·0	8·5	6·9	14·5	11·7
26	12 06·5	12 08·5	11 33·4	2·6	2·1	8·6	7·0	14·6	11·8
27	12 06·8	12 08·7	11 33·7	2·7	2·2	8·7	7·0	14·7	11·9
28	12 07·0	12 09·0	11 33·9	2·8	2·3	8·8	7·1	14·8	12·0
29	12 07·3	12 09·2	11 34·1	2·9	2·3	8·9	7·2	14·9	12·0
30	12 07·5	12 09·5	11 34·4	3·0	2·4	9·0	7·3	15·0	12·1
31	12 07·8	12 09·7	11 34·6	3·1	2·5	9·1	7·4	15·1	12·2
32	12 08·0	12 10·0	11 34·8	3·2	2·6	9·2	7·4	15·2	12·3
33	12 08·3	12 10·2	11 35·1	3·3	2·7	9·3	7·5	15·3	12·4
34	12 08·5	12 10·5	11 35·3	3·4	2·7	9·4	7·6	15·4	12·4
35	12 08·8	12 10·7	11 35·6	3·5	2·8	9·5	7·7	15·5	12·5
36	12 09·0	12 11·0	11 35·8	3·6	2·9	9·6	7·8	15·6	12·6
37	12 09·3	12 11·2	11 36·0	3·7	3·0	9·7	7·8	15·7	12·7
38	12 09·5	12 11·5	11 36·3	3·8	3·1	9·8	7·9	15·8	12·8
39	12 09·8	12 11·7	11 36·5	3·9	3·2	9·9	8·0	15·9	12·9
40	12 10·0	12 12·0	11 36·7	4·0	3·2	10·0	8·1	16·0	12·9
41	12 10·3	12 12·2	11 37·0	4·1	3·3	10·1	8·2	16·1	13·0
42	12 10·5	12 12·5	11 37·2	4·2	3·4	10·2	8·2	16·2	13·1
43	12 10·8	12 12·8	11 37·5	4·3	3·5	10·3	8·3	16·3	13·2
44	12 11·0	12 13·0	11 37·7	4·4	3·6	10·4	8·4	16·4	13·3
45	12 11·3	12 13·3	11 37·9	4·5	3·6	10·5	8·5	16·5	13·3
46	12 11·5	12 13·5	11 38·2	4·6	3·7	10·6	8·6	16·6	13·4
47	12 11·8	12 13·8	11 38·4	4·7	3·8	10·7	8·6	16·7	13·5
48	12 12·0	12 14·0	11 38·7	4·8	3·9	10·8	8·7	16·8	13·6
49	12 12·3	12 14·3	11 38·9	4·9	4·0	10·9	8·8	16·9	13·7
50	12 12·5	12 14·5	11 39·1	5·0	4·0	11·0	8·9	17·0	13·7
51	12 12·8	12 14·8	11 39·4	5·1	4·1	11·1	9·0	17·1	13·8
52	12 13·0	12 15·0	11 39·6	5·2	4·2	11·2	9·1	17·2	13·9
53	12 13·3	12 15·3	11 39·8	5·3	4·3	11·3	9·1	17·3	14·0
54	12 13·5	12 15·5	11 40·1	5·4	4·4	11·4	9·2	17·4	14·1
55	12 13·8	12 15·8	11 40·3	5·5	4·4	11·5	9·3	17·5	14·1
56	12 14·0	12 16·0	11 40·6	5·6	4·5	11·6	9·4	17·6	14·2
57	12 14·3	12 16·3	11 40·8	5·7	4·6	11·7	9·5	17·7	14·3
58	12 14·5	12 16·5	11 41·0	5·8	4·7	11·8	9·5	17·8	14·4
59	12 14·8	12 16·8	11 41·3	5·9	4·8	11·9	9·6	17·9	14·5
60	12 15·0	12 17·0	11 41·5	6·0	4·9	12·0	9·7	18·0	14·6

49m

s	SUN PLANETS	ARIES	MOON	v or d	Corrn	v or d	Corrn	v or d	Corrn
	° ′	° ′	° ′	′	′	′	′	′	′
00	12 15·0	12 17·0	11 41·5	0·0	0·0	6·0	5·0	12·0	9·9
01	12 15·3	12 17·3	11 41·8	0·1	0·1	6·1	5·0	12·1	10·0
02	12 15·5	12 17·5	11 42·0	0·2	0·2	6·2	5·1	12·2	10·1
03	12 15·8	12 17·8	11 42·2	0·3	0·2	6·3	5·2	12·3	10·1
04	12 16·0	12 18·0	11 42·5	0·4	0·3	6·4	5·3	12·4	10·2
05	12 16·3	12 18·3	11 42·7	0·5	0·4	6·5	5·4	12·5	10·3
06	12 16·5	12 18·5	11 42·9	0·6	0·5	6·6	5·4	12·6	10·4
07	12 16·8	12 18·8	11 43·2	0·7	0·6	6·7	5·5	12·7	10·5
08	12 17·0	12 19·0	11 43·4	0·8	0·7	6·8	5·6	12·8	10·6
09	12 17·3	12 19·3	11 43·7	0·9	0·7	6·9	5·7	12·9	10·6
10	12 17·5	12 19·5	11 43·9	1·0	0·8	7·0	5·8	13·0	10·7
11	12 17·8	12 19·8	11 44·1	1·1	0·9	7·1	5·9	13·1	10·8
12	12 18·0	12 20·0	11 44·4	1·2	1·0	7·2	5·9	13·2	10·9
13	12 18·3	12 20·3	11 44·6	1·3	1·1	7·3	6·0	13·3	11·0
14	12 18·5	12 20·5	11 44·9	1·4	1·2	7·4	6·1	13·4	11·1
15	12 18·8	12 20·8	11 45·1	1·5	1·2	7·5	6·2	13·5	11·1
16	12 19·0	12 21·0	11 45·3	1·6	1·3	7·6	6·3	13·6	11·2
17	12 19·3	12 21·3	11 45·6	1·7	1·4	7·7	6·4	13·7	11·3
18	12 19·5	12 21·5	11 45·8	1·8	1·5	7·8	6·4	13·8	11·4
19	12 19·8	12 21·8	11 46·1	1·9	1·6	7·9	6·5	13·9	11·5
20	12 20·0	12 22·0	11 46·3	2·0	1·7	8·0	6·6	14·0	11·6
21	12 20·3	12 22·3	11 46·5	2·1	1·7	8·1	6·7	14·1	11·6
22	12 20·5	12 22·5	11 46·8	2·2	1·8	8·2	6·8	14·2	11·7
23	12 20·8	12 22·8	11 47·0	2·3	1·9	8·3	6·8	14·3	11·8
24	12 21·0	12 23·0	11 47·2	2·4	2·0	8·4	6·9	14·4	11·9
25	12 21·3	12 23·3	11 47·5	2·5	2·1	8·5	7·0	14·5	12·0
26	12 21·5	12 23·5	11 47·7	2·6	2·1	8·6	7·1	14·6	12·0
27	12 21·8	12 23·8	11 48·0	2·7	2·2	8·7	7·2	14·7	12·1
28	12 22·0	12 24·0	11 48·2	2·8	2·3	8·8	7·3	14·8	12·2
29	12 22·3	12 24·3	11 48·4	2·9	2·4	8·9	7·3	14·9	12·3
30	12 22·5	12 24·5	11 48·7	3·0	2·5	9·0	7·4	15·0	12·4
31	12 22·8	12 24·8	11 48·9	3·1	2·6	9·1	7·5	15·1	12·5
32	12 23·0	12 25·0	11 49·2	3·2	2·6	9·2	7·6	15·2	12·5
33	12 23·3	12 25·3	11 49·4	3·3	2·7	9·3	7·7	15·3	12·6
34	12 23·5	12 25·5	11 49·6	3·4	2·8	9·4	7·8	15·4	12·7
35	12 23·8	12 25·8	11 49·9	3·5	2·9	9·5	7·8	15·5	12·8
36	12 24·0	12 26·0	11 50·1	3·6	3·0	9·6	7·9	15·6	12·9
37	12 24·3	12 26·3	11 50·3	3·7	3·1	9·7	8·0	15·7	13·0
38	12 24·5	12 26·5	11 50·6	3·8	3·1	9·8	8·1	15·8	13·0
39	12 24·8	12 26·8	11 50·8	3·9	3·2	9·9	8·2	15·9	13·1
40	12 25·0	12 27·0	11 51·1	4·0	3·3	10·0	8·3	16·0	13·2
41	12 25·3	12 27·3	11 51·3	4·1	3·4	10·1	8·3	16·1	13·3
42	12 25·5	12 27·5	11 51·5	4·2	3·5	10·2	8·4	16·2	13·4
43	12 25·8	12 27·8	11 51·8	4·3	3·5	10·3	8·5	16·3	13·4
44	12 26·0	12 28·0	11 52·0	4·4	3·6	10·4	8·6	16·4	13·5
45	12 26·3	12 28·3	11 52·3	4·5	3·7	10·5	8·7	16·5	13·6
46	12 26·5	12 28·5	11 52·5	4·6	3·8	10·6	8·7	16·6	13·7
47	12 26·8	12 28·8	11 52·7	4·7	3·9	10·7	8·8	16·7	13·8
48	12 27·0	12 29·0	11 53·0	4·8	4·0	10·8	8·9	16·8	13·9
49	12 27·3	12 29·3	11 53·2	4·9	4·0	10·9	9·0	16·9	13·9
50	12 27·5	12 29·5	11 53·4	5·0	4·1	11·0	9·1	17·0	14·0
51	12 27·8	12 29·8	11 53·7	5·1	4·2	11·1	9·2	17·1	14·1
52	12 28·0	12 30·0	11 53·9	5·2	4·3	11·2	9·2	17·2	14·2
53	12 28·3	12 30·3	11 54·2	5·3	4·4	11·3	9·3	17·3	14·3
54	12 28·5	12 30·5	11 54·4	5·4	4·5	11·4	9·4	17·4	14·4
55	12 28·8	12 30·8	11 54·6	5·5	4·5	11·5	9·5	17·5	14·4
56	12 29·0	12 31·1	11 54·9	5·6	4·6	11·6	9·6	17·6	14·5
57	12 29·3	12 31·3	11 55·1	5·7	4·7	11·7	9·7	17·7	14·6
58	12 29·5	12 31·6	11 55·4	5·8	4·8	11·8	9·7	17·8	14·7
59	12 29·8	12 31·8	11 55·6	5·9	4·9	11·9	9·8	17·9	14·8
60	12 30·0	12 32·1	11 55·8	6·0	5·0	12·0	9·9	18·0	14·9

©2009 Jack Case

TABLE 6 — ALTITUDE CORRECTION TABLES

a. Dip of the Horizon

Ht. of Eye (m)	Corr^n D	Ht. of Eye (ft.)	Ht. of Eye (m)	Corr^n D	Ht. of Eye (ft.)
0.00	−0.3	0.0	13.0	−6.4	42.8
0.03	−0.4	0.1	13.4	−6.5	44.2
0.06	−0.5	0.2	13.8	−6.6	45.5
0.09	−0.6	0.3	14.2	−6.7	47.0
0.13	−0.7	0.4	14.7	−6.8	48.4
0.18	−0.8	0.5	15.1	−6.9	49.8
0.23	−0.9	0.7	15.5	−7.0	51.3
0.29	−1.0	0.9	16.0	−7.1	52.8
0.35	−1.1	1.1	16.5	−7.2	54.3
0.42	−1.2	1.4	16.9	−7.3	55.8
0.45	−1.3	1.6	17.4	−7.4	57.4
0.5	−1.4	1.9	17.9	−7.5	58.9
0.6	−1.5	2.2	18.4	−7.6	60.5
0.7	−1.6	2.5	18.8	−7.7	62.1
0.8	−1.7	2.8	19.3	−7.8	63.8
0.9	−1.8	3.2	19.8	−7.9	65.4
1.1	−1.9	3.6	20.4	−8.0	67.1
1.2	−2.0	4.0	20.9	−8.1	68.8
1.3	−2.1	4.4	21.4	−8.2	70.5
1.4	−2.2	4.9	21.9	−8.3	72.3
1.6	−2.3	5.3	22.5	−8.4	74.1
1.7	−2.4	5.8	23.0	−8.5	75.8
1.9	−2.5	6.3	23.5	−8.6	77.6
2.0	−2.6	6.9	24.1	−8.7	79.5
2.2	−2.7	7.4	24.7	−8.8	81.3
2.4	−2.8	8.0	25.2	−8.9	83.2
2.6	−2.9	8.6	25.8	−9.0	85.1
2.8	−3.0	9.2	26.4	−9.1	87.0
3.0	−3.1	9.8	27.0	−9.2	88.9
3.2	−3.2	10.5	27.6	−9.3	90.9
3.4	−3.3	11.2	28.2	−9.4	92.9
3.6	−3.4	11.9	28.8	−9.5	94.9
3.8	−3.5	12.6	29.4	−9.6	96.9
4.0	−3.6	13.3	30.0	−9.7	98.9
4.3	−3.7	14.1	30.6	−9.8	101.0
4.5	−3.8	14.9	31.3	−9.9	103.1
4.7	−3.9	15.7	31.9	−10.0	105.2
5.0	−4.0	16.5	32.6	−10.1	107.3
5.2	−4.1	17.4	33.2	−10.2	109.4
5.5	−4.2	18.3	33.9	−10.3	111.6
5.8	−4.3	19.1	34.5	−10.4	113.8
6.1	−4.4	20.1	35.2	−10.5	116.0
6.3	−4.5	21.0	35.9	−10.6	118.2
6.6	−4.6	22.0	36.6	−10.7	120.5
6.9	−4.7	22.9	37.3	−10.8	122.8
7.2	−4.8	23.9	38.0	−10.9	125.1
7.5	−4.9	25.0	38.7	−11.0	127.4
7.9	−5.0	26.0	39.4	−11.1	129.7
8.2	−5.1	27.1	40.1	−11.2	132.1
8.5	−5.2	28.1	40.8	−11.3	134.5
8.8	−5.3	29.2	41.5	−11.4	136.9
9.2	−5.4	30.4	42.3	−11.5	139.3
9.5	−5.5	31.5	43.0	−11.6	141.7
9.9	−5.6	32.7	43.8	−11.7	144.2
10.3	−5.7	33.9	44.5	−11.8	146.7
10.6	−5.8	35.1	45.3	−11.9	149.2
11.0	−5.9	36.3	46.1	−12.0	151.7
11.4	−6.0	37.6	46.8	−12.1	154.3
11.8	−6.1	38.9	47.6	−12.2	156.8
12.2	−6.2	40.1	48.4	−12.3	159.4
12.6	−6.3	41.5	49.2	−12.4	162.1
13.0		42.8	50.0		164.7

b. Refraction for Stars & Planets

App. Alt. H_a	Corr^n R	App. Alt. H_a	Corr^n R	App. Alt. H_a	Corr^n R
0 00	−33.8	3 30	−12.9	9 55	
0 03	33.2	3 35	12.7	10 07	−5.3
0 06	32.6	3 40	12.5	10 20	−5.2
0 09	32.0	3 45	12.3	10 32	−5.1
0 12	31.5	3 50	12.1	10 46	−5.0
0 15	30.9	3 55	11.9	10 59	−4.9
0 18	−30.4	4 00	−11.7	11 14	−4.8
0 21	29.8	4 05	11.5	11 29	−4.7
0 24	29.3	4 10	11.4	11 44	−4.6
0 27	28.8	4 15	11.2	12 00	−4.5
0 30	28.3	4 20	11.0	12 17	−4.4
0 33	27.9	4 25	10.9	12 35	−4.3
0 36	−27.4	4 30	−10.7	12 53	−4.2
0 39	26.9	4 35	10.6	13 12	−4.1
0 42	26.5	4 40	10.4	13 32	−4.0
0 45	26.1	4 45	10.3	13 53	−3.9
0 48	25.7	4 50	10.1	14 16	−3.8
0 51	25.3	4 55	10.0	14 39	−3.7
0 54	−24.9	5 00	−9.8	15 03	−3.6
0 57	24.5	5 05	9.7	15 29	−3.5
1 00	24.1	5 10	9.6	15 56	−3.4
1 03	23.7	5 15	9.5	16 25	−3.3
1 06	23.4	5 20	9.3	16 55	−3.2
1 09	23.0	5 25	9.2	17 27	−3.1
1 12	−22.7	5 30	−9.1	18 01	−3.0
1 15	22.3	5 35	9.0	18 37	−2.9
1 18	22.0	5 40	8.9	19 16	−2.8
1 21	21.7	5 45	8.8	19 56	−2.7
1 24	21.4	5 50	8.7	20 40	−2.6
1 27	21.1	5 55	8.6	21 27	−2.5
1 30	−20.8	6 00	−8.5	22 17	−2.4
1 35	20.3	6 10	8.3	23 11	−2.3
1 40	19.9	6 20	8.1	24 09	−2.2
1 45	19.4	6 30	7.9	25 12	−2.1
1 50	19.0	6 40	7.7	26 20	−2.0
1 55	18.6	6 50	7.6	27 34	−1.9
2 00	−18.2	7 00	−7.4	28 54	−1.8
2 05	17.8	7 10	7.2	30 22	−1.7
2 10	17.4	7 20	7.1	31 58	−1.6
2 15	17.1	7 30	6.9	33 43	−1.5
2 20	16.7	7 40	6.8	35 38	−1.4
2 25	16.4	7 50	6.7	37 45	−1.3
2 30	−16.1	8 00	−6.6	40 06	−1.2
2 35	15.8	8 10	6.4	42 42	−1.1
2 40	15.4	8 20	6.3	45 34	−1.0
2 45	15.2	8 30	6.2	48 45	−0.9
2 50	14.9	8 40	6.1	52 16	−0.8
2 55	14.6	8 50	6.0	56 09	−0.7
3 00	−14.3	9 00	−5.9	60 26	−0.6
3 05	14.1	9 10	5.8	65 06	−0.5
3 10	13.8	9 20	5.7	70 09	−0.4
3 15	13.6	9 30	5.6	75 32	−0.3
3 20	13.4	9 40	5.5	81 12	−0.2
3 25	13.1	9 50	5.4	87 03	−0.1
3 30	−12.9	10 00	−5.3	90 00	0.0

H_a = App. Alt. = Apparent Altitude
 = Sextant altitude corrected for index error (IE) & dip (
 = H_s + IE + D

In critical cases ascend.

TABLE 6 — ALTITUDE CORRECTION TABLES

c. Additional Refraction for Non-standard Conditions

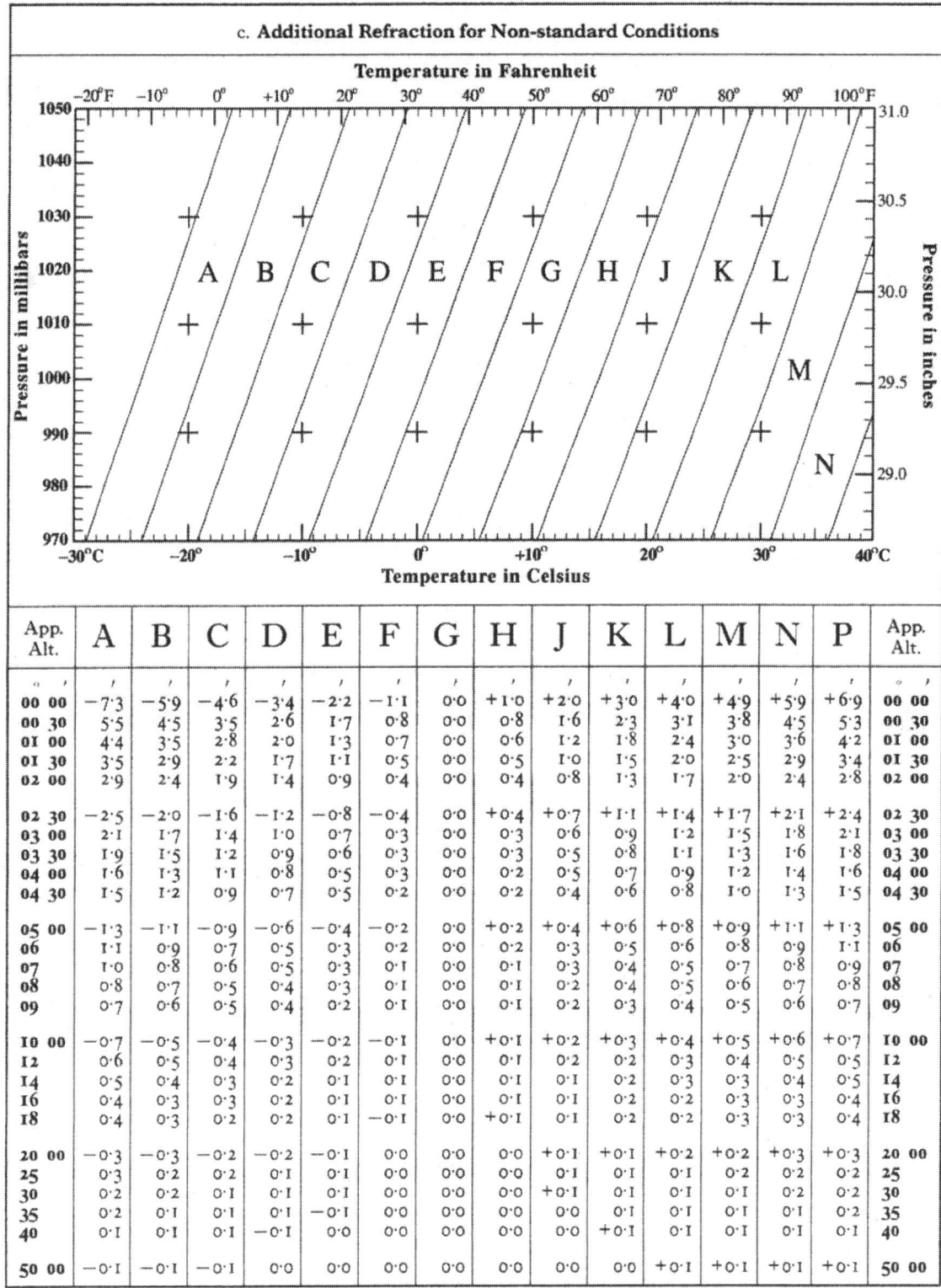

App. Alt.	A	B	C	D	E	F	G	H	J	K	L	M	N	P	App. Alt.
° ′	′	′	′	′	′	′	′	′	′	′	′	′	′	′	° ′
00 00	−7.3	−5.9	−4.6	−3.4	−2.2	−1.1	0.0	+1.0	+2.0	+3.0	+4.0	+4.9	+5.9	+6.9	00 00
00 30	5.5	4.5	3.5	2.6	1.7	0.8	0.0	0.8	1.6	2.3	3.1	3.8	4.5	5.3	00 30
01 00	4.4	3.5	2.8	2.0	1.3	0.7	0.0	0.6	1.2	1.8	2.4	3.0	3.6	4.2	01 00
01 30	3.5	2.9	2.2	1.7	1.1	0.5	0.0	0.5	1.0	1.5	2.0	2.5	2.9	3.4	01 30
02 00	2.9	2.4	1.9	1.4	0.9	0.4	0.0	0.4	0.8	1.3	1.7	2.0	2.4	2.8	02 00
02 30	−2.5	−2.0	−1.6	−1.2	−0.8	−0.4	0.0	+0.4	+0.7	+1.1	+1.4	+1.7	+2.1	+2.4	02 30
03 00	2.1	1.7	1.4	1.0	0.7	0.3	0.0	0.3	0.6	0.9	1.2	1.5	1.8	2.1	03 00
03 30	1.9	1.5	1.2	0.9	0.6	0.3	0.0	0.3	0.5	0.8	1.1	1.3	1.6	1.8	03 30
04 00	1.6	1.3	1.1	0.8	0.5	0.3	0.0	0.2	0.5	0.7	0.9	1.2	1.4	1.6	04 00
04 30	1.5	1.2	0.9	0.7	0.5	0.2	0.0	0.2	0.4	0.6	0.8	1.0	1.3	1.5	04 30
05 00	−1.3	−1.1	−0.9	−0.6	−0.4	−0.2	0.0	+0.2	+0.4	+0.6	+0.8	+0.9	+1.1	+1.3	05 00
06	1.1	0.9	0.7	0.5	0.3	0.2	0.0	0.2	0.3	0.5	0.6	0.8	0.9	1.1	06
07	1.0	0.8	0.6	0.5	0.3	0.1	0.0	0.1	0.3	0.4	0.5	0.7	0.8	0.9	07
08	0.8	0.7	0.5	0.4	0.3	0.1	0.0	0.1	0.2	0.4	0.5	0.6	0.7	0.8	08
09	0.7	0.6	0.5	0.4	0.2	0.1	0.0	0.1	0.2	0.3	0.4	0.5	0.6	0.7	09
10 00	−0.7	−0.5	−0.4	−0.3	−0.2	−0.1	0.0	+0.1	+0.2	+0.3	+0.4	+0.5	+0.6	+0.7	10 00
12	0.6	0.5	0.4	0.3	0.2	0.1	0.0	0.1	0.2	0.2	0.3	0.4	0.5	0.5	12
14	0.5	0.4	0.3	0.2	0.1	0.1	0.0	0.1	0.1	0.2	0.3	0.3	0.4	0.5	14
16	0.4	0.3	0.3	0.2	0.1	0.1	0.0	0.1	0.1	0.2	0.2	0.3	0.3	0.4	16
18	0.4	0.3	0.2	0.2	0.1	−0.1	0.0	+0.1	0.1	0.2	0.2	0.3	0.3	0.4	18
20 00	−0.3	−0.3	−0.2	−0.2	−0.1	0.0	0.0	0.0	+0.1	+0.1	+0.2	+0.2	+0.3	+0.3	20 00
25	0.3	0.2	0.2	0.1	0.1	0.0	0.0	0.0	0.1	0.1	0.1	0.2	0.2	0.2	25
30	0.2	0.2	0.1	0.1	0.1	0.0	0.0	0.0	+0.1	0.1	0.1	0.1	0.2	0.2	30
35	0.2	0.1	0.1	0.1	−0.1	0.0	0.0	0.0	0.0	0.1	0.1	0.1	0.1	0.2	35
40	0.1	0.1	0.1	−0.1	0.0	0.0	0.0	0.0	+0.1	0.1	0.1	0.1	0.1	0.1	40
50 00	−0.1	−0.1	−0.1	0.0	0.0	0.0	0.0	0.0	0.0	0.0	+0.1	+0.1	+0.1	+0.1	50 00

The graph is entered with arguments temperature and pressure to find a zone letter; using as arguments this zone letter and apparent altitude (sextant altitude corrected for index error and dip), a correction is taken from the table. This correction is to be applied to the sextant altitude in addition to the corrections for standard conditions (Table **6**b).

TABLE 6d — ALTITUDE CORRECTION TABLES FOR THE SUN

App. Alt. Ha	OCT.—MAR. Lower Limb / Upper Limb R	APR.—SEPT. Lower Limb / Upper Limb R	App. Alt. Ha	OCT.—MAR. Lower Limb / Upper Limb R	APR.—SEPT. Lower Limb / Upper Limb R	App. Alt. Ha	OCT.—MAR. Lower Limb / Upper Limb R	App. Alt. Ha	APR.—SEPT. Lower Limb / Upper Limb R
0 00	−17.5 −49.8	−17.8 −49.6	3 30	+ 3.4 −28.9	+ 3.1 −28.7	9 33	+10.8 −21.5	9 39	+10.6 −21.2
0 03	16.9 49.2	17.2 49.0	3 35	3.6 28.7	3.3 28.5	9 45	+10.9 −21.4	9 50	+10.7 −21.1
0 06	16.3 48.6	16.6 48.4	3 40	3.8 28.5	3.6 28.2	9 56	+11.0 −21.3	10 02	+10.8 −21.0
0 09	15.7 48.0	16.0 47.8	3 45	4.0 28.3	3.8 28.0	10 08	+11.1 −21.2	10 14	+10.9 −20.9
0 12	15.2 47.5	15.4 47.2	3 50	4.2 28.1	4.0 27.8	10 20	+11.2 −21.1	10 27	+11.0 −20.8
0 15	14.6 46.9	14.8 46.6	3 55	4.4 27.9	4.1 27.7	10 33	+11.3 −21.0	10 40	+11.1 −20.7
0 18	−14.1 −46.4	−14.3 −46.1	4 00	+ 4.6 −27.7	+ 4.3 −27.5	10 46	+11.4 −20.9	10 53	+11.2 −20.6
0 21	13.5 45.8	13.8 45.6	4 05	4.8 27.5	4.5 27.3	11 00	+11.5 −20.8	11 07	+11.3 −20.5
0 24	13.0 45.3	13.3 45.1	4 10	4.9 27.4	4.7 27.1	11 15	+11.6 −20.7	11 22	+11.4 −20.4
0 27	12.5 44.8	12.8 44.6	4 15	5.1 27.2	4.9 26.9	11 30	+11.7 −20.6	11 37	+11.5 −20.3
0 30	12.0 44.3	12.3 44.1	4 20	5.3 27.0	5.0 26.8	11 45	+11.8 −20.5	11 53	+11.6 −20.2
0 33	11.6 43.9	11.8 43.6	4 25	5.4 26.9	5.2 26.6	12 01	+11.9 −20.4	12 10	+11.7 −20.1
0 36	−11.1 −43.4	−11.3 −43.1	4 30	+ 5.6 −26.7	+ 5.3 −26.5	12 18	+12.0 −20.3	12 27	+11.8 −20.0
0 39	10.6 42.9	10.9 42.7	4 35	5.7 26.6	5.5 26.3	12 36	+12.1 −20.2	12 45	+11.9 −19.9
0 42	10.2 42.5	10.5 42.3	4 40	5.9 26.4	5.6 26.2	12 54	+12.2 −20.1	13 04	+12.0 −19.8
0 45	9.8 42.1	10.0 41.8	4 45	6.0 26.3	5.8 26.0	13 14	+12.3 −20.0	13 24	+12.1 −19.7
0 48	9.4 41.7	9.6 41.4	4 50	6.2 26.1	5.9 25.9	13 34	+12.4 −19.9	13 44	+12.2 −19.6
0 51	9.0 41.3	9.2 41.0	4 55	6.3 26.0	6.1 25.7	13 55	+12.5 −19.8	14 06	+12.3 −19.5
0 54	− 8.6 −40.9	− 8.8 −40.6	5 00	+ 6.4 −25.9	+ 6.2 −25.6	14 17	+12.6 −19.7	14 29	+12.4 −19.4
0 57	8.2 40.5	8.4 40.2	5 05	6.6 25.7	6.3 25.5	14 41	+12.7 −19.6	14 53	+12.5 −19.3
1 00	7.8 40.1	8.0 39.8	5 10	6.7 25.6	6.5 25.3	15 05	+12.8 −19.5	15 18	+12.6 −19.2
1 03	7.4 39.7	7.7 39.5	5 15	6.8 25.5	6.6 25.2	15 31	+12.9 −19.4	15 45	+12.7 −19.1
1 06	7.1 39.4	7.3 39.1	5 20	7.0 25.3	6.7 25.1	15 59	+13.0 −19.3	16 13	+12.8 −19.0
1 09	6.7 39.0	7.0 38.8	5 25	7.1 25.2	6.8 25.0	16 27	+13.1 −19.2	16 43	+12.9 −18.9
1 12	− 6.4 −38.7	− 6.6 −38.4	5 30	+ 7.2 −25.1	+ 6.9 −24.9	16 58	+13.2 −19.1	17 14	+13.0 −18.8
1 15	6.0 38.3	6.3 38.1	5 35	7.3 25.0	7.1 24.7	17 30	+13.3 −19.0	17 47	+13.1 −18.7
1 18	5.7 38.0	6.0 37.8	5 40	7.4 24.9	7.2 24.6	18 05	+13.4 −18.9	18 23	+13.2 −18.6
1 21	5.4 37.7	5.7 37.5	5 45	7.5 24.8	7.3 24.5	18 41	+13.5 −18.8	19 00	+13.3 −18.5
1 24	5.1 37.4	5.3 37.1	5 50	7.6 24.7	7.4 24.4	19 20	+13.6 −18.7	19 41	+13.4 −18.4
1 27	4.8 37.1	5.0 36.8	5 55	7.7 24.6	7.5 24.3	20 02	+13.7 −18.6	20 24	+13.5 −18.3
1 30	− 4.5 −36.8	− 4.7 −36.5	6 00	+ 7.8 −24.5	+ 7.6 −24.2	20 46	+13.8 −18.5	21 10	+13.6 −18.2
1 35	4.0 36.3	4.3 36.1	6 10	8.0 24.3	7.8 24.0	21 34	+13.9 −18.4	21 59	+13.7 −18.1
1 40	3.6 35.9	3.8 35.6	6 20	8.2 24.1	8.0 23.8	22 25	+14.0 −18.3	22 52	+13.8 −18.0
1 45	3.1 35.4	3.4 35.2	6 30	8.4 23.9	8.2 23.6	23 20	+14.1 −18.2	23 49	+13.9 −17.9
1 50	2.7 35.0	2.9 34.7	6 40	8.6 23.7	8.3 23.5	24 20	+14.2 −18.1	24 51	+14.0 −17.8
1 55	2.3 34.6	2.5 34.3	6 50	8.7 23.6	8.5 23.3	25 24	+14.3 −18.0	25 58	+14.1 −17.7
2 00	− 1.9 −34.2	− 2.1 −33.9	7 00	+ 8.9 −23.4	+ 8.7 −23.1	26 34	+14.4 −17.9	27 11	+14.2 −17.6
2 05	1.5 33.8	1.7 33.5	7 10	9.1 23.2	8.8 23.0	27 50	+14.5 −17.8	28 31	+14.3 −17.5
2 10	1.1 33.4	1.4 33.2	7 20	9.2 23.1	9.0 22.8	29 13	+14.6 −17.7	29 58	+14.4 −17.4
2 15	0.8 33.1	1.0 32.8	7 30	9.3 23.0	9.1 22.7	30 44	+14.7 −17.6	31 33	+14.5 −17.3
2 20	0.4 32.7	0.7 32.5	7 40	9.5 22.8	9.2 22.6	32 24	+14.8 −17.5	33 18	+14.6 −17.2
2 25	− 0.1 32.4	− 0.3 32.1	7 50	9.6 22.7	9.4 22.4	34 15	+14.9 −17.4	35 15	+14.7 −17.1
2 30	+ 0.2 −32.1	0.0 −31.8	8 00	+ 9.7 −22.6	+ 9.5 −22.3	36 17	+15.0 −17.3	37 24	+14.8 −17.0
2 35	0.5 31.8	+ 0.3 31.5	8 10	9.9 22.4	9.6 22.2	38 34	+15.1 −17.2	39 48	+14.9 −16.9
2 40	0.8 31.5	0.6 31.2	8 20	10.0 22.3	9.7 22.1	41 06	+15.2 −17.1	42 28	+15.0 −16.8
2 45	1.1 31.2	0.9 30.9	8 30	10.1 22.2	9.9 21.9	43 56	+15.3 −17.0	45 29	+15.1 −16.7
2 50	1.4 30.9	1.2 30.6	8 40	10.2 22.1	10.0 21.8	47 07	+15.4 −16.9	48 52	+15.2 −16.6
2 55	1.7 30.6	1.4 30.4	8 50	10.3 22.0	10.1 21.7	50 43	+15.5 −16.8	52 41	+15.3 −16.5
3 00	+ 2.0 −30.3	+ 1.7 −30.1	9 00	+10.4 −21.9	+10.2 −21.6	54 46	+15.6 −16.7	56 59	+15.4 −16.4
3 05	2.2 30.1	2.0 29.8	9 10	10.5 21.8	10.3 21.5	59 21	+15.7 −16.6	61 50	+15.5 −16.3
3 10	2.5 29.8	2.2 29.6	9 20	10.6 21.7	10.4 21.4	64 28	+15.8 −16.5	67 15	+15.6 −16.2
3 15	2.7 29.6	2.5 29.3	9 30	10.7 21.6	10.5 21.3	70 10	+15.9 −16.4	73 14	+15.7 −16.1
3 20	2.9 29.4	2.7 29.1	9 40	10.8 21.5	10.6 21.2	76 24	+16.0 −16.3	79 42	+15.8 −16.0
3 25	3.2 29.1	2.9 28.9	9 50	10.9 21.4	10.6 21.2	83 05	+16.1 −16.2	86 31	+15.9 −15.9
3 30	+ 3.4 −28.9	+ 3.1 −28.7	10 00	+11.0 −21.3	+10.7 −21.1	90 00		90 00	+15.9 −15.9

Extract from main table for lat. 40° (same) showing Dec 21°, LHA 268°

Extract from table 5 showing d 31' to 60'

31	32	33	34	35	36	37	38	39	40	41	42	43	44	45	46	47	48	49	50	51	52	53	54	55	56	57	58	59	60	d
'	'	'	'	'	'	'	'	'	'	'	'	'	'	'	'	'	'	'	'	'	'	'	'	'	'	'	'	'	'	'
0	0	0	0	0	0	0	0	0	0	0	0	0	0	0	0	0	0	0	0	0	0	0	0	0	0	0	0	0	0	0
1	1	1	1	1	1	1	1	1	1	1	1	1	1	1	1	1	1	1	1	1	1	1	1	1	1	1	1	1	1	1
1	1	1	1	1	1	1	1	1	1	1	1	1	1	2	2	2	2	2	2	2	2	2	2	2	2	2	2	2	2	2
2	2	2	2	2	2	2	2	2	2	2	2	2	2	2	2	2	2	2	2	3	3	3	3	3	3	3	3	3	3	3
2	2	2	2	2	2	2	3	3	3	3	3	3	3	3	3	3	3	3	3	3	3	4	4	4	4	4	4	4	4	4
3	3	3	3	3	3	3	3	3	3	3	4	4	4	4	4	4	4	4	4	4	4	4	4	5	5	5	5	5	5	5
3	3	3	3	4	4	4	4	4	4	4	4	4	5	5	5	5	5	5	5	5	5	5	5	6	6	6	6	6	6	6
4	4	4	4	4	4	4	4	5	5	5	5	5	5	5	5	5	6	6	6	6	6	6	6	6	7	7	7	7	7	7
4	4	4	5	5	5	5	5	5	5	5	6	6	6	6	6	6	6	7	7	7	7	7	7	7	7	8	8	8	8	8
5	5	5	5	5	5	6	6	6	6	6	6	6	7	7	7	7	7	7	8	8	8	8	8	8	8	9	9	9	9	9
5	5	6	6	6	6	6	6	7	7	7	7	7	7	8	8	8	8	9	9	9	9	9	10	10	10	10	10	10	10	10
6	6	6	6	6	7	7	7	7	7	8	8	8	8	8	9	9	9	9	10	10	10	10	10	11	11	11	11	11	11	11
6	7	7	7	7	7	7	8	8	8	8	8	9	9	9	9	9	10	10	10	11	11	11	11	11	11	11	12	12	12	12
7	7	7	7	8	8	8	8	8	9	9	9	9	9	10	10	10	10	11	11	11	12	12	12	12	12	12	13	13	13	13
7	7	8	8	8	8	9	9	9	9	10	10	10	10	10	11	11	11	11	12	12	12	12	13	13	13	13	14	14	14	14
8	8	8	8	9	9	9	10	10	10	10	11	11	12	12	12	12	13	13	13	14	14	14	14	14	15	15	15	15	15	15
8	9	9	9	9	10	10	10	10	11	11	12	12	12	13	13	13	14	14	14	14	14	15	15	15	15	15	16	16	16	16
9	9	9	10	10	10	11	11	11	12	12	12	13	13	14	14	14	14	15	15	15	16	16	16	16	17	17	17	17	17	17
9	10	10	10	10	11	11	11	12	12	12	13	13	13	14	14	14	15	15	15	15	16	16	16	17	17	17	17	18	18	18
10	10	10	11	11	11	12	12	12	13	13	13	14	14	14	15	15	15	16	16	16	16	17	17	17	18	18	18	19	19	19
10	11	11	11	12	12	12	13	13	13	14	14	14	15	15	15	16	16	17	17	18	18	19	19	19	20	20	19	20	20	20
11	11	12	12	12	13	13	13	14	14	14	15	15	15	16	16	16	17	17	18	18	19	19	20	20	20	20	20	21	21	21
11	12	12	12	13	13	14	14	14	15	15	15	16	16	16	17	17	18	18	18	19	20	20	21	21	22	22	21	22	22	22
12	12	13	13	13	14	14	15	15	15	16	16	16	17	17	18	18	18	19	19	20	20	20	21	21	21	22	22	23	23	23
12	13	13	14	14	14	15	15	16	16	16	17	17	18	18	18	19	19	20	20	20	21	21	22	22	22	23	23	24	24	24
13	13	14	14	15	15	15	16	16	17	17	18	18	18	19	19	20	20	20	21	21	22	22	22	23	23	24	24	25	25	25
13	14	14	15	15	16	16	17	17	18	18	19	19	19	20	20	20	21	21	22	22	23	23	23	24	24	25	25	26	26	26
14	14	15	15	16	16	17	17	18	18	19	19	20	20	21	21	21	22	22	22	23	23	24	24	25	25	26	26	27	27	27
14	15	15	16	16	17	17	18	18	19	19	20	20	21	21	21	22	22	23	23	24	24	25	25	26	26	27	27	28	28	28
15	15	16	16	17	17	18	18	19	19	20	20	21	21	22	22	23	23	24	24	25	25	26	26	27	27	28	28	29	29	29
16	16	16	17	18	18	18	19	20	20	20	21	22	22	22	23	24	24	24	25	26	26	26	27	28	28	28	29	30	30	30
16	17	17	18	18	19	19	20	20	21	21	22	22	23	23	24	24	25	25	26	26	27	27	28	28	29	29	30	30	31	31
17	17	18	18	19	19	20	20	21	21	22	22	23	23	24	25	25	26	26	27	27	28	28	29	29	30	30	31	31	32	32
17	18	18	19	19	20	20	21	21	22	22	23	23	24	25	25	26	27	27	28	28	29	29	30	30	31	31	32	32	33	33
18	18	19	19	20	20	21	22	22	23	23	24	24	25	25	26	27	27	28	28	29	29	30	31	31	32	32	33	33	34	34
18	19	19	20	20	21	22	22	23	23	24	24	25	26	26	27	27	28	29	29	30	31	32	33	33	34	35	34	34	35	35
19	19	20	20	21	22	22	23	23	24	25	25	26	26	27	28	28	29	29	30	31	31	32	32	33	34	34	35	35	36	36
19	20	20	21	22	22	23	23	24	25	25	26	26	27	28	28	29	29	30	31	31	32	33	33	34	35	35	36	36	37	37
20	20	21	22	22	23	23	24	25	25	26	27	27	28	28	29	30	30	31	32	32	33	34	34	35	35	36	37	37	38	38
20	21	21	22	23	23	24	25	25	26	27	27	28	29	29	30	31	31	32	32	33	34	34	35	36	36	37	38	38	39	39
21	21	22	23	23	24	25	25	26	27	27	28	29	29	30	31	31	32	33	33	34	35	35	36	37	37	38	39	39	40	40
21	22	23	23	24	25	25	26	27	27	28	29	29	30	31	31	32	33	33	34	35	36	36	37	38	38	39	40	40	41	41
22	22	23	24	24	25	26	27	27	28	29	29	30	31	32	32	33	34	34	35	36	36	37	38	38	39	40	41	41	42	42
22	23	24	24	25	26	27	27	28	29	29	30	31	32	32	33	34	34	35	36	37	37	38	39	39	40	41	42	42	43	43
23	23	24	25	26	26	27	28	29	29	30	31	32	32	33	34	34	35	36	37	37	38	39	40	40	41	42	43	43	44	44
23	24	25	26	26	27	28	28	29	30	31	32	32	33	34	34	35	36	37	38	38	39	40	40	41	42	43	44	44	45	45
24	25	25	26	27	28	28	29	30	31	31	32	33	34	34	35	36	37	38	38	39	40	41	41	42	43	44	44	45	46	46
24	25	26	27	27	28	29	30	31	31	32	33	34	34	35	36	37	38	38	39	40	41	42	42	43	44	45	45	46	47	47
25	26	26	27	28	29	30	30	31	32	33	34	34	35	36	37	38	38	39	40	41	42	42	43	44	45	46	46	47	48	48
25	26	27	28	29	29	30	31	32	33	33	34	35	36	37	38	38	39	40	41	42	42	43	44	45	46	47	47	48	49	49
26	27	28	28	29	30	31	32	32	33	34	35	36	37	38	38	39	40	41	42	42	43	44	45	46	47	48	48	49	50	50
26	27	28	29	30	31	31	32	33	34	35	36	37	37	38	39	40	41	42	42	43	44	45	46	47	48	48	49	50	51	51
27	28	29	29	30	31	32	33	34	34	35	36	37	38	39	40	41	42	42	43	44	45	46	47	48	49	49	50	51	52	52
27	28	29	30	31	32	32	33	34	35	36	37	38	39	40	41	42	42	43	44	45	46	47	48	49	49	50	51	52	53	53
28	29	30	31	32	32	33	34	35	36	37	38	39	40	40	41	42	43	44	45	46	47	48	49	50	50	51	52	53	54	54
28	29	30	31	32	33	34	35	36	36	37	38	39	40	41	42	43	44	45	46	47	48	49	50	50	51	52	53	54	55	55
29	30	31	32	33	34	34	35	36	37	38	39	40	41	42	43	44	45	46	47	48	49	50	51	51	52	53	54	55	56	56
29	30	31	32	33	34	35	36	37	38	39	40	41	42	43	44	45	46	47	48	49	50	51	52	53	54	55	55	56	57	57
30	31	32	33	34	35	36	37	38	39	40	41	42	43	44	45	46	47	48	49	50	51	52	53	54	55	56	56	57	58	58
30	31	32	33	34	35	36	37	38	39	40	41	42	43	44	45	46	47	48	49	50	51	52	53	54	55	56	57	58	59	59

Question 2. Calculate and plot 3 position lines to form a '3 point fix' using the information contained in the following scenario. Refer to the outline procedures in chapter 8 and the demonstration in chapter 10 as necessary.

Scenario:
D.R. Position: 47° 52'S, 57° 35'E
Date: 8 Jan 2009
Course 089° Speed: 5 knots.
Sunset: $20^h 01^m$
Nautical twilight: $21^h 37^m$
Civil twilight: $20^h 41^m$
Time of Fix: $21^h 00^m$ (-4)
DWE 25^s fast
Index error: - 0'.73
Ht. of eye: 5.0m.

Observations.
1. Using the methods described in chapter 8 for planning star and planet observations where appropriate, the following bodies were chosen for the observations at the estimated position at 1700 GMT: Moon U.L., Aldebaron, Diphda.
2. The calculated altitudes and true bearings established at steps 4 and 5 were used to help verify the identities of the observed bodies.
3. The observations below were taken in rapid succession. Movement of the ship while the observations were being taken was negligible.
4. Weather conditions at the time of the observations: Temperature: 8°C. Pressure: 980mb.

Observation 1:
DWT: $16^h 58^m 33^s$
Body observed: Moon U.L.
(lower limb obscured by cloud)
Sextant Alt: 14° 05'.35 Bearing: 012°

Observation 2:
DWT: $17^h 00^m 42^s$
Body observed: Aldebaran.
Sextant Alt: 25° 19'.25   Bearing: 008°

Observation 3.
DWT: $17^h 01^m 45^s$
Body observed: Diphda.
Sextant Alt: 39° 52'.35   Bearing: 289°

Extracts. Extracts relevant purely to this question follow the solution. Due to their sizes, some tables have been cropped to show only the required portions.

Solution To Question 2
Planning Phase

Step 1.	Approx Pos: 47° 52'S, 57° 35'E

Step 2. Calculate Greenwich Date:	
Date: 8 Jan 2009	
Time of fix:	$08^d\ 21^h\ 00^m$ (-4)
Zone corrn:	-4^h
Greenwich Date:	$08^d\ 17^h\ 00^m$

Calculate UT at time of observations.

	Moon(UL)	Aldebaran	Diphda
DWT:	$16^h\ 58^m\ 33^s$	$17^h\ 00^m\ 42^s$	$17^h\ 01^m\ 45^s$
DWE:	-25^s	-25^s	-25^s
UT:	$16^h\ 58^m\ 08^s$	$17^h\ 00^m\ 17^s$	$17^h\ 01^m\ 20^s$

Step 3. Calculate GHA and Dec.

	Moon	Aldebaran	Diphda
Dec (UT)	N26° 44'.4	N16° 31'.7	S17° 56'.3
d	(+2.7)	(n.a.)	(n.a.)
d corr.	+2'.6	0'.0	0'.0
Dec + corr.	N26° 47'.0	N16° 31'.7	S17° 56'.3
SHA (stars)		290° 52'.9	348° 59'.2
GHA ♈(UT)(for stars)		003° 22'.6	003° 22'.6
GHA ♈ + SHA (for stars)		294° 15'.5	352° 21'.8
GHA (16^h)	276° 49'.8		
Inc (m+s)	+13° 52'.3	+0° 04'.3	+0° 20'.1
v	(+1.6)	(n.a.)	(n.a.)
v corr.	+1'.6	0'.0	0'.0
GHA	290° 43'.7	294° 19'.8	352° 41'.9

Step 4. Calculate LHA

	Moon	Aldebaran	Diphda
GHA	290° 43'.7	294° 19'.8	352° 41'.9
As. long.(E)	+57° 16'.3	+57° 40'.2	+57° 18'.1
LHA	348°	352°	410°
(± 360°)			50°

Convert azimuth angle to true bearing (Zn):			
	Moon	Aldebaran	Diphda
LHA =	348°	352°	050°
Z =	169°	171°	110°
Zn =	011°	009°	290°

Rules for calculating true bearing (Zn)

	Lat. North	Lat. South
LHA>180°	Zn = Z	Zn = 180° - Z
LHA<180°	Zn = 360°-Z	Zn = 180° + Z

Please see page 161 for an explanation of the reason for calculating true bearings here. (Take the values of the Azimuth (Z) from step 5).

Step 5. Main table entry.			
	Moon	Aldebaran	Diphda
As. lat.	S48°	S48°	S48°
Dec°	N26°	N16°	S17°
Same/contr.	contr	contr	same
Dec'	47'.0	31'.7	56'.3
LHA	348°	352°	50°
Table page:	63	63	60
Tab Hc,	15° 13'	25° 36'	38° 57'
d	-59'	-60'	+45'
Z	169°	171°	110°
Correction for mins of Dec. (Table 5)			
Tab Hc,	15° 13'	25° 36'	38° 57'
Cor. for Dec'	-46'	-32'	+42'
Cal Alt(Hc)=	14° 27'	25° 04'	39° 39'

Fix Phase

Step 6. Correct sextant altitude (Hs).			
	Moon	Aldebaran	Diphda
Sex Alt(Hs)=	14° 05'.35	25° 19'.25	39° 52'.35
I.E. =	-0'.73	-0'.73	-0'.73
Dip (5m) (D)=	-3'.95	-3'.95	-3'.95
Ap Alt (Ha) =	14° 00'.67	25° 14'.57	39° 47'.67
Alt. Cor. R=		-2'.0	-1'.2
Moon Alt Cor =	+62'.8		
Moon UL Cor =	+5'.9 (Moon HP:60'.9)		
Moon s-diam. =	-30'.0		
Add'l refrac. =	+0'.1	0'.0	0'.0
Obs Alt (Ho) =	14° 39'.47	25° 12'.57	39° 46'.47

Step 7. Calculate Intercept			
	Moon	Aldebaran	Diphda
Ho =	14° 39'.47	25° 12'.57	39° 46'.47
Hc =	14° 27'.0	25° 04'.0	39° 39'.0
p(Ho-Hc) =	+12'.47	+8'.57	+7'.47

Therefore, intercepts are:
Moon: 12.47 n.m. towards 011°, Aldebaran: 8.57 n.m. towards 009°,
Dipda: 7.47 n.m. towards 290°
(Note. True bearings calculated at step 4).

Step 8. Corrections to intercept.
(No corrections required since there has been negligible movement between observations and fix).

Step 9. Plot the position lines (as shown on next page).
To revise the method, see the demonstration in chapter 10).

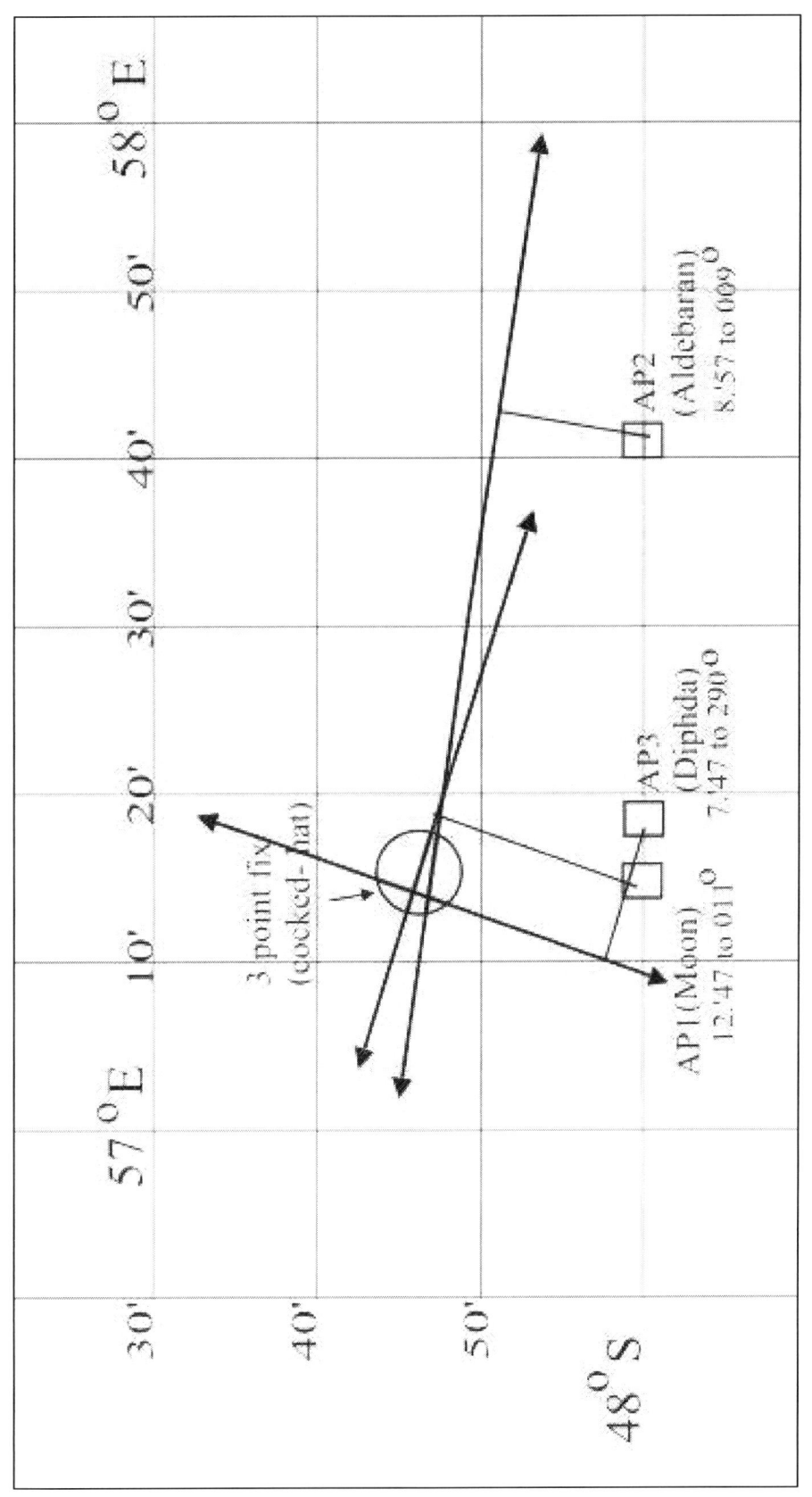

2009 JANUARY 7, 8, 9 (WED., THURS., FRI.)

UT	SUN GHA	SUN Dec	MOON GHA	MOON v	MOON Dec	MOON d	MOON HP
d h	° '	° '	° '	'	° '	'	'
7 00	178 27.6	S22 22.9	61 43.0	6.0	N22 26.1	9.5	59.7
01	193 27.3	22.6	76 08.0	5.9	22 35.6	9.5	59.7
02	208 27.0	22.3	90 32.9	5.7	22 45.1	9.2	59.7
03	223 26.7	21.9	104 57.6	5.7	22 54.3	9.2	59.8
04	238 26.5	21.6	119 22.3	5.4	23 03.5	9.0	59.8
05	253 26.2	21.3	133 46.7	5.4	23 12.5	8.8	59.8
06	268 25.9	S22 21.0	148 11.1	5.2	N23 21.3	8.8	59.9
W 07	283 25.7	20.7	162 35.3	5.1	23 30.1	8.5	59.9
E 08	298 25.4	20.3	176 59.4	5.0	23 38.6	8.5	59.9
D 09	313 25.1	20.0	191 23.4	4.8	23 47.1	8.2	60.0
N 10	328 24.9	19.7	205 47.2	4.7	23 55.3	8.1	60.0
E 11	343 24.6	19.4	220 10.9	4.6	24 03.4	8.0	60.0
S 12	358 24.3	S22 19.0	234 34.5	4.5	N24 11.4	7.8	60.1
D 13	13 24.1	18.7	248 58.0	4.3	24 19.2	7.7	60.1
A 14	28 23.8	18.4	263 21.3	4.2	24 26.9	7.5	60.1
Y 15	43 23.5	18.1	277 44.5	4.1	24 34.4	7.3	60.1
16	58 23.3	17.7	292 07.6	4.0	24 41.7	7.2	60.2
17	73 23.0	17.4	306 30.6	3.8	24 48.9	7.0	60.2
18	88 22.7	S22 17.1	320 53.4	3.8	N24 55.9	6.8	60.2
19	103 22.5	16.7	335 16.2	3.6	25 02.7	6.7	60.3
20	118 22.2	16.4	349 38.8	3.5	25 09.4	6.4	60.3
21	133 21.9	16.1	4 01.3	3.4	25 15.8	6.4	60.3
22	148 21.7	15.7	18 23.7	3.3	25 22.2	6.1	60.4
23	163 21.4	15.4	32 46.0	3.1	25 28.3	6.0	60.4
8 00	178 21.2	S22 15.1	47 08.1	3.1	N25 34.3	5.8	60.4
01	193 20.9	14.7	61 30.2	3.0	25 40.1	5.6	60.4
02	208 20.6	14.4	75 52.2	2.8	25 45.7	5.4	60.5
03	223 20.4	14.1	90 14.0	2.8	25 51.1	5.2	60.5
04	238 20.1	13.7	104 35.8	2.6	25 56.3	5.1	60.5
05	253 19.8	13.4	118 57.4	2.6	26 01.4	4.9	60.6
06	268 19.6	S22 13.0	133 19.0	2.4	N26 06.3	4.7	60.6
07	283 19.3	12.7	147 40.4	2.4	26 11.0	4.4	60.6
T 08	298 19.0	12.4	162 01.8	2.3	26 15.4	4.4	60.6
H 09	313 18.8	12.0	176 23.1	2.2	26 19.8	4.1	60.7
U 10	328 18.5	11.7	190 44.3	2.1	26 23.9	3.9	60.7
R 11	343 18.3	11.3	205 05.4	2.0	26 27.8	3.7	60.7
S 12	358 18.0	S22 11.0	219 26.4	2.0	N26 31.5	3.5	60.7
D 13	13 17.7	10.6	233 47.4	1.9	26 35.0	3.3	60.8
A 14	28 17.5	10.3	248 08.3	1.8	26 38.3	3.2	60.8
Y 15	43 17.2	09.9	262 29.1	1.7	26 41.5	2.9	60.8
16	58 17.0	09.6	276 49.8	1.6	26 44.4	2.7	60.8
17	73 16.7	09.3	291 10.4	1.6	26 47.1	2.5	60.9
18	88 16.4	S22 08.9	305 31.0	1.6	N26 49.6	2.3	60.9
19	103 16.2	08.6	319 51.6	1.4	26 51.9	2.2	60.9
20	118 15.9	08.2	334 12.0	1.4	26 54.1	1.9	60.9
21	133 15.7	07.9	348 32.4	1.4	26 56.0	1.7	60.9
22	148 15.4	07.5	2 52.8	1.3	26 57.7	1.4	61.0
23	163 15.1	07.2	17 13.1	1.3	26 59.1	1.3	61.0
9 00	178 14.9	S22 06.8	31 33.4	1.2	N27 00.4	1.1	61.0
01	193 14.6	06.4	45 53.6	1.2	27 01.5	0.9	61.0
02	208 14.4	06.1	60 13.8	1.1	27 02.4	0.6	61.0
03	223 14.1	05.7	74 33.9	1.1	27 03.0	0.4	61.1
04	238 13.8	05.4	88 54.0	1.1	27 03.4	0.3	61.1
05	253 13.6	05.0	103 14.1	1.1	27 03.7	0.0	61.1
06	268 13.3	S22 04.7	117 34.2	1.0	N27 03.7	0.2	61.1
07	283 13.1	04.3	131 54.2	1.0	27 03.5	0.4	61.1
08	298 12.8	03.9	146 14.2	1.0	27 03.1	0.6	61.1
F 09	313 12.6	03.6	160 34.2	1.0	27 02.5	0.9	61.1
R 10	328 12.3	03.2	174 54.2	0.9	27 01.6	1.0	61.2
I 11	343 12.1	02.9	189 14.1	1.0	27 00.6	1.3	61.2
D 12	358 11.8	S22 02.5	203 34.1	1.0	N26 59.3	1.4	61.2
A 13	13 11.5	02.1	217 54.1	0.9	26 57.9	1.7	61.2
Y 14	28 11.3	01.8	232 14.0	1.0	26 56.2	1.9	61.2
15	43 11.0	01.4	246 34.0	1.0	26 54.3	2.1	61.2
16	58 10.8	01.0	260 54.0	0.9	26 52.2	2.4	61.2
17	73 10.5	00.7	275 13.9	1.0	26 49.8	2.5	61.2
18	88 10.3	S22 00.3	289 33.9	1.1	N26 47.3	2.8	61.3
19	103 10.0	21 59.9	303 54.0	1.0	26 44.5	2.9	61.3
20	118 09.8	59.6	318 14.0	1.1	26 41.6	3.2	61.3
21	133 09.5	59.2	332 34.1	1.1	26 38.4	3.4	61.3
22	148 09.3	58.8	346 54.1	1.2	26 35.0	3.6	61.3
23	163 09.0	58.5	1 14.3	1.1	N26 31.4	3.8	61.3
	SD 16.3	d 0.3	SD 16.4		16.5		16.7

Lat.	Twilight Naut.	Twilight Civil	Sunrise	Moonrise 7	Moonrise 8	Moonrise 9	Moonrise 10
°	h m	h m	h m	h m	h m	h m	h m
N 72	08 15	10 18	■	□	□	□	□
N 70	07 58	09 36	■	□	□	□	□
68	07 44	09 07	11 08	□	□	□	□
66	07 32	08 46	10 12	08 58	□	□	□
64	07 23	08 28	09 40	09 59	□	□	12 32
62	07 14	08 14	09 16	10 33	10 46	11 37	13 26
60	07 07	08 02	08 57	10 59	11 25	12 23	13 59
N 58	07 00	07 51	08 41	11 19	11 52	12 53	14 23
56	06 54	07 42	08 28	11 36	12 14	13 15	14 42
54	06 48	07 34	08 16	11 50	12 32	13 34	14 58
52	06 43	07 26	08 06	12 03	12 47	13 50	15 12
50	06 38	07 19	07 57	12 14	13 00	14 04	15 24
45	06 28	07 04	07 38	12 37	13 27	14 32	15 49
N 40	06 18	06 52	07 22	12 56	13 49	14 54	16 09
35	06 09	06 41	07 09	13 12	14 06	15 12	16 26
30	06 01	06 31	06 57	13 25	14 22	15 28	16 40
20	05 46	06 13	06 37	13 49	14 48	15 54	17 05
N 10	05 31	05 57	06 20	14 09	15 11	16 17	17 25
0	05 15	05 41	06 03	14 28	15 32	16 39	17 45
S 10	04 57	05 24	05 46	14 48	15 54	17 00	18 05
20	04 35	05 04	05 29	15 08	16 16	17 23	18 25
30	04 08	04 41	05 08	15 33	16 43	17 50	18 49
35	03 50	04 26	04 56	15 47	16 59	18 06	19 03
40	03 29	04 09	04 41	16 03	17 17	18 24	19 20
45	03 00	03 48	04 25	16 23	17 40	18 46	19 39
S 50	02 19	03 21	04 04	16 48	18 08	19 14	20 03
52	01 56	03 07	03 54	17 01	18 22	19 28	20 15
54	01 23	02 51	03 42	17 14	18 38	19 43	20 28
56	////	02 31	03 29	17 31	18 57	20 02	20 43
58	////	02 06	03 14	17 50	19 21	20 25	21 00
S 60	////	01 30	02 56	18 14	19 52	20 54	21 22

Lat.	Sunset	Twilight Civil	Twilight Naut.	Moonset 7	Moonset 8	Moonset 9	Moonset 10
°	h m	h m	h m	h m	h m	h m	h m
N 72	■	13 56	16 00	□	□	□	□
N 70	■	14 38	16 17	□	□	□	□
68	13 06	15 07	16 30	□	□	□	□
66	14 02	15 28	16 42	07 15	□	□	□
64	14 34	15 46	16 52	06 14	□	□	10 33
62	14 58	16 00	17 00	05 41	07 39	09 07	09 38
60	15 17	16 12	17 07	05 16	07 00	08 21	09 05
N 58	15 33	16 23	17 14	04 56	06 33	07 51	08 41
56	15 46	16 32	17 20	04 40	06 12	07 28	08 21
54	15 58	16 40	17 26	04 26	05 54	07 10	08 04
52	16 08	16 48	17 31	04 14	05 39	06 54	07 50
50	16 17	16 55	17 36	04 03	05 27	06 40	07 38
45	16 36	17 10	17 46	03 41	05 00	06 12	07 12
N 40	16 52	17 22	17 56	03 23	04 39	05 50	06 51
35	17 05	17 33	18 04	03 09	04 21	05 31	06 34
30	17 17	17 43	18 12	02 56	04 06	05 15	06 19
20	17 37	18 00	18 28	02 34	03 41	04 49	05 53
N 10	17 54	18 17	18 43	02 15	03 19	04 25	05 31
0	18 10	18 33	18 59	01 57	02 58	04 04	05 11
S 10	18 27	18 50	19 17	01 39	02 38	03 42	04 50
20	18 45	19 09	19 38	01 21	02 16	03 19	04 28
30	19 06	19 33	20 05	00 59	01 51	02 52	04 02
35	19 18	19 47	20 23	00 46	01 36	02 36	03 46
40	19 32	20 04	20 44	00 32	01 18	02 17	03 28
45	19 49	20 25	21 13	00 15	00 58	01 55	03 07
S 50	20 09	20 52	21 53	24 32	00 32	01 26	02 39
52	20 19	21 06	22 16	24 19	00 19	01 12	02 26
54	20 31	21 22	22 48	24 05	00 05	00 56	02 10
56	20 43	21 41	////	23 49	24 37	00 37	01 52
58	20 58	22 06	////	23 29	24 13	00 13	01 29
S 60	21 16	22 41	////	23 04	23 42	25 00	01 00

Day	SUN Eqn. of Time 00ʰ	SUN Eqn. of Time 12ʰ	SUN Mer. Pass.	MOON Mer. Pass. Upper	MOON Mer. Pass. Lower	Age	Phase
d	m s	m s	h m	h m	h m	d	%
7	06 09	06 22	12 06	20 43	08 13	11	81
8	06 35	06 47	12 07	21 48	09 15	12	90
9	07 00	07 12	12 07	22 55	10 21	13	96

2009 JANUARY 7, 8, 9 (WED., THURS., FRI.)

UT	ARIES	VENUS −4.5		MARS +1.3		JUPITER −1.9		SATURN +0.9		STARS		
	GHA	GHA	Dec	GHA	Dec	GHA	Dec	GHA	Dec	Name	SHA	Dec
d h	° ′	° ′	° ′	° ′	° ′	° ′	° ′	° ′	° ′		° ′	° ′
7 00	106 41.6	130 39.5	S11 07.6	187 52.9	S23 56.7	164 05.5	S20 29.5	293 28.0	N 5 11.2	Acamar	315 20.5	S40 16.2
01	121 44.1	145 39.5	06.4	202 53.3	56.6	179 07.3	29.4	308 30.5	11.3	Achernar	335 28.9	S57 11.7
02	136 46.5	160 39.5	05.3	217 53.7	56.5	194 09.2	29.2	323 33.0	11.3	Acrux	173 13.2	S63 08.8
03	151 49.0	175 39.5 ..	04.1	232 54.1 ..	56.4	209 11.0 ..	29.1	338 35.5 ..	11.3	Adhara	255 14.8	S28 59.1
04	166 51.5	190 39.5	03.0	247 54.5	56.3	224 12.9	29.0	353 37.9	11.3	Aldebaran	290 52.9	N16 31.7
05	181 53.9	205 39.5	01.8	262 54.9	56.2	239 14.7	28.9	8 40.4	11.3			
06	196 56.4	220 39.6	S11 00.7	277 55.3	S23 56.1	254 16.6	S20 28.8	23 42.9	N 5 11.4	Alioth	166 23.3	N55 54.2
W 07	211 58.9	235 39.6	10 59.5	292 55.7	56.0	269 18.5	28.6	38 45.4	11.4	Alkaid	153 01.4	N49 15.7
E 08	227 01.3	250 39.6	58.4	307 56.1	55.9	284 20.3	28.5	53 47.9	11.4	Al Na'ir	27 48.1	S46 55.2
D 09	242 03.8	265 39.6 ..	57.2	322 56.5 ..	55.8	299 22.2 ..	28.4	68 50.4 ..	11.4	Alnilam	275 49.4	S 1 11.8
N 10	257 06.3	280 39.6	56.0	337 56.9	55.7	314 24.0	28.3	83 52.9	11.4	Alphard	217 59.1	S 8 41.9
E 11	272 08.7	295 39.6	54.9	352 57.3	55.6	329 25.9	28.1	98 55.4	11.5			
S 12	287 11.2	310 39.6	S10 53.7	7 57.7	S23 55.5	344 27.7	S20 28.0	113 57.8	N 5 11.5	Alphecca	126 14.0	N26 40.8
D 13	302 13.7	325 39.6	52.6	22 58.1	55.4	359 29.6	27.9	129 00.3	11.5	Alpheratz	357 47.1	N29 08.6
A 14	317 16.1	340 39.7	51.4	37 58.5	55.3	14 31.4	27.8	144 02.8	11.5	Altair	62 11.8	N 8 53.5
Y 15	332 18.6	355 39.7 ..	50.3	52 58.8 ..	55.2	29 33.3 ..	27.7	159 05.3 ..	11.6	Ankaa	353 18.9	S42 15.6
16	347 21.0	10 39.7	49.1	67 59.2	55.1	44 35.2	27.5	174 07.8	11.6	Antares	112 30.6	S26 27.1
17	2 23.5	25 39.7	48.0	82 59.6	55.0	59 37.0	27.4	189 10.3	11.6			
18	17 26.0	40 39.7	S10 46.8	98 00.0	S23 54.9	74 38.9	S20 27.3	204 12.8	N 5 11.6	Arcturus	145 58.8	N19 07.9
19	32 28.4	55 39.8	45.7	113 00.4	54.8	89 40.7	27.2	219 15.3	11.7	Atria	107 36.0	S69 02.5
20	47 30.9	70 39.8	44.5	128 00.8	54.7	104 42.6	27.0	234 17.8	11.7	Avior	234 18.9	S59 32.2
21	62 33.4	85 39.8 ..	43.4	143 01.2 ..	54.6	119 44.4 ..	26.9	249 20.2 ..	11.7	Bellatrix	278 35.2	N 6 21.5
22	77 35.8	100 39.8	42.2	158 01.6	54.5	134 46.3	26.8	264 22.7	11.7	Betelgeuse	271 04.5	N 7 24.5
23	92 38.3	115 39.9	41.0	173 02.0	54.4	149 48.2	26.7	279 25.2	11.7			
8 00	107 40.8	130 39.9	S10 39.9	188 02.4	S23 54.3	164 50.0	S20 26.6	294 27.7	N 5 11.8	Canopus	263 57.1	S52 42.0
01	122 43.2	145 39.9	38.7	203 02.8	54.2	179 51.9	26.4	309 30.2	11.8	Capella	280 38.9	N46 00.6
02	137 45.7	160 39.9	37.6	218 03.2	54.1	194 53.7	26.3	324 32.7	11.8	Deneb	49 34.3	N45 18.8
03	152 48.1	175 40.0 ..	36.4	233 03.6 ..	54.0	209 55.6 ..	26.2	339 35.2 ..	11.8	Denebola	182 36.9	N14 31.1
04	167 50.6	190 40.0	35.3	248 04.0	53.8	224 57.4	26.1	354 37.7	11.9	Diphda	348 59.2	S17 56.3
05	182 53.1	205 40.0	34.1	263 04.4	53.7	239 59.3	25.9	9 40.2	11.9			
06	197 55.5	220 40.1	S10 32.9	278 04.8	S23 53.6	255 01.1	S20 25.8	24 42.7	N 5 11.9	Dubhe	193 55.1	N61 41.8
T 07	212 58.0	235 40.1	31.8	293 05.2	53.5	270 03.0	25.7	39 45.2	11.9	Elnath	278 16.5	N28 37.0
H 08	228 00.5	250 40.1	30.6	308 05.6	53.4	285 04.8	25.6	54 47.6	12.0	Eltanin	90 48.2	N51 29.1
U 09	243 02.9	265 40.2 ..	29.5	323 06.0 ..	53.3	300 06.7 ..	25.4	69 50.1 ..	12.0	Enif	33 50.7	N 9 55.0
R 10	258 05.4	280 40.2	28.3	338 06.4	53.2	315 08.6	25.3	84 52.6	12.0	Fomalhaut	15 27.7	S29 34.6
S 11	273 07.9	295 40.2	27.1	353 06.8	53.1	330 10.4	25.2	99 55.1	12.0			
D 12	288 10.3	310 40.3	S10 26.0	8 07.2	S23 53.0	345 12.3	S20 25.1	114 57.6	N 5 12.1	Gacrux	172 04.7	S57 09.7
A 13	303 12.8	325 40.3	24.8	23 07.6	52.9	0 14.1	25.0	130 00.1	12.1	Gienah	175 55.7	S17 35.6
Y 14	318 15.3	340 40.3	23.7	38 08.0	52.8	15 16.0	24.8	145 02.6	12.1	Hadar	148 53.0	S60 24.8
15	333 17.7	355 40.4 ..	22.5	53 08.4 ..	52.6	30 17.8 ..	24.7	160 05.1 ..	12.1	Hamal	328 04.5	N23 30.5
16	348 20.2	10 40.4	21.3	68 08.7	52.5	45 19.7	24.6	175 07.6	12.2	Kaus Aust.	83 48.6	S34 22.9
17	3 22.6	25 40.5	20.2	83 09.1	52.4	60 21.5	24.5	190 10.1	12.2			
18	18 25.1	40 40.5	S10 19.0	98 09.5	S23 52.3	75 23.4	S20 24.3	205 12.6	N 5 12.2	Kochab	137 20.0	N74 06.7
19	33 27.6	55 40.5	17.9	113 09.9	52.2	90 25.3	24.2	220 15.1	12.2	Markab	13 41.8	N15 15.3
20	48 30.0	70 40.6	16.7	128 10.3	52.1	105 27.1	24.1	235 17.6	12.3	Menkar	314 18.3	N 4 07.6
21	63 32.5	85 40.6 ..	15.5	143 10.7 ..	52.0	120 29.0 ..	24.0	250 20.0 ..	12.3	Menkent	148 11.7	S36 24.8
22	78 35.0	100 40.7	14.4	158 11.1	51.8	135 30.8	23.8	265 22.5	12.3	Miaplacidus	221 39.9	S69 45.1
23	93 37.4	115 40.7	13.2	173 11.5	51.7	150 32.7	23.7	280 25.0	12.3			
9 00	108 39.9	130 40.8	S10 12.0	188 11.9	S23 51.6	165 34.5	S20 23.6	295 27.5	N 5 12.4	Mirfak	308 44.9	N49 53.9
01	123 42.4	145 40.8	10.9	203 12.3	51.5	180 36.4	23.5	310 30.0	12.4	Nunki	76 02.8	S26 17.2
02	138 44.8	160 40.9	09.7	218 12.7	51.4	195 38.2	23.3	325 32.5	12.4	Peacock	53 24.9	S56 42.5
03	153 47.3	175 40.9 ..	08.5	233 13.1 ..	51.3	210 40.1 ..	23.2	340 35.0 ..	12.4	Pollux	243 31.3	N28 00.2
04	168 49.8	190 41.0	07.4	248 13.5	51.1	225 42.0	23.1	355 37.5	12.5	Procyon	245 02.8	N 5 12.1
05	183 52.2	205 41.0	06.2	263 13.9	51.0	240 43.8	23.0	10 40.0	12.5			
06	198 54.7	220 41.1	S10 05.1	278 14.3	S23 50.9	255 45.7	S20 22.9	25 42.5	N 5 12.5	Rasalhague	96 09.8	N12 33.1
07	213 57.1	235 41.1	03.9	293 14.7	50.8	270 47.5	22.7	40 45.0	12.5	Regulus	207 46.7	N11 55.2
08	228 59.6	250 41.2	02.7	308 15.1	50.7	285 49.4	22.6	55 47.5	12.6	Rigel	281 14.9	S 8 11.5
F 09	244 02.1	265 41.2 ..	01.6	323 15.5 ..	50.5	300 51.2 ..	22.5	70 50.0 ..	12.6	Rigil Kent.	139 56.7	S60 52.2
R 10	259 04.5	280 41.3	10 00.4	338 15.9	50.4	315 53.1	22.4	85 52.5	12.6	Sabik	102 16.6	S15 44.2
I 11	274 07.0	295 41.4	9 59.2	353 16.3	50.3	330 54.9	22.2	100 55.0	12.6			
D 12	289 09.5	310 41.4	S 9 58.1	8 16.7	S23 50.2	345 56.8	S20 22.1	115 57.5	N 5 12.7	Schedar	349 44.6	N56 35.6
A 13	304 11.9	325 41.5	56.9	23 17.1	50.1	0 58.6	22.0	131 00.0	12.7	Shaula	96 26.8	S37 06.6
Y 14	319 14.4	340 41.5	55.7	38 17.5	49.9	16 00.5	21.9	146 02.5	12.7	Sirius	258 36.3	S16 43.7
15	334 16.9	355 41.6 ..	54.6	53 17.8 ..	49.8	31 02.4 ..	21.7	161 05.0 ..	12.7	Spica	158 34.8	S11 12.6
16	349 19.3	10 41.7	53.4	68 18.2	49.7	46 04.2	21.6	176 07.5	12.8	Suhail	222 54.6	S43 28.1
17	4 21.8	25 41.7	52.2	83 18.6	49.6	61 06.1	21.5	191 10.0	12.8			
18	19 24.3	40 41.8	S 9 51.0	98 19.0	S23 49.4	76 07.9	S20 21.4	206 12.5	N 5 12.8	Vega	80 41.6	N38 47.4
19	34 26.7	55 41.9	49.9	113 19.4	49.3	91 09.8	21.2	221 15.0	12.8	Zuben'ubi	137 09.3	S16 04.8
20	49 29.2	70 41.9	48.7	128 19.8	49.2	106 11.6	21.1	236 17.5	12.9		SHA	Mer. Pass.
21	64 31.6	85 42.0 ..	47.5	143 20.2 ..	49.1	121 13.5 ..	21.0	251 19.9 ..	12.9		° ′	h m
22	79 34.1	100 42.1	46.4	158 20.6	48.9	136 15.3	20.9	266 22.4	12.9	Venus	22 59.1	15 17
23	94 36.6	115 42.1	45.2	173 21.0	48.8	151 17.2	20.7	281 24.9	12.9	Mars	80 21.7	11 28
	h m									Jupiter	57 09.2	12 59
Mer. Pass. 16 46.5		v 0.0	d 1.2	v 0.4	d 0.1	v 1.9	d 0.1	v 2.5	d 0.0	Saturn	186 47.0	4 21

©2009 Jack Case page 253

INCREMENTS AND CORRECTIONS

0ᵐ

s	SUN PLANETS	ARIES	MOON	v or d	Corrⁿ	v or d	Corrⁿ	v or d	Corrⁿ
	° ′	° ′	° ′	′	′	′	′	′	′
00	0 00.0	0 00.0	0 00.0	0.0	0.0	6.0	0.1	12.0	0.1
01	0 00.3	0 00.3	0 00.2	0.1	0.0	6.1	0.1	12.1	0.1
02	0 00.5	0 00.5	0 00.5	0.2	0.0	6.2	0.1	12.2	0.1
03	0 00.8	0 00.8	0 00.7	0.3	0.0	6.3	0.1	12.3	0.1
04	0 01.0	0 01.0	0 01.0	0.4	0.0	6.4	0.1	12.4	0.1
05	0 01.3	0 01.3	0 01.2	0.5	0.0	6.5	0.1	12.5	0.1
06	0 01.5	0 01.5	0 01.4	0.6	0.0	6.6	0.1	12.6	0.1
07	0 01.8	0 01.8	0 01.7	0.7	0.0	6.7	0.1	12.7	0.1
08	0 02.0	0 02.0	0 01.9	0.8	0.0	6.8	0.1	12.8	0.1
09	0 02.3	0 02.3	0 02.1	0.9	0.0	6.9	0.1	12.9	0.1
10	0 02.5	0 02.5	0 02.4	1.0	0.0	7.0	0.1	13.0	0.1
11	0 02.8	0 02.8	0 02.6	1.1	0.0	7.1	0.1	13.1	0.1
12	0 03.0	0 03.0	0 02.9	1.2	0.0	7.2	0.1	13.2	0.1
13	0 03.3	0 03.3	0 03.1	1.3	0.0	7.3	0.1	13.3	0.1
14	0 03.5	0 03.5	0 03.3	1.4	0.0	7.4	0.1	13.4	0.1
15	0 03.8	0 03.8	0 03.6	1.5	0.0	7.5	0.1	13.5	0.1
16	0 04.0	0 04.0	0 03.8	1.6	0.0	7.6	0.1	13.6	0.1
17	0 04.3	0 04.3	0 04.1	1.7	0.0	7.7	0.1	13.7	0.1
18	0 04.5	0 04.5	0 04.3	1.8	0.0	7.8	0.1	13.8	0.1
19	0 04.8	0 04.8	0 04.5	1.9	0.0	7.9	0.1	13.9	0.1
20	0 05.0	0 05.0	0 04.8	2.0	0.0	8.0	0.1	14.0	0.1
21	0 05.3	0 05.3	0 05.0	2.1	0.0	8.1	0.1	14.1	0.1
22	0 05.5	0 05.5	0 05.2	2.2	0.0	8.2	0.1	14.2	0.1
23	0 05.8	0 05.8	0 05.5	2.3	0.0	8.3	0.1	14.3	0.1
24	0 06.0	0 06.0	0 05.7	2.4	0.0	8.4	0.1	14.4	0.1
25	0 06.3	0 06.3	0 06.0	2.5	0.0	8.5	0.1	14.5	0.1
26	0 06.5	0 06.5	0 06.2	2.6	0.0	8.6	0.1	14.6	0.1
27	0 06.8	0 06.8	0 06.4	2.7	0.0	8.7	0.1	14.7	0.1
28	0 07.0	0 07.0	0 06.7	2.8	0.0	8.8	0.1	14.8	0.1
29	0 07.3	0 07.3	0 06.9	2.9	0.0	8.9	0.1	14.9	0.1
30	0 07.5	0 07.5	0 07.2	3.0	0.0	9.0	0.1	15.0	0.1
31	0 07.8	0 07.8	0 07.4	3.1	0.0	9.1	0.1	15.1	0.1
32	0 08.0	0 08.0	0 07.6	3.2	0.0	9.2	0.1	15.2	0.1
33	0 08.3	0 08.3	0 07.9	3.3	0.0	9.3	0.1	15.3	0.1
34	0 08.5	0 08.5	0 08.1	3.4	0.0	9.4	0.1	15.4	0.1
35	0 08.8	0 08.8	0 08.4	3.5	0.0	9.5	0.1	15.5	0.1
36	0 09.0	0 09.0	0 08.6	3.6	0.0	9.6	0.1	15.6	0.1
37	0 09.3	0 09.3	0 08.8	3.7	0.0	9.7	0.1	15.7	0.1
38	0 09.5	0 09.5	0 09.1	3.8	0.0	9.8	0.1	15.8	0.1
39	0 09.8	0 09.8	0 09.3	3.9	0.0	9.9	0.1	15.9	0.1
40	0 10.0	0 10.0	0 09.5	4.0	0.0	10.0	0.1	16.0	0.1
41	0 10.3	0 10.3	0 09.8	4.1	0.0	10.1	0.1	16.1	0.1
42	0 10.5	0 10.5	0 10.0	4.2	0.0	10.2	0.1	16.2	0.1
43	0 10.8	0 10.8	0 10.3	4.3	0.0	10.3	0.1	16.3	0.1
44	0 11.0	0 11.0	0 10.5	4.4	0.0	10.4	0.1	16.4	0.1
45	0 11.3	0 11.3	0 10.7	4.5	0.0	10.5	0.1	16.5	0.1
46	0 11.5	0 11.5	0 11.0	4.6	0.0	10.6	0.1	16.6	0.1
47	0 11.8	0 11.8	0 11.2	4.7	0.0	10.7	0.1	16.7	0.1
48	0 12.0	0 12.0	0 11.5	4.8	0.0	10.8	0.1	16.8	0.1
49	0 12.3	0 12.3	0 11.7	4.9	0.0	10.9	0.1	16.9	0.1
50	0 12.5	0 12.5	0 11.9	5.0	0.0	11.0	0.1	17.0	0.1
51	0 12.8	0 12.8	0 12.2	5.1	0.0	11.1	0.1	17.1	0.1
52	0 13.0	0 13.0	0 12.4	5.2	0.0	11.2	0.1	17.2	0.1
53	0 13.3	0 13.3	0 12.6	5.3	0.0	11.3	0.1	17.3	0.1
54	0 13.5	0 13.5	0 12.9	5.4	0.0	11.4	0.1	17.4	0.1
55	0 13.8	0 13.8	0 13.1	5.5	0.0	11.5	0.1	17.5	0.1
56	0 14.0	0 14.0	0 13.4	5.6	0.0	11.6	0.1	17.6	0.1
57	0 14.3	0 14.3	0 13.6	5.7	0.0	11.7	0.1	17.7	0.1
58	0 14.5	0 14.5	0 13.8	5.8	0.0	11.8	0.1	17.8	0.1
59	0 14.8	0 14.8	0 14.1	5.9	0.0	11.9	0.1	17.9	0.1
60	0 15.0	0 15.0	0 14.3	6.0	0.1	12.0	0.1	18.0	0.2

1ᵐ

s	SUN PLANETS	ARIES	MOON	v or d	Corrⁿ	v or d	Corrⁿ	v or d	Corrⁿ
	° ′	° ′	° ′	′	′	′	′	′	′
00	0 15.0	0 15.0	0 14.3	0.0	0.0	6.0	0.2	12.0	0.3
01	0 15.3	0 15.3	0 14.6	0.1	0.0	6.1	0.2	12.1	0.3
02	0 15.5	0 15.5	0 14.8	0.2	0.0	6.2	0.2	12.2	0.3
03	0 15.8	0 15.8	0 15.0	0.3	0.0	6.3	0.2	12.3	0.3
04	0 16.0	0 16.0	0 15.3	0.4	0.0	6.4	0.2	12.4	0.3
05	0 16.3	0 16.3	0 15.5	0.5	0.0	6.5	0.2	12.5	0.3
06	0 16.5	0 16.5	0 15.7	0.6	0.0	6.6	0.2	12.6	0.3
07	0 16.8	0 16.8	0 16.0	0.7	0.0	6.7	0.2	12.7	0.3
08	0 17.0	0 17.0	0 16.2	0.8	0.0	6.8	0.2	12.8	0.3
09	0 17.3	0 17.3	0 16.5	0.9	0.0	6.9	0.2	12.9	0.3
10	0 17.5	0 17.5	0 16.7	1.0	0.0	7.0	0.2	13.0	0.3
11	0 17.8	0 17.8	0 16.9	1.1	0.0	7.1	0.2	13.1	0.3
12	0 18.0	0 18.0	0 17.2	1.2	0.0	7.2	0.2	13.2	0.3
13	0 18.3	0 18.3	0 17.4	1.3	0.0	7.3	0.2	13.3	0.3
14	0 18.5	0 18.6	0 17.7	1.4	0.0	7.4	0.2	13.4	0.3
15	0 18.8	0 18.8	0 17.9	1.5	0.0	7.5	0.2	13.5	0.3
16	0 19.0	0 19.1	0 18.1	1.6	0.0	7.6	0.2	13.6	0.3
17	0 19.3	0 19.3	0 18.4	1.7	0.0	7.7	0.2	13.7	0.3
18	0 19.5	0 19.6	0 18.6	1.8	0.0	7.8	0.2	13.8	0.3
19	0 19.8	0 19.8	0 18.9	1.9	0.0	7.9	0.2	13.9	0.3
20	0 20.0	0 20.1	0 19.1	2.0	0.1	8.0	0.2	14.0	0.4
21	0 20.3	0 20.3	0 19.3	2.1	0.1	8.1	0.2	14.1	0.4
22	0 20.5	0 20.6	0 19.6	2.2	0.1	8.2	0.2	14.2	0.4
23	0 20.8	0 20.8	0 19.8	2.3	0.1	8.3	0.2	14.3	0.4
24	0 21.0	0 21.1	0 20.0	2.4	0.1	8.4	0.2	14.4	0.4
25	0 21.3	0 21.3	0 20.3	2.5	0.1	8.5	0.2	14.5	0.4
26	0 21.5	0 21.6	0 20.5	2.6	0.1	8.6	0.2	14.6	0.4
27	0 21.8	0 21.8	0 20.8	2.7	0.1	8.7	0.2	14.7	0.4
28	0 22.0	0 22.1	0 21.0	2.8	0.1	8.8	0.2	14.8	0.4
29	0 22.3	0 22.3	0 21.2	2.9	0.1	8.9	0.2	14.9	0.4
30	0 22.5	0 22.6	0 21.5	3.0	0.1	9.0	0.2	15.0	0.4
31	0 22.8	0 22.8	0 21.7	3.1	0.1	9.1	0.2	15.1	0.4
32	0 23.0	0 23.1	0 22.0	3.2	0.1	9.2	0.2	15.2	0.4
33	0 23.3	0 23.3	0 22.2	3.3	0.1	9.3	0.2	15.3	0.4
34	0 23.5	0 23.6	0 22.4	3.4	0.1	9.4	0.2	15.4	0.4
35	0 23.8	0 23.8	0 22.7	3.5	0.1	9.5	0.2	15.5	0.4
36	0 24.0	0 24.1	0 22.9	3.6	0.1	9.6	0.2	15.6	0.4
37	0 24.3	0 24.3	0 23.1	3.7	0.1	9.7	0.2	15.7	0.4
38	0 24.5	0 24.6	0 23.4	3.8	0.1	9.8	0.2	15.8	0.4
39	0 24.8	0 24.8	0 23.6	3.9	0.1	9.9	0.2	15.9	0.4
40	0 25.0	0 25.1	0 23.9	4.0	0.1	10.0	0.3	16.0	0.4
41	0 25.3	0 25.3	0 24.1	4.1	0.1	10.1	0.3	16.1	0.4
42	0 25.5	0 25.6	0 24.3	4.2	0.1	10.2	0.3	16.2	0.4
43	0 25.8	0 25.8	0 24.6	4.3	0.1	10.3	0.3	16.3	0.4
44	0 26.0	0 26.1	0 24.8	4.4	0.1	10.4	0.3	16.4	0.4
45	0 26.3	0 26.3	0 25.1	4.5	0.1	10.5	0.3	16.5	0.4
46	0 26.5	0 26.6	0 25.3	4.6	0.1	10.6	0.3	16.6	0.4
47	0 26.8	0 26.8	0 25.5	4.7	0.1	10.7	0.3	16.7	0.4
48	0 27.0	0 27.1	0 25.8	4.8	0.1	10.8	0.3	16.8	0.4
49	0 27.3	0 27.3	0 26.0	4.9	0.1	10.9	0.3	16.9	0.4
50	0 27.5	0 27.6	0 26.2	5.0	0.1	11.0	0.3	17.0	0.4
51	0 27.8	0 27.8	0 26.5	5.1	0.1	11.1	0.3	17.1	0.4
52	0 28.0	0 28.1	0 26.7	5.2	0.1	11.2	0.3	17.2	0.4
53	0 28.3	0 28.3	0 27.0	5.3	0.1	11.3	0.3	17.3	0.4
54	0 28.5	0 28.6	0 27.2	5.4	0.1	11.4	0.3	17.4	0.4
55	0 28.8	0 28.8	0 27.4	5.5	0.1	11.5	0.3	17.5	0.4
56	0 29.0	0 29.1	0 27.7	5.6	0.1	11.6	0.3	17.6	0.4
57	0 29.3	0 29.3	0 27.9	5.7	0.1	11.7	0.3	17.7	0.4
58	0 29.5	0 29.6	0 28.2	5.8	0.1	11.8	0.3	17.8	0.4
59	0 29.8	0 29.8	0 28.4	5.9	0.1	11.9	0.3	17.9	0.4
60	0 30.0	0 30.1	0 28.6	6.0	0.2	12.0	0.3	18.0	0.5

©2009 Jack Case

INCREMENTS AND CORRECTIONS

58m

s	SUN PLANETS	ARIES	MOON	v or d	Corr^n	v or d	Corr^n	v or d	Corr^n
00	14 30·0	14 32·4	13 50·4	0·0	0·0	6·0	5·9	12·0	11·7
01	14 30·3	14 32·6	13 50·6	0·1	0·1	6·1	5·9	12·1	11·8
02	14 30·5	14 32·9	13 50·8	0·2	0·2	6·2	6·0	12·2	11·9
03	14 30·8	14 33·1	13 51·1	0·3	0·3	6·3	6·1	12·3	12·0
04	14 31·0	14 33·4	13 51·3	0·4	0·4	6·4	6·2	12·4	12·1
05	14 31·3	14 33·6	13 51·6	0·5	0·5	6·5	6·3	12·5	12·2
06	14 31·5	14 33·9	13 51·8	0·6	0·6	6·6	6·4	12·6	12·3
07	14 31·8	14 34·1	13 52·0	0·7	0·7	6·7	6·5	12·7	12·4
08	14 32·0	14 34·4	13 52·3	0·8	0·8	6·8	6·6	12·8	12·5
09	14 32·3	14 34·6	13 52·5	0·9	0·9	6·9	6·7	12·9	12·6
10	14 32·5	14 34·9	13 52·8	1·0	1·0	7·0	6·8	13·0	12·7
11	14 32·8	14 35·1	13 53·0	1·1	1·1	7·1	6·9	13·1	12·8
12	14 33·0	14 35·4	13 53·2	1·2	1·2	7·2	7·0	13·2	12·9
13	14 33·3	14 35·6	13 53·5	1·3	1·3	7·3	7·1	13·3	13·0
14	14 33·5	14 35·9	13 53·7	1·4	1·4	7·4	7·2	13·4	13·1
15	14 33·8	14 36·1	13 53·9	1·5	1·5	7·5	7·3	13·5	13·2
16	14 34·0	14 36·4	13 54·2	1·6	1·6	7·6	7·4	13·6	13·3
17	14 34·3	14 36·6	13 54·4	1·7	1·7	7·7	7·5	13·7	13·4
18	14 34·5	14 36·9	13 54·7	1·8	1·8	7·8	7·6	13·8	13·5
19	14 34·8	14 37·1	13 54·9	1·9	1·9	7·9	7·7	13·9	13·6
20	14 35·0	14 37·4	13 55·1	2·0	2·0	8·0	7·8	14·0	13·7
21	14 35·3	14 37·6	13 55·4	2·1	2·0	8·1	7·9	14·1	13·7
22	14 35·5	14 37·9	13 55·6	2·2	2·1	8·2	8·0	14·2	13·8
23	14 35·8	14 38·1	13 55·9	2·3	2·2	8·3	8·1	14·3	13·9
24	14 36·0	14 38·4	13 56·1	2·4	2·3	8·4	8·2	14·4	14·0
25	14 36·3	14 38·6	13 56·3	2·5	2·4	8·5	8·3	14·5	14·1
26	14 36·5	14 38·9	13 56·6	2·6	2·5	8·6	8·4	14·6	14·2
27	14 36·8	14 39·2	13 56·8	2·7	2·6	8·7	8·5	14·7	14·3
28	14 37·0	14 39·4	13 57·0	2·8	2·7	8·8	8·6	14·8	14·4
29	14 37·3	14 39·7	13 57·3	2·9	2·8	8·9	8·7	14·9	14·5
30	14 37·5	14 39·9	13 57·5	3·0	2·9	9·0	8·8	15·0	14·6
31	14 37·8	14 40·2	13 57·8	3·1	3·0	9·1	8·9	15·1	14·7
32	14 38·0	14 40·4	13 58·0	3·2	3·1	9·2	9·0	15·2	14·8
33	14 38·3	14 40·7	13 58·2	3·3	3·2	9·3	9·1	15·3	14·9
34	14 38·5	14 40·9	13 58·5	3·4	3·3	9·4	9·2	15·4	15·0
35	14 38·8	14 41·2	13 58·7	3·5	3·4	9·5	9·3	15·5	15·1
36	14 39·0	14 41·4	13 59·0	3·6	3·5	9·6	9·4	15·6	15·2
37	14 39·3	14 41·7	13 59·2	3·7	3·6	9·7	9·5	15·7	15·3
38	14 39·5	14 41·9	13 59·4	3·8	3·7	9·8	9·6	15·8	15·4
39	14 39·8	14 42·2	13 59·7	3·9	3·8	9·9	9·7	15·9	15·5
40	14 40·0	14 42·4	13 59·9	4·0	3·9	10·0	9·8	16·0	15·6
41	14 40·3	14 42·7	14 00·1	4·1	4·0	10·1	9·9	16·1	15·7
42	14 40·5	14 42·9	14 00·4	4·2	4·1	10·2	9·9	16·2	15·8
43	14 40·8	14 43·2	14 00·6	4·3	4·2	10·3	10·0	16·3	15·9
44	14 41·0	14 43·4	14 00·9	4·4	4·3	10·4	10·1	16·4	16·0
45	14 41·3	14 43·7	14 01·1	4·5	4·4	10·5	10·2	16·5	16·1
46	14 41·5	14 43·9	14 01·3	4·6	4·5	10·6	10·3	16·6	16·2
47	14 41·8	14 44·2	14 01·6	4·7	4·6	10·7	10·4	16·7	16·3
48	14 42·0	14 44·4	14 01·8	4·8	4·7	10·8	10·5	16·8	16·4
49	14 42·3	14 44·7	14 02·1	4·9	4·8	10·9	10·6	16·9	16·5
50	14 42·5	14 44·9	14 02·3	5·0	4·9	11·0	10·7	17·0	16·6
51	14 42·8	14 45·2	14 02·5	5·1	5·0	11·1	10·8	17·1	16·7
52	14 43·0	14 45·4	14 02·8	5·2	5·1	11·2	10·9	17·2	16·8
53	14 43·3	14 45·7	14 03·0	5·3	5·2	11·3	11·0	17·3	16·9
54	14 43·5	14 45·9	14 03·3	5·4	5·3	11·4	11·1	17·4	17·0
55	14 43·8	14 46·2	14 03·5	5·5	5·4	11·5	11·2	17·5	17·1
56	14 44·0	14 46·4	14 03·7	5·6	5·5	11·6	11·3	17·6	17·2
57	14 44·3	14 46·7	14 04·0	5·7	5·6	11·7	11·4	17·7	17·3
58	14 44·5	14 46·9	14 04·2	5·8	5·7	11·8	11·5	17·8	17·4
59	14 44·8	14 47·2	14 04·4	5·9	5·8	11·9	11·6	17·9	17·5
60	14 45·0	14 47·4	14 04·7	6·0	5·9	12·0	11·7	18·0	17·6

59m

s	SUN PLANETS	ARIES	MOON	v or d	Corr^n	v or d	Corr^n	v or d	Corr^n
00	14 45·0	14 47·4	14 04·7	0·0	0·0	6·0	6·0	12·0	11·9
01	14 45·3	14 47·7	14 04·9	0·1	0·1	6·1	6·0	12·1	12·0
02	14 45·5	14 47·9	14 05·2	0·2	0·2	6·2	6·1	12·2	12·1
03	14 45·8	14 48·2	14 05·4	0·3	0·3	6·3	6·2	12·3	12·2
04	14 46·0	14 48·4	14 05·6	0·4	0·4	6·4	6·3	12·4	12·3
05	14 46·3	14 48·7	14 05·9	0·5	0·5	6·5	6·4	12·5	12·4
06	14 46·5	14 48·9	14 06·1	0·6	0·6	6·6	6·5	12·6	12·5
07	14 46·8	14 49·2	14 06·4	0·7	0·7	6·7	6·6	12·7	12·6
08	14 47·0	14 49·4	14 06·6	0·8	0·8	6·8	6·7	12·8	12·7
09	14 47·3	14 49·7	14 06·8	0·9	0·9	6·9	6·8	12·9	12·8
10	14 47·5	14 49·9	14 07·1	1·0	1·0	7·0	6·9	13·0	12·9
11	14 47·8	14 50·2	14 07·3	1·1	1·1	7·1	7·0	13·1	13·0
12	14 48·0	14 50·4	14 07·5	1·2	1·2	7·2	7·1	13·2	13·1
13	14 48·3	14 50·7	14 07·8	1·3	1·3	7·3	7·2	13·3	13·2
14	14 48·5	14 50·9	14 08·0	1·4	1·4	7·4	7·3	13·4	13·3
15	14 48·8	14 51·2	14 08·3	1·5	1·5	7·5	7·4	13·5	13·4
16	14 49·0	14 51·4	14 08·5	1·6	1·6	7·6	7·5	13·6	13·5
17	14 49·3	14 51·7	14 08·7	1·7	1·7	7·7	7·6	13·7	13·6
18	14 49·5	14 51·9	14 09·0	1·8	1·8	7·8	7·7	13·8	13·7
19	14 49·8	14 52·2	14 09·2	1·9	1·9	7·9	7·8	13·9	13·8
20	14 50·0	14 52·4	14 09·5	2·0	2·0	8·0	7·9	14·0	13·9
21	14 50·3	14 52·7	14 09·7	2·1	2·1	8·1	8·0	14·1	14·0
22	14 50·5	14 52·9	14 09·9	2·2	2·2	8·2	8·1	14·2	14·1
23	14 50·8	14 53·2	14 10·2	2·3	2·3	8·3	8·2	14·3	14·2
24	14 51·0	14 53·4	14 10·4	2·4	2·4	8·4	8·3	14·4	14·3
25	14 51·3	14 53·7	14 10·6	2·5	2·5	8·5	8·4	14·5	14·4
26	14 51·5	14 53·9	14 10·9	2·6	2·6	8·6	8·5	14·6	14·5
27	14 51·8	14 54·2	14 11·1	2·7	2·7	8·7	8·6	14·7	14·6
28	14 52·0	14 54·4	14 11·4	2·8	2·8	8·8	8·7	14·8	14·7
29	14 52·3	14 54·7	14 11·6	2·9	2·9	8·9	8·8	14·9	14·8
30	14 52·5	14 54·9	14 11·8	3·0	3·0	9·0	8·9	15·0	14·9
31	14 52·8	14 55·2	14 12·1	3·1	3·1	9·1	9·0	15·1	15·0
32	14 53·0	14 55·4	14 12·3	3·2	3·2	9·2	9·1	15·2	15·1
33	14 53·3	14 55·7	14 12·6	3·3	3·3	9·3	9·2	15·3	15·2
34	14 53·5	14 55·9	14 12·8	3·4	3·4	9·4	9·3	15·4	15·3
35	14 53·8	14 56·2	14 13·0	3·5	3·5	9·5	9·4	15·5	15·4
36	14 54·0	14 56·4	14 13·3	3·6	3·6	9·6	9·5	15·6	15·5
37	14 54·3	14 56·7	14 13·5	3·7	3·7	9·7	9·6	15·7	15·6
38	14 54·5	14 56·9	14 13·8	3·8	3·8	9·8	9·7	15·8	15·7
39	14 54·8	14 57·2	14 14·0	3·9	3·9	9·9	9·8	15·9	15·8
40	14 55·0	14 57·5	14 14·2	4·0	4·0	10·0	9·9	16·0	15·9
41	14 55·3	14 57·7	14 14·5	4·1	4·1	10·1	10·0	16·1	16·0
42	14 55·5	14 58·0	14 14·7	4·2	4·2	10·2	10·1	16·2	16·1
43	14 55·8	14 58·2	14 14·9	4·3	4·3	10·3	10·2	16·3	16·2
44	14 56·0	14 58·5	14 15·2	4·4	4·4	10·4	10·3	16·4	16·3
45	14 56·3	14 58·7	14 15·4	4·5	4·5	10·5	10·4	16·5	16·4
46	14 56·5	14 59·0	14 15·7	4·6	4·6	10·6	10·5	16·6	16·5
47	14 56·8	14 59·2	14 15·9	4·7	4·7	10·7	10·6	16·7	16·6
48	14 57·0	14 59·5	14 16·1	4·8	4·8	10·8	10·7	16·8	16·7
49	14 57·3	14 59·7	14 16·4	4·9	4·9	10·9	10·8	16·9	16·8
50	14 57·5	15 00·0	14 16·6	5·0	5·0	11·0	10·9	17·0	16·9
51	14 57·8	15 00·2	14 16·9	5·1	5·1	11·1	11·0	17·1	17·0
52	14 58·0	15 00·5	14 17·1	5·2	5·2	11·2	11·1	17·2	17·1
53	14 58·3	15 00·7	14 17·3	5·3	5·3	11·3	11·2	17·3	17·2
54	14 58·5	15 01·0	14 17·6	5·4	5·4	11·4	11·3	17·4	17·3
55	14 58·8	15 01·2	14 17·8	5·5	5·5	11·5	11·4	17·5	17·4
56	14 59·0	15 01·5	14 18·0	5·6	5·6	11·6	11·5	17·6	17·5
57	14 59·3	15 01·7	14 18·3	5·7	5·7	11·7	11·6	17·7	17·6
58	14 59·5	15 02·0	14 18·5	5·8	5·8	11·8	11·7	17·8	17·7
59	14 59·8	15 02·2	14 18·8	5·9	5·9	11·9	11·8	17·9	17·8
60	15 00·0	15 02·5	14 19·0	6·0	6·0	12·0	11·9	18·0	17·9

© 2009 Jack Case

TABLE 6 — ALTITUDE CORRECTION TABLES

a. Dip of the Horizon

Ht. of Eye (m)	Corr'n D	Ht. of Eye (ft.)	Ht. of Eye (m)	Corr'n D	Ht. of Eye (ft.)
0·00	-0·3	0·0	13·0	-6·4	42·8
0·03	-0·4	0·1	13·4	-6·5	44·2
0·06	-0·5	0·2	13·8	-6·6	45·5
0·09	-0·6	0·3	14·2	-6·7	47·0
0·13	-0·7	0·4	14·7	-6·8	48·4
0·18	-0·8	0·5	15·1	-6·9	49·8
0·23	-0·9	0·7	15·5	-7·0	51·3
0·29	-1·0	0·9	16·0	-7·1	52·8
0·35	-1·1	1·1	16·5	-7·2	54·3
0·42	-1·2	1·4	16·9	-7·3	55·8
0·45	-1·3	1·6	17·4	-7·4	57·4
0·5	-1·4	1·9	17·9	-7·5	58·9
0·6	-1·5	2·2	18·4	-7·6	60·5
0·7	-1·6	2·5	18·8	-7·7	62·1
0·8	-1·7	2·8	19·3	-7·8	63·8
0·9	-1·8	3·2	19·8	-7·9	65·4
1·1	-1·9	3·6	20·4	-8·0	67·1
1·2	-2·0	4·0	20·9	-8·1	68·8
1·3	-2·1	4·4	21·4	-8·2	70·5
1·4	-2·2	4·9	21·9	-8·3	72·3
1·6	-2·3	5·3	22·5	-8·4	74·1
1·7	-2·4	5·8	23·0	-8·5	75·8
1·9	-2·5	6·3	23·5	-8·6	77·6
2·0	-2·6	6·9	24·1	-8·7	79·5
2·2	-2·7	7·4	24·7	-8·8	81·3
2·4	-2·8	8·0	25·2	-8·9	83·2
2·6	-2·9	8·6	25·8	-9·0	85·1
2·8	-3·0	9·2	26·4	-9·1	87·0
3·0	-3·1	9·8	27·0	-9·2	88·9
3·2	-3·2	10·5	27·6	-9·3	90·9
3·4	-3·3	11·2	28·2	-9·4	92·9
3·6	-3·4	11·9	28·8	-9·5	94·9
3·8	-3·5	12·6	29·4	-9·6	96·9
4·0	-3·6	13·3	30·0	-9·7	98·9
4·3	-3·7	14·1	30·6	-9·8	101·0
4·5	-3·8	14·9	31·3	-9·9	103·1
4·7	-3·9	15·7	31·9	-10·0	105·2
5·0	-4·0	16·5	32·6	-10·1	107·3
5·2	-4·1	17·4	33·2	-10·2	109·4
5·5	-4·2	18·3	33·9	-10·3	111·6
5·8	-4·3	19·1	34·5	-10·4	113·8
6·1	-4·4	20·1	35·2	-10·5	116·0
6·3	-4·5	21·0	35·9	-10·6	118·2
6·6	-4·6	22·0	36·6	-10·7	120·5
6·9	-4·7	22·9	37·3	-10·8	122·8
7·2	-4·8	23·9	38·0	-10·9	125·1
7·5	-4·9	25·0	38·7	-11·0	127·4
7·9	-5·0	26·0	39·4	-11·1	129·7
8·2	-5·1	27·1	40·1	-11·2	132·1
8·5	-5·2	28·1	40·8	-11·3	134·5
8·8	-5·3	29·2	41·5	-11·4	136·9
9·2	-5·4	30·4	42·3	-11·5	139·3
9·5	-5·5	31·5	43·0	-11·6	141·7
9·9	-5·6	32·7	43·8	-11·7	144·2
10·3	-5·7	33·9	44·5	-11·8	146·7
10·6	-5·8	35·1	45·3	-11·9	149·2
11·0	-5·9	36·3	46·1	-12·0	151·7
11·4	-6·0	37·6	46·8	-12·1	154·3
11·8	-6·1	38·9	47·6	-12·2	156·8
12·2	-6·2	40·1	48·4	-12·3	159·4
12·6	-6·3	41·5	49·2	-12·4	162·1
13·0		42·8	50·0		164·7

b. Refraction for Stars & Planets

App. Alt. H_a	Corr'n R	App. Alt. H_a	Corr'n R	App. Alt. H_a	Corr'n R
0 00	-33·8	3 30	-12·9	9 55	-5·3
0 03	33·2	3 35	12·7	10 07	-5·2
0 06	32·6	3 40	12·5	10 20	-5·1
0 09	32·0	3 45	12·3	10 32	-5·0
0 12	31·5	3 50	12·1	10 46	-4·9
0 15	30·9	3 55	11·9	10 59	-4·8
0 18	-30·4	4 00	-11·7	11 14	-4·7
0 21	29·8	4 05	11·5	11 29	-4·6
0 24	29·3	4 10	11·4	11 44	-4·5
0 27	28·8	4 15	11·2	12 00	-4·4
0 30	28·3	4 20	11·0	12 17	-4·3
0 33	27·9	4 25	10·9	12 35	-4·2
0 36	-27·4	4 30	-10·7	12 53	-4·1
0 39	26·9	4 35	10·6	13 12	-4·0
0 42	26·5	4 40	10·4	13 32	-3·9
0 45	26·1	4 45	10·3	13 53	-3·8
0 48	25·7	4 50	10·1	14 16	-3·7
0 51	25·3	4 55	10·0	14 39	-3·6
0 54	-24·9	5 00	-9·8	15 03	-3·5
0 57	24·5	5 05	9·7	15 29	-3·4
1 00	24·1	5 10	9·6	15 56	-3·3
1 03	23·7	5 15	9·5	16 25	-3·2
1 06	23·4	5 20	9·3	16 55	-3·1
1 09	23·0	5 25	9·2	17 27	-3·0
1 12	-22·7	5 30	-9·1	18 01	-2·9
1 15	22·3	5 35	9·0	18 37	-2·8
1 18	22·0	5 40	8·9	19 16	-2·7
1 21	21·7	5 45	8·8	19 56	-2·6
1 24	21·4	5 50	8·7	20 40	-2·5
1 27	21·1	5 55	8·6	21 27	-2·4
1 30	-20·8	6 00	-8·5	22 17	-2·3
1 35	20·3	6 10	8·3	23 11	-2·2
1 40	19·9	6 20	8·1	24 09	-2·1
1 45	19·4	6 30	7·9	25 12	-2·0
1 50	19·0	6 40	7·7	26 20	-1·9
1 55	18·6	6 50	7·6	27 34	-1·8
2 00	-18·2	7 00	-7·4	28 54	-1·7
2 05	17·8	7 10	7·2	30 22	-1·6
2 10	17·4	7 20	7·1	31 58	-1·5
2 15	17·1	7 30	6·9	33 43	-1·4
2 20	16·7	7 40	6·8	35 38	-1·3
2 25	16·4	7 50	6·7	37 45	-1·2
2 30	-16·1	8 00	-6·6	40 06	-1·1
2 35	15·8	8 10	6·4	42 42	-1·0
2 40	15·4	8 20	6·3	45 34	-0·9
2 45	15·2	8 30	6·2	48 45	-0·8
2 50	14·9	8 40	6·1	52 16	-0·7
2 55	14·6	8 50	6·0	56 09	-0·6
3 00	-14·3	9 00	-5·9	60 26	-0·5
3 05	14·1	9 10	5·8	65 06	-0·4
3 10	13·8	9 20	5·7	70 09	-0·3
3 15	13·6	9 30	5·6	75 32	-0·2
3 20	13·4	9 40	5·5	81 12	-0·1
3 25	13·1	9 50	5·4	87 03	0·0
3 30	-12·9	10 00	-5·3	90 00	

H_a = App. Alt. = Apparent Altitude
   = Sextant altitude corrected for index error (IE) & dip (
   = H_s + IE + D

In critical cases ascend.

TABLE 6 — ALTITUDE CORRECTION TABLES

c. Additional Refraction for Non-standard Conditions

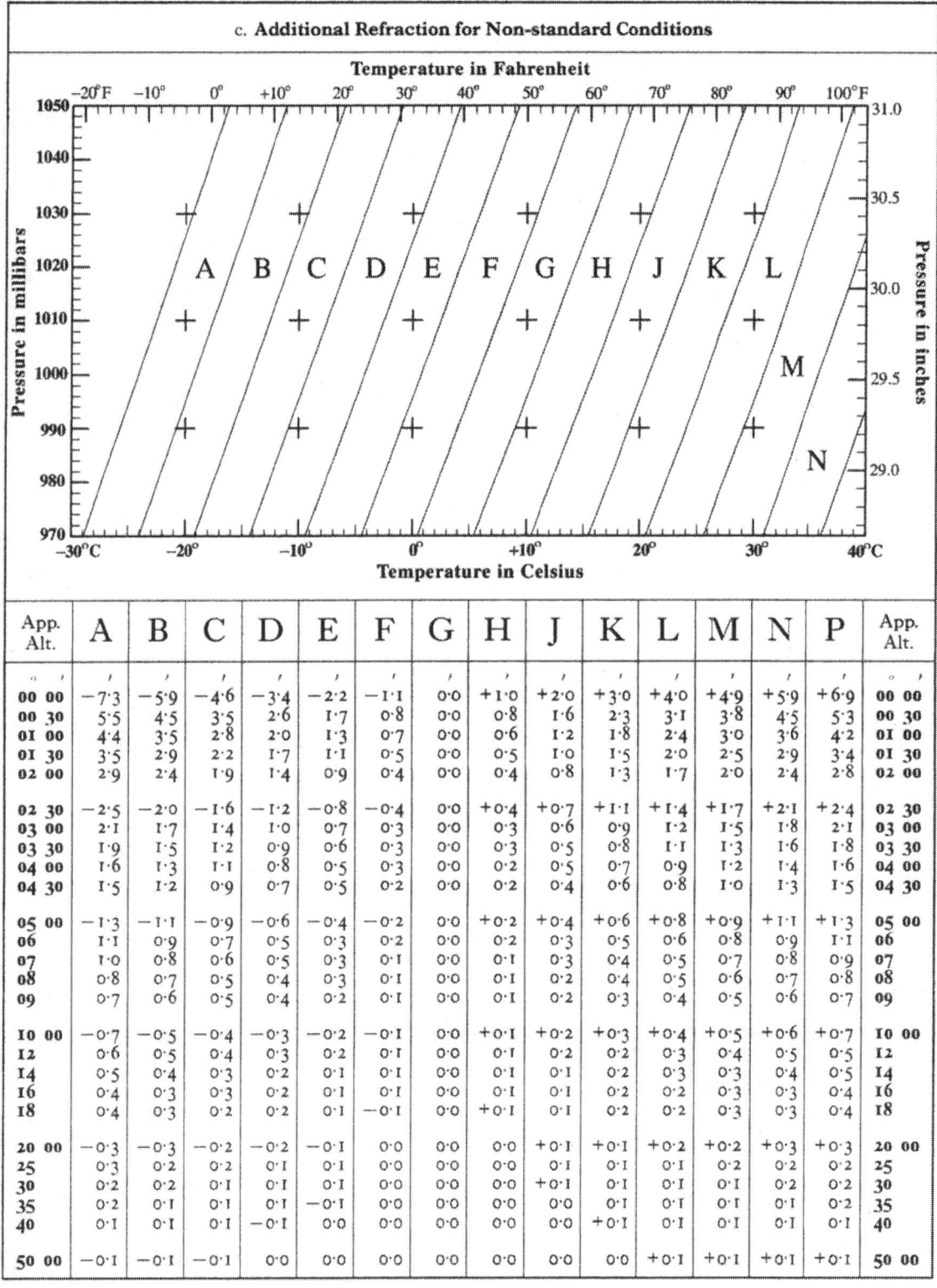

App. Alt.	A	B	C	D	E	F	G	H	J	K	L	M	N	P	App. Alt.
° ′	′	′	′	′	′	′	′	′	′	′	′	′	′	′	° ′
00 00	−7·3	−5·9	−4·6	−3·4	−2·2	−1·1	0·0	+1·0	+2·0	+3·0	+4·0	+4·9	+5·9	+6·9	00 00
00 30	5·5	4·5	3·5	2·6	1·7	0·8	0·0	0·8	1·6	2·3	3·1	3·8	4·5	5·3	00 30
01 00	4·4	3·5	2·8	2·0	1·3	0·7	0·0	0·6	1·2	1·8	2·4	3·0	3·6	4·2	01 00
01 30	3·5	2·9	2·2	1·7	1·1	0·5	0·0	0·5	1·0	1·5	2·0	2·5	2·9	3·4	01 30
02 00	2·9	2·4	1·9	1·4	0·9	0·4	0·0	0·4	0·8	1·3	1·7	2·0	2·4	2·8	02 00
02 30	−2·5	−2·0	−1·6	−1·2	−0·8	−0·4	0·0	+0·4	+0·7	+1·1	+1·4	+1·7	+2·1	+2·4	02 30
03 00	2·1	1·7	1·4	1·0	0·7	0·3	0·0	0·3	0·6	0·9	1·2	1·5	1·8	2·1	03 00
03 30	1·9	1·5	1·2	0·9	0·6	0·3	0·0	0·3	0·5	0·8	1·1	1·3	1·6	1·8	03 30
04 00	1·6	1·3	1·1	0·8	0·5	0·3	0·0	0·2	0·5	0·7	0·9	1·2	1·4	1·6	04 00
04 30	1·5	1·2	0·9	0·7	0·5	0·2	0·0	0·2	0·4	0·6	0·8	1·0	1·3	1·5	04 30
05 00	−1·3	−1·1	−0·9	−0·6	−0·4	−0·2	0·0	+0·2	+0·4	+0·6	+0·8	+0·9	+1·1	+1·3	05 00
06	1·1	0·9	0·7	0·5	0·3	0·2	0·0	0·2	0·3	0·5	0·6	0·8	0·9	1·1	06
07	1·0	0·8	0·6	0·5	0·3	0·1	0·0	0·1	0·3	0·4	0·5	0·7	0·8	0·9	07
08	0·8	0·7	0·5	0·4	0·3	0·1	0·0	0·1	0·2	0·4	0·5	0·6	0·7	0·8	08
09	0·7	0·6	0·5	0·4	0·2	0·1	0·0	0·1	0·2	0·3	0·4	0·5	0·6	0·7	09
10 00	−0·7	−0·5	−0·4	−0·3	−0·2	−0·1	0·0	+0·1	+0·2	+0·3	+0·4	+0·5	+0·6	+0·7	10 00
12	0·6	0·5	0·4	0·3	0·2	0·1	0·0	0·1	0·2	0·2	0·3	0·4	0·5	0·5	12
14	0·5	0·4	0·3	0·2	0·1	0·1	0·0	0·1	0·1	0·2	0·3	0·3	0·4	0·5	14
16	0·4	0·3	0·3	0·2	0·1	0·1	0·0	0·1	0·1	0·2	0·2	0·3	0·3	0·4	16
18	0·4	0·3	0·2	0·2	0·1	−0·1	0·0	+0·1	0·1	0·2	0·2	0·3	0·3	0·4	18
20 00	−0·3	−0·3	−0·2	−0·2	−0·1	0·0	0·0	0·0	+0·1	+0·1	+0·2	+0·2	+0·3	+0·3	20 00
25	0·3	0·2	0·2	0·1	0·1	0·0	0·0	0·0	0·1	0·1	0·1	0·2	0·2	0·2	25
30	0·2	0·2	0·1	0·1	0·1	0·0	0·0	+0·1	0·1	0·1	0·1	0·1	0·2	0·2	30
35	0·2	0·1	0·1	0·1	−0·1	0·0	0·0	0·0	0·1	0·1	0·1	0·1	0·1	0·2	35
40	0·1	0·1	0·1	−0·1	0·0	0·0	0·0	0·0	+0·1	0·1	0·1	0·1	0·1	0·1	40
50 00	−0·1	−0·1	−0·1	0·0	0·0	0·0	0·0	0·0	0·0	+0·1	+0·1	+0·1	+0·1		50 00

The graph is entered with arguments temperature and pressure to find a zone letter; using as arguments this zone letter and apparent altitude (sextant altitude corrected for index error and dip), a correction is taken from the table. This correction is to be applied to the sextant altitude in addition to the corrections for standard conditions (Table **6**b).

ALTITUDE CORRECTION TABLES 0°–35°— MOON

App. Alt.	0°–4° Corrⁿ	5°–9° Corrⁿ	10°–14° Corrⁿ	15°–19° Corrⁿ	20°–24° Corrⁿ	25°–29° Corrⁿ	30°–34° Corrⁿ	App. Alt.
00	0° 34.5	5° 58.2	10° 62.1	15° 62.8	20° 62.2	25° 60.8	30° 58.9	00
10	36.5	58.5	62.2	62.8	62.2	60.8	58.8	10
20	38.3	58.7	62.2	62.8	62.1	60.7	58.8	20
30	40.0	58.9	62.3	62.8	62.1	60.7	58.7	30
40	41.5	59.1	62.3	62.8	62.0	60.6	58.6	40
50	42.9	59.3	62.4	62.7	62.0	60.6	58.5	50
00	1° 44.2	6° 59.5	11° 62.4	16° 62.7	21° 62.0	26° 60.5	31° 58.5	00
10	45.4	59.7	62.4	62.7	61.9	60.4	58.4	10
20	46.5	59.9	62.5	62.7	61.9	60.4	58.3	20
30	47.5	60.0	62.5	62.7	61.9	60.3	58.2	30
40	48.4	60.2	62.5	62.7	61.8	60.3	58.2	40
50	49.3	60.3	62.6	62.7	61.8	60.2	58.1	50
00	2° 50.1	7° 60.5	12° 62.6	17° 62.7	22° 61.7	27° 60.1	32° 58.0	00
10	50.8	60.6	62.6	62.6	61.7	60.1	57.9	10
20	51.5	60.7	62.6	62.6	61.6	60.0	57.8	20
30	52.2	60.9	62.7	62.6	61.6	59.9	57.8	30
40	52.8	61.0	62.7	62.6	61.6	59.9	57.7	40
50	53.4	61.1	62.7	62.6	61.5	59.8	57.6	50
00	3° 53.9	8° 61.2	13° 62.7	18° 62.5	23° 61.5	28° 59.7	33° 57.5	00
10	54.4	61.3	62.7	62.5	61.4	59.7	57.4	10
20	54.9	61.4	62.7	62.5	61.4	59.6	57.4	20
30	55.3	61.5	62.8	62.5	61.3	59.5	57.3	30
40	55.7	61.6	62.8	62.4	61.3	59.5	57.2	40
50	56.1	61.6	62.8	62.4	61.2	59.4	57.1	50
00	4° 56.4	9° 61.7	14° 62.8	19° 62.4	24° 61.2	29° 59.3	34° 57.0	00
10	56.8	61.8	62.8	62.4	61.1	59.3	56.9	10
20	57.1	61.9	62.8	62.3	61.1	59.2	56.9	20
30	57.4	61.9	62.8	62.3	61.0	59.1	56.8	30
40	57.7	62.0	62.8	62.3	61.0	59.1	56.7	40
50	58.0	62.1	62.8	62.2	60.9	59.0	56.6	50

HP	L U	L U	L U	L U	L U	L U	L U	HP
54.0	0.3 0.9	0.3 0.9	0.4 1.0	0.5 1.1	0.6 1.2	0.7 1.3	0.9 1.5	54.0
54.3	0.7 1.1	0.7 1.2	0.8 1.2	0.8 1.3	0.9 1.4	1.1 1.5	1.2 1.7	54.3
54.6	1.1 1.4	1.1 1.4	1.1 1.4	1.2 1.5	1.3 1.6	1.4 1.7	1.5 1.8	54.6
54.9	1.4 1.6	1.5 1.6	1.5 1.6	1.6 1.7	1.6 1.8	1.8 1.9	1.9 2.0	54.9
55.2	1.8 1.8	1.8 1.8	1.9 1.8	1.9 1.9	2.0 2.0	2.1 2.1	2.2 2.2	55.2
55.5	2.2 2.0	2.2 2.0	2.3 2.1	2.3 2.1	2.4 2.2	2.4 2.3	2.5 2.4	55.5
55.8	2.6 2.2	2.6 2.2	2.6 2.3	2.7 2.3	2.7 2.4	2.8 2.4	2.9 2.5	55.8
56.1	3.0 2.4	3.0 2.5	3.0 2.5	3.0 2.5	3.1 2.6	3.1 2.6	3.2 2.7	56.1
56.4	3.3 2.7	3.4 2.7	3.4 2.7	3.4 2.7	3.4 2.8	3.5 2.8	3.5 2.9	56.4
56.7	3.7 2.9	3.7 2.9	3.8 2.9	3.8 2.9	3.8 3.0	3.8 3.0	3.9 3.0	56.7
57.0	4.1 3.1	4.1 3.1	4.1 3.1	4.1 3.1	4.2 3.2	4.2 3.2	4.2 3.2	57.0
57.3	4.5 3.3	4.5 3.3	4.5 3.3	4.5 3.3	4.5 3.3	4.5 3.4	4.6 3.4	57.3
57.6	4.9 3.5	4.9 3.5	4.9 3.5	4.9 3.5	4.9 3.5	4.9 3.5	4.9 3.6	57.6
57.9	5.3 3.8	5.3 3.8	5.2 3.8	5.2 3.7	5.2 3.7	5.2 3.7	5.2 3.7	57.9
58.2	5.6 4.0	5.6 4.0	5.6 4.0	5.6 4.0	5.6 3.9	5.6 3.9	5.6 3.9	58.2
58.5	6.0 4.2	6.0 4.2	6.0 4.2	6.0 4.2	6.0 4.1	5.9 4.1	5.9 4.1	58.5
58.8	6.4 4.4	6.4 4.4	6.4 4.4	6.3 4.4	6.3 4.3	6.3 4.3	6.2 4.2	58.8
59.1	6.8 4.6	6.8 4.6	6.7 4.6	6.7 4.6	6.7 4.5	6.6 4.5	6.6 4.4	59.1
59.4	7.2 4.8	7.1 4.8	7.1 4.8	7.1 4.8	7.0 4.7	7.0 4.7	6.9 4.6	59.4
59.7	7.5 5.1	7.5 5.0	7.5 5.0	7.5 5.0	7.4 4.9	7.3 4.8	7.2 4.8	59.7
60.0	7.9 5.3	7.9 5.3	7.9 5.2	7.8 5.2	7.8 5.1	7.7 5.0	7.6 4.9	60.0
60.3	8.3 5.5	8.3 5.5	8.2 5.4	8.2 5.4	8.1 5.3	8.0 5.2	7.9 5.1	60.3
60.6	8.7 5.7	8.7 5.7	8.6 5.7	8.6 5.6	8.5 5.5	8.4 5.4	8.2 5.3	60.6
60.9	9.1 5.9	9.0 5.9	9.0 5.9	8.9 5.8	8.8 5.7	8.7 5.6	8.6 5.4	60.9
61.2	9.5 6.2	9.4 6.1	9.4 6.1	9.3 6.0	9.2 5.9	9.1 5.8	8.9 5.6	61.2
61.5	9.8 6.4	9.8 6.3	9.7 6.3	9.7 6.2	9.5 6.1	9.4 5.9	9.2 5.8	61.5

DIP

Ht. of Eye (m)	Corrⁿ	Ht. of Eye (ft.)	Ht. of Eye (m)	Corrⁿ	Ht. of Eye (ft.)
2.4	−2.8	8.0	9.5	−5.5	31.5
2.6	−2.9	8.6	9.9	−5.6	32.7
2.8	−3.0	9.2	10.3	−5.7	33.9
3.0	−3.1	9.8	10.6	−5.8	35.1
3.2	−3.2	10.5	11.0	−5.9	36.3
3.4	−3.3	11.2	11.4	−6.0	37.6
3.6	−3.4	11.9	11.8	−6.1	38.9
3.8	−3.5	12.6	12.2	−6.2	40.1
4.0	−3.6	13.3	12.6	−6.3	41.5
4.3	−3.7	14.1	13.0	−6.4	42.8
4.5	−3.8	14.9	13.4	−6.5	44.2
4.7	−3.9	15.7	13.8	−6.6	45.5
5.0	−4.0	16.5	14.2	−6.7	46.9
5.2	−4.1	17.4	14.7	−6.8	48.4
5.5	−4.2	18.3	15.1	−6.9	49.8
5.8	−4.3	19.1	15.5	−7.0	51.3
6.1	−4.4	20.1	16.0	−7.1	52.8
6.3	−4.5	21.0	16.5	−7.2	54.3
6.6	−4.6	22.0	16.9	−7.3	55.8
6.9	−4.7	22.9	17.4	−7.4	57.4
7.2	−4.8	23.9	17.9	−7.5	58.9
7.5	−4.9	24.9	18.4	−7.6	60.5
7.9	−5.0	26.0	18.8	−7.7	62.1
8.2	−5.1	27.1	19.3	−7.8	63.8
8.5	−5.2	28.1	19.8	−7.9	65.4
8.8	−5.3	29.2	20.4	−8.0	67.1
9.2	−5.4	30.4	20.9	−8.1	68.8
9.5		31.5	21.4		70.5

MOON CORRECTION TABLE

The correction is in two parts; the first correction is taken from the upper part of the table with argument apparent altitude, and the second from the lower part, with argument HP, in the same column as that from which the first correction was taken. Separate corrections are given in the lower part for lower (L) and upper (U) limbs. All corrections are to be **added** to apparent altitude, *but 30′ is to be subtracted from the altitude of the upper limb.*

For corrections for pressure and temperature see page A4.

For bubble sextant observations ignore dip, take the mean of upper and lower limb corrections and subtract 15′ from the altitude.

App. Alt. = Apparent altitude = Sextant altitude corrected for index error and dip.

Extract from main table for Lat 48° (contrary), Dec 26°, LHA 348°

24°			25°			26°			27°			28°			29°			LHA
Hc	d	Z	Hc	d	Z	Hc	d	Z	Hc	d	Z	Hc	d	Z	Hc	d	Z	
° ′	′	°	° ′	′	°	° ′	′	°	° ′	′	°	° ′	′	°	° ′	′	°	°
-4 46	47	121	-5 33	47	122	-6 20	-46	122	291									
-4 12	47	122	-4 59	47	122	-5 46	47	123	292									
-3 38	47	123	-4 25	48	123	-5 13	47	124	-6 00	-47	124	293						
-3 04	48	123	-3 52	47	124	-4 39	48	125	-5 27	47	125	-6 14	-47	126	294			
-2 31	-48	124	-3 19	-47	125	-4 06	-48	125	-4 54	-48	126	-5 42	-47	126	295			
-1 58	48	125	-2 46	48	125	-3 34	48	126	-4 22	47	127	-5 09	48	127	-5 57	-48	128	296
-1 25	48	125	-2 13	48	126	-3 01	49	127	-3 50	48	127	-4 38	48	128	-5 26	48	128	297
-0 53	48	126	-1 41	48	127	-2 29	49	127	-3 18	48	128	-4 06	48	129	-4 54	49	129	298
-0 20	49	127	-1 09	49	128	-1 58	48	128	-2 46	49	129	-3 35	48	129	-4 23	49	130	299
00 12	-49	128	-0 37	-49	128	-1 26	-49	129	-2 15	-49	129	-3 04	-49	130	-3 53	-49	131	300
00 43	49	128	-0 06	49	129	-0 55	49	130	-1 44	49	130	-2 33	49	131	-3 22	50	131	301
01 14	49	129	00 25	49	130	-0 24	50	130	-1 14	49	131	-2 03	49	131	-2 52	50	132	302
01 45	49	130	00 56	50	131	00 06	49	132	-0 43	50	132	-1 33	49	132	-2 23	49	133	303
02 16	50	131	01 26	50	131	00 36	50	132	-0 14	50	132	-1 04	50	133	-1 54	50	133	304
02 46	-50	131	01 56	-50	132	01 06	-50	133	00 16	-50	133	-0 34	-51	134	-1 25	-50	134	305
03 16	50	132	02 26	51	133	01 35	50	133	00 45	51	134	-0 06	50	134	-0 56	50	135	306
03 46	51	133	02 55	51	134	02 04	50	134	01 14	51	135	00 23	51	135	-0 28	50	136	307
04 15	51	134	03 24	51	134	02 33	51	135	01 42	51	135	00 51	51	136	00 00	51	136	308
04 44	51	134	03 52	51	135	03 01	51	136	02 10	51	136	01 19	51	137	00 28	52	137	309
05 12	-51	135	04 21	-52	136	03 29	-51	136	02 38	-52	137	01 46	-51	137	00 55	-52	138	310
05 40	52	136	04 48	51	137	03 57	52	137	03 05	52	138	02 13	52	138	01 21	51	139	311
06 08	52	137	05 16	52	137	04 24	52	138	03 32	52	138	02 40	52	139	01 48	52	139	312
06 35	52	138	05 43	53	138	04 50	52	139	03 58	52	139	03 06	52	140	02 14	53	140	313
07 02	53	139	06 09	52	139	05 17	53	140	04 24	52	140	03 32	53	140	02 39	53	141	314
07 28	-53	139	06 35	-52	140	05 43	-53	140	04 50	-53	141	03 57	-53	141	03 04	-53	142	315
07 54	53	140	07 01	53	141	06 08	53	141	05 15	53	142	04 22	53	142	03 29	53	143	316
08 20	54	141	07 26	53	141	06 33	53	142	05 40	54	142	04 46	53	143	03 53	53	143	317
08 45	54	142	07 51	53	142	06 58	54	143	06 04	54	143	05 10	53	144	04 17	54	144	318
09 09	54	143	08 15	53	143	07 22	54	144	06 28	54	144	05 34	54	144	04 40	54	145	319
09 33	-54	143	08 39	-54	144	07 45	-54	144	06 51	-54	145	05 57	-54	145	05 03	-54	146	320
09 57	54	144	09 03	55	145	08 08	54	145	07 14	54	146	06 20	55	146	05 25	54	146	321
10 20	54	145	09 26	55	146	08 31	54	146	07 37	55	146	06 42	55	147	05 47	54	147	322
10 43	55	146	09 48	55	146	08 53	54	147	07 59	55	147	07 04	55	148	06 09	55	148	323
11 05	55	147	10 10	55	147	09 15	55	148	08 20	55	148	07 25	55	148	06 30	55	149	324
11 27	-55	148	10 32	-56	148	09 36	-55	148	08 41	-55	149	07 46	-56	149	06 50	-55	150	325
11 48	55	149	10 53	56	149	09 57	55	149	09 02	56	150	08 06	56	150	07 10	55	150	326
12 09	56	149	11 13	56	150	10 17	55	150	09 22	56	151	08 26	56	151	07 30	56	151	327
12 29	56	150	11 33	56	151	10 37	56	151	09 41	56	151	08 45	56	152	07 49	56	152	328
12 49	57	151	11 52	56	152	10 56	56	152	10 00	56	152	09 04	56	153	08 08	57	153	329
13 08	-57	152	12 11	-56	152	11 15	-57	153	10 18	-56	153	09 22	-56	153	08 26	-57	154	330
13 26	56	153	12 30	57	153	11 33	57	154	10 36	56	154	09 40	57	154	08 43	57	155	331
13 44	57	154	12 47	56	154	11 51	57	154	10 54	57	155	09 57	57	155	09 00	57	155	332
14 02	57	155	13 05	57	155	12 08	57	155	11 11	57	156	10 14	58	156	09 16	57	156	333
14 19	58	156	13 21	57	156	12 24	57	156	11 27	57	156	10 30	58	157	09 32	57	157	334
14 35	-58	156	13 37	-57	157	12 40	-57	157	11 43	-58	157	10 45	-57	158	09 48	-58	158	335
14 51	58	157	13 53	58	158	12 55	57	158	11 58	58	158	11 00	58	159	10 02	57	159	336
15 06	58	158	14 08	58	159	13 10	58	159	12 12	57	159	11 15	58	159	10 17	58	160	337
15 20	58	159	14 22	58	159	13 24	58	160	12 26	58	160	11 28	58	160	10 30	58	161	338
15 34	58	160	14 36	58	160	13 38	58	161	12 40	58	161	11 42	59	161	10 43	58	161	339
15 48	-59	161	14 49	-58	161	13 51	-58	162	12 53	-59	162	11 54	-58	162	10 56	-59	162	340
16 00	58	162	15 02	59	162	14 03	58	162	13 05	59	163	12 06	58	163	11 08	59	163	341
16 12	58	163	15 14	59	163	14 15	58	163	13 17	59	164	12 18	59	164	11 19	58	164	342
16 24	59	164	15 25	59	164	14 26	58	164	13 28	59	164	12 29	59	165	11 30	59	165	343
16 35	59	165	15 36	59	165	14 37	59	165	13 38	59	165	12 39	59	166	11 40	59	166	344
16 45	-59	166	15 46	-59	166	14 47	-59	166	13 48	-59	166	12 49	-59	166	11 50	-59	167	345
16 55	60	167	15 55	59	167	14 56	59	167	13 57	59	167	12 58	59	167	11 59	59	168	346
17 04	60	168	16 04	59	168	15 05	59	168	14 06	60	168	13 06	59	168	12 07	59	168	347
17 12	60	169	16 12	59	169	15 13	59	169	14 14	60	169	13 14	59	169	12 15	59	169	348
17 19	59	169	16 20	59	170	15 21	60	170	14 21	59	170	13 22	60	170	12 22	59	170	349
17 26	-59	170	16 27	-59	171	15 27	-59	171	14 28	-60	171	13 28	-59	171	12 29	-60	171	350
17 33	60	171	16 33	59	171	15 34	60	172	14 34	60	172	13 34	59	172	12 35	60	172	351
17 39	60	172	16 39	60	172	15 39	60	173	14 39	59	173	13 40	60	173	12 40	60	173	352
17 44	60	173	16 44	60	173	15 44	60	173	14 44	60	174	13 44	59	174	12 45	60	174	353
17 48	60	174	16 48	60	174	15 48	60	174	14 48	59	174	13 49	60	175	12 49	60	175	354
17 52	60	175	16 52	60	175	15 52	60	175	14 52	-60	175	13 52	-60	176	12 52	-60	176	355
17 55	60	176	16 55	60	176	15 55	60	176	14 55	60	176	13 55	60	176	12 55	60	176	356
17 57	60	177	16 57	60	177	15 57	60	177	14 57	60	177	13 57	60	177	12 57	60	177	357
17 59	60	178	16 59	60	178	15 59	60	178	14 59	60	178	13 59	60	178	12 59	60	178	358
18 00	60	179	17 00	60	179	16 00	60	179	15 00	60	179	14 00	60	179	13 00	60	179	359
18 00	-60	180	17 00	-60	180	16 00	-60	180	15 00	-60	180	14 00	-60	180	13 00	-60	180	360

©2009 Jack Case

Extract from main table for Lat 48° (contrary), Dec 16°, LHA 352°

LHA	15° Hc d Z	16° Hc d Z	17° Hc d Z	18° Hc d Z	19° Hc d Z	28° Hc d Z	29° Hc d Z	LHA
69	02 15 47 116	01 28 47 116	00 41 46 117	−0 05 47 117	−0 52 47 118			
68	02 51 47 116	02 04 47 117	01 17 47 118	00 30 47 118	−0 17 47 119			
67	03 27 47 117	02 40 47 118	01 53 48 118	01 05 47 119	00 18 47 119	293		
66	04 03 48 118	03 15 47 118	02 28 48 119	01 40 47 120	00 53 48 120	−6 14 −47 126	294	
65	04 38 −48 119	03 50 −47 119	03 03 −48 120	02 15 −48 120	01 27 −47 121	−5 42 −47 126	295	
64	05 13 48 119	04 25 47 120	03 38 48 121	02 50 48 121	02 02 48 122	−5 09 48 127	−5 57 −48 128	296
63	05 48 48 120	05 00 48 121	04 12 48 121	03 24 48 122	02 36 48 123	−4 38 48 128	−5 26 48 128	297
62	06 23 48 121	05 34 48 121	04 46 48 122	03 58 49 123	03 09 48 123	−4 06 48 129	−4 54 49 129	298
61	06 57 48 122	06 09 49 122	05 20 49 123	04 31 48 123	03 43 49 124	−3 35 48 129	−4 23 49 130	299
60	07 31 −49 122	06 42 −48 123	05 54 −49 124	05 05 −49 124	04 16 −49 125	−3 04 −49 130	−3 53 −49 131	300
59	08 05 49 123	07 16 49 124	06 27 49 124	05 38 49 125	04 49 49 126	−2 33 49 131	−3 22 50 131	301
58	08 38 49 124	07 49 49 125	07 00 49 125	06 11 50 126	05 21 49 126	−2 03 49 131	−2 52 50 132	302
57	09 11 49 125	08 22 49 125	07 32 49 126	06 43 50 127	05 53 49 127	−1 33 50 132	−2 23 49 133	303
56	09 44 50 126	08 54 49 126	08 05 50 127	07 15 50 127	06 25 50 128	−1 04 51 133	−1 54 49 133	304
55	10 17 −50 126	09 27 −50 127	08 37 −50 128	07 47 −50 128	06 57 −50 129	−0 34 −51 134	−1 25 −50 134	305
54	10 49 50 127	09 59 51 128	09 08 50 128	08 18 50 129	07 28 50 130	−0 06 50 134	−0 56 50 135	306
53	11 20 50 128	10 30 50 129	09 40 51 129	08 49 50 130	07 59 51 130	00 23 51 135	−0 28 50 136	307
52	11 52 51 129	11 01 50 129	10 11 51 130	09 20 51 131	08 29 51 131	00 51 51 136	00 00 51 136	308
51	12 23 51 130	11 32 51 130	10 41 51 131	09 50 51 131	08 59 51 132	01 19 51 137	00 28 52 137	309
50	12 54 −52 131	12 02 −51 131	11 11 −51 132	10 20 −51 132	09 29 −51 133	01 46 −51 137	00 55 −51 138	310
49	13 24 52 131	12 32 51 132	11 41 51 133	10 50 52 133	09 58 51 134	02 13 52 138	01 21 51 139	311
48	13 54 52 132	13 02 52 133	12 10 51 133	11 19 52 134	10 27 52 134	02 40 52 139	01 48 52 139	312
47	14 23 52 133	13 31 52 134	12 39 52 134	11 48 52 135	10 56 52 135	03 06 52 140	02 14 53 140	313
46	14 52 52 134	14 00 52 135	13 08 52 135	12 16 52 136	11 24 52 136	03 32 53 140	02 39 53 141	314
45	15 21 −52 135	14 29 −53 135	13 36 −52 136	12 44 −53 136	11 51 −52 137	03 57 −53 141	03 04 −53 142	315
44	15 49 52 136	14 57 53 136	14 04 53 137	13 11 53 137	12 18 52 138	04 22 53 142	03 29 53 143	316
43	16 17 53 137	15 24 53 137	14 31 53 138	13 38 53 138	12 45 53 139	04 46 53 143	03 53 53 143	317
42	16 44 53 138	15 51 53 138	14 58 53 139	14 05 53 139	13 12 54 139	05 10 53 144	04 17 54 144	318
41	17 11 53 138	16 18 54 139	15 24 53 139	14 31 54 140	13 37 53 140	05 34 54 144	04 40 54 145	319
40	17 37 −53 139	16 44 −54 140	15 50 −53 140	14 57 −54 141	14 03 −54 141	05 57 −54 145	05 03 −54 146	320
39	18 03 53 140	17 10 54 141	16 16 54 141	15 22 54 142	14 28 54 142	06 20 55 146	05 25 54 146	321
38	18 29 54 141	17 35 54 142	16 41 55 142	15 46 54 143	14 52 54 143	06 42 55 147	05 47 54 147	322
37	18 54 55 142	17 59 54 143	17 05 54 143	16 11 55 143	15 16 54 144	07 04 55 148	06 09 55 148	323
36	19 18 54 143	18 24 55 143	17 29 55 144	16 34 54 144	15 40 55 145	07 25 55 148	06 30 55 149	324
35	19 42 −55 144	18 47 −55 144	17 52 −55 145	16 57 −55 145	16 02 −55 146	07 46 −56 149	06 50 −55 150	325
34	20 05 55 145	19 10 55 145	18 15 55 146	17 20 55 146	16 25 55 147	08 06 56 150	07 10 55 150	326
33	20 28 55 146	19 33 55 146	18 38 56 147	17 42 55 147	16 47 56 147	08 26 55 151	07 30 56 151	327
32	20 50 55 147	19 55 56 147	18 59 56 148	18 04 56 148	17 08 56 148	08 45 56 152	07 49 56 152	328
31	21 12 56 148	20 16 55 148	19 21 56 149	18 25 56 149	17 29 56 149	09 04 56 153	08 08 57 153	329
30	21 33 −56 149	20 37 −56 149	19 41 −56 149	18 45 −56 150	17 49 −56 150	09 22 −56 153	08 26 −57 154	330
29	21 54 56 150	20 58 57 150	20 01 56 150	19 05 56 151	18 09 57 151	09 40 57 154	08 43 57 155	331
28	22 14 57 151	21 17 56 151	20 21 57 151	19 24 57 152	18 28 57 152	09 57 57 155	09 00 57 155	332
27	22 33 56 152	21 37 57 152	20 40 57 152	19 43 57 153	18 46 57 153	10 14 58 156	09 16 57 156	333
26	22 52 57 153	21 55 57 153	20 58 57 153	20 01 57 154	19 04 57 154	10 30 58 157	09 32 57 157	334
25	23 10 −57 154	22 13 −57 154	21 16 −57 154	20 19 −58 155	19 21 −57 155	10 45 −57 158	09 48 −58 158	335
24	23 28 58 155	22 30 57 155	21 33 57 155	20 36 58 156	19 38 57 156	11 00 58 159	10 02 57 159	336
23	23 44 57 156	22 47 58 156	21 49 57 156	20 52 58 157	19 54 57 157	11 15 58 159	10 17 58 160	337
22	24 01 58 157	23 03 58 157	22 05 58 157	21 07 58 158	20 10 58 158	11 28 58 160	10 30 58 161	338
21	24 16 58 158	23 18 58 158	22 20 58 158	21 22 57 159	20 25 59 159	11 42 59 161	10 43 58 161	339
20	24 31 −58 159	23 33 −58 159	22 35 −58 159	21 37 −58 160	20 39 −58 160	11 54 −58 162	10 56 −58 162	340
19	24 45 58 160	23 47 58 160	22 49 58 160	21 51 59 161	20 52 58 161	12 06 58 163	11 08 59 163	341
18	24 59 58 161	24 01 58 161	23 02 58 161	22 04 59 162	21 05 58 162	12 18 59 164	11 19 59 164	342
17	25 12 59 162	24 13 58 162	23 15 59 162	22 16 59 163	21 17 59 163	12 29 59 165	11 30 59 165	343
16	25 24 59 163	24 25 58 163	23 27 59 163	22 28 59 164	21 29 59 164	12 39 59 166	11 40 59 166	344
15	25 36 −59 164	24 37 −59 164	23 38 −59 164	22 39 −59 165	21 40 −59 165	12 49 −59 166	11 50 −59 167	345
14	25 46 59 165	24 47 59 165	23 48 59 165	22 49 59 166	21 50 59 166	12 58 59 167	11 59 59 168	346
13	25 56 59 166	24 57 59 166	23 58 59 166	22 59 59 167	22 00 60 167	13 06 59 168	12 07 59 168	347
12	26 06 60 167	25 06 59 167	24 07 59 167	23 08 59 168	22 09 60 168	13 14 59 169	12 15 59 169	348
11	26 14 59 168	25 15 59 168	24 16 60 168	23 16 59 169	22 17 60 169	13 22 60 170	12 22 59 170	349
10	26 22 −59 169	25 23 −60 169	24 23 −59 169	23 24 −60 170	22 24 −59 170	13 28 −59 171	12 29 −60 171	350
9	26 29 59 170	25 30 60 170	24 30 59 171	23 31 60 171	22 31 60 171	13 34 59 172	12 35 60 172	351
8	26 36 60 171	25 36 59 172	24 36 59 172	23 37 60 172	22 37 60 172	13 40 60 173	12 40 60 173	352
7	26 41 59 172	25 42 60 173	24 42 60 173	23 42 60 173	22 42 59 173	13 44 60 174	12 45 60 174	353
6	26 46 59 174	25 47 60 174	24 47 60 174	23 47 60 174	22 47 60 174	13 49 60 175	12 49 60 175	354
5	26 51 −60 175	25 51 −60 175	24 51 −60 175	23 51 −60 175	22 51 −60 175	13 52 −60 175	12 52 −60 176	355
4	26 54 60 176	25 54 60 176	24 54 60 176	23 54 60 176	22 54 60 176	13 55 60 176	12 55 60 176	356
3	26 57 60 177	25 57 60 177	24 57 60 177	23 57 60 177	22 57 60 177	13 57 60 177	12 57 60 177	357
2	26 58 59 178	25 59 60 178	24 59 60 178	23 59 60 178	22 59 60 178	13 59 60 178	12 59 60 178	358
1	27 00 60 179	26 00 60 179	25 00 60 179	24 00 60 179	23 00 60 179	14 00 60 179	13 00 60 179	359
0	27 00 −60 180	26 00 −60 180	25 00 −60 180	24 00 −60 180	23 00 −60 180	14 00 −60 180	13 00 −60 180	360

©2009 Jack Case

Extract from main table for Lat 48° (same), Dec 17°, LHA 50°

LHA	15° Hc	d	Z	16° Hc	d	Z	17° Hc	d	Z	18° Hc	d	Z	19° Hc	d	Z	20° Hc	d	Z
0	57 00	+60	180	58 00	+60	180	59 00	+60	180	60 00	+60	180	61 00	+60	180	62 00	+60	180
1	56 59	60	178	57 59	60	178	58 59	60	178	59 59	60	178	60 59	60	178	61 59	60	178
2	56 58	59	176	57 57	60	176	58 57	60	176	59 57	60	176	60 57	60	176	61 57	60	176
3	56 54	60	175	57 54	60	175	58 54	60	174	59 54	60	174	60 54	60	174	61 54	60	174
4	56 50	60	173	57 50	60	173	58 50	59	173	59 49	60	172	60 49	60	172	61 49	60	172
5	56 45	+59	171	57 44	+60	171	58 44	+59	171	59 43	+60	171	60 43	+60	170	61 43	+59	170
6	56 38	59	169	57 37	60	169	58 37	59	169	59 36	60	169	60 36	59	168	61 35	59	168
7	56 30	59	168	57 29	59	167	58 28	60	167	59 28	59	167	60 27	59	166	61 26	59	166
8	56 21	59	166	57 20	59	166	58 19	59	165	59 18	59	165	60 17	59	165	61 16	59	164
9	56 10	59	164	57 09	59	164	58 08	59	164	59 07	59	163	60 06	58	163	61 04	59	162
10	55 59	+59	163	56 58	+58	162	57 56	+59	162	58 55	+58	161	59 53	+58	161	60 51	+59	160
11	55 46	59	161	56 45	58	160	57 43	58	160	58 41	58	160	59 39	58	159	60 37	58	159
12	55 33	58	159	56 31	58	159	57 29	58	158	58 27	57	158	59 24	58	157	60 22	57	157
13	55 18	58	158	56 16	57	157	57 13	58	157	58 11	57	156	59 08	58	156	60 06	57	155
14	55 02	57	156	55 59	58	155	56 57	57	155	57 54	57	154	58 51	57	154	59 48	57	153
15	54 45	+57	154	55 42	+57	154	56 39	+57	153	57 36	+57	153	58 33	+56	152	59 29	+57	151
16	54 27	57	153	55 24	57	152	56 21	56	152	57 17	56	151	58 13	57	150	59 10	55	150
17	54 08	57	151	55 05	56	151	56 01	56	150	56 57	56	149	57 53	56	149	58 49	55	148
18	53 48	57	150	54 45	55	149	55 40	56	148	56 36	56	148	57 32	55	147	58 27	55	146
19	53 28	55	148	54 23	56	147	55 19	55	147	56 14	55	146	57 09	55	145	58 04	55	145
20	53 06	+55	147	54 01	+56	146	54 57	+54	145	55 51	+55	145	56 46	+54	144	57 40	+55	143
21	52 44	55	145	53 39	54	144	54 33	55	144	55 28	54	143	56 22	54	142	57 16	54	141
22	52 20	55	144	53 15	54	143	54 09	54	142	55 03	54	142	55 57	54	141	56 51	53	140
23	51 56	54	142	52 50	54	142	53 44	54	141	54 38	53	140	55 31	53	139	56 24	53	138
24	51 31	54	141	52 25	53	140	53 18	54	139	54 12	53	139	55 05	52	138	55 57	52	137
25	51 05	+54	139	51 59	+53	139	52 52	+53	138	53 45	+52	137	54 37	+52	136	55 29	+52	135
26	50 39	53	138	51 32	53	137	52 25	52	137	53 17	52	136	54 09	52	135	55 01	51	134
27	50 12	52	137	51 04	53	136	51 57	52	135	52 49	51	134	53 40	52	134	54 32	51	133
28	49 44	52	135	50 36	52	135	51 28	52	134	52 20	51	133	53 11	51	132	54 02	50	131
29	49 15	52	134	50 07	52	133	50 59	51	133	51 50	51	132	52 41	50	131	53 31	51	130
30	48 46	+52	133	49 38	+51	132	50 29	+51	131	51 20	+50	130	52 10	+50	130	53 00	+50	129
31	48 17	51	132	49 08	51	131	49 59	50	130	50 49	50	129	51 39	50	128	52 29	49	127
32	47 46	51	130	48 37	51	130	49 28	50	129	50 18	49	128	51 07	50	127	51 57	48	126
33	47 15	51	129	48 06	50	128	48 56	50	128	49 46	49	127	50 35	49	126	51 24	48	125
34	46 44	50	128	47 34	50	127	48 24	49	126	49 13	49	125	50 02	49	125	50 51	48	124
35	46 12	+50	127	47 02	+49	126	47 51	+49	125	48 40	+49	124	49 29	+48	123	50 17	+48	122
36	45 40	49	126	46 29	49	125	47 18	49	124	48 07	48	123	48 55	48	122	49 43	47	121
37	45 07	49	125	45 56	49	124	46 45	48	123	47 33	48	122	48 21	48	121	49 09	47	120
38	44 34	48	123	45 22	49	123	46 11	48	122	46 59	47	121	47 46	48	120	48 34	46	119
39	44 00	47	122	44 48	48	122	45 36	48	121	46 24	47	120	47 11	47	119	47 58	47	118
40	43 26	+48	121	44 14	+48	120	45 02	+47	120	45 49	+47	119	46 36	+47	118	47 23	+46	117
41	42 51	48	120	43 39	48	119	44 27	47	119	45 14	46	118	46 00	47	117	46 47	46	116
42	42 16	48	119	43 04	47	118	43 51	47	117	44 38	46	117	45 24	46	116	46 10	46	115
43	41 41	47	118	42 28	47	117	43 15	47	116	44 02	46	116	44 48	46	115	45 34	45	114
44	41 06	47	117	41 53	46	116	42 39	47	115	43 26	45	115	44 11	46	114	44 57	45	113
45	40 30	+46	116	41 16	+47	115	42 03	+46	114	42 49	+46	114	43 35	+45	113	44 20	+45	112
46	39 53	47	115	40 40	46	114	41 26	46	113	42 12	45	113	42 57	45	112	43 42	45	111
47	39 17	46	114	40 03	46	113	40 49	46	112	41 35	45	112	42 20	45	111	43 05	44	110
48	38 40	46	113	39 26	46	112	40 12	45	111	40 57	45	111	41 42	45	110	42 27	44	109
49	38 03	46	112	38 49	45	111	39 34	46	111	40 20	44	110	41 04	45	109	41 49	44	108
50	37 26	+45	111	38 11	+46	110	38 57	+45	110	39 42	+44	109	40 26	+44	108	41 10	+44	107
51	36 48	46	110	37 34	45	110	38 19	45	109	39 04	44	108	39 48	44	107	40 32	43	106
52	36 11	45	109	36 56	45	109	37 41	44	108	38 25	44	107	39 09	44	106	39 53	44	105
53	35 33	45	109	36 18	44	108	37 02	45	107	37 47	44	106	38 31	43	105	39 14	44	104
54	34 54	45	108	35 39	45	107	36 24	44	106	37 08	44	105	37 52	43	104	38 35	44	103
55	34 16	+45	107	35 01	+44	106	35 45	+44	105	36 29	+44	104	37 13	+43	103	37 56	+43	103
56	33 38	44	106	34 22	44	105	35 06	44	104	35 50	44	103	36 34	43	103	37 17	43	102
57	32 59	44	105	33 43	44	104	34 27	44	103	35 11	44	103	35 55	43	102	36 38	42	101
58	32 20	44	104	33 04	44	103	33 48	44	103	34 32	43	102	35 15	43	101	35 58	43	100
59	31 41	44	103	32 25	44	103	33 09	44	102	33 53	43	101	34 36	43	100	35 19	42	99
60	31 02	+44	103	31 46	+44	102	32 30	+43	101	33 13	+43	100	33 56	+43	99	34 39	+42	98
61	30 23	44	102	31 07	43	101	31 50	44	100	32 34	43	99	33 17	42	98	33 59	43	98
62	29 43	44	101	30 27	44	100	31 11	43	99	31 54	43	98	32 37	42	98	33 19	43	97
63	29 04	44	100	29 48	43	99	30 31	43	98	31 14	43	98	31 57	43	97	32 40	42	96
64	28 24	44	99	29 08	43	98	29 51	43	98	30 34	43	97	31 17	43	96	32 00	42	95
65	27 45	+43	98	28 28	+43	98	29 11	+43	97	29 54	+43	96	30 37	+43	95	31 20	+42	94
66	27 05	43	98	27 48	44	97	28 32	42	96	29 14	43	95	29 57	42	94	30 39	43	94
67	26 25	43	97	27 08	44	96	27 52	42	95	28 34	43	95	29 17	42	94	29 59	42	93
68	25 45	43	96	26 28	44	95	27 12	42	95	27 54	43	94	28 37	42	93	29 19	42	92
69	25 05	43	95	25 48	44	95	26 32	42	94	27 14	43	93	27 57	42	92	28 39	42	91

Extract from table 5 showing values of d from 34' to 60'

34	35	36	37	38	39	40	41	42	43	44	45	46	47	48	49	50	51	52	53	54	55	56	57	58	59	60	d
'	'	'	'	'	'	'	'	'	'	'	'	'	'	'	'	'	'	'	'	'	'	'	'	'	'	'	'
0	0	0	0	0	0	0	0	0	0	0	0	0	0	0	0	0	0	0	0	0	0	0	0	0	0	0	0
1	1	1	1	1	1	1	1	1	1	1	1	1	1	1	1	1	1	1	1	1	1	1	1	1	1	1	1
1	1	1	1	1	1	1	1	1	1	1	2	2	2	2	2	2	2	2	2	2	2	2	2	2	2	2	2
2	2	2	2	2	2	2	2	2	2	2	2	2	2	2	2	2	3	3	3	3	3	3	3	3	3	3	3
2	2	2	2	3	3	3	3	3	3	3	3	3	3	3	3	3	3	3	4	4	4	4	4	4	4	4	4
3	3	3	3	3	3	3	3	4	4	4	4	4	4	4	4	4	4	4	4	4	5	5	5	5	5	5	5
3	4	4	4	4	4	4	4	4	4	4	4	5	5	5	5	5	5	5	5	5	6	6	6	6	6	6	6
4	4	4	4	4	5	5	5	5	5	5	5	5	5	6	6	6	6	6	6	6	6	7	7	7	7	7	7
5	5	5	5	5	5	5	5	6	6	6	6	6	6	6	7	7	7	7	7	7	7	7	8	8	8	8	8
5	5	5	6	6	6	6	6	6	6	7	7	7	7	7	7	8	8	8	8	8	8	8	9	9	9	9	9
6	6	6	6	6	6	7	7	7	7	7	8	8	8	8	8	8	8	9	9	9	9	9	10	10	10	10	10
6	6	7	7	7	7	7	8	8	8	8	8	8	9	9	9	9	9	10	10	10	10	10	10	11	11	11	11
7	7	7	7	8	8	8	8	8	9	9	9	9	9	10	10	10	10	10	11	11	11	11	11	12	12	12	12
7	8	8	8	8	8	9	9	9	9	10	10	10	10	10	11	11	11	11	12	12	12	13	13	13	13	13	13
8	8	8	9	9	9	9	10	10	10	10	10	11	11	11	11	12	12	12	12	13	13	13	13	14	14	14	14
8	9	9	9	10	10	10	10	10	11	11	11	12	12	12	12	12	13	13	13	14	14	14	14	14	15	15	15
9	9	10	10	10	10	11	11	11	11	12	12	12	13	13	13	13	14	14	14	14	15	15	15	15	16	16	16
10	10	10	10	11	11	11	12	12	12	13	13	13	13	14	14	14	15	15	15	15	16	16	16	16	17	17	17
10	10	11	11	11	12	12	12	13	13	13	14	14	14	14	15	15	15	16	16	16	16	17	17	17	18	18	18
11	11	11	12	12	12	13	13	13	14	14	14	15	15	15	16	16	16	16	17	17	17	18	18	18	19	19	19
11	12	12	12	13	13	13	14	14	14	15	15	15	16	16	16	17	17	17	18	18	18	19	19	19	20	20	20
12	12	13	13	13	14	14	14	15	15	15	16	16	16	17	17	18	18	18	19	19	19	20	20	20	21	21	21
12	13	13	14	14	14	15	15	15	16	16	16	17	17	18	18	18	19	19	19	20	20	21	21	21	22	22	22
13	13	14	14	15	15	15	16	16	16	17	17	18	18	18	19	19	20	20	20	21	21	21	22	22	23	23	23
14	14	14	15	15	16	16	16	17	17	18	18	18	19	19	20	20	20	21	21	22	22	22	23	23	24	24	24
14	15	15	15	16	16	17	17	18	18	18	19	19	20	20	20	21	21	22	22	22	23	23	24	24	25	25	25
15	15	16	16	16	17	17	18	18	19	19	20	20	20	21	21	22	22	23	23	23	24	25	25	25	26	26	26
15	16	16	17	17	18	18	19	19	20	20	20	21	21	22	22	22	23	23	24	24	25	25	26	26	27	27	27
16	16	17	17	18	18	19	19	20	20	21	21	21	22	22	23	23	24	24	25	25	26	26	27	27	28	28	28
16	17	17	18	18	19	19	20	20	21	21	22	22	23	23	24	24	25	25	26	26	27	27	28	28	29	29	29
17	18	18	18	19	20	20	20	21	22	22	22	23	24	24	24	25	26	26	26	27	28	28	28	29	30	30	30
18	18	19	19	20	20	21	21	22	22	23	23	24	24	25	25	26	26	27	27	28	28	29	29	30	30	31	31
18	19	19	20	20	21	21	22	22	23	23	24	25	25	26	26	27	27	28	28	29	29	30	30	31	31	32	32
19	19	20	20	21	21	22	23	23	24	24	25	25	26	26	27	27	28	29	29	30	30	31	31	32	32	33	33
19	20	20	21	22	22	23	23	24	24	25	26	26	27	27	28	28	29	29	30	31	31	32	32	33	33	34	34
20	20	21	22	22	23	23	24	24	25	26	26	27	27	28	29	29	30	30	31	32	32	33	33	34	34	35	35
20	21	22	22	23	23	24	25	25	26	26	27	28	28	29	29	30	31	31	32	32	33	34	34	35	35	36	36
21	22	22	23	23	24	25	25	26	27	27	28	28	29	30	30	31	31	32	33	33	34	35	35	36	36	37	37
22	22	23	23	24	25	25	26	27	27	28	28	29	30	30	31	32	32	33	34	34	35	35	36	37	37	38	38
22	23	23	24	25	25	26	27	27	28	29	29	30	31	31	32	32	33	34	34	35	36	36	37	38	38	39	39
23	23	24	25	25	26	27	27	28	29	29	30	31	31	32	33	33	34	35	35	36	37	37	38	39	39	40	40
23	24	25	25	26	27	27	28	29	29	30	31	31	32	33	33	34	35	36	36	37	38	38	39	40	40	41	41
24	24	25	26	27	27	28	29	29	30	31	32	32	33	34	34	35	36	36	37	38	38	39	40	41	41	42	42
24	25	26	27	27	28	29	29	30	31	32	32	33	34	34	35	36	37	37	38	39	39	40	41	42	42	43	43
25	26	26	27	28	29	29	30	31	32	32	33	34	34	35	36	37	37	38	39	40	40	41	42	43	43	44	44
26	26	27	28	28	29	30	31	32	32	33	34	34	35	36	37	38	38	39	40	40	41	42	43	44	44	45	45
26	27	28	28	29	30	31	31	32	33	34	34	35	36	37	38	38	39	40	41	41	42	43	44	44	45	46	46
27	27	28	29	30	31	31	32	33	34	34	35	36	37	38	38	39	40	41	42	42	43	44	45	45	46	47	47
27	28	29	30	30	31	32	33	34	34	35	36	37	38	38	39	40	41	42	42	43	44	45	46	46	47	48	48
28	29	29	30	31	32	33	33	34	35	36	37	38	38	39	40	41	42	42	43	44	45	46	47	47	48	49	49
28	29	30	31	32	32	33	34	35	36	37	38	38	39	40	41	42	42	43	44	45	46	47	48	48	49	50	50
29	30	31	31	32	33	34	35	36	37	37	38	39	40	41	42	42	43	44	45	46	47	48	48	49	50	51	51
29	30	31	32	33	34	35	36	36	37	38	39	40	41	42	42	43	44	45	46	47	48	49	49	50	51	52	52
30	31	32	33	34	34	35	36	37	38	39	40	41	42	42	43	44	45	46	47	48	49	49	50	51	52	53	53
31	32	32	33	34	35	36	37	38	39	40	40	41	42	43	44	45	46	47	48	49	50	50	51	52	53	54	54
31	32	33	34	35	36	37	38	38	39	40	41	42	43	44	45	46	47	48	49	50	50	51	52	53	54	55	55
32	33	34	35	35	36	37	38	39	40	41	42	43	44	45	46	47	48	49	50	50	51	52	53	54	55	56	56
32	33	34	35	36	37	38	39	40	41	42	43	44	45	46	47	48	49	50	51	51	52	53	54	55	56	57	57
33	34	35	36	37	38	39	40	41	42	43	44	45	46	47	48	49	50	51	52	53	53	54	55	56	57	58	58
33	34	35	36	37	38	39	40	41	42	43	44	45	46	47	48	49	50	51	52	53	54	55	56	57	58	59	59

Question 3. Calculate and plot 3 position lines to form a '3 point fix' using the information contained in the following scenario. Refer to the outline procedures in chapter 8 and the demonstration in chapter 11 as necessary.

Scenario:
At 1200 on 28 May 2009, the D.R. position of a sailing vessel is 50° 03'.4N, 18° 35'W.
Three observations of the Sun's lower limb were made during the day's run and were used to fix the position of the vessel at $12^h\ 00^m$ (+1).
The course and speed made good throughout the day were as follows: course 270° speed 5 knots.
For all three observations, the following details applied:
DWE 12^s slow
Index error: +0'.62
Ht. of eye: 10m.

Task: Calculate the fix at $12^h\ 00^m$(+1) using position lines obtained from the following 3 observations of the Sun's lower limb. Remember to take account of the movement of the observer (MOO) between each observation.

Observation 1:
DWT: $06^h\ 20^m\ 08^s$
Body observed: Sun L.L.
Sextant Alt: 08° 13'.8   Bearing: 068°
Temperature: 8°C. Pressure: 1008mb

Observation 2:
DWT: $13^h\ 00^m\ 01^s$
Body observed: Sun L.L.
Sextant Alt: 61° 17'.49   Bearing: 173°
Temperature: 12°C. Pressure: 1006mb

Observation 3.
DWT: $18^h\ 05^m\ 09^s$
Body observed: Sun L.L.
Sextant Alt: 27° 59'.66   Bearing: 270°
Temperature: 18°C. Pressure: 1002mb

Solution. A suggested solution to this question can be found on the following pages.

Extracts. Extracts relevant purely to this question follow the solution. Due to their sizes, some tables have been cropped to show only the required portions.

Solution Question 3.

Planning Phase

Step 1.	Approx Pos: 50° 03'.4N, 18° 35'W

Step 2. Calculate UT of observations:

Date: 28 May 2009

Time of fix:	$28^d\ 12^h\ 00^m$
Zone corrn:	$+1^h$
Greenwich Date:	$28^d\ 13^h\ 00^m$

Calculate UT of observations.

	Obs.1	Obs.2	Obs.3
DWT:	$06^h\ 20^m\ 08^s$	$13^h\ 00^m\ 01^s$	$18^h\ 05^m\ 09^s$
DWE:	$+12^s$	$+12^s$	$+12^s$
a. UT obs:	$06^h\ 20^m\ 20^s$	$13^h\ 00^m\ 13^s$	$18^h\ 05^m\ 21^s$
b. UT fix:	$13^h\ 00^m\ 00^s$	$13^h\ 00^m\ 00^s$	$13^h\ 00^m\ 00^s$
b~a:	$+06^h\ 39^m\ 40^s$	$-00^h\ 00^m\ 13^s$	$-05^h\ 05^m\ 21^s$
Fix is	later	earlier	earlier

Step 3. Calculate GHA and Dec. at UT of observations.

UT obs:	$06^h\ 20^m\ 20^s$	$13^h\ 00^m\ 13^s$	$18^h\ 05^m\ 21^s$
Dec UT^h	N21° 29'.3	N21° 32'.1	N21° 34'.0
d	(0'.4)	(0'.4)	(0'.4)
d corr	+0'.1	0'.0	0'.0
Dec+corr	N21° 29'.4	N21° 32'.1	N21° 34'.0
GHA UT^h	270° 41'.1	15° 40'.6	90° 40'.2
Inc (m+s)	+5° 05'.0	+0° 03'.3	+1° 20'.3
GHA =	275° 46'.1	15° 43'.9	92° 00'.5

Step 4. Calculate LHA

GHA	275° 46'.1	15° 43'.9	92° 00'.5
As Long(W)	-18° 46'.1	-18° 43'.9	-19° 00'.5
LHA (W-)	257°	-3°	73°
±360°		357°	

Convert azimuth angle to true bearing (Zn):

LHA	257°	357°	073°
Z	67°	174°	089°
Bearing (Zn)	067°	174°	271°

Please see page 161 for an explanation of the reason for calculating true bearings here. (Take the values of the Azimuth (Z) from step 5).

Rules for calculating true bearing (Zn)

	Lat. North	Lat. South
LHA > 180°	Zn = Z	Zn = 180° - Z
LHA < 180°	Zn = 360° - Z	Zn = 180° + Z

Step 5. Main table entry.

	Obs.1	Obs.2	Obs.3
As. lat.	N50°	N50°	N50°
Dec°	N21°(same)	N21°(same)	N21°(same)
Dec'	29'.4	32'.1	34'.0
LHA	257°	357°	73°
Table page:	77	76	76
Tab Hc,	08° 01'	60° 54'	26° 45'
d	+47'	+60'	+43'
Z	67°	174°	089°

Correction for mins of Dec. (Table 5)

	Obs.1	Obs.2	Obs.3
Tab Hc,	08° 01'	60° 54'	26° 45'
Cor. for Dec'	+23'	+32'	+24'
Cal. Alt. Hc (Hc+Cor.Dec')	08° 24'	61° 26'	27° 09'

Fix Phase

Step 6. Correct sextant altitude (Hs).

	Obs.1	Obs.2	Obs.3
Sext. Alt. (Hs)	08° 13'.80	61° 17'.49	27° 59'.66
I.E.	+0'.62	+0'.62	+0'.62
Dip (10m) (D)	-5'.60	-5'.60	-5'.60
Ap. Alt. (Ha)	08° 08'.82	61° 12'.51	27° 54'.68
Alt. Cor. (R)	+9'.6	+15'.4	+14'.2
Add'l refrac.	+0'.0	0'.0	+0'.1
Obs. Alt. (Ho)	08° 18'.42	61° 27'.91	28° 08.98

Step 7. Calculate Intercept

	Obs.1	Obs.2	Obs.3
Ho	08° 18'.42	61° 27'.91	28° 08'.98
Hc	08° 24'.00	61° 26'.00	27° 09'.00
p(Ho-Hc)	-5'.58	+1'.91	+59'.98

Therefore, intercepts are: Obs.1: 5.58 n.m. from 067°,
Obs.2: 1.91 n.m. towards 174°, Obs.3: 59.98 n.m. towards 271°.

(Note. True bearings calculated at end of step 4).

Step 8. Corrections to intercept.(MOO) (Table 1)			
Movement of vessel:			
Distance made good between observations 1 and 2: 6.66hrs. @ 5 knots = 33.3n.m., course 270°			
Distance made good between observations 2 and 3: 5.1 hrs. @ 5 knots = 25.5n.m., course 270°.			
	Obs.1	Obs.2	Obs.3
Bearing (Zn)	067°	174°	271°
±360°	360°	360°	
(to make Zn > C)	427°	534°	
Course (C)	270°	270°	270°
Rel Zn (Zn – C)	157°	264°	001°
Dist. (DMG)	33'.3	negligible	25'.50
Table 1 corn.	-30'.4	0	-25'.00
Fix is	later		earlier
Intercept (p)	-5'.58	+1'.91	+59'.98
Correction	-30'.4	0	-25'.00
Corrected (p)	-35'.98 (from)	+1'.91 (to)	+34'.98 (to)
Azimuth (Zn)	067°	174°	271°

Therefore, corrected intercepts are: Obs.1: 35.98 n.m. from 067°, Obs.2: 1.91 n.m. towards 174°, Obs.3: 34.98 n.m. towards 271°.

Note. Interpolate as necessary when using table 1

Rules for application of sign from table 1 & 2		
Time of fix	Sign from table	To intercept
Later than observation	+	Add
	-	Subtract
Earlier than observation	+	Subtract
	-	Add

Step 9. Plot the position lines. (To revise the method, see the demonstration in chapter 11).
As in the previous example, a small 'cocked-hat' is produced

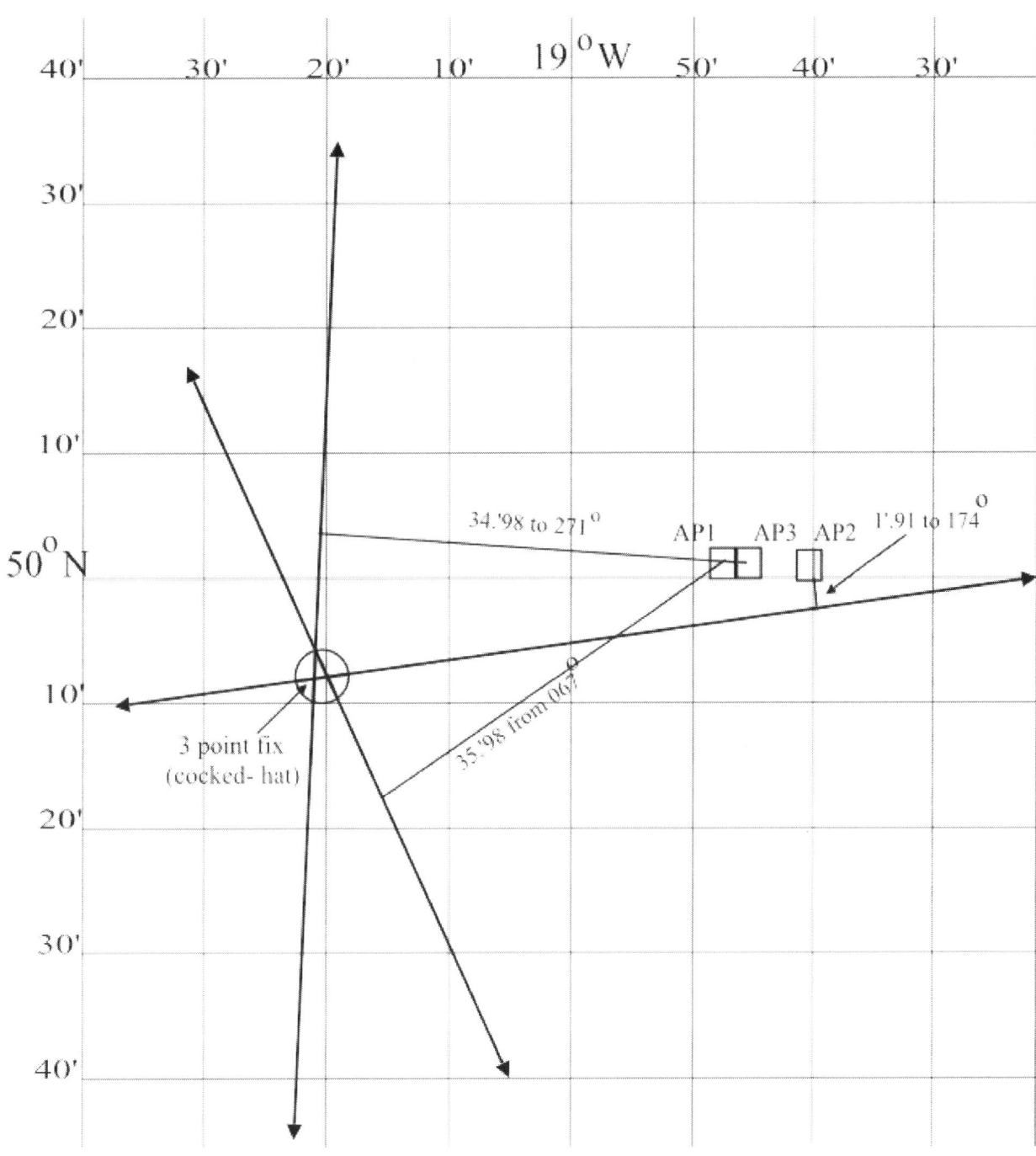

2009 MAY 28, 29, 30 (THURS., FRI., SAT.)

UT	SUN GHA	SUN Dec	MOON GHA	MOON v	MOON Dec	MOON d	MOON HP
d h	° '	° '	° '	'	° '	'	'
28 00	180 41.6	N21 26.9	128 18.4	5.5	N21 37.8	10.0	60.3
01	195 41.5	27.3	142 42.9	5.6	21 27.8	10.1	60.3
02	210 41.4	27.7	157 07.5	5.7	21 17.7	10.2	60.3
03	225 41.4	28.1	171 32.2	5.8	21 07.5	10.4	60.2
04	240 41.3	28.5	185 57.0	6.0	20 57.1	10.4	60.2
05	255 41.2	28.9	200 22.0	6.0	20 46.7	10.7	60.2
06	270 41.1	N21 29.3	214 47.0	6.2	N20 36.0	10.7	60.2
07	285 41.0	29.7	229 12.2	6.3	20 25.3	10.9	60.2
T 08	300 41.0	30.1	243 37.5	6.4	20 14.4	11.0	60.1
H 09	315 40.9	30.5	258 02.9	6.5	20 03.4	11.1	60.1
U 10	330 40.8	30.9	272 28.4	6.6	19 52.3	11.2	60.1
R 11	345 40.7	31.3	286 54.0	6.7	19 41.1	11.3	60.1
S 12	0 40.6	N21 31.7	301 19.7	6.9	N19 29.8	11.5	60.1
D 13	15 40.6	32.1	315 45.6	6.9	19 18.3	11.6	60.0
A 14	30 40.5	32.5	330 11.5	7.1	19 06.7	11.7	60.0
Y 15	45 40.4	32.8	344 37.6	7.2	18 55.0	11.7	60.0
16	60 40.3	33.2	359 03.8	7.3	18 43.3	11.9	60.0
17	75 40.2	33.6	13 30.1	7.5	18 31.4	12.0	59.9
18	90 40.2	N21 34.0	27 56.6	7.5	N18 19.4	12.1	59.9
19	105 40.1	34.4	42 23.1	7.6	18 07.3	12.2	59.9
20	120 40.0	34.8	56 49.7	7.8	17 55.1	12.3	59.9
21	135 39.9	35.2	71 16.5	7.9	17 42.8	12.4	59.8
22	150 39.8	35.6	85 43.4	8.0	17 30.4	12.5	59.8
23	165 39.8	36.0	100 10.4	8.1	17 17.9	12.6	59.8
29 00	180 39.7	N21 36.4	114 37.5	8.2	N17 05.3	12.7	59.8
01	195 39.6	36.8	129 04.7	8.3	16 52.6	12.8	59.8
02	210 39.5	37.1	143 32.0	8.4	16 39.8	12.8	59.7
03	225 39.4	37.5	157 59.4	8.6	16 27.0	13.0	59.7
04	240 39.3	37.9	172 27.0	8.6	16 14.0	13.0	59.7
05	255 39.3	38.3	186 54.6	8.8	16 01.0	13.1	59.7
06	270 39.2	N21 38.7	201 22.4	8.8	N15 47.9	13.2	59.6
07	285 39.1	39.1	215 50.2	9.0	15 34.7	13.2	59.6
08	300 39.0	39.4	230 18.2	9.1	15 21.5	13.4	59.6
F 09	315 38.9	39.8	244 46.3	9.2	15 08.1	13.4	59.5
R 10	330 38.8	40.2	259 14.5	9.2	14 54.7	13.5	59.5
I 11	345 38.8	40.6	273 42.7	9.4	14 41.2	13.5	59.5
D 12	0 38.7	N21 41.0	288 11.1	9.5	N14 27.7	13.6	59.5
A 13	15 38.6	41.4	302 39.6	9.6	14 14.1	13.7	59.4
Y 14	30 38.5	41.7	317 08.2	9.7	14 00.4	13.8	59.4
15	45 38.4	42.1	331 36.9	9.7	13 46.6	13.8	59.4
16	60 38.3	42.5	346 05.6	9.9	13 32.8	13.9	59.4
17	75 38.2	42.9	0 34.5	10.0	13 18.9	13.9	59.3
18	90 38.2	N21 43.2	15 03.5	10.1	N13 05.0	14.0	59.3
19	105 38.1	43.6	29 32.6	10.1	12 51.0	14.1	59.3
20	120 38.0	44.0	44 01.7	10.3	12 36.9	14.1	59.3
21	135 37.9	44.4	58 31.0	10.4	12 22.8	14.2	59.2
22	150 37.8	44.7	73 00.4	10.4	12 08.6	14.2	59.2
23	165 37.7	45.1	87 29.8	10.5	11 54.4	14.3	59.2
30 00	180 37.6	N21 45.5	101 59.3	10.6	N11 40.1	14.3	59.1
01	195 37.6	45.9	116 28.9	10.8	11 25.8	14.3	59.1
02	210 37.5	46.2	130 58.7	10.8	11 11.5	14.4	59.1
03	225 37.4	46.6	145 28.5	10.8	10 57.1	14.5	59.1
04	240 37.3	47.0	159 58.3	11.0	10 42.6	14.5	59.1
05	255 37.2	47.3	174 28.3	11.0	10 28.1	14.5	59.0
06	270 37.1	N21 47.7	188 58.3	11.2	N10 13.6	14.6	59.0
07	285 37.0	48.1	203 28.5	11.2	9 59.0	14.6	58.9
S 08	300 37.0	48.5	217 58.7	11.3	9 44.4	14.6	58.9
A 09	315 36.9	48.8	232 29.0	11.3	9 29.8	14.7	58.9
T 10	330 36.8	49.2	246 59.3	11.5	9 15.1	14.7	58.9
U 11	345 36.7	49.5	261 29.8	11.5	9 00.4	14.8	58.8
R 12	0 36.6	N21 49.9	276 00.3	11.6	N 8 45.6	14.8	58.8
D 13	15 36.5	50.3	290 30.9	11.7	8 30.9	14.8	58.8
A 14	30 36.4	50.6	305 01.6	11.7	8 16.1	14.9	58.7
Y 15	45 36.3	51.0	319 32.3	11.8	8 01.2	14.8	58.7
16	60 36.2	51.4	334 03.1	11.9	7 46.4	14.9	58.7
17	75 36.2	51.7	348 34.0	11.9	7 31.5	14.9	58.7
18	90 36.1	N21 52.1	3 04.9	12.0	N 7 16.6	14.9	58.6
19	105 36.0	52.4	17 35.9	12.1	7 01.7	15.0	58.6
20	120 35.9	52.8	32 07.0	12.2	6 46.7	14.9	58.6
21	135 35.8	53.2	46 38.2	12.2	6 31.8	15.0	58.5
22	150 35.7	53.5	61 09.4	12.2	6 16.8	15.0	58.5
23	165 35.6	53.9	75 40.6	12.4	N 6 01.8	15.0	58.5
	SD 15.8	d 0.4	SD 16.4		16.2		16.0

Lat.	Twilight Naut.	Twilight Civil	Sunrise	Moonrise 28	Moonrise 29	Moonrise 30	Moonrise 31
°	h m	h m	h m	h m	h m	h m	h m
N 72	▭	▭	▭	▭	06 42	09 31	11 48
N 70	▭	▭	▭	▭	07 22	09 47	11 52
68	▭	▭	▭	05 05	07 49	09 59	11 56
66	////	////	01 25	05 53	08 09	10 08	11 58
64	////	////	02 05	06 24	08 25	10 17	12 00
62	////	00 38	02 33	06 47	08 39	10 24	12 02
60	////	01 35	02 54	07 05	08 50	10 30	12 04
N 58	////	02 06	03 11	07 20	09 00	10 35	12 05
56	00 34	02 29	03 25	07 33	09 08	10 40	12 07
54	01 26	02 47	03 38	07 44	09 16	10 44	12 08
52	01 55	03 03	03 49	07 54	09 22	10 47	12 09
50	02 17	03 16	03 58	08 03	09 28	10 51	12 10
45	02 56	03 42	04 18	08 21	09 41	10 58	12 12
N 40	03 23	04 03	04 35	08 36	09 52	11 04	12 14
35	03 44	04 19	04 48	08 49	10 01	11 10	12 15
30	04 01	04 33	05 00	09 00	10 09	11 14	12 17
20	04 27	04 56	05 20	09 19	10 22	11 22	12 19
N 10	04 48	05 15	05 38	09 35	10 34	11 29	12 21
0	05 06	05 32	05 54	09 50	10 45	11 36	12 23
S 10	05 21	05 47	06 10	10 06	10 56	11 42	12 25
20	05 36	06 03	06 27	10 22	11 08	11 49	12 27
30	05 51	06 20	06 46	10 40	11 21	11 57	12 30
35	05 58	06 30	06 57	10 51	11 29	12 02	12 31
40	06 07	06 40	07 10	11 03	11 38	12 07	12 32
45	06 16	06 52	07 25	11 18	11 48	12 12	12 34
S 50	06 26	07 07	07 44	11 35	12 00	12 19	12 36
52	06 31	07 13	07 53	11 43	12 06	12 23	12 37
54	06 36	07 21	08 03	11 53	12 12	12 26	12 38
56	06 41	07 29	08 14	12 03	12 19	12 30	12 40
58	06 47	07 38	08 27	12 14	12 26	12 34	12 41
S 60	06 53	07 48	08 42	12 28	12 35	12 39	12 42

Lat.	Sunset	Twilight Civil	Twilight Naut.	Moonset 28	Moonset 29	Moonset 30	Moonset 31
°	h m	h m	h m	h m	h m	h m	h m
N 72	▭	▭	▭	▭	02 45	01 44	01 08
N 70	▭	▭	▭	▭	02 04	01 26	01 01
68	▭	▭	▭	02 22	01 35	01 12	00 55
66	22 34	////	////	01 33	01 13	01 00	00 50
64	21 52	////	////	01 01	00 56	00 51	00 46
62	21 24	23 25	////	00 38	00 41	00 42	00 42
60	21 02	22 23	////	00 19	00 29	00 35	00 39
N 58	20 45	21 51	////	00 03	00 18	00 28	00 36
56	20 30	21 27	23 28	24 09	00 09	00 22	00 33
54	20 18	21 09	22 31	24 00	00 00	00 17	00 31
52	20 07	20 53	22 01	23 53	24 12	00 12	00 28
50	19 57	20 40	21 39	23 46	24 08	00 08	00 26
45	19 37	20 13	21 00	23 32	23 59	24 22	00 22
N 40	19 21	19 53	20 33	23 19	23 51	24 18	00 18
35	19 07	19 36	20 12	23 09	23 44	24 15	00 15
30	18 55	19 22	19 54	23 00	23 38	24 12	00 12
20	18 35	18 59	19 28	22 44	23 28	24 07	00 07
N 10	18 17	18 40	19 07	22 30	23 18	24 03	00 03
0	18 01	18 23	18 49	22 17	23 10	23 59	24 45
S 10	17 45	18 08	18 34	22 04	23 01	23 54	24 45
20	17 28	17 52	18 19	21 50	22 51	23 50	24 46
30	17 09	17 34	18 04	21 33	22 40	23 44	24 46
35	16 57	17 25	17 56	21 23	22 34	23 41	24 46
40	16 44	17 14	17 48	21 12	22 27	23 38	24 46
45	16 29	17 02	17 39	20 59	22 18	23 34	24 47
S 50	16 10	16 48	17 24	20 43	22 08	23 29	24 47
52	16 02	16 41	17 24	20 36	22 03	23 27	24 47
54	15 52	16 34	17 19	20 27	21 58	23 24	24 47
56	15 41	16 26	17 14	20 18	21 52	23 21	24 47
58	15 28	16 17	17 08	20 07	21 45	23 17	24 47
S 60	15 13	16 07	17 01	19 54	21 38	23 15	24 48

Day	SUN Eqn. of Time 00h	SUN Eqn. of Time 12h	SUN Mer. Pass.	MOON Mer. Pass. Upper	MOON Mer. Pass. Lower	MOON Age	MOON Phase
d	m s	m s	h m	h m	h m	d	%
28	02 47	02 43	11 57	16 04	03 35	04	22
29	02 39	02 35	11 57	16 58	04 31	05	32
30	02 31	02 27	11 58	17 47	05 23	06	43

INCREMENTS AND CORRECTIONS

0ᵐ

s	SUN PLANETS	ARIES	MOON	v or d	Corrⁿ	v or d	Corrⁿ	v or d	Corrⁿ
	° ′	° ′	° ′	′	′	′	′	′	′
00	0 00·0	0 00·0	0 00·0	0·0	0·0	6·0	0·1	12·0	0·1
01	0 00·3	0 00·3	0 00·2	0·1	0·0	6·1	0·1	12·1	0·1
02	0 00·5	0 00·5	0 00·5	0·2	0·0	6·2	0·1	12·2	0·1
03	0 00·8	0 00·8	0 00·7	0·3	0·0	6·3	0·1	12·3	0·1
04	0 01·0	0 01·0	0 01·0	0·4	0·0	6·4	0·1	12·4	0·1
05	0 01·3	0 01·3	0 01·2	0·5	0·0	6·5	0·1	12·5	0·1
06	0 01·5	0 01·5	0 01·4	0·6	0·0	6·6	0·1	12·6	0·1
07	0 01·8	0 01·8	0 01·7	0·7	0·0	6·7	0·1	12·7	0·1
08	0 02·0	0 02·0	0 01·9	0·8	0·0	6·8	0·1	12·8	0·1
09	0 02·3	0 02·3	0 02·1	0·9	0·0	6·9	0·1	12·9	0·1
10	0 02·5	0 02·5	0 02·4	1·0	0·0	7·0	0·1	13·0	0·1
11	0 02·8	0 02·8	0 02·6	1·1	0·0	7·1	0·1	13·1	0·1
12	0 03·0	0 03·0	0 02·9	1·2	0·0	7·2	0·1	13·2	0·1
13	0 03·3	0 03·3	0 03·1	1·3	0·0	7·3	0·1	13·3	0·1
14	0 03·5	0 03·5	0 03·3	1·4	0·0	7·4	0·1	13·4	0·1
15	0 03·8	0 03·8	0 03·6	1·5	0·0	7·5	0·1	13·5	0·1
16	0 04·0	0 04·0	0 03·8	1·6	0·0	7·6	0·1	13·6	0·1
17	0 04·3	0 04·3	0 04·1	1·7	0·0	7·7	0·1	13·7	0·1
18	0 04·5	0 04·5	0 04·3	1·8	0·0	7·8	0·1	13·8	0·1
19	0 04·8	0 04·8	0 04·5	1·9	0·0	7·9	0·1	13·9	0·1
20	0 05·0	0 05·0	0 04·8	2·0	0·0	8·0	0·1	14·0	0·1
21	0 05·3	0 05·3	0 05·0	2·1	0·0	8·1	0·1	14·1	0·1
22	0 05·5	0 05·5	0 05·2	2·2	0·0	8·2	0·1	14·2	0·1
23	0 05·8	0 05·8	0 05·5	2·3	0·0	8·3	0·1	14·3	0·1
24	0 06·0	0 06·0	0 05·7	2·4	0·0	8·4	0·1	14·4	0·1
25	0 06·3	0 06·3	0 06·0	2·5	0·0	8·5	0·1	14·5	0·1
26	0 06·5	0 06·5	0 06·2	2·6	0·0	8·6	0·1	14·6	0·1
27	0 06·8	0 06·8	0 06·4	2·7	0·0	8·7	0·1	14·7	0·1
28	0 07·0	0 07·0	0 06·7	2·8	0·0	8·8	0·1	14·8	0·1
29	0 07·3	0 07·3	0 06·9	2·9	0·0	8·9	0·1	14·9	0·1
30	0 07·5	0 07·5	0 07·2	3·0	0·0	9·0	0·1	15·0	0·1
31	0 07·8	0 07·8	0 07·4	3·1	0·0	9·1	0·1	15·1	0·1
32	0 08·0	0 08·0	0 07·6	3·2	0·0	9·2	0·1	15·2	0·1
33	0 08·3	0 08·3	0 07·9	3·3	0·0	9·3	0·1	15·3	0·1
34	0 08·5	0 08·5	0 08·1	3·4	0·0	9·4	0·1	15·4	0·1
35	0 08·8	0 08·8	0 08·4	3·5	0·0	9·5	0·1	15·5	0·1
36	0 09·0	0 09·0	0 08·6	3·6	0·0	9·6	0·1	15·6	0·1
37	0 09·3	0 09·3	0 08·8	3·7	0·0	9·7	0·1	15·7	0·1
38	0 09·5	0 09·5	0 09·1	3·8	0·0	9·8	0·1	15·8	0·1
39	0 09·8	0 09·8	0 09·3	3·9	0·0	9·9	0·1	15·9	0·1
40	0 10·0	0 10·0	0 09·5	4·0	0·0	10·0	0·1	16·0	0·1
41	0 10·3	0 10·3	0 09·8	4·1	0·0	10·1	0·1	16·1	0·1
42	0 10·5	0 10·5	0 10·0	4·2	0·0	10·2	0·1	16·2	0·1
43	0 10·8	0 10·8	0 10·3	4·3	0·0	10·3	0·1	16·3	0·1
44	0 11·0	0 11·0	0 10·5	4·4	0·0	10·4	0·1	16·4	0·1
45	0 11·3	0 11·3	0 10·7	4·5	0·0	10·5	0·1	16·5	0·1
46	0 11·5	0 11·5	0 11·0	4·6	0·0	10·6	0·1	16·6	0·1
47	0 11·8	0 11·8	0 11·2	4·7	0·0	10·7	0·1	16·7	0·1
48	0 12·0	0 12·0	0 11·5	4·8	0·0	10·8	0·1	16·8	0·1
49	0 12·3	0 12·3	0 11·7	4·9	0·0	10·9	0·1	16·9	0·1
50	0 12·5	0 12·5	0 11·9	5·0	0·0	11·0	0·1	17·0	0·1
51	0 12·8	0 12·8	0 12·2	5·1	0·0	11·1	0·1	17·1	0·1
52	0 13·0	0 13·0	0 12·4	5·2	0·0	11·2	0·1	17·2	0·1
53	0 13·3	0 13·3	0 12·6	5·3	0·0	11·3	0·1	17·3	0·1
54	0 13·5	0 13·5	0 12·9	5·4	0·0	11·4	0·1	17·4	0·1
55	0 13·8	0 13·8	0 13·1	5·5	0·0	11·5	0·1	17·5	0·1
56	0 14·0	0 14·0	0 13·4	5·6	0·0	11·6	0·1	17·6	0·1
57	0 14·3	0 14·3	0 13·6	5·7	0·0	11·7	0·1	17·7	0·1
58	0 14·5	0 14·5	0 13·8	5·8	0·0	11·8	0·1	17·8	0·1
59	0 14·8	0 14·8	0 14·1	5·9	0·0	11·9	0·1	17·9	0·1
60	0 15·0	0 15·0	0 14·3	6·0	0·1	12·0	0·1	18·0	0·2

1ᵐ

s	SUN PLANETS	ARIES	MOON	v or d	Corrⁿ	v or d	Corrⁿ	v or d	Corrⁿ
	° ′	° ′	° ′	′	′	′	′	′	′
00	0 15·0	0 15·0	0 14·3	0·0	0·0	6·0	0·2	12·0	0·3
01	0 15·3	0 15·3	0 14·6	0·1	0·0	6·1	0·2	12·1	0·3
02	0 15·5	0 15·5	0 14·8	0·2	0·0	6·2	0·2	12·2	0·3
03	0 15·8	0 15·8	0 15·0	0·3	0·0	6·3	0·2	12·3	0·3
04	0 16·0	0 16·0	0 15·3	0·4	0·0	6·4	0·2	12·4	0·3
05	0 16·3	0 16·3	0 15·5	0·5	0·0	6·5	0·2	12·5	0·3
06	0 16·5	0 16·5	0 15·7	0·6	0·0	6·6	0·2	12·6	0·3
07	0 16·8	0 16·8	0 16·0	0·7	0·0	6·7	0·2	12·7	0·3
08	0 17·0	0 17·0	0 16·2	0·8	0·0	6·8	0·2	12·8	0·3
09	0 17·3	0 17·3	0 16·5	0·9	0·0	6·9	0·2	12·9	0·3
10	0 17·5	0 17·5	0 16·7	1·0	0·0	7·0	0·2	13·0	0·3
11	0 17·8	0 17·8	0 16·9	1·1	0·0	7·1	0·2	13·1	0·3
12	0 18·0	0 18·0	0 17·2	1·2	0·0	7·2	0·2	13·2	0·3
13	0 18·3	0 18·3	0 17·4	1·3	0·0	7·3	0·2	13·3	0·3
14	0 18·5	0 18·6	0 17·7	1·4	0·0	7·4	0·2	13·4	0·3
15	0 18·8	0 18·8	0 17·9	1·5	0·0	7·5	0·2	13·5	0·3
16	0 19·0	0 19·1	0 18·1	1·6	0·0	7·6	0·2	13·6	0·3
17	0 19·3	0 19·3	0 18·4	1·7	0·0	7·7	0·2	13·7	0·3
18	0 19·5	0 19·6	0 18·6	1·8	0·0	7·8	0·2	13·8	0·3
19	0 19·8	0 19·8	0 18·9	1·9	0·0	7·9	0·2	13·9	0·3
20	0 20·0	0 20·1	0 19·1	2·0	0·1	8·0	0·2	14·0	0·4
21	0 20·3	0 20·3	0 19·3	2·1	0·1	8·1	0·2	14·1	0·4
22	0 20·5	0 20·6	0 19·6	2·2	0·1	8·2	0·2	14·2	0·4
23	0 20·8	0 20·8	0 19·8	2·3	0·1	8·3	0·2	14·3	0·4
24	0 21·0	0 21·1	0 20·0	2·4	0·1	8·4	0·2	14·4	0·4
25	0 21·3	0 21·3	0 20·3	2·5	0·1	8·5	0·2	14·5	0·4
26	0 21·5	0 21·6	0 20·5	2·6	0·1	8·6	0·2	14·6	0·4
27	0 21·8	0 21·8	0 20·8	2·7	0·1	8·7	0·2	14·7	0·4
28	0 22·0	0 22·1	0 21·0	2·8	0·1	8·8	0·2	14·8	0·4
29	0 22·3	0 22·3	0 21·2	2·9	0·1	8·9	0·2	14·9	0·4
30	0 22·5	0 22·6	0 21·5	3·0	0·1	9·0	0·2	15·0	0·4
31	0 22·8	0 22·8	0 21·7	3·1	0·1	9·1	0·2	15·1	0·4
32	0 23·0	0 23·1	0 22·0	3·2	0·1	9·2	0·2	15·2	0·4
33	0 23·3	0 23·3	0 22·2	3·3	0·1	9·3	0·2	15·3	0·4
34	0 23·5	0 23·6	0 22·4	3·4	0·1	9·4	0·2	15·4	0·4
35	0 23·8	0 23·8	0 22·7	3·5	0·1	9·5	0·2	15·5	0·4
36	0 24·0	0 24·1	0 22·9	3·6	0·1	9·6	0·2	15·6	0·4
37	0 24·3	0 24·3	0 23·1	3·7	0·1	9·7	0·2	15·7	0·4
38	0 24·5	0 24·6	0 23·4	3·8	0·1	9·8	0·2	15·8	0·4
39	0 24·8	0 24·8	0 23·6	3·9	0·1	9·9	0·2	15·9	0·4
40	0 25·0	0 25·1	0 23·9	4·0	0·1	10·0	0·3	16·0	0·4
41	0 25·3	0 25·3	0 24·1	4·1	0·1	10·1	0·3	16·1	0·4
42	0 25·5	0 25·6	0 24·3	4·2	0·1	10·2	0·3	16·2	0·4
43	0 25·8	0 25·8	0 24·6	4·3	0·1	10·3	0·3	16·3	0·4
44	0 26·0	0 26·1	0 24·8	4·4	0·1	10·4	0·3	16·4	0·4
45	0 26·3	0 26·3	0 25·1	4·5	0·1	10·5	0·3	16·5	0·4
46	0 26·5	0 26·6	0 25·3	4·6	0·1	10·6	0·3	16·6	0·4
47	0 26·8	0 26·8	0 25·5	4·7	0·1	10·7	0·3	16·7	0·4
48	0 27·0	0 27·1	0 25·8	4·8	0·1	10·8	0·3	16·8	0·4
49	0 27·3	0 27·3	0 26·0	4·9	0·1	10·9	0·3	16·9	0·4
50	0 27·5	0 27·6	0 26·2	5·0	0·1	11·0	0·3	17·0	0·4
51	0 27·8	0 27·8	0 26·5	5·1	0·1	11·1	0·3	17·1	0·4
52	0 28·0	0 28·1	0 26·7	5·2	0·1	11·2	0·3	17·2	0·4
53	0 28·3	0 28·3	0 27·0	5·3	0·1	11·3	0·3	17·3	0·4
54	0 28·5	0 28·6	0 27·2	5·4	0·1	11·4	0·3	17·4	0·4
55	0 28·8	0 28·8	0 27·4	5·5	0·1	11·5	0·3	17·5	0·4
56	0 29·0	0 29·1	0 27·7	5·6	0·1	11·6	0·3	17·6	0·4
57	0 29·3	0 29·3	0 27·9	5·7	0·1	11·7	0·3	17·7	0·4
58	0 29·5	0 29·6	0 28·2	5·8	0·1	11·8	0·3	17·8	0·4
59	0 29·8	0 29·8	0 28·4	5·9	0·1	11·9	0·3	17·9	0·4
60	0 30·0	0 30·1	0 28·6	6·0	0·2	12·0	0·3	18·0	0·5

INCREMENTS AND CORRECTIONS

4ᵐ

s	SUN PLANETS	ARIES	MOON	v or d	Corrⁿ	v or d	Corrⁿ	v or d	Corrⁿ
	° ′	° ′	° ′	′	′	′	′	′	′
00	1 00·0	1 00·2	0 57·3	0·0	0·0	6·0	0·5	12·0	0·9
01	1 00·3	1 00·4	0 57·5	0·1	0·0	6·1	0·5	12·1	0·9
02	1 00·5	1 00·7	0 57·7	0·2	0·0	6·2	0·5	12·2	0·9
03	1 00·8	1 00·9	0 58·0	0·3	0·0	6·3	0·5	12·3	0·9
04	1 01·0	1 01·2	0 58·2	0·4	0·0	6·4	0·5	12·4	0·9
05	1 01·3	1 01·4	0 58·5	0·5	0·0	6·5	0·5	12·5	0·9
06	1 01·5	1 01·7	0 58·7	0·6	0·0	6·6	0·5	12·6	0·9
07	1 01·8	1 01·9	0 58·9	0·7	0·1	6·7	0·5	12·7	1·0
08	1 02·0	1 02·2	0 59·2	0·8	0·1	6·8	0·5	12·8	1·0
09	1 02·3	1 02·4	0 59·4	0·9	0·1	6·9	0·5	12·9	1·0
10	1 02·5	1 02·7	0 59·7	1·0	0·1	7·0	0·5	13·0	1·0
11	1 02·8	1 02·9	0 59·9	1·1	0·1	7·1	0·5	13·1	1·0
12	1 03·0	1 03·2	1 00·1	1·2	0·1	7·2	0·5	13·2	1·0
13	1 03·3	1 03·4	1 00·4	1·3	0·1	7·3	0·5	13·3	1·0
14	1 03·5	1 03·7	1 00·6	1·4	0·1	7·4	0·6	13·4	1·0
15	1 03·8	1 03·9	1 00·8	1·5	0·1	7·5	0·6	13·5	1·0
16	1 04·0	1 04·2	1 01·1	1·6	0·1	7·6	0·6	13·6	1·0
17	1 04·3	1 04·4	1 01·3	1·7	0·1	7·7	0·6	13·7	1·0
18	1 04·5	1 04·7	1 01·6	1·8	0·1	7·8	0·6	13·8	1·0
19	1 04·8	1 04·9	1 01·8	1·9	0·1	7·9	0·6	13·9	1·0
20	1 05·0	1 05·2	1 02·0	2·0	0·2	8·0	0·6	14·0	1·1
21	1 05·3	1 05·4	1 02·3	2·1	0·2	8·1	0·6	14·1	1·1
22	1 05·5	1 05·7	1 02·5	2·2	0·2	8·2	0·6	14·2	1·1
23	1 05·8	1 05·9	1 02·8	2·3	0·2	8·3	0·6	14·3	1·1
24	1 06·0	1 06·2	1 03·0	2·4	0·2	8·4	0·6	14·4	1·1
25	1 06·3	1 06·4	1 03·2	2·5	0·2	8·5	0·6	14·5	1·1
26	1 06·5	1 06·7	1 03·5	2·6	0·2	8·6	0·6	14·6	1·1
27	1 06·8	1 06·9	1 03·7	2·7	0·2	8·7	0·7	14·7	1·1
28	1 07·0	1 07·2	1 03·9	2·8	0·2	8·8	0·7	14·8	1·1
29	1 07·3	1 07·4	1 04·2	2·9	0·2	8·9	0·7	14·9	1·1
30	1 07·5	1 07·7	1 04·4	3·0	0·2	9·0	0·7	15·0	1·1
31	1 07·8	1 07·9	1 04·7	3·1	0·2	9·1	0·7	15·1	1·1
32	1 08·0	1 08·2	1 04·9	3·2	0·2	9·2	0·7	15·2	1·1
33	1 08·3	1 08·4	1 05·1	3·3	0·2	9·3	0·7	15·3	1·1
34	1 08·5	1 08·7	1 05·4	3·4	0·3	9·4	0·7	15·4	1·2
35	1 08·8	1 08·9	1 05·6	3·5	0·3	9·5	0·7	15·5	1·2
36	1 09·0	1 09·2	1 05·9	3·6	0·3	9·6	0·7	15·6	1·2
37	1 09·3	1 09·4	1 06·1	3·7	0·3	9·7	0·7	15·7	1·2
38	1 09·5	1 09·7	1 06·3	3·8	0·3	9·8	0·7	15·8	1·2
39	1 09·8	1 09·9	1 06·6	3·9	0·3	9·9	0·7	15·9	1·2
40	1 10·0	1 10·2	1 06·8	4·0	0·3	10·0	0·8	16·0	1·2
41	1 10·3	1 10·4	1 07·0	4·1	0·3	10·1	0·8	16·1	1·2
42	1 10·5	1 10·7	1 07·3	4·2	0·3	10·2	0·8	16·2	1·2
43	1 10·8	1 10·9	1 07·5	4·3	0·3	10·3	0·8	16·3	1·2
44	1 11·0	1 11·2	1 07·8	4·4	0·3	10·4	0·8	16·4	1·2
45	1 11·3	1 11·4	1 08·0	4·5	0·3	10·5	0·8	16·5	1·2
46	1 11·5	1 11·7	1 08·2	4·6	0·3	10·6	0·8	16·6	1·2
47	1 11·8	1 11·9	1 08·5	4·7	0·4	10·7	0·8	16·7	1·3
48	1 12·0	1 12·2	1 08·7	4·8	0·4	10·8	0·8	16·8	1·3
49	1 12·3	1 12·4	1 09·0	4·9	0·4	10·9	0·8	16·9	1·3
50	1 12·5	1 12·7	1 09·2	5·0	0·4	11·0	0·8	17·0	1·3
51	1 12·8	1 12·9	1 09·4	5·1	0·4	11·1	0·8	17·1	1·3
52	1 13·0	1 13·2	1 09·7	5·2	0·4	11·2	0·8	17·2	1·3
53	1 13·3	1 13·5	1 09·9	5·3	0·4	11·3	0·8	17·3	1·3
54	1 13·5	1 13·7	1 10·2	5·4	0·4	11·4	0·9	17·4	1·3
55	1 13·8	1 14·0	1 10·4	5·5	0·4	11·5	0·9	17·5	1·3
56	1 14·0	1 14·2	1 10·6	5·6	0·4	11·6	0·9	17·6	1·3
57	1 14·3	1 14·5	1 10·9	5·7	0·4	11·7	0·9	17·7	1·3
58	1 14·5	1 14·7	1 11·1	5·8	0·4	11·8	0·9	17·8	1·3
59	1 14·8	1 15·0	1 11·3	5·9	0·4	11·9	0·9	17·9	1·3
60	1 15·0	1 15·2	1 11·6	6·0	0·5	12·0	0·9	18·0	1·4

5ᵐ

s	SUN PLANETS	ARIES	MOON	v or d	Corrⁿ	v or d	Corrⁿ	v or d	Corrⁿ
	° ′	° ′	° ′	′	′	′	′	′	′
00	1 15·0	1 15·2	1 11·6	0·0	0·0	6·0	0·6	12·0	1·1
01	1 15·3	1 15·5	1 11·8	0·1	0·0	6·1	0·6	12·1	1·1
02	1 15·5	1 15·7	1 12·1	0·2	0·0	6·2	0·6	12·2	1·1
03	1 15·8	1 16·0	1 12·3	0·3	0·0	6·3	0·6	12·3	1·1
04	1 16·0	1 16·2	1 12·5	0·4	0·0	6·4	0·6	12·4	1·1
05	1 16·3	1 16·5	1 12·8	0·5	0·0	6·5	0·6	12·5	1·1
06	1 16·5	1 16·7	1 13·0	0·6	0·1	6·6	0·6	12·6	1·2
07	1 16·8	1 17·0	1 13·3	0·7	0·1	6·7	0·6	12·7	1·2
08	1 17·0	1 17·2	1 13·5	0·8	0·1	6·8	0·6	12·8	1·2
09	1 17·3	1 17·5	1 13·7	0·9	0·1	6·9	0·6	12·9	1·2
10	1 17·5	1 17·7	1 14·0	1·0	0·1	7·0	0·6	13·0	1·2
11	1 17·8	1 18·0	1 14·2	1·1	0·1	7·1	0·7	13·1	1·2
12	1 18·0	1 18·2	1 14·4	1·2	0·1	7·2	0·7	13·2	1·2
13	1 18·3	1 18·5	1 14·7	1·3	0·1	7·3	0·7	13·3	1·2
14	1 18·5	1 18·7	1 14·9	1·4	0·1	7·4	0·7	13·4	1·2
15	1 18·8	1 19·0	1 15·2	1·5	0·1	7·5	0·7	13·5	1·2
16	1 19·0	1 19·2	1 15·4	1·6	0·1	7·6	0·7	13·6	1·2
17	1 19·3	1 19·5	1 15·6	1·7	0·2	7·7	0·7	13·7	1·3
18	1 19·5	1 19·7	1 15·9	1·8	0·2	7·8	0·7	13·8	1·3
19	1 19·8	1 20·0	1 16·1	1·9	0·2	7·9	0·7	13·9	1·3
20	1 20·0	1 20·2	1 16·4	2·0	0·2	8·0	0·7	14·0	1·3
21	1 20·3	1 20·5	1 16·6	2·1	0·2	8·1	0·7	14·1	1·3
22	1 20·5	1 20·7	1 16·8	2·2	0·2	8·2	0·8	14·2	1·3
23	1 20·8	1 21·0	1 17·1	2·3	0·2	8·3	0·8	14·3	1·3
24	1 21·0	1 21·2	1 17·3	2·4	0·2	8·4	0·8	14·4	1·3
25	1 21·3	1 21·5	1 17·5	2·5	0·2	8·5	0·8	14·5	1·3
26	1 21·5	1 21·7	1 17·8	2·6	0·2	8·6	0·8	14·6	1·3
27	1 21·8	1 22·0	1 18·0	2·7	0·2	8·7	0·8	14·7	1·3
28	1 22·0	1 22·2	1 18·3	2·8	0·3	8·8	0·8	14·8	1·4
29	1 22·3	1 22·5	1 18·5	2·9	0·3	8·9	0·8	14·9	1·4
30	1 22·5	1 22·7	1 18·7	3·0	0·3	9·0	0·8	15·0	1·4
31	1 22·8	1 23·0	1 19·0	3·1	0·3	9·1	0·8	15·1	1·4
32	1 23·0	1 23·2	1 19·2	3·2	0·3	9·2	0·8	15·2	1·4
33	1 23·3	1 23·5	1 19·5	3·3	0·3	9·3	0·9	15·3	1·4
34	1 23·5	1 23·7	1 19·7	3·4	0·3	9·4	0·9	15·4	1·4
35	1 23·8	1 24·0	1 19·9	3·5	0·3	9·5	0·9	15·5	1·4
36	1 24·0	1 24·2	1 20·2	3·6	0·3	9·6	0·9	15·6	1·4
37	1 24·3	1 24·5	1 20·4	3·7	0·3	9·7	0·9	15·7	1·4
38	1 24·5	1 24·7	1 20·7	3·8	0·3	9·8	0·9	15·8	1·4
39	1 24·8	1 25·0	1 20·9	3·9	0·4	9·9	0·9	15·9	1·5
40	1 25·0	1 25·2	1 21·1	4·0	0·4	10·0	0·9	16·0	1·5
41	1 25·3	1 25·5	1 21·4	4·1	0·4	10·1	0·9	16·1	1·5
42	1 25·5	1 25·7	1 21·6	4·2	0·4	10·2	0·9	16·2	1·5
43	1 25·8	1 26·0	1 21·8	4·3	0·4	10·3	0·9	16·3	1·5
44	1 26·0	1 26·2	1 22·1	4·4	0·4	10·4	1·0	16·4	1·5
45	1 26·3	1 26·5	1 22·3	4·5	0·4	10·5	1·0	16·5	1·5
46	1 26·5	1 26·7	1 22·6	4·6	0·4	10·6	1·0	16·6	1·5
47	1 26·8	1 27·0	1 22·8	4·7	0·4	10·7	1·0	16·7	1·5
48	1 27·0	1 27·2	1 23·0	4·8	0·4	10·8	1·0	16·8	1·5
49	1 27·3	1 27·5	1 23·3	4·9	0·4	10·9	1·0	16·9	1·5
50	1 27·5	1 27·7	1 23·5	5·0	0·5	11·0	1·0	17·0	1·6
51	1 27·8	1 28·0	1 23·8	5·1	0·5	11·1	1·0	17·1	1·6
52	1 28·0	1 28·2	1 24·0	5·2	0·5	11·2	1·0	17·2	1·6
53	1 28·3	1 28·5	1 24·2	5·3	0·5	11·3	1·0	17·3	1·6
54	1 28·5	1 28·7	1 24·5	5·4	0·5	11·4	1·0	17·4	1·6
55	1 28·8	1 29·0	1 24·7	5·5	0·5	11·5	1·1	17·5	1·6
56	1 29·0	1 29·2	1 24·9	5·6	0·5	11·6	1·1	17·6	1·6
57	1 29·3	1 29·5	1 25·2	5·7	0·5	11·7	1·1	17·7	1·6
58	1 29·5	1 29·7	1 25·4	5·8	0·5	11·8	1·1	17·8	1·6
59	1 29·8	1 30·0	1 25·7	5·9	0·5	11·9	1·1	17·9	1·6
60	1 30·0	1 30·2	1 25·9	6·0	0·6	12·0	1·1	18·0	1·7

©2009 Jack Case

INCREMENTS AND CORRECTIONS

20m

s	SUN PLANETS	ARIES	MOON	v or d	Corrn	v or d	Corrn	v or d	Corrn
	° ′	° ′	° ′	′	′	′	′	′	′
00	5 00·0	5 00·8	4 46·3	0·0	0·0	6·0	2·1	12·0	4·1
01	5 00·3	5 01·1	4 46·6	0·1	0·0	6·1	2·1	12·1	4·1
02	5 00·5	5 01·3	4 46·8	0·2	0·1	6·2	2·1	12·2	4·2
03	5 00·8	5 01·6	4 47·0	0·3	0·1	6·3	2·2	12·3	4·2
04	5 01·0	5 01·8	4 47·3	0·4	0·1	6·4	2·2	12·4	4·2
05	5 01·3	5 02·1	4 47·5	0·5	0·2	6·5	2·2	12·5	4·3
06	5 01·5	5 02·3	4 47·8	0·6	0·2	6·6	2·3	12·6	4·3
07	5 01·8	5 02·6	4 48·0	0·7	0·2	6·7	2·3	12·7	4·3
08	5 02·0	5 02·8	4 48·2	0·8	0·3	6·8	2·3	12·8	4·4
09	5 02·3	5 03·1	4 48·5	0·9	0·3	6·9	2·4	12·9	4·4
10	5 02·5	5 03·3	4 48·7	1·0	0·3	7·0	2·4	13·0	4·4
11	5 02·8	5 03·6	4 49·0	1·1	0·4	7·1	2·4	13·1	4·5
12	5 03·0	5 03·8	4 49·2	1·2	0·4	7·2	2·5	13·2	4·5
13	5 03·3	5 04·1	4 49·4	1·3	0·4	7·3	2·5	13·3	4·5
14	5 03·5	5 04·3	4 49·7	1·4	0·5	7·4	2·5	13·4	4·6
15	5 03·8	5 04·6	4 49·9	1·5	0·5	7·5	2·6	13·5	4·6
16	5 04·0	5 04·8	4 50·2	1·6	0·5	7·6	2·6	13·6	4·6
17	5 04·3	5 05·1	4 50·4	1·7	0·6	7·7	2·6	13·7	4·7
18	5 04·5	5 05·3	4 50·6	1·8	0·6	7·8	2·7	13·8	4·7
19	5 04·8	5 05·6	4 50·9	1·9	0·6	7·9	2·7	13·9	4·7
20	5 05·0	5 05·8	4 51·1	2·0	0·7	8·0	2·7	14·0	4·8
21	5 05·3	5 06·1	4 51·3	2·1	0·7	8·1	2·8	14·1	4·8
22	5 05·5	5 06·3	4 51·6	2·2	0·8	8·2	2·8	14·2	4·9
23	5 05·8	5 06·6	4 51·8	2·3	0·8	8·3	2·8	14·3	4·9
24	5 06·0	5 06·8	4 52·1	2·4	0·8	8·4	2·9	14·4	4·9
25	5 06·3	5 07·1	4 52·3	2·5	0·9	8·5	2·9	14·5	5·0
26	5 06·5	5 07·3	4 52·5	2·6	0·9	8·6	2·9	14·6	5·0
27	5 06·8	5 07·6	4 52·8	2·7	0·9	8·7	3·0	14·7	5·0
28	5 07·0	5 07·8	4 53·0	2·8	1·0	8·8	3·0	14·8	5·1
29	5 07·3	5 08·1	4 53·3	2·9	1·0	8·9	3·0	14·9	5·1
30	5 07·5	5 08·3	4 53·5	3·0	1·0	9·0	3·1	15·0	5·1
31	5 07·8	5 08·6	4 53·7	3·1	1·1	9·1	3·1	15·1	5·2
32	5 08·0	5 08·8	4 54·0	3·2	1·1	9·2	3·1	15·2	5·2
33	5 08·3	5 09·1	4 54·2	3·3	1·1	9·3	3·2	15·3	5·2
34	5 08·5	5 09·3	4 54·4	3·4	1·2	9·4	3·2	15·4	5·3
35	5 08·8	5 09·6	4 54·7	3·5	1·2	9·5	3·2	15·5	5·3
36	5 09·0	5 09·8	4 54·9	3·6	1·2	9·6	3·3	15·6	5·3
37	5 09·3	5 10·1	4 55·2	3·7	1·3	9·7	3·3	15·7	5·4
38	5 09·5	5 10·3	4 55·4	3·8	1·3	9·8	3·3	15·8	5·4
39	5 09·8	5 10·6	4 55·6	3·9	1·3	9·9	3·4	15·9	5·4
40	5 10·0	5 10·8	4 55·9	4·0	1·4	10·0	3·4	16·0	5·5
41	5 10·3	5 11·1	4 56·1	4·1	1·4	10·1	3·5	16·1	5·5
42	5 10·5	5 11·4	4 56·4	4·2	1·4	10·2	3·5	16·2	5·5
43	5 10·8	5 11·6	4 56·6	4·3	1·5	10·3	3·5	16·3	5·6
44	5 11·0	5 11·9	4 56·8	4·4	1·5	10·4	3·6	16·4	5·6
45	5 11·3	5 12·1	4 57·1	4·5	1·5	10·5	3·6	16·5	5·6
46	5 11·5	5 12·4	4 57·3	4·6	1·6	10·6	3·6	16·6	5·7
47	5 11·8	5 12·6	4 57·5	4·7	1·6	10·7	3·7	16·7	5·7
48	5 12·0	5 12·9	4 57·8	4·8	1·6	10·8	3·7	16·8	5·7
49	5 12·3	5 13·1	4 58·0	4·9	1·7	10·9	3·7	16·9	5·8
50	5 12·5	5 13·4	4 58·3	5·0	1·7	11·0	3·8	17·0	5·8
51	5 12·8	5 13·6	4 58·5	5·1	1·7	11·1	3·8	17·1	5·8
52	5 13·0	5 13·9	4 58·7	5·2	1·8	11·2	3·8	17·2	5·9
53	5 13·3	5 14·1	4 59·0	5·3	1·8	11·3	3·9	17·3	5·9
54	5 13·5	5 14·4	4 59·2	5·4	1·8	11·4	3·9	17·4	5·9
55	5 13·8	5 14·6	4 59·5	5·5	1·9	11·5	3·9	17·5	6·0
56	5 14·0	5 14·9	4 59·7	5·6	1·9	11·6	4·0	17·6	6·0
57	5 14·3	5 15·1	4 59·9	5·7	1·9	11·7	4·0	17·7	6·0
58	5 14·5	5 15·4	5 00·2	5·8	2·0	11·8	4·0	17·8	6·1
59	5 14·8	5 15·6	5 00·4	5·9	2·0	11·9	4·1	17·9	6·1
60	5 15·0	5 15·9	5 00·7	6·0	2·1	12·0	4·1	18·0	6·2

21m

s	SUN PLANETS	ARIES	MOON	v or d	Corrn	v or d	Corrn	v or d	Corrn
	° ′	° ′	° ′	′	′	′	′	′	′
00	5 15·0	5 15·9	5 00·7	0·0	0·0	6·0	2·2	12·0	4·3
01	5 15·3	5 16·1	5 00·9	0·1	0·0	6·1	2·2	12·1	4·3
02	5 15·5	5 16·4	5 01·1	0·2	0·1	6·2	2·2	12·2	4·4
03	5 15·8	5 16·6	5 01·4	0·3	0·1	6·3	2·3	12·3	4·4
04	5 16·0	5 16·9	5 01·6	0·4	0·1	6·4	2·3	12·4	4·4
05	5 16·3	5 17·1	5 01·8	0·5	0·2	6·5	2·3	12·5	4·5
06	5 16·5	5 17·4	5 02·1	0·6	0·2	6·6	2·4	12·6	4·5
07	5 16·8	5 17·6	5 02·3	0·7	0·3	6·7	2·4	12·7	4·6
08	5 17·0	5 17·9	5 02·6	0·8	0·3	6·8	2·4	12·8	4·6
09	5 17·3	5 18·1	5 02·8	0·9	0·3	6·9	2·5	12·9	4·6
10	5 17·5	5 18·4	5 03·0	1·0	0·4	7·0	2·5	13·0	4·7
11	5 17·8	5 18·6	5 03·3	1·1	0·4	7·1	2·5	13·1	4·7
12	5 18·0	5 18·9	5 03·5	1·2	0·4	7·2	2·6	13·2	4·7
13	5 18·3	5 19·1	5 03·8	1·3	0·5	7·3	2·6	13·3	4·8
14	5 18·5	5 19·4	5 04·0	1·4	0·5	7·4	2·7	13·4	4·8
15	5 18·8	5 19·6	5 04·2	1·5	0·5	7·5	2·7	13·5	4·8
16	5 19·0	5 19·9	5 04·5	1·6	0·6	7·6	2·7	13·6	4·9
17	5 19·3	5 20·1	5 04·7	1·7	0·6	7·7	2·8	13·7	4·9
18	5 19·5	5 20·4	5 04·9	1·8	0·6	7·8	2·8	13·8	4·9
19	5 19·8	5 20·6	5 05·2	1·9	0·7	7·9	2·8	13·9	5·0
20	5 20·0	5 20·9	5 05·4	2·0	0·7	8·0	2·9	14·0	5·0
21	5 20·3	5 21·1	5 05·7	2·1	0·8	8·1	2·9	14·1	5·1
22	5 20·5	5 21·4	5 05·9	2·2	0·8	8·2	2·9	14·2	5·1
23	5 20·8	5 21·6	5 06·1	2·3	0·8	8·3	3·0	14·3	5·1
24	5 21·0	5 21·9	5 06·4	2·4	0·9	8·4	3·0	14·4	5·2
25	5 21·3	5 22·1	5 06·6	2·5	0·9	8·5	3·0	14·5	5·2
26	5 21·5	5 22·4	5 06·9	2·6	0·9	8·6	3·1	14·6	5·2
27	5 21·8	5 22·6	5 07·1	2·7	1·0	8·7	3·1	14·7	5·3
28	5 22·0	5 22·9	5 07·3	2·8	1·0	8·8	3·2	14·8	5·3
29	5 22·3	5 23·1	5 07·6	2·9	1·0	8·9	3·2	14·9	5·3
30	5 22·5	5 23·4	5 07·8	3·0	1·1	9·0	3·2	15·0	5·4
31	5 22·8	5 23·6	5 08·0	3·1	1·1	9·1	3·3	15·1	5·4
32	5 23·0	5 23·9	5 08·3	3·2	1·1	9·2	3·3	15·2	5·4
33	5 23·3	5 24·1	5 08·5	3·3	1·2	9·3	3·3	15·3	5·5
34	5 23·5	5 24·4	5 08·8	3·4	1·2	9·4	3·4	15·4	5·5
35	5 23·8	5 24·6	5 09·0	3·5	1·3	9·5	3·4	15·5	5·6
36	5 24·0	5 24·9	5 09·2	3·6	1·3	9·6	3·4	15·6	5·6
37	5 24·3	5 25·1	5 09·5	3·7	1·3	9·7	3·5	15·7	5·6
38	5 24·5	5 25·4	5 09·7	3·8	1·4	9·8	3·5	15·8	5·7
39	5 24·8	5 25·6	5 10·0	3·9	1·4	9·9	3·5	15·9	5·7
40	5 25·0	5 25·9	5 10·2	4·0	1·4	10·0	3·6	16·0	5·7
41	5 25·3	5 26·1	5 10·4	4·1	1·5	10·1	3·6	16·1	5·8
42	5 25·5	5 26·4	5 10·7	4·2	1·5	10·2	3·7	16·2	5·8
43	5 25·8	5 26·6	5 10·9	4·3	1·5	10·3	3·7	16·3	5·8
44	5 26·0	5 26·9	5 11·1	4·4	1·6	10·4	3·7	16·4	5·9
45	5 26·3	5 27·1	5 11·4	4·5	1·6	10·5	3·8	16·5	5·9
46	5 26·5	5 27·4	5 11·6	4·6	1·6	10·6	3·8	16·6	5·9
47	5 26·8	5 27·6	5 11·9	4·7	1·7	10·7	3·8	16·7	6·0
48	5 27·0	5 27·9	5 12·1	4·8	1·7	10·8	3·9	16·8	6·0
49	5 27·3	5 28·1	5 12·3	4·9	1·8	10·9	3·9	16·9	6·1
50	5 27·5	5 28·4	5 12·6	5·0	1·8	11·0	3·9	17·0	6·1
51	5 27·8	5 28·6	5 12·8	5·1	1·8	11·1	4·0	17·1	6·1
52	5 28·0	5 28·9	5 13·1	5·2	1·9	11·2	4·0	17·2	6·2
53	5 28·3	5 29·1	5 13·3	5·3	1·9	11·3	4·1	17·3	6·2
54	5 28·5	5 29·4	5 13·5	5·4	1·9	11·4	4·1	17·4	6·2
55	5 28·8	5 29·7	5 13·8	5·5	2·0	11·5	4·1	17·5	6·3
56	5 29·0	5 29·9	5 14·0	5·6	2·0	11·6	4·2	17·6	6·3
57	5 29·3	5 30·2	5 14·3	5·7	2·0	11·7	4·2	17·7	6·3
58	5 29·5	5 30·4	5 14·5	5·8	2·1	11·8	4·2	17·8	6·4
59	5 29·8	5 30·7	5 14·7	5·9	2·1	11·9	4·3	17·9	6·4
60	5 30·0	5 30·9	5 15·0	6·0	2·2	12·0	4·3	18·0	6·5

© 2009 Jack Case

TABLE 6 — ALTITUDE CORRECTION TABLES

a. Dip of the Horizon

Ht. of Eye (m)	Corr^n D	Ht. of Eye (ft.)	Ht. of Eye (m)	Corr^n D	Ht. of Eye (ft.)
0·00	−0·3	0·0	13·0	−6·4	42·8
0·03	−0·4	0·1	13·4	−6·5	44·2
0·06	−0·5	0·2	13·8	−6·6	45·5
0·09	−0·6	0·3	14·2	−6·7	47·0
0·13	−0·7	0·4	14·7	−6·8	48·4
0·18	−0·8	0·5	15·1	−6·9	49·8
0·23	−0·9	0·7	15·5	−7·0	51·3
0·29	−1·0	0·9	16·0	−7·1	52·8
0·35	−1·1	1·1	16·5	−7·2	54·3
0·42	−1·2	1·4	16·9	−7·3	55·8
0·45	−1·3	1·6	17·4	−7·4	57·4
0·5	−1·4	1·9	17·9	−7·5	58·9
0·6	−1·5	2·2	18·4	−7·6	60·5
0·7	−1·6	2·5	18·8	−7·7	62·1
0·8	−1·7	2·8	19·3	−7·8	63·8
0·9	−1·8	3·2	19·8	−7·9	65·4
1·1	−1·9	3·6	20·4	−8·0	67·1
1·2	−2·0	4·0	20·9	−8·1	68·8
1·3	−2·1	4·4	21·4	−8·2	70·5
1·4	−2·2	4·9	21·9	−8·3	72·3
1·6	−2·3	5·3	22·5	−8·4	74·1
1·7	−2·4	5·8	23·0	−8·5	75·8
1·9	−2·5	6·3	23·5	−8·6	77·6
2·0	−2·6	6·9	24·1	−8·7	79·5
2·2	−2·7	7·4	24·7	−8·8	81·3
2·4	−2·8	8·0	25·2	−8·9	83·2
2·6	−2·9	8·6	25·8	−9·0	85·1
2·8	−3·0	9·2	26·4	−9·1	87·0
3·0	−3·1	9·8	27·0	−9·2	88·9
3·2	−3·2	10·5	27·6	−9·3	90·9
3·4	−3·3	11·2	28·2	−9·4	92·9
3·6	−3·4	11·9	28·8	−9·5	94·9
3·8	−3·5	12·6	29·4	−9·6	96·9
4·0	−3·6	13·3	30·0	−9·7	98·9
4·3	−3·7	14·1	30·6	−9·8	101·0
4·5	−3·8	14·9	31·3	−9·9	103·1
4·7	−3·9	15·7	31·9	−10·0	105·2
5·0	−4·0	16·5	32·6	−10·1	107·3
5·2	−4·1	17·4	33·2	−10·2	109·4
5·5	−4·2	18·3	33·9	−10·3	111·6
5·8	−4·3	19·1	34·5	−10·4	113·8
6·1	−4·4	20·1	35·2	−10·5	116·0
6·3	−4·5	21·0	35·9	−10·6	118·2
6·6	−4·6	22·0	36·6	−10·7	120·5
6·9	−4·7	22·9	37·3	−10·8	122·8
7·2	−4·8	23·9	38·0	−10·9	125·1
7·5	−4·9	25·0	38·7	−11·0	127·4
7·9	−5·0	26·0	39·4	−11·1	129·7
8·2	−5·1	27·1	40·1	−11·2	132·1
8·5	−5·2	28·1	40·8	−11·3	134·5
8·8	−5·3	29·2	41·5	−11·4	136·9
9·2	−5·4	30·4	42·3	−11·5	139·3
9·5	−5·5	31·5	43·0	−11·6	141·7
9·9	−5·6	32·7	43·8	−11·7	144·2
10·3	−5·7	33·9	44·5	−11·8	146·7
10·6	−5·8	35·1	45·3	−11·9	149·2
11·0	−5·9	36·3	46·1	−12·0	151·7
11·4	−6·0	37·6	46·8	−12·1	154·3
11·8	−6·1	38·9	47·6	−12·2	156·8
12·2	−6·2	40·1	48·4	−12·3	159·4
12·6	−6·3	41·5	49·2	−12·4	162·1
13·0		42·8	50·0		164·7

b. Refraction for Stars & Planets

App. Alt. H_a	Corr^n R	App. Alt. H_a	Corr^n R	App. Alt. H_a	Corr^n R
0 00	−33·8	3 30	−12·9	9 55	−5·3
0 03	33·2	3 35	12·7	10 07	−5·2
0 06	32·6	3 40	12·5	10 20	−5·1
0 09	32·0	3 45	12·3	10 32	−5·0
0 12	31·5	3 50	12·1	10 46	−4·9
0 15	30·9	3 55	11·9	10 59	−4·8
0 18	−30·4	4 00	−11·7	11 14	−4·7
0 21	29·8	4 05	11·5	11 29	−4·6
0 24	29·3	4 10	11·4	11 44	−4·5
0 27	28·8	4 15	11·2	12 00	−4·4
0 30	28·3	4 20	11·0	12 17	−4·3
0 33	27·9	4 25	10·9	12 35	−4·2
0 36	−27·4	4 30	−10·7	12 53	−4·1
0 39	26·9	4 35	10·6	13 12	−4·0
0 42	26·5	4 40	10·4	13 32	−3·9
0 45	26·1	4 45	10·3	13 53	−3·8
0 48	25·7	4 50	10·1	14 16	−3·7
0 51	25·3	4 55	10·0	14 39	−3·6
0 54	−24·9	5 00	−9·8	15 03	−3·5
0 57	24·5	5 05	9·7	15 29	−3·4
1 00	24·1	5 10	9·6	15 56	−3·3
1 03	23·7	5 15	9·5	16 25	−3·2
1 06	23·4	5 20	9·3	16 55	−3·1
1 09	23·0	5 25	9·2	17 27	−3·0
1 12	−22·7	5 30	−9·1	18 01	−2·9
1 15	22·3	5 35	9·0	18 37	−2·8
1 18	22·0	5 40	8·9	19 16	−2·7
1 21	21·7	5 45	8·8	19 56	−2·6
1 24	21·4	5 50	8·7	20 40	−2·5
1 27	21·1	5 55	8·6	21 27	−2·4
1 30	−20·8	6 00	−8·5	22 17	−2·3
1 35	20·3	6 10	8·3	23 11	−2·2
1 40	19·9	6 20	8·1	24 09	−2·1
1 45	19·4	6 30	7·9	25 12	−2·0
1 50	19·0	6 40	7·7	26 20	−1·9
1 55	18·6	6 50	7·6	27 34	−1·8
2 00	−18·2	7 00	−7·4	28 54	−1·7
2 05	17·8	7 10	7·2	30 22	−1·6
2 10	17·4	7 20	7·1	31 58	−1·5
2 15	17·1	7 30	6·9	33 43	−1·4
2 20	16·7	7 40	6·8	35 38	−1·3
2 25	16·4	7 50	6·7	37 45	−1·2
2 30	−16·1	8 00	−6·6	40 06	−1·1
2 35	15·8	8 10	6·4	42 42	−1·0
2 40	15·4	8 20	6·3	45 34	−0·9
2 45	15·2	8 30	6·2	48 45	−0·8
2 50	14·9	8 40	6·1	52 16	−0·7
2 55	14·6	8 50	6·0	56 09	−0·6
3 00	−14·3	9 00	−5·9	60 26	−0·5
3 05	14·1	9 10	5·8	65 06	−0·4
3 10	13·8	9 20	5·7	70 09	−0·3
3 15	13·6	9 30	5·6	75 32	−0·2
3 20	13·4	9 40	5·5	81 12	−0·1
3 25	13·1	9 50	5·4	87 03	0·0
3 30	−12·9	10 00	−5·3	90 00	

H_a = App. Alt. = Apparent Altitude
 = Sextant altitude corrected for index error (IE) & dip (
 = H_s + IE + D

In critical cases ascend.

TABLE 6 — ALTITUDE CORRECTION TABLES

c. Additional Refraction for Non-standard Conditions

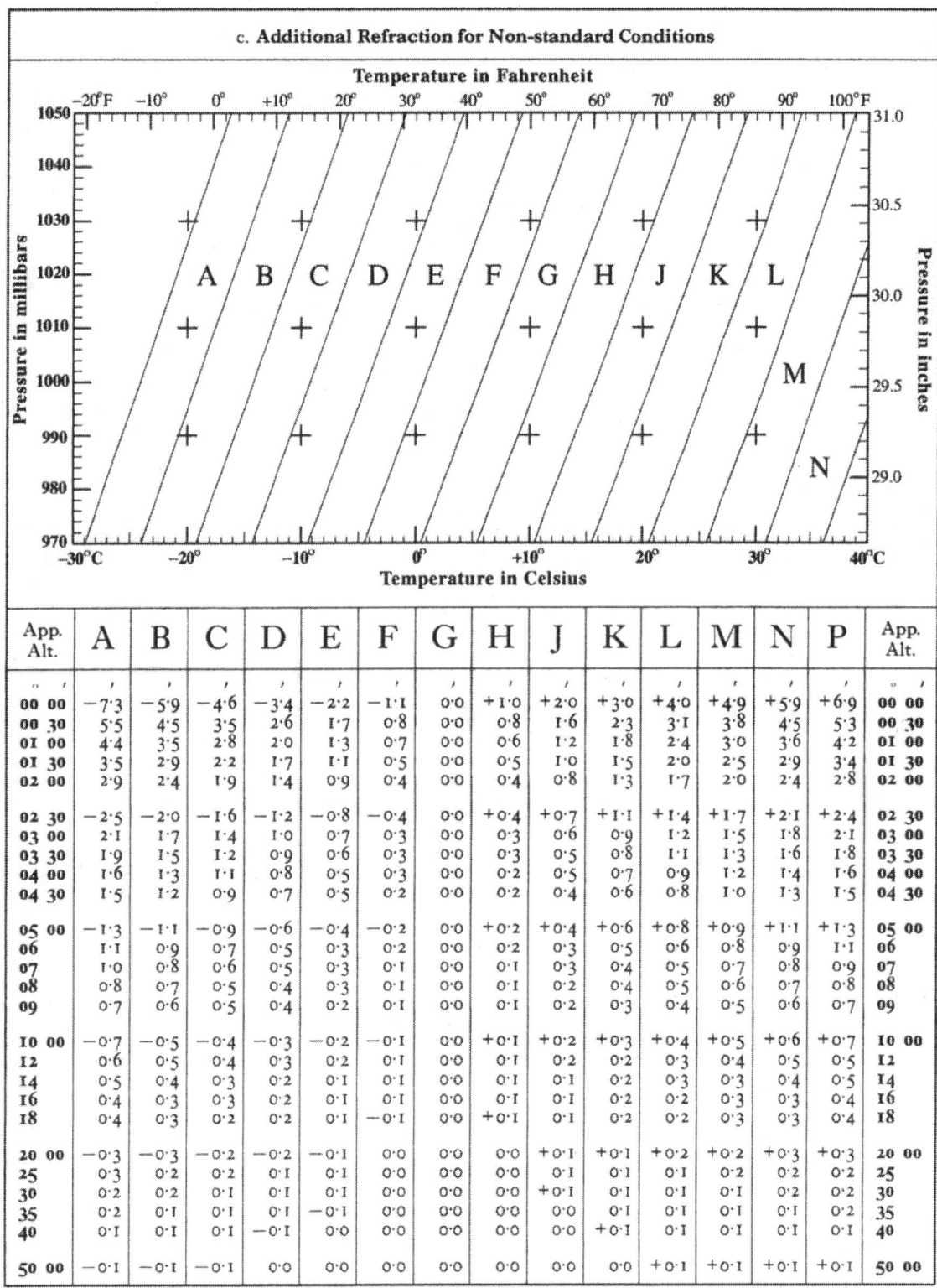

App. Alt.	A	B	C	D	E	F	G	H	J	K	L	M	N	P	App. Alt.
° ′	′	′	′	′	′	′	′	′	′	′	′	′	′	′	° ′
00 00	−7·3	−5·9	−4·6	−3·4	−2·2	−1·1	0·0	+1·0	+2·0	+3·0	+4·0	+4·9	+5·9	+6·9	00 00
00 30	5·5	4·5	3·5	2·6	1·7	0·8	0·0	0·8	1·6	2·3	3·1	3·8	4·5	5·3	00 30
01 00	4·4	3·5	2·8	2·0	1·3	0·7	0·0	0·6	1·2	1·8	2·4	3·0	3·6	4·2	01 00
01 30	3·5	2·9	2·2	1·7	1·1	0·5	0·0	0·5	1·0	1·5	2·0	2·5	2·9	3·4	01 30
02 00	2·9	2·4	1·9	1·4	0·9	0·4	0·0	0·4	0·8	1·3	1·7	2·0	2·4	2·8	02 00
02 30	−2·5	−2·0	−1·6	−1·2	−0·8	−0·4	0·0	+0·4	+0·7	+1·1	+1·4	+1·7	+2·1	+2·4	02 30
03 00	2·1	1·7	1·4	1·0	0·7	0·3	0·0	0·3	0·6	0·9	1·2	1·5	1·8	2·1	03 00
03 30	1·9	1·5	1·2	0·9	0·6	0·3	0·0	0·3	0·5	0·8	1·1	1·3	1·6	1·8	03 30
04 00	1·6	1·3	1·1	0·8	0·5	0·3	0·0	0·2	0·5	0·7	0·9	1·2	1·4	1·6	04 00
04 30	1·5	1·2	0·9	0·7	0·5	0·2	0·0	0·2	0·4	0·6	0·8	1·0	1·3	1·5	04 30
05 00	−1·3	−1·1	−0·9	−0·6	−0·4	−0·2	0·0	+0·2	+0·4	+0·6	+0·8	+0·9	+1·1	+1·3	05 00
06	1·1	0·9	0·7	0·5	0·3	0·2	0·0	0·2	0·3	0·5	0·6	0·8	0·9	1·1	06
07	1·0	0·8	0·6	0·5	0·3	0·1	0·0	0·1	0·3	0·4	0·5	0·7	0·8	0·9	07
08	0·8	0·7	0·5	0·4	0·3	0·1	0·0	0·1	0·2	0·4	0·5	0·6	0·7	0·8	08
09	0·7	0·6	0·5	0·4	0·2	0·1	0·0	0·1	0·2	0·3	0·4	0·5	0·6	0·7	09
10 00	−0·7	−0·5	−0·4	−0·3	−0·2	−0·1	0·0	+0·1	+0·2	+0·3	+0·4	+0·5	+0·6	+0·7	10 00
12	0·6	0·5	0·4	0·3	0·2	0·1	0·0	0·1	0·2	0·2	0·3	0·4	0·5	0·5	12
14	0·5	0·4	0·3	0·2	0·1	0·1	0·0	0·1	0·1	0·2	0·3	0·3	0·4	0·5	14
16	0·4	0·3	0·3	0·2	0·1	0·1	0·0	0·1	0·1	0·2	0·2	0·3	0·3	0·4	16
18	0·4	0·3	0·2	0·2	0·1	−0·1	0·0	+0·1	0·1	0·2	0·2	0·3	0·3	0·4	18
20 00	−0·3	−0·3	−0·2	−0·2	−0·1	0·0	0·0	0·0	+0·1	+0·1	+0·2	+0·2	+0·3	+0·3	20 00
25	0·3	0·2	0·2	0·1	0·1	0·0	0·0	0·0	0·1	0·1	0·1	0·2	0·2	0·2	25
30	0·2	0·2	0·1	0·1	0·1	0·0	0·0	0·0	+0·1	0·1	0·1	0·1	0·2	0·2	30
35	0·2	0·1	0·1	0·1	−0·1	0·0	0·0	0·0	0·0	0·1	0·1	0·1	0·1	0·2	35
40	0·1	0·1	0·1	−0·1	0·0	0·0	0·0	0·0	+0·1	0·1	0·1	0·1	0·1	0·1	40
50 00	−0·1	−0·1	−0·1	0·0	0·0	0·0	0·0	0·0	0·0	0·0	+0·1	+0·1	+0·1	+0·1	50 00

The graph is entered with arguments temperature and pressure to find a zone letter; using as arguments this zone letter and apparent altitude (sextant altitude corrected for index error and dip), a correction is taken from the table. This correction is to be applied to the sextant altitude in addition to the corrections for standard conditions (Table **6**b).

TABLE 6d — ALTITUDE CORRECTION TABLES FOR THE SUN

App. Alt. H_a	OCT.–MAR. Lower Limb R	OCT.–MAR. Upper Limb R	APR.–SEPT. Lower Limb R	APR.–SEPT. Upper Limb R	App. Alt. H_a	OCT.–MAR. Lower Limb R	OCT.–MAR. Upper Limb R	APR.–SEPT. Lower Limb R	APR.–SEPT. Upper Limb R	App. Alt. H_a	OCT.–MAR. Lower Limb R	OCT.–MAR. Upper Limb R	App. Alt. H_a	APR.–SEPT. Lower Limb R	APR.–SEPT. Upper Limb R
° ′	′	′	′	′	° ′	′	′	′	′	° ′	′	′	° ′	′	′
0 00	−17.5	−49.8	−17.8	−49.6	3 30	+ 3.4	−28.9	+ 3.1	−28.7	9 33	+10.8	−21.5	9 39	+10.6	−21.2
0 03	16.9	49.2	17.2	49.0	3 35	3.6	28.7	3.3	28.5	9 45	+10.9	−21.4	9 50	+10.7	−21.1
0 06	16.3	48.6	16.6	48.4	3 40	3.8	28.5	3.6	28.2	9 56	+11.0	−21.3	10 02	+10.8	−21.0
0 09	15.7	48.0	16.0	47.8	3 45	4.0	28.3	3.8	28.0	10 08	+11.1	−21.2	10 14	+10.9	−20.9
0 12	15.2	47.5	15.4	47.2	3 50	4.2	28.1	4.0	27.8	10 20	+11.2	−21.1	10 27	+11.0	−20.8
0 15	14.6	46.9	14.8	46.6	3 55	4.4	27.9	4.1	27.7	10 33	+11.3	−21.0	10 40	+11.1	−20.7
0 18	−14.1	−46.4	−14.3	−46.1	4 00	+ 4.6	−27.7	+ 4.3	−27.5	10 46	+11.4	−20.9	10 53	+11.2	−20.6
0 21	13.5	45.8	13.8	45.6	4 05	4.8	27.5	4.5	27.3	11 00	+11.5	−20.8	11 07	+11.3	−20.5
0 24	13.0	45.3	13.3	45.1	4 10	4.9	27.4	4.7	27.1	11 15	+11.6	−20.7	11 22	+11.4	−20.4
0 27	12.5	44.8	12.8	44.6	4 15	5.1	27.2	4.9	26.9	11 30	+11.7	−20.6	11 37	+11.5	−20.3
0 30	12.0	44.3	12.3	44.1	4 20	5.3	27.0	5.0	26.8	11 45	+11.8	−20.5	11 53	+11.6	−20.2
0 33	11.6	43.9	11.8	43.6	4 25	5.4	26.9	5.2	26.6	12 01	+11.9	−20.4	12 10	+11.7	−20.1
0 36	−11.1	−43.4	−11.3	−43.1	4 30	+ 5.6	−26.7	+ 5.3	−26.5	12 18	+12.0	−20.3	12 27	+11.8	−20.0
0 39	10.6	42.9	10.9	42.7	4 35	5.7	26.6	5.5	26.3	12 36	+12.1	−20.2	12 45	+11.9	−19.9
0 42	10.2	42.5	10.5	42.3	4 40	5.9	26.4	5.6	26.2	12 54	+12.2	−20.1	13 04	+12.0	−19.8
0 45	9.8	42.1	10.0	41.8	4 45	6.0	26.3	5.8	26.0	13 14	+12.3	−20.0	13 24	+12.1	−19.7
0 48	9.4	41.7	9.6	41.4	4 50	6.2	26.1	5.9	25.9	13 34	+12.4	−19.9	13 44	+12.2	−19.6
0 51	9.0	41.3	9.2	41.0	4 55	6.3	26.0	6.1	25.7	13 55	+12.5	−19.8	14 06	+12.3	−19.5
0 54	− 8.6	−40.9	− 8.8	−40.6	5 00	+ 6.4	−25.9	+ 6.2	−25.6	14 17	+12.6	−19.7	14 29	+12.4	−19.4
0 57	8.2	40.5	8.4	40.2	5 05	6.6	25.7	6.3	25.5	14 41	+12.7	−19.6	14 53	+12.5	−19.3
1 00	7.8	40.1	8.0	39.8	5 10	6.7	25.6	6.5	25.3	15 05	+12.8	−19.5	15 18	+12.6	−19.2
1 03	7.4	39.7	7.7	39.5	5 15	6.8	25.5	6.6	25.2	15 31	+12.9	−19.4	15 45	+12.7	−19.1
1 06	7.1	39.4	7.3	39.1	5 20	7.0	25.3	6.7	25.1	15 59	+13.0	−19.3	16 13	+12.8	−19.0
1 09	6.7	39.0	7.0	38.8	5 25	7.1	25.2	6.8	25.0	16 27	+13.1	−19.2	16 43	+12.9	−18.9
1 12	− 6.4	−38.7	− 6.6	−38.4	5 30	+ 7.2	−25.1	+ 6.9	−24.9	16 58	+13.2	−19.1	17 14	+13.0	−18.8
1 15	6.0	38.3	6.3	38.1	5 35	7.3	25.0	7.1	24.7	17 30	+13.3	−19.0	17 47	+13.1	−18.7
1 18	5.7	38.0	6.0	37.8	5 40	7.4	24.9	7.2	24.6	18 05	+13.4	−18.9	18 23	+13.2	−18.6
1 21	5.4	37.7	5.7	37.5	5 45	7.5	24.8	7.3	24.5	18 41	+13.5	−18.8	19 00	+13.3	−18.5
1 24	5.1	37.4	5.3	37.1	5 50	7.6	24.7	7.4	24.4	19 20	+13.6	−18.7	19 41	+13.4	−18.4
1 27	4.8	37.1	5.0	36.8	5 55	7.7	24.6	7.5	24.3	20 02	+13.7	−18.6	20 24	+13.5	−18.3
1 30	− 4.5	−36.8	− 4.7	−36.5	6 00	+ 7.8	−24.5	+ 7.6	−24.2	20 46	+13.8	−18.5	21 10	+13.6	−18.2
1 35	4.0	36.3	4.3	36.1	6 10	8.0	24.3	7.8	24.0	21 34	+13.9	−18.4	21 59	+13.7	−18.1
1 40	3.6	35.9	3.8	35.6	6 20	8.2	24.1	8.0	23.8	22 25	+14.0	−18.3	22 52	+13.8	−18.0
1 45	3.1	35.4	3.4	35.2	6 30	8.4	23.9	8.2	23.6	23 20	+14.1	−18.2	23 49	+13.9	−17.9
1 50	2.7	35.0	2.9	34.7	6 40	8.6	23.7	8.3	23.5	24 20	+14.2	−18.1	24 51	+14.0	−17.8
1 55	2.3	34.6	2.5	34.3	6 50	8.7	23.6	8.5	23.3	25 24	+14.3	−18.0	25 58	+14.1	−17.7
2 00	− 1.9	−34.2	− 2.1	−33.9	7 00	+ 8.9	−23.4	+ 8.7	−23.1	26 34	+14.4	−17.9	27 11	+14.2	−17.6
2 05	1.5	33.8	1.7	33.5	7 10	9.1	23.2	8.8	23.0	27 50	+14.5	−17.8	28 31	+14.3	−17.5
2 10	1.1	33.4	1.4	33.2	7 20	9.2	23.1	9.0	22.8	29 13	+14.6	−17.7	29 58	+14.4	−17.4
2 15	0.8	33.1	1.0	32.8	7 30	9.3	23.0	9.1	22.7	30 44	+14.7	−17.6	31 33	+14.5	−17.3
2 20	0.4	32.7	0.7	32.5	7 40	9.5	22.8	9.2	22.6	32 24	+14.8	−17.5	33 18	+14.6	−17.2
2 25	− 0.1	32.4	− 0.3	32.1	7 50	9.6	22.7	9.4	22.4	34 15	+14.9	−17.4	35 15	+14.7	−17.1
2 30	+ 0.2	−32.1	0.0	−31.8	8 00	+ 9.7	−22.6	+ 9.5	−22.3	36 17	+15.0	−17.3	37 24	+14.8	−17.0
2 35	0.5	31.8	+ 0.3	31.5	8 10	9.9	22.4	9.6	22.2	38 34	+15.1	−17.2	39 48	+14.9	−16.9
2 40	0.8	31.5	0.6	31.2	8 20	10.0	22.3	9.7	22.1	41 06	+15.2	−17.1	42 28	+15.0	−16.8
2 45	1.1	31.2	0.9	30.9	8 30	10.1	22.2	9.9	21.9	43 56	+15.3	−17.0	45 29	+15.1	−16.7
2 50	1.4	30.9	1.2	30.6	8 40	10.2	22.1	10.0	21.8	47 07	+15.4	−16.9	48 52	+15.2	−16.6
2 55	1.7	30.6	1.4	30.4	8 50	10.3	22.0	10.1	21.7	50 43	+15.5	−16.8	52 41	+15.3	−16.5
3 00	+ 2.0	−30.3	+ 1.7	−30.1	9 00	+10.4	−21.9	+10.2	−21.6	54 46	+15.6	−16.7	56 59	+15.4	−16.4
3 05	2.2	30.1	2.0	29.8	9 10	10.5	21.8	10.3	21.5	59 21	+15.7	−16.6	61 50	+15.5	−16.3
3 10	2.5	29.8	2.2	29.6	9 20	10.6	21.7	10.4	21.4	64 28	+15.8	−16.5	67 15	+15.6	−16.2
3 15	2.7	29.6	2.5	29.3	9 30	10.7	21.6	10.5	21.3	70 10	+15.9	−16.4	73 14	+15.7	−16.1
3 20	2.9	29.4	2.7	29.1	9 40	10.8	21.5	10.6	21.2	76 24	+16.0	−16.3	79 42	+15.8	−16.0
3 25	3.2	29.1	2.9	28.9	9 50	10.9	21.4	10.6	21.2	83 05	+16.1	−16.2	86 31	+15.9	−15.9
3 30	+ 3.4	−28.9	+ 3.1	−28.7	10 00	+11.0	−21.3	+10.7	−21.1	90 00			90 00		

Extract from main table for Lat 50° (same) Dec 21° LHA 73°

N. Lat. { LHA greater than 180° $Z_n = Z$
         LHA less than 180° $Z_n = 360° - Z$

DECLINATION (15°- 2

LHA	15° Hc / d / Z	16° Hc / d / Z	17° Hc / d / Z	18° Hc / d / Z	19° Hc / d / Z	20° Hc / d / Z	21° Hc / d / Z
70	24 15 +44 95	24 59 +45 95	25 44 +44 94	26 28 +45 93	27 13 +44 92	27 57 +43 92	28 40 +44 91
71	23 36 45 95	24 21 45 94	25 06 44 93	25 50 44 92	26 34 44 92	27 18 44 91	28 02 43 90
72	22 58 45 94	23 43 44 93	24 27 44 92	25 11 45 92	25 56 43 91	26 39 44 90	27 23 43 89
73	22 19 45 93	23 04 45 92	23 49 44 92	24 33 44 91	25 17 44 90	26 01 44 89	26 45 43 89
74	21 41 44 92	22 25 45 92	23 10 44 91	23 54 44 90	24 38 44 89	25 22 44 89	26 06 43 88
75	21 02 +45 92	21 47 +44 91	22 31 +45 90	23 16 +44 89	24 00 +44 89	24 44 +43 88	25 27 +44 87
76	20 24 44 91	21 08 45 90	21 53 44 89	22 37 44 89	23 21 44 88	24 05 44 87	24 49 43 86
77	19 45 45 90	20 30 44 89	21 14 45 89	21 59 44 88	22 43 44 87	23 27 43 86	24 10 44 86
78	19 07 44 89	19 51 45 89	20 36 44 88	21 20 44 87	22 04 44 86	22 48 44 86	23 32 44 85
79	18 28 45 88	19 13 44 88	19 57 45 87	20 42 44 86	21 26 44 86	22 10 44 85	22 54 43 84
80	17 49 +45 88	18 34 +45 87	19 19 +44 86	20 03 +44 86	20 47 +44 85	21 31 +44 84	22 15 +44 83
81	17 11 45 87	17 56 44 86	18 40 45 86	19 25 44 85	20 09 44 84	20 53 44 83	21 37 44 83
82	16 32 45 86	17 17 45 86	18 02 44 85	18 46 45 84	19 31 44 83	20 15 44 83	20 59 44 82
83	15 54 45 85	16 39 44 85	17 23 45 84	18 08 44 83	18 52 45 83	19 37 44 82	20 21 44 81
84	15 15 45 85	16 00 45 84	16 45 45 83	17 30 44 83	18 14 44 82	18 58 45 81	19 43 44 80
85	14 37 +45 84	15 22 +45 83	16 07 +44 83	16 51 +45 82	17 36 +44 81	18 20 +45 80	19 05 +44 80
86	13 59 45 83	14 44 45 83	15 29 44 82	16 13 45 81	16 58 44 80	17 42 45 80	18 27 44 79
87	13 21 45 82	14 06 44 82	14 50 45 81	15 35 45 80	16 20 44 80	17 04 45 79	17 49 44 78
88	12 42 45 82	13 27 45 81	14 12 45 80	14 57 45 80	15 42 45 79	16 27 44 78	17 11 44 78
89	12 04 45 81	12 49 45 80	13 34 45 80	14 19 45 79	15 04 45 78	15 49 45 78	16 34 44 77
90	11 26 +45 80	12 11 +46 80	12 57 +45 79	13 42 +45 78	14 27 +44 78	15 11 +45 77	15 56 +45 76
91	10 48 46 79	11 34 45 79	12 19 45 78	13 04 45 77	13 49 45 77	14 34 45 76	15 19 44 75
92	10 10 46 79	10 56 45 78	11 41 45 77	12 26 45 77	13 11 45 76	13 56 45 75	14 41 45 75
93	09 33 45 78	10 18 45 77	11 03 46 77	11 49 45 76	12 34 45 75	13 19 45 75	14 04 45 74
94	08 55 45 77	09 40 46 77	10 26 45 76	11 11 46 75	11 57 45 75	12 42 45 74	13 27 45 73
95	08 17 +46 77	09 03 +46 76	09 49 +45 75	10 34 +46 75	11 20 +45 74	12 05 +45 73	12 50 +46 73
96	07 40 46 76	08 26 45 75	09 11 46 74	09 57 46 74	10 43 45 73	11 28 46 72	12 14 45 72
97	07 03 45 75	07 48 46 74	08 34 46 74	09 20 46 73	10 06 46 72	10 52 45 72	11 37 46 71
98	06 25 46 74	07 11 46 74	07 57 46 73	08 43 46 72	09 29 46 72	10 15 46 71	11 01 45 70
99	05 48 46 74	06 34 47 73	07 21 46 72	08 07 46 72	08 53 46 71	09 39 45 70	10 24 46 70
100	05 11 +47 73	05 58 +46 72	06 44 +46 71	07 30 +46 71	08 16 +46 70	09 02 +46 70	09 48 +46 69
101	04 35 46 72	05 21 46 71	06 07 47 71	06 54 46 70	07 40 46 69	08 26 46 69	09 12 47 68
102	03 58 47 71	04 45 46 71	05 31 47 70	06 18 46 69	07 04 46 69	07 50 47 68	08 37 46 67
103	03 22 46 71	04 08 47 70	04 55 47 69	05 42 46 69	06 28 47 68	07 15 46 67	08 01 47 67
104	02 45 47 70	03 32 47 69	04 19 47 69	05 06 47 68	05 53 46 67	06 39 47 67	07 26 47 66
105	02 09 +47 69	02 56 +47 68	03 43 +47 68	04 30 +47 67	05 17 +47 67	06 04 +47 66	06 51 +47 65
106	01 33 46 68	02 20 48 68	03 08 47 67	03 55 47 66	04 42 47 66	05 29 47 65	06 16 47 65
107	00 58 47 67	01 45 47 67	02 32 47 66	03 19 48 66	04 07 47 65	04 54 47 64	05 41 47 64
108	00 22 47 67	01 09 48 66	01 57 47 66	02 44 48 65	03 32 47 64	04 19 48 64	05 07 47 63
109	–0 13 47 66	00 34 48 65	01 22 48 65	02 10 47 64	02 57 48 64	03 45 47 63	04 32 48 62
110	–0 48 +47 65	–0 01 +48 65	00 47 +48 64	01 35 +48 63	02 23 +48 63	03 11 +47 62	03 58 +48 62
111	–1 23 48 64	–0 35 48 64	00 13 48 63	01 01 48 63	01 49 48 62	02 37 48 61	03 25 48 61
112	–1 58 48 64	–1 10 48 63	–0 22 49 62	00 27 48 62	01 15 48 61	02 03 48 61	02 51 48 60
113	–2 32 48 63	–1 44 48 62	–0 56 49 62	–0 07 48 61	00 41 48 61	01 29 49 60	02 18 48 59
114	–3 07 49 62	–2 18 48 62	–1 30 49 61	–0 41 49 60	00 08 48 60	00 56 49 59	01 45 48 59
115	–3 41 +49 61	–2 52 +49 61	–2 03 +49 60	–1 14 +48 60	–0 26 +49 59	00 23 +49 58	01 12 +49 58
116	–4 14 49 61	–3 25 49 60	–2 36 48 59	–1 48 49 59	–0 59 49 58	–0 10 49 58	00 39 49 57
117	–4 48 49 60	–3 59 49 59	–3 10 50 59	–2 20 49 58	–1 31 49 57	–0 42 49 57	00 07 49 56
118	–5 21 49 59	–4 32 50 58	–3 42 49 58	–2 53 49 57	–2 04 50 57	–1 14 49 56	–0 25 50 56
119	–5 54 +50 58	–5 04 49 58	–4 15 50 57	–3 25 49 56	–2 36 50 56	–1 46 50 55	–0 56 49 55
120		–5 37 +50 57	–4 47 +50 56	–3 57 +50 56	–3 07 +49 55	–2 18 +50 55	–1 28 +50 54
121		–6 09 +50 56	–5 19 50 55	–4 29 50 55	–3 39 50 54	–2 49 50 54	–1 59 50 53
122			–5 50 +50 55	–5 00 50 54	–4 10 50 54	–3 20 50 53	–2 30 51 52
123				–5 31 50 53	–4 41 51 53	–3 50 50 52	–3 00 51 52
124				–6 02 +51 52	–5 11 50 52	–4 21 51 51	–3 30 51 51
125					–5 42 +51 51	–4 51 +51 51	–4 00 +51 50
126					–6 11 +51 50	–5 20 51 50	–4 29 51 49
127						–5 50 52 49	–4 58 51 48
128						–6 18 +51 48	–5 27 52 48
129							–5 55 +52 47
130							
131							

©2009 Jack Case

Extract from main table for Lat 50° (same) Dec 21° LHA 257°

20° Hc	d	Z	21° Hc	d	Z	22° Hc	d	Z	23° Hc	d	Z	24° Hc	d	Z	25° Hc	d	LHA
° ′	′	°	° ′	′	°	° ′	′	°	° ′	′	°	° ′	′	°	° ′	′	°
27 57	+43	92	28 40	+44	91	29 24	+43	90	30 07	+43	89	30 50	+42	89	31 32	+42	290
27 18	44	91	28 02	43	90	28 45	43	89	29 28	43	89	30 11	42	88	30 53	43	289
26 39	44	90	27 23	43	89	28 06	44	89	28 50	42	88	29 32	43	87	30 15	42	288
26 01	44	89	26 45	43	89	27 28	43	88	28 11	43	87	28 54	43	86	29 37	42	287
25 22	44	89	26 06	43	88	26 49	44	87	27 33	42	86	28 15	43	86	28 58	42	286
24 44	+43	88	25 27	+44	87	26 11	+43	86	26 54	+43	86	27 37	+43	85	28 20	+42	285
24 05	44	87	24 49	43	86	25 32	44	86	26 16	43	85	26 59	42	84	27 41	43	284
23 27	43	86	24 10	44	86	24 54	43	85	25 37	43	84	26 20	43	83	27 03	43	283
22 48	44	86	23 32	44	85	24 16	43	84	24 59	43	83	25 42	43	83	26 25	42	282
22 10	44	85	22 54	43	84	23 37	44	83	24 21	43	83	25 04	43	82	25 47	42	281
21 31	+44	84	22 15	+44	83	22 59	+43	83	23 42	+44	82	24 26	+43	81	25 09	+42	280
20 53	44	83	21 37	44	83	22 21	43	82	23 04	44	81	23 48	43	80	24 31	43	279
20 15	44	83	20 59	44	82	21 43	43	81	22 26	44	80	23 10	43	80	23 53	43	278
19 37	44	82	20 21	44	81	21 05	43	80	21 48	44	80	22 32	43	79	23 15	43	277
18 58	45	81	19 43	44	80	20 27	43	80	21 10	44	79	21 54	43	78	22 37	43	276
18 20	+45	80	19 05	+44	80	19 49	+43	79	20 32	+44	78	21 16	+44	78	22 00	+43	275
17 42	45	80	18 27	44	79	19 11	44	78	19 55	44	78	20 39	43	77	21 22	44	274
17 04	45	79	17 49	44	78	18 33	44	78	19 17	44	77	20 01	44	76	20 45	43	273
16 27	44	78	17 11	44	78	17 55	45	77	18 40	44	76	19 24	44	75	20 08	43	272
15 49	45	78	16 34	44	77	17 18	44	76	18 02	44	75	18 46	44	75	19 30	44	271
15 11	+45	77	15 56	+45	76	16 41	+44	75	17 25	+44	75	18 09	+44	74	18 53	+44	270
14 34	45	76	15 19	44	75	16 03	45	75	16 48	44	74	17 32	44	73	18 16	45	269
13 56	45	75	14 41	45	75	15 26	45	74	16 11	44	73	16 55	45	73	17 40	44	268
13 19	45	75	14 04	45	74	14 49	45	73	15 34	45	73	16 19	44	72	17 03	45	267
12 42	45	74	13 27	45	73	14 12	45	73	14 57	45	72	15 42	45	71	16 27	44	266
12 05	+45	73	12 50	+46	73	13 36	+45	72	14 21	+45	71	15 06	+44	70	15 50	+45	265
11 28	46	72	12 14	45	72	12 59	45	71	13 44	45	70	14 29	45	70	15 14	45	264
10 52	45	72	11 37	46	71	12 23	45	70	13 08	45	70	13 53	45	69	14 38	45	263
10 15	46	71	11 01	45	70	11 46	46	70	12 32	45	69	13 17	46	68	14 03	45	262
09 39	45	70	10 24	46	70	11 10	46	69	11 56	46	68	12 42	45	68	13 27	45	261
09 02	+46	70	09 48	+46	69	10 34	+46	68	11 20	+46	68	12 06	+46	67	12 52	+45	260
08 26	46	69	09 12	47	68	09 59	46	68	10 45	46	67	11 31	45	66	12 16	46	259
07 50	47	68	08 37	46	67	09 23	46	67	10 09	46	66	10 55	46	66	11 41	46	258
07 15	46	67	08 01	47	67	08 48	46	66	09 34	46	65	10 20	47	65	11 07	46	257
06 39	47	67	07 26	47	66	08 13	46	65	08 59	47	65	09 46	46	64	10 32	46	256
06 04	+47	66	06 51	+47	65	07 38	+46	65	08 24	+47	64	09 11	+47	63	09 58	+46	255
05 29	47	65	06 16	47	65	07 03	47	64	07 50	47	63	08 37	46	63	09 23	47	254
04 54	47	64	05 41	47	64	06 28	47	63	07 15	47	63	08 02	48	62	08 50	46	253
04 19	48	64	05 07	47	63	05 54	47	62	06 41	48	62	07 29	47	61	08 16	47	252
03 45	47	63	04 32	48	62	05 20	47	62	06 07	48	61	06 55	47	60	07 42	48	251
03 11	+47	62	03 58	+48	62	04 46	+48	61	05 34	+47	60	06 21	+48	60	07 09	+48	250
02 37	48	61	03 25	48	61	04 13	47	60	05 00	48	60	05 48	48	59	06 36	48	249
02 03	48	61	02 51	48	60	03 39	48	59	04 27	48	59	05 15	48	58	06 03	48	248
01 29	49	60	02 18	48	59	03 06	48	59	03 54	49	58	04 43	48	58	05 31	48	247
00 56	49	59	01 45	48	59	02 33	49	58	03 22	48	57	04 10	49	57	04 59	48	246
00 23	+49	58	01 12	+49	58	02 01	+48	57	02 49	+49	57	03 38	+49	56	04 27	+49	245
−0 10	49	58	00 39	49	57	01 28	49	56	02 17	49	56	03 06	49	55	03 55	49	244
−0 42	49	57	00 07	49	56	00 56	50	56	01 46	49	55	02 35	49	55	03 24	49	243
−1 14	49	56	−0 25	50	56	00 25	49	55	01 14	49	54	02 03	50	54	02 53	49	242
−1 46	50	55	−0 56	49	55	−0 07	50	54	00 43	49	54	01 32	50	53	02 22	50	241
−2 18	+50	55	−1 28	+50	54	−0 38	+50	53	00 12	+50	53	01 02	+50	52	01 52	+49	240
−2 49	50	54	−1 59	50	53	−1 09	50	53	−0 19	50	52	00 31	50	52	01 21	51	239
−3 20	50	53	−2 30	51	52	−1 39	50	52	−0 49	50	51	00 01	51	51	00 52	50	238
−3 50	50	52	−3 00	51	52	−2 09	50	51	−1 19	51	51	−0 28	50	50	00 22	51	237
−4 21	51	51	−3 30	51	51	−2 39	51	50	−1 48	50	50	−0 58	51	49	−0 07	51	236
−4 51	+51	51	−4 00	+51	50	−3 09	+51	50	−2 18	+51	49	−1 27	+51	48	−0 36	+51	235
−5 20	51	50	−4 29	51	49	−3 38	51	49	−2 47	52	48	−1 55	51	48	−1 04	51	234
−5 50	52	49	−4 58	51	48	−4 07	52	48	−3 15	51	47	−2 24	52	47	−1 32	51	233
−6 18	+51	48	−5 27	52	48	−4 35	52	47	−3 43	51	47	−2 52	52	46	−2 00	52	232
		129	−5 55	+52	47	−5 03	52	46	−4 11	52	46	−3 19	52	45	−2 27	52	231
					130	−5 31	+52	46	−4 39	+52	45	−3 47	+53	45	−2 54	+52	230
					131	−5 58	+52	45	−5 06	53	44	−4 13	52	44	−3 21	52	229

Extract from main table for Lat 50° (same) Dec 21° LHA 357°

DECLINATION (15°- 29°) **SAME** NAME AS LATITUDE — LAT 50°

20° Hc d Z	21° Hc d Z	22° Hc d Z	23° Hc d Z	24° Hc d Z	25° Hc d Z	LHA
60 00 +60 180	61 00 +60 180	62 00 +60 180	63 00 +60 180	64 00 +60 180	65 00 +60 180	360
59 59 60 178	60 59 60 178	61 59 60 178	62 59 60 178	63 59 60 178	64 59 60 178 3	359
59 57 60 176	60 57 60 176	61 57 60 176	62 57 60 176	63 57 60 176	64 57 60 176 5	358
59 54 60 174	60 54 60 174	61 54 60 174	62 54 60 174	63 54 60 174	64 54 59 174 3	357
59 50 60 173	60 50 59 172	61 49 60 172	62 49 60 172	63 49 59 172	64 48 60 171 0	356
59 44 +60 171	60 44 +59 170	61 43 +60 170	62 43 +60 170	63 43 +59 170	64 42 +60 169 3	355
59 37 60 169	60 37 59 169	61 36 60 168	62 36 59 168	63 35 59 168	64 34 59 167 5	354
59 29 60 167	60 29 59 167	61 28 59 166	62 27 59 166	63 26 59 166	64 25 59 165 3	353
59 20 59 165	60 19 59 165	61 18 59 164	62 17 59 164	63 16 59 164	64 15 58 163 1	352
59 10 58 163	60 08 59 163	61 07 59 163	62 06 58 162	63 04 59 162	64 03 58 161 9	351
58 58 +58 162	59 56 +59 161	60 55 +58 161	61 53 +58 160	62 51 +59 160	63 50 +58 159 5	350
58 45 58 160	59 43 58 159	60 41 59 159	61 40 57 158	62 37 58 158	63 35 58 157 4	349
58 31 58 158	59 29 58 158	60 27 58 157	61 25 57 156	62 22 58 156	63 20 57 155 2	348
58 16 58 156	59 14 57 156	60 11 58 155	61 09 57 155	62 06 57 154	63 03 57 153 0	347
58 00 58 155	58 58 57 154	59 55 57 153	60 52 56 153	61 48 57 152	62 45 56 151 8	346
57 43 +57 153	58 40 +57 152	59 37 +56 152	60 33 +57 151	61 30 +56 150	62 26 +55 150 6	345
57 25 57 151	58 22 56 151	59 18 56 150	60 14 56 149	61 10 56 149	62 06 55 148 4	344
57 06 56 150	58 02 56 149	58 58 56 148	59 54 56 148	60 50 55 147	61 45 55 146 2	343
56 46 56 148	57 42 56 147	58 38 55 147	59 33 55 146	60 28 55 145	61 23 54 144 0	342
56 25 56 146	57 21 55 145	58 16 55 145	59 11 54 144	60 05 55 143	61 00 54 143 9	341
56 03 +55 145	56 58 +55 144	57 53 +55 143	58 48 +54 143	59 42 +54 142	60 36 +53 141 7	340
55 41 54 143	56 35 55 143	57 30 54 142	58 24 54 141	59 18 53 140	60 11 53 139 5	339
55 17 55 142	56 12 54 141	57 06 53 140	57 59 53 139	58 52 53 139	59 45 53 138 3	338
54 53 54 140	55 47 54 140	56 41 53 139	57 34 53 138	58 27 52 137	59 19 52 136 2	337
54 28 54 139	55 22 53 138	56 15 53 137	57 07 53 136	58 00 52 135	58 52 51 135 0	336
54 02 +53 137	54 55 +53 137	55 48 +53 136	56 41 +51 135	57 32 +52 134	58 24 +51 133 9	335
53 36 53 136	54 29 52 135	55 21 52 134	56 13 51 133	57 04 51 133	57 55 51 132 7	334
53 09 52 135	54 01 52 134	54 53 52 133	55 45 51 132	56 36 50 131	57 26 50 130 6	333
52 41 52 133	53 33 52 132	54 25 51 132	55 16 50 131	56 06 51 130	56 57 49 129 4	332
52 13 51 132	53 04 51 131	53 55 51 130	54 46 50 129	55 36 50 128	56 26 49 127 3	331
51 44 +51 131	52 35 +51 130	53 26 +50 129	54 16 +50 128	55 06 +49 127	55 55 +49 126 2	330
51 14 51 129	52 05 50 129	52 55 50 128	53 45 50 127	54 35 49 126	55 24 48 125 0	329
50 44 51 128	51 35 50 127	52 25 49 126	53 14 49 125	54 03 49 124	54 52 48 123 9	328
50 14 50 127	51 04 49 126	51 53 49 125	52 42 49 124	53 31 48 123	54 19 48 122 8	327
49 42 50 126	50 32 49 125	51 21 49 124	52 10 49 123	52 59 48 122	53 47 47 121 6	326
49 11 +49 124	50 00 +49 124	50 49 +49 123	51 38 +48 122	52 26 +47 121	53 13 +47 120 5	325
48 39 49 123	49 28 49 122	50 17 48 121	51 05 47 121	51 52 48 120	52 40 46 119 4	324
48 06 49 122	48 55 48 121	49 43 48 120	50 31 48 119	51 19 47 118	52 06 46 117 3	323
47 34 48 121	48 22 48 120	49 10 47 119	49 57 48 118	50 45 46 117	51 31 46 116 2	322
47 00 48 120	47 48 48 119	48 36 47 118	49 23 47 117	50 10 46 116	50 56 46 115 1	321
46 27 +48 119	47 15 +47 118	48 02 +47 117	48 49 +46 116	49 35 +46 115	50 21 +46 114 0	320
45 53 47 118	46 40 47 117	47 27 47 116	48 14 46 115	49 00 46 114	49 46 45 113 9	319
45 18 48 117	46 06 47 116	46 53 46 115	47 39 46 114	48 25 45 113	49 10 45 112 8	318
44 44 47 116	45 31 46 115	46 17 47 114	47 04 45 113	47 49 45 112	48 34 45 111 7	317
44 09 47 115	44 56 46 114	45 42 46 113	46 28 45 112	47 13 45 111	47 58 45 110 6	316
43 34 +46 114	44 20 +46 113	45 06 +46 112	45 52 +45 111	46 37 +45 110	47 22 +44 109 5	315
42 58 46 113	43 44 46 112	44 30 46 111	45 16 45 110	46 01 44 109	46 45 44 108 4	314
42 22 46 112	43 08 46 111	43 54 45 110	44 39 45 109	45 24 44 108	46 08 44 107 3	313
41 46 46 111	42 32 45 110	43 18 45 109	44 03 44 108	44 47 44 107	45 31 44 106 2	312
41 10 46 110	41 56 45 109	42 41 45 108	43 26 44 107	44 10 44 106	44 54 44 105 1	311
40 34 +45 109	41 19 +45 108	42 04 +45 107	42 49 +44 106	43 33 +44 105	44 17 +43 104 0	310
39 57 45 108	40 42 45 107	41 27 45 106	42 12 44 105	42 56 44 104	43 40 43 103 9	309
39 20 45 107	40 05 45 106	40 50 44 105	41 34 44 104	42 18 44 103	43 02 43 102 3	308
38 43 45 106	39 28 45 105	40 13 44 104	40 57 44 103	41 41 43 102	42 24 43 101 7	307
38 06 45 105	38 51 44 104	39 35 44 103	40 19 44 102	41 03 43 101	41 46 43 101 7	306
37 29 +44 104	38 13 +45 103	38 58 +44 102	39 42 +43 101	40 25 +43 101	41 08 +43 100 6	305
36 51 45 103	37 36 44 102	38 20 44 101	39 04 43 101	39 47 43 100	40 30 43 99 5	304
36 14 44 102	36 58 44 101	37 42 44 101	38 26 43 100	39 09 43 99	39 52 43 98 4	303
35 36 44 101	36 20 44 101	37 04 44 100	37 48 43 99	38 31 43 98	39 14 42 97 3	302
34 58 44 101	35 42 44 100	36 26 44 99	37 10 43 98	37 53 43 97	38 36 42 96 3	301
34 20 +44 100	35 04 +44 99	35 48 +43 98	36 31 +44 97	37 15 +42 96	37 57 +43 95 2	300
33 42 44 99	34 26 44 98	35 10 43 97	35 53 43 96	36 36 43 96	37 19 42 95 1	299
33 04 44 98	33 48 43 97	34 31 44 96	35 15 43 96	35 58 42 95	36 40 43 94 0	298
32 26 44 97	33 10 43 96	33 53 43 96	34 36 43 95	35 19 42 94	36 02 42 93 9	297
31 47 44 96	32 31 44 96	33 15 43 95	33 58 43 94	34 41 42 93	35 23 42 92 9	296
31 09 +44 96	31 53 +43 95	32 36 +43 94	33 19 +43 93	34 02 +43 92	34 45 +42 92 3	295
30 31 43 95	31 14 44 94	31 58 43 93	32 41 43 92	33 24 42 92	34 06 42 91 7	294
29 52 44 94	30 36 43 93	31 19 43 92	32 02 43 92	32 45 43 91	33 28 42 90 6	293
29 14 43 93	29 57 44 92	30 41 43 92	31 24 43 91	32 07 42 90	32 49 42 89 5	292
28 35 44 92	29 19 43 92	30 02 43 91	30 45 43 90	31 28 43 89	32 11 42 88 5	291
20°	21°	22°	23°	24°	25°	

© 2009 Jack Case

Extract from table 6 showing values of d from 34' to 60'

34	35	36	37	38	39	40	41	42	43	44	45	46	47	48	49	50	51	52	53	54	55	56	57	58	59	60	d
'	'	'	'	'	'	'	'	'	'	'	'	'	'	'	'	'	'	'	'	'	'	'	'	'	'	'	'
0	0	0	0	0	0	0	0	0	0	0	0	0	0	0	0	0	0	0	0	0	0	0	0	0	0	0	0
1	1	1	1	1	1	1	1	1	1	1	1	1	1	1	1	1	1	1	1	1	1	1	1	1	1	1	1
1	1	1	1	1	1	1	1	1	1	1	2	2	2	2	2	2	2	2	2	2	2	2	2	2	2	2	2
2	2	2	2	2	2	2	2	2	2	2	2	2	2	2	2	2	3	3	3	3	3	3	3	3	3	3	3
2	2	2	2	3	3	3	3	3	3	3	3	3	3	3	3	3	3	3	4	4	4	4	4	4	4	4	4
3	3	3	3	3	3	3	3	4	4	4	4	4	4	4	4	4	4	4	4	4	5	5	5	5	5	5	5
3	4	4	4	4	4	4	4	4	4	4	4	5	5	5	5	5	5	5	5	5	6	6	6	6	6	6	6
4	4	4	4	4	4	4	5	5	5	5	5	5	5	6	6	6	6	6	6	6	6	7	7	7	7	7	7
5	5	5	5	5	5	5	5	6	6	6	6	6	6	6	7	7	7	7	7	7	7	7	8	8	8	8	8
5	5	5	6	6	6	6	6	6	6	7	7	7	7	7	7	8	8	8	8	8	8	8	9	9	9	9	9
6	6	6	6	6	6	7	7	7	7	7	8	8	8	8	8	8	8	9	9	9	9	9	10	10	10	10	10
6	6	7	7	7	7	7	8	8	8	8	8	8	9	9	9	9	9	10	10	10	10	10	10	11	11	11	11
7	7	7	7	8	8	8	8	8	9	9	9	9	10	10	10	10	11	11	11	12	12	12	12	13	13	13	12
7	8	8	8	8	8	9	9	9	9	10	10	10	10	11	11	11	12	12	12	13	13	13	13	14	14	14	13
8	8	8	9	9	9	9	10	10	10	10	10	11	11	11	11	12	12	12	13	13	13	13	14	14	14	14	14
8	9	9	9	10	10	10	10	10	11	11	11	12	12	12	12	13	14	13	13	14	14	14	15	14	15	15	15
9	9	10	10	10	10	11	11	11	11	12	12	12	13	13	13	13	14	14	14	14	15	15	15	15	16	16	16
10	10	10	10	11	11	11	12	12	12	12	13	13	13	14	14	14	14	15	15	15	16	16	16	16	17	17	17
10	10	11	11	11	12	12	12	13	13	13	14	14	14	14	15	15	15	16	16	16	16	17	17	17	18	18	18
11	11	11	12	12	12	13	13	13	14	14	14	15	15	15	16	16	16	16	17	17	17	18	18	18	19	19	19
11	12	12	12	13	13	13	14	14	14	15	15	15	16	16	16	17	17	17	18	18	18	19	19	19	20	20	20
12	12	13	13	13	14	14	14	15	15	15	16	16	16	17	17	18	18	18	19	19	19	20	20	20	21	21	21
12	13	13	14	14	14	15	15	15	16	16	16	17	17	18	18	18	19	19	19	20	20	21	21	21	22	22	22
13	13	14	14	15	15	15	16	16	16	17	17	18	18	18	19	19	20	20	20	21	21	21	22	22	23	23	23
14	14	14	15	15	16	16	16	17	17	18	18	18	19	19	20	20	20	21	21	22	22	22	23	23	24	24	24
14	15	15	15	16	16	17	17	18	18	18	19	19	20	20	20	21	21	22	22	22	23	23	24	24	25	25	25
15	15	16	16	16	17	17	18	18	19	19	20	20	20	21	21	22	22	23	23	23	24	24	25	25	26	26	26
15	16	16	17	17	18	18	18	19	19	20	20	21	21	22	22	22	23	23	24	24	25	25	26	26	27	27	27
16	16	17	17	18	18	19	19	20	20	21	21	21	22	22	23	23	24	24	24	25	25	26	26	27	28	28	28
16	17	17	18	18	19	19	20	20	21	21	22	22	23	23	24	24	25	25	26	26	27	27	28	28	29	29	29
17	18	18	18	19	20	20	20	21	22	22	22	23	24	24	24	25	26	26	26	27	28	28	28	29	30	30	30
18	18	19	19	20	20	21	21	22	22	23	23	24	24	25	25	26	26	27	27	28	28	29	29	30	30	31	31
18	19	19	20	20	21	21	22	22	23	23	24	24	25	26	26	27	27	28	28	29	29	30	30	31	31	32	32
19	19	20	20	21	21	22	23	23	24	24	25	25	26	26	27	28	28	29	30	30	30	31	31	32	32	33	33
19	20	20	21	22	22	23	23	24	24	25	26	26	27	27	28	28	29	29	30	31	31	32	32	33	33	34	34
20	20	21	22	22	23	23	24	24	25	26	26	27	27	28	29	29	30	30	31	32	32	33	33	34	34	35	35
20	21	22	22	23	23	24	25	25	26	26	27	28	28	29	29	30	31	31	32	32	33	34	34	35	35	36	36
21	22	22	23	23	24	25	25	26	27	27	28	28	29	30	30	31	31	32	33	33	34	35	35	36	36	37	37
22	22	23	23	24	25	25	26	27	27	28	28	29	30	30	31	32	32	33	34	34	35	35	36	37	37	38	38
22	23	23	24	25	25	26	27	27	28	29	29	30	31	31	32	32	33	34	34	35	36	36	37	38	38	39	39
23	23	24	25	25	26	27	27	28	29	29	30	31	31	32	33	33	34	35	35	36	37	37	38	39	39	40	40
23	24	25	25	26	27	27	28	29	29	30	31	31	32	33	33	34	35	36	36	37	38	38	39	40	40	41	41
24	24	25	26	27	27	28	29	29	30	31	32	32	33	34	34	35	36	36	37	38	39	40	40	41	41	42	42
24	25	26	27	27	28	29	29	30	31	32	32	33	34	34	35	36	37	37	38	39	39	40	41	42	42	43	43
25	26	26	27	28	29	29	30	31	32	32	33	34	34	35	36	37	37	38	39	40	40	41	42	43	43	44	44
26	26	27	28	28	29	30	31	32	32	33	34	34	35	36	37	38	38	39	40	40	41	42	43	44	44	45	45
26	27	28	28	29	30	31	31	32	33	34	34	35	36	37	38	38	39	40	41	41	42	43	44	44	45	46	46
27	27	28	29	30	31	31	32	33	34	34	35	36	37	38	38	39	40	41	42	42	43	44	45	45	46	47	47
27	28	29	30	30	31	32	33	34	34	35	36	37	38	38	39	40	41	42	42	43	44	45	46	46	47	48	48
28	29	29	30	31	32	33	33	34	35	36	37	38	38	39	40	41	42	42	43	44	45	46	47	47	48	49	49
28	29	30	31	32	32	33	34	35	36	37	38	38	39	40	41	42	42	43	44	45	46	47	48	48	49	50	50
29	30	31	31	32	33	34	35	36	37	37	38	39	40	41	42	43	43	44	45	46	47	48	48	49	50	51	51
29	30	31	32	33	34	34	35	36	37	38	39	40	41	42	42	43	44	45	46	47	48	49	49	50	51	52	52
30	31	32	33	34	34	35	36	37	38	39	40	41	42	42	43	44	45	46	47	48	49	49	50	51	52	53	53
31	32	32	33	34	35	36	37	38	39	40	40	41	42	43	44	45	46	47	48	49	50	50	51	52	53	54	54
31	32	33	34	35	36	37	38	38	39	40	41	42	43	44	45	46	47	48	49	50	50	51	52	53	54	55	55
32	33	34	35	35	36	37	38	39	40	41	42	43	44	45	46	47	48	49	50	51	51	52	53	54	55	56	56
32	33	34	35	36	37	38	39	40	41	42	43	44	45	46	47	48	48	49	50	51	52	53	54	55	56	57	57
33	34	35	36	37	38	39	40	41	42	43	44	44	45	46	47	48	49	50	51	52	53	54	55	56	57	58	58
33	34	35	36	37	38	39	40	41	42	43	44	45	46	47	48	49	50	51	52	53	54	55	56	57	58	59	59

Extract from table 1 showing distance 33n.m. Rel Zn 157

MOO TABLE 1 — ALTITUDE CORRECTION FOR CHANGE IN POSITION OF OBSERV

Rel. Zn	1	2	3	4	5	6	7	8	10	15	20	25	30	35	40	45	50	75	100
000	+1.0	+2.0	+3.0	+4.0	+5.0	+6.0	+7.0	+8.0	+10.0	+15.0	+20.0	+25.0	+30.0	+35.0	+40.0	+45.0	+50.0	+75.0	+100.0
002	1.0	2.0	3.0	4.0	5.0	6.0	7.0	8.0	10.0	15.0	20.0	25.0	30.0	35.0	40.0	45.0	50.0	75.0	99.9
004	1.0	2.0	3.0	4.0	5.0	6.0	7.0	8.0	10.0	15.0	20.0	24.9	29.9	34.9	39.9	44.9	49.9	74.8	99.8
006	1.0	2.0	3.0	4.0	5.0	6.0	7.0	8.0	9.9	14.9	19.9	24.9	29.8	34.8	39.8	44.8	49.7	74.6	99.5
008	1.0	2.0	3.0	4.0	5.0	5.9	6.9	7.9	9.9	14.9	19.8	24.8	29.7	34.7	39.6	44.6	49.5	74.3	99.0
010	+1.0	+2.0	+3.0	+3.9	+4.9	+5.9	+6.9	+7.9	+9.8	+14.8	+19.7	+24.6	+29.5	+34.5	+39.4	+44.3	+49.2	+73.9	+98.5
012	1.0	2.0	2.9	3.9	4.9	5.9	6.8	7.8	9.8	14.7	19.6	24.5	29.3	34.2	39.1	44.0	48.9	73.4	97.8
014	1.0	1.9	2.9	3.9	4.9	5.8	6.8	7.8	9.7	14.6	19.4	24.3	29.1	34.0	38.8	43.7	48.5	72.8	97.0
016	1.0	1.9	2.9	3.8	4.8	5.8	6.7	7.7	9.6	14.4	19.2	24.0	28.8	33.6	38.5	43.3	48.1	72.1	96.1
018	1.0	1.9	2.9	3.8	4.8	5.7	6.7	7.6	9.5	14.3	19.0	23.8	28.5	33.3	38.0	42.8	47.6	71.3	95.1
020	+0.9	+1.9	+2.8	+3.8	+4.7	+5.6	+6.6	+7.5	+9.4	+14.1	+18.8	+23.5	+28.2	+32.9	+37.6	+42.3	+47.0	+70.5	+94.0
022	0.9	1.9	2.8	3.7	4.6	5.6	6.5	7.4	9.3	13.9	18.5	23.2	27.8	32.5	37.1	41.7	46.4	69.5	92.7
024	0.9	1.8	2.7	3.7	4.6	5.5	6.4	7.3	9.1	13.7	18.3	22.8	27.4	32.0	36.5	41.1	45.7	68.5	91.4
026	0.9	1.8	2.7	3.6	4.5	5.4	6.3	7.2	9.0	13.5	18.0	22.5	27.0	31.5	36.0	40.4	44.9	67.4	89.9
028	0.9	1.8	2.6	3.5	4.4	5.3	6.2	7.1	8.8	13.2	17.7	22.1	26.5	30.9	35.3	39.7	44.1	66.2	88.3
030	+0.9	+1.7	+2.6	+3.5	+4.3	+5.2	+6.1	+6.9	+8.7	+13.0	+17.3	+21.7	+26.0	+30.3	+34.6	+39.0	+43.3	+65.0	+86.6
032	0.8	1.7	2.5	3.4	4.2	5.1	5.9	6.8	8.5	12.7	17.0	21.2	25.4	29.7	33.9	38.2	42.4	63.6	84.8
034	0.8	1.7	2.5	3.3	4.1	5.0	5.8	6.6	8.3	12.4	16.6	20.7	24.9	29.0	33.2	37.3	41.5	62.2	82.9
036	0.8	1.6	2.4	3.2	4.0	4.9	5.7	6.5	8.1	12.1	16.2	20.2	24.3	28.3	32.4	36.4	40.5	60.7	80.9
038	0.8	1.6	2.4	3.2	3.9	4.7	5.5	6.3	7.9	11.8	15.8	19.7	23.6	27.6	31.5	35.5	39.4	59.1	78.8
040	+0.8	+1.5	+2.3	+3.1	+3.8	+4.6	+5.4	+6.1	+7.7	+11.5	+15.3	+19.2	+23.0	+26.8	+30.6	+34.5	+38.3	+57.5	+76.6
042	0.7	1.5	2.2	3.0	3.7	4.5	5.2	5.9	7.4	11.1	14.9	18.6	22.3	26.0	29.7	33.4	37.2	55.7	74.3
044	0.7	1.4	2.2	2.9	3.6	4.3	5.0	5.8	7.2	10.8	14.4	18.0	21.6	25.2	28.8	32.4	36.0	54.0	71.9
046	0.7	1.4	2.1	2.8	3.5	4.2	4.9	5.6	6.9	10.4	13.9	17.4	20.8	24.3	27.8	31.3	34.7	52.1	69.5
048	0.7	1.3	2.0	2.7	3.3	4.0	4.7	5.4	6.7	10.0	13.4	16.7	20.1	23.4	26.8	30.1	33.5	50.2	66.9
050	+0.6	+1.3	+1.9	+2.6	+3.2	+3.9	+4.5	+5.1	+6.4	+9.6	+12.9	+16.1	+19.3	+22.5	+25.7	+28.9	+32.1	+48.2	+64.3
052	0.6	1.2	1.8	2.5	3.1	3.7	4.3	4.9	6.2	9.2	12.3	15.4	18.5	21.5	24.6	27.7	30.8	46.2	61.6
054	0.6	1.2	1.8	2.4	2.9	3.5	4.1	4.7	5.9	8.8	11.8	14.7	17.6	20.6	23.5	26.5	29.4	44.1	58.8
056	0.6	1.1	1.7	2.2	2.8	3.4	3.9	4.5	5.6	8.4	11.2	14.0	16.8	19.6	22.4	25.2	28.0	41.9	55.9
058	0.5	1.1	1.6	2.1	2.6	3.2	3.7	4.2	5.3	7.9	10.6	13.2	15.9	18.5	21.2	23.8	26.5	39.7	53.0
060	+0.5	+1.0	+1.5	+2.0	+2.5	+3.0	+3.5	+4.0	+5.0	+7.5	+10.0	+12.5	+15.0	+17.5	+20.0	+22.5	+25.0	+37.5	+50.0
062	0.5	0.9	1.4	1.9	2.3	2.8	3.3	3.8	4.7	7.0	9.4	11.7	14.1	16.4	18.8	21.1	23.5	35.2	46.9
064	0.4	0.9	1.3	1.8	2.2	2.6	3.1	3.5	4.4	6.6	8.8	11.0	13.2	15.3	17.5	19.7	21.9	32.9	43.8
066	0.4	0.8	1.2	1.6	2.0	2.4	2.8	3.3	4.1	6.1	8.1	10.2	12.2	14.2	16.3	18.3	20.3	30.5	40.7
068	0.4	0.7	1.1	1.5	1.9	2.2	2.6	3.0	3.7	5.6	7.5	9.4	11.2	13.1	15.0	16.9	18.7	28.1	37.5
070	+0.3	+0.7	+1.0	+1.4	+1.7	+2.1	+2.4	+2.7	+3.4	+5.1	+6.8	+8.6	+10.3	+12.0	+13.7	+15.4	+17.1	+25.7	+34.2
072	0.3	0.6	0.9	1.2	1.5	1.9	2.2	2.5	3.1	4.6	6.2	7.7	9.3	10.8	12.4	13.9	15.5	23.2	30.9
074	0.3	0.6	0.8	1.1	1.4	1.7	1.9	2.2	2.8	4.1	5.5	6.9	8.3	9.6	11.0	12.4	13.8	20.7	27.6
076	0.2	0.5	0.7	1.0	1.2	1.5	1.7	1.9	2.4	3.6	4.8	6.0	7.3	8.5	9.7	10.9	12.1	18.1	24.2
078	0.2	0.4	0.6	0.8	1.0	1.2	1.5	1.7	2.1	3.1	4.2	5.2	6.2	7.3	8.3	9.4	10.4	15.6	20.8
080	+0.2	+0.3	+0.5	+0.7	+0.9	+1.0	+1.2	+1.4	+1.7	+2.6	+3.5	+4.3	+5.2	+6.1	+6.9	+7.8	+8.7	+13.0	+17.4
082	0.1	0.3	0.4	0.6	0.7	0.8	1.0	1.1	1.4	2.1	2.8	3.5	4.2	4.9	5.6	6.3	7.0	10.4	13.9
084	0.1	0.2	0.3	0.4	0.5	0.6	0.7	0.8	1.0	1.6	2.1	2.6	3.1	3.7	4.2	4.7	5.2	7.8	10.5
086	0.1	0.1	0.2	0.3	0.3	0.4	0.5	0.6	0.7	1.0	1.4	1.7	2.1	2.4	2.8	3.1	3.5	5.2	7.0
088	0.0	0.1	0.1	0.1	0.2	0.2	0.2	0.3	0.3	0.5	0.7	0.9	1.0	1.2	1.4	1.6	1.7	2.6	3.5
090	0.0	0.0	0.0	0.0	0.0	0.0	0.0	0.0	0.0	0.0	0.0	0.0	0.0	0.0	0.0	0.0	0.0	0.0	0.0
092	0.0	−0.1	−0.1	−0.1	−0.2	−0.2	−0.2	−0.3	− 0.3	− 0.5	− 0.7	− 0.9	− 1.0	− 1.2	− 1.4	− 1.6	− 1.7	− 2.6	− 3.5
094	0.1	0.1	0.2	0.3	0.3	0.4	0.5	0.6	0.7	1.0	1.4	1.7	2.1	2.4	2.8	3.1	3.5	5.2	7.0
096	0.1	0.2	0.3	0.4	0.5	0.6	0.7	0.8	1.0	1.6	2.1	2.6	3.1	3.7	4.2	4.7	5.2	7.8	10.5
098	0.1	0.3	0.4	0.6	0.7	0.8	1.0	1.1	1.4	2.1	2.8	3.5	4.2	4.9	5.6	6.3	7.0	10.4	13.9
100	0.2	0.3	0.5	0.7	0.9	1.0	1.2	1.4	1.7	2.6	3.5	4.3	5.2	6.1	6.9	7.8	8.7	13.0	17.4
102	−0.2	−0.4	−0.6	−0.8	−1.0	−1.2	−1.5	−1.7	− 2.1	− 3.1	− 4.2	− 5.2	− 6.2	− 7.3	− 8.3	− 9.4	−10.4	−15.6	− 20.8
104	0.2	0.5	0.7	1.0	1.2	1.5	1.7	1.9	2.4	3.6	4.8	6.0	7.3	8.5	9.7	10.9	12.1	18.1	24.2
106	0.3	0.6	0.8	1.1	1.4	1.7	1.9	2.2	2.8	4.1	5.5	6.9	8.3	9.6	11.0	12.4	13.8	20.7	27.6
108	0.3	0.6	0.9	1.2	1.5	1.9	2.2	2.5	3.1	4.6	6.2	7.7	9.3	10.8	12.4	13.9	15.5	23.2	30.9
110	0.3	0.7	1.0	1.4	1.7	2.1	2.4	2.7	3.4	5.1	6.8	8.6	10.3	12.0	13.7	15.4	17.1	25.7	34.2
112	−0.4	−0.7	−1.1	−1.5	−1.9	−2.2	−2.6	−3.0	− 3.7	− 5.6	− 7.5	− 9.4	−11.2	−13.1	−15.0	−16.9	−18.7	−28.1	− 37.5
114	0.4	0.8	1.2	1.6	2.0	2.4	2.8	3.3	4.1	6.1	8.1	10.2	12.2	14.2	16.3	18.3	20.3	30.5	40.7
116	0.4	0.9	1.3	1.8	2.2	2.6	3.1	3.5	4.4	6.6	8.8	11.0	13.2	15.3	17.5	19.7	21.9	32.9	43.8
118	0.5	0.9	1.4	1.9	2.3	2.8	3.3	3.8	4.7	7.0	9.4	11.7	14.1	16.4	18.8	21.1	23.5	35.2	46.9
120	0.5	1.0	1.5	2.0	2.5	3.0	3.5	4.0	5.0	7.5	10.0	12.5	15.0	17.5	20.0	22.5	25.0	37.5	50.0
122	−0.5	−1.1	−1.6	−2.1	−2.6	−3.2	−3.7	−4.2	− 5.3	− 7.9	−10.6	−13.2	−15.9	−18.5	−21.2	−23.8	−26.5	−39.7	− 53.0
124	0.6	1.1	1.7	2.2	2.8	3.4	3.9	4.5	5.6	8.4	11.2	14.0	16.8	19.6	22.4	25.2	28.0	41.9	55.9
126	0.6	1.2	1.8	2.4	2.9	3.5	4.1	4.7	5.9	8.8	11.8	14.7	17.6	20.6	23.5	26.5	29.4	44.1	58.8
128	0.6	1.2	1.8	2.5	3.1	3.7	4.3	4.9	6.2	9.2	12.3	15.4	18.5	21.5	24.6	27.7	30.8	46.2	61.6
130	0.6	1.3	1.9	2.6	3.2	3.9	4.5	5.1	6.4	9.6	12.9	16.1	19.3	22.5	25.7	28.9	32.1	48.2	64.3
132	−0.7	−1.3	−2.0	−2.7	−3.3	−4.0	−4.7	−5.4	− 6.7	−10.0	−13.4	−16.7	−20.1	−23.4	−26.8	−30.1	−33.5	−50.2	− 66.9
134	0.7	1.4	2.1	2.8	3.5	4.2	4.9	5.6	6.9	10.4	13.9	17.4	20.8	24.3	27.8	31.3	34.7	52.1	69.5
136	0.7	1.4	2.2	2.9	3.6	4.3	5.0	5.8	7.2	10.8	14.4	18.0	21.6	25.2	28.8	32.4	36.0	54.0	71.9
138	0.7	1.5	2.2	3.0	3.7	4.5	5.2	5.9	7.4	11.1	14.9	18.6	22.3	26.0	29.7	33.4	37.2	55.7	74.3
140	0.8	1.5	2.3	3.1	3.8	4.6	5.4	6.1	7.7	11.5	15.3	19.2	23.0	26.8	30.6	34.5	38.3	57.5	76.6
142	−0.8	−1.6	−2.4	−3.2	−3.9	−4.7	−5.5	−6.3	− 7.9	−11.8	−15.8	−19.7	−23.6	−27.6	−31.5	−35.5	−39.4	−59.1	− 78.8
144	0.8	1.6	2.4	3.2	4.0	4.9	5.7	6.5	8.1	12.1	16.2	20.2	24.3	28.3	32.4	36.4	40.5	60.7	80.9
146	0.8	1.7	2.5	3.3	4.1	5.0	5.8	6.6	8.3	12.4	16.6	20.7	24.9	29.0	33.2	37.3	41.5	62.2	82.9
148	0.8	1.7	2.5	3.4	4.2	5.1	5.9	6.8	8.5	12.7	17.0	21.2	25.4	29.7	33.9	38.2	42.4	63.6	84.8
150	0.9	1.7	2.6	3.5	4.3	5.2	6.1	6.9	8.7	13.0	17.3	21.7	26.0	30.3	34.6	39.0	43.3	65.0	86.6
152	−0.9	−1.8	−2.6	−3.5	−4.4	−5.3	−6.2	−7.1	− 8.8	−13.2	−17.7	−22.1	−26.5	−30.9	−35.3	−39.7	−44.1	−66.2	− 88.3
154	0.9	1.8	2.7	3.6	4.5	5.4	6.3	7.2	9.0	13.5	18.0	22.5	27.0	31.5	36.0	40.4	44.9	67.4	89.9
156	0.9	1.8	2.7	3.7	4.6	5.5	6.4	7.3	9.1	13.7	18.3	22.8	27.4	32.0	36.5	41.1	45.7	68.5	91.4
158	0.9	1.9	2.8	3.7	4.6	5.6	6.5	7.4	9.3	13.9	18.5	23.2	27.8	32.5	37.1	41.7	46.4	69.5	92.7
160	0.9	1.9	2.8	3.8	4.7	5.6	6.6	7.5	9.4	14.1	18.8	23.5	28.2	32.9	37.6	42.3	47.0	70.5	94.0

Extract from table 1 showing distance 25 n.m. Rel Zn 001

— ALTITUDE CORRECTION FOR CHANGE IN POSITION OF OBSERVER

4	5	6	7	8	10	15	20	25	30	35	40	45	50	75	100	150	Rel. Zn
+4.0	+5.0	+6.0	+7.0	+8.0	+10.0	+15.0	+20.0	+25.0	+30.0	+35.0	+40.0	+45.0	+50.0	+75.0	+100.0	+150.0	000
4.0	5.0	6.0	7.0	8.0	10.0	15.0	20.0	25.0	30.0	35.0	40.0	45.0	50.0	75.0	99.9	149.9	358
4.0	5.0	6.0	7.0	8.0	10.0	15.0	20.0	24.9	29.9	34.9	39.9	44.9	49.9	74.8	99.8	149.6	356
4.0	5.0	6.0	7.0	8.0	9.9	14.9	19.9	24.9	29.8	34.8	39.8	44.8	49.7	74.6	99.5	149.2	354
4.0	5.0	5.9	6.9	7.9	9.9	14.9	19.8	24.8	29.7	34.7	39.6	44.6	49.5	74.3	99.0	148.5	352
+3.9	+4.9	+5.9	+6.9	+7.9	+ 9.8	+14.8	+19.7	+24.6	+29.5	+34.5	+39.4	+44.3	+49.2	+73.9	+ 98.5	+147.7	350
3.9	4.9	5.9	6.8	7.8	9.8	14.7	19.6	24.5	29.3	34.2	39.1	44.0	48.9	73.4	97.8	146.7	348
3.9	4.9	5.8	6.8	7.8	9.7	14.6	19.4	24.3	29.1	34.0	38.8	43.7	48.5	72.8	97.0	145.5	346
3.8	4.8	5.8	6.7	7.7	9.6	14.4	19.2	24.0	28.8	33.6	38.5	43.3	48.1	72.1	96.1	144.2	344
3.8	4.8	5.7	6.7	7.6	9.5	14.3	19.0	23.8	28.5	33.3	38.0	42.8	47.6	71.3	95.1	142.7	342
+3.8	+4.7	+5.6	+6.6	+7.5	+ 9.4	+14.1	+18.8	+23.5	+28.2	+32.9	+37.6	+42.3	+47.0	+70.5	+ 94.0	+141.0	340
3.7	4.6	5.6	6.5	7.4	9.3	13.9	18.5	23.2	27.8	32.5	37.1	41.7	46.4	69.5	92.7	139.1	338
3.7	4.6	5.5	6.4	7.3	9.1	13.7	18.3	22.8	27.4	32.0	36.5	41.1	45.7	68.5	91.4	137.0	336
3.6	4.5	5.4	6.3	7.2	9.0	13.5	18.0	22.5	27.0	31.5	36.0	40.4	44.9	67.4	89.9	134.8	334
3.5	4.4	5.3	6.2	7.1	8.8	13.2	17.7	22.1	26.5	30.9	35.3	39.7	44.1	66.2	88.3	132.4	332
+3.5	+4.3	+5.2	+6.1	+6.9	+ 8.7	+13.0	+17.3	+21.7	+26.0	+30.3	+34.6	+39.0	+43.3	+65.0	+ 86.6	+129.9	330
3.4	4.2	5.1	5.9	6.8	8.5	12.7	17.0	21.2	25.4	29.7	33.9	38.2	42.4	63.6	84.8	127.2	328
3.3	4.1	5.0	5.8	6.6	8.3	12.4	16.6	20.7	24.9	29.0	33.2	37.3	41.5	62.2	82.9	124.4	326
3.2	4.0	4.9	5.7	6.5	8.1	12.1	16.2	20.2	24.3	28.3	32.4	36.4	40.5	60.7	80.9	121.4	324
3.2	3.9	4.7	5.5	6.3	7.9	11.8	15.8	19.7	23.6	27.6	31.5	35.5	39.4	59.1	78.8	118.2	322
+3.1	+3.8	+4.6	+5.4	+6.1	+ 7.7	+11.5	+15.3	+19.2	+23.0	+26.8	+30.6	+34.5	+38.3	+57.5	+ 76.6	+114.9	320
3.0	3.7	4.5	5.2	5.9	7.4	11.1	14.9	18.6	22.3	26.0	29.7	33.4	37.2	55.7	74.3	111.5	318
2.9	3.6	4.3	5.0	5.8	7.2	10.8	14.4	18.0	21.6	25.2	28.8	32.4	36.0	54.0	71.9	107.9	316
2.8	3.5	4.2	4.9	5.6	6.9	10.4	13.9	17.4	20.8	24.3	27.8	31.3	34.7	52.1	69.5	104.2	314
2.7	3.3	4.0	4.7	5.4	6.7	10.0	13.4	16.7	20.1	23.4	26.8	30.1	33.5	50.2	66.9	100.4	312
+2.6	+3.2	+3.9	+4.5	+5.1	+ 6.4	+ 9.6	+12.9	+16.1	+19.3	+22.5	+25.7	+28.9	+32.1	+48.2	+ 64.3	+ 96.4	310
2.5	3.1	3.7	4.3	4.9	6.2	9.2	12.3	15.4	18.5	21.5	24.6	27.7	30.8	46.2	61.6	92.3	308
2.4	2.9	3.5	4.1	4.7	5.9	8.8	11.8	14.7	17.6	20.6	23.5	26.5	29.4	44.1	58.8	88.2	306
2.2	2.8	3.4	3.9	4.5	5.6	8.4	11.2	14.0	16.8	19.6	22.4	25.2	28.0	41.9	55.9	83.9	304
2.1	2.6	3.2	3.7	4.2	5.3	7.9	10.6	13.2	15.9	18.5	21.2	23.8	26.5	39.7	53.0	79.5	302
+2.0	+2.5	+3.0	+3.5	+4.0	+ 5.0	+ 7.5	+10.0	+12.5	+15.0	+17.5	+20.0	+22.5	+25.0	+37.5	+ 50.0	+ 75.0	300
1.9	2.3	2.8	3.3	3.8	4.7	7.0	9.4	11.7	14.1	16.4	18.8	21.1	23.5	35.2	46.9	70.4	298
1.8	2.2	2.6	3.1	3.5	4.4	6.6	8.8	11.0	13.2	15.3	17.5	19.7	21.9	32.9	43.8	65.8	296
1.6	2.0	2.4	2.8	3.3	4.1	6.1	8.1	10.2	12.2	14.2	16.3	18.3	20.3	30.5	40.7	61.0	294
1.5	1.9	2.2	2.6	3.0	3.7	5.6	7.5	9.4	11.2	13.1	15.0	16.9	18.7	28.1	37.5	56.2	292
+1.4	+1.7	+2.1	+2.4	+2.7	+ 3.4	+ 5.1	+ 6.8	+ 8.6	+10.3	+12.0	+13.7	+15.4	+17.1	+25.7	+ 34.2	+ 51.3	290
1.2	1.5	1.9	2.2	2.5	3.1	4.6	6.2	7.7	9.3	10.8	12.4	13.9	15.5	23.2	30.9	46.4	288
1.1	1.4	1.7	1.9	2.2	2.8	4.1	5.5	6.9	8.3	9.6	11.0	12.4	13.8	20.7	27.6	41.3	286
1.0	1.2	1.5	1.7	1.9	2.4	3.6	4.8	6.0	7.3	8.5	9.7	10.9	12.1	18.1	24.2	36.3	284
0.8	1.0	1.2	1.5	1.7	2.1	3.1	4.2	5.2	6.2	7.3	8.3	9.4	10.4	15.6	20.8	31.2	282
+0.7	+0.9	+1.0	+1.2	+1.4	+ 1.7	+ 2.6	+ 3.5	+ 4.3	+ 5.2	+ 6.1	+ 6.9	+ 7.8	+ 8.7	+13.0	+ 17.4	+ 26.0	280
0.6	0.7	0.8	1.0	1.1	1.4	2.1	2.8	3.5	4.2	4.9	5.6	6.3	7.0	10.4	13.9	20.9	278
0.4	0.5	0.6	0.7	0.8	1.0	1.6	2.1	2.6	3.1	3.7	4.2	4.7	5.2	7.8	10.5	15.7	276
0.3	0.3	0.4	0.5	0.6	0.7	1.0	1.4	1.7	2.1	2.4	2.8	3.1	3.5	5.2	7.0	10.5	274
0.1	0.2	0.2	0.2	0.3	0.3	0.5	0.7	0.9	1.0	1.2	1.4	1.6	1.7	2.6	3.5	5.2	272
0.0	0.0	0.0	0.0	0.0	0.0	0.0	0.0	0.0	0.0	0.0	0.0	0.0	0.0	0.0	0.0	0.0	270
−0.1	−0.2	−0.2	−0.2	−0.3	− 0.3	− 0.5	− 0.7	− 0.9	− 1.0	− 1.2	− 1.4	− 1.6	− 1.7	− 2.6	− 3.5	− 5.2	268
0.3	0.3	0.4	0.5	0.6	0.7	1.0	1.4	1.7	2.1	2.4	2.8	3.1	3.5	5.2	7.0	10.5	266
0.4	0.5	0.6	0.7	0.8	1.0	1.6	2.1	2.6	3.1	3.7	4.2	4.7	5.2	7.8	10.5	15.7	264
0.6	0.7	0.8	1.0	1.1	1.4	2.1	2.8	3.5	4.2	4.9	5.6	6.3	7.0	10.4	13.9	20.9	262
0.7	0.9	1.0	1.2	1.4	1.7	2.6	3.5	4.3	5.2	6.1	6.9	7.8	8.7	13.0	17.4	26.0	260
−0.8	−1.0	−1.2	−1.5	−1.7	− 2.1	− 3.1	− 4.2	− 5.2	− 6.2	− 7.3	− 8.3	− 9.4	−10.4	−15.6	− 20.8	− 31.2	258
1.0	1.2	1.5	1.7	1.9	2.4	3.6	4.8	6.0	7.3	8.5	9.7	10.9	12.1	18.1	24.2	36.3	256
1.1	1.4	1.7	1.9	2.2	2.8	4.1	5.5	6.9	8.3	9.6	11.0	12.4	13.8	20.7	27.6	41.3	254
1.2	1.5	1.9	2.2	2.5	3.1	4.6	6.2	7.7	9.3	10.8	12.4	13.9	15.5	23.2	30.9	46.4	252
1.4	1.7	2.1	2.4	2.7	3.4	5.1	6.8	8.6	10.3	12.0	13.7	15.4	17.1	25.7	34.2	51.3	250
−1.5	−1.9	−2.2	−2.6	−3.0	− 3.7	− 5.6	− 7.5	− 9.4	−11.2	−13.1	−15.0	−16.9	−18.7	−28.1	− 37.5	− 56.2	248
1.6	2.0	2.4	2.8	3.3	4.1	6.1	8.1	10.2	12.2	14.2	16.3	18.3	20.3	30.5	40.7	61.0	246
1.8	2.2	2.6	3.1	3.5	4.4	6.6	8.8	11.0	13.2	15.3	17.5	19.7	21.9	32.9	43.8	65.8	244
1.9	2.3	2.8	3.3	3.8	4.7	7.0	9.4	11.7	14.1	16.4	18.8	21.1	23.5	35.2	46.9	70.4	242
2.0	2.5	3.0	3.5	4.0	5.0	7.5	10.0	12.5	15.0	17.5	20.0	22.5	25.0	37.5	50.0	75.0	240
−2.1	−2.6	−3.2	−3.7	−4.2	− 5.3	− 7.9	−10.6	−13.2	−15.9	−18.5	−21.2	−23.8	−26.5	−39.7	− 53.0	− 79.5	238
2.2	2.8	3.4	3.9	4.5	5.6	8.4	11.2	14.0	16.8	19.6	22.4	25.2	28.0	41.9	55.9	83.9	236
2.4	2.9	3.5	4.1	4.7	5.9	8.8	11.8	14.7	17.6	20.6	23.5	26.5	29.4	44.1	58.8	88.2	234
2.5	3.1	3.7	4.3	4.9	6.2	9.2	12.3	15.4	18.5	21.5	24.6	27.7	30.8	46.2	61.6	92.3	232
2.6	3.2	3.9	4.5	5.1	6.4	9.6	12.9	16.1	19.3	22.5	25.7	28.9	32.1	48.2	64.3	96.4	230
−2.7	−3.3	−4.0	−4.7	−5.4	− 6.7	−10.0	−13.4	−16.7	−20.1	−23.4	−26.8	−30.1	−33.5	−50.2	− 66.9	−100.4	228
2.8	3.5	4.2	4.9	5.6	6.9	10.4	13.9	17.4	20.8	24.3	27.8	31.3	34.7	52.1	69.5	104.2	226

©2009 Jack Case

Question 4.

Planning phase.
Basing your calculations on the scenario below, use the special technique for planning observations to fix the position of the ship at $18^h\ 00^m$ (-1). The procedure was demonstrated in chapter 12.

Fix phase.
Use the results of the observations shown in the scenario to fix the position of the ship at $18^h\ 00^m$ (-1). Remember to take account of the movement of the body being observed (MOB) between each observation.

Scenario.
D.R. Position: 39° 59'.6S, 15° 33'E.
Zone: -1
Date: 23 July 2009
Sunset: $17^h\ 00^m$ (nautical almanac daily page)
Civil twilight: $17^h\ 29^m$
Nautical twilight: $18^h\ 02^m$
Time of planning: $15^h\ 00^m$
Course 054° speed 12 knots.
Time of Fix: $18^h\ 00^m$ (this is within the optimum time period for star and planet observations (i.e. between civil and nautical twilight)).
E.P. at time of fix: 39° 55'.3S, 16° 08'E.
DWE 35^s fast.
Index error: -1'.2
Ht. of eye: 4.5m.

Observations.
1. Bodies selected: Where appropriate, the method described in chapter 8 for planning star and planet observations was used to select the following celestial bodies for the observations: The Moon, Saturn, and Alphard.
2. The calculated altitudes and true bearings established at steps 4 and 5 were used to help verify the identities of the bodies observed at the fix stage (see page 161).

The solution to this question begins on the next page.
Relevant extracts of navigational data can be found after the solution.

Solution Question 4.

Planning phase

Remember that, at this stage, calculations are based on the UT of the planned fix. Details of the actual observations are added later, when they have been made.

Step 1. Approx Pos: $18^h\ 00^m$ (-1) $39°\ 55'.3S,\ 16°\ 08'E$.

Step 2. Calculate UT.

Z.T. of fix:	$23^d\ 18^h\ 00^m$
Zone corrn:	-1^h
Greenwich Date:	$23^d\ 17^h\ 00^m$
UT of fix:	$17^h\ 00^m$

Step 3. Calculate GHA and Dec.

	Moon LL	Saturn	Alphard
Dec 17^h	N11° 40'.3	N6° 23'.0	S8° 42'.0
d	(15'.2)	(0'.1)	n.a.
d cor at 00^m	-0'.1	0'.0	n.a.
Dec + corr.	N11° 40'.2	N6° 23'.0	S8° 42'.0
SHA (stars only)			217° 59'.2
GHAΥ(17^h) (stars only)			196° 33'.9
(GHAΥ+SHA) (stars only)			414° 33'.1
GHA 17^h	51° 22'.1	26° 20'.2	
Inc (m+s)	0'.0	0'.0	0'.0
v	(8.6)	(2.2)	n.a.
v corr.	+0'.1	0'.0	n.a.
Sum = GHA	51° 22'.2	26° 20'.2	414° 33'.1

Step 4. Calculate LHA

	Moon LL	Saturn	Alphard
GHA	51° 22'.2	26° 20'.2	414° 33'.1
As.long(E)	+16° 37'.8	+16° 39'.8	+16° 26'.9
LHA	68°	43°	431°
±360°			71°
Convert azimuth angle to true bearing (Zn):			
LHA =	68°	43°	71°
Z =	113°	129°	96°
Bear'g(Zn)	293°	309°	276°

Rules for calculating true bearing(Zn):		
	Lat. North	Lat. South
LHA>180°	Zn = Z	Zn = 180° - Z
LHA<180°	Zn = 360°-Z	Zn = 180° + Z

Step 5. Main table entry.

	Moon LL	Saturn	Alphard
As. lat.	S40°	S40°	S40°
Dec°	N11°	N6°	S8°
	contr.	contr.	same
Dec'	40'.2	23'.0	42'.0
LHA	68°	43°	71°
Table page:	10	10	9
Tab Hc,	09° 09'	29° 20'	19° 40'
d	-42'	-48'	+38'
Z	113°	129°	96°
Correction for mins of Dec. (Table 5)			
Dec'	40'.2	23'.0	42'.0
d	-42'	-48	+38
Tab Hc,	09° 09'	29° 20'	19° 40'
Cor. for Dec'	-28'	-18 '	+27'
Cal. Alt. Hc (Hc+Cor.Dec'	08° 41'	29° 02 '	20° 07'

The fix phase

Details of the observations made:

Observation 1:
DWT: 16h 59m 45s
Body observed: Moon LL.
Sextant Alt: 07° 40.'35 Bearing: 292°

Observation 2:
DWT: 17h 00m 45s
Body observed: Saturn.
Sextant Alt: 28° 57'.65 Bearing: 310°

Observation 3.
DWT: 17h 01m 05s
Body observed: Alphard.
Sextant Alt: 19° 58'.25 Bearing: 275°
Weather conditions at $18^h 00^m$ (-1): Temp: 21°C. Pressure: 988mb.

The UT of each observation is calculated to determine whether it is earlier or later than the UT of the fix and by how much.

	Moon LL	Saturn	Alphard
DWT:	$16^h\ 59^m\ 45^s$	$17^h\ 00^m\ 45^s$	$17^h\ 01^m\ 05^s$
DWE:	-35^s	-35^s	-35^s
a. UT obs:	$16^h\ 59^m\ 10^s$	$17^h\ 00^m\ 10^s$	$17^h\ 00^m\ 30^s$
b. UT fix:	$17^h\ 00^m\ 00^s$	$17^h\ 00^m\ 00^s$	$17^h\ 00^m\ 00^s$
b ~ a	$+50^s$	-10^s	-30^s
∴ Fix is	50^s later	10^s earlier	30^s earlier
than the observation			

Now continue from step 6:

Step 6. Correct sextant altitude (Hs).			
Sextant altitudes are now added to the previous calculations.			
	Moon LL	Saturn	Alphard
Sext. Alt. (Hs)	07° 40'.35	28° 57'.65	19° 58'.25
IE	-1'.2	-1'.2	-1'.2
Dip (4.5m) (D)	-3'.75	-3'.75	-3'.75
Ap. Alt. (Ha)	07° 35'.40	28° 52'.70	19° 53'.3
Alt. Cor. (R	+60'.95	-1'.8	-2'.7
HPMoon: (60.8)			
Corr. For HP	+8'.8		
Add'l refrac.	+0'.35	+0'.1	+0'.1
Obs.Alt. (Ho)	08° 45'.5	28° 51'.0	19° 50'.7
Step 7. Calculate Intercept			
Ho	08° 45'.5	28° 51'.0	19° 50'.7
Hc	08° 41'.0	29° 02'.0	20° 07'.0
p(Ho-Hc)	+4'.5	-11'.0	-16'.3

Therefore, intercepts are as follows: Moon: 4.5 n.m. towards 293°, 11 n.m. from 309°, 16.3 n.m. from 276°.
(Note. True bearings calculated at step 4).

Step 8

	Moon LL	**Saturn**	**Alphard**
(a) MOO:	nil	nil	nil
(b) MOB (table 2)			
Lat =	S40°	S40°	S40°
Zn =	293°	309°	276°
Fix is	50^s(0.83^m)	10^s(0.17^m)	30^s(0.5^m)
	later	earlier	earlier
corr for 1^m :	-10'.6	-9'.0	-11'.5
corr for fraction:	-8'.8	-1'.53	-5'.75
Apply rules below:	-8'.8	+1'.53	+5'.75
intercept:	+4'.5	-11'.0	-16'.3
Corrected 'cpt	-4'.3 from 293°	-9'.47 from 309°	-10'.55 from 276°

Rules for applying the sign from table 1 & 2		
Time of fix	Sign from table	To intercept
Later than observation	+	Add
	-	Subtract
Earlier than observation	+	Subtract
	-	Add

9. Plot the position lines. (See chapter 12 for method). As in the previous example, a small 'cocked-hat' is produced.

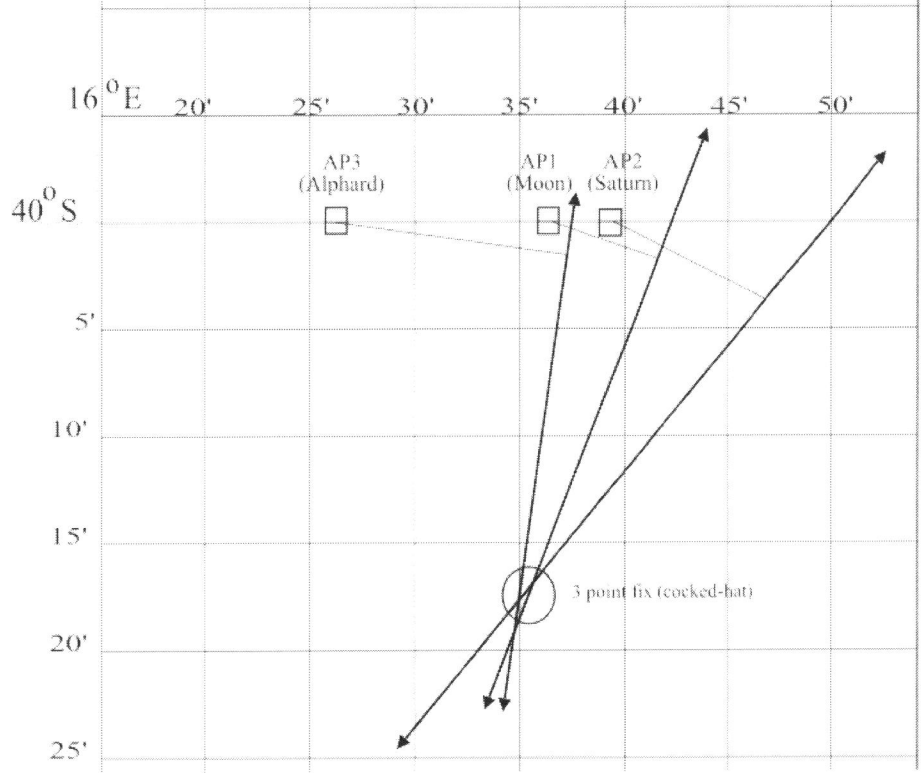

2009 JULY 21, 22, 23 (TUES., WED., THURS.)

UT	SUN GHA	Dec	MOON GHA	v	Dec	d	HP	Lat.	Twilight Naut.	Civil	Sunrise	Moonrise 21	22	23	24
d h	° '	° '	° '	'	° '	'	'	°	h m	h m	h m	h m	h m	h m	h m
21 00	178 24.0	N20 29.1	194 59.4	2.5	N24 26.0	7.0	61.2	N 72	▭	▭	▭	▭	▭	03 27	06 20
01	193 23.9	28.6	209 20.9	2.6	24 19.0	7.2	61.2	N 70	▭	▭	▭	▭	▭	04 02	06 32
02	208 23.9	28.1	223 42.5	2.6	24 11.8	7.4	61.2	68	////	////	01 15	▭	01 38	04 27	06 43
03	223 23.9	27.6	238 04.1	2.7	24 04.4	7.6	61.3	66	////	////	02 06	▭	02 23	04 46	06 51
04	238 23.8	27.1	252 25.8	2.8	23 56.8	7.8	61.3	64	////	////	02 37	00 33	02 53	05 01	06 58
05	253 23.8	26.7	266 47.6	2.8	23 49.0	7.9	61.3	62	////	01 31	03 00	01 14	03 15	05 13	07 04
								60	////	02 08	03 18	01 42	03 33	05 24	07 10
06	268 23.8	N20 26.2	281 09.4	2.9	N23 41.1	8.2	61.3	N 58	////	02 33	03 33	02 04	03 48	05 33	07 14
07	283 23.8	25.7	295 31.3	3.0	23 32.9	8.3	61.3	56	01 23	02 53	03 46	02 21	04 00	05 41	07 18
08	298 23.7	25.2	309 53.3	3.0	23 24.6	8.4	61.3	54	01 57	03 09	03 57	02 36	04 11	05 48	07 22
T 09	313 23.7	24.7	324 15.3	3.2	23 16.2	8.7	61.3	52	02 21	03 23	04 07	02 49	04 21	05 55	07 25
U 10	328 23.7	24.2	338 37.5	3.2	23 07.5	8.8	61.3	50	02 40	03 35	04 16	03 01	04 30	06 00	07 28
E 11	343 23.6	23.7	352 59.7	3.3	22 58.7	9.0	61.3	45	03 14	03 59	04 34	03 24	04 48	06 13	07 34
S 12	358 23.6	N20 23.2	7 22.0	3.3	N22 49.7	9.2	61.3	N 40	03 39	04 18	04 49	03 43	05 03	06 23	07 40
D 13	13 23.6	22.8	21 44.3	3.5	22 40.5	9.3	61.3	35	03 59	04 34	05 02	03 59	05 15	06 31	07 44
A 14	28 23.5	22.3	36 06.8	3.5	22 31.2	9.5	61.3	30	04 15	04 47	05 13	04 12	05 26	06 39	07 48
Y 15	43 23.5	21.8	50 29.3	3.6	22 21.7	9.7	61.3	20	04 40	05 08	05 32	04 36	05 45	06 52	07 55
16	58 23.5	21.3	64 51.9	3.7	22 12.0	9.8	61.3	N 10	04 59	05 25	05 48	04 56	06 01	07 03	08 02
17	73 23.5	20.8	79 14.6	3.8	22 02.2	10.0	61.3	0	05 15	05 41	06 03	05 14	06 16	07 14	08 07
18	88 23.4	N20 20.3	93 37.4	3.9	N21 52.2	10.1	61.3	S 10	05 30	05 55	06 18	05 33	06 31	07 24	08 13
19	103 23.4	19.8	108 00.3	4.0	21 42.1	10.3	61.3	20	05 43	06 10	06 33	05 53	06 47	07 35	08 19
20	118 23.4	19.3	122 23.3	4.1	21 31.8	10.5	61.3	30	05 57	06 26	06 51	06 16	07 05	07 48	08 26
21	133 23.4	18.8	136 46.4	4.2	21 21.3	10.6	61.3	35	06 04	06 35	07 02	06 29	07 16	07 55	08 30
22	148 23.3	18.3	151 09.6	4.2	21 10.7	10.8	61.3	40	06 11	06 46	07 14	06 44	07 28	08 04	08 34
23	163 23.3	17.8	165 32.8	4.4	N20 59.9	10.9	61.3	45	06 19	06 55	07 28	07 03	07 42	08 13	08 39
22 00	178 23.3	N20 17.3						S 50	06 28	07 08	07 45	07 25	07 59	08 25	08 45
01	193 23.2	16.8	A total eclipse of					52	06 32	07 14	07 52	07 36	08 07	08 30	08 48
02	208 23.2	16.3	the Sun occurs on this					54	06 36	07 21	08 01	07 48	08 16	08 36	08 51
03	223 23.2	15.8	date. See page 5.					56	06 41	07 28	08 11	08 02	08 26	08 43	08 55
04	238 23.2	15.3						58	06 46	07 36	08 23	08 18	08 38	08 50	08 58
05	253 23.1	14.8						S 60	06 51	07 45	08 36	08 38	08 51	08 58	09 03
06	268 23.1	N20 14.3	266 18.5	5.0	N19 40.6	11.9	61.3								
W 07	283 23.1	13.8	280 42.5	5.2	19 28.7	12.0	61.3	Lat.	Sunset	Twilight Civil	Naut.	Moonset 21	22	23	24
E 08	298 23.1	13.3	295 06.7	5.3	19 16.7	12.2	61.3								
D 09	313 23.1	12.8	309 31.0	5.4	19 04.5	12.3	61.3	°	h m	h m	h m	h m	h m	h m	h m
N 10	328 23.0	12.3	323 55.4	5.4	18 52.2	12.4	61.3	N 72	▭	▭	▭	▭	22 58	22 00	21 24
E 11	343 23.0	11.8	338 19.8	5.6	18 39.8	12.5	61.3	N 70	▭	▭	▭	▭	22 21	21 44	21 19
S 12	358 23.0	N20 11.3	352 44.4	5.7	N18 27.3	12.7	61.3	68	22 52	////	////	22 40	21 55	21 32	21 14
D 13	13 23.0	10.8	7 09.1	5.8	18 14.6	12.7	61.3	66	22 04	////	////	21 54	21 34	21 21	21 10
A 14	28 22.9	10.3	21 33.9	5.9	18 01.9	12.9	61.3	64	21 34	////	////	21 23	21 17	21 12	21 07
Y 15	43 22.9	09.8	35 58.8	6.0	17 49.0	13.0	61.2	62	21 11	22 37	////	21 00	21 03	21 04	21 04
16	58 22.9	09.3	50 23.8	6.1	17 36.0	13.1	61.2	60	20 53	22 02	////	20 41	20 52	20 58	21 02
17	73 22.9	08.8	64 48.9	6.2	17 22.9	13.2	61.2								
18	88 22.8	N20 08.3	79 14.1	6.3	N17 09.7	13.4	61.2	N 58	20 38	21 37	////	20 26	20 41	20 52	21 00
19	103 22.8	07.8	93 39.4	6.4	16 56.3	13.4	61.2	56	20 26	21 18	22 46	20 12	20 32	20 47	20 58
20	118 22.8	07.3	108 04.8	6.5	16 42.9	13.5	61.2	54	20 15	21 02	22 13	20 00	20 24	20 42	20 56
21	133 22.8	06.8	122 30.3	6.6	16 29.4	13.7	61.2	52	20 05	20 49	21 50	19 50	20 17	20 37	20 54
22	148 22.8	06.3	136 55.9	6.8	16 15.7	13.7	61.2	50	19 56	20 37	21 32	19 41	20 11	20 33	20 53
23	163 22.7	05.8	151 21.7	6.8	16 02.0	13.9	61.1	45	19 38	20 13	20 58	19 21	19 56	20 25	20 49
23 00	178 22.7	N20 05.3	165 47.5	6.9	N15 48.1	13.9	61.1	N 40	19 23	19 54	20 33	19 05	19 45	20 18	20 46
01	193 22.7	04.7	180 13.4	7.1	15 34.2	14.0	61.1	35	19 10	19 39	20 14	18 51	19 35	20 11	20 44
02	208 22.7	04.2	194 39.5	7.1	15 20.2	14.1	61.1	30	19 00	19 26	19 58	18 40	19 26	20 06	20 42
03	223 22.7	03.7	209 05.6	7.3	15 06.1	14.2	61.1	20	18 41	19 05	19 33	18 19	19 10	19 56	20 38
04	238 22.6	03.2	223 31.9	7.3	14 51.9	14.3	61.1	N 10	18 25	18 47	19 14	18 01	18 57	19 48	20 35
05	253 22.6	02.7	237 58.2	7.5	14 37.6	14.4	61.0	0	18 10	18 32	18 58	17 44	18 44	19 40	20 31
06	268 22.6	N20 02.2	252 24.7	7.5	N14 23.2	14.4	61.0	S 10	17 55	18 18	18 43	17 27	18 31	19 31	20 28
07	283 22.6	01.7	266 51.2	7.7	14 08.8	14.6	61.0	20	17 40	18 03	18 30	17 09	18 17	19 23	20 24
T 08	298 22.6	01.2	281 17.9	7.7	13 54.2	14.6	61.0	30	17 22	17 47	18 17	16 47	18 01	19 12	20 20
H 09	313 22.6	00.6	295 44.6	7.9	13 39.6	14.7	61.0	35	17 11	17 39	18 10	16 35	17 52	19 07	20 18
U 10	328 22.5	20 00.1	310 11.5	7.9	13 24.9	14.7	60.9	40	17 00	17 29	18 02	16 20	17 41	19 00	20 15
R 11	343 22.5	19 59.6	324 38.4	8.0	13 10.2	14.8	60.9	45	16 46	17 18	17 54	16 03	17 28	18 52	20 12
S 12	358 22.5	N19 59.1	339 05.4	8.2	N12 55.4	14.9	60.9	S 50	16 29	17 05	17 45	15 42	17 13	18 42	20 08
D 13	13 22.5	58.6	353 32.6	8.2	12 40.5	15.0	60.9	52	16 21	16 59	17 41	15 31	17 05	18 38	20 07
A 14	28 22.5	58.1	7 59.8	8.3	12 25.5	15.0	60.9	54	16 12	16 53	17 37	15 20	16 57	18 33	20 05
Y 15	43 22.5	57.5	22 27.1	8.5	12 10.5	15.1	60.8	56	16 02	16 46	17 32	15 06	16 48	18 27	20 02
16	58 22.4	57.0	36 54.6	8.5	11 55.4	15.1	60.8	58	15 51	16 38	17 28	14 51	16 37	18 21	20 00
17	73 22.4	56.5	51 22.1	8.6	11 40.3	15.2	60.8	S 60	15 38	16 29	17 22	14 32	16 25	18 14	19 57
18	88 22.4	N19 56.0	65 49.7	8.7	N11 25.1	15.3	60.8		SUN			MOON			
19	103 22.4	55.5	80 17.4	8.8	11 09.8	15.3	60.7	Day	Eqn. of Time		Mer. Pass.	Mer. Pass.		Age	Phase
20	118 22.4	54.9	94 45.2	8.9	10 54.5	15.3	60.7		00^h	12^h		Upper	Lower		
21	133 22.4	54.4	109 13.1	9.0	10 39.2	15.5	60.7	d	m s	m s	h m	h m	h m	d	%
22	148 22.3	53.9	123 41.1	9.0	10 23.7	15.4	60.7	21	06 24	06 26	12 06	11 29	24 00	29	1	●
23	163 22.3	53.4	138 09.1	9.2	N10 08.3	15.5	60.6	22	06 27	06 28	12 06	12 30	00 00	00	0
	SD 15.8	d 0.5	SD 16.7		16.7		16.6	23	06 29	06 30	12 06	13 27	00 59	01	3

2009 JULY 21, 22, 23 (TUES., WED., THURS.)

UT	ARIES GHA	VENUS −4.1 GHA	Dec	MARS +1.1 GHA	Dec	JUPITER −2.8 GHA	Dec	SATURN +1.1 GHA	Dec	STARS Name	SHA	Dec
d h	° ′	° ′	° ′	° ′	° ′	° ′	° ′	° ′	° ′		° ′	° ′
21 00	298 53.7	222 23.0	N20 42.4	234 29.2	N20 54.7	331 11.8	S14 07.1	128 54.4	N 6 29.4	Acamar	315 20.5	S40 15.6
01	313 56.2	237 22.5	42.8	249 29.8	55.0	346 14.5	07.2	143 56.6	29.3	Achernar	335 28.7	S57 10.9
02	328 58.7	252 21.9	43.2	264 30.5	55.3	1 17.2	07.2	158 58.9	29.2	Acrux	173 13.0	S63 09.5
03	344 01.1	267 21.4	.. 43.5	279 31.1	.. 55.7	16 19.9	.. 07.3	174 01.1	.. 29.1	Adhara	255 15.2	S28 59.0
04	359 03.6	282 20.9	43.9	294 31.7	56.0	31 22.6	07.4	189 03.4	29.0	Aldebaran	290 52.9	N16 31.8
05	14 06.1	297 20.4	44.3	309 32.4	56.3	46 25.3	07.5	204 05.6	28.9			
06	29 08.5	312 19.8	N20 44.6	324 33.0	N20 56.6	61 28.0	S14 07.6	219 07.9	N 6 28.8	Alioth	166 23.1	N55 54.7
07	44 11.0	327 19.3	45.0	339 33.7	57.0	76 30.7	07.7	234 10.1	28.7	Alkaid	153 01.0	N49 16.1
T 08	59 13.4	342 18.8	45.3	354 34.3	57.3	91 33.4	07.8	249 12.4	28.6	Al Na'ir	27 46.8	S46 54.6
U 09	74 15.9	357 18.2	.. 45.7	9 35.0	.. 57.6	106 36.2	.. 07.9	264 14.6	.. 28.5	Alnilam	275 49.5	S 1 11.6
E 10	89 18.4	12 17.7	46.1	24 35.6	57.9	121 38.9	08.0	279 16.8	28.4	Alphard	217 59.2	S 8 42.0
S 11	104 20.8	27 17.2	46.4	39 36.3	58.3	136 41.6	08.1	294 19.1	28.3			
D 12	119 23.3	42 16.6	N20 46.8	54 36.9	N20 58.6	151 44.3	S14 08.2	309 21.3	N 6 28.2	Alphecca	126 13.2	N26 41.1
A 13	134 25.8	57 16.1	47.1	69 37.5	58.9	166 47.0	08.3	324 23.6	28.1	Alpheratz	357 46.4	N29 08.6
Y 14	149 28.2	72 15.5	47.5	84 38.2	59.3	181 49.7	08.4	339 25.8	28.0	Altair	62 10.7	N 8 53.7
15	164 30.7	87 15.0	.. 47.8	99 38.8	.. 59.6	196 52.4	.. 08.5	354 28.1	.. 27.9	Ankaa	353 18.2	S42 14.9
16	179 33.2	102 14.5	48.2	114 39.5	20 59.9	211 55.2	08.6	9 30.3	27.8	Antares	112 29.6	S26 27.3
17	194 35.6	117 13.9	48.5	129 40.1	21 00.2	226 57.9	08.7	24 32.6	27.7			
18	209 38.1	132 13.4	N20 48.9	144 40.8	N21 00.5	242 00.6	S14 08.8	39 34.8	N 6 27.6	Arcturus	145 58.3	N19 08.0
19	224 40.5	147 12.9	49.2	159 41.4	00.9	257 03.3	08.9	54 37.1	27.5	Atria	107 33.6	S69 02.9
20	239 43.0	162 12.3	49.6	174 42.0	01.2	272 06.0	09.0	69 39.3	27.4	Avior	234 20.0	S59 32.4
21	254 45.5	177 11.8	.. 49.9	189 42.7	.. 01.5	287 08.7	.. 09.1	84 41.5	.. 27.3	Bellatrix	278 35.3	N 6 21.6
22	269 47.9	192 11.2	50.3	204 43.3	01.8	302 11.4	09.2	99 43.8	27.2	Betelgeuse	271 04.7	N 7 24.6
23	284 50.4	207 10.7	50.6	219 44.0	02.2	317 14.2	09.3	114 46.0	27.1			
22 00	299 52.9	222 10.2	N20 51.0	234 44.6	N21 02.5	332 16.9	S14 09.4	129 48.3	N 6 27.0	Canopus	263 57.9	S52 41.9
01	314 55.3	237 09.6	51.3	249 45.3	02.8	347 19.6	09.5	144 50.5	26.9	Capella	280 39.1	N46 00.4
02	329 57.8	252 09.1	51.7	264 45.9	03.1	2 22.3	09.6	159 52.8	26.9	Deneb	49 33.1	N45 18.9
03	345 00.3	267 08.5	.. 52.0	279 46.6	.. 03.4	17 25.0	.. 09.6	174 55.0	.. 26.8	Denebola	182 36.7	N14 31.2
04	0 02.7	282 08.0	52.3	294 47.2	03.8	32 27.7	09.7	189 57.3	26.7	Diphda	348 58.6	S17 55.8
05	15 05.2	297 07.5	52.7	309 47.8	04.1	47 30.5	09.8	204 59.5	26.6			
06	30 07.7	312 06.9	N20 53.0	324 48.5	N21 04.4	62 33.2	S14 09.9	220 01.7	N 6 26.5	Dubhe	193 55.4	N61 42.1
W 07	45 10.1	327 06.4	53.4	339 49.1	04.7	77 35.9	10.0	235 04.0	26.4	Elnath	278 16.6	N28 36.9
E 08	60 12.6	342 05.8	53.7	354 49.8	05.0	92 38.6	10.1	250 06.2	26.3	Eltanin	90 47.0	N51 29.4
D 09	75 15.0	357 05.3	.. 54.0	9 50.4	.. 05.4	107 41.3	.. 10.2	265 08.5	.. 26.2	Enif	33 49.7	N 9 55.2
N 10	90 17.5	12 04.7	54.4	24 51.1	05.7	122 44.0	10.3	280 10.7	26.1	Fomalhaut	15 26.8	S29 34.0
E 11	105 20.0	27 04.2	54.7	39 51.7	06.0	137 46.8	10.4	295 13.0	26.0			
S 12	120 22.4	42 03.6	N20 55.0	54 52.3	N21 06.3	152 49.5	S14 10.5	310 15.2	N 6 25.9	Gacrux	172 04.5	S57 10.3
D 13	135 24.9	57 03.1	55.4	69 53.0	06.6	167 52.2	10.6	325 17.4	25.8	Gienah	175 55.4	S17 35.8
A 14	150 27.4	72 02.6	55.7	84 53.6	06.9	182 54.9	10.7	340 19.7	25.7	Hadar	148 52.1	S60 25.5
Y 15	165 29.8	87 02.0	.. 56.0	99 54.3	.. 07.3	197 57.6	.. 10.8	355 21.9	.. 25.6	Hamal	328 04.1	N23 30.5
16	180 32.3	102 01.5	56.4	114 54.9	07.6	213 00.4	10.9	10 24.2	25.5	Kaus Aust.	83 47.2	S34 22.8
17	195 34.8	117 00.9	56.7	129 55.6	07.9	228 03.1	11.0	25 26.4	25.4			
18	210 37.2	132 00.4	N20 57.0	144 56.2	N21 08.2	243 05.8	S14 11.1	40 28.7	N 6 25.3	Kochab	137 19.2	N74 07.2
19	225 39.7	146 59.8	57.4	159 56.9	08.5	258 08.5	11.2	55 30.9	25.2	Markab	13 41.0	N15 15.5
20	240 42.2	161 59.3	57.7	174 57.5	08.8	273 11.2	11.3	70 33.1	25.1	Menkar	314 18.2	N 4 07.8
21	255 44.6	176 58.7	.. 58.1	189 58.1	.. 09.2	288 14.0	.. 11.4	85 35.4	.. 25.0	Menkent	148 11.0	S36 25.3
22	270 47.1	191 58.2	58.4	204 58.8	09.5	303 16.7	11.5	100 37.6	24.9	Miaplacidus	221 41.5	S69 45.5
23	285 49.5	206 57.6	58.7	219 59.4	09.8	318 19.4	11.6	115 39.9	24.8			
23 00	300 52.0	221 57.1	N20 59.0	235 00.1	N21 10.1	333 22.1	S14 11.7	130 42.1	N 6 24.7	Mirfak	308 44.8	N49 53.6
01	315 54.5	236 56.5	59.3	250 00.7	10.4	348 24.8	11.8	145 44.4	24.6	Nunki	76 01.5	S26 17.1
02	330 56.9	251 56.0	20 59.7	265 01.4	10.7	3 27.6	11.9	160 46.6	24.5	Peacock	53 23.0	S56 42.1
03	345 59.4	266 55.4	21 00.0	280 02.0	.. 11.0	18 30.3	.. 12.0	175 48.8	.. 24.4	Pollux	243 31.6	N28 00.2
04	1 01.9	281 54.9	00.3	295 02.7	11.3	33 33.0	12.1	190 51.1	24.3	Procyon	245 03.0	N 5 12.1
05	16 04.3	296 54.3	00.6	310 03.3	11.7	48 35.7	12.2	205 53.3	24.2			
06	31 06.8	311 53.7	N21 00.9	325 03.9	N21 12.0	63 38.4	S14 12.3	220 55.6	N 6 24.1	Rasalhague	96 08.8	N12 33.2
07	46 09.3	326 53.2	01.3	340 04.6	12.3	78 41.2	12.4	235 57.8	24.0	Regulus	207 46.8	N11 55.3
T 08	61 11.7	341 52.6	01.6	355 05.2	12.6	93 43.9	12.5	251 00.0	23.9	Rigel	281 15.1	S 8 11.3
H 09	76 14.2	356 52.1	.. 01.9	10 05.9	.. 12.9	108 46.6	.. 12.6	266 02.3	.. 23.8	Rigil Kent.	139 55.7	S60 52.8
U 10	91 16.7	11 51.5	02.2	25 06.5	13.2	123 49.3	12.7	281 04.5	23.7	Sabik	102 15.6	S15 44.2
R 11	106 19.1	26 51.0	02.5	40 07.2	13.5	138 52.1	12.8	296 06.8	23.6			
S 12	121 21.6	41 50.4	N21 02.8	55 07.8	N21 13.8	153 54.8	S14 12.9	311 09.0	N 6 23.5	Schedar	349 43.8	N56 35.3
D 13	136 24.0	56 49.9	03.2	70 08.5	14.1	168 57.5	13.0	326 11.3	23.4	Shaula	96 25.5	S37 06.7
A 14	151 26.5	71 49.3	03.5	85 09.1	14.4	184 00.2	13.1	341 13.5	23.3	Sirius	258 36.6	S16 43.6
Y 15	166 29.0	86 48.7	.. 03.8	100 09.8	.. 14.8	199 03.0	.. 13.2	356 15.7	.. 23.2	Spica	158 34.3	S11 12.8
16	181 31.4	101 48.2	04.1	115 10.4	15.1	214 05.7	13.3	11 18.0	23.1	Suhail	222 55.1	S43 28.3
17	196 33.9	116 47.6	04.4	130 11.0	15.4	229 08.4	13.4	26 20.2	23.0			
18	211 36.4	131 47.1	N21 04.7	145 11.7	N21 15.7	244 11.1	S14 13.5	41 22.5	N 6 22.9	Vega	80 40.5	N38 47.6
19	226 38.8	146 46.5	05.0	160 12.3	16.0	259 13.9	13.6	56 24.7	22.8	Zuben'ubi	137 08.5	S16 05.0
20	241 41.3	161 46.0	05.3	175 13.0	16.3	274 16.6	13.7	71 26.9	22.7		SHA	Mer.Pass.
21	256 43.8	176 45.4	.. 05.6	190 13.6	.. 16.6	289 19.3	.. 13.8	86 29.2	.. 22.6		° ′	h m
22	271 46.2	191 44.8	05.9	205 14.3	16.9	304 22.0	13.9	101 31.4	22.5	Venus	282 17.3	9 12
23	286 48.7	206 44.3	06.2	220 14.9	17.2	319 24.8	14.0	116 33.7	22.4	Mars	294 51.7	8 21
	h m									Jupiter	32 24.0	1 51
Mer.Pass. 3 59.8	v −0.5	d 0.3	v 0.6	d 0.3	v 2.7	d 0.1	v 2.2	d 0.1	Saturn	189 55.4	15 18

INCREMENTS AND CORRECTIONS

0ᵐ

s	SUN PLANETS	ARIES	MOON	v or d Corrⁿ	v or d Corrⁿ	v or d Corrⁿ
	° ′	° ′	° ′	′ ′	′ ′	′ ′
00	0 00·0	0 00·0	0 00·0	0·0 0·0	6·0 0·1	12·0 0·1
01	0 00·3	0 00·3	0 00·2	0·1 0·0	6·1 0·1	12·1 0·1
02	0 00·5	0 00·5	0 00·5	0·2 0·0	6·2 0·1	12·2 0·1
03	0 00·8	0 00·8	0 00·7	0·3 0·0	6·3 0·1	12·3 0·1
04	0 01·0	0 01·0	0 01·0	0·4 0·0	6·4 0·1	12·4 0·1
05	0 01·3	0 01·3	0 01·2	0·5 0·0	6·5 0·1	12·5 0·1
06	0 01·5	0 01·5	0 01·4	0·6 0·0	6·6 0·1	12·6 0·1
07	0 01·8	0 01·8	0 01·7	0·7 0·0	6·7 0·1	12·7 0·1
08	0 02·0	0 02·0	0 01·9	0·8 0·0	6·8 0·1	12·8 0·1
09	0 02·3	0 02·3	0 02·1	0·9 0·0	6·9 0·1	12·9 0·1
10	0 02·5	0 02·5	0 02·4	1·0 0·0	7·0 0·1	13·0 0·1
11	0 02·8	0 02·8	0 02·6	1·1 0·0	7·1 0·1	13·1 0·1
12	0 03·0	0 03·0	0 02·9	1·2 0·0	7·2 0·1	13·2 0·1
13	0 03·3	0 03·3	0 03·1	1·3 0·0	7·3 0·1	13·3 0·1
14	0 03·5	0 03·5	0 03·3	1·4 0·0	7·4 0·1	13·4 0·1
15	0 03·8	0 03·8	0 03·6	1·5 0·0	7·5 0·1	13·5 0·1
16	0 04·0	0 04·0	0 03·8	1·6 0·0	7·6 0·1	13·6 0·1
17	0 04·3	0 04·3	0 04·1	1·7 0·0	7·7 0·1	13·7 0·1
18	0 04·5	0 04·5	0 04·3	1·8 0·0	7·8 0·1	13·8 0·1
19	0 04·8	0 04·8	0 04·5	1·9 0·0	7·9 0·1	13·9 0·1
20	0 05·0	0 05·0	0 04·8	2·0 0·0	8·0 0·1	14·0 0·1
21	0 05·3	0 05·3	0 05·0	2·1 0·0	8·1 0·1	14·1 0·1
22	0 05·5	0 05·5	0 05·2	2·2 0·0	8·2 0·1	14·2 0·1
23	0 05·8	0 05·8	0 05·5	2·3 0·0	8·3 0·1	14·3 0·1
24	0 06·0	0 06·0	0 05·7	2·4 0·0	8·4 0·1	14·4 0·1
25	0 06·3	0 06·3	0 06·0	2·5 0·0	8·5 0·1	14·5 0·1
26	0 06·5	0 06·5	0 06·2	2·6 0·0	8·6 0·1	14·6 0·1
27	0 06·8	0 06·8	0 06·4	2·7 0·0	8·7 0·1	14·7 0·1
28	0 07·0	0 07·0	0 06·7	2·8 0·0	8·8 0·1	14·8 0·1
29	0 07·3	0 07·3	0 06·9	2·9 0·0	8·9 0·1	14·9 0·1
30	0 07·5	0 07·5	0 07·2	3·0 0·0	9·0 0·1	15·0 0·1
31	0 07·8	0 07·8	0 07·4	3·1 0·0	9·1 0·1	15·1 0·1
32	0 08·0	0 08·0	0 07·6	3·2 0·0	9·2 0·1	15·2 0·1
33	0 08·3	0 08·3	0 07·9	3·3 0·0	9·3 0·1	15·3 0·1
34	0 08·5	0 08·5	0 08·1	3·4 0·0	9·4 0·1	15·4 0·1
35	0 08·8	0 08·8	0 08·4	3·5 0·0	9·5 0·1	15·5 0·1
36	0 09·0	0 09·0	0 08·6	3·6 0·0	9·6 0·1	15·6 0·1
37	0 09·3	0 09·3	0 08·8	3·7 0·0	9·7 0·1	15·7 0·1
38	0 09·5	0 09·5	0 09·1	3·8 0·0	9·8 0·1	15·8 0·1
39	0 09·8	0 09·8	0 09·3	3·9 0·0	9·9 0·1	15·9 0·1
40	0 10·0	0 10·0	0 09·5	4·0 0·0	10·0 0·1	16·0 0·1
41	0 10·3	0 10·3	0 09·8	4·1 0·0	10·1 0·1	16·1 0·1
42	0 10·5	0 10·5	0 10·0	4·2 0·0	10·2 0·1	16·2 0·1
43	0 10·8	0 10·8	0 10·3	4·3 0·0	10·3 0·1	16·3 0·1
44	0 11·0	0 11·0	0 10·5	4·4 0·0	10·4 0·1	16·4 0·1
45	0 11·3	0 11·3	0 10·7	4·5 0·0	10·5 0·1	16·5 0·1
46	0 11·5	0 11·5	0 11·0	4·6 0·0	10·6 0·1	16·6 0·1
47	0 11·8	0 11·8	0 11·2	4·7 0·0	10·7 0·1	16·7 0·1
48	0 12·0	0 12·0	0 11·5	4·8 0·0	10·8 0·1	16·8 0·1
49	0 12·3	0 12·3	0 11·7	4·9 0·0	10·9 0·1	16·9 0·1
50	0 12·5	0 12·5	0 11·9	5·0 0·0	11·0 0·1	17·0 0·1
51	0 12·8	0 12·8	0 12·2	5·1 0·0	11·1 0·1	17·1 0·1
52	0 13·0	0 13·0	0 12·4	5·2 0·0	11·2 0·1	17·2 0·1
53	0 13·3	0 13·3	0 12·6	5·3 0·0	11·3 0·1	17·3 0·1
54	0 13·5	0 13·5	0 12·9	5·4 0·0	11·4 0·1	17·4 0·1
55	0 13·8	0 13·8	0 13·1	5·5 0·0	11·5 0·1	17·5 0·1
56	0 14·0	0 14·0	0 13·4	5·6 0·0	11·6 0·1	17·6 0·1
57	0 14·3	0 14·3	0 13·6	5·7 0·0	11·7 0·1	17·7 0·1
58	0 14·5	0 14·5	0 13·8	5·8 0·0	11·8 0·1	17·8 0·1
59	0 14·8	0 14·8	0 14·1	5·9 0·0	11·9 0·1	17·9 0·1
60	0 15·0	0 15·0	0 14·3	6·0 0·1	12·0 0·1	18·0 0·2

1ᵐ

s	SUN PLANETS	ARIES	MOON	v or d Corrⁿ	v or d Corrⁿ	v or d Corrⁿ
	° ′	° ′	° ′	′ ′	′ ′	′ ′
00	0 15·0	0 15·0	0 14·3	0·0 0·0	6·0 0·2	12·0 0·3
01	0 15·3	0 15·3	0 14·6	0·1 0·0	6·1 0·2	12·1 0·3
02	0 15·5	0 15·5	0 14·8	0·2 0·0	6·2 0·2	12·2 0·3
03	0 15·8	0 15·8	0 15·0	0·3 0·0	6·3 0·2	12·3 0·3
04	0 16·0	0 16·0	0 15·3	0·4 0·0	6·4 0·2	12·4 0·3
05	0 16·3	0 16·3	0 15·5	0·5 0·0	6·5 0·2	12·5 0·3
06	0 16·5	0 16·5	0 15·7	0·6 0·0	6·6 0·2	12·6 0·3
07	0 16·8	0 16·8	0 16·0	0·7 0·0	6·7 0·2	12·7 0·3
08	0 17·0	0 17·0	0 16·2	0·8 0·0	6·8 0·2	12·8 0·3
09	0 17·3	0 17·3	0 16·5	0·9 0·0	6·9 0·2	12·9 0·3
10	0 17·5	0 17·5	0 16·7	1·0 0·0	7·0 0·2	13·0 0·3
11	0 17·8	0 17·8	0 16·9	1·1 0·0	7·1 0·2	13·1 0·3
12	0 18·0	0 18·0	0 17·2	1·2 0·0	7·2 0·2	13·2 0·3
13	0 18·3	0 18·3	0 17·4	1·3 0·0	7·3 0·2	13·3 0·3
14	0 18·5	0 18·6	0 17·7	1·4 0·0	7·4 0·2	13·4 0·3
15	0 18·8	0 18·8	0 17·9	1·5 0·0	7·5 0·2	13·5 0·3
16	0 19·0	0 19·1	0 18·1	1·6 0·0	7·6 0·2	13·6 0·3
17	0 19·3	0 19·3	0 18·4	1·7 0·0	7·7 0·2	13·7 0·3
18	0 19·5	0 19·6	0 18·6	1·8 0·0	7·8 0·2	13·8 0·3
19	0 19·8	0 19·8	0 18·9	1·9 0·0	7·9 0·2	13·9 0·3
20	0 20·0	0 20·1	0 19·1	2·0 0·1	8·0 0·2	14·0 0·4
21	0 20·3	0 20·3	0 19·3	2·1 0·1	8·1 0·2	14·1 0·4
22	0 20·5	0 20·6	0 19·6	2·2 0·1	8·2 0·2	14·2 0·4
23	0 20·8	0 20·8	0 19·8	2·3 0·1	8·3 0·2	14·3 0·4
24	0 21·0	0 21·1	0 20·0	2·4 0·1	8·4 0·2	14·4 0·4
25	0 21·3	0 21·3	0 20·3	2·5 0·1	8·5 0·2	14·5 0·4
26	0 21·5	0 21·6	0 20·5	2·6 0·1	8·6 0·2	14·6 0·4
27	0 21·8	0 21·8	0 20·8	2·7 0·1	8·7 0·2	14·7 0·4
28	0 22·0	0 22·1	0 21·0	2·8 0·1	8·8 0·2	14·8 0·4
29	0 22·3	0 22·3	0 21·2	2·9 0·1	8·9 0·2	14·9 0·4
30	0 22·5	0 22·6	0 21·5	3·0 0·1	9·0 0·2	15·0 0·4
31	0 22·8	0 22·8	0 21·7	3·1 0·1	9·1 0·2	15·1 0·4
32	0 23·0	0 23·1	0 22·0	3·2 0·1	9·2 0·2	15·2 0·4
33	0 23·3	0 23·3	0 22·2	3·3 0·1	9·3 0·2	15·3 0·4
34	0 23·5	0 23·6	0 22·4	3·4 0·1	9·4 0·2	15·4 0·4
35	0 23·8	0 23·8	0 22·7	3·5 0·1	9·5 0·2	15·5 0·4
36	0 24·0	0 24·1	0 22·9	3·6 0·1	9·6 0·2	15·6 0·4
37	0 24·3	0 24·3	0 23·1	3·7 0·1	9·7 0·2	15·7 0·4
38	0 24·5	0 24·6	0 23·4	3·8 0·1	9·8 0·2	15·8 0·4
39	0 24·8	0 24·8	0 23·6	3·9 0·1	9·9 0·2	15·9 0·4
40	0 25·0	0 25·1	0 23·9	4·0 0·1	10·0 0·3	16·0 0·4
41	0 25·3	0 25·3	0 24·1	4·1 0·1	10·1 0·3	16·1 0·4
42	0 25·5	0 25·6	0 24·3	4·2 0·1	10·2 0·3	16·2 0·4
43	0 25·8	0 25·8	0 24·6	4·3 0·1	10·3 0·3	16·3 0·4
44	0 26·0	0 26·1	0 24·8	4·4 0·1	10·4 0·3	16·4 0·4
45	0 26·3	0 26·3	0 25·1	4·5 0·1	10·5 0·3	16·5 0·4
46	0 26·5	0 26·6	0 25·3	4·6 0·1	10·6 0·3	16·6 0·4
47	0 26·8	0 26·8	0 25·5	4·7 0·1	10·7 0·3	16·7 0·4
48	0 27·0	0 27·1	0 25·8	4·8 0·1	10·8 0·3	16·8 0·4
49	0 27·3	0 27·3	0 26·0	4·9 0·1	10·9 0·3	16·9 0·4
50	0 27·5	0 27·6	0 26·2	5·0 0·1	11·0 0·3	17·0 0·4
51	0 27·8	0 27·8	0 26·5	5·1 0·1	11·1 0·3	17·1 0·4
52	0 28·0	0 28·1	0 26·7	5·2 0·1	11·2 0·3	17·2 0·4
53	0 28·3	0 28·3	0 27·0	5·3 0·1	11·3 0·3	17·3 0·4
54	0 28·5	0 28·6	0 27·2	5·4 0·1	11·4 0·3	17·4 0·4
55	0 28·8	0 28·8	0 27·4	5·5 0·1	11·5 0·3	17·5 0·4
56	0 29·0	0 29·1	0 27·7	5·6 0·1	11·6 0·3	17·6 0·4
57	0 29·3	0 29·3	0 27·9	5·7 0·1	11·7 0·3	17·7 0·4
58	0 29·5	0 29·6	0 28·2	5·8 0·1	11·8 0·3	17·8 0·4
59	0 29·8	0 29·8	0 28·4	5·9 0·1	11·9 0·3	17·9 0·4
60	0 30·0	0 30·1	0 28·6	6·0 0·2	12·0 0·3	18·0 0·5

©2009 Jack Case

TABLE 6 — ALTITUDE CORRECTION TABLES

a. Dip of the Horizon

Ht. of Eye (m)	Corr^n D	Ht. of Eye (ft.)	Ht. of Eye (m)	Corr^n D	Ht. of Eye (ft.)
0.00	−0.3	0.0	13.0	−6.4	42.8
0.03	−0.4	0.1	13.4	−6.5	44.2
0.06	−0.5	0.2	13.8	−6.6	45.5
0.09	−0.6	0.3	14.2	−6.7	47.0
0.13	−0.7	0.4	14.7	−6.8	48.4
0.18	−0.8	0.5	15.1	−6.9	49.8
0.23	−0.9	0.7	15.5	−7.0	51.3
0.29	−1.0	0.9	16.0	−7.1	52.8
0.35	−1.1	1.1	16.5	−7.2	54.3
0.42	−1.2	1.4	16.9	−7.3	55.8
0.45	−1.3	1.6	17.4	−7.4	57.4
0.5	−1.4	1.9	17.9	−7.5	58.9
0.6	−1.5	2.2	18.4	−7.6	60.5
0.7	−1.6	2.5	18.8	−7.7	62.1
0.8	−1.7	2.8	19.3	−7.8	63.8
0.9	−1.8	3.2	19.8	−7.9	65.4
1.1	−1.9	3.6	20.4	−8.0	67.1
1.2	−2.0	4.0	20.9	−8.1	68.8
1.3	−2.1	4.4	21.4	−8.2	70.5
1.4	−2.2	4.9	21.9	−8.3	72.3
1.6	−2.3	5.3	22.5	−8.4	74.1
1.7	−2.4	5.8	23.0	−8.5	75.8
1.9	−2.5	6.3	23.5	−8.6	77.6
2.0	−2.6	6.9	24.1	−8.7	79.5
2.2	−2.7	7.4	24.7	−8.8	81.3
2.4	−2.8	8.0	25.2	−8.9	83.2
2.6	−2.9	8.6	25.8	−9.0	85.1
2.8	−3.0	9.2	26.4	−9.1	87.0
3.0	−3.1	9.8	27.0	−9.2	88.9
3.2	−3.2	10.5	27.6	−9.3	90.9
3.4	−3.3	11.2	28.2	−9.4	92.9
3.6	−3.4	11.9	28.8	−9.5	94.9
3.8	−3.5	12.6	29.4	−9.6	96.9
4.0	−3.6	13.3	30.0	−9.7	98.9
4.3	−3.7	14.1	30.6	−9.8	101.0
4.5	−3.8	14.9	31.3	−9.9	103.1
4.7	−3.9	15.7	31.9	−10.0	105.2
5.0	−4.0	16.5	32.6	−10.1	107.3
5.2	−4.1	17.4	33.2	−10.2	109.4
5.5	−4.2	18.3	33.9	−10.3	111.6
5.8	−4.3	19.1	34.5	−10.4	113.8
6.1	−4.4	20.1	35.2	−10.5	116.0
6.3	−4.5	21.0	35.9	−10.6	118.2
6.6	−4.6	22.0	36.6	−10.7	120.5
6.9	−4.7	22.9	37.3	−10.8	122.8
7.2	−4.8	23.9	38.0	−10.9	125.1
7.5	−4.9	25.0	38.7	−11.0	127.4
7.9	−5.0	26.0	39.4	−11.1	129.7
8.2	−5.1	27.1	40.1	−11.2	132.1
8.5	−5.2	28.1	40.8	−11.3	134.5
8.8	−5.3	29.2	41.5	−11.4	136.9
9.2	−5.4	30.4	42.3	−11.5	139.3
9.5	−5.5	31.5	43.0	−11.6	141.7
9.9	−5.6	32.7	43.8	−11.7	144.2
10.3	−5.7	33.9	44.5	−11.8	146.7
10.6	−5.8	35.1	45.3	−11.9	149.2
11.0	−5.9	36.3	46.1	−12.0	151.7
11.4	−6.0	37.6	46.8	−12.1	154.3
11.8	−6.1	38.9	47.6	−12.2	156.8
12.2	−6.2	40.1	48.4	−12.3	159.4
12.6	−6.3	41.5	49.2	−12.4	162.1
13.0		42.8	50.0		164.7

b. Refraction for Stars & Planets

App. Alt. H_a ° '	Corr^n R '	App. Alt. H_a ° '	Corr^n R '	App. Alt. H_a ° '	Corr^n R '
0 00	−33.8	3 30	−12.9	9 55	−5.3
0 03	33.2	3 35	12.7	10 07	−5.2
0 06	32.6	3 40	12.5	10 20	−5.1
0 09	32.0	3 45	12.3	10 32	−5.0
0 12	31.5	3 50	12.1	10 46	−4.9
0 15	30.9	3 55	11.9	10 59	−4.8
0 18	−30.4	4 00	−11.7	11 14	−4.7
0 21	29.8	4 05	11.5	11 29	−4.6
0 24	29.3	4 10	11.4	11 44	−4.5
0 27	28.8	4 15	11.2	12 00	−4.4
0 30	28.3	4 20	11.0	12 17	−4.3
0 33	27.9	4 25	10.9	12 35	−4.2
0 36	−27.4	4 30	−10.7	12 53	−4.1
0 39	26.9	4 35	10.6	13 12	−4.0
0 42	26.5	4 40	10.4	13 32	−3.9
0 45	26.1	4 45	10.3	13 53	−3.8
0 48	25.7	4 50	10.1	14 16	−3.7
0 51	25.3	4 55	10.0	14 39	−3.6
0 54	−24.9	5 00	−9.8	15 03	−3.5
0 57	24.5	5 05	9.7	15 29	−3.4
1 00	24.1	5 10	9.6	15 56	−3.3
1 03	23.7	5 15	9.5	16 25	−3.2
1 06	23.4	5 20	9.3	16 55	−3.1
1 09	23.0	5 25	9.2	17 27	−3.0
1 12	−22.7	5 30	−9.1	18 01	−2.9
1 15	22.3	5 35	9.0	18 37	−2.8
1 18	22.0	5 40	8.9	19 16	−2.7
1 21	21.7	5 45	8.8	19 56	−2.6
1 24	21.4	5 50	8.7	20 40	−2.5
1 27	21.1	5 55	8.6	21 27	−2.4
1 30	−20.8	6 00	−8.5	22 17	−2.3
1 35	20.3	6 10	8.3	23 11	−2.2
1 40	19.9	6 20	8.1	24 09	−2.1
1 45	19.4	6 30	7.9	25 12	−2.0
1 50	19.0	6 40	7.7	26 20	−1.9
1 55	18.6	6 50	7.6	27 34	−1.8
2 00	−18.2	7 00	−7.4	28 54	−1.7
2 05	17.8	7 10	7.2	30 22	−1.6
2 10	17.4	7 20	7.1	31 58	−1.5
2 15	17.1	7 30	6.9	33 43	−1.4
2 20	16.7	7 40	6.8	35 38	−1.3
2 25	16.4	7 50	6.7	37 45	−1.2
2 30	−16.1	8 00	−6.6	40 06	−1.1
2 35	15.8	8 10	6.4	42 42	−1.0
2 40	15.4	8 20	6.3	45 34	−0.9
2 45	15.2	8 30	6.2	48 45	−0.8
2 50	14.9	8 40	6.1	52 16	−0.7
2 55	14.6	8 50	6.0	56 09	−0.6
3 00	−14.3	9 00	−5.9	60 26	−0.5
3 05	14.1	9 10	5.8	65 06	−0.4
3 10	13.8	9 20	5.7	70 09	−0.3
3 15	13.6	9 30	5.6	75 32	−0.2
3 20	13.4	9 40	5.5	81 12	−0.1
3 25	13.1	9 50	5.4	87 03	0.0
3 30	−12.9	10 00	−5.3	90 00	

H_a = App. Alt. = Apparent Altitude
 = Sextant altitude corrected for index error (IE) & dip (*d*)
 = H_s + IE + D

In critical cases ascend.

TABLE 6 — ALTITUDE CORRECTION TABLES

c. Additional Refraction for Non-standard Conditions

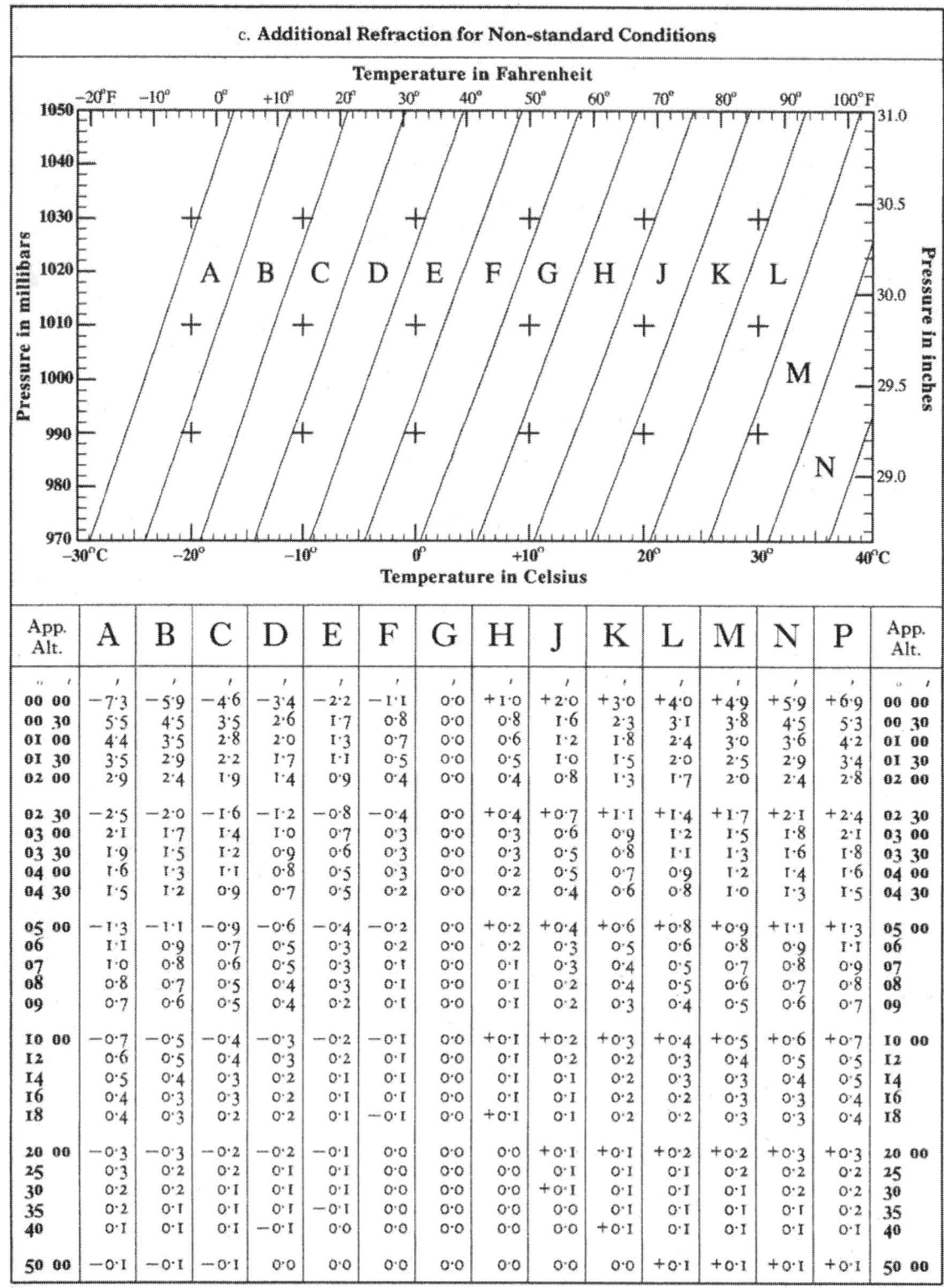

App. Alt.	A	B	C	D	E	F	G	H	J	K	L	M	N	P	App. Alt.
° ′	′	′	′	′	′	′	′	′	′	′	′	′	′	′	° ′
00 00	−7·3	−5·9	−4·6	−3·4	−2·2	−1·1	0·0	+1·0	+2·0	+3·0	+4·0	+4·9	+5·9	+6·9	00 00
00 30	5·5	4·5	3·5	2·6	1·7	0·8	0·0	0·8	1·6	2·3	3·1	3·8	4·5	5·3	00 30
01 00	4·4	3·5	2·8	2·0	1·3	0·7	0·0	0·6	1·2	1·8	2·4	3·0	3·6	4·2	01 00
01 30	3·5	2·9	2·2	1·7	1·1	0·5	0·0	0·5	1·0	1·5	2·0	2·5	2·9	3·4	01 30
02 00	2·9	2·4	1·9	1·4	0·9	0·4	0·0	0·4	0·8	1·3	1·7	2·0	2·4	2·8	02 00
02 30	−2·5	−2·0	−1·6	−1·2	−0·8	−0·4	0·0	+0·4	+0·7	+1·1	+1·4	+1·7	+2·1	+2·4	02 30
03 00	2·1	1·7	1·4	1·0	0·7	0·3	0·0	0·3	0·6	0·9	1·2	1·5	1·8	2·1	03 00
03 30	1·9	1·5	1·2	0·9	0·6	0·3	0·0	0·3	0·5	0·8	1·1	1·3	1·6	1·8	03 30
04 00	1·6	1·3	1·1	0·8	0·5	0·3	0·0	0·2	0·5	0·7	0·9	1·2	1·4	1·6	04 00
04 30	1·5	1·2	0·9	0·7	0·5	0·2	0·0	0·2	0·4	0·6	0·8	1·0	1·3	1·5	04 30
05 00	−1·3	−1·1	−0·9	−0·6	−0·4	−0·2	0·0	+0·2	+0·4	+0·6	+0·8	+0·9	+1·1	+1·3	05 00
06	1·1	0·9	0·7	0·5	0·3	0·2	0·0	0·2	0·3	0·5	0·6	0·8	0·9	1·1	06
07	1·0	0·8	0·6	0·5	0·3	0·1	0·0	0·1	0·3	0·4	0·5	0·7	0·8	0·9	07
08	0·8	0·7	0·5	0·4	0·3	0·1	0·0	0·1	0·2	0·4	0·5	0·6	0·7	0·8	08
09	0·7	0·6	0·5	0·4	0·2	0·1	0·0	0·1	0·2	0·3	0·4	0·5	0·6	0·7	09
10 00	−0·7	−0·5	−0·4	−0·3	−0·2	−0·1	0·0	+0·1	+0·2	+0·3	+0·4	+0·5	+0·6	+0·7	10 00
12	0·6	0·5	0·4	0·3	0·2	0·1	0·0	0·1	0·2	0·2	0·3	0·4	0·5	0·5	12
14	0·5	0·4	0·3	0·2	0·1	0·1	0·0	0·1	0·1	0·2	0·3	0·3	0·4	0·5	14
16	0·4	0·3	0·3	0·2	0·1	0·1	0·0	0·1	0·1	0·2	0·2	0·3	0·3	0·4	16
18	0·4	0·3	0·2	0·2	0·1	−0·1	0·0	+0·1	0·1	0·2	0·2	0·3	0·3	0·4	18
20 00	−0·3	−0·3	−0·2	−0·2	−0·1	0·0	0·0	0·0	+0·1	+0·1	+0·2	+0·2	+0·3	+0·3	20 00
25	0·3	0·2	0·2	0·1	0·1	0·0	0·0	0·0	0·1	0·1	0·1	0·2	0·2	0·2	25
30	0·2	0·2	0·1	0·1	0·1	0·0	0·0	0·0	+0·1	0·1	0·1	0·1	0·2	0·2	30
35	0·2	0·1	0·1	0·1	−0·1	0·0	0·0	0·0	0·0	0·1	0·1	0·1	0·1	0·2	35
40	0·1	0·1	0·1	−0·1	0·0	0·0	0·0	0·0	0·0	+0·1	0·1	0·1	0·1	0·1	40
50 00	−0·1	−0·1	−0·1	0·0	0·0	0·0	0·0	0·0	0·0	0·0	+0·1	+0·1	+0·1	+0·1	50 00

The graph is entered with arguments temperature and pressure to find a zone letter; using as arguments this zone letter and apparent altitude (sextant altitude corrected for index error and dip), a correction is taken from the table. This correction is to be applied to the sextant altitude in addition to the corrections for standard conditions (Table 6b).

ALTITUDE CORRECTION TABLES 0°–35°— MOON

App. Alt.	0°–4° Corrⁿ	5°–9° Corrⁿ	10°–14° Corrⁿ	15°–19° Corrⁿ	20°–24° Corrⁿ	25°–29° Corrⁿ	30°–34° Corrⁿ	App. Alt.
00	0° 34.5	5° 58.2	10° 62.1	15° 62.8	20° 62.2	25° 60.8	30° 58.9	00
10	36.5	58.5	62.2	62.8	62.2	60.8	58.8	10
20	38.3	58.7	62.2	62.8	62.1	60.7	58.8	20
30	40.0	58.9	62.3	62.8	62.1	60.7	58.7	30
40	41.5	59.1	62.3	62.8	62.0	60.6	58.6	40
50	42.9	59.3	62.4	62.7	62.0	60.6	58.5	50
00	1° 44.2	6° 59.5	11° 62.4	16° 62.7	21° 62.0	26° 60.5	31° 58.5	00
10	45.4	59.7	62.4	62.7	61.9	60.4	58.4	10
20	46.5	59.9	62.5	62.7	61.9	60.4	58.3	20
30	47.5	60.0	62.5	62.7	61.9	60.3	58.2	30
40	48.4	60.2	62.5	62.7	61.8	60.3	58.2	40
50	49.3	60.3	62.6	62.7	61.8	60.2	58.1	50
00	2° 50.1	7° 60.5	12° 62.6	17° 62.7	22° 61.7	27° 60.1	32° 58.0	00
10	50.8	60.6	62.6	62.7	61.7	60.1	57.9	10
20	51.5	60.7	62.6	62.6	61.6	60.0	57.8	20
30	52.2	60.9	62.7	62.6	61.6	59.9	57.8	30
40	52.8	61.0	62.7	62.6	61.6	59.9	57.7	40
50	53.4	61.1	62.7	62.6	61.5	59.8	57.6	50
00	3° 53.9	8° 61.2	13° 62.7	18° 62.5	23° 61.5	28° 59.7	33° 57.5	00
10	54.4	61.3	62.7	62.5	61.4	59.7	57.4	10
20	54.9	61.4	62.7	62.5	61.4	59.6	57.4	20
30	55.3	61.5	62.8	62.5	61.3	59.5	57.3	30
40	55.7	61.6	62.8	62.4	61.3	59.5	57.2	40
50	56.1	61.6	62.8	62.4	61.2	59.4	57.1	50
00	4° 56.4	9° 61.7	14° 62.8	19° 62.4	24° 61.2	29° 59.3	34° 57.0	00
10	56.8	61.8	62.8	62.4	61.1	59.3	56.9	10
20	57.1	61.9	62.8	62.3	61.1	59.2	56.9	20
30	57.4	61.9	62.8	62.3	61.0	59.1	56.8	30
40	57.7	62.0	62.8	62.3	61.0	59.1	56.7	40
50	58.0	62.1	62.8	62.2	60.9	59.0	56.6	50

HP	L U	L U	L U	L U	L U	L U	L U	HP
54.0	0.3 0.9	0.3 0.9	0.4 1.0	0.5 1.1	0.6 1.2	0.7 1.3	0.9 1.5	54.0
54.3	0.7 1.1	0.7 1.2	0.8 1.2	0.8 1.3	0.9 1.4	1.1 1.5	1.2 1.7	54.3
54.6	1.1 1.4	1.1 1.4	1.1 1.4	1.2 1.5	1.3 1.6	1.4 1.7	1.5 1.8	54.6
54.9	1.4 1.6	1.5 1.6	1.5 1.6	1.6 1.7	1.6 1.8	1.8 1.9	1.9 2.0	54.9
55.2	1.8 1.8	1.8 1.8	1.9 1.8	1.9 1.9	2.0 2.0	2.1 2.1	2.2 2.2	55.2
55.5	2.2 2.0	2.2 2.0	2.3 2.1	2.3 2.1	2.4 2.2	2.4 2.3	2.5 2.4	55.5
55.8	2.6 2.2	2.6 2.2	2.6 2.3	2.7 2.3	2.7 2.4	2.8 2.4	2.9 2.5	55.8
56.1	3.0 2.4	3.0 2.5	3.0 2.5	3.0 2.5	3.1 2.6	3.1 2.6	3.2 2.7	56.1
56.4	3.3 2.7	3.4 2.7	3.4 2.7	3.4 2.7	3.4 2.8	3.5 2.8	3.5 2.9	56.4
56.7	3.7 2.9	3.7 2.9	3.8 2.9	3.8 2.9	3.8 3.0	3.8 3.0	3.9 3.0	56.7
57.0	4.1 3.1	4.1 3.1	4.1 3.1	4.1 3.1	4.2 3.2	4.2 3.2	4.2 3.2	57.0
57.3	4.5 3.3	4.5 3.3	4.5 3.3	4.5 3.3	4.5 3.3	4.5 3.4	4.6 3.4	57.3
57.6	4.9 3.5	4.9 3.5	4.9 3.5	4.9 3.5	4.9 3.5	4.9 3.5	4.9 3.6	57.6
57.9	5.3 3.8	5.3 3.8	5.2 3.8	5.2 3.7	5.2 3.7	5.2 3.7	5.2 3.7	57.9
58.2	5.6 4.0	5.6 4.0	5.6 4.0	5.6 4.0	5.6 3.9	5.6 3.9	5.6 3.9	58.2
58.5	6.0 4.2	6.0 4.2	6.0 4.2	6.0 4.2	6.0 4.1	5.9 4.1	5.9 4.1	58.5
58.8	6.4 4.4	6.4 4.4	6.4 4.4	6.3 4.4	6.3 4.3	6.3 4.2	6.2 4.2	58.8
59.1	6.8 4.6	6.8 4.6	6.7 4.6	6.7 4.6	6.7 4.5	6.6 4.5	6.4 4.4	59.1
59.4	7.2 4.8	7.1 4.8	7.1 4.8	7.1 4.8	7.0 4.7	7.0 4.7	6.9 4.6	59.4
59.7	7.5 5.1	7.5 5.0	7.5 5.0	7.5 5.0	7.4 4.9	7.3 4.8	7.2 4.8	59.7
60.0	7.9 5.3	7.9 5.3	7.9 5.2	7.8 5.2	7.8 5.1	7.7 5.0	7.6 4.9	60.0
60.3	8.3 5.5	8.3 5.5	8.2 5.4	8.2 5.4	8.1 5.3	8.0 5.2	7.9 5.1	60.3
60.6	8.7 5.7	8.7 5.7	8.6 5.7	8.6 5.6	8.5 5.5	8.4 5.4	8.2 5.3	60.6
60.9	9.1 5.9	9.0 5.9	9.0 5.9	8.9 5.8	8.8 5.7	8.7 5.6	8.6 5.4	60.9
61.2	9.5 6.2	9.4 6.1	9.4 6.1	9.3 6.0	9.2 5.9	9.1 5.8	8.9 5.6	61.2
61.5	9.8 6.4	9.8 6.3	9.7 6.3	9.7 6.2	9.5 6.1	9.4 5.9	9.2 5.8	61.5

DIP

Ht. of Eye	Corrⁿ	Ht. of Eye	Ht. of Eye	Corrⁿ	Ht. of Eye
m		ft.	m		ft.
2.4	−2.8	8.0	9.5	−5.5	31.5
2.6	−2.9	8.6	9.9	−5.6	32.7
2.8	−3.0	9.2	10.3	−5.7	33.9
3.0	−3.1	9.8	10.6	−5.8	35.1
3.2	−3.2	10.5	11.0	−5.9	36.3
3.4	−3.3	11.2	11.4	−6.0	37.6
3.6	−3.4	11.9	11.8	−6.1	38.9
3.8	−3.5	12.6	12.2	−6.2	40.1
4.0	−3.6	13.3	12.6	−6.3	41.5
4.3	−3.7	14.1	13.0	−6.4	42.8
4.5	−3.8	14.9	13.4	−6.5	44.2
4.7	−3.9	15.7	13.8	−6.6	45.5
5.0	−4.0	16.5	14.2	−6.7	46.9
5.2	−4.1	17.4	14.7	−6.8	48.4
5.5	−4.2	18.3	15.1	−6.9	49.8
5.8	−4.3	19.1	15.5	−7.0	51.3
6.1	−4.4	20.1	16.0	−7.1	52.8
6.3	−4.5	21.0	16.5	−7.2	54.3
6.6	−4.6	22.0	16.9	−7.3	55.8
6.9	−4.7	22.9	17.4	−7.4	57.4
7.2	−4.8	23.9	17.9	−7.5	58.9
7.5	−4.9	24.9	18.4	−7.6	60.5
7.9	−5.0	26.0	18.8	−7.7	62.1
8.2	−5.1	27.1	19.3	−7.8	63.8
8.5	−5.2	28.1	19.8	−7.9	65.4
8.8	−5.3	29.2	20.4	−8.0	67.1
9.2	−5.4	30.4	20.9	−8.1	68.8
9.5		31.5	21.4		70.5

MOON CORRECTION TABLE

The correction is in two parts; the first correction is taken from the upper part of the table with argument apparent altitude, and the second from the lower part, with argument HP, in the same column as that from which the first correction was taken. Separate corrections are given in the lower part for lower (L) and upper (U) limbs. All corrections are to be **added** to apparent altitude, *but 30′ is to be subtracted from the altitude of the upper limb.*

For corrections for pressure and temperature see page A4.

For bubble sextant observations ignore dip, take the mean of upper and lower limb corrections and subtract 15′ from the altitude.

App. Alt. = Apparent altitude = Sextant altitude corrected for index error and dip.

Extract of main table for Lat 40° (contrary) Dec 6°, LHA 43°

LHA	0° Hc d Z	1° Hc d Z	2° Hc d Z	3° Hc d Z	4° Hc d Z	5° Hc d Z	6° Hc d Z
69	15 56 40 104	15 16 41 105	14 35 40 105	13 55 41 106	13 14 40 107	12 34 41 108	11 53 41 108
68	16 41 41 105	16 00 40 105	15 20 41 106	14 39 41 107	13 58 41 108	13 17 41 108	12 36 41 109
67	17 25 41 105	16 44 40 106	16 04 41 107	15 23 41 108	14 42 41 108	14 01 41 109	13 20 42 110
66	18 09 40 106	17 29 41 107	16 48 41 108	16 07 41 108	15 26 42 109	14 44 41 110	14 03 42 111
65	18 53 -40 107	18 13 -42 107	17 31 -41 108	16 50 -41 109	16 09 -42 110	15 27 -41 110	14 46 -42 111
64	19 37 41 107	18 56 41 108	18 15 41 109	17 34 42 110	16 52 42 110	16 10 42 111	15 28 42 112
63	20 21 41 108	19 40 42 109	18 58 41 110	18 17 42 110	17 35 42 111	16 53 42 112	16 11 42 113
62	21 05 42 109	20 23 41 110	19 42 42 110	19 00 42 111	18 18 42 112	17 36 43 113	16 53 42 113
61	21 48 42 110	21 06 41 110	20 25 42 111	19 43 43 112	19 00 42 113	18 18 43 113	17 35 43 114
60	22 31 -42 110	21 49 -42 111	21 07 -42 112	20 25 -43 113	19 43 -43 113	19 00 -43 114	18 17 -43 115
59	23 14 42 111	22 32 42 112	21 50 43 113	21 07 42 113	20 25 43 114	19 42 43 115	18 59 44 116
58	23 57 42 112	23 15 43 113	22 32 43 113	21 49 43 114	21 06 43 115	20 23 43 116	19 40 44 116
57	24 40 43 113	23 57 43 113	23 14 43 114	22 31 43 115	21 48 43 116	21 05 44 116	20 21 44 117
56	25 22 43 113	24 39 43 114	23 56 43 115	23 13 44 116	22 29 43 116	21 46 44 117	21 02 44 118
55	26 04 -43 114	25 21 -43 115	24 38 -44 116	23 54 -44 117	23 10 -44 117	22 26 -44 118	21 42 -44 119
54	26 46 44 115	26 02 43 116	25 19 44 117	24 35 44 117	23 51 44 118	23 07 45 119	22 22 44 120
53	27 27 43 116	26 44 44 117	26 00 44 117	25 16 45 118	24 31 44 119	23 47 45 120	23 02 45 120
52	28 08 43 117	27 25 45 117	26 40 44 118	25 56 45 119	25 11 44 120	24 27 45 120	23 42 46 121
51	28 49 44 117	28 05 44 118	27 21 45 119	26 36 45 120	25 51 45 121	25 06 45 121	24 21 46 122
50	29 30 -45 118	28 45 -44 119	28 01 -45 120	27 16 -45 121	26 31 -46 121	25 45 -45 122	25 00 -46 123
49	30 10 45 119	29 25 45 120	28 40 45 121	27 55 45 121	27 10 46 122	26 24 46 123	25 38 46 124
48	30 50 45 120	30 05 45 121	29 20 46 122	28 34 45 122	27 48 46 123	27 02 46 124	26 16 46 124
47	31 30 46 121	30 44 45 122	29 59 46 122	29 13 46 123	28 27 47 124	27 40 46 125	26 54 47 125
46	32 09 46 122	31 23 46 123	30 37 46 123	29 51 46 124	29 05 47 125	28 18 47 126	27 31 47 126
45	32 48 -46 123	32 02 -46 123	31 16 -47 124	30 29 -47 125	29 42 -47 126	28 55 -47 126	28 08 -47 127
44	33 26 46 124	32 40 47 124	31 53 47 125	31 06 47 126	30 19 47 127	29 32 48 127	28 44 47 128
43	34 04 46 125	33 18 47 125	32 31 48 126	31 43 47 127	30 56 48 128	30 08 48 128	29 20 48 129
42	34 42 47 126	33 55 47 126	33 08 47 127	32 20 48 128	31 32 48 128	30 44 48 129	29 56 49 130
41	35 19 47 126	34 32 48 127	33 44 48 128	32 56 48 129	32 08 48 129	31 20 49 130	30 31 49 131
40	35 56 -48 127	35 08 -48 128	34 20 -48 129	33 32 -49 130	32 43 -48 130	31 55 -49 131	31 06 -50 132
39	36 32 48 128	35 44 48 129	34 56 49 130	34 07 49 131	33 18 49 131	32 29 49 132	31 40 50 133
38	37 08 49 129	36 19 48 130	35 31 49 131	34 42 50 132	33 52 49 132	33 03 50 133	32 13 50 134
37	37 43 49 130	36 54 49 131	36 05 49 132	35 16 50 133	34 26 50 133	33 36 50 134	32 46 50 135
36	38 18 49 131	37 29 50 132	36 39 50 133	35 49 50 134	34 59 50 134	34 09 50 135	33 19 51 136
35	38 52 -50 133	38 02 -50 133	37 12 -50 134	36 22 -50 135	35 32 -51 135	34 41 -51 136	33 50 -51 137
34	39 26 51 134	38 35 50 134	37 45 50 135	36 55 51 136	36 04 51 136	35 13 51 137	34 22 52 138
33	39 59 51 135	39 08 51 135	38 17 51 136	37 26 51 137	36 35 51 137	35 44 52 138	34 52 51 139
32	40 31 51 136	39 40 51 137	38 49 51 137	37 58 52 138	37 06 52 138	36 14 52 139	35 22 52 140
31	41 03 52 137	40 11 51 138	39 20 52 138	38 28 52 139	37 36 52 140	36 44 52 140	35 52 53 141
30	41 34 -52 138	40 42 -52 139	39 50 -52 139	38 58 -52 140	38 06 -53 141	37 13 -53 141	36 20 -52 142
29	42 04 52 139	41 12 52 140	40 20 53 141	39 27 53 141	38 34 52 142	37 42 54 142	36 48 53 143
28	42 34 54 140	41 41 52 141	40 49 53 142	39 56 53 142	39 03 54 143	38 09 53 144	37 16 54 144
27	43 03 53 142	42 10 53 142	41 17 54 143	40 23 53 143	39 30 54 144	38 36 54 145	37 42 54 145
26	43 31 54 143	42 37 53 143	41 44 54 144	40 50 54 145	39 56 54 145	39 02 54 146	38 08 54 146
25	43 58 -54 144	43 04 -53 145	42 11 -54 145	41 17 -55 146	40 22 -54 146	39 28 -55 147	38 33 -54 147
24	44 25 54 145	43 31 55 146	42 36 54 146	41 42 55 147	40 47 54 148	39 53 55 148	38 58 55 149
23	44 51 55 147	43 56 55 147	43 01 54 148	42 07 55 148	41 12 56 149	40 16 55 149	39 21 55 150
22	45 15 54 148	44 21 56 148	43 25 55 149	42 30 55 150	41 35 56 150	40 39 55 151	39 44 56 151
21	45 39 55 149	44 44 55 150	43 49 56 150	42 53 56 151	41 57 55 151	41 02 56 152	40 06 56 152
20	46 03 -56 150	45 07 -56 151	44 11 -56 152	43 15 -56 152	42 19 -56 153	41 23 -56 153	40 27 -57 153
19	46 25 56 152	45 29 56 152	44 33 57 153	43 36 56 153	42 40 57 154	41 43 56 154	40 47 57 155
18	46 46 56 153	45 50 57 154	44 53 57 154	43 56 56 155	43 00 57 155	42 03 57 156	41 06 57 156
17	47 06 57 155	46 09 56 155	45 13 57 155	44 16 57 156	43 19 58 156	42 21 57 157	41 24 57 157
16	47 25 57 156	46 28 57 156	45 31 57 157	44 34 57 157	43 37 57 158	42 39 57 158	41 42 58 158
15	47 44 -58 157	46 46 -57 158	45 49 -58 158	44 51 -57 159	43 54 -58 159	42 56 -58 159	41 58 -58 160
14	48 01 58 159	47 03 58 159	46 05 58 160	45 07 58 160	44 09 58 160	43 11 58 161	42 13 58 161
13	48 17 58 160	47 19 58 161	46 21 58 161	45 23 59 161	44 24 58 162	43 26 58 162	42 28 59 162
12	48 32 58 162	47 34 59 162	46 35 58 162	45 37 59 163	44 38 58 163	43 40 59 163	42 41 58 164
11	48 46 59 163	47 47 58 164	46 49 59 164	45 50 59 164	44 51 59 164	43 52 58 165	42 54 59 165
10	48 58 -58 165	48 00 -59 165	47 01 -59 165	46 02 -59 166	45 03 -59 166	44 04 -59 166	43 05 -59 166
9	49 10 59 166	48 11 59 166	47 12 59 167	46 13 59 167	45 14 59 167	44 15 60 167	43 15 59 168
8	49 20 59 168	48 21 59 168	47 22 59 168	46 23 60 168	45 23 59 169	44 24 59 169	43 25 60 169
7	49 30 60 169	48 30 59 169	47 31 60 170	46 31 59 170	45 32 60 170	44 32 59 170	43 33 60 170
6	49 38 60 171	48 38 59 171	47 39 60 171	46 39 60 171	45 39 59 171	44 40 60 172	43 40 60 172
5	49 44 -59 172	48 45 -60 172	47 45 -60 173	46 45 -59 173	45 46 -60 173	44 46 -60 173	43 46 -60 173
4	49 50 60 174	48 50 60 174	47 50 59 174	46 51 60 174	45 51 60 174	44 51 60 174	43 51 60 174
3	49 54 59 175	48 55 60 175	47 55 60 176	46 55 60 176	45 55 60 176	44 55 60 176	43 55 60 176
2	49 58 60 177	48 58 60 177	47 58 60 177	46 58 60 177	45 58 60 177	44 58 60 177	43 58 60 177
1	49 59 60 178	48 59 60 178	47 59 60 179	46 59 60 179	45 59 60 179	44 59 60 179	43 59 60 179
0	50 00 -60 180	49 00 -60 180	48 00 -60 180	47 00 -60 180	46 00 -60 180	45 00 -60 180	44 00 -60 180

©2009 Jack Case

Extract of main table for Lat 40° (contrary) Dec 11°, LHA 68°

LHA	d	10° Hc d Z	11° Hc d Z	12° Hc d Z	13° Hc d Z	14° Hc d Z	LHA
69	11	09 08 41 111	08 27 42 112	07 45 41 113	07 04 42 114	06 22 42 114	291
68	10	09 51 42 112	09 09 42 113	08 27 41 114	07 46 42 114	07 04 42 115	292
67	12	10 33 42 113	09 51 42 113	09 09 42 114	08 27 42 115	07 45 42 116	293
66	13	11 15 42 113	10 33 42 114	09 51 42 115	09 09 42 116	08 27 43 116	294
65	13	11 58 −43 114	11 15 −42 115	10 33 −43 116	09 50 −42 116	09 08 −43 117	295
64	19	12 39 42 115	11 57 43 116	11 14 43 116	10 31 43 117	09 48 43 118	296
63	20	13 21 43 116	12 38 43 116	11 55 43 117	11 12 43 118	10 29 43 118	297
62	23	14 02 43 116	13 19 43 117	12 36 43 118	11 53 44 118	11 09 43 119	298
61	23	14 43 43 117	14 00 43 118	13 17 44 118	12 33 44 119	11 49 44 120	299
60	27	15 24 −44 118	14 40 −43 119	13 57 −44 119	13 13 −44 120	12 29 −44 121	300
59	23	16 05 44 119	15 21 44 119	14 37 44 120	13 53 45 121	13 08 44 121	301
58	29	16 45 44 119	16 01 45 120	15 16 44 121	14 32 45 121	13 47 44 122	302
57	29	17 25 45 120	16 40 44 121	15 56 45 121	15 11 45 122	14 26 45 123	303
56	25	18 04 44 121	17 20 45 122	16 35 45 122	15 50 45 123	15 05 46 124	304
55	20	18 44 −45 122	17 59 −45 122	17 14 −46 123	16 28 −45 124	15 43 −46 124	305
54	20	19 23 46 122	18 37 45 123	17 52 46 124	17 06 45 124	16 21 46 125	306
53	22	20 01 45 123	19 16 46 124	18 30 46 125	17 44 46 125	16 58 46 126	307
52	28	20 40 46 124	19 54 46 125	19 08 47 125	18 21 46 126	17 35 46 127	308
51	24	21 18 47 125	20 31 46 125	19 45 47 126	18 58 46 127	18 12 47 127	309
50	25	21 55 −46 126	21 09 −47 126	20 22 −47 127	19 35 −47 128	18 48 −47 128	310
49	36	22 32 47 126	21 45 47 127	20 58 47 128	20 11 47 128	19 24 48 129	311
48	37	23 09 47 127	22 22 47 128	21 35 48 129	20 47 48 129	19 59 47 130	312
47	37	23 46 48 128	22 58 48 129	22 10 48 129	21 22 48 130	20 34 48 131	313
46	38	24 21 47 129	23 34 48 130	22 46 49 130	21 57 48 131	21 09 48 132	314
45	39	24 57 −48 130	24 09 −49 130	23 20 −48 131	22 32 −49 132	21 43 −49 132	315
44	30	25 32 48 131	24 44 49 131	23 55 49 132	23 06 49 133	22 17 49 133	316
43	31	26 07 49 132	25 18 49 132	24 29 49 133	23 40 50 133	22 50 49 134	317
42	32	26 41 49 132	25 52 50 133	25 02 49 134	24 13 50 134	23 23 50 135	318
41	33	27 14 49 133	26 25 50 134	25 35 50 135	24 45 50 135	23 55 50 136	319
40	34	27 48 −50 134	26 58 −50 135	26 08 −51 136	25 17 −50 136	24 27 −51 137	320
39	36	28 20 50 135	27 30 50 136	26 40 51 136	25 49 51 137	24 58 51 138	321
38	36	28 52 50 136	28 02 51 137	27 11 51 137	26 20 51 138	25 29 51 139	322
37	37	29 24 51 137	28 33 51 138	27 42 52 138	26 50 51 139	25 59 52 139	323
36	37	29 55 51 138	29 04 52 139	28 12 52 139	27 20 51 140	26 29 52 140	324
35	38	30 25 −51 139	29 34 −52 140	28 42 −52 140	27 50 −52 141	26 58 −53 141	325
34	39	30 55 52 140	30 03 52 141	29 11 53 141	28 18 52 142	27 26 53 142	326
33	30	31 24 52 141	30 32 53 142	29 39 52 142	28 47 53 143	27 54 53 143	327
32	40	31 53 53 142	31 00 53 143	30 07 53 143	29 14 53 144	28 21 53 144	328
31	43	32 21 53 143	31 28 54 144	30 34 53 144	29 41 54 145	28 47 53 145	329
30	44	32 48 −53 144	31 55 −54 145	31 01 −54 145	30 07 −54 146	29 13 −54 146	330
29	45	33 15 54 145	32 21 54 146	31 27 54 146	30 33 54 147	29 39 55 147	331
28	45	33 41 55 146	32 46 54 147	31 52 54 147	30 58 55 148	30 03 55 148	332
27	47	34 06 55 147	33 11 54 148	32 17 55 148	31 22 55 149	30 27 55 149	333
26	43	34 30 55 148	33 35 55 149	32 40 55 149	31 45 55 150	30 50 55 150	334
25	49	34 54 −55 150	33 59 −56 150	33 03 −55 150	32 08 −56 151	31 12 −55 151	335
24	40	35 17 56 151	34 21 55 151	33 26 56 152	32 30 56 152	31 34 56 152	336
23	41	35 39 56 152	34 43 56 152	33 47 56 153	32 51 56 153	31 55 56 153	337
22	42	36 00 56 153	35 04 56 153	34 08 56 154	33 12 57 154	32 15 56 155	338
21	44	36 21 57 154	35 24 56 154	34 28 57 155	33 31 56 155	32 35 57 156	339
20	46	36 41 −57 155	35 44 −57 156	34 47 −57 156	33 50 −57 156	32 53 −57 157	340
19	46	36 59 57 156	36 02 57 157	35 05 57 157	34 08 57 157	33 11 57 158	341
18	47	37 17 57 158	36 20 57 158	35 23 58 158	34 25 57 159	33 28 58 159	342
17	48	37 35 58 159	36 37 58 159	35 39 57 159	34 42 58 160	33 44 58 160	343
16	40	37 51 58 160	36 53 58 160	35 55 58 161	34 57 58 161	33 59 58 161	344
15	41	38 06 −58 161	37 08 −58 161	36 10 −58 162	35 12 −58 162	34 14 −59 162	345
14	42	38 21 59 162	37 22 58 163	36 24 58 163	35 26 59 163	34 27 58 163	346
13	48	38 34 58 164	37 36 59 164	36 37 59 164	35 38 58 164	34 40 59 165	347
12	45	38 47 59 165	37 48 59 165	36 49 59 165	35 50 59 166	34 51 58 166	348
11	46	38 58 59 166	37 59 59 166	37 00 59 166	36 01 59 167	35 02 59 167	349
10	47	39 09 −59 167	38 10 −59 167	37 11 −60 168	36 11 −59 168	35 12 −59 168	350
9	48	39 19 60 169	38 19 59 169	37 20 59 169	36 21 60 169	35 21 59 169	351
8	49	39 27 59 170	38 28 60 170	37 28 59 170	36 29 60 170	35 29 59 170	352
7	41	39 35 60 171	38 35 59 171	37 36 59 171	36 36 59 171	35 37 60 172	353
6	42	39 41 59 172	38 42 60 172	37 42 60 173	36 42 59 173	35 43 60 173	354
5	48	39 47 −60 174	38 47 −59 174	37 48 −60 174	36 48 −60 174	35 48 −60 174	355
4	45	39 52 60 175	38 52 60 175	37 52 60 175	36 52 60 175	35 52 60 175	356
3	45	39 55 60 176	38 55 59 176	37 56 60 176	36 56 60 176	35 56 60 176	357
2	47	39 58 60 178	38 58 60 178	37 58 60 178	36 58 60 178	35 58 60 178	358
1	49	39 59 60 179	38 59 59 179	38 00 60 179	37 00 60 179	36 00 60 179	359
0	50	40 00 −60 180	39 00 −60 180	38 00 −60 180	37 00 −60 180	36 00 −60 180	360

LAT 40°

Extract of main table for Lat 40° (same) Dec 8°, LHA 71°

LHA	7° Hc	d	Z	8° Hc	d	Z	9° Hc	d	Z	10° Hc	d	Z	11° Hc	d	Z	12° Hc	d	Z	
70	19 47	+38	98	20 25	+39	97	21 04	+38	96	21 42	+37	95	22 19	+38	94	22 57	+37	93	2
71	19 01	39	97	19 40	38	96	20 18	38	95	20 56	38	94	21 34	37	94	22 11	37	93	2
72	18 15	39	96	18 54	38	95	19 32	38	95	20 10	38	94	20 48	37	93	21 25	37	92	2
73	17 30	38	96	18 08	38	95	18 46	38	94	19 24	38	93	20 02	37	92	20 39	37	92	2
74	16 44	38	95	17 22	38	94	18 00	38	93	18 38	38	93	19 16	37	92	19 53	37	91	2
75	15 58	+38	94	16 36	+38	93	17 14	+38	93	17 52	+38	92	18 30	+37	91	19 07	+38	90	1
76	15 12	38	94	15 50	38	93	16 28	38	92	17 06	38	91	17 44	37	90	18 21	38	90	1
77	14 26	39	93	15 05	38	92	15 43	37	91	16 20	38	91	16 58	37	90	17 35	38	89	1
78	13 41	38	92	14 19	38	92	14 57	37	91	15 34	38	90	16 12	37	89	16 49	38	88	1
79	12 55	38	92	13 33	38	91	14 11	37	90	14 48	38	89	15 26	38	88	16 04	37	88	1
80	12 09	+38	91	12 47	+38	90	13 25	+37	89	14 02	+38	89	14 40	+38	88	15 18	+37	87	1
81	11 23	38	90	12 01	38	90	12 39	38	89	13 17	37	88	13 54	38	87	14 32	37	86	1
82	10 37	38	90	11 15	38	89	11 53	38	88	12 31	37	87	13 08	38	87	13 46	37	86	1
83	09 51	38	89	10 29	38	88	11 07	38	88	11 45	37	87	12 22	38	86	13 00	37	85	1
84	09 05	38	88	09 43	38	88	10 21	38	87	10 59	38	86	11 37	37	85	12 14	38	85	1
85	08 19	+38	88	08 57	+38	87	09 35	+38	86	10 13	+38	85	10 51	+38	85	11 29	+37	84	1
86	07 33	38	87	08 11	38	86	08 49	38	86	09 27	38	85	10 05	38	84	10 43	38	83	1
87	06 47	38	87	07 25	38	86	08 03	38	85	08 41	38	84	09 19	38	83	09 57	38	83	1
88	06 01	38	86	06 39	39	85	07 18	38	84	07 56	38	84	08 34	38	83	09 12	38	82	0
89	05 15	39	85	05 54	38	84	06 32	38	84	07 10	38	83	07 48	38	82	08 26	38	81	0
90	04 30	+38	85	05 08	+38	84	05 46	+39	83	06 25	+38	82	07 03	+38	82	07 41	+38	81	0
91	03 44	38	84	04 22	39	83	05 01	38	82	05 39	38	82	06 17	38	81	06 55	39	80	0
92	02 58	39	83	03 37	38	83	04 15	39	82	04 54	38	81	05 32	38	80	06 10	38	79	0
93	02 13	38	83	02 51	39	82	03 30	38	81	04 08	39	80	04 47	38	80	05 25	38	79	0
94	01 27	39	82	02 06	38	81	02 44	39	81	03 23	39	80	04 02	38	79	04 40	39	78	0
95	00 41	+39	81	01 20	+39	81	01 59	+39	80	02 38	+38	79	03 16	+39	78	03 55	+39	78	0
96	-0 04	39	81	00 35	39	80	01 14	39	79	01 53	38	79	02 31	39	78	03 10	39	77	0
97	-0 49	39	80	-0 10	39	79	00 29	39	79	01 08	39	78	01 47	39	77	02 26	38	76	0
98	-1 34	39	80	-0 55	39	79	-0 16	39	78	00 23	39	77	01 02	39	76	01 41	39	76	0
	-2 20	40	79	-1 40	39	78	-1 01	39	77	-0 22	39	77	00 17	39	76	00 56	40	75	0
	-3 05	+40	78	-2 25	+39	77	-1 46	+39	77	-1 07	+40	76	-0 27	+39	75	00 12	+40	74	0
	-3 50	40	78	-3 10	39	77	-2 31	40	76	-1 51	39	75	-1 12	40	75	-0 32	39	74	0
	-4 34	39	77	-3 55	40	76	-3 15	39	75	-2 36	40	75	-1 56	40	74	-1 16	40	73	-
	-5 19	40	76	-4 39	39	75	-4 00	40	75	-3 20	40	74	-2 40	40	73	-2 00	40	72	-
	-6 04	+40	76	-5 24	40	75	-4 44	40	74	-4 04	40	73	-3 24	40	73	-2 44	40	72	-
			105	-6 08	+40	74	-5 28	+40	73	-4 48	+40	73	-4 08	+41	72	-3 27	+40	71	-
						106	-6 12	+40	73	-5 32	41	72	-4 51	40	71	-4 11	41	71	-
									107	-6 15	+40	72	-5 35	41	71	-4 54	41	70	-
												108	-6 18	+41	70	-5 37	41	69	-
															109	-6 20	+41	69	-
																		110	-

LHA	Hc	d	Z	Hc	d	Z	Hc	d	Z	Hc	d	Z	Hc	d	Z	Hc	d	Z	
97																			
96																			
95																			
94																			
93	**267**																		
92	-6 01	-38	94	**268**															
91	-5 15	39	95	-5 54	-38	96	**269**												
90	-4 30	-38	95	-5 08	-38	96	-5 46	-39	97	**270**									
89	-3 44	38	96	-4 22	39	97	-5 01	38	98	-5 39	-38	98	-6 17	-38	99	**271**			
88	-2 58	39	97	-3 37	38	97	-4 15	39	98	-4 54	38	99	-5 32	38	100	-6 10	-38	101	?
87	-2 13	38	97	-2 51	39	98	-3 30	38	99	-4 08	39	100	-4 47	38	100	-5 25	38	101	-
86	-1 27	39	98	-2 06	38	99	-2 44	39	99	-3 23	39	100	-4 02	38	101	-4 40	39	102	-
85	-0 41	-39	99	-1 20	-39	99	-1 59	-39	100	-2 38	-38	101	-3 16	-39	102	-3 55	-39	102	-
84	00 04	39	99	-0 35	39	100	-1 14	39	101	-1 53	38	101	-2 31	39	102	-3 10	39	103	-
83	00 49	39	100	00 10	39	101	-0 29	39	101	-1 08	39	102	-1 47	39	103	-2 26	38	104	-
82	01 34	39	100	00 55	39	101	00 16	39	102	-0 23	39	103	-1 02	39	104	-1 41	39	104	-
81	02 20	40	101	01 40	39	102	01 01	39	103	00 22	39	103	-0 17	39	104	-0 56	40	105	-
80	03 05	-40	102	02 25	-39	103	01 46	-39	103	01 07	-40	104	00 27	-39	105	-0 12	-40	106	-
79	03 50	40	102	03 10	39	103	02 31	40	104	01 51	39	105	01 12	40	105	00 32	39	106	-
78	04 34	39	103	03 55	40	104	03 15	39	105	02 36	40	105	01 56	40	106	01 16	40	107	0
77	05 19	40	104	04 39	39	105	04 00	40	105	03 20	40	106	02 40	40	107	02 00	40	108	0
76	06 04	40	104	05 24	40	105	04 44	40	106	04 04	40	107	03 24	40	107	02 44	40	108	0
75	06 48	-40	105	06 08	-40	106	05 28	-40	107	04 48	-40	107	04 08	-41	108	03 27	-40	109	0
74	07 32	40	106	06 52	40	107	06 12	40	107	05 32	41	108	04 51	40	109	04 11	41	109	0
73	08 17	41	106	07 36	40	107	06 56	41	108	06 15	40	109	05 35	41	109	04 54	41	110	0
72	09 01	41	107	08 20	41	108	07 39	40	109	06 59	41	109	06 18	41	110	05 37	41	111	0
71	09 44	40	108	09 04	41	109	08 23	41	109	07 42	41	110	07 01	41	111	06 20	41	111	0
70	10 28	-41	108	09 47	-41	109	09 06	-41	110	08 25	-41	111	07 44	-41	111	07 03	-42	112	0

©2009 Jack Case

Extract from table 5 for values of d from 34' to 60'

34	35	36	37	38	39	40	41	42	43	44	45	46	47	48	49	50	51	52	53	54	55	56	57	58	59	60	d
'	'	'	'	'	'	'	'	'	'	'	'	'	'	'	'	'	'	'	'	'	'	'	'	'	'	'	'
0	0	0	0	0	0	0	0	0	0	0	0	0	0	0	0	0	0	0	0	0	0	0	0	0	0	0	0
1	1	1	1	1	1	1	1	1	1	1	1	1	1	1	1	1	1	1	1	1	1	1	1	1	1	1	1
1	1	1	1	1	1	1	1	1	1	1	2	2	2	2	2	2	2	2	2	2	2	2	2	2	2	2	2
2	2	2	2	2	2	2	2	2	2	2	2	2	2	2	2	2	3	3	3	3	3	3	3	3	3	3	3
2	2	2	2	3	3	3	3	3	3	3	3	3	3	3	3	3	3	3	4	4	4	4	4	4	4	4	4
3	3	3	3	3	3	3	3	4	4	4	4	4	4	4	4	4	4	4	4	4	5	5	5	5	5	5	5
3	4	4	4	4	4	4	4	4	4	4	4	5	5	5	5	5	5	5	5	5	6	6	6	6	6	6	6
4	4	4	4	4	5	5	5	5	5	5	5	5	5	6	6	6	6	6	6	6	6	7	7	7	7	7	7
5	5	5	5	5	5	5	5	6	6	6	6	6	6	6	7	7	7	7	7	7	7	7	8	8	8	8	8
5	5	5	6	6	6	6	6	6	6	7	7	7	7	7	7	8	8	8	8	8	8	8	9	9	9	9	9
6	6	6	6	6	6	7	7	7	7	7	8	8	8	8	8	8	8	9	9	9	9	9	10	10	10	10	10
6	6	7	7	7	7	7	8	8	8	8	8	8	9	9	9	9	9	10	10	10	10	10	10	11	11	11	11
7	7	7	7	8	8	8	8	8	9	9	9	9	9	10	10	10	10	10	11	11	11	11	11	12	12	12	12
7	8	8	8	8	8	9	9	9	9	10	10	10	10	10	11	11	11	11	11	12	12	12	12	13	13	13	13
8	8	8	9	9	9	9	10	10	10	10	10	11	11	11	11	12	12	12	12	13	13	13	13	14	14	14	14
8	9	9	9	10	10	10	10	10	11	11	11	12	12	12	12	12	13	13	13	14	14	14	14	14	15	15	15
9	9	10	10	10	10	11	11	11	11	12	12	12	13	13	13	13	14	14	14	14	15	15	15	15	16	16	16
10	10	10	10	11	11	11	12	12	12	12	13	13	13	14	14	14	14	15	15	15	16	16	16	16	17	17	17
10	10	11	11	11	12	12	12	13	13	13	14	14	14	14	15	15	15	16	16	16	16	17	17	17	18	18	18
11	11	11	12	12	12	13	13	13	14	14	14	15	15	15	16	16	16	16	17	17	17	18	18	18	19	19	19
11	12	12	12	13	13	13	14	14	14	15	15	15	16	16	16	17	17	17	18	18	18	19	19	19	20	20	20
12	12	13	13	13	14	14	14	15	15	15	16	16	16	17	17	18	18	18	19	19	19	20	20	20	21	21	21
12	13	13	14	14	14	15	15	15	16	16	16	17	17	18	18	18	19	19	19	20	20	21	21	21	22	22	22
13	13	14	14	15	15	15	16	16	16	17	17	18	18	18	19	19	20	20	20	21	21	21	22	22	23	23	23
14	14	14	15	15	16	16	16	17	17	18	18	18	19	19	20	20	20	21	21	22	22	22	23	23	24	24	24
14	15	15	15	16	16	17	17	18	18	18	19	19	20	20	20	21	21	22	22	22	23	23	24	24	25	25	25
15	15	16	16	16	17	17	18	18	19	19	20	20	20	21	21	22	22	23	23	23	24	24	25	25	26	26	26
15	16	16	17	17	18	18	18	19	19	20	20	21	21	22	22	22	23	23	24	24	25	25	26	26	27	27	27
16	16	17	17	18	18	19	19	20	20	21	21	21	22	22	23	23	24	24	25	25	26	26	27	27	28	28	28
16	17	17	18	18	19	19	20	20	21	21	22	22	23	23	24	24	25	25	26	26	27	27	28	28	29	29	29
17	18	18	18	19	20	20	20	21	22	22	22	23	24	24	24	25	26	26	26	27	28	28	28	29	30	30	30
18	18	19	19	20	20	21	21	22	22	23	23	24	24	25	25	26	26	27	27	28	28	29	29	30	30	31	31
18	19	19	20	20	21	21	22	22	23	23	24	25	25	26	26	27	27	28	28	29	29	30	30	31	31	32	32
19	19	20	20	21	21	22	23	23	24	24	25	25	26	26	27	28	28	29	29	30	30	31	31	32	32	33	33
19	20	20	21	22	22	23	23	24	24	25	26	26	27	27	28	28	29	29	30	31	31	32	32	33	33	34	34
20	20	21	22	22	23	23	24	24	25	26	26	27	27	28	29	29	30	30	31	32	32	33	33	34	34	35	35
20	21	22	22	23	23	24	25	25	26	26	27	28	28	29	29	30	31	31	32	32	33	34	34	35	35	36	36
21	22	22	23	23	24	25	25	26	27	27	28	28	29	30	30	31	31	32	33	33	34	35	35	36	36	37	37
22	22	23	23	24	25	25	26	27	27	28	28	29	30	30	31	32	32	33	34	34	35	35	36	37	37	38	38
22	23	23	24	25	25	26	27	27	28	29	29	30	31	31	32	32	33	34	34	35	36	36	37	38	38	39	39
23	23	24	25	25	26	27	27	28	29	29	30	31	31	32	33	33	34	35	35	36	37	37	38	39	39	40	40
23	24	25	25	26	27	27	28	29	29	30	31	31	32	33	33	34	35	36	36	37	38	38	39	40	40	41	41
24	24	25	26	27	27	28	29	29	30	31	32	32	33	34	34	35	36	36	37	38	38	39	40	41	41	42	42
24	25	26	27	27	28	29	29	30	31	32	32	33	34	34	35	36	37	37	38	39	39	40	41	42	42	43	43
25	26	26	27	28	29	29	30	31	32	32	33	34	34	35	36	37	37	38	39	40	40	41	42	43	43	44	44
26	26	27	28	28	29	30	31	32	32	33	34	34	35	36	37	38	38	39	40	40	41	42	43	44	44	45	45
26	27	28	28	29	30	31	31	32	33	34	34	35	36	37	38	38	39	40	41	41	42	43	44	44	45	46	46
27	27	28	29	30	31	31	32	33	34	34	35	36	37	38	38	39	40	41	42	42	43	44	45	45	46	47	47
27	28	29	30	30	31	32	33	34	34	35	36	37	38	38	39	40	41	42	42	43	44	45	46	46	47	48	48
28	29	29	30	31	32	33	33	34	35	36	37	38	38	39	40	41	42	42	43	44	45	46	47	47	48	49	49
28	29	30	31	32	32	33	34	35	36	37	38	38	39	40	41	42	42	43	44	45	46	47	48	48	49	50	50
29	30	31	31	32	33	34	35	36	37	37	38	39	40	41	42	42	43	44	45	46	47	48	48	49	50	51	51
29	30	31	32	33	34	35	35	36	37	38	39	40	41	42	42	43	44	45	46	47	48	49	49	50	51	52	52
30	31	32	33	34	34	35	36	37	38	39	40	41	42	42	43	44	45	46	47	48	49	49	50	51	52	53	53
31	32	32	33	34	35	36	37	38	39	40	40	41	42	43	44	45	46	47	48	49	50	50	51	52	53	54	54
31	32	33	34	35	36	37	38	38	39	40	41	42	43	44	45	46	47	48	49	50	50	51	52	53	54	55	55
32	33	33	35	35	36	37	38	39	40	41	42	43	44	45	46	47	48	49	50	50	51	52	53	54	55	56	56
32	33	34	35	36	37	38	39	40	41	42	43	44	45	46	47	48	48	49	50	51	52	53	54	55	56	57	57
33	34	35	36	37	38	39	40	41	42	43	44	45	46	47	48	49	50	50	51	52	53	54	55	56	57	58	58
33	34	35	36	37	38	39	40	41	42	43	44	45	46	47	48	49	50	51	52	53	54	55	56	57	58	59	59

TABLE 2 — ALTITUDE CORRECTION FOR CHANGE IN POSITION OF BODY

Correction for 1 Minute of Time

MOB

True Zn	Latitude																		True Zn
	0°	5°	10°	15°	20°	25°	30°	35°	40°	45°	50°	55°	60°	65°	70°	75°	80°	85°	
°	′	′	′	′	′	′	′	′	′	′	′	′	′	′	′	′	′	′	°
090	+15.0	+15.0	+14.8	+14.5	+14.1	+13.6	+13.0	+12.3	+11.5	+10.6	+9.7	+8.6	+7.5	+6.4	+5.1	+3.9	+2.6	+1.3	090
093	15.0	15.0	14.8	14.5	14.1	13.6	13.0	12.3	11.5	10.6	9.7	8.6	7.5	6.3	5.1	3.9	2.6	1.3	087
096	15.0	14.9	14.7	14.4	14.1	13.6	13.0	12.3	11.5	10.6	9.6	8.6	7.5	6.3	5.1	3.9	2.6	1.3	084
099	14.9	14.8	14.6	14.3	14.0	13.5	12.9	12.2	11.4	10.5	9.5	8.5	7.4	6.3	5.1	3.8	2.6	1.3	081
102	14.7	14.7	14.5	14.2	13.8	13.3	12.7	12.1	11.3	10.4	9.5	8.4	7.4	6.2	5.0	3.8	2.6	1.3	078
105	+14.5	+14.5	+14.3	+14.0	+13.7	+13.2	+12.6	+11.9	+11.1	+10.3	+9.3	+8.3	+7.3	+6.1	+5.0	+3.8	+2.5	+1.3	075
108	14.3	14.3	14.1	13.8	13.4	13.0	12.4	11.7	11.0	10.1	9.2	8.2	7.2	6.0	4.9	3.7	2.5	1.2	072
111	14.0	14.0	13.8	13.6	13.2	12.7	12.2	11.5	10.8	9.9	9.0	8.1	7.0	5.9	4.8	3.6	2.4	1.2	069
114	13.7	13.7	13.5	13.3	12.9	12.5	11.9	11.3	10.5	9.7	8.8	7.9	6.9	5.8	4.7	3.6	2.4	1.2	066
117	13.4	13.4	13.2	12.9	12.6	12.1	11.6	11.0	10.3	9.5	8.6	7.7	6.7	5.7	4.6	3.5	2.3	1.2	063
120	+13.0	+13.0	+12.8	+12.6	+12.2	+11.8	+11.3	+10.7	+10.0	+9.2	+8.4	+7.5	+6.5	+5.5	+4.5	+3.4	+2.3	+1.1	060
123	12.6	12.6	12.4	12.2	11.9	11.4	10.9	10.3	9.7	8.9	8.1	7.2	6.3	5.3	4.3	3.3	2.2	1.1	057
126	12.2	12.1	12.0	11.8	11.4	11.0	10.5	10.0	9.3	8.6	7.8	7.0	6.1	5.1	4.2	3.1	2.1	1.1	054
129	11.7	11.6	11.5	11.3	11.0	10.6	10.1	9.6	9.0	8.3	7.5	6.7	5.8	4.9	4.0	3.0	2.0	1.0	051
132	11.2	11.1	11.0	10.8	10.5	10.1	9.7	9.2	8.6	7.9	7.2	6.4	5.6	4.7	3.8	2.9	1.9	1.0	048
135	+10.6	+10.6	+10.5	+10.3	+10.0	+9.6	+9.2	+8.7	+8.1	+7.5	+6.8	+6.1	+5.3	+4.5	+3.6	+2.8	+1.8	+0.9	045
138	10.1	10.0	9.9	9.7	9.5	9.1	8.7	8.2	7.7	7.1	6.5	5.8	5.0	4.3	3.4	2.6	1.7	0.9	042
141	9.5	9.4	9.3	9.1	8.9	8.6	8.2	7.8	7.3	6.7	6.1	5.4	4.7	4.0	3.2	2.4	1.6	0.8	039
144	8.8	8.8	8.7	8.5	8.3	8.0	7.7	7.2	6.8	6.3	5.7	5.1	4.4	3.7	3.0	2.3	1.5	0.8	036
147	8.2	8.2	8.1	7.9	7.7	7.4	7.1	6.7	6.3	5.8	5.3	4.7	4.1	3.5	2.8	2.1	1.4	0.7	033
150	+7.5	+7.5	+7.4	+7.3	+7.1	+6.8	+6.5	+6.2	+5.8	+5.3	+4.8	+4.3	+3.8	+3.2	+2.6	+1.9	+1.3	+0.7	030
153	6.8	6.8	6.7	6.6	6.4	6.2	5.9	5.6	5.2	4.8	4.4	3.9	3.4	2.9	2.3	1.8	1.2	0.6	027
156	6.1	6.1	6.0	5.9	5.7	5.5	5.3	5.0	4.7	4.3	3.9	3.5	3.1	2.6	2.1	1.6	1.1	0.5	024
159	5.4	5.4	5.3	5.2	5.1	4.9	4.7	4.4	4.1	3.8	3.5	3.1	2.7	2.3	1.8	1.4	0.9	0.5	021
162	4.6	4.6	4.6	4.5	4.4	4.2	4.0	3.8	3.6	3.3	3.0	2.7	2.3	2.0	1.6	1.2	0.8	0.4	018
165	+3.9	+3.9	+3.8	+3.8	+3.7	+3.5	+3.4	+3.2	+3.0	+2.8	+2.5	+2.2	+1.9	+1.6	+1.3	+1.0	+0.7	+0.3	015
168	3.1	3.1	3.1	3.0	2.9	2.8	2.7	2.6	2.4	2.2	2.0	1.8	1.6	1.3	1.1	0.8	0.5	0.3	012
171	2.4	2.3	2.3	2.3	2.2	2.1	2.0	1.9	1.8	1.7	1.5	1.3	1.2	1.0	0.8	0.6	0.4	0.2	009
174	1.6	1.6	1.5	1.5	1.5	1.4	1.4	1.3	1.2	1.1	1.0	0.9	0.8	0.7	0.5	0.4	0.3	0.1	006
177	0.8	0.8	0.8	0.8	0.7	0.7	0.7	0.6	0.6	0.6	0.5	0.5	0.4	0.3	0.3	0.2	0.1	0.1	003
180	0.0	0.0	0.0	0.0	0.0	0.0	0.0	0.0	0.0	0.0	0.0	0.0	0.0	0.0	0.0	0.0	0.0	0.0	000
183	−0.8	−0.8	−0.8	−0.8	−0.7	−0.7	−0.7	−0.6	−0.6	−0.6	−0.5	−0.5	−0.4	−0.3	−0.3	−0.2	−0.1	−0.1	357
186	1.6	1.6	1.5	1.5	1.5	1.4	1.4	1.3	1.2	1.1	1.0	0.9	0.8	0.7	0.5	0.4	0.3	0.1	354
189	2.4	2.3	2.3	2.3	2.2	2.1	2.0	1.9	1.8	1.7	1.5	1.3	1.2	1.0	0.8	0.6	0.4	0.2	351
192	3.1	3.1	3.1	3.0	2.9	2.8	2.7	2.6	2.4	2.2	2.0	1.8	1.6	1.3	1.1	0.8	0.5	0.3	348
195	3.9	3.9	3.8	3.8	3.7	3.5	3.4	3.2	3.0	2.8	2.5	2.2	1.9	1.6	1.3	1.0	0.7	0.3	345
198	−4.6	−4.6	−4.6	−4.5	−4.4	−4.2	−4.0	−3.8	−3.6	−3.3	−3.0	−2.7	−2.3	−2.0	−1.6	−1.2	−0.8	−0.4	342
201	5.4	5.4	5.3	5.2	5.1	4.9	4.7	4.4	4.1	3.8	3.5	3.1	2.7	2.3	1.8	1.4	0.9	0.5	339
204	6.1	6.1	6.0	5.9	5.7	5.5	5.3	5.0	4.7	4.3	3.9	3.5	3.1	2.6	2.1	1.6	1.1	0.5	336
207	6.8	6.8	6.7	6.6	6.4	6.2	5.9	5.6	5.2	4.8	4.4	3.9	3.4	2.9	2.3	1.8	1.2	0.6	333
210	7.5	7.5	7.4	7.3	7.1	6.8	6.5	6.2	5.8	5.3	4.8	4.3	3.8	3.2	2.6	1.9	1.3	0.7	330
213	−8.2	−8.2	−8.1	−7.9	−7.7	−7.4	−7.1	−6.7	−6.3	−5.8	−5.3	−4.7	−4.1	−3.5	−2.8	−2.1	−1.4	−0.7	327
216	8.8	8.8	8.7	8.5	8.3	8.0	7.7	7.2	6.8	6.3	5.7	5.1	4.4	3.7	3.0	2.3	1.5	0.8	324
219	9.5	9.4	9.3	9.1	8.9	8.6	8.2	7.8	7.3	6.7	6.1	5.4	4.7	4.0	3.2	2.4	1.6	0.8	321
222	10.1	10.0	9.9	9.7	9.5	9.1	8.7	8.2	7.7	7.1	6.5	5.8	5.0	4.3	3.4	2.6	1.7	0.9	318
225	10.6	10.6	10.5	10.3	10.0	9.6	9.2	8.7	8.1	7.5	6.8	6.1	5.3	4.5	3.6	2.8	1.8	0.9	315
228	−11.2	−11.1	−11.0	−10.8	−10.5	−10.1	−9.7	−9.2	−8.6	−7.9	−7.2	−6.4	−5.6	−4.7	−3.8	−2.9	−1.9	−1.0	312
231	11.7	11.6	11.5	11.3	11.0	10.6	10.1	9.6	9.0	8.3	7.5	6.7	5.8	4.9	4.0	3.0	2.0	1.0	309
234	12.2	12.1	12.0	11.8	11.4	11.0	10.5	10.0	9.3	8.6	7.8	7.0	6.1	5.1	4.2	3.1	2.1	1.1	306
237	12.6	12.6	12.4	12.2	11.9	11.4	10.9	10.3	9.7	8.9	8.1	7.2	6.3	5.3	4.3	3.3	2.2	1.1	303
240	13.0	13.0	12.8	12.6	12.2	11.8	11.3	10.7	10.0	9.2	8.4	7.5	6.5	5.5	4.5	3.4	2.3	1.1	300
243	−13.4	−13.4	−13.2	−12.9	−12.6	−12.1	−11.6	−11.0	−10.3	−9.5	−8.6	−7.7	−6.7	−5.7	−4.6	−3.5	−2.3	−1.2	297
246	13.7	13.7	13.5	13.3	12.9	12.5	11.9	11.3	10.5	9.7	8.8	7.9	6.9	5.8	4.7	3.6	2.4	1.2	294
249	14.0	14.0	13.8	13.6	13.2	12.7	12.2	11.5	10.8	9.9	9.0	8.1	7.0	5.9	4.8	3.6	2.4	1.2	291
252	14.3	14.3	14.1	13.8	13.4	13.0	12.4	11.7	11.0	10.1	9.2	8.2	7.2	6.0	4.9	3.7	2.5	1.2	288
255	14.5	14.5	14.3	14.0	13.7	13.2	12.6	11.9	11.1	10.3	9.3	8.3	7.3	6.1	5.0	3.8	2.5	1.3	285
258	−14.7	−14.7	−14.5	−14.2	−13.8	−13.3	−12.7	−12.1	−11.3	−10.4	−9.5	−8.4	−7.4	−6.2	−5.0	−3.8	−2.6	−1.3	282
261	14.9	14.8	14.6	14.3	14.0	13.5	12.9	12.2	11.4	10.5	9.5	8.5	7.4	6.3	5.1	3.8	2.6	1.3	279
264	15.0	14.9	14.7	14.4	14.1	13.6	13.0	12.3	11.5	10.6	9.6	8.6	7.5	6.3	5.1	3.9	2.6	1.3	276
267	15.0	15.0	14.8	14.5	14.1	13.6	13.0	12.3	11.5	10.6	9.7	8.6	7.5	6.3	5.1	3.9	2.6	1.3	273
270	15.0	15.0	14.8	14.5	14.1	13.6	13.0	12.3	11.5	10.6	9.7	8.6	7.5	6.4	5.1	3.9	2.6	1.3	270

©2009 Jack Case

Appendix
A Brief Revision of Mathematical Topics Relevant To Astro Navigation.

To avoid making demonstrations and techniques overly complicated, certain mathematical formulae have so far been given without explanation throughout this book. This appendix is devoted to revising some of the mathematical principles that underpin astro navigation. For those who are fully conversant with this area of mathematics and for those who are only interested in knowing how to apply the necessary techniques without understanding how they were developed, nothing will be lost by skipping this appendix.

Revising Angles of Elevation and Trigonometric Ratios.

The activity of measuring the height of an object by shadow comparison provides a good starting point for investigating the uses that can be made of angles of elevation.

As shown in Diagram 45, if we put an upright pole in the ground at the tip of the shadow of a tree, the pole, its shadow and a ray of Sunlight make a triangle with angles of 90°, ø, and 90°- ø. In the same way, the tree, its shadow and a ray of Sunlight make another triangle with the same angles (i.e. the triangles are similar).

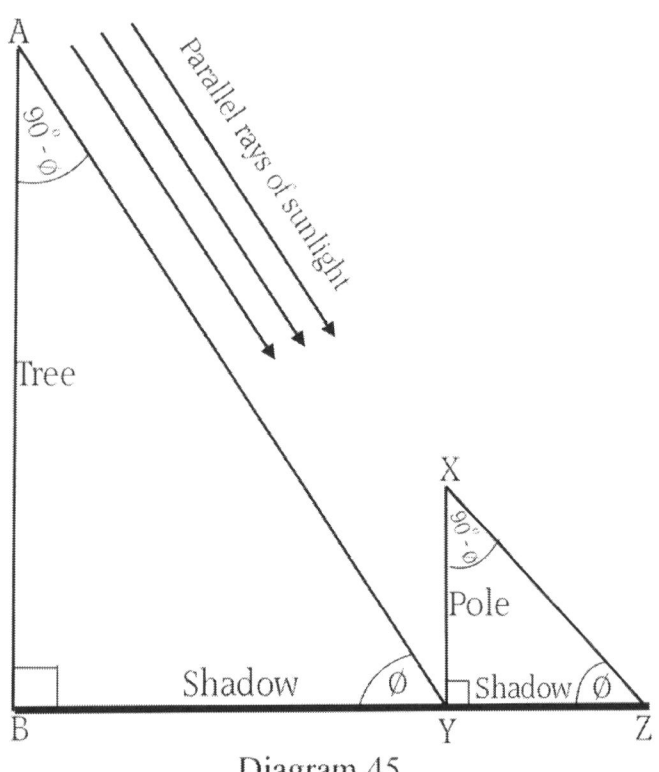

Diagram 45

It follows that since the triangles are similar, the ratios of corresponding sides must be the same.

i.e. $\dfrac{AB}{BY} = \dfrac{XY}{YZ}$

The angle ø is the Sun's inclination to the horizon (known as altitude) and is the same for both triangles. This leads us to suppose that there must be a relationship between this angle and the ratios of the sides of the triangles. If we were to examine the ratios of the sides of the triangles for all possible angles of inclination, and put the results in a table we would have a method of determining the heights and distances of all inaccessible objects.

The sides of right-angle triangles, are named according to their position in relation to angle ø, as shown in Diagram 46.

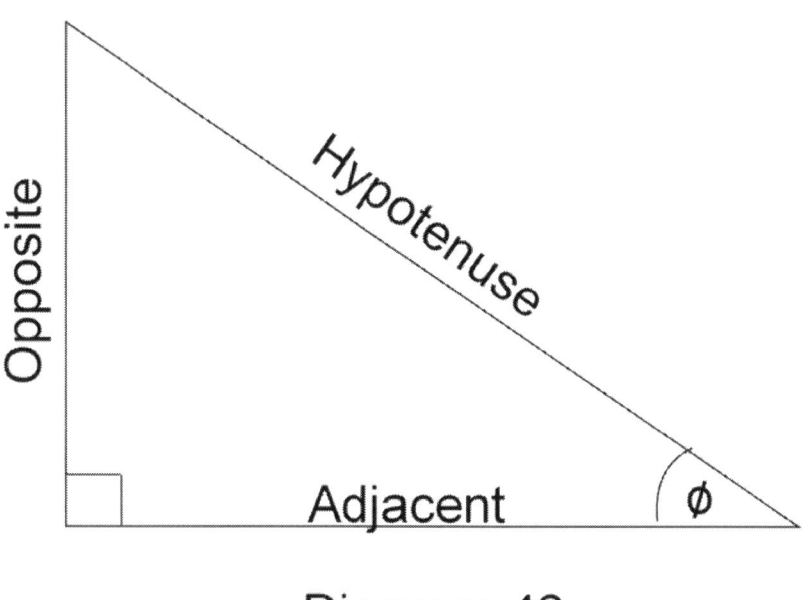

Diagram 46

The various ratios of the sides of a right-angle triangle are given names. The names of the ratios of the various sides in relation to angle ø are shown in the following table:

RATIO	NAME
$\dfrac{\text{OPPOSITE}}{\text{ADJACENT}}$	TANGENT
$\dfrac{\text{OPPOSITE}}{\text{HYPOTENUSE}}$	SINE
$\dfrac{\text{ADJACENT}}{\text{HYPOTENUSE}}$	COSINE

The reciprocals of these ratios also have names and these are listed below:

RATIO	NAME
ADJACENT / OPPOSITE	COTANGENT
HYPOTENUSE / OPPOSITE	COSECANT
HYPOTENUSE / ADJACENT	SECANT

Using Trigonometry to Solve Problems in Navigation.

Let us consider a simple problem that might be faced by a sailor and investigate how trigonometry could be used to solve that problem:

Suppose a yachtsman wishes to find the distance of his boat from a lighthouse marking a dangerous rock and suppose also that he has no radar or GPS equipment. The problem is similar to that of finding the height of the tree when we know our distance from it. The difference with this problem however, is that the sailor can find out the height of the lighthouse from a chart of the area but doesn't know the boat's distance from it.

Consider diagram 47,

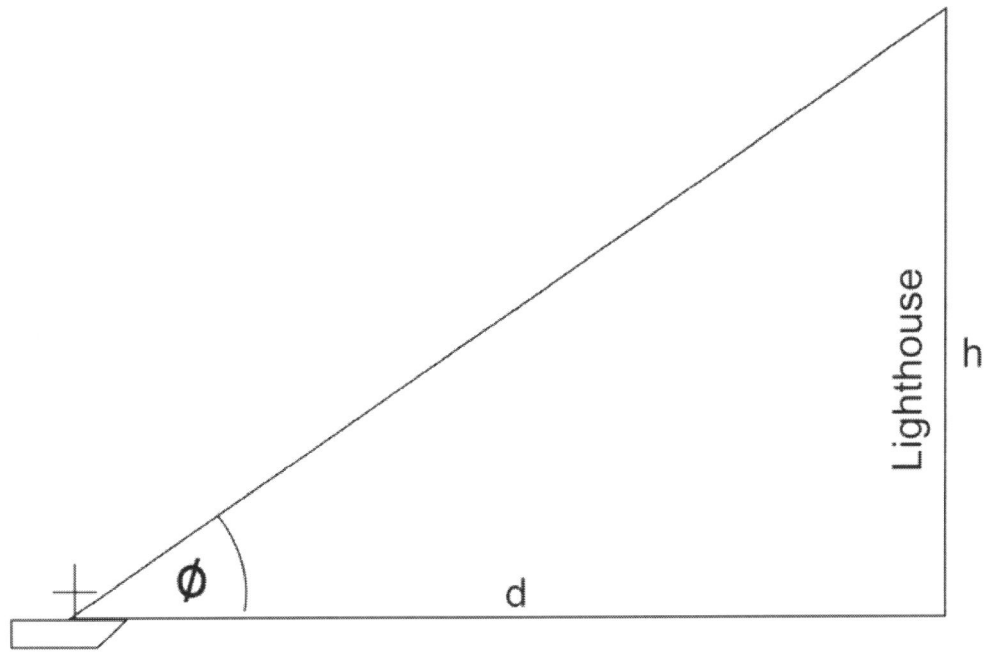

Diagram 47

In diagram 47,
h is the height of the lighthouse, which is 30m above sea level.
d is the distance of the boat from the base of the lighthouse.
Angle ø is the angle of elevation of the top of the lighthouse which the sailor has measured and found to be 20°.
We can see that the problem can be expressed in the form of a right-angle triangle in which (in relation to angle ø) side h is the opposite, side d is the adjacent and the line joining the boat to the top of the lighthouse is the hypotenuse. It follows that we can solve the problem by using one of the trigonometric ratios discussed earlier.

Since we know the values of side h and angle ø and wish to find the value of side d, the problem can be expressed as follows:

$$h = 30, \quad ø = 20°$$

$$\tan \alpha = \frac{\text{opposite}}{\text{adjacent}} = \frac{h}{d}$$

$$\therefore \tan 20° = \frac{h}{d}$$

$$\therefore d = \frac{h}{(\tan 20°)}$$

$$\therefore d = \frac{30}{0.364}$$

$$= 82.4 \text{ m}$$

∴ The boat is 82.4 metres from the base of the lighthouse.
However, accuracy is of the utmost importance, especially if the sailor is anxious to avoid rocks and there is one measurement of which we have not taken account. The height of the sailor's eye above water level is obviously important since his sextant or other measuring device will have been held at eye level. If his height of eye is 3m. then side d will be 3m. above sea level and so side h will be 27m. instead of 30m.

The solution to the problem now becomes:

$$d = \frac{27}{0.364} = 74.2$$

This puts the boat 8.2m. closer to the lighthouse and in rock strewn waters, this could make the difference between life and death. This illustrates the importance of taking the height of the observer's eye into account when making calculations involving altitude.

Spherical Trigonometry.

In practical astro navigation, we mostly rely on the use of tables of computed data and rote-learned procedures. We can operate quite efficiently in this way for most of the time; however, for those who wish to develop a thorough understanding of the subject, it is important to study its underlying principles of spherical trigonometry. Consider diagram 48:

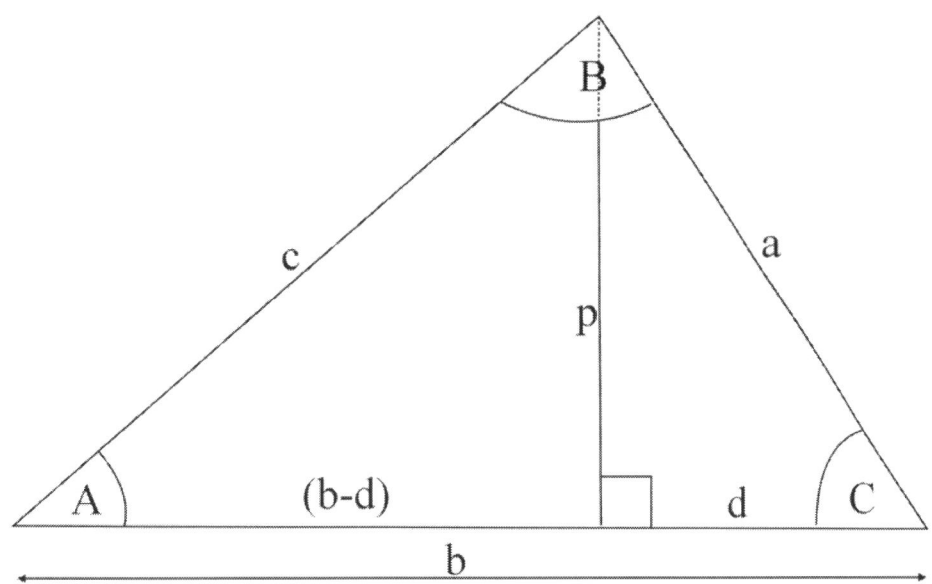

Diagram 48

Cos C = d/a and d = a Cos C
$P^2 = c^2 - (b-d)^2$ and $p^2 = a^2 - d^2$
☐ $a^2 - d^2 = c^2 - (b-d)^2$
☐ $a^2 - d^2 = c^2 - b^2 + 2bd - d^2$
☐ $a^2 = c^2 - b^2 + 2bd$
☐ $a^2 = c^2 - b^2 + 2b.a\ Cos\ C$ (since d = a Cos C)
☐ $c^2 = a^2 + b^2 - 2b.a\ Cos\ C$

This is the **cosine rule** for 'flat' triangles but does this rule also apply to spherical triangles?

Diagram 49 shows a spherical triangle ABC formed by the intersection of three circles with their common centre O at the centre of the sphere.

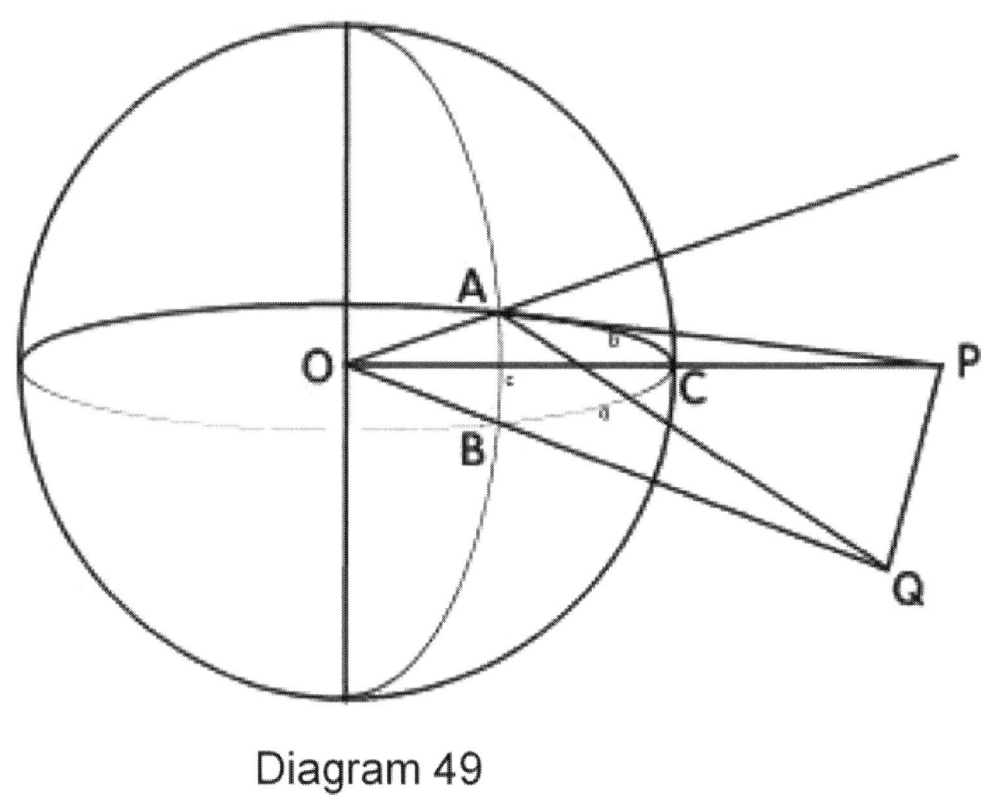

Diagram 49

The edges of the 'flat' planes in which the sides a, b, & c lie, meet along OA, OQ and OP.

The edges AQ and AP just graze the great circles of the arcs c and b at A; that is, AQ is the tangent to c and AP is the tangent to b. So angles OAQ and OAP are really right angles although it is impossible to draw them as such in the 'flat'.

The edges of the three planes in which a, b, & c lie form a flat (two-dimensional) triangle PAQ, of which the apical angle PAQ is equivalent to the angle A of the spherical triangle.

Now, by the rule for 'flat' triangles, we have:
$$PQ^2 = PO^2 + QO^2 - 2PO \cdot QO \cdot \cos(a)$$

and $PQ^2 = PA^2 + QA^2 - 2PA \cdot QA \cdot Cos(A)$
$\Rightarrow$ $[PO^2 - PA^2] + [QO^2 - QA^2] - [2PO \cdot QO \cdot Cos(a)] + [2PA \cdot QA \cdot Cos(A)] = 0$
We also have:
$PO^2 - PA^2 = AO^2$ and $QO^2 - QA^2 = AO^2$
$\Rightarrow$ $PO^2 - PA^2] + [QO^2 - QA^2] = 2AO^2$
Substituting $2AO^2$ in the equation gives us:
$2PO \cdot QO \cdot Cos(a) = 2AO^2 + 2PA \cdot QA \cdot Cos(A)$
Dividing through by $2PO \cdot QO$, we have:
$Cos(a) = \dfrac{AO}{PO} \cdot \dfrac{AO}{QO} + \dfrac{PA}{PO} \cdot \dfrac{QA}{QO} Cos(A)$
$= [Cos(POA) \cdot Cos(QOA)] + [Sin(POA) \cdot Sin(QOA) \cdot Cos(A)]$
$= [Cos(b) \cdot Cos(c)] + [Sin(b) \cdot Sin(c) \cdot Cos(A)]$

Hence, the formula for finding the third side (a) of a spherical triangle when the other two sides (b and c) are known together with the included angle (A) is: $Cos(a) = [Cos(b) \cdot Cos(c)] + [Sin(b) \cdot Sin(c) \cdot Cos(A)]$
(This is the cosine rule for spherical triangles).
We apply this rule when solving the spherical triangle PZX (see diagram 3). For example, to find the side ZX in the spherical triangle PZX, we have:
$Cos\ ZX = [Cos\ PZ \cdot Cos\ PX] + [Sin\ PZ \cdot Sin\ PX \cdot Cos\ ZPX]$

Revision of Pythagoras' Theorem.

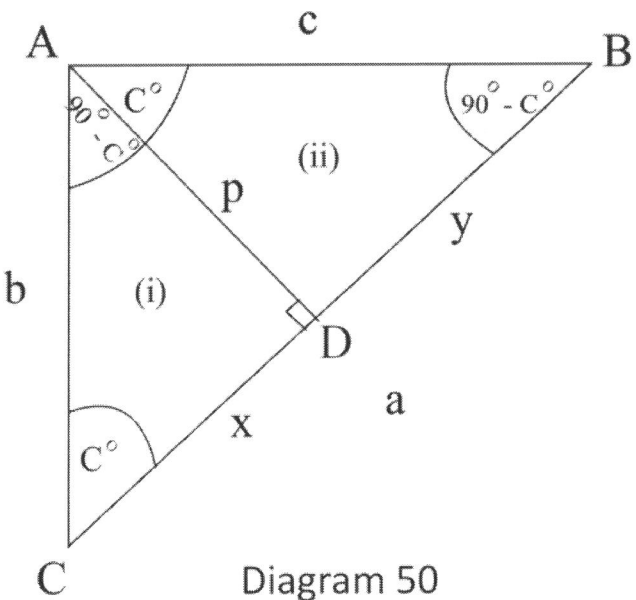

Diagram 50

Consider Diagram 50.

We can see that, in the triangle ABC, if a perpendicular is dropped from point A to side a at point D, ABC is divided into 2 smaller triangles which are both similar to ABC.

Consider triangle (i). Since this triangle is similar to ABC, it follows that:

$$\frac{a}{b} = \frac{b}{x}$$

$\Rightarrow b^2 = ax$ ....(i)

Now consider triangle (ii); by similar argument:

$$\frac{a}{c} = \frac{c}{y}$$

$\Rightarrow c^2 = ay$ ....(ii)

Combining (i) and (ii) we get:

$b^2 + c^2 = ax + ay$

$\Rightarrow a(x+y) = b^2 + c^2$

But $a = x + y$

$\Rightarrow a^2 = b^2 + c^2$ **This is Pythagoras' Theorem**

Practical Application of Pythagoras' Theorem.
(Calculating the Earth's Radius).

Early sailors would have been puzzled by the mystery of why an object, such as a point of land, would gradually disappear over the horizon as a ship sailed further out to sea. Let us use this phenomenon to devise a method of calculating the Earth's radius.

Suppose a boat sails out to sea from the vicinity of a lighthouse. If the height of the lighthouse is 100 metres, the top of it will disappear over the horizon at approximately 35.7 Km.

What can be deduced about the radius of the Earth from this?

Consider diagram 51:

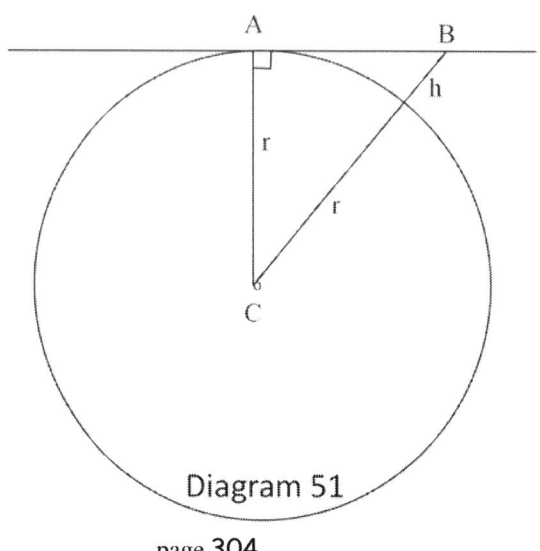

Diagram 51

In diagram 51, AB represents the line of sight from the boat to the top of the lighthouse (assuming the sailor's height of eye is at sea level).

AB is a tangent to the Earth's surface and is at right-angles to a line joining the boat to the Earth's centre (AC).

BC represents a line joining the lighthouse to the Earth's centre.

ABC forms a right-angle triangle and therefore, we can use Pythagoras' Theorem to calculate the Earth's radius as follows:

If, A is the boat's position,
B is the top of the lighthouse,
C is the Earth's centre,
r is the radius of the Earth
h is the height of the lighthouse and h = 100m. (0.1 Km.)
AB = 35.7 Km.

then, by Pythagoras:

$$BC^2 = AB^2 + AC^2$$
$$\Rightarrow (r + h)^2 = AB^2 + r^2$$
$$\Rightarrow r^2 + 2rh + h^2 = AB^2 + r^2$$
$$\Rightarrow 2rh + h^2 = AB^2$$
$$\Rightarrow 2rh = AB^2 - h^2$$
$$\Rightarrow r = \frac{AB^2 - h^2}{2h}$$

$$\therefore r = \frac{35.7^2 - 0.1^2}{2 \times 0.1} \text{ Km.}$$
$$= \frac{1274.49 - 0.01}{0.2}$$
$$= \frac{1274.48}{0.2}$$

$$\therefore r = 6372.4 \text{ Km.}$$

Note. This calculation is based on the assumption that the Earth is a perfect sphere. However, this is not the case; the Earth is in fact an oblate spheroid and so there can be no single value for its radius. The value generally accepted as the mean radius is 6367.45 Km. which is approximately 3438 n.m.

Revision of Circular Measurement involving Pi.

We know that the ratio of the circumference of a circle to its diameter is the same for all circles. We call this ratio pi (π) and its value is 3.1416 (to 4 dp.). So if c is the circumference of a circle, d is its diameter, and r is its radius, then $c \div d = \pi$

$\Rightarrow \quad c = \pi d$

$\Rightarrow \quad c = 2\pi r$

Practical Application of Pi (Calculating the Earth's circumference).

Given that the Earth's radius is 3438 n.m. and that the value of Pi is 3.1416, we can use the formula $c = 2\pi r$ to calculate the Earth's circumference as follows:

$c = 2\pi r = 2\pi \, 6372.4$ Km.

$\therefore \; c = 40039.1$ Km.

Note. Since this calculation is based on the mean radius of the Earth, it follows that what we have calculated is the mean circumference of the Earth. However, the calculation assumes that the Earth is a perfect sphere and if that were the case, the formula for the circumference of a circle ($C = 2\pi r$) would be appropriate for calculating the circumference of any great circle contained within that sphere. However, as discussed above, the Earth is in fact an oblate spheroid and so there can be no single value for its circumference; however, the important values are as follows:

Equatorial Circumference = 40075.16 Km.
Polar (or Meridional) Circumference = 40008 Km.
Mean circumference = 40041.58 Km.

Calculating the Circumference of a Parallel of Latitude.

We know the Equator is a great circle and since we know the equatorial radius, it is a simple matter to calculate the equatorial circumference. Parallels of latitude, on the other hand, are small circles; so how do we calculate the circumference of a parallel of latitude?

Consider diagram 52:

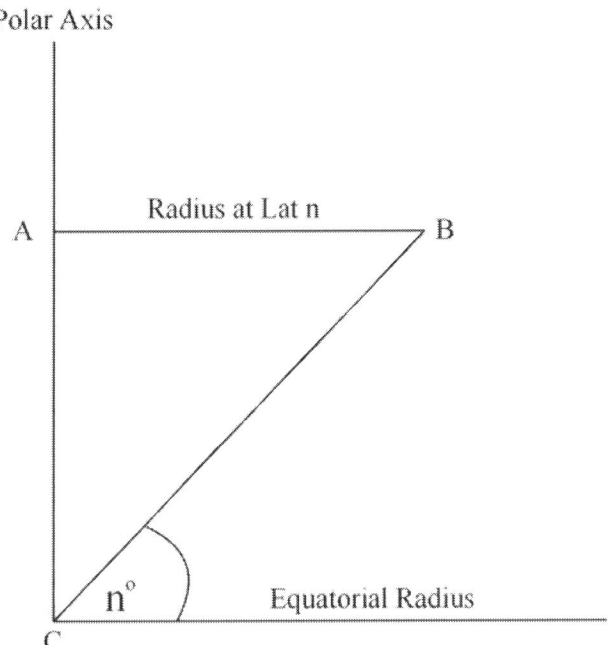

Diagram 52

In diagram 52,
ABC is a right angle triangle.
C   represents the Earth's centre.
A   represents a point on the Polar axis.
B   represents a point on the Earth's surface at latitude $n°$.
AB is perpendicular to the polar axis and represents the radius of the small circle described by latitude $n°$.
Angle ABC = $n°$ (alternate angles).
BC = the mean radius of the Earth (since B is a point on the Earth's surface and C is the Earth's centre).
By trigonometry, $\dfrac{AB}{BC} = \cos n°$

$\Rightarrow$ AB = BC $\times \cos(n°)$

The circumference at latitude ($n°$) equals $2\pi r$ (the circumference of a circle).
∴ circumference at latitude ($n°$) = 2 $\times \pi \times$ AB = 2 $\times \pi \times$ BC $\times \cos(n°)$
Now, as discussed above, BC is equal to the Earth's mean radius.
So 2 $\times \pi \times$ BC must equal the Earth's mean circumference (40041.58 Km.).
Therefore, circumference at latitude $n°$ = mean circumference $\times \cos(n°)$
= 40041.58 $\cos(n°)$

For example, the circumference at latitude 50°N. = 40041.58 cos(50°)
= 40041.58 x 0.64279
= 25738.3 Km.

Proof For The Derivation Of The Formula To Calculate The Difference In Distance Along A Parallel Of Latitude (Dlist) Corresponding To A Difference In Longitude (Dlong)

Consider diagram 53:
Let DC be an arc of n° of longitude measured along the circumference of the great circle which is the Equator.
Let O be the centre of the plane of the great circle which is the Equator.
Then CO will be equal to the radius of the Earth.
Let AB be an arc of n° of longitude measured along a parallel of latitude. Let Q be the centre of the plane of the small circle which is the parallel of latitude.
Then angle A Q B = angle D O C

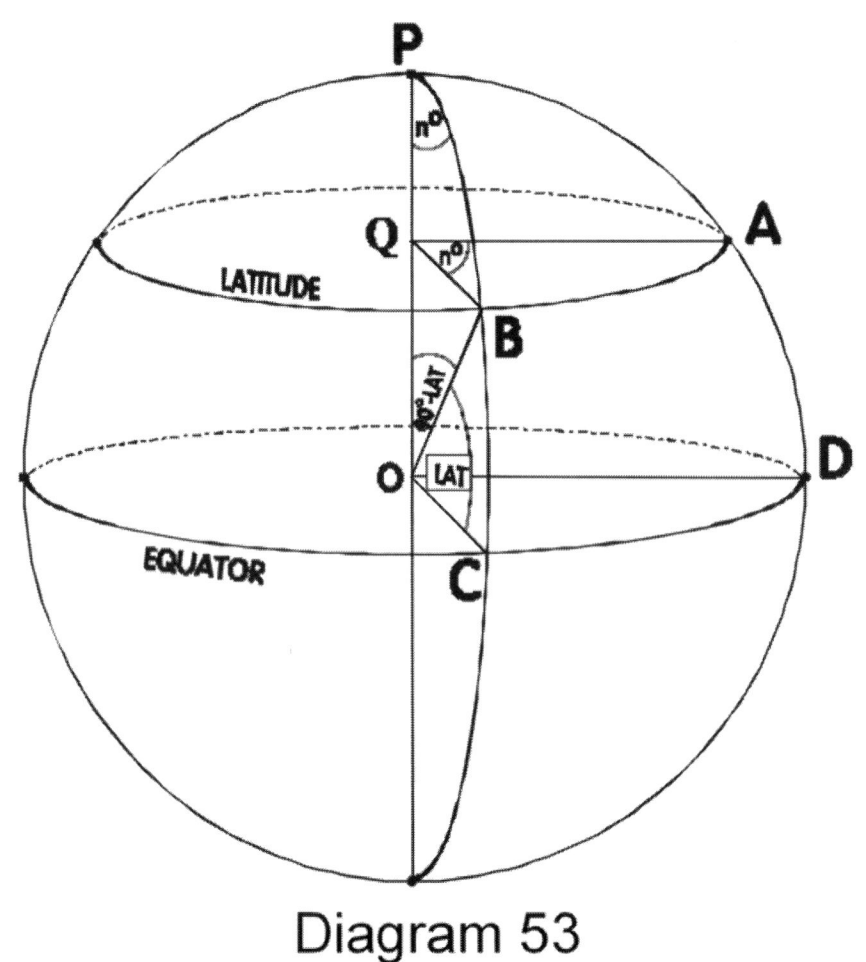

Diagram 53

Using the formula $C = 2\pi r$, we can deduce that the circumference of the Equator equals $2\pi CO$.

Therefore it follows that 1° of longitude corresponds to an arc of $\dfrac{2\pi \cdot CO}{360}$

$\Rightarrow \quad DC = \dfrac{2\pi n \cdot CO}{360}$

$\Rightarrow \quad CO = \dfrac{360 \cdot DC}{2\pi n}$

by similar argument:

$\quad AB = \dfrac{2\pi n \cdot BQ}{360}$

$\Rightarrow \quad BQ = \dfrac{360 \cdot AB}{2\pi n}$

$\therefore$ angle BOC represents the latitude

and angle QOB = 90° - Lat

and angle QBO = Lat. (alternate angles)

Angle OQB = 90° (since the plane of QAB is at right-angles to the polar axis)

$\Rightarrow \quad$ QBO is a right angle triangle

$\therefore \quad \text{Sin (QOB)} = \dfrac{BQ}{BO}$

$\Rightarrow \quad \text{Sin (90° - Lat.)} = \dfrac{BQ}{BO}$

Also CO = BO = r (since r = Earth's radius)

$\Rightarrow \text{Sin (90°-Lat)} = \dfrac{BQ}{CO}$ (angle QBO = Lat)

Also Cos Lat. = $\dfrac{BQ}{CO}$ (since CO = BO)

$\Rightarrow \quad BQ = CO \text{ Cos Lat.}$

$\Rightarrow \quad \dfrac{360 \cdot AB}{2\pi n} = \dfrac{360 \cdot DC \text{ Cos Lat.}}{2\pi n}$

$\Rightarrow \quad$ Arc AB = arc DC Cos Lat.

$\Rightarrow$ Distance AB = DC x Cos Lat (since DC is the difference in longitude (Dlong) and 1 n.m. along the Equator equals 1 minute of arc).

Therefore, to calculate the difference in distance along a parallel of latitude (Ddist) corresponding to a difference in longitude (Dlong) we have the formula:

Ddist = Dlong x Cos Lat.

and

Dlong = $\dfrac{\text{Ddist}}{\text{Cos Lat.}}$

Since the secant is the inverse of the cosine, the formula for Dlong can be simplified to:

Dlong = Ddist x Sec Lat.

Eratosthenes' Proof that the Earth is a Sphere.

In approximately 300 B.C., Eratosthenes, a Greek mathematician living in Alexandria, devised a method of calculating the circumference of the Earth and in doing so, provided evidence to support his long held belief that the Earth is a sphere.

Eratosthenes reasoned that if two sticks of equal height are placed on the Earth's surface and aligned with the direction of the Sun, then at any given point in time, assuming the Earth is flat, they will cast shadows of equal length. In other words, the angle between the sticks and rays of Sunlight would be the same. However, when he conducted an experiment with the help of friends in Assouan, he found that this was not so.

In his experiment, sticks were placed vertically in the ground at Alexandria and at Assouan. At noon the angle made by the Sun's rays and the top of the sticks was measured it was found that there was a difference of 7.2° between the two measurements. Now, Alexandria and Assouan lie on the same meridian of longitude and so noon always occurs at exactly the same instant in those places, which means that they must be in line with the direction of the Sun at that moment.

From the results of the experiment, Eratosthenes deduced that the towns could not be on a flat surface and that they must, instead be on the surface of a sphere. He went on to calculate the circumference of the Earth by the following simple reasoning:

He knew that Assouan lay due south of Alexandria at a distance of 580 statute miles. He reasoned that if the Earth were indeed a sphere, then a line drawn around its surface would describe a circle. Since there are 360° degrees in a circle and since 7.2° is one fiftieth of 360°, the distance from Alexandria to Assouan must be one fiftieth of the circumference of the Earth. Therefore, the circumference of the Earth, according to Eratosthenes, must be 50 x 580, that is 29000 statute miles.

The latest measurements show that the correct distance from Alexandria to Assouan is 500 miles and if Eratosthenes had known this, the result of his calculation of the Earth's circumference would have been: 50 x 500 = 25000 statute miles.

Eratosthenes believed that the Earth is a perfect sphere and the above calculations were based on that assumption. However, we now know that the Earth is in fact an **oblate spheroid** (i.e. it bulges at the Equator). For this reason, there can be no single value for its circumference but the value generally accepted as its mean circumference is 24825.8 statute miles and when we consider that this measurement was made with the use of modern technology, it is quite remarkable that it is not too far removed from the findings of Eratosthenes whose only form of technology was the 'shadow stick'.

Eratosthenes' Proof To Support His Calculation Of The Circumference of the Earth.

Consider diagram 54 below:

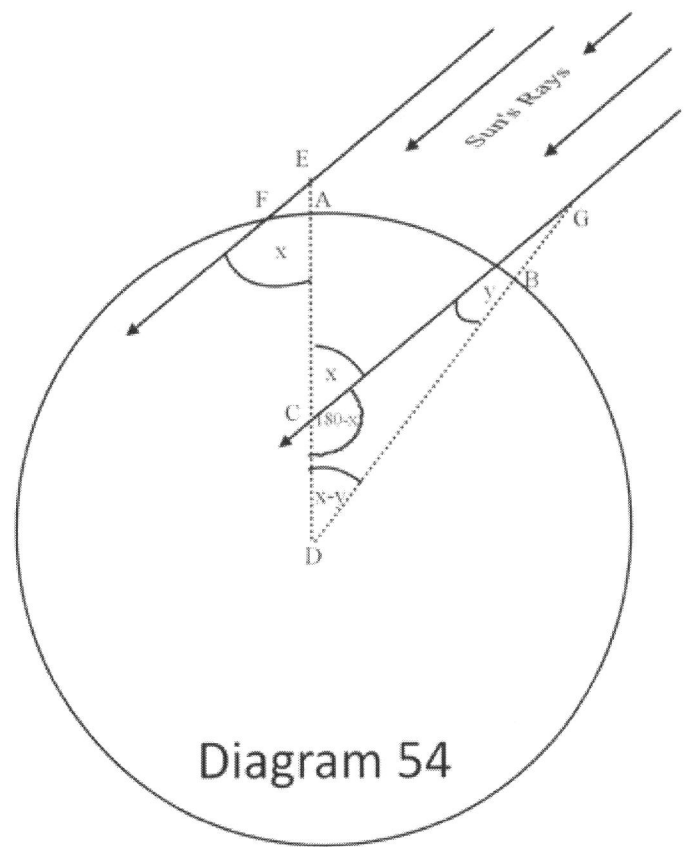

Diagram 54

Point A represents Alexandria.
Point B represents Assouan.
Point D represents the Earths centre.
x represent the angle made by the Sun's rays and a stick at point A.
y represent the angle made by the Sun's rays and a stick at point B.
Arc AB = 580 miles (*by Eratosthenes' reckoning*)
Angle FEA = angle ECG (*alternate angles*).
Angle GCD = 180° - x (*supplementary angles*)
∴ angle CDB = 180° - (angle GCD + angle CGD) (Angles in a triangle)
∴ angle CDB = 180° - (180° - x + y) = x - y
∴ arc AB subtends angle (x - y) at the Earth's centre.
∴ 1° subtends arc of $\frac{AB}{(x-y)}$ miles

∴ Earth's circumference = $360 \times \frac{AB}{(x-y)}$ miles

= $360 \times \frac{580}{7.2}$ miles

= 29000 miles (statute miles).

INDEX

Actual Position	52
Additional Correction	85
Altitude	77
Apparent Altitude	83
Apparent Noon	17
Apparent Solar Time	16
Approximate Position	52
Assumed Position	157
Astronomical Day	19
Astronomical Position Line	135
Atomic Radio Clocks	23
Autumnal Equinox	7
Azimuth	76
Celestial Body	5
Celestial Sphere	2
Chronometer	33
Civil Day	19
Civil Twilight	17
Co-Latitude	50
Combined Correction Tables	88
Crescent Moon	87
D.Lat	40
Day	16
D.Dist	36
Deck Watch	23
Declination	6
Departure	37
Difference In Latitude	40
Dip	82
D.Long	36
Earth Dimensions Table	32
Ecliptic	4
Equation Of Time	24
Equator	30
Equatorial Circumference	306
Equinoxes	6
Eratosthenes	310

First Point Of Aries	6
Geographical Mile	31
Geographical Position	5
GHA	10
Gibbous Moon	87
GMT	20
Gnomic Charts	50
GP	5
Great Circles	29
Great Circle Sailing	49
Greenwich Hour Angle	10
Greenwich Mean Time	20
Horizontal Parallax	88
Index Error	81
Intercept Method	143
Kilometre	31
Latitude	30
LHA	9
LMT	19
Local Hour Angle	9
Local Hour Angle Of The Sun	17
Local Mean Time	19
Longitude	29
Lower Limb (LL)	86
Marcq St. Hilaire	143
Mean Circumference	306
Mean Latitude	38
Mean Noon	17
Mean Solar Day	16
Mean Solar Time	16
Mean Sun	16
Measurement Conversion Table	31
Mer. Pas.	102
Meridian Passage	102
Meridional Circumference	306
Meridians Of Longitude	30
Middle Latitude	37
Movement of Body (MOB)	170

Movement of Observer (MOO)	168
Nautical Almanac	52
Nautical Mile	31
Nautical Twilight	17
Observed Altitude	77
Off The Arc	82
On The Arc	82
Parallax	77, 88
Polar Circumference	306
Position Circle	135
Position Line	3
Ptolemy	2
Pythagoras Theorem	303
RA	7
Rapid Sight Reduction Tables	156
Refraction	83
Rhumb Line	36
Rhumb Line Formulae	39
Right Ascension	7
Semi-Diameter	86
Sextant	80
Sextant Altitude	81
SHA	8
Short Distance Sailing	40
Sidereal Hour Angle	8
Small Circles	29
Spherical Triangle	5
Spherical Trigonometry	301
Standard Time	22
Star And Planet Observations	17
Statute Mile	31
Summer Solstice	7
Time Zones	22
Time Zone System	22
Time-Keeping At Sea	23
Traverse Tables	45
Triangle PZX	5, 135
True Altitude	77

True Bearing	161
True Sun	16
Twilight	17
Units Of Length	31
Universal Time	21
Universal Time Signal	23
Upper Limb (UL)	86
UT	21
v Correction	58
Vernal Equinox	7
Winter Solstice	7
Zenith	4 ,17
Zenith Distance	79
Zone Time	22

Made in the USA
Lexington, KY
28 December 2015